Introductory Algebra

FOURTH EDITION

Introductory Algebra

ARNOLD R. STEFFENSEN
L. MURPHY JOHNSON
Northern Arizona University

HarperCollins*Publishers*

To Barbara, Barbara, Becky, Cindy, and Pam

To the Student

A *Student's Solutions Manual,* by Joseph Mutter, provides complete, worked-out solutions to all of the exercises in set A, the chapter review exercises, and the chapter tests. Your college bookstore either has this book or can order it for you.

Sponsoring Editor:	Bill Poole
Development Editor:	Linda Youngman
Project Editor:	Janet Tilden/Randee Wire
Art Direction:	Julie Anderson
Text Design:	Lucy Lesiak Design/Lucy Lesiak
Cover Design:	Lucy Lesiak Design/Lucy Lesiak
Cover Photo:	Chicago Photographic Company ©
Photo Research:	Judy Ladendorf
Production:	Linda Murray
Compositor:	York Graphic Services
Printer and Binder:	Courier Corporation

Introductory Algebra, Fourth Edition

ISBN 0-673-46281-1

90 91 92 93 9 8 7 6 5 4 3 2 1

PREFACE

Introductory Algebra, Fourth Edition, is designed for college students who have not been exposed to algebra, or who require a review before taking further mathematics or computer science courses. Informal yet carefully worded explanations, detailed examples with accompanying practice exercises, pedagogical second color, abundant exercises, and comprehensive chapter reviews are hallmarks of the book. The text has been written for maximum instructor flexibility. Both core and peripheral topics can be selected to fit individual course needs. An annotated instructor's edition, testing manual, solutions manual, and test generator are provided for the instructor. A solutions manual is available for student purchase. Interactive tutorial software and a set of instructional videotapes are also available.

FEATURES

STUDENT GUIDEPOSTS ● Designed to help students locate important concepts as they study or review, student guideposts specify the major topics, rules, and procedures in each section. The guideposts are listed at the beginning of the section, then each is repeated as the corresponding material is discussed in the section.

EXAMPLES   More than 650 carefully selected examples include detailed, step-by-step solutions and side annotations in color. Each example is headed by a brief descriptive title to help students focus on the concept being developed and to aid in review.

PRACTICE EXERCISES   These parallel each example and keep students involved with the presentation by allowing them to check immediately their understanding of ideas. Answers immediately follow these exercises.

CAUTIONS   This feature calls students' attention to common mistakes and special problems to avoid.

COLOR   Pedagogical color highlights important information throughout the book. Key definitions, rules, and procedures are set off in colored boxes for increased emphasis. Figures and graphs utilize color to clarify the concepts presented. Examples present important steps and helpful side comments in color.

EXERCISES   As a key feature of the text, more than 5800 exercises, including about 800 applied problems, are provided. Two parallel exercise sets (A and B) and a collection of extension exercises (C) follow each section and offer a wealth of practice for students and flexibility for instructors. Exercises ranging from the routine to the more challenging, including application and calculator problems, are provided.

Exercises A   This set of exercises includes space for working the problems, with answers immediately following the exercises. Some of these problems, identified by a colored circle 4, have their solutions worked out at the back of the book. Many of these solutions are to exercises that students frequently have difficulty solving.

Exercises B   This set matches the exercises in set A problem for problem but is presented without work space or answers.

Exercises C   This set is designed to give students an extra challenge. These problems extend the concepts of the section or demand more thought than exercises in sets A and B. Answers or hints are given for selected exercises in this set.

REVIEW EXERCISES   To provide ample opportunities for review, the text features a variety of review exercises.

For Review exercises are located at the end of most A and B exercise sets. They not only encourage continuous review of previously covered material, but also often provide special review preparation for topics covered in the upcoming section.

Chapter Review Exercises and a practice **Chapter Test** conclude each chapter. The Chapter Review Exercises are divided into two parts: the problems in Part I are ordered and marked by section. Those in Part II are not referenced to the source section, and are thoroughly mixed to help students prepare for examinations. Answers to all review and test exercises are provided in the text.

Final Review Exercises, referenced to each chapter and with answers supplied, are located at the back of the book.

CHAPTER REVIEWS   In addition to the Chapter Review Exercises and Chapter Tests, comprehensive chapter reviews also include **Key Words** and **Key Concepts**. Key Words, listed by section, have brief definitions. Key Concepts summarize the major points of each section.

CALCULATORS   It is assumed that most students have a hand-held scientific calculator. However, there is disagreement among mathematics educators as to the use of a calculator at this level. When working exercises that involve dividing decimals or approximating irrational numbers, a calculator can be an invaluable tool, and it is important for students to learn to judge for themselves when a calculator should or should not be used. Some problems are marked with this calculator symbol: 🖩.

INSTRUCTIONAL FLEXIBILITY

Introductory Algebra, Fourth Edition, offers proven flexibility for a variety of teaching situations such as individualized instruction, lab instruction, lecture classes, or a combination of methods.

Material in each section of the book is presented in a well-paced, easy-to-follow sequence. Students in a tutorial or lab instruction setting, aided by the student guideposts, can work through a section completely by reading the explanation, following the detailed steps in the examples, working the practice exercises, and then doing the exercises in set A.

The book can also serve as the basis for, or as a supplement to, classroom lectures. The straightforward presentation of material, numerous examples, practice exercises, and three sets of exercises offer the traditional lecture class an alternative approach within the convenient workbook format.

NEW IN THIS EDITION

In a continuing effort to make this text even better suited to the needs of instructors and students, the following are some of the enhanced features of the new edition.

- The student guideposts are more visible.
- Explanations have been polished, reworded, and streamlined where appropriate. More figures and illustrations are used in discussions.
- Exercise sets have been reviewed for grading and balance of coverage. The number and variety of practical applications are increased. Additional challenging exercises and the use of calculators have been incorporated into Exercises C.
- Additional For Review exercises review topics in preparation for the next section.
- Chapter Review Exercises are presented in two parts, one with sectional references and one without, to help students recognize problem types and better prepare for tests.
- Key Words given in the Chapter Review are expanded to include brief definitions.
- Greater emphasis has been given to the use of figures and illustrations in examples and exercises, particularly applications.
- A thorough review of geometry has been added as an appendix.

SUPPLEMENTS

An expanded supplemental package is available for use with *Introductory Algebra, Fourth Edition*.

For the Instructor

The **ANNOTATED INSTRUCTOR'S EDITION** provides instructors with immediate access to the answers to every exercise in the text; each answer is printed in color next to the corresponding text exercise.

The **INSTRUCTOR'S TESTING MANUAL** contains a series of **ready-to-duplicate tests,** including a Placement Test, six different but equivalent tests for each chapter (four open-response and two multiple-choice), and two final exams, all with answers supplied. Section-by-section **teaching tips** provide suggestions for content implementation that an instructor, tutor, or teaching assistant might find helpful.

The **INSTRUCTOR'S SOLUTIONS MANUAL** contains complete, worked-out solutions to every exercise in sets B and C of the text.

HARPERCOLLINS TEST GENERATOR FOR MATHEMATICS Available in Apple, IBM, and Macintosh versions, the test generator enables instructors to select questions by objective, section, or chapter, or to use a ready-made test for each chapter. Instructors may generate tests in multiple-choice or open-response formats, scramble the order of questions while printing, and produce multiple versions of each test (up to 9 with Apple, up to 25 with IBM and Macintosh). The system features printed graphics and accurate mathematics symbols. It also features a preview option that allows instructors to view questions before printing, to regenerate variables, and to replace or skip questions if desired. The IBM version includes an editor that allows instructors to add their own problems to existing data disks.

VIDEOTAPES A new videotape series, ALGEBRA CONNECTION: *The Introductory Algebra Course,* has been developed to accompany *Introductory Algebra, Fourth Edition.* Produced by an Emmy Award–winning team in consultation with a task force of academicians from both two-year and four-year colleges, the tapes cover all objectives, topics, and problem-solving techniques within the text. In addition, each lesson is preceded by motivational ''launchers'' that connect classroom activity to real-world applications.

For the Student

The **STUDENT'S SOLUTIONS MANUAL** contains complete, worked-out solutions to each practice exercise in the text, to every exercise in set A, and to all Chapter Review Exercises and Chapter Tests.

INTERACTIVE TUTORIAL SOFTWARE   This innovative package is also available in Apple, IBM, and Macintosh versions. It offers interactive modular units, specifically linked to the text, for reinforcement of selected topics. The tutorial is self-paced and provides unlimited opportunities to review lessons and to practice problem solving. When students give a wrong answer, they can request to see the problem worked out. The program is menu-driven for ease of use, and on-screen help can be obtained at any time with a single keystroke. Students' scores are automatically recorded and can be printed for a permanent record.

ACKNOWLEDGMENTS

We extend our sincere gratitude to the students and instructors who used the previous editions of this book and offered many suggestions for improvement. Special thanks go to the instructors at Northern Arizona University and Yavapai Community College. In particular, the assistance given over the years by James Kirk and Michael Ratliff is most appreciated. Also, we sincerely appreciate the support and encouragement of the Northern Arizona University administration, especially President Eugene M. Hughes. It is a pleasure and privilege to serve on the faculty of a university that recognizes quality teaching as its primary role.

We also express our thanks to the following instructors who responded to a questionnaire sent out by HarperCollins: Gregg Cox, College of Boca Raton; Robert Dunker, Metropolitan Technical Community College; James Head, Dalton Junior College; Virginia Licata, Camden County College; and Virginia Singer, Northwest Community College.

We are also indebted to the following reviewers for their countless beneficial suggestions at various stages of the book's revision:

Sharon Abramson, *Nassau Community College*

LaVerne Blagmon-Earl, *University of the District of Columbia*

June A. Barrett, *California State University–Sacramento*

Douglas J. Campbell, *West Valley College*

Elizabeth J. Collins, *Glassboro State College*

Sarah R. Evangelista, *Temple University*

Margaret H. Finster, *Erie Community College, South Campus*

Harry Lassiter, *Craven Community College*

Marion Littlepage, *Clark County Community College*

Lee H. Marsh, *Kalamazoo Valley Community College*

Kenneth R. Ohm, *Sheridan College*

Thomas J. Ribley, *Valencia Community College West*

David Sack, *Lincoln Land Community College*

Mark Serebransky, *Camden County College*

Virginia Singer, *Northwest Community College*

Gayle Smith, *Eastern Washington University*

Helen J. Smith, *South Mountain Community College*

Michael White, *Jefferson Community College*

We extend special appreciation to Joseph Mutter for writing the *Student's Solutions Manual* and for the countless suggestions and support given over the years. Thanks go to Diana Denlinger for typing this edition and the *Instructor's Solutions Manual*.

We thank our editors, Jack Pritchard, Bill Poole, Leslie Borns, Sarah Joseph, and Randee Wire, whose support has been most appreciated.

Finally, we are indebted to our families and in particular our wives, Barbara and Barbara, whose encouragement over the years cannot be measured.

Arnold R. Steffensen
L. Murphy Johnson

CONTENTS

5 SYSTEMS OF LINEAR EQUATIONS

6 EXPONENTS AND POLYNOMIALS

7 FACTORING POLYNOMIALS

8 RATIONAL EXPRESSIONS

During the past few years we have taught Introductory Algebra to more than 1400 students and have heard the following comments more than just a few times. "I've always been afraid of math and have avoided it as much as possible. Now my major requires algebra and I'm petrified." "I don't like math, but it's required to graduate." "I can't do word problems!" If you have ever made a similar statement, now is the time to think positively, stop making negative comments, and start down the path toward success in mathematics. Don't worry about this course as a whole. The material in the text is presented in a way that enables you to take one small step at a time. Here are some general and specific guidelines that are necessary and helpful.

GENERAL GUIDELINES

1. Mastering algebra requires motivation and dedication. A successful athlete does not become a champion without commitment to his or her goal. The same is true for the successful student of algebra. Be prepared to work hard and spend time studying.

2. Algebra is not learned simply by watching, listening, or reading; *it is learned by doing*. Use your pencil and practice. When your thoughts are organized and written in a neat and orderly fashion, you have taken a giant step toward success. Be complete and write out all details. The following are samples of two students' work on a word problem. Can you tell which one was the most successful in the course?

STUDENT A

Let x = marked price
0.05x = tax on marked price
x + 0.05x = price plus tax
x + 0.05x = 37.59
x(1 + 0.05) = 37.59
x(1.05) = 37.59
$$x = \frac{37.59}{1.05}$$
$$= \$35.80$$

STUDENT F

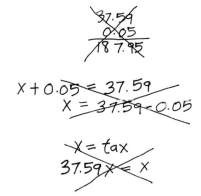

3. If the use of a calculator is permitted in your course, become familiar with its features by consulting your owner's manual. When you are computing with decimals or approximating irrational numbers, a calculator can be a time-saving device. On the other hand, you should not be so dependent on a calculator that you use it for simple calculations that can be done mentally. For example, it would be ridiculous to use a calculator to solve an equation such as $2x = 8$, while it would be helpful when solving $1.12x = 7343.84$. It is important that you learn when to use and when not to use your calculator.

SPECIFIC GUIDELINES

1. As you begin to study each section, glance through the material and obtain a preview of what is coming.

2. Return to the beginning of the section and start reading slowly. The STUDENT GUIDEPOSTS specify the important terms, definitions, and rules that must be understood. They also will help you locate important concepts as you progress or as you review.

3. Read through each EXAMPLE and make sure that you understand each step. The side comments in color will help you if something is not quite clear.

4. After reading each example, work the parallel PRACTICE EXERCISE. This will reinforce what you have just read and start the process of practice.

5. Periodically you will encounter a CAUTION. These warn you of common mistakes and special problems that should be avoided.

6. After you have completed the material in the body of the section, you must test your mastery of the skills and practice, practice, practice! Begin with the exercises in set A. Answers to all of the problems are placed at the end of the set for easy reference. Some of the problems are identified as having complete solutions at the back of the book. After trying these problems, refer to the step-by-step solutions if you have difficulty. You may want to practice more by doing the exercises in set B or to challenge yourself by trying the exercises in set C. (Note: Solutions to every problem in set A and to all Review Exercises and Chapter Tests are available in the *Student's Solutions Manual*.)

7. After you have completed all of the sections in the chapter, read the CHAPTER REVIEW that contains key words and key concepts. The review exercises provide additional practice before you take the CHAPTER TEST. Answers to the tests are at the back of the book.

8. To aid you in studying for your final examination, we have concluded the book with a comprehensive set of FINAL REVIEW EXERCISES.

If you follow these steps and work closely with your instructor, you will greatly improve your chance of success in algebra.

Best of luck, and remember that you can do it!

Fundamentals

1.1 FRACTIONS

1 SETS OF NUMBERS

Two terms that are used repeatedly in algebra are *set* and *element*. A **set** is a collection of objects. These objects, the **elements** of the set, are often listed within braces { }.

Most sets we deal with are sets of *numbers*. A **number** is a human invention, an idea formed when questions about "how many" or "in what order" are asked. Symbols for numbers are called **numerals.** For example, "4" is a symbol or numeral that represents the number four. The most basic numbers are the **natural** or **counting numbers.**

$$N = \{1, 2, 3, 4, \ldots\} \quad \text{Natural (counting) numbers}$$

We use three dots to mean that a sequence or pattern continues on in the same manner. When zero is included with the natural numbers, we obtain the set of **whole numbers.**

$$W = \{0, 1, 2, 3, 4, \ldots\} \quad \text{Whole numbers}$$

2 PROPERTIES OF FRACTIONS

Numbers written as the quotient of a whole number and a natural number, such as

$$\frac{1}{2}, \frac{3}{4}, \frac{7}{3}, \frac{5}{5}, \frac{5}{1}, \frac{0}{1}, \frac{15}{1}, \frac{61}{21},$$

are called **fractions.** The **numerator** of a fraction is the number above the fraction bar, and the **denominator** is the number below the fraction bar. In the fraction $\frac{3}{4}$, 3 is the numerator and 4 is the denominator.

When the numerator is a smaller number than the denominator, the fraction is called a **proper fraction;** otherwise it is an **improper fraction.** For example,

$\frac{1}{2}$ and $\frac{3}{4}$ are proper fractions, while $\frac{7}{5}$, $\frac{5}{5}$, $\frac{5}{1}$, and $\frac{61}{27}$ are improper fractions.

Every whole number can be written as a fraction by letting the whole number be the numerator and 1 be the denominator. For example,

$$5 = \frac{5}{1}, \qquad 2 = \frac{2}{1}, \qquad 0 = \frac{0}{1}, \qquad \text{and} \qquad 15 = \frac{15}{1}.$$

Five could also be written as $\frac{10}{2}$, $\frac{15}{3}$, $\frac{20}{4}$, etc. Similarly, the fraction $\frac{1}{4}$ could be expressed as $\frac{1}{4}$, $\frac{2}{8}$, $\frac{3}{12}$, $\frac{4}{16}$, $\frac{5}{20}$, and so on. We can always find another name for a fraction by multiplying both the numerator and the denominator by the same nonzero number. Fractions that are names for the same number or have the same value are called **equivalent fractions.**

Fundamental Principle of Fractions

If both the numerator and denominator of a fraction are multiplied or divided by the same nonzero number, the resulting fraction is equivalent to the original fraction.

❸ REDUCING TO LOWEST TERMS

We **build up** fractions when we multiply both the numerator and denominator by the same natural number other than 1 and **reduce** fractions when we divide both by the same natural number other than 1. For example,

$$\frac{3}{4} = \frac{3 \cdot 2}{4 \cdot 2} = \frac{6}{8}, \qquad \frac{3}{4} = \frac{3 \cdot 5}{4 \cdot 5} = \frac{15}{20}, \qquad \frac{3}{4} = \frac{3 \cdot 10}{4 \cdot 10} = \frac{30}{40}.$$

Thus, $\frac{3}{4}$ has been built up to the equivalent fractions $\frac{6}{8}$, $\frac{15}{20}$, $\frac{30}{40}$. Also,

$$\frac{12}{18} = \frac{12 \div 2}{18 \div 2} = \frac{6}{9}, \qquad \frac{12}{18} = \frac{12 \div 3}{18 \div 3} = \frac{4}{6}, \qquad \frac{12}{18} = \frac{12 \div 6}{18 \div 6} = \frac{2}{3}.$$

Thus, $\frac{12}{18}$ has been reduced to the equivalent fractions $\frac{6}{9}$, $\frac{4}{6}$, and $\frac{2}{3}$.

Reducing Fractions

A fraction is **reduced to lowest terms** when 1 is the only natural number that divides both numerator and denominator.

❹ PRIME NUMBERS AND FACTORS

We reduced $\frac{12}{18}$ to $\frac{6}{9}$, $\frac{4}{6}$, and $\frac{2}{3}$ above. Notice that $\frac{6}{9}$ and $\frac{4}{6}$ can each be reduced further, while $\frac{2}{3}$ cannot. Thus, $\frac{2}{3}$ is reduced to lowest terms. The concept of a *prime number* is useful for reducing fractions to lowest terms. A **prime number** is a natural number greater than 1 whose only divisors are 1 and itself. Since 1 and 5 are the only divisors of 5, 5 is prime; since 2 and 3 are divisors of 6 along with 1 and 6, 6 is not prime.

The first few primes are

$$2, \quad 3, \quad 5, \quad 7, \quad 11, \quad 13, \quad 17, \quad 19, \quad 23, \ldots.$$

Prime Numbers and Factors

Every natural number greater than 1 either is prime or can be expressed as a product of primes.

The primes in the product are called **prime factors,** and the process of expressing a number as a product of primes is called **factoring into primes.**

| EXAMPLE 1 FACTORING INTO PRIMES | PRACTICE EXERCISE 1 |

Express as a product of primes.

(a) $15 = 3 \cdot 5$ Why not $15 = 15 \cdot 1$?

(b) $28 = 2 \cdot 2 \cdot 7$ Why not $28 = 4 \cdot 7$?

(c) $41 = 41$ 41 is a prime

Express as a product of primes.

(a) 21 $7 \cdot 3$

(b) 36

(c) 53

Answers: (a) $3 \cdot 7$ (b) $2 \cdot 2 \cdot 3 \cdot 3$
(c) 53 is a prime

To Reduce a Fraction to Lowest Terms

1. Factor the numerator and denominator into products of primes.
2. Divide both the numerator and denominator by all common factors.
3. Multiply remaining factors in the numerator and multiply the remaining factors in the denominator.

Dividing both the numerator and denominator of a fraction by common factors is sometimes called **canceling factors.** This is shown by crossing out the common factors. For example,

$$\frac{9 \cdot \cancel{5}}{\cancel{5}} = \frac{9}{1} \quad \text{and} \quad \frac{\cancel{2} \cdot \cancel{7}}{\cancel{2} \cdot 3 \cdot \cancel{7}} = \frac{1}{3}.$$

Notice that when we cancel *all* of the factors out of a numerator or a denominator, it makes the numerator or denominator 1, *not* 0.

| EXAMPLE 2 REDUCING FRACTIONS | PRACTICE EXERCISE 2 |

Reduce the fractions to lowest terms.

(a) $\dfrac{6}{15} = \dfrac{2 \cdot \cancel{3}}{\cancel{3} \cdot 5} = \dfrac{2}{5}$ Cancel common factor 3 from both numerator and denominator

(b) $\dfrac{12}{42} = \dfrac{\cancel{2} \cdot 2 \cdot \cancel{3}}{\cancel{2} \cdot \cancel{3} \cdot 7} = \dfrac{2}{7}$ Cancel common factors 2 and 3 from both numerator and denominator

(c) $\dfrac{14}{15} = \dfrac{2 \cdot 7}{3 \cdot 5}$ There are no common factors, so $\frac{14}{15}$ is already reduced to lowest terms

Reduce the fractions to lowest terms.

(a) $\dfrac{2}{4}$ $1/2$

(b) $\dfrac{70}{25}$ $\dfrac{2 \cdot 5 \cdot 7}{5 \cdot 5}$

(c) $\dfrac{15}{77}$ $\dfrac{3 \cdot 5}{7 \cdot 11}$

Answers: (a) $\frac{1}{2}$ (b) $\frac{14}{5}$
(c) already in lowest terms

CAUTION

Divide out or cancel only *factors* (numbers that are multiplied) and never cross out parts of sums. For example,

$$\frac{2 \cdot 7}{2} = \frac{\cancel{2} \cdot 7}{\cancel{2}} = 7, \quad \text{but} \quad \frac{2 + 7}{2} \text{ is } not \text{ equal to } \frac{\cancel{2} + 7}{\cancel{2}}.$$

With practice, we often shorten the process of reducing fractions by dividing out common factors without first factoring into primes. For example, if we notice that 6 is a common factor of both 12 and 42, we can write

$$\frac{12}{42} = \frac{2 \cdot \cancel{6}}{7 \cdot \cancel{6}} = \frac{2}{7}.$$

Fractions describe parts of quantities in many applied situations. If a baseball player got three hits in five times at bat, we would say he had a hit $\frac{3}{5}$ (three-fifths) of the time. Thus, fractions are used to indicate what part one number is of another.

EXAMPLE 3 FRACTIONAL PARTS

A group of 100 students was made up of 60 girls and 40 boys. Exactly 75 students were taking math, and exactly 35 students were taking history.

(a) What fractional part of the total number of students were girls?

There were 60 girls among 100 students for a fractional part of $\frac{60}{100}$, or $\frac{3}{5}$, reduced to lowest terms.

(b) What fraction of the students were taking math?

There were 75 math students out of the 100 total for a fraction of $\frac{75}{100}$, or $\frac{3}{4}$, reduced to lowest terms.

(c) What fraction of the students were *not* taking history?

Since 35 students were taking history, 65 were not. Thus, $\frac{65}{100}$, or $\frac{13}{20}$, of the students were not taking history.

PRACTICE EXERCISE 3

In a sample of 200 pens, 120 are red and 80 are green. Exactly 65 of the pens have a fine point, and exactly 185 have retractable points.

(a) What fractional part of the total number of pens is green?

(b) What fraction of the total have fine points?

(c) What fraction do not have retractable points?

Answers: (a) $\frac{2}{5}$ (b) $\frac{13}{40}$ (c) $\frac{3}{40}$

1.1 EXERCISES A

Answer true *or* false *in Exercises 1–10. If the statement is false, tell why.*

1. A collection of objects is also called a set of objects.

2. The name or symbol for a number is a numeral.

3. $\{1, 2, 3, \ldots\}$ is the set of whole numbers.

4. In the fraction $\frac{5}{7}$, the number 7 is called the numerator.

5. If the numerator of a fraction is a smaller number than the denominator, it is a proper fraction.

6. Two fractions with the same value are called equivalent fractions.

7. We build up a fraction when we divide both numerator and denominator by a natural number other than 1.

8. A natural number other than 1 that is divisible only by itself and 1 is called prime.

9. The process of canceling factors from a numerator and denominator is really the process of adding common factors.

10. An idea that comes to mind when considering questions such as "how many" or "in what order" is called a number.

11. Consider the fractions $\frac{1}{5}$, $\frac{3}{8}$, $\frac{9}{9}$, $\frac{10}{2}$, $\frac{6}{6}$, $\frac{13}{15}$, and $\frac{8}{1}$.
  (a) Which of these are proper fractions?   **(b)** Which of these are improper fractions?

$\frac{1}{5}$ $\frac{3}{8}$ $\frac{13}{15}$ $\frac{10}{2}$ $\frac{6}{6}$ $\frac{8}{1}$ $\frac{9}{9}$

Write three fractions that are equivalent to the following.

12. $\frac{2}{5}$ $\frac{4}{10}$ $\frac{6}{15}$ $\frac{8}{20}$

13. $\frac{5}{2}$ $\frac{10}{4}$ $\frac{15}{6}$

14. 4 $\frac{4}{1}$ $\frac{8}{2}$ $\frac{20}{5}$

15. 0

16. $\frac{105}{2}$ $\frac{210}{4}$

17. $\frac{1}{100}$ $\frac{3}{300}$

18. List all prime numbers between 1 and 20.

2, 3, 5, 7, 11, 13, 17, 19

Express as a product of primes.

19. 70 $3 \cdot 5 \cdot 7 \cdot 2$

20. 45 $3 \cdot 5 \cdot 3$

21. 47

22. 81

23. 180

24. 1470

Reduce to lowest terms.

25. $\frac{9}{36}$ $\frac{3}{12}$ $\frac{1}{4}$

26. $\frac{8}{28}$ $\frac{2}{7}$

27. $\frac{25}{30}$ $\frac{5}{6}$

28. $\frac{12}{35}$

29. $\frac{48}{20}$ $\frac{24}{10}$ $\frac{12}{5}$

30. $\frac{72}{16}$ $\frac{36}{8}$ $\frac{13}{4}$ $\frac{9}{2}$

31. $\frac{72}{72}$ 1

32. $\frac{72}{114}$ $\frac{36}{57}$ $\frac{12}{19}$

33. $\frac{120}{162}$ $\frac{60}{81}$ $\frac{20}{27}$

34. Write an equivalent fraction for $\frac{2}{3}$ with 15 as the denominator. $\frac{10}{15}$

35. Write an equivalent fraction for $\frac{7}{8}$ with 32 as the denominator. $\frac{28}{32}$

36. Write an equivalent fraction for $\frac{5}{4}$ with 15 as the numerator. $\frac{15}{12}$

37. Write an equivalent fraction for $\frac{6}{5}$ with 36 as the numerator. $\frac{36}{30}$

Supply the missing numerator.

38. $\frac{3}{4} = \frac{9}{12}$

39. $\frac{5}{7} = \frac{20}{28}$

40. $\frac{10}{3} = \frac{80}{24}$

Supply the missing denominator.

41. $\dfrac{3}{7} = \dfrac{6}{14}$

42. $\dfrac{2}{11} = \dfrac{10}{55}$

43. $\dfrac{21}{2} = \dfrac{63}{6}$

44. A softball pitcher struck out 12 of the 21 batters she faced. What fractional part of the total number of batters **(a)** did she strike out? **(b)** did she not strike out?

45. Marvin received $150 from his dad. He spent $55 for a coat and deposited the rest into his savings account. What fractional part of the $150 went for **(a)** the coat? **(b)** savings?

ANSWERS: 1. true 2. true 3. false (natural numbers) 4. false (denominator) 5. true 6. true 7. false (reduce) 8. true 9. false (dividing) 10. true 11. (a) $\frac{1}{5}$, $\frac{3}{8}$, and $\frac{13}{15}$ (b) $\frac{9}{9}$, $\frac{10}{2}$, $\frac{6}{6}$, and $\frac{8}{1}$ *Answers to 12–17 will vary; some possibilities are given.* 12. $\frac{4}{10}$, $\frac{6}{15}$, $\frac{8}{20}$ 13. $\frac{10}{4}$, $\frac{15}{6}$, $\frac{20}{8}$ 14. $\frac{4}{1}$, $\frac{8}{2}$, $\frac{12}{3}$ 15. $\frac{0}{1}$, $\frac{0}{2}$, $\frac{0}{3}$ 16. $\frac{210}{4}$, $\frac{315}{6}$, $\frac{420}{8}$ 17. $\frac{2}{200}$, $\frac{3}{300}$, $\frac{4}{400}$ 18. 2, 3, 5, 7, 11, 13, 17, 19 19. $2 \cdot 5 \cdot 7$ 20. $3 \cdot 3 \cdot 5$ 21. 47 (prime) 22. $3 \cdot 3 \cdot 3 \cdot 3$ 23. $2 \cdot 2 \cdot 3 \cdot 3 \cdot 5$ 24. $2 \cdot 3 \cdot 5 \cdot 7 \cdot 7$ 25. $\frac{1}{4}$ 26. $\frac{2}{7}$ 27. $\frac{5}{6}$ 28. $\frac{12}{35}$ (already reduced to lowest terms) 29. $\frac{12}{5}$ 30. $\frac{9}{2}$ 31. 1 32. $\frac{12}{19}$ 33. $\frac{20}{27}$ 34. $\frac{10}{15}$ 35. $\frac{28}{32}$ 36. $\frac{15}{12}$ 37. $\frac{36}{30}$ 38. 9 39. 20 40. 80 41. 14 42. 55 43. 6 44. (a) $\frac{12}{21}$, or $\frac{4}{7}$ when reduced to lowest terms (b) $\frac{9}{21}$, or $\frac{3}{7}$ 45. (a) $\frac{55}{150}$, or $\frac{11}{30}$ when reduced (b) $\frac{95}{150}$, or $\frac{19}{30}$ when reduced

1.1 EXERCISES B

Answer true or false in Exercises 1–10. If the statement is false, tell why.

1. The objects that belong to a set are called the elements of the set.

2. A numeral is the name or symbol for a set.

3. $\{0, 1, 2, 3, \ldots\}$ is the set of natural numbers.

4. The denominator of the fraction $\frac{5}{7}$ is 5.

5. If the numerator of a fraction is equal to or larger than the denominator, it is a proper fraction.

6. Equivalent fractions name the same number or have the same value.

7. We reduce a fraction when we divide both numerator and denominator by a natural number other than 1.

8. Prime numbers are divisible only by themselves and 1.

9. The process of dividing common factors from a numerator and denominator is sometimes called reducing common factors.

10. A number expressed as a quotient of a whole number and a natural number is called a fraction.

11. Consider the fractions $\frac{1}{7}$, $\frac{3}{11}$, $\frac{8}{8}$, $\frac{15}{3}$, $\frac{17}{17}$, $\frac{21}{23}$, and $\frac{5}{1}$.
(a) Which of these are proper fractions? **(b)** Which of these are improper fractions?

Write three fractions that are equivalent to the following.

12. $\dfrac{3}{5}$

13. $\dfrac{7}{2}$

14. 8

15. 2

16. $\dfrac{95}{3}$

17. $\dfrac{1}{1000}$

18. List all prime numbers between 20 and 40.   *23, 29, 31, 37, 39*

Express as a product of primes.

19. 42

20. 20

21. 53

22. 625

23. 1000

24. 2541

Reduce to lowest terms.

25. $\dfrac{7}{42}$ *1/6*

26. $\dfrac{12}{28}$ *3/7*

27. $\dfrac{20}{32}$ *5/8*

28. $\dfrac{15}{77}$

29. $\dfrac{60}{35}$ *12/7*

30. $\dfrac{88}{16}$ *11/2*

31. $\dfrac{54}{54}$ *1*

32. $\dfrac{90}{147}$ *30/49*

33. $\dfrac{168}{228}$ *84/114 42/57*

34. Write an equivalent fraction for $\frac{3}{4}$ with 20 as the denominator.

35. Write an equivalent fraction for $\frac{5}{8}$ with 48 as the denominator.

36. Write an equivalent fraction for $\frac{7}{4}$ with 21 as the numerator.

37. Write an equivalent fraction for $\frac{8}{5}$ with 32 as the numerator.

Supply the missing numerator.

38. $\dfrac{2}{3} = \dfrac{?}{12}$

39. $\dfrac{3}{7} = \dfrac{?}{35}$

40. $\dfrac{12}{5} = \dfrac{?}{45}$

Supply the missing denominator.

41. $\dfrac{4}{7} = \dfrac{28}{?}$

42. $\dfrac{8}{11} = \dfrac{24}{?}$

43. $\dfrac{52}{3} = \dfrac{104}{?}$

44. A basketball player hit 21 free throws in 28 attempts during a recent tournament. What fractional part of the attempts **(a)** did he make? **(b)** did he miss?

45. Bill received $250 from his grandmother. He spent $120 to repair his car and the rest went for gasoline. What fractional part of the $250 went for **(a)** car repairs? **(b)** gasoline?

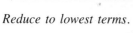

1.1 EXERCISES C

Reduce to lowest terms.

1. $\dfrac{632}{492}$ *316/224 = 158/112 = 79/56*

2. $\dfrac{246}{1230}$ *82/410   41/25*

3. $\dfrac{780}{5005}$ [Answer: $\frac{12}{77}$]

1.2 OPERATIONS ON FRACTIONS

STUDENT GUIDEPOSTS

1. Division by Zero
2. Multiplying Fractions
3. Reciprocals
4. Dividing Fractions

5. Least Common Denominator
6. Adding Fractions
7. Mixed Numbers
8. Subtracting Fractions

1 DIVISION BY ZERO

The four basic operations on fractions are addition ($+$), subtraction ($-$), multiplication ($\cdot$), and division ($\div$ or a fraction bar). We assume that these four operations applied to whole numbers are well understood. Division by zero, however, causes a problem. Why? The quotient of two numbers such as 8 and 2, written $8 \div 2$ or $\frac{8}{2}$, is the number which, when multiplied by 2, equals 8.

That is, $\dfrac{8}{2} = 4$ because $8 = 2 \cdot 4$.

Similarly, $\dfrac{20}{4} = 5$ because $20 = 4 \cdot 5$.

If $\dfrac{20}{0} = \text{(a number)},$ then $20 = 0 \cdot \text{(a number)}.$

But zero times any number is zero, not 20, so there is no number equal to $\frac{20}{0}$. Since we could have used any number in our example, no number can be divided by zero (except possibly zero itself). What would $\frac{0}{0}$ equal?

$\dfrac{0}{0}$ could be 5   since $0 = 0 \cdot 5$

$\dfrac{0}{0}$ could be 259   since $0 = 0 \cdot 259$

$\dfrac{0}{0}$ could be (any number)   since $0 \cdot \text{(any number)} = 0$

Since $\frac{0}{0}$ cannot be determined, we agree never to divide 0 by 0. Thus, *any division by 0 is said to be undefined* and is excluded from mathematics. However, $0 \div 5$ or $\frac{0}{5}$ does make sense, and in fact $\frac{0}{5} = 0$ (why?).

2 MULTIPLYING FRACTIONS

We now review the four operations on fractions starting with multiplication.

To Multiply Two (or More) Fractions

1. Factor all numerators and denominators into products of primes.
2. Place all numerator factors over all denominator factors.
3. Divide out or cancel all common factors from both numerator and denominator.
4. Multiply the remaining factors in the numerator and multiply the remaining factors in the denominator.
5. The resulting fraction is the product (reduced to lowest terms) of the original fractions.

EXAMPLE 1   MULTIPLYING FRACTIONS

Multiply.

(a) $\dfrac{3}{8} \cdot \dfrac{20}{9} = \dfrac{3}{2 \cdot 2 \cdot 2} \cdot \dfrac{2 \cdot 2 \cdot 5}{3 \cdot 3}$ Factor all numerators and denominators

$\qquad\qquad = \dfrac{3 \cdot 2 \cdot 2 \cdot 5}{2 \cdot 2 \cdot 2 \cdot 3 \cdot 3}$ All numerator factors over all denominator factors

PRACTICE EXERCISE 1

Multiply.

(a) $\dfrac{2}{15} \cdot \dfrac{5}{14}$

$$= \frac{\cancel{3} \cdot \cancel{2} \cdot 2 \cdot 5}{\cancel{2} \cdot 2 \cdot 2 \cdot \cancel{3} \cdot 3}$$ Divide out common factors

$$= \frac{5}{6}$$ Multiply the remaining factors

(b) $\dfrac{3}{7} \cdot 14 = \dfrac{3}{7} \cdot \dfrac{14}{1} = \dfrac{3}{7} \cdot \dfrac{2 \cdot 7}{1} = \dfrac{3 \cdot 2 \cdot \cancel{7}}{\cancel{7} \cdot 1} = \dfrac{6}{1} = 6$$

(b) $\dfrac{3}{8} \cdot 24$ $\; = \dfrac{3}{8} \cdot \dfrac{24}{1} = \dfrac{3}{2 \cdot 2 \cdot 2} \cdot \dfrac{2 \cdot 2 \cdot 3}{1} =$

$$\frac{3 \cdot \cancel{2} \cdot \cancel{2} \cdot 2 \cdot 3}{\cancel{2} \cdot \cancel{2} \cdot 2 \cdot 1} \quad \frac{9}{1}$$

Answers: (a) $\frac{1}{21}$ (b) 9

With practice we can shorten our work by not factoring completely to primes when larger common factors can be recognized. For example, in Example 1 (a) we might multiply as follows:

$$\frac{\overset{1}{\cancel{3}}}{\underset{2}{\cancel{8}}} \cdot \frac{\overset{5}{\cancel{20}}}{\underset{3}{\cancel{9}}} = \frac{5}{6}.$$ Divide out 4 and 3

③ RECIPROCALS

To divide fractions, we need to understand the idea of a *reciprocal*. The **reciprocal** of a fraction is the fraction formed by interchanging its numerator and denominator. Some fractions and their reciprocals are given in the following table.

Fraction	Reciprocal
$\dfrac{2}{3}$	$\dfrac{3}{2}$
$\dfrac{7}{8}$	$\dfrac{8}{7}$
$5 \left(\text{or } \dfrac{5}{1} \right)$	$\dfrac{1}{5}$
$\dfrac{1}{3}$	$3 \left(\text{or } \dfrac{3}{1} \right)$

Any fraction multiplied by its reciprocal gives us the number 1. For example,

$$\frac{2}{3} \cdot \frac{3}{2} = 1, \qquad \frac{7}{8} \cdot \frac{8}{7} = 1, \qquad \frac{5}{1} \cdot \frac{1}{5} = 1, \qquad \frac{1}{3} \cdot \frac{3}{1} = 1.$$

④ DIVIDING FRACTIONS

When one number is divided by a second, the first is the **dividend,** the second is the **divisor,** and the result is the **quotient.** In $10 \div 2 = 5$, 10 is the dividend, 2 the divisor, and 5 the quotient.

To Divide Two Fractions

1. Replace the divisor with its reciprocal and change the division sign to multiplication.
2. Multiply the fractions.

EXAMPLE 2 DIVIDING FRACTIONS

Divide.

(a) $\dfrac{13}{15} \div \dfrac{39}{5} = \dfrac{13}{15} \cdot \dfrac{5}{39}$ Replace divisor with its reciprocal and multiply

$= \dfrac{13}{3 \cdot 5} \cdot \dfrac{5}{3 \cdot 13}$ Factor

$= \dfrac{13 \cdot 5}{3 \cdot 5 \cdot 3 \cdot 13}$ Indicate products and divide out common factors

$= \dfrac{1}{9}$ A 1, not 0, remains in the numerator

(b) $\dfrac{4}{5} \div 44 = \dfrac{4}{5} \cdot \dfrac{1}{44} = \dfrac{2 \cdot 2}{5} \cdot \dfrac{1}{2 \cdot 2 \cdot 11} = \dfrac{2 \cdot 2 \cdot 1}{5 \cdot 2 \cdot 2 \cdot 11} = \dfrac{1}{55}$

(c) $\dfrac{3}{7} \div \dfrac{3}{7} = \dfrac{3}{7} \cdot \dfrac{7}{3} = \dfrac{3 \cdot 7}{7 \cdot 3} = \dfrac{1}{1} = 1$ Does this answer seem reasonable?

PRACTICE EXERCISE 2

Divide.

(a) $\dfrac{11}{6} \div \dfrac{44}{21} =$

(b) $\dfrac{3}{7} \div 6$

(c) $\dfrac{9}{13} \div \dfrac{9}{13}$

Answers: (a) $\frac{7}{8}$ (b) $\frac{1}{14}$ (c) 1

To add or subtract fractions with the same denominators, simply add numerators and place the result over the common denominator. For example,

$$\frac{3}{7} + \frac{5}{7} = \frac{3 + 5}{7} = \frac{8}{7}.$$

CAUTION

When adding fractions, we add numerators but *not* denominators.

5 LEAST COMMON DENOMINATOR

To add or subtract fractions with different denominators, change the fractions to equivalent fractions having a common denominator. For this it is a good idea to find the **least common denominator (LCD)** of the fractions, that is, the *smallest* number that has both denominators as factors. For example,

$$\frac{1}{2} + \frac{1}{3} \quad \text{can be changed to} \quad \frac{3}{6} + \frac{2}{6}.$$

Here 6 is the least common denominator of $\frac{1}{2}$ and $\frac{1}{3}$ since there is no smaller number that has both 2 and 3 as factors. Notice we could have used 12 or 18 as a common denominator, but 6 is the *least* common denominator.

To Find the LCD of Two (or More) Fractions

1. Factor each denominator into a product of primes.
2. If there are no common factors in the denominators, the LCD is the product of *all* denominators.
3. If there are common factors in the denominators, each factor must appear in the LCD as many times as it appears in the denominator where it is found the greatest number of times.

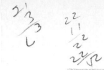

| EXAMPLE 3 FINDING THE LCD | PRACTICE EXERCISE 3 |

Find the LCD of the fractions.

(a) $\dfrac{1}{6}$ and $\dfrac{4}{15}$

Factor the denominators: $6 = 2 \cdot 3$ and $15 = 3 \cdot 5$. The LCD must consist of one 2, one 3, and one 5. Thus, the LCD $= 2 \cdot 3 \cdot 5 = 30$.

(b) $\dfrac{13}{90}$ and $\dfrac{7}{24}$

Factor the denominators: $90 = 2 \cdot 3 \cdot 3 \cdot 5$ and $24 = 2 \cdot 2 \cdot 2 \cdot 3$. The LCD must consist of three 2's, two 3's, and one 5. Thus, the LCD $= 2 \cdot 2 \cdot 2 \cdot 3 \cdot 3 \cdot 5 = 360$.

(c) $\dfrac{3}{10}$ and $\dfrac{5}{21}$

Factor the denominators: $10 = 2 \cdot 5$ and $21 = 3 \cdot 7$. Since there are no common factors, the LCD is the product of all factors. Thus, the LCD $= 2 \cdot 3 \cdot 5 \cdot 7 = 210$.

(d) $\dfrac{1}{2}, \dfrac{5}{6}$, and $\dfrac{7}{25}$

Factor the denominators: $2 = 2$, $6 = 2 \cdot 3$, and $75 = 3 \cdot 5 \cdot 5$. Thus, the LCD $= 2 \cdot 3 \cdot 5 \cdot 5 = 150$.

Find the LCD of the fractions.

(a) $\dfrac{5}{6}$ and $\dfrac{1}{21}$

(b) $\dfrac{1}{45}$ and $\dfrac{7}{20}$

(c) $\dfrac{2}{33}$ and $\dfrac{5}{14}$

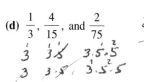

(d) $\dfrac{1}{3}, \dfrac{4}{15}$, and $\dfrac{2}{75}$

Answers: (a) 42   (b) 180
(c) 462   (d) 75

To see why the rule above works, let us look again at the fractions $\frac{1}{6}$ and $\frac{4}{15}$ in Example 3(a). By definition, each denominator must be a factor of the LCD. Since 6 is a factor of the LCD, the LCD must contain the factors 2 and 3. Also, with 15 a factor of the LCD, the LCD must contain the factors 3 and 5. Since we already have 3 as a factor from the denominator 6, we only need to supply the factor 5. Thus,

$$LCD = 2 \cdot 3 \cdot 5 = 30,$$

which is what we found in Example 3(a) using the rule above.

6 ADDING FRACTIONS

We now use the LCD to add fractions with different denominators.

> **To Add Two (or More) Fractions with Different Denominators**
>
> 1. Rewrite the sum with each denominator expressed as a product of primes.
> 2. Find the LCD.
> 3. Multiply the numerators and denominators of each fraction by all those factors present in the LCD but missing in the denominator of the particular fraction.
> 4. Place the sum of all numerators over the LCD.
> 5. Simplify the resulting numerator and reduce the fraction to lowest terms.

EXAMPLE 4 ADDING FRACTIONS

PRACTICE EXERCISE 4

Add.

(a) $\dfrac{5}{6} + \dfrac{7}{15} = \dfrac{5}{2\cdot3} + \dfrac{7}{3\cdot5}$ Factor denominators
the LCD is $2\cdot3\cdot5$

$= \dfrac{5\cdot5}{2\cdot3\cdot5} + \dfrac{7\cdot2}{3\cdot5\cdot2}$ Multiply numerators
and denominators by
missing factors

$= \dfrac{25+14}{2\cdot3\cdot5}$ Simplify and add over LCD

$= \dfrac{39}{2\cdot3\cdot5}$ Simplify

$= \dfrac{3\cdot13}{2\cdot3\cdot5}$ Factor numerator and divide out
common factors

$= \dfrac{13}{10}$ The desired sum

(b) $\dfrac{1}{2} + \dfrac{5}{6} + \dfrac{7}{25}$

$= \dfrac{1}{2} + \dfrac{5}{2\cdot3} + \dfrac{7}{5\cdot5}$ Factor: the LCD $= 2\cdot3\cdot5\cdot5$

$= \dfrac{1\cdot3\cdot5\cdot5}{2\cdot3\cdot5\cdot5} + \dfrac{5\cdot5\cdot5}{2\cdot3\cdot5\cdot5} + \dfrac{7\cdot2\cdot3}{5\cdot5\cdot2\cdot3}$ Multiply numerators
and denominators
by missing factors

$= \dfrac{75+125+42}{2\cdot3\cdot5\cdot5}$ Simplify and add over LCD

$= \dfrac{242}{2\cdot3\cdot5\cdot5}$ Simplify

$= \dfrac{2\cdot11\cdot11}{2\cdot3\cdot5\cdot5}$ Factor numerator and divide out common factors

$= \dfrac{121}{75}$ The desired sum

Add.

(a) $\dfrac{7}{15} + \dfrac{8}{21}$

(b) $\dfrac{2}{3} + \dfrac{1}{5} + \dfrac{5}{6}$

Answers: **(a)** $\frac{89}{105}$ **(b)** $\frac{17}{10}$

7 MIXED NUMBERS

It is sometimes helpful to write an improper fraction, such as $\frac{121}{75}$, as a *mixed number*. A **mixed number** is the sum of a whole number and a proper fraction. For example,

$$\dfrac{121}{75} \quad \text{can be written as} \quad 1 + \dfrac{46}{75} \quad \text{or} \quad 1\dfrac{46}{75}.$$

This is read ''one and forty-six seventy-fifths.'' To change an improper fraction to a mixed number, reduce the fraction to lowest terms and divide the denominator into the numerator. The quotient is the whole number part of the mixed number, and the fractional part is the remainder over the divisor.

EXAMPLE 5 IMPROPER FRACTION TO MIXED NUMBER	PRACTICE EXERCISE 5

Change each improper fraction to a mixed number.

(a) $\dfrac{13}{4}$

Divide 4 into 13.

$$4\overline{)13} \;\; \begin{array}{r} 3 \\ 12 \\ \hline 1 \end{array}$$

The mixed number is $3\frac{1}{4}$, read "three and one-fourth."

(b) $\dfrac{27}{6}$

Reduce $\frac{27}{6}$ to lowest terms and then divide.

$$\frac{27}{6} = \frac{9 \cdot 3}{2 \cdot 3} = \frac{9}{2} \qquad 2\overline{)9} \;\; \begin{array}{r} 4 \\ 8 \\ \hline 1 \end{array}$$

Thus, $\frac{27}{6} = 4\frac{1}{2}$.

Change each improper fraction to a mixed number.

(a) $\dfrac{25}{8}$

(b) $\dfrac{40}{15}$

Answers: (a) $3\frac{1}{8}$ (b) $2\frac{2}{3}$

$$\frac{2}{1} + \frac{11}{18} =$$

To change a mixed number to an improper fraction, we can think of the mixed number as a sum of two fractions. For example, the mixed number $2\frac{11}{18} = 2 + \frac{11}{18}$. Adding, we have

$$\frac{2}{1} + \frac{11}{18} = \frac{2}{1} + \frac{11}{2 \cdot 3 \cdot 3} = \frac{2 \cdot 2 \cdot 3 \cdot 3}{2 \cdot 3 \cdot 3} + \frac{11}{2 \cdot 3 \cdot 3} = \frac{36 + 11}{2 \cdot 3 \cdot 3} = \frac{47}{18}.$$

Rather than going through all of these steps, we can use a shortcut. Consider $2\frac{11}{18}$ again. Notice that if we multiply the denominator, 18, by the whole number, 2, and add the numerator, 11, we obtain

$$(18 \cdot 2) + 11 = 47.$$

When we put this result over the denominator, 18, we have the desired improper fraction $\frac{47}{18}$. We can show this as

$$\frac{(\text{denominator} \times \text{whole number}) + \text{numerator}}{\text{denominator}}.$$

EXAMPLE 6 MIXED NUMBER TO IMPROPER FRACTION	PRACTICE EXERCISE 6

Change each mixed number to an improper fraction.

(a) $4\dfrac{3}{5} = \dfrac{(5 \cdot 4) + 3}{5}$

$\quad = \dfrac{20 + 3}{5} = \dfrac{23}{5}$

(b) $6\dfrac{7}{10} = \dfrac{(10 \cdot 6) + 7}{10}$

$\quad = \dfrac{60 + 7}{10} = \dfrac{67}{10}$

Change each mixed number to an improper fraction.

(a) $6\dfrac{2}{7}$

(b) $3\dfrac{23}{100}$

Answers: (a) $\frac{44}{7}$ (b) $\frac{323}{100}$

To add two or more mixed numbers or mixed numbers and proper fractions, convert all mixed numbers to improper fractions and proceed as before.

EXAMPLE 7   ADDING MIXED NUMBERS

Add.

$$3\frac{1}{4} + 6\frac{7}{10} + \frac{1}{3} = \frac{13}{4} + \frac{67}{10} + \frac{1}{3} \qquad \text{Convert to improper fractions}$$

$$= \frac{13}{2 \cdot 2} + \frac{67}{2 \cdot 5} + \frac{1}{3} \qquad \text{The LCD} = 2 \cdot 2 \cdot 3 \cdot 5$$

$$= \frac{13 \cdot 3 \cdot 5}{2 \cdot 2 \cdot 3 \cdot 5} + \frac{67 \cdot 2 \cdot 3}{2 \cdot 5 \cdot 2 \cdot 3} + \frac{1 \cdot 2 \cdot 2 \cdot 5}{3 \cdot 2 \cdot 2 \cdot 5}$$

$$= \frac{195 + 402 + 20}{2 \cdot 2 \cdot 3 \cdot 5}$$

$$= \frac{617}{60} \text{ or } 10\frac{17}{60}$$

PRACTICE EXERCISE 7

Add.

$$2\frac{2}{3} + 5\frac{1}{10} + \frac{1}{5}$$

Answer: $\frac{239}{30}$, or $7\frac{29}{30}$

---

8 SUBTRACTING FRACTIONS

The fourth operation we consider is subtraction.

> ### To Subtract Two Fractions
>
> 1. Rewrite the difference with each denominator expressed as a product of prime factors.
> 2. Find the LCD.
> 3. Supply missing factors just as when adding.
> 4. Place the difference of the numerators over the LCD.
> 5. Simplify the resulting numerator and reduce the fraction to lowest terms.

EXAMPLE 8   SUBTRACTING FRACTIONS

Subtract.

(a) $\dfrac{7}{12} - \dfrac{5}{9} = \dfrac{7}{2 \cdot 2 \cdot 3} - \dfrac{5}{3 \cdot 3}$ Factor; LCD is $2 \cdot 2 \cdot 3 \cdot 3$

$$= \frac{7 \cdot 3}{2 \cdot 2 \cdot 3 \cdot 3} - \frac{5 \cdot 2 \cdot 2}{3 \cdot 3 \cdot 2 \cdot 2} \qquad \text{Supply missing factors}$$

$$= \frac{21 - 20}{2 \cdot 2 \cdot 3 \cdot 3} \qquad \text{Subtract and simplify}$$

$$= \frac{1}{36} \qquad \text{The difference in reduced form}$$

PRACTICE EXERCISE 8

Subtract.

(a) $\dfrac{8}{35} - \dfrac{1}{21}$

(b) $4\dfrac{1}{2} - 3\dfrac{7}{8}$

First convert the mixed numbers to improper fractions.

$$4\dfrac{1}{2} - 3\dfrac{7}{8} = \dfrac{9}{2} - \dfrac{31}{8} = \dfrac{9 \cdot 4}{2 \cdot 4} - \dfrac{31}{8} = \dfrac{36 - 31}{8} = \dfrac{5}{8}$$

(b) $6\dfrac{1}{4} - 2\dfrac{6}{7}$

Answers: (a) $\frac{19}{105}$ (b) $\frac{95}{28}$, or $3\frac{11}{28}$

Notice in Example 8(b) that we took a shortcut and did not factor denominators into primes. It was clear that the LCD of the two fractions is 8, so we simply supplied the factor of 4 to both numerator and denominator of $\frac{9}{2}$. With practice, shortcuts such as this can save time.

The rules for adding and subtracting fractions can be extended to include combinations of these operations, as shown in the next example.

EXAMPLE 9   COMBINATION OF OPERATIONS

Perform the indicated operations.

$$2\dfrac{1}{15} + 4\dfrac{1}{5} - 3\dfrac{1}{2}$$

$$= \dfrac{31}{15} + \dfrac{21}{5} - \dfrac{7}{2} \qquad \text{Convert to improper fractions}$$

$$= \dfrac{31}{3 \cdot 5} + \dfrac{21}{5} - \dfrac{7}{2} \qquad \text{The LCD} = 2 \cdot 3 \cdot 5$$

$$= \dfrac{31 \cdot 2}{3 \cdot 5 \cdot 2} + \dfrac{21 \cdot 2 \cdot 3}{5 \cdot 2 \cdot 3} - \dfrac{7 \cdot 3 \cdot 5}{2 \cdot 3 \cdot 5} \qquad \text{Supply missing factors}$$

$$= \dfrac{62 + 126 - 105}{2 \cdot 3 \cdot 5}$$

$$= \dfrac{83}{30} \text{ or } 2\dfrac{23}{30}$$

PRACTICE EXERCISE 9

Perform the indicated operations.

$$7\dfrac{1}{15} - 2\dfrac{2}{3} + 1\dfrac{1}{2}$$

Answers: $\frac{59}{10}$, or $5\frac{9}{10}$

1.2   EXERCISES A

1. $\dfrac{0}{3} =$ _____

2. $\dfrac{3}{0} =$ _____

3. $\dfrac{0}{0} =$ _____

Multiply.

4. $\dfrac{1}{3} \cdot \dfrac{6}{5}$

5. $\dfrac{1}{2} \cdot \dfrac{3}{7}$

6. $\dfrac{6}{35} \cdot \dfrac{20}{12}$

7. $\dfrac{12}{5} \cdot 10$

* **8** $4\dfrac{2}{5} \cdot \dfrac{3}{11}$

9. $2\dfrac{2}{3} \cdot 1\dfrac{1}{8} \cdot \dfrac{1}{6}$

---

*A colored number means there is a complete solution in the back of the text. See *To the Student* for more details.

Divide.

10. $\dfrac{3}{7} \div \dfrac{9}{28}$

11. $\dfrac{\frac{20}{9}}{\frac{2}{3}}$

12. $\dfrac{20}{27} \div \dfrac{35}{36}$

13. $\dfrac{8}{9} \div 4$

14 $3\dfrac{1}{5} \div \dfrac{2}{5}$

15. $3\dfrac{1}{3} \div 1\dfrac{3}{5}$

Find the LCD of the following fractions.

16. $\dfrac{2}{7}$ and $\dfrac{4}{33}$

17. $\dfrac{7}{12}$ and $\dfrac{1}{45}$

18. $\dfrac{3}{16}$ and $\dfrac{7}{20}$

19. 7 and $\dfrac{9}{2}$

20. $\dfrac{2}{3}, \dfrac{3}{4}$, and $\dfrac{1}{5}$

21. $\dfrac{1}{9}, \dfrac{5}{6}$, and $\dfrac{1}{75}$

Add.

22. $\dfrac{1}{2} + \dfrac{1}{2}$

23. $\dfrac{1}{6} + \dfrac{3}{10}$

24. $\dfrac{3}{28} + \dfrac{13}{70}$

25. $4 + \dfrac{3}{5}$

26. $2\dfrac{1}{5} + 1\dfrac{2}{3}$

27. $\dfrac{2}{3} + \dfrac{1}{6} + \dfrac{3}{4}$

Change each improper fraction to a mixed number.

28. $\dfrac{23}{8}$

29. $\dfrac{135}{4}$

30. $\dfrac{142}{11}$

Change each mixed number to an improper fraction.

31. $4\dfrac{2}{5}$

32. $66\dfrac{2}{3}$

33. $9\dfrac{7}{11}$

Subtract.

34. $\dfrac{3}{4} - \dfrac{3}{4}$

35. $\dfrac{7}{11} - \dfrac{2}{7}$

36. $\dfrac{7}{15} - \dfrac{13}{35}$

37. $\dfrac{19}{2} - 3$

38. $5\dfrac{1}{3} - 2\dfrac{3}{4}$

39. $11\dfrac{4}{5} - 5\dfrac{2}{3}$

Perform the indicated operations.

40. $\dfrac{7}{20} + \dfrac{3}{8} - \dfrac{1}{4}$

41. $\dfrac{14}{15} - \dfrac{2}{5} - \dfrac{1}{3}$

42. $\dfrac{7}{3} + \dfrac{1}{7} - 2$

Solve.

43 If $\frac{5}{6}$ of a class consists of girls and there are 72 students in the class, how many students are girls? How many are boys?

44. On a map, 1 inch represents $\frac{1}{2}$ of a mile. How many miles are represented by $4\frac{1}{4}$ inches?

45. A piece of rope 20 m long is to be cut into pieces, each of whose length is $\frac{2}{3}$ m. How many pieces can be cut?

46 A fuel tank holds $12\frac{1}{2}$ gallons when it is $\frac{3}{4}$ full. What is the capacity of the tank?

47. Alphonso hiked from Whiskey Creek to Bald Mountain, a distance of $2\frac{1}{3}$ mi. From there he hiked to Spook Hollow, a distance of $4\frac{1}{5}$ mi. How far did he hike?

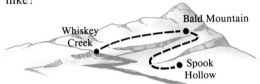

48. To get the right shade of paint for her living room, Tracy Bell mixed $\frac{7}{8}$ of a gallon of white paint with $\frac{2}{3}$ of a gallon of light blue paint. How much paint did she have?

49. The weight of one cubic foot of water is $62\frac{1}{2}$ lb. How much do $3\frac{1}{5}$ cubic feet of water weigh?

50. Fahrenheit temperature can be found by multiplying Celsius temperature by $\frac{9}{5}$ and adding $32°$. If Celsius temperature is $15°$, what is Fahrenheit temperature?

FOR REVIEW

51. Write an equivalent fraction for $\frac{7}{8}$ with
(a) 21 for the numerator.
(b) 72 for the denominator.

52. Max received $125 for doing a particular job. If he had to pay $15 for income tax, what fractional part of the $125 went for tax?

ANSWERS: **1.** 0   **2.** undefined   **3.** undefined   **4.** $\frac{2}{5}$ **5.** $\frac{3}{14}$ **6.** $\frac{2}{7}$ **7.** 24   **8.** $\frac{6}{5}$ **9.** $\frac{1}{2}$ **10.** $\frac{4}{3}$ **11.** $\frac{10}{3}$ **12.** $\frac{16}{21}$ **13.** $\frac{2}{9}$
14. 8   **15.** $\frac{25}{12}$ **16.** 231   **17.** 180   **18.** 80   **19.** 2   **20.** 60   **21.** 450   **22.** 1   **23.** $\frac{7}{15}$ **24.** $\frac{41}{140}$ **25.** $\frac{23}{5}$ **26.** $\frac{58}{15}$ **27.** $\frac{19}{12}$
28. $2\frac{7}{8}$ **29.** $33\frac{3}{4}$ **30.** $12\frac{10}{11}$ **31.** $\frac{22}{5}$ **32.** $\frac{200}{3}$ **33.** $\frac{106}{11}$ **34.** 0   **35.** $\frac{27}{77}$ **36.** $\frac{2}{21}$ **37.** $\frac{13}{2}$ **38.** $2\frac{7}{12}$ **39.** $6\frac{2}{15}$ **40.** $\frac{19}{40}$
41. $\frac{1}{5}$ **42.** $\frac{10}{21}$ **43.** 60 girls, 12 boys   **44.** $2\frac{1}{8}$ mi   **45.** 30   **46.** $16\frac{2}{3}$ gallons   **47.** $6\frac{8}{15}$ mi   **48.** $1\frac{13}{24}$ gallons   **49.** 200 lb
50. $59°$ **51.** (a) $\frac{21}{24}$ (b) $\frac{63}{72}$ **52.** $\frac{3}{25}$

1.2 EXERCISES B

1. $\dfrac{7}{0} = $ ___?___

2. $\dfrac{0}{0} = $ ___?___

3. $\dfrac{0}{7} = $ ___?___

Multiply.

4. $\dfrac{2}{3} \cdot \dfrac{6}{5}$

5. $\dfrac{3}{8} \cdot \dfrac{1}{2}$

6. $\dfrac{7}{35} \cdot \dfrac{20}{14}$

7. $\dfrac{11}{4} \cdot 8$

8. $4\dfrac{4}{5} \cdot \dfrac{1}{12}$

9. $1\dfrac{1}{3} \cdot 3\dfrac{3}{8} \cdot \dfrac{2}{3}$

Divide.

10. $\dfrac{2}{7} \div \dfrac{6}{28}$

11. $\dfrac{\frac{15}{8}}{\frac{3}{4}}$

12. $\dfrac{10}{27} \div \dfrac{15}{36}$

13. $6 \div \dfrac{8}{11}$

14. $2\dfrac{3}{5} \div \dfrac{26}{5}$

15. $2\dfrac{2}{3} \div 1\dfrac{1}{5}$

Find the LCD of the following fractions.

16. $\dfrac{1}{9}$ and $\dfrac{3}{22}$

17. $\dfrac{5}{12}$ and $\dfrac{2}{27}$

18. $\dfrac{1}{18}$ and $\dfrac{3}{20}$

19. $\dfrac{2}{9}$ and 5

20. $\dfrac{1}{3}, \dfrac{3}{4},$ and $\dfrac{2}{7}$

21. $\dfrac{1}{6}, \dfrac{5}{9},$ and $\dfrac{7}{30}$

Add.

22. $\dfrac{1}{4} + \dfrac{3}{4}$

23. $\dfrac{1}{8} + \dfrac{5}{12}$

24. $\dfrac{5}{24} + \dfrac{7}{60}$

25. $\dfrac{1}{8} + 2$

26. $4\dfrac{1}{5} + 2\dfrac{1}{3}$

27. $\dfrac{1}{3} + \dfrac{5}{6} + \dfrac{1}{5}$

Change each improper fraction to a mixed number.

28. $\dfrac{81}{10}$

29. $\dfrac{245}{9}$

30. $\dfrac{167}{11}$

Change each mixed number to an improper fraction.

31. $10\dfrac{1}{5}$

32. $42\dfrac{3}{10}$

33. $8\dfrac{4}{11}$

Subtract.

34. $\dfrac{5}{8} - \dfrac{1}{4}$

35. $\dfrac{8}{15} - \dfrac{12}{35}$

36. $5 - \dfrac{1}{9}$

37. $\dfrac{9}{2} - 1\dfrac{1}{4}$

38. $6\dfrac{2}{3} - 1\dfrac{3}{4}$

39. $11\dfrac{3}{5} - 4\dfrac{1}{3}$

Perform the indicated operations.

40. $\dfrac{9}{20} + \dfrac{5}{8} - \dfrac{3}{4}$

41. $\dfrac{13}{15} - \dfrac{1}{5} - \dfrac{1}{3}$

42. $\dfrac{8}{3} + \dfrac{2}{7} - 2$

Solve.

43. Raoul took out a loan for $\frac{2}{3}$ of the cost of a motorbike. If the bike cost \$744, how much did he borrow?

44. Sandi can hike $3\frac{1}{2}$ mi in one hour. How many miles can she hike in $\frac{3}{4}$ of an hour?

45. When Fred Vickers pounds a nail into a board, he sinks it $1\frac{1}{4}$ inches with each blow. How many times will he have to hit a 5-in nail to drive it completely into the board?

46. After driving 285 miles, the Stotts had completed $\frac{5}{8}$ of their trip. What was the total length of the trip?

47. In July, Memphis had two rain storms, one that dropped $1\frac{1}{4}$ in of rain and another that dropped $\frac{3}{5}$ in. How much rain did Memphis receive in July?

48. It took Burford $8\frac{1}{2}$ hr to fix his car himself. If he had the money, he could have hired a mechanic to do the job in $1\frac{1}{3}$ hr. How many hours of work could Burford have saved?

49. Cindy Brown is paid $4\frac{1}{2}$ dollars per hr to type a manuscript. If she types for a period of $20\frac{2}{3}$ hr, how much is she paid?

50. Celsius temperature can be found by subtracting 32° from Fahrenheit temperature and multiplying the result by $\frac{5}{9}$. If Fahrenheit temperature is 122°, what is the corresponding Celsius temperature?

FOR REVIEW

51. Write an equivalent fraction for $\frac{8}{9}$ with **(a)** 32 as the numerator, and **(b)** 81 as the denominator.

52. A farmer collected 280 bushels of grain from his field. If he had to give 40 bushels to his landlord for rent, what fractional part of the 280 bushels went for rent?

1.2 EXERCISES C

Perform the indicated operations.

1. $22\dfrac{5}{8} + 17\dfrac{3}{10} - 8\dfrac{3}{4}$

$\left[\text{Answer: } 31\frac{7}{40}\right]$

2. $42\dfrac{7}{15} - 18\dfrac{5}{6} - 11\dfrac{3}{10}$

3. $125\dfrac{16}{25} + 37\dfrac{2}{15} - 83\dfrac{7}{20}$

1.3 DECIMALS

> **STUDENT GUIDEPOSTS**
> 1. Place-Value Number System
> 2. Decimal Notation
> 3. Operations on Decimals

1 PLACE-VALUE NUMBER SYSTEM

Remember that our number system is a **place-value system.** That is, each **digit** (0, 1, 2, 3, 4, 5, 6, 7, 8, or 9) in a numeral has a particular value determined by its location or place in the symbol. For example, the numeral

$$125.378$$

is a shorthand symbol for the **expanded numeral**

$$100 + 20 + 5 + \frac{3}{10} + \frac{7}{100} + \frac{8}{1000},$$

or $$1 \cdot 100 + 2 \cdot 10 + 5 \cdot 1 + 3 \cdot \frac{1}{10} + 7 \cdot \frac{1}{100} + 8 \cdot \frac{1}{1000},$$

where each of the digits, 1, 2, 5, 3, 7, and 8 is multiplied by 100, 10, 1, $\frac{1}{10}$, $\frac{1}{100}$, or $\frac{1}{1000}$, depending upon the location of the digit in the symbol 125.378. We read 125.378 as "one hundred twenty-five *and* three hundred seventy-eight thousandths." This results from thinking of the number as a mixed number

$$125 + \frac{378}{1000}, \quad \text{or} \quad 125\frac{378}{1000}.$$

2 DECIMAL NOTATION

The numeral 125.378, representing the mixed number $125\frac{378}{1000}$, is in **decimal notation,** and numbers expressed in this manner are referred to simply as **decimals.** Decimal notation is based on the number 10, and the period used in a decimal is the **decimal point.** If there are no digits to the right of the decimal point we usually omit it; we write 53 rather than 53., for example. The diagram below gives the value of each position. Notice that the decimal point separates the ones digit from the tenths digit.

Hundreds	Tens	Ones	Decimal Point	Tenths	Hundredths	Thousandths
1	2	5	.	3	7	8

The following table expresses each decimal in expanded notation, as a mixed number, and as a proper or improper fraction.

Decimal	Expanded notation	Mixed number	Fraction
1.25	$1 + \dfrac{2}{10} + \dfrac{5}{100}$	$1\dfrac{25}{100}$	$\dfrac{125}{100}$
23.46	$20 + 3 + \dfrac{4}{10} + \dfrac{6}{100}$	$23\dfrac{46}{100}$	$\dfrac{2346}{100}$
43.278	$40 + 3 + \dfrac{2}{10} + \dfrac{7}{100} + \dfrac{8}{1000}$	$43\dfrac{278}{1000}$	$\dfrac{43{,}278}{1000}$
562.34	$500 + 60 + 2 + \dfrac{3}{10} + \dfrac{4}{100}$	$562\dfrac{34}{100}$	$\dfrac{56{,}234}{100}$
0.23	$\dfrac{2}{10} + \dfrac{3}{100}$	$0\dfrac{23}{100}$	$\dfrac{23}{100}$
47 or 47.0	$40 + 7$ or $40 + 7 + \dfrac{0}{10}$	47 or $47\dfrac{0}{10}$	$\dfrac{47}{1}$ or $\dfrac{470}{10}$

③ OPERATIONS ON DECIMALS

We have already reviewed the operations of adding, subtracting, multiplying, and dividing numbers in fractional notation. Now we review these operations when the numbers are in decimal notation.

To Add Two (or More) Decimals

1. Arrange the numbers in a column so that the decimal points line up vertically.
2. Add in columns from right to left as if there were no decimal points.
3. Place the decimal point in the sum in line with the other decimal points.

EXAMPLE 1   ADDING DECIMALS

Add.

(a)
```
  21.3
+  4.2
  25.5
```

(b)
```
 123.417
  45.
+  1.14
 169.557
```

PRACTICE EXERCISE 1

Add.

(a)
```
  43.8
+  6.1
```

(b)
```
  10.001
   0.52
+129.007
```

Answers: (a) **49.9** (b) **139.528**

To Subtract Two Decimals

1. Arrange the numbers in a column so that the decimal points line up vertically.
2. Subtract in columns from right to left as if there were no decimal points. (Additional zeros may be placed to the right of the decimal points if desired.)
3. Place the decimal point in the difference in line with the other decimal points.

EXAMPLE 2 SUBTRACTING DECIMALS	PRACTICE EXERCISE 2

Subtract.

(a) 27.5
   − 2.4
    25.1

(b) 421.3000
   −   8.9999
    412.3001

(In (b), the three zeros to the right of 3 have been supplied.)

Subtract.

(a) 51.2
   − 9.8

(b) 654.2
   − 26.888

Answers: (a) 41.4 (b) 627.312

We will agree to say that 42.7 has *one decimal place*, 12.16 has *two decimal places*, 5.006 has *three decimal places,* and so forth.

To Multiply Two Decimals

1. Ignore the decimal points and multiply as if the numbers were whole numbers.
2. Place the decimal point so the product has the same number of decimal places as the *sum* of the number of decimal places in the factors.

EXAMPLE 3 MULTIPLYING DECIMALS	PRACTICE EXERCISE 3

Multiply.

(a) 1.23   2 decimal places
  × 2.1   1 decimal place
    123
    246
    2.583   2 + 1 = 3 decimal places

(b) 42.103   3 decimal places
  ×   2.15   2 decimal places
    210515
    42103
    84206
    90.52145   3 + 2 = 5 decimal places

(c) (0.02)(0.013) = 0.00026   5 decimal places

Multiply.

(a) 41.3
  × 2.11

(b) 6.003
  × 1.09

(c) (0.004)(0.0255)

Answers: (a) 87.143 (b) 6.54327
(c) 0.000102

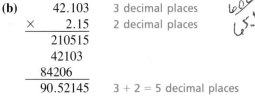

To Divide One Decimal into Another

1. Move the decimal point in the divisor (the dividing number) to the right until the divisor becomes a whole number.
2. Move the decimal point in the dividend (the number divided into) to the right the same number of places adding zeros as necessary.
3. Divide as if the numbers were whole numbers.
4. Place the decimal point in the quotient directly above the decimal point in the dividend.

EXAMPLE 4 DIVIDING DECIMALS

Divide.

(a) $0.12\overline{)14.4}$

$$0.12\underset{2}{\overbrace{}}\overline{)14.\underset{2}{\underbrace{40}}} \quad \text{becomes} \quad 12\overline{)1440}$$

$$\begin{array}{r} 120. \\ 12\overline{)1440} \\ \underline{12} \\ 24 \\ \underline{24} \\ 0 \end{array}$$

(handwritten)
$$\begin{array}{r}120.\\12\,\overline{)1440}\\\underline{12}\\240\end{array}$$

(b) $0.003\overline{)42.3}$

$$0.003\underset{3}{\underbrace{}}\overline{)42.\underset{3}{\underbrace{300}}} \quad \text{becomes} \quad 3\overline{)42,300} = 14,100.$$

PRACTICE EXERCISE 4

Divide.

(a) $2.1\overline{)28.35}$

(b) $0.0002\overline{)1.5448}$

Answers: (a) 13.5 (b) 7724

We need to know how to operate on decimals since these techniques are used often in algebra. However, using a calculator can eliminate much of the tedious, time-consuming work with decimal operations. If you have a calculator, be sure to read the instruction manual that comes with your model. You might use a calculator to check part of your work in some of the following exercises.

1.3 EXERCISES A

Express in expanded notation.

1. 3.47

2. 107.28

3. 203.005

Express in decimal notation.

4. $400 + 20 + 7 + \dfrac{3}{10}$

5. $500 + 3 + \dfrac{2}{10} + \dfrac{8}{1000}$

6. $4000 + 30 + \dfrac{2}{100} + \dfrac{5}{1000}$

Add.

7. $\begin{array}{r}31.4\\+\ 1.07\\\hline 32.47\end{array}$

8. $\begin{array}{r}403.1\\+\ 21.08\\\hline 424.18\end{array}$

9. $\begin{array}{r}201.005\\30.2\\+\ 1.07\\\hline 232.275\end{array}$

10. $23.4 + 9.75 = 33.15$

11. $512.3 + 2.009 + 13 = 514.309$

12. $15.0001 + 9.3 + 0.004 = 24.3041$

Subtract.

13. $\begin{array}{r}4.0003\\-\ 1.1000\\\hline 2.9003\end{array}$

14. $\begin{array}{r}427.006\\-\ 135.040\\\hline 291.966\end{array}$

15. $\begin{array}{r}36.5000\\-\ 0.0007\\\hline 36.4993\end{array}$

16. $46.01 - 13.4$

17. $923.006 - 23$

18. $47.2 - 0.0003$

Multiply.

19. $\begin{array}{r} 3.42 \\ \times\ 2.1 \\ \hline \end{array}$

342
684
7.182

20. $\begin{array}{r} 15.207 \\ \times\ 3.12 \\ \hline \end{array}$

21. $\begin{array}{r} 0.0034 \\ \times\ 0.209 \\ \hline \end{array}$

00216
00000
00068
00000

22. $(0.003)(0.003)$

23. $(0.02)(0.02)(0.02)$

24. $(0.0001)(0.00001)$

00007016

Divide.

25. $2.1\overline{)70.14}$

33.4
21)7014
63
71
63
84

26. $0.007\overline{)38.717}$

5534.8
7)38717.
35
37
35
27
27
60

27. $20.4\overline{)0.687072}$

28. $8\overline{)7.0}$

29. $2\overline{)1.0}$

30. $9\overline{)1.0}$

Solve.

31. Before leaving on a trip, Harley's odometer read 23,411.8. If he drove 723.9 mi on the trip, what did it read when he returned?

23411.8
723.9
24135.7

32. The Northern Arizona four-mile relay team members recorded times of 4.23, 4.17, 3.99, and 4.08 minutes. What was their combined time for the race?

33. Roseanne Strauch buys a pair of hose for $3.95. If she pays with a $20 bill and the sales tax amounts to $0.20, how much change should she receive?

3.95
.20
4.15

20.00
- 4.15
15.85

34. When Kimmie was sick, she ran a temperature of 103.2° Fahrenheit. If normal body temperature is 98.6° Fahrenheit, how many degrees above normal was her temperature?

103.2
98.6
4.6

35. Jose is paid $7.35 per hr. If he worked a total of 35.2 hr last week, how much was he paid?

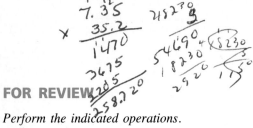

36. How many months will it take to pay off a loan of $6562.80 if the monthly payments are $182.30?

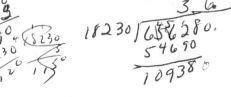

FOR REVIEW

Perform the indicated operations.

37. $\dfrac{2}{9} + \dfrac{5}{12}$

38. $\dfrac{23}{36} - \dfrac{4}{9}$

39. $\dfrac{3}{4} \cdot \dfrac{16}{27}$

40. $1\dfrac{1}{12} \div 9\dfrac{3}{4}$

Solve.

41 Nancy Chandler plans a two-day hike from Gopher Gulch to Aspen Glen by way of North Face. It is $7\frac{3}{4}$ mi from Gopher Gulch to North Face and $9\frac{1}{8}$ mi from there to Aspen Glen. If she plans to travel half the total distance on each day, how far will she hike on the first day?

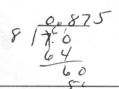

42. Julio has a rope $26\frac{2}{3}$ meters in length. He wishes to cut the rope into 3 pieces of equal length. What will be the length of each piece?

ANSWERS: 1. $3 + \frac{4}{10} + \frac{7}{100}$ 2. $100 + 7 + \frac{2}{10} + \frac{8}{100}$ 3. $200 + 3 + \frac{5}{1000}$ 4. 427.3 5. 503.208 6. 4030.025 7. 32.47 8. 424.18 9. 232.275 10. 33.15 11. 527.309 12. 24.3041 13. 2.9003 14. 291.966 15. 36.4993 16. 32.61 17. 900.006 18. 47.1997 19. 7.182 20. 47.44584 21. 0.0007106 22. 0.000009 23. 0.000008 24. 0.000000001 25. 33.4 26. 5531 27. 0.03368 28. 0.875 29. 0.5 30. 0.111 . . . 31. 24,135.7 32. 16.47 minutes 33. $15.85 34. 4.6° 35. $258.72 36. 36 months 37. $\frac{23}{36}$ 38. $\frac{7}{36}$ 39. $\frac{4}{9}$ 40. $\frac{1}{9}$ 41. $8\frac{7}{16}$ mi 42. $8\frac{8}{9}$ meters

1.3 EXERCISES B

Express in expanded notation.

1. 35.6

2. 257.663

3. 3215.802

Express in decimal notation.

4. $700 + 30 + 8 + \frac{6}{10}$

5. $4000 + 6 + \frac{7}{100}$

6. $300 + 40 + \frac{2}{100} + \frac{7}{1000}$

Add.

7. 52.3
  $+\ \ 2.09$

8. 602.9
  $+\ \ 43.07$

9. 421.544
  2.8
  $+\ \ 43.09$

10. $42.6 + 8.41$

11. $631.9 + 5.007 + 21$

12. $12.0003 + 7.2 + 0.003$

Subtract.

13. 6.0004
  $-\ 2.3$

14. 637.008
  $-\ 123.05$

15. 51.6
  $-\ \ 0.0004$

16. $43.02 - 17.1$

17. $843.002 - 51$

18. $39.4 - 0.0006$

Multiply.

19. 2.91
  $\times\ \ 3.4$

20. 17.603
  $\times\ \ 4.15$

21. 0.0053
  $\times\ 0.402$

22. (0.002)(0.002)   **23.** (0.03)(0.03)(0.03)   **24.** (0.0002)(0.00002)

Divide.

25. $4.2\overline{)94.5}$ **26.** $0.003\overline{)13.014}$ **27.** $30.1\overline{)6.52267}$

28. $8\overline{)5.0}$ **29.** $4\overline{)1.0}$ **30.** $9\overline{)4.0}$

Solve.

31. The Saturday and Sunday receipts at Waldo's Waffle Palace were $621.13 and $743.19, respectively. How much money did Waldo take in on that weekend?

32. What is the **perimeter** of (the distance around) a four-sided figure with sides of 6.031 m, 5.411 m, 4.003 m, and 5.772 m?

33. Sam Passamonte had a balance of $635.86 in his checking account when he wrote a check for $249.93. What was his new balance?

34. When George filled the gas tank on his Bronco, the odometer read 62,421.8. The next time he filled it, the odometer read 62,914.2. How far had he driven?

35. Betty Sue bought 13.4 pounds of peaches at $0.55 per pound. How much did she pay for the peaches?

36. Kate Williams paid $58.00 to rent a car for two days. If the rental fee amounts to $12.50 per day plus $0.15 per mile driven, how far did she drive the car?

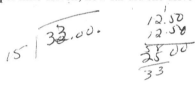

FOR REVIEW

Perform the indicated operations.

37. $\dfrac{12}{15} + \dfrac{8}{35}$ **38.** $\dfrac{13}{24} - \dfrac{3}{8}$ **39.** $1\dfrac{2}{3} \cdot 4\dfrac{1}{5}$ **40.** $\dfrac{6}{5} \div \dfrac{24}{45}$

Solve.

41. A wheel on a child's car is turning at a rate of $40\frac{1}{2}$ revolutions per minute. If the car is driven for $3\frac{1}{9}$ minutes, how many revolutions does the wheel make?

42. Marguerite is proofing a computer program that is $14\frac{1}{3}$ pages in length. If she has completed $10\frac{1}{5}$ pages, how much remains to be done?

1.3 EXERCISES C

Perform the indicated operation. A calculator would be useful for these problems.

1.   374.26
      × 8.603

2. $0.0257\overline{)0.02584392}$ [Answer: 1.0056]

1.4 CONVERTING FRACTIONS, DECIMALS, AND PERCENTS

STUDENT GUIDEPOSTS

1 Converting Fractions to Decimals
2 Rational and Irrational Numbers
3 Converting Terminating Decimals to Fractions
4 Conversions Involving Percent

1 CONVERTING FRACTIONS TO DECIMALS

Every fraction can be converted to a decimal. Several problems in the previous section gave a preview of this process.

To Convert a Fraction to a Decimal

1. Divide the denominator of the fraction into the numerator.
2. Place the decimal point as in any decimal division problem.

| EXAMPLE 1 CONVERTING A FRACTION TO A DECIMAL | PRACTICE EXERCISE 1 |

Convert each fraction to a decimal.

(a) $\dfrac{3}{20}$ We divide 20 into 3 as follows:

$$\begin{array}{r} .15 \\ 20\overline{)3.00} \\ \underline{2\ 0} \\ 1\ 00 \\ \underline{1\ 00} \\ 0 \end{array}$$ $\frac{3}{20} = 0.15$

(b) $\dfrac{1}{3}$

$$\begin{array}{r} .333\ \dots \\ 3\overline{)1.000} \\ \underline{9} \\ 10 \\ \underline{9} \\ 10 \\ \underline{9} \\ 1 \end{array}$$

It is clear that we will continue to obtain "3" on each division. When this happens we write $\frac{1}{3} = 0.333 \dots = 0.\overline{3}$, where the bar denotes the digit (sometimes digits) that continues to repeat. For the mixed decimal number $3\frac{1}{3}$ we write $3.\overline{3}$ (not $\overline{3}$).

(c) $\dfrac{2}{7}$

$$\begin{array}{r} .285714285714\ \dots \\ 7\overline{)2.000000000000} \end{array}$$

$$\frac{2}{7} = 0.285714\mathbf{285714} \dots$$

$$= 0.\overline{285714}$$

Convert each fraction to a decimal.

(a) $\dfrac{1}{2}$

(b) $\dfrac{13}{40}$

(c) $\dfrac{2}{3}$

Answers: (a) 0.5   (b) 0.325
(c) 0.$\overline{6}$

It is a property of every fraction that in decimal notation, the digits after the decimal either terminate, as in

$$\frac{1}{2} = 0.5, \qquad \frac{3}{20} = 0.15, \qquad \frac{13}{40} = 0.325,$$

or contain a block of digits that repeats without ending, as in

$$\frac{1}{3} = 0.\overline{3}, \qquad \frac{2}{7} = 0.\overline{285714}, \qquad \frac{2}{3} = 0.\overline{6}.$$

❷ RATIONAL AND IRRATIONAL NUMBERS

It is also true that if a number has a decimal representation that terminates or repeats, the number can always be expressed as a fraction. This fact is often used to define the set of **rational numbers** or fractions. The rational numbers include all of the numbers we have considered thus far. Numbers that are not rational numbers, called **irrational numbers,** have decimal representations that neither terminate nor repeat. One of the most famous irrational numbers is π (the ratio of the circumference of a circle to its diameter).

❸ CONVERTING TERMINATING DECIMALS TO FRACTIONS

We now turn our attention to converting a terminating decimal to a fraction. For example,

$$0.5 \quad \text{is} \quad \frac{5}{10},$$

$$0.23 \quad \text{is} \quad \frac{2}{10} + \frac{3}{100} = \frac{20}{100} + \frac{3}{100} = \frac{23}{100},$$

$$0.537 \quad \text{is} \quad \frac{5}{10} + \frac{3}{100} + \frac{7}{1000} = \frac{500}{1000} + \frac{30}{1000} + \frac{7}{1000} = \frac{537}{1000}.$$

To Convert a Terminating Decimal to a Fraction
1. Express the decimal in expanded notation.
2. Add the whole number terms and add the fractional terms to obtain a mixed number.
3. Convert the mixed number to a single fraction by addition.

EXAMPLE 2   CONVERTING A DECIMAL TO A FRACTION

Convert each decimal to a fraction.

(a) $0.47 = \dfrac{4}{10} + \dfrac{7}{100} = \dfrac{40}{100} + \dfrac{7}{100} = \dfrac{47}{100}$

(b) $2.35 = 2 + \dfrac{3}{10} + \dfrac{5}{100} = 2 + \dfrac{30}{100} + \dfrac{5}{100}$

$\qquad = 2 + \dfrac{35}{100} = \dfrac{2 \cdot 100}{100} + \dfrac{35}{100} = \dfrac{235}{100}$

(c) $43.619 = 40 + 3 + \dfrac{6}{10} + \dfrac{1}{100} + \dfrac{9}{1000} = 43 + \dfrac{619}{1000} = \dfrac{43{,}619}{1000}$

PRACTICE EXERCISE 2

Convert each decimal to a fraction.

(a) 0.11

(b) 3.86

(c) 51.015

Answers: (a) $\frac{11}{100}$ (b) $\frac{193}{50}$
(c) $\frac{10{,}203}{200}$

❹ CONVERSIONS INVOLVING PERCENT

The concept *percent* is closely related to decimals and is used in a variety of applied problems. The word **percent** literally means *per hundred;* it refers to the number of parts in one hundred parts. A five percent tax means five parts per one hundred parts, or a tax of 5¢ on every 100¢. Five percent would be written

$$5\%.$$

Since the % symbol means "divide by 100" or "multiply by $\frac{1}{100} = 0.01$,"

$$5\% = 5(0.01) = 0.05 \quad \text{or} \quad 5\% = \frac{5}{100} = \frac{1}{20}.$$

To Convert a Percent to a Decimal

Remove the % symbol and multiply by 0.01. Or remove the % symbol and move the decimal point two places to the *left*, adding zeros as necessary.

To Convert a Percent to a Fraction

Remove the % symbol and divide by 100. Reduce the resulting fraction to lowest terms.

EXAMPLE 3   CONVERTING PERCENT TO A DECIMAL

Convert each percent to a decimal.

(a) $43\% = (43)(0.01) = 0.43$ $43\% = 0.43$

(b) $325.5\% = (325.5)(0.01) = 3.255$ $325.5\% = 3.255$

(c) $0.05\% = (0.05)(0.01) = 0.0005$ $0.05\% = 0.0005$

(d) $0.2\% = (0.2)(0.01) = 0.002$ $0.2\% = 0.002$

PRACTICE EXERCISE 3

Convert each percent to a decimal.

(a) 79%

(b) 640.5%

(c) 0.09%

(d) 0.8%

Answers: (a) **0.79**   (b) **6.405**
(c) **0.0009**   (d) **0.008**

EXAMPLE 4   CONVERTING PERCENT TO A FRACTION

Convert each percent to a fraction.

(a) $28\% = \frac{28}{100} = \frac{4 \cdot 7}{4 \cdot 25} = \frac{7}{25}$

(b) $37.5\% = \frac{37.5}{100} = \frac{(37.5)(10)}{(100)(10)}$ Clear decimal by multiplying by 10

$\qquad = \frac{375}{1000}$

$\qquad = \frac{3 \cdot 125}{8 \cdot 125}$

$\qquad = \frac{3}{8}$ Reduce fraction

PRACTICE EXERCISE 4

Convert each percent to a fraction.

(a) 65%

(b) 42.5%

(c) $66\frac{2}{3}\% = \left(66\frac{2}{3}\right)\left(\frac{1}{100}\right)$ Dividing by 100 is the same as multiplying by $\frac{1}{100}$

$= \left(\frac{200}{3}\right)\left(\frac{1}{100}\right)$ $66\frac{2}{3} = \frac{(3 \cdot 66) + 2}{3} = \frac{200}{3}$

$= \frac{2 \cdot \cancel{100}}{3 \cdot \cancel{100}} = \frac{2}{3}$ Reduce fraction

(d) $100\% = \frac{100}{100} = 1$

(c) $16\frac{1}{6}\%$

(d) 500%

Answers: (a) $\frac{13}{20}$ (b) $\frac{17}{40}$ (c) $\frac{97}{600}$ (d) 5

> ### To Convert a Decimal to a Percent
>
> Multiply by 100 and attach the % symbol. Or move the decimal point two places to the *right*, adding zeros as necessary, and attach the % symbol.

> ### To Convert a Fraction to a Percent
>
> Convert the fraction to a decimal by dividing the numerator by the denominator. Then change the resulting decimal to a percent.

EXAMPLE 5   CONVERTING A DECIMAL TO PERCENT

Convert each decimal to a percent.
(a) $0.37 = (0.37)(100\%) = 37\%$ $0.37 = 37.\% = 37\%$
(b) $5.81 = (5.81)(100)\% = 581\%$ $5.81 = 581.\% = 581\%$
(c) $0.0004 = (0.0004)(100)\% = 0.04\%$ $0.0004 = 000.04\% = 0.04\%$
(d) $0.\overline{3} = (0.\overline{3})(100)\% = (0.333 \ldots)(100)\%$

$= 33.33 \ldots \%$

$= 33.\overline{3}\%$ or $33\frac{1}{3}\%$ $0.\overline{3} = \frac{1}{3}$

PRACTICE EXERCISE 5

Convert each decimal to a percent.
(a) 0.81

(b) 9.05

(c) 0.003

(d) $0.\overline{7}$

Answers: (a) 81% (b) 905% (c) 0.3% (d) $77.\overline{7}\%$

EXAMPLE 6   CONVERTING A FRACTION TO PERCENT

Convert each fraction to a percent.
(a) $\frac{3}{4} = 0.75$ Divide 4 into 3

$= (0.75)(100)\% = 75\%$ $0.75 = 075.\% = 75\%$

(b) $\frac{1}{8} = 0.125$ Divide 8 into 1

$= (0.125)(100)\% = 12.5\%$ $0.125 = 012.5\% = 12.5\%$

PRACTICE EXERCISE 6

Convert each fraction to a percent.
(a) $\frac{4}{5}$

(b) $\frac{7}{10}$

(c) $\dfrac{5}{3} = 1.\overline{6}$ Divide 3 into 5   **(c)** $\dfrac{1}{9}$

$\quad = (1.\overline{6})(100)\%$

$\quad = (1.666\ldots)(100)\%$

$\quad = 166.666\ldots\%$

$\quad = 166.\overline{6}\%$ or $166\dfrac{2}{3}\%$ $0.\overline{6} = \tfrac{2}{3}$ Answers: **(a)** 80%   **(b)** 70%
                                                                     (c) $11.\overline{1}\%$

Keeping a simple example in mind will help you remember whether you should multiply by 0.01 or by 100 when making percent conversions. 50% of some quantity is 0.5 or $\frac{1}{2}$ of that quantity. Thus, to convert

$$50\% \text{ to } 0.5, \quad \text{multiply by } 0.01$$

and to convert $\qquad 0.5 \text{ to } 50\%, \quad \text{multiply by } 100.$

Many applied problems involve finding a percent of some number. For example, we might ask

What is 5% of $40.00?

To solve such problems we change the percent to a decimal or fraction and then multiply it by the given number. The word *of* means "multiply" or "times" in this setting.

$$\begin{array}{ccc} 5\% & \text{of} & \$40.00 \\ \downarrow & \downarrow & \downarrow \\ (0.05) & \cdot & (40.00) \end{array}$$

Thus, 5% of $40.00 is $(0.05)(40.00) = \$2.00$.

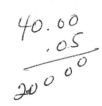

1.4 EXERCISES A

Convert each fraction to a decimal.

1. $\dfrac{1}{8}$ **2.** $\dfrac{7}{20}$ **3.** $\dfrac{5}{9}$

4. $\dfrac{8}{3}$ **5.** $\dfrac{5}{6}$ **6.** $2\dfrac{1}{7}$

Convert each decimal to a fraction.

7. 0.3   **8.** 0.007   **9.** 0.48

10. 27.95   **11.** 3.207   **12.** 25.001

Convert each percent to a decimal.

13. 65.5%

14. $\frac{1}{2}$%

15. 0.6%

16. 200%

17. $8\frac{1}{4}$%

18. $\frac{1}{3}$%

Convert each percent to a fraction.

19. 57%

20. 100%

21. $\frac{1}{4}$%

22. $3\frac{2}{3}$%

23. 1000%

24. $\frac{1}{10}$%

Convert each decimal to a percent.

25. 2.37

26. 0.06

27. 0.005

28. $0.62\overline{3}$

29. $0.1\overline{6}$

30. 1.1

Convert each fraction to a percent.

31. $\frac{1}{2}$

32. $\frac{6}{5}$

33. $\frac{10}{3}$

34. $\frac{7}{50}$

35. $\frac{5}{6}$

36. $\frac{50}{3}$

Solve.

 A basketball player made 15 shots in 20 attempts during a recent game. What fractional part of his shots did he make? What percent of his shots did he make?

38. Bill Schulz earned $24,000 last year and gave $3000 to charity. What fractional part of his income was given to charity? What percent of his income was given to charity?

39 There were 5000 votes cast in an election. Mr. Gomez received 55% of the votes. How many votes did he receive?

40. Snow, Montana has a normal annual snowfall of 120 in. This year the city received 150% of normal. How many inches of snow fell this year?

FOR REVIEW

41. Express 32.51 in expanded notation.

42. Add: 403.2
                   1.006
            +   0.04

43. Subtract: $3.21 - 0.0005$

44. Multiply: 0.0014
              $\times$ 1.06

45. Divide: $0.004\overline{)0.086}$

46. John Hagood rented an Escort for 3 days. If the rental fee was $8.95 per day plus $0.14 per mi, how much was he charged if he drove a total of 185.5 mi?

ANSWERS: 1. 0.125   2. 0.35   3. $0.\overline{5}$ 4. $2.\overline{6}$ 5. $0.8\overline{3}$ 6. $2.\overline{142857}$ 7. $\frac{3}{10}$ 8. $\frac{7}{1000}$ 9. $\frac{12}{25}$ 10. $\frac{559}{20}$ 11. $\frac{3207}{1000}$
12. $\frac{25,001}{1000}$ 13. 0.655   14. 0.005   15. 0.006   16. 2   17. 0.0825   18. $0.00\overline{3}$ 19. $\frac{57}{100}$ 20. 1   21. $\frac{1}{400}$ 22. $\frac{11}{300}$ 23. 10
24. $\frac{1}{1000}$ 25. 237%   26. 6%   27. 0.5%   28. $62.\overline{3}$% or $62\frac{1}{3}$%   29. $16.\overline{6}$% or $16\frac{2}{3}$%   30. 110%   31. 50%
32. 120%   33. $333\frac{1}{3}$%   34. 14%   35. $83\frac{1}{3}$%   36. $1666\frac{2}{3}$%   37. $\frac{3}{4}$, 75%   38. $\frac{1}{8}$, 12.5%   39. 2750 votes   40. 180 in
41. $30 + 2 + \frac{5}{10} + \frac{1}{100}$ 42. 404.246   43. 3.2095   44. 0.001484   45. 21.5   46. $52.82

1.4   EXERCISES B

Convert each fraction to a decimal.

1. $\dfrac{5}{8}$

2. $\dfrac{5}{16}$

3. $\dfrac{11}{15}$

4. $\dfrac{4}{5}$

5. $\dfrac{1}{6}$

6. $2\dfrac{3}{7}$

Convert each decimal to a fraction.

7. 0.09

8. 0.002

9. 2.6

10. 31.85

11. 20.02

12. 35.001

Convert each percent to a decimal.

13. 35.5%

14. $3\dfrac{1}{2}$%

15. 26.54%

16. 300%

17. $7\dfrac{3}{4}$%

18. $\dfrac{2}{3}$%

Convert each percent to a fraction.

19. 90%

20. 500%

21. 275%

22. $4\frac{1}{3}\%$ **23.** 2000%   **24.** $\frac{1}{100}\%$

Convert each decimal to a percent.

25. 4.29   **26.** 0.01   **27.** 0.007

28. $0.57\overline{3}$ **29.** 7   **30.** 2.2

Convert each fraction to a percent.

31. $\frac{3}{5}$ **32.** $\frac{11}{3}$ **33.** $\frac{9}{50}$

34. $\frac{3}{10}$ **35.** $\frac{5}{8}$ **36.** $\frac{25}{3}$

Solve.

37. A football quarterback completed 21 passes in 35 attempts during a game last fall. What fraction of his attempts did he complete? What percent of his attempts did he complete?

38. Amy received $12,000 from her grandfather. She spent $8000 on a new car. What fraction of the money went for the car? What percent of the money went for the car?

39. There are 0.2% impurities in a laboratory solution weighing 485 grams. What is the weight of the impurities?

40. Diane received a raise giving her a salary equal to 120% of her previous salary. If she earned $16,000 last year, how much will she earn this year?

FOR REVIEW

41. Express 29.65 in expanded notation.

42. Add: 209.6
             4.007
       +   0.01

43. Subtract: $4.81 - 0.0002$

44. Multiply: 0.0015
             $\times$ 1.08

45. Divide: $0.003\overline{)0.1488}$

46. Ben Whitney rented a midsize car for two days. If the rental fee was $11.95 per day plus $0.18 per mi, how much was he charged if he drove a total of 230.5 mi?

1.4   EXERCISES C

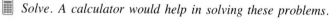 *Solve. A calculator would help in solving these problems.*

1. A retailer paid $92.00 for a dress and marked it up 30% to sell in her store. When the dress did not sell, she sold it at a 25% discount on the marked price. What was the percent profit or loss based on the original cost?   [Answer: 2.5% loss]

2. A salesman receives a 3% commission on all sales of $50,000 or less and a 4% commission on the amount he sells over $50,000. What is his commission on sales of $85,000? What is the effective total commission rate on the $85,000 sales?

1.5   VARIABLES, EXPONENTS, AND ORDER OF OPERATIONS

========= STUDENT GUIDEPOSTS =========

1 Variables   **4** Symbols of Grouping
2 Exponential Notation   **5** Evaluating Algebraic Expressions
3 Order of Operations   **6** Formulas

1 VARIABLES

In algebra, we often use letters such as a, b, x, y, A and B to represent numbers. A letter that can be replaced by various numbers is called a **variable.** Lengthy verbal expressions can often be symbolized by brief algebraic expressions using variables, as in the following examples.

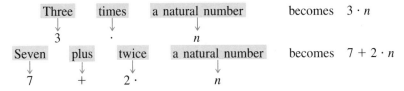

In each example, the variable n represents a natural number. Other notations for $3 \cdot n$ are $3(n)$, $(3)(n)$, and $3n$, with the last form preferred. The product of two natural numbers m and n would most likely be represented by mn, but could also be expressed by $m \cdot n$, $m(n)$, and $(m)(n)$. We usually avoid using the symbol $\times$ for multiplication since it can be confused with the letter x used as a variable.

2 EXPONENTIAL NOTATION

When a variable or number is multiplied by itself several times, as in

$$2 \cdot 2 \cdot 2 \cdot 2 \text{ or } x \cdot x \cdot x,$$

we can use **exponential notation** to avoid writing long strings of **factors** (the individual numbers in the expressed product). For example, we write $2 \cdot 2 \cdot 2 \cdot 2$ as 2^4,

$$\underbrace{2 \cdot 2 \cdot 2 \cdot 2}_{4 \text{ factors}} = 2\underset{\uparrow \atop \text{base}}{^{4,}} {}^{\text{exponent}}$$

where 2 is called the **base,** 4 the **exponent,** and 2^4 the **exponential expression** (read 2 to the **fourth power**). Similarly, $x \cdot x \cdot x = x^3$ is called the **third power** or **cube** of x. The **square** or **second power** of a is a^2. The **first power** of a is a^1, which we write simply as a.

Exponential Notation

If a is any number and n is a natural number,

$$a^n = \underbrace{a \cdot a \cdot a \cdots a.}_{n \text{ factors}}$$

EXAMPLE 1 USING EXPONENTS	PRACTICE EXERCISE 1

Write in exponential notation.

(a) $\underbrace{7 \cdot 7 \cdot 7}_{3 \text{ factors}} = 7^3$

(b) $\underbrace{a \cdot a \cdot a \cdot a \cdot a \cdot a}_{6 \text{ factors}} = a^6$

(c) $\underbrace{3 \cdot 3}_{\substack{2 \\ \text{factors}}} \cdot \underbrace{x \cdot x \cdot x \cdot x \cdot x}_{5 \text{ factors}} = 3^2 x^5$

Write in exponential notation.

(a) $4 \cdot 4 \cdot 4 \cdot 4 \cdot 4 \cdot 4 \cdot 4$

(b) $z \cdot z \cdot z \cdot z$

(c) $6 \cdot 6 \cdot 6 \cdot y \cdot y \cdot y \cdot y$

Answers: (a) 4^7 (b) z^4
(c) $6^3 y^4$

---

③ ORDER OF OPERATIONS

When exponents or powers are used together with the operations of addition, subtraction, multiplication, and division, the resulting numerical expressions can be confusing unless we agree to an order of operations. For example,

$2 \cdot 5^2$ *could* equal 10^2 or 100, If we first multiply then square

or $2 \cdot 5^2$ *could* equal $2 \cdot 25$ or 50. If we first square then multiply

Similarly,

$2 \cdot 3 + 4$ *could* equal $2 \cdot 7$ or 14, If we first add then multiply

or $2 \cdot 3 + 4$ *could* equal $6 + 4$ or 10. If we first multiply then add

According to the following rule, we see that the second procedure in each of these examples is the correct one to use.

To Evaluate a Numerical Expression

1. Evaluate all powers, in any order, first.
2. Do all multiplications and divisions in order from left to right.
3. Do all additions and subtractions in order from left to right.

EXAMPLE 2 EVALUATING NUMERICAL EXPRESSIONS	PRACTICE EXERCISE 2

Evaluate each numerical expression.

(a) $2 + 3 \cdot 4 = 2 + 12 = 14$ Multiply first, then add

(b) $20 - 3^2 = 20 - 9 = 11$ Square first, then subtract

(c) $5 \cdot 6 - 12 \div 3 = 30 - 4 = 26$ Multiply and divide first, then subtract

(d) $25 \div 5 + 3 \cdot 2^3 = 25 \div 5 + 3 \cdot 8$ Cube first

$ = 5 + 24$ Divide and multiply second

$ = 29$ Add last

Evaluate each numerical expression.

(a) $5 \cdot 2 + 1$

(b) $6 + 2^3$

(c) $18 \div 9 + 3 \cdot 2$

(d) $16 \cdot 2 - 3 \cdot 2^2$

Answers: (a) 11 (b) 14 (c) 8
(d) 20

④ SYMBOLS OF GROUPING

Suppose that we want to evaluate three times the sum of 2 and 5. If we write $3 \cdot 2 + 5$ and use the above rule, we will obtain $6 + 5$ or 11. However, it is clear from the first sentence, that we want 3 times 7 or 21 for the result. **Symbols of grouping,** such as parentheses ( ), square brackets [ ], or braces { }, can help us symbolize the problem correctly as $3 \cdot (2 + 5)$. The grouping symbols contain the expression that must be evaluated first. In this case, we must add before multiplying.

$$3 \cdot (2 + 5) = 3 \cdot (7) = 21$$

Generally, we omit the dot and write $3(2 + 5)$ for $3 \cdot (2 + 5)$. Also, instead of writing $3 \cdot (7)$, we will write $(3)(7)$.

One other method of grouping is to use a fraction bar. For example, in

$$\frac{4 \cdot 5}{7 + 3},$$

the fraction bar acts like parentheses since the expression is the same as

$$(4 \cdot 5) \div (7 + 3).$$

We first multiply 4 and 5, then add 7 and 3, and finally we divide the results. Thus,

$$\frac{4 \cdot 5}{7 + 3} = \frac{20}{10} = 2.$$

> ### To Evaluate a Numerical Expression Involving Grouping Symbols
>
> Evaluate all expressions within the grouping symbols first, and begin with the innermost symbols of grouping if more than one set of symbols is present.

EXAMPLE 3 EVALUATING NUMERICAL EXPRESSIONS	PRACTICE EXERCISE 3

Evaluate each numerical expression.

(a) $3 + (2 \cdot 5) - 3 + 10 = 13$ Evaluate inside the parentheses first

(b) $(4 + 5)2 + 3 = (9)2 + 3$ Work inside the parentheses first

$\qquad\qquad = 18 + 3$ Multiply before adding

$\qquad\qquad = 21$

(c) $[3(8 + 2) + 1]4 = [3(10) + 1]4$ Innermost grouping symbol first

$\qquad\qquad\qquad = [30 + 1]4$ Multiply before adding inside brackets

$\qquad\qquad\qquad = [31]4$ Combine numbers inside brackets before multiplying

$\qquad\qquad\qquad = 124$

Evaluate each numerical expression.

(a) $3 \cdot (2 + 5)$

(b) $(6 - 1)3 - 2$

(c) $2[2(7 + 2) - 1]$

Answers: (a) 21 (b) 13 (c) 34

We now summarize the order to follow when evaluating a numerical expression.

Order of Operations

1. Evaluate within grouping symbols first, beginning with the innermost set if more than one set is used.
2. Evaluate all powers.
3. Perform all multiplications and divisions in order from left to right.
4. Perform all additions and subtractions in order from left to right.

EXAMPLE 4 EVALUATING USING ALL RULES

Evaluate each expression.

(a) $2^3 + 4^3 = 8 + 64 = 72$ Cube first, then add

(b) $(2 + 4)^3 = 6^3 = 216$ Add first inside parentheses, then cube

(c) $(3 \cdot 5)^2 = 15^2 = 225$ Multiply first, then square

(d) $3 \cdot 5^2 = 3 \cdot 25 = 75$ Square first, then multiply

(e) $5^2 - 4^2 = 25 - 16 = 9$ Square first, then subtract

(f) $(5 - 4)^2 = 1^2 = 1$ Subtract first, then square

(g) $3^2 - 15 \div 5 + 5 \cdot 2 = 9 - 15 \div 5 + 5 \cdot 2$ Square first

             $= 9 - 3 + 10$ Divide and multiply in order

             $= 16$ Subtract and add in order

(h) The square of seven minus three
    $(7 - 3)^2 = 4^2 = 16$ Subtract first, then square

(i) Seven squared minus three squared
    $7^2 - 3^2 = 49 - 9 = 40$ Square first, then subtract

PRACTICE EXERCISE 4

Evaluate each expression.

(a) $1^2 + 6^2$

(b) $(1 + 6)^2$

(c) $(2 \cdot 7)^2$

(d) $2 \cdot 7^2$

(e) $8^2 - 6^2$

(f) $(8 - 6)^2$

(g) $2^3 - 20 \div 4 + 4 \cdot 3$

(h) The square of nine plus two

(i) Nine squared plus two squared

Answers: (a) 37   (b) 49
(c) 196   (d) 98   (e) 28   (f) 4
(g) 15   (h) 121   (i) 85

CAUTION

Example 4 points out common errors to avoid when working with exponents. For example,

$$2^3 + 4^3 \neq (2 + 4)^3, \quad (3 \cdot 5)^2 \neq 3 \cdot 5^2, \quad 5^2 - 4^2 \neq (5 - 4)^2.$$

The symbol $\neq$ means "is *not* equal to." In general, if a, b, and n are numbers,

$$a^n + b^n \neq (a + b)^n, \ (ab)^n \neq ab^n, \ a^n - b^n \neq (a - b)^n.$$

---

5 EVALUATING ALGEBRAIC EXPRESSIONS

An **algebraic expression** contains variables as well as numbers. We can use the order of operations for evaluating numerical expressions to evaluate algebraic expressions when specific values for the variables are given.

To Evaluate an Algebraic Expression

1. Replace each variable (letter) with the specified value.
2. Proceed as in evaluating numerical expressions.

EXAMPLE 5   EVALUATING AN ALGEBRAIC EXPRESSION

Evaluate $2(a + b) - c$ when $a = 3$, $b = 4$, and $c = 5$.

$2(a + b) - c = 2(3 + 4) - 5$ Replace each letter with given value

$= 2(7) - 5$ Evaluate inside parentheses first

$= 14 - 5 = 9$

PRACTICE EXERCISE 5

Evaluate $3(x - y) + z$, when $x = 4$, $y = 2$, and $z = 1$.

Answer: 7

EXAMPLE 6   EVALUATING AN ALGEBRAIC EXPRESSION

Evaluate $5[12 - 3(a + 1) + b] - c$ when $a = 2$, $b = 7$, and $c = 4$.

$5[12 - 3(a + 1) + b] - c = 5[12 - 3(2 + 1) + 7] - 4$ Replace variables with numbers

$= 5[12 - 3(3) + 7] - 4$

$= 5[12 - 9 + 7] - 4$ Multiply before adding or subtracting

$= 5[10] - 4 = 50 - 4 = 46$

PRACTICE EXERCISE 6

Evaluate $w - 2[3(u - 1) + v]$, when $u = 3$, $v = 1$, and $w = 14$.

Answer: 0

EXAMPLE 7   EVALUATING AN ALGEBRAIC EXPRESSION

Evaluate $\dfrac{ab - 1}{c}$ for the given values.

(a) $a = 2$, $b = 3$, and $c = 0$

$\dfrac{ab - 1}{c} = \dfrac{(2)(3) - 1}{0}$ Division by zero undefined

That is, this expression is undefined when $c = 0$.

(b) $a = 3$, $b = \dfrac{1}{3}$, and $c = 5$

$\dfrac{ab - 1}{c} = \dfrac{(3)\left(\frac{1}{3}\right) - 1}{5} = \dfrac{1 - 1}{5} = \dfrac{0}{5} = 0$ $\frac{0}{5} = 0$

PRACTICE EXERCISE 7

Evaluate $\dfrac{3x}{yz - x}$ for the given values.

(a) $x = 0$, $y = 5$, $z = 8$

(b) $x = 6$, $y = 2$, $z = 3$

Answers: (a) 0   (b) undefined

EXAMPLE 8   EVALUATING AN ALGEBRAIC EXPRESSION

Evaluate the following when $a = 2$, $b = 1$, $c = 3$.

(a) $3a^2 = 3(2)^2$ Not $(3 \cdot 2)^2 = 6^2 = 36$

$= 3 \cdot 4 = 12$

(b) $2ab^2c^3 = 2(2)(1)^2(3)^3$

$= 2(2)(1)(27) = 4 \cdot 27 = 108$

(c) $(2c)^2 - 2c^2 = (2 \cdot 3)^2 - 2(3)^2$ Watch the substitution

$= 6^2 - 2 \cdot 9$

$= 36 - 18 = 18$

(d) $a^a + c^c = 2^2 + 3^3$

$= 4 + 27 = 31$

PRACTICE EXERCISE 8

Evaluate the following when $u = 5$, $v = 3$, and $w = 1$.

(a) $2u^2$

(b) $3uv^2w$

(c) $(3u)^2 - 3u^2$

(d) $v^v - w^w$

Answers: (a) 50   (b) 135
(c) 150   (d) 26

⑥ FORMULAS

A **formula** is an algebraic statement that relates two or more quantities. For example, the formula

$$d = rt$$

relates the distance, d, traveled by an object moving at an average rate, r, for a period of time, t. Other familiar formulas related to geometric figures, such as

$A = lw$ Area of a rectangle in terms of its length and width
and $P = 2l + 2w$ Perimeter of a rectangle in terms of its length and width

are summarized on the inside back cover of the text. To use a particular formula in an applied problem, we must first understand the meaning of the variables, and next, evaluate the algebraic expression (formula) for the given values.

EXAMPLE 9   EVALUATING FORMULAS

If we invest P dollars (called the principal) at an interest rate, r, compounded annually for a period of t years, it will grow to an amount, A, given by

$$A = P(1 + r)^t.$$

If a principal of \$1000 is invested at 12% interest, compounded annually, how much will be in the account at the end of two years?

$A = P(1 + r)^t$ Start with the formula

$\quad = 1000(1 + 0.12)^2$ Substitute 1000 for P, 0.12 for r, and 2 for t

$\quad = 1000(1.12)^2$

$\quad = 1000(1.2544)$

$\quad = 1254.4$

Thus, there will be \$1254.40 in the account. Notice that 12% was converted to the decimal 0.12, in order to solve this problem.

PRACTICE EXERCISE 9

The perimeter of a rectangle, P, is given by $P = 2l + 2w$, where l is its length and w is its width. What is the perimeter of a rectangle with length 15 in and width 8 in?

Answer: 46 in

1.5 EXERCISES A

1. What is a letter used to represent a number called?

2. Given the exponential expression x^7, **(a)** what is the base?   **(b)** what is the exponent?

3. In the expression $3x$, what is the exponent on x?

Write in exponential notation.

4. $8 \cdot 8 \cdot 8 \cdot 8 \cdot 8$ **5.** $(2x)(2x)$ **6.** $2 \cdot x \cdot x$

7. $(b + c)(b + c)(b + c)$ **8.** $7aaaaaaa$ **9.** $6 \cdot 6 \cdot 6 \cdot y \cdot y \cdot z \cdot z \cdot z \cdot z$

Write without using exponents.

10. $a^2b^3c^4$

11. $3y^3$

12. $(3y)^3$

13. 1^{51}

14. $a^3 - c^3$

15. $\left(\dfrac{2}{3}\right)^3 x^2$

Evaluate.

16. $3 \cdot 2^2$

17. $(3 \cdot 2)^2$

18. $(3 + 2)^2$

19. $3^2 + 2^2$

20. $(5 - 2)^3$

21. $5^3 - 2^3$

22. $8 - 2^3$

23. $(8 - 2)^3$

24. $(7 - 7)^3$

25. $9 - 2 \cdot 3 + 1$

26. $(9 - 2) \cdot 3 + 4 \cdot 0$

27. $\dfrac{2(3 - 1) - 4}{5}$

28. $\dfrac{4(3 + 5) - 10}{0}$

29. $12 \div 4 + 2 \cdot 3 - 1$

30. $2[8 - 2(4 - 1) + 5]$

31. $15 - 3\{2(5 - 4) + 3\}$

32. $10 - 2[8 - 2(7 - 3)]$

33. $16 \div 4 \cdot 2 \div 4 - 2$

34. The cube of the difference five minus two

35. Five cubed minus two cubed

36. Ten minus three squared

37. The square of the difference ten minus three

Evaluate when $a = 2$, $b = 3$, $c = 5$, $d = 0$, and $x = 12$.

38. $6a + b$

39. $6(a + b)$

40. $3ab \div d$

41. $2b^2$

42. $(2b)^2$

43. $(c - b)^2$

44. $c^2 - b^2$

45. $(a + b)^3$

46. $a^3 + b^3$

47. $3a^3 + 2$

48. $(3a)^3 + 2$

49. $a^a + b^b$

50. $\dfrac{2abcd - 3d}{x}$

51 $x - [a(b + 1) - c]$

52. $5a + 2[c + x \div b]$

Solve.

53. The perimeter of a rectangle, P, is given by $P = 2l + 2w$, where l is its length and w is its width. What is the perimeter of a rectangle of length 20 ft and width 13 ft?

54. The distance an automobile travels, d, at an average rate of speed, r, for a period of time, t, is given by $d = rt$. How far does a car travel in 11 hr at an average speed of 55 mph?

55. Use $A = P(1 + r)^t$ to find the amount of money in an account at the end of two years if a principal of $500 is invested at 14% interest, compounded annually.

56. The area of a square, A, is given by $A = s^2$, where s is the length of a side. What is the area of a square with side 3.5 cm?

57. The area of a triangle, A, is given by $A = \frac{1}{2} bh$, where b is the length of its base and h is its height. Find the area of a triangle with height 15 m and base 8 m.

58. The temperature measured in degrees Celsius, °C, can be obtained from degrees Fahrenheit, °F, by using $C = \frac{5}{9}(F - 32)$. Find C when F is 200°.

59. The surface area of a cube, A, is given by $A = 6e^2$ where e is the length of an edge. Find the surface area of a cube with an edge of $2\frac{1}{2}$ in.

60 The surface area of a cylinder, A, with height, h, and base radius, r, is given by $A = 2\pi rh + 2\pi r^2$. Use 3.14 for π and find the surface area of a cylinder with a radius of 2 cm and a height of 10 cm.

FOR REVIEW

61. Convert 3.7 to a fraction.

62. Convert $\frac{5}{9}$ to a decimal.

63. Convert 0.085 to a percent

64. Convert $\frac{5}{11}$ to a percent.

65. Convert $7\frac{3}{4}\%$ to a decimal.

66. A chemist has a solution that is 0.6% salt. If the solution weighs 650 g, what is the weight of the salt?

ANSWERS: 1. variable 2. (a) x (b) 7 3. 1 4. 8^5 5. $(2x)^2$ 6. $2x^2$ 7. $(b+c)^3$, *not* b^3+c^3 8. $7a^7$ 9. $6^3y^2z^4$
10. $aabbbcccc$ 11. $3yyy$ 12. $(3y)(3y)(3y)$ or $27yyy$ 13. 1 14. $aaa-ccc$ 15. $\left(\frac{2}{3}\right)\left(\frac{2}{3}\right)\left(\frac{2}{3}\right)xx$ or $\frac{8}{27}xx$ 16. 12 17. 36
18. 25 19. 13 20. 27 21. 117 22. 0 23. 216 24. 0 25. 4 26. 21 27. 0 28. undefined 29. 8 30. 14
31. 0 32. 10 33. 0 34. 27 35. 117 36. 1 37. 49 38. 15 39. 30 40. undefined 41. 18 42. 36 43. 4
44. 16 45. 125 46. 35 47. 26 48. 218 49. 31 50. 0 51. 9 52. 28 53. 66 ft 54. 605 mi 55. $649.80
56. 12.25 cm² 57. 60 m² 58. 93.$\overline{3}$° 59. 37.5 in² 60. 150.72 cm² 61. $\frac{37}{10}$ 62. 0.$\overline{5}$ 63. 8.5% 64. 45.$\overline{45}$%
65. 0.0775 66. 3.9 g

1.5 EXERCISES B

1. When evaluating a numerical expression with more than one set of grouping symbols, always evaluate within which set of symbols first?

2. In the exponential expression a^6, **(a)** what is the base? **(b)** what is the exponent?

3. In the expression $8y$, what is the exponent on y?

Write in exponential notation.

4. $c \cdot c \cdot c \cdot c \cdot c$

5. $3 \cdot 3 \cdot y \cdot y \cdot y \cdot y \cdot y \cdot y$

6. $(3a)(3a)$

7. $3 \cdot a \cdot a$

8. $(x-y)(x-y)(x-y)$

9. $5 \cdot 5 \cdot 5 \cdot a \cdot a \cdot c \cdot c \cdot c$

Write without using exponents.

10. $a^3b^4c^2$

11. $2z^3$

12. $(2z)^3$

13. $x^2 - y^2$

14. $a^3 + b^3$

15. $\left(\frac{1}{2}\right)^3 a^2$

Evaluate.

16. $2 \cdot 7^2$

17. $(2 \cdot 7)^2$

18. $(2+7)^2$

19. $2^2 + 7^2$

20. $(4-3)^2$

21. $4^2 - 3^2$

22. $4 - 2^2$

23. $(4-2)^2$

24. $(3-3)^3$

25. $14 - 5 + 2 \cdot 6$

26. $0 \cdot (5-2) + 0$

27. $\dfrac{2(5+1)-3}{3}$

28. $\dfrac{5(2+8)-19}{0}$

29. $25 \div 5 + 2 \cdot 4 - 1$

30. $3[10 - 2(3 - 1) + 6]$

31. $20 - 2[3(6 - 5) + 4]$

32. $30 - 5[7 - 2(6 - 3)]$

33. $20 \div 5 \cdot 2 \div 4 - 2$

34. The square of the sum of eleven and three

35. Eleven squared plus three squared

36. Twelve minus two cubed

37. The cube of the difference twelve minus two

Evaluate when $x = 3$, $y = 2$, $z = 4$, $w = 0$, and $a = 24$.

38. $5x + y$

39. $5(x + y)$

40. $4xy \div w$

41. $3y^2$

42. $(3y)^2$

43. $(z - y)^2$

44. $z^2 - y^2$

45. $(x + z)^3$

46. $x^3 + z^3$

47. $2z^3 + 1$

48. $(2z)^3 + 1$

49. $x^y + y^x$

50. $\dfrac{3wxyz - 5w}{a}$

51. $a - [x(y + 3) - z]$

52. $3x + 4[x + z \div y]$

Solve.

53. The perimeter of a square, P, with side of length s is given by $P = 4s$. What is the perimeter of a square with side of length 15 ft?

54. The area of a rectangle, A, is given by $A = lw$, where l is its length and w its width. If a rectangle is 2 cm wide and 7 cm long, what is its area?

55. Use the formula $A = P(1 + r)^t$ to find the amount of money in an account at the end of two years if a principal of $2000 is invested at 8% interest, compounded annually.

56. The area of a circle, A, with radius r, is given by $A = \pi r^2$. Use 3.14 for π and find the area of a circle with radius 20 m.

57. The volume, V, of a box with length l, width w, and height h, is given by $V = lwh$. If the box is 9 in long, 6 in wide, and 11 in high, what is its volume?

58. The temperature measured in degrees Fahrenheit, °F, can be obtained from degrees Celsius, °C, by using $F = \frac{9}{5}C + 32$. Find F when C is 20°.

59. The surface area of a sphere, A, with radius r, is given by $A = 4\pi r^2$. Use 3.14 for π and find the surface area of a sphere with radius 50 cm.

60. The surface area of a cylinder, A, is given by $A = 2\pi rh + 2\pi r^2$, where r is the radius of its base and h is its height. Find A if $r = 10$ in, $h = 8$ in, and $\pi = 3.14$.

FOR REVIEW

61. Convert 8.3 to a fraction.

62. Convert $\frac{6}{11}$ to a decimal.

63. Convert 0.017 to a percent.

64. Convert $\frac{8}{9}$ to a percent.

65. Convert $10\frac{1}{4}\%$ to a decimal.

66. It has been estimated that 32% of the population of Northern, Minnesota is of Danish ancestry. If 19,270 people live in Northern, approximately how many of them are of Danish background?

1.5 EXERCISES C

Evaluate.

1. $2(5 - 3)^3 - [7 - (8 - 6)^2]$

2. $13 - [11 - (7 - 4)^2]^3$

3. $[(6 \div 3)^2 + 9 \cdot (3 - 1)^2]^2$
[Answer: 1600]

Evaluate when a = 1.35, b = 2.75, and c = 6.2. A calculator would be helpful for these problems.

4. $abc + b^2$

5. $c^2 - a^2 + b^2$

6. $2c^2 - (a + b)$

CHAPTER 1 REVIEW

KEY WORDS

1.1   A **numeral** is the name or symbol for a number.

The **natural** or **counting numbers** are 1, 2, 3, 4, ...

The **whole numbers** are 0, 1, 2, 3, ...

Fractions that are names for the same number are called **equivalent fractions.**

A fraction is **reduced to lowest terms** when 1 is the only natural number that divides both numerator and denominator.

A **prime number** is a natural number greater than 1 whose only divisors are 1 and itself.

1.2   The **reciprocal** of a fraction is the fraction formed by interchanging its numerator and denominator.

The **least common denominator (LCD)** of two fractions is the smallest number that has both denominators as factors.

1.3   A **place-value system** is a number system in which a digit has a particular value determined by its place in the numeral.

The **digits** are 0, 1, 2, 3, 4, 5, 6, 7, 8, and 9.

1.4   A **rational number** has a decimal representation that terminates or repeats.

An **irrational number** has a decimal representation that neither terminates nor repeats.

The word **percent** means per hundred.

1.5   A **variable** is a letter that can be replaced by various numbers.

An **exponent** is a power on a number or variable indicating how many times the number or variable is used as a factor.

The **base** of an exponential expression is the number or variable that is raised to a power.

KEY CONCEPTS

1.1   **1.** When reducing fractions by canceling common factors, do not make the numerator 0 when it is actually 1. For example,

$$\frac{5}{5 \cdot 3} = \frac{\cancel{5}}{\cancel{5} \cdot 3} = \frac{1}{3}, \; not \; \frac{0}{3}.$$

2. Factors common to the numerator and denominator of a fraction may be divided or canceled. However, do not cancel numbers that are not factors. For example,

$$\frac{2 \cdot 3}{2} = \frac{\cancel{2} \cdot 3}{\cancel{2}} = 3,$$

but $\frac{2 + 3}{2}$ is *not* equal to $\frac{\cancel{2} + 3}{\cancel{2}}$.

1.2 **1.** To add or subtract fractions with different denominators, find the LCD. However, *do not* find the LCD to multiply or divide fractions.

2. For any nonzero number a, $\frac{0}{a} = 0$, but $\frac{a}{0}$ is undefined.

1.4 **1.** To convert **a percent to a decimal,** remove the % symbol and multiply by 0.01 (move the decimal point two places to the *left*). For example,

$$48\% = 48\,(0.01) = 0.48.$$

2. To convert **a decimal to a percent,** multiply by 100 (move the decimal point two places to the *right*) and attach the % symbol. For example,

$$0.172 = 0.172\,(100\%) = 17.2\%.$$

1.5 **1.** In an expression such as $2y^3$, only y is cubed—not $2y$. That is, $2y^3$ is not the same as $(2y)^3$.

2. When simplifying a numerical expression,
First, evaluate within grouping symbols.
Second, evaluate all powers.
Third, perform all multiplications and divisions in order from left to right.
Fourth, perform all additions and subtractions in order from left to right.

REVIEW EXERCISES

Part I

1.1 **1.** Express 490 as a product of primes.

2. Reduce $\frac{110}{385}$ to lowest terms.

3. Write an equivalent fraction for $\frac{3}{11}$ having 44 for the denominator.

4. Supply the missing numerator. $\frac{9}{13} = \frac{}{39}$

1.2 *Perform the indicated operations.*

5. $\frac{3}{14} \cdot \frac{21}{9}$

6. $3\frac{1}{3} \div \frac{20}{9}$

7. $\frac{5}{9} + \frac{7}{12}$

8. $\frac{17}{18} - \frac{2}{3}$

9. Change $\frac{219}{4}$ to a mixed number.

10. Change $5\frac{2}{7}$ to an improper fraction.

11. On a map, one inch represents $\frac{3}{4}$ mi. How many miles are represented by $2\frac{1}{3}$ in?

12. Julie Simpson bought two pieces of fabric, one $3\frac{1}{8}$ yd long and the other $4\frac{3}{4}$ yd long. How many yards of fabric did she purchase?

1.3 **13.** Express 29.34 in expanded notation.

14. Express $2000 + 4 + \frac{4}{10} + \frac{2}{1000}$ in decimal notation.

15. Add.
14.5
0.007
+ 325.16

16. Subtract.
307.2
− 12.009

17. Multiply.
1.31
× 0.02

18. Divide.
0.005)36.15

1.4 **19.** Convert $\frac{3}{8}$ to a decimal.

20. Convert 27.235 to a fraction.

21. Convert 35.2% to a decimal.

22. Convert $\frac{4}{3}$ to a percent.

23. Sherman's Shoes has an inventory consisting of 1980 pairs of men's and women's shoes. If 55% of the inventory consists of women's shoes, how many pairs of women's shoes are in stock?

1.5 *Write in exponential notation.*

24. *aaaaa*

25. $(3z)(3z)(3z)$

26. $(a + w)(a + w)$

Write without using exponents.

27. b^7

28. $2x^3$

29. $(2x)^3$

Evaluate.

30. $2 \cdot 5 - 9 \div 3 + 1$

31. $2 \cdot 3^3 - 3 \cdot 2^3$

Evaluate when $a = 2$, $b = 5$, and $x = 4$.

32. $3a^2$

33. $(3a)^2$

34. 3^2a

35. $2 + 5[x + (b - a)]$

36. $2a + 3[x \div a \cdot b - 4]$

Part II

Evaluate when $a = 3$, $b = 6$, and $x = 1$.

37. $(b + x)^2$

38. $b^2 + x^2$

39. $abx - 10x$

40. $x(ab - 10)$

41. In a test of 560 video games, 420 were working. What fractional part were working? Reduce the fraction.

42. Graydon Bell works in Perko's Delicatessen as a registered cheese cutter. If he is paid $9.85 per hour and worked a total of 35.6 hr last week, what did he earn?

Perform the indicated operations.

43. $0 \div \dfrac{5}{4}$

44. $4.7 - 0.001$

45. $1\dfrac{1}{5} + \dfrac{9}{10} - \dfrac{1}{3}$

46. $(1.04)(1.04)(1.04)$

47. Convert 0.5% to a decimal.

48. Convert $\frac{3}{11}$ to a decimal.

49. Convert 0.03 to a percent.

50. Convert 25 to a percent.

51. Change $\frac{31}{9}$ to a mixed number.

52. Convert 1.85 to a fraction.

53. Change $5\frac{23}{27}$ to an improper fraction.

54. Write an equivalent fraction for $\frac{2}{13}$ with 52 as the denominator.

55. There were 942 books in a shipment. If $\frac{1}{6}$ of these were science books, how many science books were in the shipment?

56. Supply the missing term. $\frac{5}{6} = \frac{35}{\underline{}}$

57. Use $A = P(1 + r)^t$ to find the amount of money in an account at the end of two years if a principal of $5000 is invested at 15% interest, compounded annually.

58. Kelly received 250 of the 600 votes cast for the two candidates in an election. What fractional part of the total number of votes did she receive?

ANSWERS: 1. $2 \cdot 5 \cdot 7 \cdot 7$ 2. $\frac{2}{7}$ 3. $\frac{12}{44}$ 4. 27 5. $\frac{1}{2}$ 6. $\frac{3}{2}$ 7. $\frac{41}{36}$ 8. $\frac{5}{18}$ 9. $54\frac{3}{4}$ 10. $\frac{37}{7}$ 11. $1\frac{3}{4}$ mi 12. $7\frac{7}{8}$ yd
13. $20 + 9 + \frac{3}{10} + \frac{4}{100}$ 14. 2004.402 15. 339.667 16. 295.191 17. 0.0262 18. 7230 19. 0.375 20. $\frac{5447}{200}$
21. 0.352 22. $133\frac{1}{3}\%$ 23. 1089 24. a^5 25. $(3z)^3$ 26. $(a + w)^2$ 27. *bbbbbb* 28. $2xxx$ 29. $(2x)(2x)(2x)$ or $8xxx$
30. 8 31. 30 32. 12 33. 36 34. 18 35. 37 36. 22 37. 49 38. 37 39. 8 40. 8 41. $\frac{3}{4}$ 42. $350.66 43. 0
44. 4.699 45. $1\frac{23}{30}$ 46. 1.124864 47. 0.005 48. $0.\overline{27}$ 49. 3% 50. 2500% 51. $3\frac{4}{9}$ 52. $\frac{37}{20}$ 53. $\frac{158}{27}$ 54. $\frac{8}{52}$
55. 157 56. 42 57. $6612.50 58. $\frac{5}{12}$

1. Reduce $\frac{322}{350}$ to lowest terms.

1. _____

2. Write an equivalent fraction for $\frac{8}{25}$ with 75 as the denominator.

2. _____

3. Supply the missing term. $\frac{4}{15} = \frac{}{90}$

3. _____

4. In a contest, Michelle received 385 of the 525 possible points. What fractional part of the points did Michelle get? Reduce the fraction.

4. _____

Perform the indicated operation.

5. $\frac{2}{5} \cdot \frac{15}{22}$

5. _____

6. $\frac{9}{14} \div \frac{2}{7}$

6. _____

7. $\frac{5}{6} + \frac{7}{9}$

7. _____

8. $\frac{17}{24} - \frac{2}{3}$

8. _____

9. Change $\frac{28}{5}$ to a mixed number.

9. _____

10. Change $2\frac{14}{15}$ to an improper fraction.

10. _____

11. There were 720 tires in a shipment. If $\frac{3}{5}$ of these were whitewalls, how many whitewall tires were in the shipment?

11. _____

12. A bottle contains $7\frac{2}{5}$ liters of solution. If $2\frac{7}{10}$ liters are used, how many liters are left in the bottle?

12. _____

Perform the indicated operation.

13. $14.06 + 3.057$

13. _____

14. $26.3 - 7.441$

14. _____

15. $(0.42)(5.06)$

15. _____

16. $46.2 \div 1.05$

16. _____

17. Convert $\frac{1}{5}$ to a percent.

17. _____

18. Convert 3.15 to a fraction.

18. _____

19. The tax rate on purchases is 4%. What is the tax on a $27 purchase?

19. _____

20. Write $a \cdot a \cdot a$ in exponential notation.

20. _____

21. Evaluate. $42 \div 6 - (4 - 2) \cdot 3$

21. _____

Evaluate when $x = 3$, $y = 2$, and $z = 5$.

22. $3z^2$

22. _____

23. $(3z)^2$

23. _____

24. $(x + y)^3$

24. _____

25. $x^3 + y^3$

25. _____

26. $z + 2[(x + y) - z]$

26. _____

27. Use $A = P(1 + r)^t$ to find the amount of money in an account at the end of two years if a principal of $500 is invested at 9% interest, compounded annually.

27. _____

Rational and Real Numbers

2.1 RATIONAL NUMBERS AND THE NUMBER LINE

STUDENT GUIDEPOSTS

1. Number Line
2. Less Than and Greater Than
3. Integers
4. Equal and Unequal Numbers
5. Rational Numbers
6. Absolute Value

1 NUMBER LINE

Many ideas in algebra can be better understood if we ''picture'' them in some way. Numbers are often pictured using a *number line*. Consider the set of whole numbers. We draw a line like the one in Figure 2.1, and select some unit of length. Then, starting at an arbitrary point that we label 0 and call the **origin,** we mark off unit lengths to the right, labeling the points 1, 2, 3, ... The result is a **number line** displaying the whole numbers, shown in Figure 2.1. Every whole number is paired with a point on this line. The arrowhead points to the direction in which the whole numbers continue.

Figure 2.1

EXAMPLE 1   PAIRING NUMBERS WITH POINTS

What whole numbers are paired with *a*, *b*, and *c* on the number line in Figure 2.2?

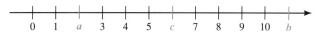

Figure 2.2

Point *a* is paired with 2, point *b* with 11, and point *c* with 6.

PRACTICE EXERCISE 1

What whole numbers are paired with *x*, *y*, and *z* on the following number line?

Answer: *x* paired with 4, *y* paired with 3, *z* paired with 1

❷ LESS THAN AND GREATER THAN

The whole numbers occur in a natural order. For example, we know that 3 is less than (has a smaller value than) 8, and that 8 is greater than (has a larger value than) 3. Note on the number line in Figure 2.2, that 3 is to the left of 8, while 8 is to the right of 3. The symbol $<$ means **"is less than."** We write "3 is less than 8" as

$$3 < 8.$$

Similarly, the symbol $>$ means **"is greater than."** We write "8 is greater than 3" as

$$8 > 3.$$

Notice that $3 < 8$ and $8 > 3$ have the same meaning even though they are read differently.

Order of Whole Numbers

Suppose a and b are any two whole numbers.

1. If a is to the left of b on a number line, then $a < b$.

2. If a is to the right of b on a number line, then $a > b$.

EXAMPLE 2   ORDERING WHOLE NUMBERS

Place the correct symbol ($<$ or $>$) between the numbers in the given pairs. Use part of a number line if necessary.

(a) 2   7

$2 < 7$. Notice that 2 is to the left of 7 on the number line in Figure 2.3.

Figure 2.3

(b) 12   9

$12 > 9$. Notice that 12 is to the right of 9 on the number line in Figure 2.4.

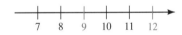

Figure 2.4

PRACTICE EXERCISE 2

Place the correct symbol ($<$ or $>$) between the given pairs of numbers.

(a) 3   11

(b) 15   7

Answers: (a) $3 < 11$ (b) $15 > 7$

   Notice in Example 2 that when $<$ and $>$ are used, the symbol *always points to the smaller of the two numbers.*

   The number lines we have looked at so far display no numbers to the left of 0. What meaning could we give to a number to the left of 0? One example of such a number is a temperature of 5° below zero on a cold day, which is sometimes represented as $-5°$ (read "negative 5°"). Others are shown in the following table.

Measurement	Number
5° above zero	5° (or +5°)
5° below zero	−5°
100 ft above sea level	100 ft (or +100 ft)
100 ft below sea level	−100 ft
$16 deposit into an account	$16 (or +$16)
$16 check written on an account	−$16

❸ INTEGERS

When we put a negative sign in front of each of the counting numbers, we obtain the **negative integers.** The collection of negative integers together with the counting numbers (sometimes called **positive integers**) and zero is called the set of **integers.**

$$I = \{. \, . \, . \, , \; -3, \; -2, \; -1, \; 0, \; 1, \; 2, \; 3, \; . \, . \, .\} \qquad \text{Integers}$$

The whole numbers are often called **nonnegative integers.** The number line in Figure 2.5 shows negative integers, positive integers, and zero. Note that we can write 1 as +1, 2 as +2, and so on. Also, we read −1 as "negative 1," for example.

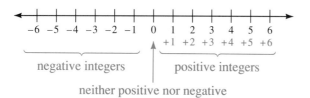

Figure 2.5

❹ EQUAL AND UNEQUAL NUMBERS

We say that two numbers are **equal** (for example, $3 + 2 = 5$) if they correspond to the same point on a number line. If two numbers correspond to different points on a number line, they are **unequal.** We use $\neq$ to represent "is not equal to." We can expand our definitions of *less than* and *greater than* for whole numbers to include the integers. That is, one integer is **less than** a second if it is to the left of the second on a number line, and one integer is **greater than** a second if it is to the right of the second on a number line. If one number is **less than or equal to** another number, we use the symbol $\leq$ (similarly $\geq$ represents **greater than or equal to**). For example,

$$-4 \leq 4, \quad -2 \leq 0, \quad 0 \leq 3, \quad 2 \leq 2, \quad \text{and} \quad -1 \leq 1.$$

All negative integers are less than zero (left of zero), all positive integers are greater than zero (right of zero), and any positive integer is greater than any negative integer.

| **EXAMPLE 3** ORDER RELATIONS BETWEEN NUMBERS | **PRACTICE EXERCISE 3** |

Some relationships between numbers on the number line in Figure 2.6 follow.

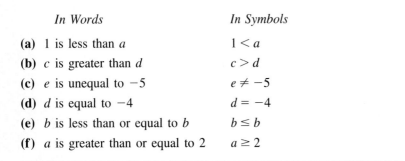

Figure 2.6

	In Words	*In Symbols*
(a)	1 is less than *a*	$1 < a$
(b)	*c* is greater than *d*	$c > d$
(c)	*e* is unequal to -5	$e \neq -5$
(d)	*d* is equal to -4	$d = -4$
(e)	*b* is less than or equal to *b*	$b \leq b$
(f)	*a* is greater than or equal to 2	$a \geq 2$

Use Figure 2.6 and write the following in symbols.

(a) *d* is less than 0

(b) *a* is greater than or equal to -1

(c) *b* is equal to 7

(d) *e* is less than or equal to *e*

(e) *a* is greater than *d*

(f) *c* is unequal to -8

Answers: (a) $d < 0$ (b) $a \geq -1$
(c) $b = 7$ (d) $e \leq e$ (e) $a > d$
(f) $c \neq -8$

⑤ RATIONAL NUMBERS

In addition to the integers, the fractions we studied in Chapter 1, together with their negatives, can be shown or **plotted** on a number line. All of the fractions to the right of zero are called **positive rational numbers.** Those to the left of zero are **negative rational numbers,** and together, along with zero, they form the set of **rational numbers.** We can think of the rational numbers as

$Q = \{$all numbers that can be written as a quotient of two integers$\}$.

Remember from Chapter 1 that any integer can be written as a quotient of two integers, for example $-2 = -\frac{2}{1}$ and $1 = \frac{1}{1}$, so all integers are also rational numbers. Some examples of rational numbers are plotted in Figure 2.7.

Figure 2.7

Just as with integers, we can define the order relationships of *less than* and *greater than* for rational numbers. For example, we see that

$$-\frac{1}{2} < \frac{1}{2}, \qquad \frac{3}{2} < \frac{9}{4}, \qquad 2\frac{7}{8} > -\frac{3}{2}, \qquad \text{and} \qquad -\frac{9}{4} > -3\frac{1}{3}$$

by looking at the number line in Figure 2.7.

A very important property of our number system states that if two numbers are unequal, then one of them must be less than the other. With two integers, it is easy to see if they are equal and, if not, which is less than the other. But this is not quite so easy for two rational numbers.

Ordering Rational Numbers

Given two positive rational numbers $\frac{a}{b}$ and $\frac{c}{d}$,

1. $\dfrac{a}{b} = \dfrac{c}{d}$ whenever $ad = bc$.

2. $\dfrac{a}{b} > \dfrac{c}{d}$ whenever $ad > bc$.

3. $\dfrac{a}{b} < \dfrac{c}{d}$ whenever $ad < bc$.

The product ad is called the **first cross product,** and bc is the **second cross product.** Thus, two positive fractions are equal if the cross products are equal, and the order of two unequal fractions is the same as the order of the first and second cross products. Similar results apply to negative rational numbers.

EXAMPLE 4 ORDERING RATIONAL NUMBERS	PRACTICE EXERCISE 4

Place the correct symbol, $=$, $>$, or $<$, between the given pairs of fractions.

(a) $\dfrac{4}{6}$ $\dfrac{28}{42}$

Since $4 \cdot 42 = 168$ and $6 \cdot 28 = 168$, the fractions are equal.

$$\frac{4}{6} = \frac{28}{42}.$$

(b) $\dfrac{24}{7}$ $\dfrac{39}{11}$

Since $24 \cdot 11 = 264$ and $7 \cdot 39 = 273$, and $264 < 273$,

$$\frac{24}{7} < \frac{39}{11}.$$

(c) $\dfrac{15}{9}$ $\dfrac{31}{20}$

Since $15 \cdot 20 = 300$ and $9 \cdot 31 = 279$, and $300 > 279$,

$$\frac{15}{9} > \frac{31}{20}.$$

Place the correct symbol, $=$, $>$, or $<$, between the given pairs of fractions.

(a) $\dfrac{14}{3}$ $\dfrac{23}{5}$

(b) $\dfrac{21}{13}$ $\dfrac{18}{11}$

(c) $\dfrac{9}{28}$ $\dfrac{36}{112}$

Answers: (a) $>$ (b) $<$ (c) $=$

---

⑥ ABSOLUTE VALUE

In Chapter 1 we reviewed the four basic operations of addition, subtraction, multiplication, and division on the nonnegative rational numbers. When operating on *all* rational numbers we use the idea of *absolute value*.

Absolute Value of a Number

The **absolute value** of a number x is the distance from zero to x on a number line. We symbolize the absolute value of x by $|x|$.

| EXAMPLE 5   ABSOLUTE VALUE USING A NUMBER LINE | PRACTICE EXERCISE 5 |

(a) As shown in Figure 2.8, 3 is 3 units from 0. Thus, $|3| = 3$.

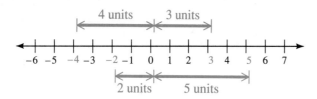

Figure 2.8

(b) 5 is 5 units from 0. Thus, $|5| = 5$.
(c) 0 is 0 units from 0. Thus, $|0| = 0$.
(d) -2 is 2 units from 0. Thus, $|-2| = 2$.
(e) -4 is 4 units from 0. Thus, $|-4| = 4$.

Find the following absolute values.

(a) $|-8|$

(b) $|8|$

(c) $|0.05|$

(d) $|-0.05|$

(e) $\left| -\dfrac{3}{8} \right|$

Answers: **(a) 8**   **(b) 8**   **(c) 0.05**
(d) 0.05   **(e)** $\frac{3}{8}$

Remember that the absolute value of a positive number or zero is the number itself. The absolute value of a negative number is the positive number formed by removing the minus sign. Thus, *the absolute value of a number is always greater than or equal to zero, never negative.*

2.1 EXERCISES A

Answer true *or* false. *If the statement is false, tell why.*

1. $\{0, 1, 2, 3, \ldots\}$ is the set of nonnegative integers.

2. If a number x is located to the left of a number y on a number line, then $x > y$.

3. The numeral -7 is read "negative seven."

4. A temperature of 32° below zero could be written as $+32°$.

5. The distance from zero to a number y on a number line is called the absolute value of y.

6. An elevation of 4500 feet above sea level would be denoted by -4500.

7. A withdrawal of $200 from a savings account could be denoted by $-\$200$.

Place the correct symbol, =, >, *or* <, *between the given pairs of numbers.*

8. 7   1   **9.** 0   3   **10.** -2 0   **11.** 2   9

12. -3 2   **13.** $\dfrac{1}{2}$ $-\dfrac{1}{2}$ **14.** $-\dfrac{3}{4}$ $\dfrac{1}{4}$ **15.** -5 -5

Is the given statement true or false? If the statement is false, tell why.

16. $-12 \le 0$ **17.** $6 \le 6$ **18.** $0 \le -3$ **19.** $-5 \le -5$

20. $-2 \le -7$ **21.** $\dfrac{2}{3} \le -\dfrac{2}{3}$ **22.** $4\dfrac{1}{2} \ge 4.5$ **23.** $-0.5 \le -\dfrac{1}{2}$

24. Plot the numbers $\dfrac{2}{3}$, $-\dfrac{7}{8}$, $\dfrac{5}{4}$, $-\dfrac{7}{4}$, $\dfrac{5}{2}$, and $-\dfrac{10}{3}$ on the given number line.

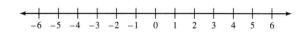

25. Insert the correct symbol, $=$, $>$, or $<$, between the two positive fractions.

    (a) If $ad = bc$, then $\dfrac{a}{b}$ $\dfrac{c}{d}$.   **(b)** If $ad < bc$, then $\dfrac{a}{b}$ $\dfrac{c}{d}$.   **(c)** If $ad > bc$, then $\dfrac{a}{b}$ $\dfrac{c}{d}$.

    (d) The products ad and bc are called _____.

Place the correct symbol, $=$, $>$, or $<$, between the given pairs of fractions.

26 $\dfrac{17}{43}$ $\dfrac{28}{79}$ **27.** $\dfrac{35}{11}$ $\dfrac{41}{13}$ **28.** $\dfrac{4}{11}$ $\dfrac{7}{19}$

29. $\dfrac{13}{50}$ $\dfrac{39}{150}$ **30.** $\dfrac{5}{7}$ $\dfrac{60}{83}$ **31.** $\dfrac{2}{9}$ $\dfrac{21}{91}$

Evaluate the absolute values.

32. $|17|$ **33.** $|-1|$ **34.** $|0|$ **35.** $\left|-\dfrac{3}{4}\right|$

36. $|3.1|$ **37.** $\left|-\dfrac{5}{2}\right|$ **38.** $|45|$ **39.** $|-0.8|$

ANSWERS: 1. true   2. false $(x < y)$ 3. true   4. false $(-32°)$ 5. true   6. false (4500)   7. true   8. $>$
9. $<$ 10. $<$ 11. $<$ 12. $<$ 13. $>$ 14. $<$ 15. $=$ 16. true   17. true   18. false $(0 > -3)$ 19. true

20. false $(-2 > -7)$ 21. false $\left(\frac{2}{3} > -\frac{2}{3}\right)$ 22. true   23. true   24.

25. (a) $=$ (b) $<$ (c) $>$ (d) cross products   26. $>$ 27. $>$ 28. $<$ 29. $=$ 30. $<$ 31. $<$ 32. 17   33. 1   34. 0
35. $\frac{3}{4}$ 36. 3.1   37. $\frac{5}{2}$ 38. 45   39. 0.8

2.1 EXERCISES B

Answer true *or* false. *If the statement is false, tell why.*

1. $\{\ldots, -3, -2, -1\}$ is called the set of negative integers.

2. If a number w is located to the right of a number v on a number line, then $w < v$.

3. The numeral -11 is read "negative eleven."

4. The low temperature last winter in Twin Falls, Idaho of $38°$ below zero could be written $-38°$.

5. The absolute value of z, denoted by $|z|$, is the distance from zero to z on a number line.

6. An airplane flying at an altitude of 25,000 feet could be described as flying at an altitude of $-25,000$ feet.

7. A check in the amount of \$35.00 written on an account could be denoted by $-\$35.00$.

Place the correct symbol, $=$, $>$, or $<$, between the given pairs of numbers.

8. 4   9

9. 5   0

10. 0 -7

11. 3   12

12. -5 1

13. $\dfrac{2}{3}$ $-\dfrac{2}{3}$

14. -0.2 0.2

15. -6 -6

Is the given statement true or false? If the statement is false, tell why.

16. $-3 \geq -4$

17. $-1 \geq -1$

18. $0 \leq -1$

19. $-8 \leq -8$

20. $-5 \leq -40$

21. $\dfrac{1}{4} \geq -\dfrac{1}{4}$

22. $3\dfrac{1}{5} \leq 3.2$

23. $-0.1 \leq -\dfrac{1}{10}$

24. Plot the numbers $\dfrac{1}{3}$, $-\dfrac{3}{4}$, $\dfrac{7}{4}$, $-\dfrac{9}{4}$, $\dfrac{7}{2}$, and $-\dfrac{11}{3}$ on a number line.

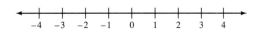

25. Insert the correct symbol, $=$, $>$, or $<$, between the two positive fractions.

(a) If $xw < yv$ then $\dfrac{x}{y}$ $\dfrac{v}{w}$.

(b) If $xw = yv$ then $\dfrac{x}{y}$ $\dfrac{v}{w}$.

(c) If $xw > yv$ then $\dfrac{x}{y}$ $\dfrac{v}{w}$.

(d) The cross products used to determine order on the two rational numbers $\dfrac{x}{y}$ and $\dfrac{v}{w}$ are _____ and _____.

Place the correct symbol, $=$, $>$, or $<$, between the given pairs of fractions.

26. $\dfrac{3}{5}$ $\dfrac{7}{10}$

27. $\dfrac{36}{11}$ $\dfrac{77}{20}$

28. $\dfrac{19}{43}$ $\dfrac{57}{129}$

29. $\dfrac{6}{7}$ $\dfrac{47}{55}$

30. $\dfrac{121}{9}$ $\dfrac{55}{3}$

31. $\dfrac{27}{201}$ $\dfrac{13}{121}$

Evaluate the absolute values.

32. $|25|$

33. $|-2|$

34. $\left|-\dfrac{1}{2}\right|$

35. $|-0|$

36. $|8.7|$

37. $\left|-\dfrac{9}{4}\right|$

38. $|112|$

39. $|-0.03|$

2.1 EXERCISES C

Place the correct symbol, =, >, or <, between the pairs of fractions.

1. $-\dfrac{14}{19}$ ___ $-\dfrac{2}{3}$

2. $-\dfrac{21}{8}$ ___ $-\dfrac{31}{11}$

3. $-\dfrac{77}{14}$ ___ $-\dfrac{11}{2}$

2.2   ADDITION AND SUBTRACTION OF RATIONAL NUMBERS

STUDENT GUIDEPOSTS

1 Adding Numbers with Like Signs

2 Adding Numbers with Unlike Signs

3 Additive Identity

4 Additive Inverse (Negative)

5 Commutative Law of Addition

6 Associative Law of Addition

7 Adding More Than Two Numbers

8 Subtracting Numbers

In Chapter 1 we reviewed the four basic operations on the nonnegative rational numbers. Now we extend our treatment to addition and subtraction of all rational numbers. Consider the addition problem $3 + 2$ on the number line in Figure 2.9.

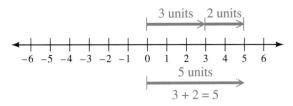

Figure 2.9

To find $3 + 2$, we start at 0, move 3 units to the *right,* and then move 2 more units to the *right*. We are then 5 units to the *right* of zero. Thus, $3 + 2 = 5$.

We can also find $3 + (-2)$ on a number line. We first move 3 units to the *right* and then 2 units to the *left*. This puts us 1 unit to the *right* of 0, as shown in Figure 2.10. Thus, $3 + (-2) = 1$.

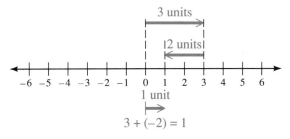

Figure 2.10

If we think of numbers as temperatures, an increase of $3°$ $(+3)$ followed by a second increase of $2°$ $(+2)$ results in a total increase of $5°$ $(+5)$. Similarly, an increase of $3°$ $(+3)$ followed by a decrease of $2°$ (-2) results in a net increase of $1°$ $(+1)$.

Notice that when adding numbers using a number line, we move to the *right when a number is positive* and to the *left when it is negative.*

| **EXAMPLE 1   ADDING NUMBERS WITH LIKE SIGNS** | **PRACTICE EXERCISE 1** |

Find $(-3) + (-2)$.

Draw a number line. Move 3 units to the left for -3 and then 2 more units to the left for -2. As seen in Figure 2.11, we end up 5 units to the left of 0, at -5. Thus, $(-3) + (-2) = -5$.

Find $(-11) + (-8)$.

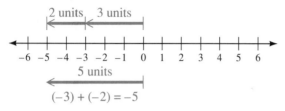

Figure 2.11

Answer: -19

| **EXAMPLE 2   ADDING NUMBERS WITH UNLIKE SIGNS** | **PRACTICE EXERCISE 2** |

Find $(-3) + 2$.

Draw a number line. Move 3 units to the left and then 2 units to the right. As Figure 2.12 shows, we are then at -1. Thus, $(-3) + 2 = -1$.

Find $(-11) + 8$.

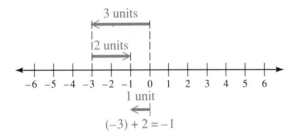

Figure 2.12

Answer: -3

| **EXAMPLE 3   ADDING NUMBERS WITH UNLIKE SIGNS** | **PRACTICE EXERCISE 3** |

Find $2 + (-3)$.

Draw a number line. Move 2 units to the right and then 3 units to the left. (See Figure 2.13.) This puts us at -1. Thus, $2 + (-3) = -1$.

Find $8 + (-11)$.

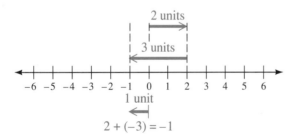

Figure 2.13

Answer: -3

➊ ADDING NUMBERS WITH LIKE SIGNS

The number-line method of adding numbers shows us what addition means, but it takes too much time. For example, we found $(-3) + (-2)$ to be -5 using a number line (Figure 2.11). However, we could have added $3 + 2$ and attached a minus sign:

$$(-3) + (-2) = -(3 + 2) = -5.$$

This is an example of the following rule.

> ### Adding Numbers Having Like Signs
>
> To add two numbers having **like signs,** add their absolute values. The sum has the same sign as the numbers being added.

➋ ADDING NUMBERS WITH UNLIKE SIGNS

When the signs of the numbers being added are not the same, we need another rule. For $3 + (-2) = 1$ (Figure 2.10), we could have found the difference between 3 and 2 and attached a plus sign:

$$3 + (-2) = +(3 - 2) = +1 = 1.$$

Also, for $(-3) + 2 = 2 + (-3) = -1$ (Figures 2.12 and 2.13), we could have subtracted 2 from 3 and attached a minus sign:

$$(-3) + 2 = 2 + (-3) = -(3 - 2) = -1.$$

These are examples of the following rule.

> ### Adding Numbers Having Unlike Signs
>
> To add two numbers having **unlike signs,** subtract the smaller absolute value from the larger absolute value. The result has the same sign as the number with the larger absolute value. If the absolute values are the same, the sum is zero.

Remember that if a number (other than zero) has no sign, this is the same as having a plus sign attached. For example, 5 and +5 are the same.

EXAMPLE 4 ADDING RATIONAL NUMBERS	PRACTICE EXERCISE 4

Add using the preceding rules.

(a) $6 + 7 = 13$

(b) $(-6) + (-7) = -(6 + 7) = -13$

(c) $6 + (-7) = -(7 - 6) = -1$ $|-7| = 7 > 6 = |6|$

(d) $(-6) + 7 = +(7 - 6) = +1 = 1$ $|-6| = 6 < 7 = |7|$

(e) $12 + (-12) = 0$ $|12| = 12 = |-12|$

Add.

(a) $5 + 14$

(b) $(-5) + (-14)$

(c) $5 + (-14)$

(d) $(-5) + 14$

(e) $(-14) + 14$

Answers: (a) **19**   (b) **−19**
(c) **−9**   (d) **9**   (e) **0**

❸ ADDITIVE IDENTITY

The number 0 plays a special role in our number system since for any number a,

$$a + 0 = 0 + a = a.$$

Intuitively, we see any number a remains identically the same when added to 0. For this reason, 0 is called the **additive identity.**

❹ ADDITIVE INVERSE (NEGATIVE)

Example 4(e) illustrates another property of our number system. If a is any number, we call $-a$ the **additive inverse,** or **negative** of a, which means

$$a + (-a) = (-a) + a = 0.$$

EXAMPLE 5 ADDITIVE IDENTITY AND INVERSES	PRACTICE EXERCISE 5

Add.

(a) $0 + 7 = 7 + 0 = 7$ 0 is the additive identity

(b) $0 + (-7) = (-7) + 0 = -7$ 0 is the additive identity

(c) $(-2.3) + (-4.1) = -(2.3 + 4.1) = -6.4$

(d) $(-3.5) + 1.7 = -(3.5 - 1.7) = -1.8$

(e) $\frac{1}{2} + \left(-\frac{3}{4}\right) = -\left(\frac{3}{4} - \frac{1}{2}\right) = -\left(\frac{3}{4} - \frac{2}{4}\right) = -\frac{1}{4}$

Add.

(a) $13 + 0$

(b) $(-13) + 0$

(c) $(-4.02) + (-1.35)$

(d) $(-6.35) + 2.11$

(e) $\left(-\frac{2}{9}\right) + \frac{4}{11}$

Answers: (a) 13 (b) -13 (c) -5.37 (d) -4.24 (e) $\frac{14}{99}$

❺ COMMUTATIVE LAW OF ADDITION

With two more properties of addition, we can change the order of addition, and rearrange grouping symbols in sums of three (or more) numbers. If a and b are two numbers, the **commutative law of addition** states that

$$a + b = b + a.$$

That is, the order of addition can be changed by the commutative law. For example, $16 + 5 = 5 + 16$.

❻ ASSOCIATIVE LAW OF ADDITION

Given another number c, the **associative law of addition** states that

$$(a + b) + c = a + (b + c).$$

That is, the grouping symbols can be rearranged by the associative law. For example, $(5 + 2) + 3 = 5 + (2 + 3)$.

EXAMPLE 6 COMMUTATIVE AND ASSOCIATIVE LAWS	PRACTICE EXERCISE 6

(a) Use the commutative law to complete the equation $5 + \frac{1}{4} =$ _____ . The commutative law says the order of addition can be changed, so

$$5 + \frac{1}{4} = \frac{1}{4} + 5.$$

(a) Complete the following using the commutative law.
$\left(-\frac{1}{2}\right) + 9 =$ _____

(b) Use the associative law to complete the equation $(7 + \frac{1}{2}) + (-2) =$ _____ .

The associative law says the grouping symbols can be rearranged, so

$$\left(7 + \frac{1}{2}\right) + (-2) = 7 + \left[\frac{1}{2} + (-2)\right].$$

(b) Complete the following using the associative law.
$$(7 + 0.5) + 2 = \underline{\hphantom{xxx}}$$

Answers: (a) $9 + \left(-\frac{1}{2}\right)$
(b) $7 + (0.5 + 2)$

7 ADDING MORE THAN TWO NUMBERS

Taken together, the commutative and associative laws of addition allow us to add several signed numbers by reordering and regrouping the numbers to obtain the sum in the most convenient manner.

Adding More Than Two Numbers

1. Add all positive numbers.
2. Add all negative numbers.
3. Now use the rule for adding numbers with unlike signs.

EXAMPLE 7   ADDING SEVERAL NUMBERS

Add.

$$(-2) + (-3) + 2 + 7 + (-5) + 6 + 9 + (-1)$$
$$= [2 + 7 + 6 + 9] + [(-2) + (-3) + (-5) + (-1)]$$
$$= [2 + 7 + 6 + 9] + [-(2 + 3 + 5 + 1)]$$
$$= [24] + [-11] = +[24 - 11] = 13$$

PRACTICE EXERCISE 7

Add.

$$5 + (-2) + (-1) + 7 + (-9) + 6$$

Answer: 6

With practice we should be able to skip many of the intermediate steps in the preceding examples and make computations mentally.

8 SUBTRACTING NUMBERS

Subtraction of numbers can be defined as addition of an additive inverse or a negative. For example, we can think of $5 - 3 = 2$ as $5 + (-3) = 2$.

Subtracting Numbers

To subtract one number from another, change the sign of the number being subtracted and use the rule for adding numbers. Thus, for rational numbers a and b,

$$a - b = a + (-b).$$

EXAMPLE 8   SUBTRACTING NUMBERS

Subtract.

(a) $5 - (+3) = 5 + (-3)$ Change sign and add
$$= 5 - 3 = 2$$

(b) $5 - (-3) = 5 + 3 = 8$ Change sign and add

PRACTICE EXERCISE 8

Subtract.

(a) $17 - (+2)$

(b) $17 - (-2)$

(c) $(-5) - 3 = (-5) + (-3)$ Change sign and add
$= -(5 + 3) = -8$

(d) $(-5) - (-3) = (-5) + 3$ Change sign and add
$= -(5 - 3) = -2$

(e) $(-5) - 0 = (-5) + (-0)$ Change sign and add
$= (-5) + 0$ The negative of zero is zero: $-0 = 0$
$= -5$

(c) $(-17) - 2$

(d) $(-17) - (-2)$

(e) $0 - (-17)$

Answers: (a) **15**   (b) **19**
(c) **−19**   (d) **−15**   (e) **17**

| EXAMPLE 9   Subtracting Numbers | PRACTICE EXERCISE 9 |

Subtract.

(a) $5 - (+5) = 5 + (-5) = 0$

(b) $5 - (-5) = 5 + 5 = 10$

(c) $(-5) - 5 = (-5) + (-5) = -(5 + 5) = -10$

(d) $(-5) - (-5) = (-5) + 5 = 0$

(e) $(-5.8) - (-3.2) = (-5.8) + 3.2 = -(5.8 - 3.2) = -2.6$

(f) $\left(-\dfrac{2}{3}\right) - \dfrac{1}{6} = \left(-\dfrac{2}{3}\right) + \left(-\dfrac{1}{6}\right) = \left(-\dfrac{4}{6}\right) + \left(-\dfrac{1}{6}\right) = -\dfrac{5}{6}$

Subtract.

(a) $12 - (+12)$

(b) $12 - (-12)$

(c) $(-12) - 12$

(d) $(-12) - (-12)$

(e) $(-1.2) - (-5.7)$

(f) $\left(-\dfrac{1}{8}\right) - \dfrac{3}{5}$

Answers: (a) **0**   (b) **24**   (c) **−24**
(d) **0**   (e) **4.5**   (f) $-\frac{29}{40}$

CAUTION

Although addition satisfies both the commutative and associative laws, subtraction does not. For example,

$3 = 5 - 2 \neq 2 - 5 = -3.$ Subtraction is *not* commutative

Likewise,

$(8 - 4) - 3 = 4 - 3 = 1,$

but $8 - (4 - 3) = 8 - 1 = 7.$ Subtraction is *not* associative

---

| EXAMPLE 10   Application to Personal Banking | PRACTICE EXERCISE 10 |

John had \$932 in his bank account on May 1. He deposited \$326 and \$791 and wrote checks for \$816, \$315, and \$940 during the month. What was his balance at the end of May?

We can consider the deposits as positive numbers and the checks written as negative numbers. We need to add 932, 326, 791, −816, −315, and −940.

932	−816		
326	−315	−2071	
+ 791	−940	2049	
2049 Sum of deposits	− 2071 Sum of checks	− 22 Final sum	

Thus, John is overdrawn (in the red) \$22 on his account.

Marlo had a temperature of 100.1° at noon. It rose 2.3° before dropping 3.5° by 3:00 P.M. What was her temperature at 3:00 P.M.?

Answer: **98.9°**

2.2 EXERCISES A

Perform the indicated operation.

1. $4 + 3$

2. $(-4) + (-3)$

3. $4 + (-3)$

4. $(-4) + 3$

5. $4 + 0$

6. $0 + (-4)$

7. $7 + (-7)$

8. $(-7) + 7$

9. $(-7) + (-7)$

10. $7 + 7$

11. $(-8) + (-5)$

12. $9 + (-1)$

13. $19 + (-25)$

14. $(-16) + 14$

15. $4 - 3$

16. $(-4) - (-3)$

17. $4 - (-3)$

18. $(-4) - 3$

19. $0 - 4$

20. $(-4) - 0$

21. $7 - (-7)$

22. $(-7) - 7$

23. $(-7) - (-7)$

24. $7 - 7$

25. $(-8) - (-5)$

26. $9 - (-1)$

27. $(-1) - 9$

28. $14 - 21$

29. $19 - (-25)$

30. $(-16) - 14$

31. $1.3 + (-2.5)$

32. $(-1.3) + 2.5$

33. $(-1.3) + (-2.5)$

34. $1.3 - (-2.5)$

35. $(-1.3) - 2.5$

36. $(-1.3) - (-2.5)$

37. $\dfrac{1}{2} + \left(-\dfrac{2}{5}\right)$

38. $\left(-\dfrac{1}{2}\right) + \dfrac{2}{5}$

39. $\left(-\dfrac{1}{2}\right) + \left(-\dfrac{2}{5}\right)$

40. $\dfrac{1}{2} - \left(-\dfrac{2}{5}\right)$

41. $\left(-\dfrac{1}{2}\right) - \dfrac{2}{5}$

42. $\left(-\dfrac{1}{2}\right) - \left(-\dfrac{2}{5}\right)$

43. $(-3) + 2 + (-7) + (-8) + 9$

44. $5 + 8 + 9 + (-3) + (-7)$

45. $(-8) + (-6) + 9 + (-3) + 5 + (-6)$

46. $(-4) + (-2) + 6 + 7 + (-1)$

47. Show by example that subtraction does not satisfy the commutative law. Can you find numbers a and b for which $a - b$ does equal $b - a$?

Add.

48. 23
-16
-36
5
$\underline{81}$

49. -16
-81
-14
-70
$\underline{-93}$

50. -64
25
38
-17
$\underline{-83}$

51 When a cold front came through Cut Bank, Montana, the temperature dropped from 35° above zero to 13° below zero (−13°). What was the total change in temperature?

52. On March 1, Allen had $64 in his bank account. During the month, he wrote checks for $81, $108, and $192 and made deposits of $123 and $140. What was the value of his account at the end of the month?

53. A helicopter is 600 ft above sea level and a submarine directly below it is 250 ft below sea level (−250 ft). How far apart are they?

54. At 5:30 A.M., it was −42°F. By 2:00 P.M. the same day, the temperature had risen 57° to the recorded high temperature. Shortly thereafter, a cold front passed through dropping the temperature 41° by 5:00 P.M. What was the temperature at 5:00 P.M.?

FOR REVIEW

Place the correct symbol, =, <, or >, between the given pairs of numbers.

55. −5   −4

56. $\dfrac{2}{3}$ $\dfrac{77}{110}$

57. $\dfrac{5}{12}$ $\dfrac{30}{72}$

58. −4.7   −1.2

Evaluate the absolute values.

59. $|-81|$

60. $\left|-\dfrac{5}{4}\right|$

61. $|17|$

62. $|-7.25|$

ANSWERS: 1. 7   2. −7   3. 1   4. −1   5. 4   6. −4   7. 0   8. 0   9. −14   10. 14   11. −13   12. 8   13. −6   14. −2   15. 1   16. −1   17. 7   18. −7   19. −4   20. −4   21. 14   22. −14   23. 0   24. 0   25. −3   26. 10   27. −10   28. −7   29. 44   30. −30   31. −1.2   32. 1.2   33. −3.8   34. 3.8   35. −3.8   36. 1.2   37. $\frac{1}{10}$ 38. $-\frac{1}{10}$ 39. $-\frac{9}{10}$ 40. $\frac{9}{10}$ 41. $-\frac{9}{10}$ 42. $-\frac{1}{10}$ 43. −7   44. 12   45. −9   46. 6   47. $2 - 1 = 1 \neq -1 = 1 - 2$; if $a = b$ then $a - b = 0 = b - a$ 48. 57   49. −274   50. −101   51. −48°   52. −$54   53. 850 ft   54. −26°   55. <   56. <   57. =   58. <   59. 81   60. $\frac{5}{4}$ 61. 17   62. 7.25

2.2 EXERCISES B

Perform the indicated operations.

1. $5 + 2$

2. $(-5) + (-2)$

3. $5 + (-2)$

4. $(-5) + 2$

5. $0 + 5$

6. $(-5) + 0$

7. $5 + (-5)$

8. $(-5) + 5$

9. $(-5) + (-5)$

10. $5 + 5$

11. $(-9) + (-3)$

12. $(-2) + 11$

13. $18 + (-23)$

14. $(-15) + 19$

15. $8 - 7$

16. $(-8) - (-7)$

17. $8 - (-7)$

18. $(-8) - 7$

19. $10 - 0$

20. $0 - (-10)$

21. $6 - (-6)$

22. $(-6) - 6$

23. $(-6) - (-6)$

24. $6 - 6$

25. $(-11) - (-4)$

26. $(-13) - (-1)$

27. $(-1) - 7$

28. $(-12) - 24$

29. $18 - (-27)$

30. $(-17) - 11$

31. $1.5 + (-3.8)$

32. $(-1.5) + 3.8$

33. $(-1.5) + (-3.8)$

34. $1.5 - (-3.8)$

35. $(-1.5) - 3.8$

36. $(-1.5) - (-3.8)$

37. $\dfrac{1}{4} + \left(-\dfrac{3}{8}\right)$

38. $\left(-\dfrac{1}{4}\right) + \dfrac{3}{8}$

39. $\left(-\dfrac{1}{4}\right) + \left(-\dfrac{3}{8}\right)$

40. $\dfrac{1}{4} - \left(-\dfrac{3}{8}\right)$

41. $\left(-\dfrac{1}{4}\right) - \dfrac{3}{8}$

42. $\left(-\dfrac{1}{4}\right) - \left(-\dfrac{3}{8}\right)$

43. $(-5) + 4 + (-9) + (-10) + 6$

44. $5 + 9 + (-3) + 1 + (-10)$

45. $(-5) + (-7) + 10 + (-3) + 4 + (-7)$

46. $(-6) + (-3) + 6 + 10 + (-2)$

47. Show by example that subtraction does not satisfy the associative law. Can you find numbers, a, b, and c for which $(a - b) - c$ does equal $a - (b - c)$?

Add.

48.
```
  26
 -14
 -51
   3
  92
```

49.
```
 -17
 -20
 -43
  -5
-110
```

50.
```
 -78
  32
  51
 -60
 -92
```

51. In a five-hour period of time, the temperature in Empire, N.Y. dropped from 41° to 12° below zero. What was the total temperature change?

52. On July 4, Uncle Sam had $97 in his checking account. During the next week he wrote checks for $15, $102, and $312, and made deposits of $145 and $170. What was the value of his account at the end of the week?

53. A balloon is 1200 ft above sea level, and a diving bell directly below it is 340 ft beneath the surface of the water. How far apart are the two?

54. At 8:00 A.M., Leah Johnson had a temperature of 99.8°. It rose another 3.1° before falling 4.3° by noon. What was her temperature at noon?

FOR REVIEW

Place the correct symbol, =, <, or >, between the given pairs of numbers.

55. $-6 \quad -7$

56. $\dfrac{3}{4} \quad \dfrac{96}{120}$

57. $\dfrac{7}{8} \quad \dfrac{77}{88}$

58. $0 \quad -11.5$

Evaluate the absolute values.

59. $\left| -\dfrac{9}{10} \right|$

60. $|-210|$

61. $|42|$

62. $|-0.01|$

2.2 EXERCISES C

Perform the indicated operations.

1. $(-962) - (-508)$

2. $-48.62 - 40.002$

3. $64\frac{7}{24} - 98\frac{13}{16}$

[Answer: $-34\frac{25}{48}$]

2.3 MULTIPLICATION AND DIVISION OF RATIONAL NUMBERS

STUDENT GUIDEPOSTS

1 Multiplying Numbers

2 Multiplicative Identity

3 Multiplicative Inverse (Reciprocal)

4 Commutative Law of Multiplication

5 Associative Law of Multiplication

6 Dividing Numbers

7 Double Sign Properties

1 MULTIPLYING NUMBERS

In multiplying and dividing numbers, we need to decide what sign to give the product or quotient. To help us, we look for patterns in several products.

Decreases by 1 each time ⟶   Decreases by 2 each time ⟵

$$
\begin{aligned}
4 \cdot 2 &= 8 \\
3 \cdot 2 &= 6 \quad \text{6 is 2 less than 8}\\
2 \cdot 2 &= 4 \\
1 \cdot 2 &= 2 \\
0 \cdot 2 &= 0 \\
(-1) \cdot 2 &= -2 \quad \text{-2 is 2 less than 0}\\
(-2) \cdot 2 &= -4 \quad \text{-4 is 2 less than -2}\\
(-3) \cdot 2 &= -6 \\
(-4) \cdot 2 &= -8
\end{aligned}
$$

We know the products $4 \cdot 2$, $3 \cdot 2$, $2 \cdot 2$, $1 \cdot 2$, and $0 \cdot 2$, and we can see that these products decrease by 2 each time. For this pattern to continue, a negative number times a positive number must be negative. For example,

$$(-1) \cdot 2 = -2, \quad (-2) \cdot 2 = -4, \quad (-3) \cdot 2 = -6, \quad (-4) \cdot 2 = -8.$$

Similarly, any positive number times any negative number must be negative. We use this to find a pattern in the following products.

Decreases by 1 each time ⟶   Increases by 2 each time ⟵

$$
\begin{aligned}
4 \cdot (-2) &= -8 \\
3 \cdot (-2) &= -6 \quad \text{-6 is 2 more than -8}\\
2 \cdot (-2) &= -4 \\
1 \cdot (-2) &= -2 \\
0 \cdot (-2) &= 0 \quad \text{0 is 2 more than -2}\\
(-1) \cdot (-2) &= 2 \\
(-2) \cdot (-2) &= 4 \\
(-3) \cdot (-2) &= 6 \\
(-4) \cdot (-2) &= 8
\end{aligned}
$$

For this pattern to continue, a negative number times a negative number must be a positive number. For example,

$$(-1) \cdot (-2) = 2, \quad (-2) \cdot (-2) = 4, \quad (-3) \cdot (-2) = 6, \quad (-4) \cdot (-2) = 8.$$

Our results are summarized in the following rule.

Multiplying Numbers

1. Multiply the absolute values of the numbers.
2. If both signs are positive or both negative, the product is positive.
3. If one sign is positive and one is negative, the product is negative.
4. If one number (or both numbers) is zero, the product is zero.

When multiplying numbers, remember that *like signs have a positive product and unlike signs have a negative product.*

EXAMPLE 1 MULTIPLYING NUMBERS	PRACTICE EXERCISE 1

Multiply.

(a) $5 \cdot 2 = 10$ Like signs

(b) $(-5)(-2) = (5 \cdot 2) = 10$ Like signs

(c) $5 \cdot (-2) = -(5 \cdot 2) = -10$ Unlike signs

(d) $(-5) \cdot 2 = -(5 \cdot 2) = -10$ Unlike signs

(e) $0 \cdot (-8) = 0$

(f) $0 \cdot 0 = 0$

(g) $(3)\left(\dfrac{1}{3}\right) = 1$

Multiply.

(a) $4 \cdot 11$

(b) $(-4)(-11)$

(c) $4 \cdot (-11)$

(d) $(-4) \cdot 11$

(e) $(12) \cdot 0$

(f) $0 \cdot (-12)$

(g) $\left(\dfrac{3}{7}\right)\left(\dfrac{7}{3}\right)$

Answers: (a) 44   (b) 44
(c) −44   (d) −44   (e) 0   (f) 0
(g) 1

② MULTIPLICATIVE IDENTITY

The number 1 plays a role in multiplication similar to the one 0 plays in addition. Since for any number a,

$$a \cdot 1 = 1 \cdot a = a,$$

we see that a number remains identically the same when multiplied by 1. For this reason, 1 is called the **multiplicative identity.**

③ MULTIPLICATIVE INVERSE (RECIPROCAL)

Example 1(g) illustrates a property of reciprocals that we have seen before. If a is any number except 0, we call $\frac{1}{a}$ the **multiplicative inverse** (or **reciprocal**) of a, which means that

$$a \cdot \frac{1}{a} = \frac{1}{a} \cdot a = 1.$$

EXAMPLE 2 MULTIPLICATIVE IDENTITY AND INVERSES	PRACTICE EXERCISE 2

Multiply.

(a) $(-5)\left(-\dfrac{1}{5}\right) = 1$ -5 and $-\frac{1}{5}$ are reciprocals or multiplicative inverses

(b) $1 \cdot (-7) = (-7) \cdot 1 = -7$ 1 is the multiplicative identity

(c) $(-2.1) \cdot 3.5 = -(2.1)(3.5) = -7.35$

(d) $(-2.1) \cdot (-3.5) = (2.1)(3.5) = 7.35$

(e) $\left(\dfrac{1}{4}\right)\left(-\dfrac{3}{5}\right) = -\left(\dfrac{1}{4}\right)\left(\dfrac{3}{5}\right) = -\dfrac{3}{20}$

(f) $\left(-\dfrac{1}{4}\right)\left(-\dfrac{3}{5}\right) = \left(\dfrac{1}{4}\right)\left(\dfrac{3}{5}\right) = \dfrac{3}{20}$

Multiply.

(a) $\left(-\dfrac{1}{9}\right)(-9)$

(b) $1 \cdot (-13)$

(c) $(-4.1) \cdot (-2.9)$

(d) $(-4.1) \cdot (2.9)$

(e) $\left(\dfrac{1}{7}\right)\left(-\dfrac{3}{8}\right)$

(f) $\left(-\dfrac{1}{7}\right)\left(-\dfrac{3}{8}\right)$

Answers: (a) **1**   (b) **−13**
(c) **11.89**   (d) **−11.89**   (e) $-\frac{3}{56}$
(f) $\frac{3}{56}$

④ COMMUTATIVE LAW OF MULTIPLICATION

With two more properties of multiplication, like the ones for addition, we can change the order of multiplication and rearrange grouping symbols in products of three (or more) numbers. If a and b are two numbers, the **commutative law of multiplication** states that

$$a \cdot b = b \cdot a.$$

That is, the order of multiplication can be changed by the commutative law. For example, $4 \cdot 7 = 7 \cdot 4$.

⑤ ASSOCIATIVE LAW OF MULTIPLICATION

Given another number c, the **associative law of multiplication** states that

$$(a \cdot b) \cdot c = a \cdot (b \cdot c).$$

That is, the grouping symbols can be rearranged by the associative law. For example, $(2 \cdot 5) \cdot 3 = 2 \cdot (5 \cdot 3)$.

EXAMPLE 3 COMMUTATIVE AND ASSOCIATIVE LAWS	PRACTICE EXERCISE 3

(a) Use the commutative law to complete the equation $7 \cdot \frac{3}{4} =$ _____.
The commutative law says the order of multiplication can be changed, so

$$7 \cdot \frac{3}{4} = \frac{3}{4} \cdot 7.$$

(b) Use the associative law to complete the equation $(3 \cdot \frac{1}{5}) \cdot (-5) =$ _____.
The associative law says the grouping symbols can be rearranged, so

$$\left(3 \cdot \frac{1}{5}\right) \cdot (-5) = 3 \cdot \left(\frac{1}{5} \cdot (-5)\right).$$

(a) Use the commutative law to complete the following.
$(0.09) \cdot (1000) =$ _____

(b) Use the associative law to complete the following.
$\left(\dfrac{1}{2} \cdot \dfrac{3}{10}\right) \cdot (-6) =$ _____

Answers: (a) $(1000) \cdot (0.09)$
(b) $\left(\frac{1}{2}\right) \cdot \left(\frac{3}{10} \cdot (-6)\right)$

Often we will need to multiply more than two numbers. To do this, multiply in pairs and keep track of the sign. Notice in Example 4 that *products involving an odd number of minus signs are negative while those with an even number are positive.*

EXAMPLE 4   MULTIPLYING MORE THAN TWO NUMBERS

Multiply.

(a) $(3)(-2)(-4) = (-6)(-4) = 24$ Multiply 3 times -2 first

(b) $(-1)(-1)(-2)(-3)(-4) = (1)(-2)(-3)(-4)$ $(-1)(-1) = 1$
$$= (-2)(-3)(-4) \qquad (1)(-2) = -2$$
$$= (6)(-4) \qquad (-2)(-3) = 6$$
$$= -24$$

(c) $(-2)^3 = (-2)(-2)(-2) = (4)(-2)$ $(-2)(-2) = 4$
$$= -8$$

PRACTICE EXERCISE 4

Multiply.

(a) $(2)(-1)(-4)(3)$

(b) $(-6)(-1)(-5)(-2)(-1)$

(c) $(-3)^4$

Answers: (a) **24**   (b) **−60**
(c) **81**

⑥ DIVIDING NUMBERS

We have seen that the only difference between multiplying positive numbers and multiplying rational numbers is finding the sign of the product. Recall that division can be thought of as multiplication by a reciprocal. For example,

$$15 \div 3 \quad \text{or} \quad \frac{15}{3} \qquad \text{is the same as} \qquad 15 \cdot \frac{1}{3}.$$

Thus, we can use the same rules for signs in division that we used for multiplication.

Dividing Numbers

1. Divide the absolute values of the numbers.

2. If both signs are positive or both negative, the quotient is positive.

3. If one sign is positive and one negative, the quotient is negative.

4. Zero divided by any number (except zero) is zero, and division of any number by 0 is undefined.

Remember that the rule of signs for division is the same as for multiplication: *like signs have a positive quotient and unlike signs have a negative quotient.*

EXAMPLE 5   DIVIDING NUMBERS

Divide.

(a) $6 \div 3 = 2$ Like signs

(b) $(-6) \div (-3) = 6 \div 3 = 2$ Like signs

(c) $6 \div (-3) = -(6 \div 3) = -2$ Unlike signs

(d) $(-6) \div 3 = -(6 \div 3) = -2$ Unlike signs

(e) $0 \div (-3) = 0$

(f) $(-3) \div 0$ is undefined

(g) $\dfrac{1}{2} \div \left(-\dfrac{3}{5}\right) = \dfrac{1}{2} \cdot \left(-\dfrac{5}{3}\right) = -\left(\dfrac{1}{2} \cdot \dfrac{5}{3}\right) = -\dfrac{5}{6}$

PRACTICE EXERCISE 5

Divide.

(a) $12 \div 4$

(b) $(-12) \div (-4)$

(c) $12 \div (-4)$

(d) $(-12) \div 4$

(e) $(-0.3) \div 0$

(f) $0 \div (-0.3)$

(g) $\left(-\dfrac{3}{5}\right) \div \dfrac{9}{10}$

Answers: (a) **3**   (b) **3**   (c) **−3**
(d) **−3**   (e) **undefined**   (f) **0**
(g) $-\dfrac{2}{3}$

///////////////////| **CAUTION** |///////////////////

Although multiplication satisfies both the commutative and associative laws, division does not. For example,

$$3 = 6 \div 2 \neq 2 \div 6 = \frac{2}{6} = \frac{1}{3}.$$ Division is not commutative

Likewise,

$$(8 \div 4) \div 2 = 2 \div 2 = 1$$

but $8 \div (4 \div 2) = 8 \div 2 = 4.$ Division is not associative

///////////////////|

| **EXAMPLE 6** APPLICATION TO PERSONAL BANKING | **PRACTICE EXERCISE 6** |

Willie wrote 3 checks for $75 each and 4 checks for $105 each. By how much did this change his bank balance?

The 3 checks for $75 could be calculated as follows:

$$(3)(-75) = -225.$$

There are 4 checks for $105, so

$$(4)(-105) = -420.$$

Thus, his account is changed by

$$-225 + (-420) = -225 - 420 = -645 \text{ dollars}.$$

His balance is $645 less.

A diver descended below the surface of the ocean by diving 25 ft each minute for 11 minutes. How far below sea level was he at this time?

Answer: 275 ft (-275)

| **EXAMPLE 7** APPLICATION TO PERSONAL FINANCE | **PRACTICE EXERCISE 7** |

Mary, Sue, and Maria share an apartment with total rent of $720. By how much is each woman's bank account changed if they each write a check for an equal share of the rent?

Since the rent is a decrease in their accounts, we use -720. To find the shares, we divide -720 by 3:

$$-720 \div 3 = -240.$$

Thus, each woman's account is decreased by $240.

During a 5-hr period, the temperature decreased 42°. What was the average decrease per hour?

Answer: 8.4° ($-8.4°$)

When evaluating numerical expressions involving rational numbers, follow the order of operations and rules of grouping given in Section 1.5. That is, evaluate within grouping symbols first, beginning with the innermost, evaluate powers next, and then perform multiplication and division left to right followed by addition and subtraction.

EXAMPLE 8 ORDER OF OPERATIONS	PRACTICE EXERCISE 8

Evaluate each numerical expression.

(a) $(-3) + 5 \cdot (-1) - (-4)$

$= (-3) + (-5) - (-4)$ Multiply first

$= (-8) - (-4)$ Add first from left

$= (-8) + 4$ Then subtract

$= -4$

(b) $(-2)^3 + 1 = (-2)(-2)(-2) + 1$ $a^3 = aaa$

$= (4)(-2) + 1$ $(-2)(-2) = 4$

$= (-8) + 1$ Multiply first

$= -7$ Then add

(c) $\dfrac{1}{2} - \left\{ \dfrac{1}{4} - \left[\left(-\dfrac{3}{4} \right) - \left(-\dfrac{1}{4} \right) \right] \right\}$

$= \dfrac{1}{2} - \left\{ \dfrac{1}{4} - \left[\left(-\dfrac{3}{4} \right) + \dfrac{1}{4} \right] \right\}$

$= \dfrac{1}{2} - \left\{ \dfrac{1}{4} - \left[-\dfrac{2}{4} \right] \right\}$ Innermost parentheses first

$= \dfrac{1}{2} - \left\{ \dfrac{1}{4} + \dfrac{2}{4} \right\}$

$= \dfrac{1}{2} - \left(\dfrac{3}{4} \right)$

$= \dfrac{2}{4} - \dfrac{3}{4} = -\dfrac{1}{4}$

Evaluate each numerical expression.

(a) $5 + (-3) \cdot (-2) - 6$

(b) $(-4)^2 + (-1)^3$

(c) $-0.15 + [3.2 - (5.65 - 1.6)]$

Answers: (a) **5**   (b) **15**   (c) **−1**

7 DOUBLE SIGN PROPERTIES

Notice in Example 8 that we use parentheses to avoid writing plus and minus signs right next to each other. The parentheses may be removed and the two signs replaced by one according to the following rule, part 4 of which is called the **double negative property.**

Double Sign Properties

If a is any number,

1. $+(+a) = +a$ 2. $+(-a) = -a$

3. $-(+a) = -a$ 4. $-(-a) = +a$.

Notice that when a plus sign is before parentheses, we *do not* change the sign inside when the parentheses and plus sign are removed. However, when a minus sign is before parentheses, we *do* change the sign inside when the parentheses and minus sign are removed.

EXAMPLE 9 DOUBLE SIGN PROPERTIES

Write without parentheses.
(a) $+(+6) = +6 = 6$

(b) $+(-8) = -8$

(c) $-(+7) = -7$

(d) $-(-5) = +5 = 5$

(e) $-[-(-2)] = -[+2] = -2$ Remove innermost parentheses first

---

PRACTICE EXERCISE 9

Write without parentheses.
(a) $+(+11)$

(b) $+(-19)$

(c) $-(+17)$

(d) $-(-15)$

(e) $-[+(-12)]$

Answers: (a) 11   (b) -19
(c) -17 (d) 15   (e) 12

EXAMPLE 10 EVALUATING EXPRESSIONS

Evaluate when $a = -3$ and $b = -2$.
(a) $a - b = (-3) - (-2) = (-3) + 2 = -1$ Use parentheses to substitute

(b) $b - [-a] = (-2) - [-(-3)] = (-2) - [3]$
$\qquad = -2 - 3 = -5$

(c) $2a - 3b = 2(-3) - 3(-2) = -6 - (-6) = -6 + 6 = 0$

(d) $|a - 2b| = |(-3) - 2(-2)| = |(-3) - (-4)|$
$\qquad = |(-3) + 4| = |1| = 1$

PRACTICE EXERCISE 10

Evaluate when $x = -5$ and $y = -1$.
(a) $x - y$

(b) $y - (-x)$

(c) $2x - 6y$

(d) $|y - 3x|$

Answers: (a) -4 (b) -6
(c) -4 (d) 14

EXAMPLE 11 EVALUATING EXPRESSIONS

Evaluate $3[2(a + 2) - 3(b + c)]$ when $a = -2$, $b = 4$, and $c = -8$.

$3[2(a + 2) - 3(b + c)] = 3[2(-2 + 2) - 3(4 + (-8))]$ Replace variables with numbers

$\qquad = 3[2(0) - 3(-4)]$ Add inside parentheses first

$\qquad = 3[0 + 12]$

$\qquad = 3[12]$

$\qquad = 36$

PRACTICE EXERCISE 11

Evaluate $-4[(u - v) - (w - u)]$ when $u = -3$, $v = 5$, and $w = -1$.

Answer: 40

2.3 EXERCISES A

Perform the indicated operations.

1. $(2)(6)$ **2.** $(-2)(-6)$ **3.** $(2)(-6)$ **4.** $(-2)(6)$

5. $0 \cdot (-6)$ **6.** $(-5) \cdot 7$ **7.** $(-3)(-9)$ **8.** $(-10)(8)$

9. $(9)(-12)$ **10.** $(12)(-12)$ **11.** $(-12)(-12)$ **12.** $(-15) \cdot 20$

13. $6 \div 2$

14. $(-6) \div (-2)$

15. $6 \div (-2)$

16. $(-6) \div 2$

17. $0 \div (-6)$

18. $(-6) \div 0$

19. $\dfrac{99}{-11}$

20. $\dfrac{-36}{-4}$

21. $\dfrac{-12}{12}$

22. $\dfrac{0}{-38}$

23. $\dfrac{-48}{-16}$

24. $\dfrac{52}{-13}$

25. $(1.2)(-2.3)$

26. $(-1.2)(-2.3)$

27. $(-1.2)(2.3)$

28. $\left(\dfrac{2}{5}\right)\left(-\dfrac{15}{4}\right)$

29. $\left(-\dfrac{2}{5}\right)\left(-\dfrac{15}{4}\right)$

30. $\left(-\dfrac{2}{5}\right)\left(\dfrac{15}{4}\right)$

31. $6.4 \div (-0.8)$

32. $(-6.4) \div (-0.8)$

33. $\dfrac{-6.4}{0.8}$

34. $\dfrac{3}{5} \div \left(-\dfrac{9}{15}\right)$

35. $\left(-\dfrac{3}{5}\right) \div \left(-\dfrac{9}{15}\right)$

36. $\dfrac{-\frac{3}{5}}{\frac{9}{15}}$

37. $(-1)(-1)(-1)$

38. $(-1)(2)(-3)$

39. $(-1)(-2)(-3)(-4)$

40. $(-1)(-2)(-3)(4)(-5)(6)(-1)$

41. $(-1)(-1)(-1)(-1)(-1)(-1)(-1)$

42. Show by example that division does not satisfy the associative law. Can you find numbers a, b, and c for which $(a \div b) \div c$ does equal $a \div (b \div c)$?

43. Tony had to write four $12 checks during the month of June. By how much did this change his bank balance?

44 Martha had 423 points in a contest, but then she received 20 penalty points 3 times. What was her total after the 3 penalties?

Evaluate each numerical expression.

45. $(-3)^3 + 27$

46. $(-1)^5 + (-1)^3$

47. $2 - [3 - (4 - 1)]$

48. $4 + 2[(-1) - (-5)]$

49. $(3 - 8) \cdot 4 + 1$

50 $3 - \dfrac{2 + (-4)}{1 - 5}$

Write without parentheses using only one sign (or no sign).

51. $-(-3)$ **52.** $-(+4)$ **53.** $+(-x)$ **54.** $-(-x)$

55. $-[+(-2)]$ **56.** $-[-(+3)]$ **57.** $-[-(-4)]$ **58.** $+[-(-10)]$

Evaluate the following expressions when $a = -1$, $b = 2$, and $c = -4$.

59. $a^2 - 1$ **60.** $abc + b^2$ **61.** $3(a + b) + c$

62. $2(b - a) - c$ **63** $2[3(a + 1) + 2(4 + c)]$ **64.** $a^2 + b^2 + c$

FOR REVIEW

Perform the following operations.

65. $(-12) + 9$ **66.** $(-12) + (-9)$ **67.** $15 + (-8)$

68. $(-12) - 9$ **69.** $(-12) - (-9)$ **70.** $15 - (-8)$

Add.

71. -13 **72.** -28 **73.** -92 **74.** -102
12 -29 81 321
26 -42 27 -421
-18 -36 -61 -60

75 The Brisebois family owed the credit union $3000. They repaid $1200 and later borrowed $650. What number represents the status of their account?

2.3 EXERCISES B

Perform the indicated operations.

1. $(5)(9)$ **2.** $(-5)(-9)$ **3.** $(5)(-9)$ **4.** $(-5)(9)$

5. $(-7)(0)$ **6.** $(-2)(-7)$ **7.** $(12)(-2)$ **8.** $(-12)(2)$

9. $(-6)(-13)$ **10.** $(-6)(13)$ **11.** $(-11)(-11)$ **12.** $(-15)(30)$

13. $9 \div 3$ **14.** $(-9) \div (-3)$ **15.** $9 \div (-3)$ **16.** $(-9) \div 3$

17. $0 \div (-5)$ **18.** $(-5) \div 0$ **19.** $\dfrac{88}{-11}$ **20.** $\dfrac{-24}{-8}$

21. $\dfrac{-15}{-15}$ **22.** $\dfrac{0}{-25}$ **23.** $\dfrac{-100}{10}$ **24.** $\dfrac{52}{-26}$

25. $(1.5)(-2.2)$ **26.** $(-1.5)(-2.2)$ **27.** $(-1.5)(2.2)$ **28.** $\left(\dfrac{3}{5}\right)\left(-\dfrac{10}{9}\right)$

29. $\left(-\dfrac{3}{5}\right)\left(-\dfrac{10}{9}\right)$ **30.** $\left(-\dfrac{3}{5}\right)\left(\dfrac{10}{9}\right)$ **31.** $4.8 \div (-0.6)$ **32.** $(-4.8) \div (-0.6)$

33. $\dfrac{-4.8}{-0.6}$ **34.** $\dfrac{2}{5} \div \left(-\dfrac{6}{15}\right)$ **35.** $\left(-\dfrac{2}{5}\right) \div \left(-\dfrac{6}{15}\right)$ **36.** $\dfrac{-\frac{2}{5}}{\frac{6}{15}}$

37. $(-1)(-1)(-1)(-1)$ **38.** $(-1)(-2)(-3)$ **39.** $(-1)(2)(-3)(4)$

40. $(-1)(-2)(3)(-1)(5)(-4)$ **41.** $(-1)(-1)(-1)(-1)(-1)(-1)(-1)(-1)(-1)$

42. Show by example that division does not satisfy the commutative law. Can you find numbers a and b for which $a \div b$ does equal $b \div a$?

43. During each of 5 consecutive hours, the temperature dropped 7°. By how much did this change the temperature?

44. Hank Anderson received 330 points for a performance, but then was penalized 25 points each for four rule infractions. What was his point total after the penalties were imposed?

Evaluate each numerical expression.

45. $(-3) + 2 \cdot (-4)$ **46.** $(-2)^2 + 3$ **47.** $4 - [2 - (5 - 3)]$

48. $6 + 3[(-2) - (-4)]$ **49.** $(4 - 7) \cdot 5 + 2$ **50.** $2 - \dfrac{4 + (-5)}{3 - 5}$

Write without parentheses using only one sign (or no sign).

51. $-(-5)$ **52.** $+(-2)$ **53.** $-(+a)$ **54.** $-(-a)$

55. $-[+(-3)]$ **56.** $-[-(+2)]$ **57.** $-[-(-7)]$ **58.** $+[-(-8)]$

Evaluate the following expressions when $x = -2$, $y = 1$, and $z = -3$.

59. $3xy + z$ **60.** $xyz + z^2$ **61.** $3(y - x) - z$

62. $x + \dfrac{y}{z}$ **63.** $2[4(x + 2) + 7(z + 3)]$ **64.** $x^2 + y^2 + z$

FOR REVIEW

Perform the following operations.

65. $(-17) + 6$ **66.** $(-17) + (-6)$ **67.** $17 + (-6)$

68. $(-17) - 6$ **69.** $(-17) - (-6)$ **70.** $17 - (-6)$

Add.

71.	-16	**72.**	-40	**73.**	-97	**74.**	-205
	15		-27		27		420
	3		-19		13		-318
	$\underline{-27}$		$\underline{-52}$		$\underline{-47}$		$\underline{-\ 65}$

75. A football team lost 6 yards on first down, gained 11 yards on second down, and lost 4 yards on third down. If the series started on their 25 yard line, where was the ball placed on fourth down?

2.3 EXERCISES C

Evaluate each numerical expression.

1. $-2 - [3 - (4 - 2)^2]^2$

2. $[17 - (-5 + 8)^2 - 6 \div 2]^2$

3. $-[2^3 - (2 - 4)]^2 - \dfrac{16 - (-6)}{15 - 4}$

[Answer: -102]

A calculator would be helpful in Exercises 4–6.

4. $(3.65)^2 - (2.57 - 1.22)^2$

5. $0.0123[7.2 - (5.5 - 4.6)]$

6. $\dfrac{1.63 - 0.7}{4.8} + \dfrac{(0.76)^2}{1 - (2.71 - 2.11)}$

2.4 THE DISTRIBUTIVE LAWS AND SIMPLIFYING EXPRESSIONS

STUDENT GUIDEPOSTS

1 Distributive Laws   **3** Collecting Like Terms
2 Terms, Factors, and Coefficients   **4** Simplified Expressions

1 DISTRIBUTIVE LAWS

The operations of addition (or subtraction) and multiplication are related by two important properties called the **distributive laws.** Consider the numerical expression $2(3 + 5)$. Since the order of operations tells us to work within the parentheses first, we evaluate it as follows:

$$2 (3 + 5) = 2 (8) = 16.$$

However, in this case if we were to "distribute" the product of 2 over the sum of $3 + 5$,

$$2(3 + 5) = 2 \cdot 3 + 2 \cdot 5 = 6 + 10 = 16,$$

we obtain the same result. Similarly,

$$4(8 - 3) = 4(5) = 20,$$
$$\text{and}\quad 4(8 - 3) = 4 \cdot 8 - 4 \cdot 3 = 32 - 12 = 20.$$

Thus multiplication can be "distributed over" addition and subtraction. These examples illustrate the following laws.

Distributive Laws of Multiplication Over Addition and Subtraction

If a, b, and c are numbers,

1. $a(b + c) = ab + ac,$ and
2. $a(b - c) = ab - ac.$

Since multiplication is commutative, products are not affected by changing the order of multiplication. Thus we also have

$$(b + c)a = ba + ca \quad \text{and} \quad (b - c)a = ba - ca.$$

Also, multiplication distributes over sums with more than two terms.

$$a(b + c + d) = ab + ac + ad$$

EXAMPLE 1   EVALUATING USING THE DISTRIBUTIVE LAW

Evaluate each numerical expression in two ways.

(a) $7(3 + 2) = 7 \cdot 3 + 7 \cdot 2 = 21 + 14 = 35.$ Also,
$7(3 + 2) = 7(5) = 35.$

(b) $5(8 - 3) = 5 \cdot 8 - 5 \cdot 3 = 40 - 15 = 25.$ Also,
$5(8 - 3) = 5(5) = 25.$

(c) $6(7 + 3 - 5) = 6 \cdot 7 + 6 \cdot 3 - 6 \cdot 5 = 42 + 18 - 30 = 30.$ Also,
$6(7 + 3 - 5) = 6(5) = 30.$

(d) $-7(3 + 2) = (-7)(3) + (-7)(2) = (-21) + (-14) = -35.$ Also,
$-7(3 + 2) = (-7)(5) = -35.$

(e) $-2(-3 - 6) = (-2)(-3) - (-2)(6) = 6 - (-12) = 6 + 12 = 18.$
Also, $-2(-3 - 6) = -2(-9) = (2)(9) = 18.$

PRACTICE EXERCISE 1

Evaluate each numerical expression in two ways.

(a) $4(1 + 9)$

(b) $3(2 - 7)$

(c) $5(4 - 6 + 3)$

(d) $-3(5 - 1)$

(e) $-9(-2 - 8)$

Answers: (a) 40   (b) -15 (c) 5
(d) -12 (e) 90

The distributive laws are useful in computation, but they are more useful when we work with algebraic expressions. In Section 1.5, we introduced algebraic expressions involving sums, differences, products, or quotients of numbers and variables.

❷ TERMS, FACTORS, AND COEFFICIENTS

A **term** is a part of an expression that is a product of numbers and variables and that is separated from the rest of the expression by plus or minus signs. The numbers and letters that are multiplied in a term are **factors** of the term. The numerical factor is the **(numerical) coefficient** of the term. For example,

$$3x, \quad 4a + 7, \quad 3x + 7y - z, \quad 4x + 5a + 3 - 8x$$

are algebraic expressions with one, two, three, and four terms, respectively. In $4a + 7$, the term $4a$ has factors 4 and a, and 4 is the coefficient of the term.

Two terms are **similar** or **like terms** if they contain the same variables to the same powers. In $4x + 5a + 3 - 8x$, the terms $4x$ and $-8x$ are like terms. Note that the minus sign goes with the term and thus the coefficient of $-8x$ is -8.

If the terms of an expression have a common factor, the distributive laws can be used in reverse to **remove the common factor** by a process called **factoring.** We use the distributive law in the following way.

$$ab + ac = a(b + c)$$

EXAMPLE 2 FACTORING USING THE DISTRIBUTIVE LAW	**PRACTICE EXERCISE 2**

Use the distributive laws to factor.

(a) $3\,x + 3\,y = 3\,(x + y)$ The distributive law in reverse order

(b) $4\,a - 4\,b = 4\,(a - b)$ 4 and $(a - b)$ are factors of $4a - 4b$

(c) $5\,u + 5\,v - 5\,w = 5\,(u + v - w)$

(d) $3x + 6 = 3 \cdot x + 3 \cdot 2$ 6 is $3 \cdot 2$

            $= 3\,(x + 2)$ Factor out 3

(e) $8a - 8 = 8 \cdot a - 8 \cdot 1$ Express 8 as $8 \cdot 1$

            $= 8\,(a - 1)$ Factor out 8

(f) $-3x - 3y = (-3)\,x + (-3)\,y$ Factor out -3

            $= (-3)\,(x + y)$

Note that $-3x - 3y$ is not $-3(x - y)$ since $-3(x - y) = -3x + 3y$.

Use the distributive laws to factor.

(a) $7a + 7b$

(b) $3u - 3w$

(c) $11c - 11d + 11v$

(d) $6u + 6$

(e) $3w - 15$

(f) $-12a - 12b$

Answers: (a) $7(a + b)$
(b) $3(u - w)$ (c) $11(c - d + v)$
(d) $6(u + 1)$ (e) $3(w - 5)$
(f) $-12(a + b)$

In any factoring problem, we check by multiplying. For example, since

$$3(x + y) = 3x + 3y \quad \text{and} \quad 4(a - b) = 4a - 4b,$$

our factoring in the first two parts of Example 2 is correct.

③ COLLECTING LIKE TERMS

When an expression contains like terms, it can be simplified by **collecting like terms.** This process is illustrated in the next example.

EXAMPLE 3 COLLECTING LIKE TERMS	**PRACTICE EXERCISE 3**

Use the distributive laws to collect like terms.

(a) $5\,x + 7\,x = (5 + 7)\,x = 12x$ Factor out x

(b) $-8x + 2x = (-8 + 2)x = -6x$ Factor out x

(c) $-2a + 7a - 9a = (-2 + 7 - 9)a = -4a$ Factor out a

Use the distributive laws to collect like terms.

(a) $3u + 10u$

(b) $-9v + 5v$

(d) $6y + 2y - y + 5 = 6 \cdot y + 2 \cdot y - 1 \cdot y + 5$ $-y = -1 \cdot y$

$\qquad\qquad\quad = (6 + 2 - 1)y + 5$ Factor y out of first three terms

$\qquad\qquad\quad = 7y + 5$

The terms $7y$ and 5 *cannot* be collected since they are not like terms. Thus, $7y + 5$ *is not* $12y$. You can see this more easily if you replace y by some number. For example, when $y = 2$,

$\qquad 7y + 5 = 7(2) + 5 = 14 + 5 = 19,$ but $12y = 12(2) = 24.$

(e) $4a + 7b - a + 6b = 4a - a + 7b + 6b$ Commutative law

$\qquad\qquad\qquad\quad = 4 \cdot a - 1 \cdot a + 7 \cdot b + 6 \cdot b$ $-a = -1 \cdot a$

$\qquad\qquad\qquad\quad = (4 - 1)a + (7 + 6)b$ Distributive law

$\qquad\qquad\qquad\quad = 3a + 13b$

With practice some steps can be left out. We should be able to see, for example, that $4a - a = 3a$ and $7b + 6b = 13b$.

(f) $0.07x + x = (0.07)x + 1 \cdot x$ $x = 1 \cdot x$

$\qquad\qquad\;\; = (0.07 + 1)x$ Distributive law

$\qquad\qquad\;\; = 1.07x$

(c) $-w + 2w - 5w$

(d) $3z - z + 2z + 13$

(e) $6x + 6y - x - 7y$

(f) $p + (0.14)p$

Answers: (a) **$13u$**   (b) **$-4v$**
(c) **$-4w$**   (d) **$4z + 13$**
(e) **$5x - y$**   (f) **$(1.14)p$**

④ SIMPLIFIED EXPRESSIONS

When all like terms of an expression have been collected, we say that the expression has been **simplified.**

Using the distributive law and the fact that $-x = (-1) \cdot x$, we get

$$-(a + b) = (-1)(a + b) = (-1)(a) + (-1)(b) = -a - b,$$
$$-(a - b) = (-1)(a - b) = (-1)(a) - (-1)(b) = -a + b,$$
$$-(-a - b) = (-1)(-a - b) = (-1)(-a) - (-1)(b) = a + b.$$

These observations lead us to the next rule.

To Simplify an Expression by Removing Parentheses

1. When a negative sign is before parentheses, remove the parentheses (and the negative sign in front of the parentheses) by changing the sign of every term within the parentheses.

2. When a plus sign is before parentheses, remove the parentheses without changing any of the signs of the terms.

EXAMPLE 4   REMOVING PARENTHESES

Simplify by removing parentheses.

(a) $-(x + 1) = -x - 1$ Change all signs

(b) $-(x - 1) = -x + 1$ Change all signs

(c) $-(-x + y + 5) = +x - y - 5$
$\qquad\qquad\qquad = x - y - 5$ Change all signs

(d) $+(-x + y + 5) = -x + y + 5$ Change *no* signs

(e) $x - (y - 3) = x - y + 3$ Change all signs within parentheses

PRACTICE EXERCISE 4

Simplify by removing parentheses.

(a) $-(2 + y)$

(b) $-(-2 - y)$

(c) $-(5 - u - w)$

(d) $+(5 - u - w)$

(e) $a - (1 - b)$

Answers: (a) $-2 - y$ (b) $2 + y$
(c) $-5 + u + w$ (d) $5 - u - w$
(e) $a - 1 + b$

The process of removing parentheses can be called **clearing parentheses.**

| EXAMPLE 5 CLEARING PARENTHESES AND COLLECTING LIKE TERMS | PRACTICE EXERCISE 5 |

Clear parentheses and collect like terms.

(a) $3x + (2x - 7) = 3x + 2x - 7$ — Do not change signs
$= (3 + 2)x - 7$ — Collect like terms
$= 5x - 7$

(b) $y - (4y - 4) = y - 4y + 4$ — Change all signs within parentheses
$= (1 - 4)y + 4$ — Collect like terms
$= -3y + 4$

(c) $3 - (5a + 2) + 7a = 3 - 5a - 2 + 7a$ — Change all signs within parentheses
$= 7a - 5a + 3 - 2$ — Commutative law
$= 2a + 1$ — Collect like terms

(d) $3x - (-2x - 7) + 5 = 3x + 2x + 7 + 5$ — Change all signs within parentheses
$= (3 + 2)x + 7 + 5$
$= 5x + 12$

Clear parentheses and collect like terms.

(a) $a + (2 - 3a)$

(b) $b - (1 - b)$

(c) $2x - (3x + 4) - x$

(d) $5 - (-8 - 2w) + 8$

Answers: (a) $-2a + 2$ (b) $2b - 1$
(c) $-2x - 4$ (d) $2w + 21$

CAUTION

Change *all* signs within the parentheses.

$$-(-2x - 7) = 2x + 7, \quad not \quad 2x - 7.$$

| EXAMPLE 6 EVALUATING EXPRESSIONS | PRACTICE EXERCISE 6 |

Evaluate the expressions when $a = -2$ and $b = -1$.

(a) $3a + 7 = 3(-2) + 7 = -6 + 7 = 1.$

Note how using parentheses at the substitution step helps us avoid ambiguous statements. Without parentheses we would have $3 \cdot -2 + 7$, which is confusing.

(b) $4a^2 = 4(-2)^2 = 4 \cdot 4 = 16$ — Only -2 is squared. Compare this example with the next one

(c) $(4a)^2 = (4 \cdot (-2))^2 = (-8)^2 = 64$ — From (b) and (c) we see that $4a^2 \neq (4a)^2$

(d) $-a^2 = -(-2)^2 = -(4) = -4$ — Compare this with the next example

(e) $(-a)^2 = (-(-2))^2 = (2)^2 = 4$ — From (d) and (e) we see that $-a^2 \neq (-a)^2$

(f) $-a - b = -(-2) - (-1) = 2 + 1 = 3$

(g) $-a^3 = -(-2)^3 = -(-8) = +8 = 8$

Evaluate the expressions when $u = -3$ and $w = -1$.

(a) $2u - 3$

(b) $3w^2$

(c) $(3w)^2$

(d) $-u^3$

(e) $(-u)^3$

(f) $-u - w$

(g) $-(-u + w^2)$

Answers: (a) -9 (b) 3 (c) 9
(d) 27 (e) 27 (f) 4 (g) -4

///////////////// **CAUTION** ///////////////

Two common mistakes to avoid are: (1) forgetting to change *all* signs when a minus sign appears in front of a set of parentheses (for example, $-(x - 2)$ is not $-x - 2$), and (2) making sign errors when evaluating expressions such as $-a^2$ (for example, $-a^2$ is not the same as $(-a)^2$).

///////////////

2.4 EXERCISES A

1. (a) Compute $-2(3 - 8)$ **(b)** Compute $(-2)(3) - (-2)(8)$ **(c)** Why are these two equal?

2. (a) Compute $-3(4 + 2)$ **(b)** Compute $(-3)(4) + (-3)(2)$ **(c)** Why are these two equal?

3. In $2x - 3y + 7 - 9x$ the terms $2x$ and $-9x$ are called similar or _____ terms.

How many terms does each expression have?

4. $2a + b - 3$ **5.** $-2x - 3y + 4 - z$ **6.** $3w$

Multiply.

7. $4(x + y)$ **8.** $5(x - y)$ **9.** $10(x + 2y)$

10. $10(x + 2y + 3z)$ **11.** $2(2a + 1 + 6b)$ **12.** $-5(x + y)$

Use the distributive laws to factor.

13. $4x + 4y$ **14.** $5x - 5y$ **15.** $10x + 20y$

16. $10x + 20y + 30z$ **17.** $4a + 2 + 12b$ **18.** $-5x - 5y$

Simplify by removing parentheses.

19. $-(y + 2)$ **20.** $-(-a - b)$ **21.** $-(-x + 2)$

22. $-(2y - 2)$ **23.** $x - (3 - b)$ **24.** $-(x - y - 4)$

25. $+(-x - y - z)$ **26.** $-(1 + x) + (a - b)$ **27.** $+(u - v) - (w - 3)$

Use the distributive laws and collect like terms.

28. $3x - 8x$ **29.** $-4z - 9z$ **30.** $1 - 2x + x$

31. $y - 3 - 5y$

32. $2x - y - 5x + y$

33. $a - b + a - b$

34. $1 - x + 2y - 3x - y$

35. $-a + 3 + b - 2a + b$

36. $u - 3 + 3v - 2u + 1$

37 $\dfrac{1}{2}x - \dfrac{2}{3}y - \dfrac{5}{2}x - \dfrac{1}{3}y$

38. $-\dfrac{3}{4}a + \dfrac{1}{4} + \dfrac{3}{4}a - \dfrac{1}{4}b$

39. $2.1u - 3 - 5.8u$

40. $x - 3(x + 1)$

41. $2a - (1 - 3a)$

42. $2(1 - 2u) + u$

43 $2x - (-x + 1) + 3$

44. $a + (2 - 5a) - 3$

45. $7u - (-3u - 1) + 5$

46. $-2[x - 3(x + 1)]$

47 $a - [3a - (1 - 2a)]$

48. $2 - [3u - (-2 + 3u)]$

49. $3(x - 2) - 2[5y - (2x - y)]$

50. $2(3a - 4b) - (-4a + b) - (-4b - 5a)$

51. $-[2u - (u - v)] - 3[(u - 2v) - 3v]$

Evaluate the following expressions when $x = -3$, $y = -1$ and $z = 2$.

52. $-2x - y$

53. $2x^2$

54. $(2x)^2$

55 $-2x^2$

56 $(-2x)^2$

57. $-x - y + z$

58. $x^2 - y^2 - z^2$

59. $-3y^2 - (x + z)$

60. $z^3 - x^3$

61. $x^2 - 4yz$

62. $(x - y)^3$

63. $x^3 - y^3$

64. $x^2y^2z^2$

65. $2x + 3z + y + 1$

66. $-(-x)$

67. $|x + y|$

68 $|x - y|$

69. $|x^2 + y^2|$

FOR REVIEW

Perform the indicated operations.

70. $(3)(-11)$

71. $(-3)(-11)$

72. $(-12) \div 6$

73. $(-12) \div (-6)$

74. $\left(-\dfrac{1}{3}\right)\left(-\dfrac{3}{4}\right)$

75. $(-6.3) \div (-0.9)$

ANSWERS: 1. (a) 10 (b) 10 (c) distributive law 2. (a) −18 (b) −18 (c) distributive law 3. like 4. 3 5. 4
6. 1 7–12. answers given in exercises 13–18 13–18. answers given in exercises 7–12 19. −y − 2 20. a + b
21. x − 2 22. −2y + 2 23. x − 3 + b 24. −x + y + 4 25. −x − y − z 26. −1 − x + a − b 27. u − v − w + 3
28. −5x 29. −13z 30. 1 − x 31. −4y − 3 32. −3x 33. 2a − 2b 34. 1 − 4x + y 35. −3a + 3 + 2b
36. −u − 2 + 3v 37. −2x − y 38. $\frac{1}{4} - \frac{1}{4}b$ 39. −3 − 3.7u 40. −2x − 3 41. 5a − 1 42. 2 − 3u 43. 3x + 2
44. −4a − 1 45. 10u + 6 46. 4x + 6 47. −4a + 1 48. 0 49. 7x − 12y − 6 50. 15a − 5b 51. −4u + 14v
52. 7 53. 18 54. 36 55. −18 56. 36 57. 6 58. 4 59. −2 60. 35 61. 17 62. −8 63. −26 64. 36 65. 0
66. −3 67. 4 68. 2 69. 10 70. −33 71. 33 72. −2 73. 2 74. $\frac{1}{4}$ 75. 7

2.4 EXERCISES B

1. (a) Compute −3(5 − 2).   **(b)** Compute (−3)(5) − (−3)(2).   **(c)** Why are these two equal?

2. (a) Compute −4(6 + 2).   **(b)** Compute (−4)(6) + (−4)(2).   **(c)** Why are these two equal?

3. In $2a + b + 8 − 10a$ the terms $2a$ and $−10a$ are called like or _____ terms.

How many terms does each expression have?

4. $2a + 7 + 6w − 3$ **5.** $4a + 8b$ **6.** $2w − 3 + 7b + a + c$

Multiply.

7. $4(x − y)$ **8.** $a(x + y)$ **9.** $10(a + 4b)$

10. $11(x + 2y + 3z)$ **11.** $2(3x + 1 + 4y)$ **12.** $−6(a + w)$

Use the distributive laws to factor.

13. $4x − 4y$ **14.** $ax + ay$ **15.** $10a + 40b$

16. $11x + 22y + 33z$ **17.** $6x + 2 + 8y$ **18.** $−6a − 6w$

Simplify by removing parentheses.

19. $−(x + 3)$ **20.** $−(−x − y)$ **21.** $−(−w + 5)$

22. $−(4a − 4)$ **23.** $a − (2 − y)$ **24.** $−(a − b − 9)$

25. $+(−a − b − w)$ **26.** $−(1 + a) + (w − c)$ **27.** $+(x − y) − (a − 8)$

Use the distributive laws and collect like terms.

28. $−3x + 8x$ **29.** $−3w − 10w$ **30.** $1 − 3x + 2x$

31. $5y + 2 − 3y$ **32.** $x + y − x − y$ **33.** $2y − u + y − 3u$

34. $1 − 2a + b − a − 3b$ **35.** $w + 3 − 2w + b − 4$ **36.** $z − 4 + 3z − a + 2$

37. $\frac{1}{4}a - \frac{1}{5}w + \frac{3}{4}a + \frac{2}{5}w$ **38.** $-\frac{1}{3}x + \frac{1}{2} - \frac{2}{3}x + \frac{1}{3}y$ **39.** $2.5w − 3 + 4.1w$

40. $a − 2(a + 5)$ **41.** $2w − (4 − 5w)$ **42.** $3(1 − 5x) + x$

43. $2b - (-b + 2) + 5$ **44.** $y + (3 - 2y) - 5$ **45.** $2y - (-5y - 2) + 7$

46. $-3[a - 2(a + 2)]$ **47.** $x - [4x - (3 - 4x)]$ **48.** $4 - [2w - (-3 - 2w)]$

49. $-6(x + 2y) - 5[8x - (-2x + y)]$ **50.** $-3(-a + 3b) + 5(-2a - 2b) - 2(3a - b)$

51. $-[(6u - v) - 5v] - 4[-u - (2u - v)]$

Evaluate the following expressions when $a = -2$, $b = -1$, and $c = 3$.

52. $-3a - b$ **53.** $3a^2$ **54.** $(3a)^2$ **55.** $-3a^2$

56. $(-3a)^2$ **57.** $-a - b + c$ **58.** $a^2 - b^2 - c^2$ **59.** $-2b^2 - (a + c)$

60. $c^3 - a^3$ **61.** $a^2 - 3bc$ **62.** $(a - b)^3$ **63.** $a^3 - b^3$

64. $a^2b^2c^2$ **65.** $2a + b - 3c + 5$ **66.** $-(-a)$ **67.** $|a + c|$

68. $|a - c|$ **69.** $|b^2 + c^2|$

FOR REVIEW

Perform the indicated operations.

70. $(4)(-10)$ **71.** $(-4)(-10)$ **72.** $(-18) \div 9$

73. $(-18) \div (-9)$ **74.** $\left(-\dfrac{1}{5}\right)\left(-\dfrac{5}{6}\right)$ **75.** $(-5.4) \div (-0.9)$

2.4 EXERCISES C

Use the distributive laws and collect like terms.

1. $3\{x - 2[x - (y - x)] - (2x - y)\}$

2. $4\{[5(a - 2b) - 3b] - 2[6a - (3b - a)]\}$
[Answer: $-36a - 28b$]

3. $7x - \{[6x - 2y - 3(2x - y)] - 3[-x - (3y - x)]\}$

4. $-5\{[-4(a + 5b) - 2b] - 7[a - (-a - b)]\}$
[Answer: $90a + 145b$]

2.5 IRRATIONAL AND REAL NUMBERS

STUDENT GUIDEPOSTS

1 Irrational and Real Numbers
2 Perfect Squares and Square Roots
3 Principal Square Roots, Radicals, and Radicands

So far we have concentrated on working with the rational numbers. Remember that a **ratio**nal number is the quotient or **ratio** of two integers. Equivalently a rational number is a number with a decimal form that either terminates or repeats a block of digits.

❶ IRRATIONAL AND REAL NUMBERS

There are many numbers that cannot be expressed this way. One of the most familiar of these is π, the number equal to the ratio of the circumference of any circle to its diameter. Such numbers are called *irrational numbers*. In decimal notation **irrational numbers** do not terminate nor do they have a repeating block of digits. The set of **real numbers** includes both the rational numbers and the irrational numbers.

❷ PERFECT SQUARES AND SQUARE ROOTS

Before identifying some of the more familiar irrational numbers, we look at the idea of perfect squares and their *square roots*.

Perfect Square and Square Root

When an integer is squared, the result is called a **perfect square.** Either of the identical factors of a perfect square number is called a **square root** of the number.

EXAMPLE 1 SQUARE ROOTS

(a) 4 is a perfect square since $4 = 2^2$. We call 2 a square root of 4. Since $4 = (-2)^2$, -2 is also a square root of 4.

(b) 25 is a perfect square since $25 = 5^2$. The two square roots of 25 are 5 and -5.

(c) 0 is a perfect square since $0 = 0^2$. Unlike other perfect squares, 0 has only one square root, namely itself, 0.

CAUTION

Do not confuse the terms *square* and *square root!*

5 is a **square** root of 25 ⟶ ⌐ 25 is the **square** of 5
$$5^2 = 5 \cdot 5 = 25$$

---

❸ PRINCIPAL SQUARE ROOTS, RADICALS, AND RADICANDS

All perfect squares other than 0 have two square roots, one positive and one negative. When we refer to the square root of a number, do we mean the positive root or the negative root? To avoid confusion, we refer to the positive root as the **principal square root** and we use the symbol $\sqrt{}$, called a **radical,** to represent it. The principal or positive square root of 4 is $\sqrt{4}$, or 2. The number under the radical (4 in this case) is the **radicand.** Since the radical symbol only designates the principal or positive (possibly zero) square root of a number, we designate the negative square root by $-\sqrt{}$. For example, $-\sqrt{4} = -2$.

The first sixteen perfect squares and their square roots are listed in the following table.

Perfect square	Positive square root of N	Negative square root of N
N	$\sqrt{N}$	$-\sqrt{N}$
0	0	$-0 = 0$
1	1	-1
4	2	-2
9	3	-3
16	4	-4
25	5	-5
36	6	-6
49	7	-7
64	8	-8
81	9	-9
100	10	-10
121	11	-11
144	12	-12
169	13	-13
196	14	-14
225	15	-15

In addition to whole number perfect squares, there are also fractional perfect squares. A fraction that can be factored into the product of two identical fractional factors is called a **perfect square,** and each factor is called a **square root** of the fraction.

EXAMPLE 2   FRACTIONAL PERFECT SQUARES

(a) $\dfrac{4}{9}$ is a perfect square, with $\sqrt{\dfrac{4}{9}} = \dfrac{2}{3}$ and $-\sqrt{\dfrac{4}{9}} = -\dfrac{2}{3}$.

(b) $\dfrac{25}{81}$ is a perfect square, with $\sqrt{\dfrac{25}{81}} = \dfrac{5}{9}$ and $-\sqrt{\dfrac{25}{81}} = -\dfrac{5}{9}$.

PRACTICE EXERCISE 2

(a) Is $\dfrac{36}{121}$ a perfect square?

(b) Is $\dfrac{50}{200}$ a perfect square?

Answers: (a) yes; $\dfrac{36}{121} = \left(\dfrac{6}{11}\right)^2$
(b) Yes; first reduce the fraction to $\dfrac{1}{4}$, and $\dfrac{1}{4} = \left(\dfrac{1}{2}\right)^2$.

Example 2 shows that a fraction is a perfect square if its numerator and denominator are whole number perfect squares. However, a fraction can be a perfect square without this feature. For example, if we reduce $\frac{8}{18}$ to lowest terms, we change it to a ratio of whole number perfect squares:

$$\sqrt{\frac{8}{18}} = \sqrt{\frac{\cancel{2} \cdot 4}{\cancel{2} \cdot 9}} = \sqrt{\frac{4}{9}} = \frac{2}{3}.$$

Thus far we have discussed only square roots of perfect squares. What about square roots of other numbers? This problem can be divided into two parts, square roots of positive numbers and square roots of negative numbers. Square roots of negative numbers are not real numbers and will not be discussed in this text. However, square roots of positive numbers that are not perfect squares supply us with many examples of irrational numbers. These include $\sqrt{2}$, $\sqrt{3}$, $\sqrt{5}$, $\sqrt{6}$, and $\sqrt{7}$ among many, many more.

Suppose we consider $\sqrt{2}$, the number which when squared is 2. We know that

$$\sqrt{1} = 1 \text{ and } \sqrt{4} = 2.$$

Since $1 < 2 < 4$, we would expect $\sqrt{1} < \sqrt{2} < \sqrt{4}$ so that
$$1 < \sqrt{2} < 2.$$

Since there is no integer between 1 and 2, $\sqrt{2}$ cannot be an integer. In more advanced work it can be shown that $\sqrt{2}$ is not a rational number either.

Most of our work in this text will be with rational numbers until Chapter 9 when we consider operations with radicals.

EXAMPLE 3   NUMBER SYSTEMS

Find the numbers in the set $\{-5, 0, 4, \sqrt{7}, 5, -\pi, \frac{3}{2}, -1.8\}$ that belong to the specified set.

(a) Natural numbers: 4 and 5

(b) Whole numbers: 0, 4, and 5

(c) Integers: 0, 4, 5, and -5

(d) Rational numbers: 0, 4, 5, -5, $\frac{3}{2}$, and -1.8

(e) Irrational numbers: $\sqrt{7}$, $-\pi$

(f) Real numbers: all the numbers in the set are real numbers

PRACTICE EXERCISE 3

Find the numbers in the set $\{-7, -\frac{3}{5}, -\sqrt{17}, 0, 11, \pi, \frac{11}{2}, 7.22\}$ that belong to the specified set.

(a) Natural numbers

(b) Whole numbers

(c) Integers

(d) Rational numbers

(e) Irrational numbers

(f) Real numbers

Answers: (a) 11   (b) 0, 11
(c) $-7, 0, 11$ (d) $-7, -\frac{3}{5}, 0$,
$11, \frac{11}{2}, 7.22$ (e) $-\sqrt{17}, \pi$
(f) all the numbers

We conclude with the diagram in Figure 2.14 that shows the relationships among the number systems we have discussed.

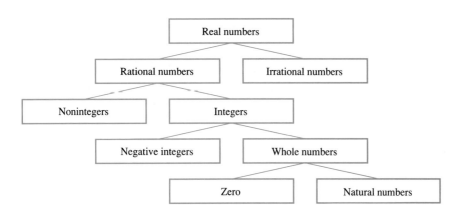

Figure 2.14

2.5 EXERCISES A

Evaluate the square roots.

1. $\sqrt{36}$ **2.** $-\sqrt{49}$ **3.** $\sqrt{225}$ **4.** $-\sqrt{196}$

5. $\sqrt{0}$ **6.** $-\sqrt{1}$ **7.** $-\sqrt{169}$ **8.** $\sqrt{81}$

9. $\sqrt{\dfrac{9}{25}}$ 　　　**10.** $\sqrt{\dfrac{64}{196}}$ 　　　**11.** $-\sqrt{\dfrac{49}{81}}$ 　　　**12.** $-\sqrt{\dfrac{36}{225}}$

13. $\sqrt{\dfrac{1}{144}}$ 　　　**14.** $\sqrt{\dfrac{4}{169}}$ 　　　**15.** $\sqrt{\dfrac{49}{196}}$ 　　　**16.** $\sqrt{\dfrac{9}{81}}$

17 $\sqrt{\dfrac{8}{18}}$ 　　　**18.** $-\sqrt{\dfrac{2}{50}}$ 　　　**19.** $\sqrt{\dfrac{12}{48}}$ 　　　**20.** $\sqrt{\dfrac{75}{363}}$

21. (a) Compute $(16)^2$　**(b)** $\sqrt{256} =$ 　　　　**22. (a)** Compute $(17)^2$　**(b)** $\sqrt{289} =$

23. (a) Compute $(18)^2$　**(b)** $\sqrt{324} =$ 　　　　**24. (a)** Compute $(19)^2$　**(b)** $\sqrt{361} =$

25. If x is any whole number, what is $\sqrt{x^2}$? 　　　**26.** If x is any natural number, what is $-\sqrt{x^2}$?

27. (a) What is the square of 9?　**(b)** What is the principal square root of 9?

28. (a) What is the square of $\frac{1}{4}$?　**(b)** What is the principal square root of $\frac{1}{4}$?

29 Between what two positive integers is $\sqrt{15}$ located?

30. Between what two positive integers is $\sqrt{150}$ located?

In Exercises 31–36, list the numbers in the set $\left\{-\frac{7}{2},\, -1,\, 0,\, -\sqrt{11},\, -3,\, 2.66,\, 2\pi,\, \frac{\sqrt{2}}{2}\right\}$ *that belong to the specified set.*

31. Natural numbers 　　　**32.** Whole numbers 　　　**33.** Integers

34. Rational numbers 　　　**35.** Irrational numbers 　　　**36.** Real numbers

Of the natural numbers, the whole numbers, the integers, the rational numbers, and the real numbers, which is the smallest set that contains each of the following numbers?

37. -3 　　　**38.** 0 　　　**39.** $\sqrt{13}$ 　　　**40.** -10

41. 22 　　　**42.** -1.326 　　　**43.** $-\dfrac{11}{5}$ 　　　**44.** $3\sqrt{3}$

FOR REVIEW

Remove parentheses and collect like terms.

45. $2a - (4a - 1)$

46. $-(1 - a) + (a + 1)$

47. $-[-(1 - a)]$

48. $2[y - (3y - 2)]$

49. $-3[-y + (1 - y)]$

50. $-[-(-y)] + y$

ANSWERS: 1. 6 2. -7 3. 15 4. -14 5. 0 6. -1 7. -13 8. 9 9. $\frac{3}{5}$ 10. $\frac{4}{7}$ 11. $-\frac{7}{9}$ 12. $-\frac{2}{5}$ 13. $\frac{1}{12}$ 14. $\frac{2}{13}$ 15. $\frac{1}{2}$ 16. $\frac{1}{3}$ 17. $\frac{2}{3}$ 18. $-\frac{1}{5}$ 19. $\frac{1}{2}$ 20. $\frac{5}{11}$ 21. (a) 256 (b) 16 22. (a) 289 (b) 17 23. (a) 324 (b) 18 24. (a) 361 (b) 19 25. x 26. $-x$ 27. (a) 81 (b) 3 28. (a) $\frac{1}{16}$ (b) $\frac{1}{2}$ 29. 3 and 4 30. 12 and 13 31. none 32. 0 33. $-1, 0, -3$ 34. $-\frac{7}{2}, -1, 0, -3, 2.66$ 35. $-\sqrt{11}, 2\pi, \frac{\sqrt{2}}{2}$ 36. all the numbers 37. integers 38. whole numbers 39. real numbers 40. integers 41. natural numbers 42. rational numbers 43. rational numbers 44. real numbers 45. $-2a + 1$ 46. $2a$ 47. $1 - a$ 48. $-4y + 4$ 49. $6y - 3$ 50. 0

2.5 EXERCISES B

Evaluate the square roots.

1. $\sqrt{196}$

2. $-\sqrt{36}$

3. $\sqrt{49}$

4. $-\sqrt{121}$

5. $\sqrt{144}$

6. $-\sqrt{64}$

7. $-\sqrt{100}$

8. $\sqrt{169}$

9. $\sqrt{\dfrac{4}{9}}$

10. $\sqrt{\dfrac{25}{196}}$

11. $-\sqrt{\dfrac{81}{121}}$

12. $-\sqrt{\dfrac{36}{49}}$

13. $\sqrt{\dfrac{1}{225}}$

14. $\sqrt{\dfrac{4}{36}}$

15. $\sqrt{\dfrac{49}{121}}$

16. $\sqrt{\dfrac{4}{16}}$

17. $\sqrt{\dfrac{8}{32}}$

18. $-\sqrt{\dfrac{50}{98}}$

19. $\sqrt{\dfrac{12}{75}}$

20. $\sqrt{\dfrac{147}{363}}$

21. (a) Compute $(20)^2$ **(b)** $\sqrt{400} =$

22. (a) Compute $(21)^2$ **(b)** $\sqrt{441} =$

23. (a) Compute $(24)^2$ **(b)** $\sqrt{576} =$

24. (a) Compute $(25)^2$ **(b)** $\sqrt{625} =$

25. If a is any whole number, what is $\sqrt{a^2}$?

26. If a is any natural number what is $-\sqrt{a^2}$?

27. (a) What is the square of 16?   **(b)** What is the principal square root of 16?

28. (a) What is the square of $\frac{1}{9}$?   **(b)** What is the principal square root of $\frac{1}{9}$?

29. Between what two positive integers is $\sqrt{30}$ located?

30. Between what two positive integers is $\sqrt{200}$ located?

In Exercises 31–36, list the numbers in the set $\left\{6, \frac{5}{4}, -0.235, 0, \sqrt{22}, -4, \frac{\pi}{2}\right\}$ that belong to the specified set.

31. Natural numbers

32. Whole numbers

33. Integers

34. Rational numbers

35. Irrational numbers

36. Real numbers

Of the natural numbers, the whole numbers, the integers, the rational numbers, and the real numbers, which is the smallest set that contains each of the following numbers?

37. 0   **38.** -16 **39.** 45   **40.** $\sqrt{23}$

41. 63.21   **42.** 4π **43.** $6\sqrt{5}$ **44.** $-\frac{21}{5}$

FOR REVIEW

Remove parentheses and collect like terms.

45. $4y - (6y - 3)$ **46.** $-(2 - y) + (2 + y)$ **47.** $-[-(3 - y)]$

48. $3[w - (2w - 4)]$ **49.** $-4[-w + (2 - w)]$ **50.** $-[-(-w)] - w$

2.5   EXERCISES C

Evaluate the square roots where a is a natural number.

1. $\sqrt{25a^2}$ **2.** $\sqrt{\dfrac{16}{a^4}}$ **3.** $-\sqrt{\dfrac{625}{576}}$ **4.** $-\sqrt{\dfrac{847}{1008}}$
[Answer: $-\frac{11}{12}$]

Use a calculator to find the square root of each number to two decimal places.

5. $\sqrt{2}$ **6.** $\sqrt{7}$ **7.** $\sqrt{26}$ **8.** $\sqrt{191}$
[Answer: 13.82]

CHAPTER 2 REVIEW

KEY WORDS

2.1   The **integers** are . . . , $-3, -2, -1, 0, 1, 2,$ $3, \ldots$.

The **rational numbers** are the numbers that can be written as the ratio of two integers.

2.2   The **additive identity** is 0.

The **additive inverse** or **negative** of a is $-a$.

2.3   The **multiplicative identity** is 1.

The **multiplicative inverse** or **reciprocal** of a ($a \neq 0$) is $\frac{1}{a}$.

2.4   A **term** is part of an expression that is a product of numbers and variables and that is separated from the rest of the expression by plus or minus signs.

The **factors** of a term are numbers or letters that are multiplied in the term.

The **coefficient** of a term is the numerical factor.

2.5   An **irrational number** is a number that is not rational and has a decimal representation that does not terminate or repeat.

The **real numbers** are the rational numbers together with the irrational numbers.

A **perfect square** results when an integer or a rational number is squared.

Either of the identical factors of a perfect square number is a **square root** of the number.

KEY CONCEPTS

2.1 **1.** If a and b are any numbers such that a is to the left of b on a number line, then $a < b$ and $b > a$.

  2. The absolute value of a number x, denoted by $|x|$, is the distance from x to zero on a number line and as a result, is never negative.

2.2 If a, b, and c are any numbers,

  (a) $a + 0 = 0 + a = a$,   Additive identity

  (b) $a + (-a) = (-a) + a = 0$,
    Additive inverses (negatives)

  (c) $a + b = b + a$,
    Commutative law of addition

  (d) $(a + b) + c = a + (b + c)$.
    Associative law of addition

2.3 If a, b, and c are any numbers,

  (a) $a \cdot 1 = 1 \cdot a = a$,   Multiplicative identity

  (b) $a \cdot \dfrac{1}{a} = \dfrac{1}{a} \cdot a = 1$ $(a \neq 0)$,
    Multiplicative inverses (reciprocals)

  (c) $ab = ba$,
    Commutative law of multiplication

  (d) $(ab)c = a(bc)$.
    Associative law of multiplication

2.4 **1.** The distributive laws,

$$a(b + c) = ab + ac$$
$$\text{and} \quad a(b - c) = ab - ac,$$

  are used to factor and to collect like terms in algebraic expressions.

  2. Only like terms can be combined. For example, $4y + 5 \neq 9y$ since $4y$ and 5 are *not* like terms.

  3. A minus sign before a set of parentheses changes the sign of *every* term inside when the parentheses are removed. For example, $-(2 - a) = -2 + a$, *not* $-2 - a$.

2.5 The radical symbol by itself only designates the principal (positive or zero) square root of a number. For example, $\sqrt{9} = 3$, not -3. We must write $-\sqrt{9}$ to represent -3.

REVIEW EXERCISES

Part I

2.1 *Find the absolute values.*

  1. $|12|$ **2.** $|-7|$ **3.** $|-2.51|$

Are the following true or false? If the statement is false, tell why.

  4. $-11 < -1$ **5.** $0 \leq -4$ **6.** $5 \leq 5$

Place the correct symbol, =, >, or <, between the pairs of fractions.

  7. $\dfrac{2}{17}$ $\dfrac{10}{85}$ **8.** $\dfrac{13}{5}$ $\dfrac{28}{11}$ **9.** $\dfrac{6}{7}$ $\dfrac{11}{12}$

  10. The lowest temperature ever recorded in Hawley Lake, Arizona was 41° below zero. How could this be written?

2.2 *Perform the indicated operations.*

  11. $(-2) + (-5)$ **12.** $2 + (-5)$ **13.** $(-2) - (-5)$

  14. $(-2) - 5$ **15.** $2 - (-5)$ **16.** $0 - (-2)$

  17. $(-2.3) + (-1.5)$ **18.** $\left(\dfrac{1}{5}\right) - \left(-\dfrac{3}{10}\right)$ **19.** $\left(-\dfrac{1}{5}\right) - \left(-\dfrac{3}{10}\right)$

20. $(-5) + (-3) + 5 + (-8) + (-4) + (-6)$

21. On first down the LSU football team made 7 yards. It lost 9 yards on second down, then picked up 11 yards on third down. If the series started at the LSU 30 yard line, where was the ball spotted on fourth down?

2.3 *Perform the indicated operations.*

22. $(-4)(2)$ **23.** $(-14) \div 7$ **24.** $(-14) \div (-7)$

25. $\dfrac{0}{-12}$ **26.** $\left(-\dfrac{1}{3}\right)\left(\dfrac{15}{8}\right)$ **27.** $(-2.5)(-1.6)$

28. $(-2)(-2)(-1)(2)(-3)(-1)(-1)(-1)$ **29.** When Bertha was on a crash diet, she lost 7 pounds in each of 6 consecutive weeks. Use a number to express her total weight loss.

Evaluate the following:

30. $(-2) + (-8) \div (-4)$ **31.** $(-1)^3 + 5$

32. $3 - 2[1 - (4 - 3)]$ **33.** $(6 - 8) \cdot 2 - (-4)$

Write without parentheses using only one sign (or no sign).

34. $-(-9)$ **35.** $-(-y)$ **36.** $-[-(-8)]$ **37.** $-[+(-x)]$

2.4 *Factor.*

38. $-4x + 12$ **39.** $-2 - 6x - 10y$ **40.** $-5x - 25y$

Collect like terms.

41. $3a - 2 - 7a + 4 - a$ **42.** $-x + 2 - 3y + 4x + y$

Clear parentheses and collect like terms.

43. $y - (2y - 3) + y - (-y + 2)$ **44.** $-3[2a - (4 - a)] - (-a - 3)$

Evaluate when $b = -2$ and $c = 4$.

45. $7b^2$ **46.** $-7b^2$ **47.** $(-7b)^2$

48. $(c - b)^2$ **49.** $-b^3$ **50.** $|b^2 - c|$

2.5 *Evaluate the square roots.*

51. $-\sqrt{169}$ **52.** $\sqrt{\dfrac{48}{75}}$ **53.** $-\sqrt{\dfrac{27}{3}}$

54. **(a)** What is the square of 25?   **(b)** What is the principal square root of 25?

Of the natural numbers, the whole numbers, the integers, the rational numbers, and the real numbers, which is the smallest set that contains the given number?

55. 11

56. −13

57. $\dfrac{4}{5}$

58. $-\sqrt{17}$

59. −3.2

60. 0

Part II

Perform the indicated operations.

61. $\left(\dfrac{3}{4}\right) \div \left(-\dfrac{15}{8}\right)$

62. $(0)(24)$

63. $(-1)(-2)(-4)(-6)$

64. $\left(\dfrac{1}{8}\right) + \left(-\dfrac{3}{4}\right)$

65. $(6.8) - (9.2)$

66. $\left(-\dfrac{2}{7}\right) - \left(-\dfrac{3}{14}\right)$

Evaluate.

67. $(-3) - 14 \div 7$

68. $(2 - 8) \cdot 5 + 4 \div 2$

69. $|-1.7|$

70. $|6 - 8|$

71. $2^3 - (6 - 3) \cdot 5$

72. $-[-(-8)]$

73. $-\sqrt{64}$

74. $\sqrt{\dfrac{32}{50}}$

75. $-\sqrt{\dfrac{169}{144}}$

Evaluate when $x = -2$ and $y = -5$.

76. $9x^2$

77. $-y^3$

78. $|x - y^2|$

Factor.

79. $-8x + 2$

80. $-6x - 12$

81. $3a - 6b + 9c$

Place the correct symbol, =, >, or <, between the pairs of fractions.

82. $\dfrac{6}{7}\quad \dfrac{4}{5}$

83. $\dfrac{21}{11}\quad \dfrac{19}{10}$

84. $\dfrac{36}{26}\quad \dfrac{18}{13}$

Clear parentheses and collect like terms.

85. $x + (x - 3) - (2x + 1)$

86. $-2[5 - (y - 2)] - 3(y + 1)$

ANSWERS: 1. 12 2. 7 3. 2.51 4. true 5. false (0 > −4) 6. true 7. = 8. > 9. < 10. −41° 11. −7 12. −3 13. 3 14. −7 15. 7 16. 2 17. −3.8 18. $\frac{1}{2}$ 19. $\frac{1}{10}$ 20. −21 21. 39 yard line 22. −8 23. −2 24. 2 25. 0 26. $-\frac{5}{8}$ 27. 4 28. −24 29. −42 pounds 30. 0 31. 4 32. 3 33. 0 34. 9 35. y 36. −8 37. x 38. $-4(x - 3)$ 39. $-2(1 + 3x + 5y)$ 40. $-5(x + 5y)$ 41. $-5a + 2$ 42. $3x - 2y + 2$ 43. $y + 1$ 44. $-8a + 15$ 45. 28 46. −28 47. 196 48. 36 49. 8 50. 0 51. −13 52. $\frac{4}{5}$ 53. −3 54. (a) 625 (b) 5 55. natural numbers 56. integers 57. rational numbers 58. real numbers 59. rational numbers 60. whole numbers 61. $-\frac{2}{5}$ 62. 0 63. 48 64. $-\frac{5}{8}$ 65. −2.4 66. $-\frac{1}{14}$ 67. −5 68. −28 69. 1.7 70. 2 71. −7 72. −8 73. −8 74. $\frac{4}{5}$ 75. $-\frac{13}{12}$ 76. 36 77. 125 78. 27 79. $-2(4x - 1)$ 80. $-6(x + 2)$ 81. $3(a - 2b + 3c)$ 82. > 83. > 84. = 85. −4 86. $-y - 17$

1. True or false: the number $\frac{2}{7}$ is called the negative of $\frac{7}{2}$.

1. _____

2. Find the absolute value. $|-13|$

2. _____

3. Is the following true or false? $-15 < -5$

3. _____

4. Place the correct symbol, $=$, $<$, or $>$, between the pair of fractions. $\frac{6}{13}$ $\frac{3}{5}$

4. _____

5. The lowest temperature recorded in Polar, Idaho was 49° below zero. If the highest temperature was 92°, what is the difference between these extremes?

5. _____

Perform the indicated operations.

6. $(-8) + (-14)$

6. _____

7. $5 - (-3)$

7. _____

8. $\left(\frac{1}{3}\right)\left(-\frac{3}{8}\right)$

8. _____

9. $(-4.5) \div (-1.5)$

9. _____

10. $(-3) + 8 - (-7) - 9 - (-2)$

10. _____

11. $(-2)(-1)(-1)(-1)(3)(-3)$

11. _____

12. Evaluate.
$(2 - 7) \cdot 5 - (-6)$

12. _____

13. Write $-[-(-6)]$ without parentheses using only one sign (or no sign).

13. _____

14. Factor.
  $-3y + 15$

14. _____

15. Collect like terms.
  $3y - 2b + 5 + b - 5y$

15. _____

16. Clear parentheses and collect like terms.
  $y - (2y - 3) + y$

16. _____

17. Evaluate $x^2 - y^2$ when $x = 4$ and $y = 1$.

17. _____

18. Multiply. $-3(4y - 2)$

18. _____

19. Evaluate. $\sqrt{121}$

19. _____

20. Evaluate. $-\sqrt{\dfrac{50}{32}}$

20. _____

Are the following true or false?

21. $\sqrt{81}$ is an integer.

21. _____

22. 4π is a rational number.

22. _____

Linear Equations and Inequalities

3.1 LINEAR EQUATIONS AND THE ADDITION-SUBTRACTION RULE

1 EQUATIONS

An **equation** is a statement that two quantities are equal. The two quantities are written with an equal sign ($=$) between them. Some equations are true, some are false, and for some the truth value cannot be determined. For example,

$1 + 2 = 3$ is true,
$2 + 5 = 3 - 7$ is false, and
$x + 2 = 9$ is neither true nor false since the value of x is not known.

Equations such as $x + 2 = 9$, which involve variables, are our primary concern in this chapter. If the variable can be replaced by a number that makes the resulting equation true, that number is called a **solution** of the equation. The process of finding all solutions is called **solving the equation.**

2 LINEAR EQUATIONS

In this chapter we concentrate on **linear equations** in one variable in which the exponent on the variable is 1. For this reason, a linear equation is often called a **first-degree equation.** Every linear equation in one variable x can be written in the form

$$ax + b = 0$$

where a and b are known real numbers and $a \neq 0$. Some simple linear equations can be solved directly by inspection or observation. Others require more sophisticated techniques which are discussed in the material that follows. For example, it is easy to see that 5 is a solution to the equation $x = 5$, whereas finding the solution of $2x + 3 = 8 - x$ is not so obvious and will require something beyond simple inspection. The equation

$$x + 2 = 9$$

has 7 as a solution because when x is replaced by 7,

$$7 + 2 = 9 \quad \text{True}$$

is a true equation. Note that 3 is not a solution since

$$3 + 2 = 9 \quad \text{False}$$

is false. An equation such as this, which is true for some replacements of the variable and false for others, is a **conditional equation.** The equation

$$x + 1 = 1 + x$$

has many solutions. In fact every number is a solution of this equation. An equation like this, called an **identity,** has the entire set of real numbers for its solution set. The equation

$$x + 1 = x$$

has no solutions. An equation like this is called a **contradiction.**

❸ EQUIVALENT EQUATIONS

Consider the two equations

$$x + 1 = 3 \quad \text{and} \quad x = 2.$$

Both equations have 2 as a solution. Although it is easier to see that 2 is a solution to the second, both can be solved by inspection. When two equations have exactly the same solutions, they are called **equivalent equations.** To solve an equation, we try to change it into an equivalent equation that can be solved by direct inspection. This usually means that the variable is isolated on one side, as in $x = 2$.

❹ ADDITION-SUBTRACTION RULE

Following is the first rule for solving equations.

Addition-Subtraction Rule

An equivalent equation is obtained if the same quantity is added to or subtracted from both sides of an equation.
Suppose a, b, and c are real numbers.

If $a = b$, then $a + c = b + c$ and $a - c = b - c.$

For example, if we start with the true equation $5 = 5$ and then add 2 to both sides or subtract 3 from both sides,

$$5 + 2 = 5 + 2 \quad \text{and} \quad 5 - 3 = 5 - 3,$$

the resulting equations are also true since $7 = 7$ and $2 = 2$.

EXAMPLE 1 USING THE ADDITION-SUBTRACTION RULE	**PRACTICE EXERCISE 1**
Solve. $x - 2 = 7$	Solve. $y - 8 = -1$
$x - 2 \boxed{+ 2} = 7 \boxed{+ 2}$ Add 2 to both sides to isolate x on the left side	
$x + 0 = 9$	
$x = 9$	
The solution is 9.	Answer: 7

EXAMPLE 2 Using the Addition-Subtraction Rule

Solve. $y + 3 = 17$

$y + 3 - 3 = 17 - 3$ Subtract 3 from both sides to isolate y on the left side
$y + 0 = 14$
$y = 14$

The solution is 14.

PRACTICE EXERCISE 2

Solve. $z + 9 = 4$

Answer: -5

Get in the habit of checking all indicated solutions to an equation. To do this, replace the variable throughout the original equation with the indicated solution to see whether the resulting equation is true.

EXAMPLE 3 Using the Addition-Subtraction Rule

Solve. $3 = x + 5$

$3 - 5 = x + 5 - 5$ Subtract 5 from both sides to isolate x on the right side
$-2 = x + 0$
$-2 = x$

The solution is -2.
Check: $3 \stackrel{?}{=} -2 + 5$ Replace x with -2 in the original equation
$3 = 3$

The solution -2 does check.

PRACTICE EXERCISE 3

Solve. $21 = z + 15$

Answer: 6

In Example 3, we isolated the variable on the right side of the equation instead of the left side as in the first two examples. Since $-2 = x$ and $x = -2$ are equivalent, we see that it makes no difference whether the isolated variable is on the left or on the right.

Equations involving fractions or decimals are solved in the same manner.

EXAMPLE 4 Equation with Fractions

Solve. $\dfrac{1}{2} + y = -3$

$\dfrac{1}{2} - \dfrac{1}{2} + y = -3 - \dfrac{1}{2}$ Subtract $\frac{1}{2}$ from both sides

$0 + y = -\dfrac{6}{2} - \dfrac{1}{2}$ The LCD of -3 and $-\frac{1}{2}$ is 2

$y = \dfrac{-6 - 1}{2}$

$y = -\dfrac{7}{2}$ Subtract fractions

PRACTICE EXERCISE 4

Solve. $\dfrac{3}{4} + x = \dfrac{1}{8}$

The solution is $-\frac{7}{2}$. To check this, substitute $-\frac{7}{2}$ for y in $\frac{1}{2} + y = -3$.

$$\frac{1}{2} + \left(-\frac{7}{2}\right) \stackrel{?}{=} -3$$

$$\frac{1}{2} - \frac{7}{2} \stackrel{?}{=} -3$$

$$\frac{1-7}{2} \stackrel{?}{=} -3$$

$$-\frac{6}{2} \stackrel{?}{=} -3$$

$$-3 = -3$$

Answer: $-\frac{5}{8}$

3.1 EXERCISES A

Give an example of each of the following.

1. A true equation

2. A contradiction

3. A linear equation

Solve the following equations by direct observation or inspection.

4. $x + 3 = 5$

5. $2y = 8$

6. $2 + z = 2 - z$

7. $2 + x = 2 + x$

8. $y + 5 = y + 1$

9. $2z = 5z$

Use the equation $y - 1 = 5$ to answer Exercises 10–12.

10. What is the variable?

11. What is the left side of the equation?

12. Is it an identity?

Use the addition-subtraction rule to solve. Check all solutions.

13. $x + 3 = 7$

14. $y - 4 = 9$

15. $z + 12 = -4$

16. $x - 9 = -6$

17. $y + 16 = 15$

18. $z + 11 = 13$

19. $4 = x + 1$

20. $-7 = y - 3$

21. $-3 = z + 2$

22. $\frac{1}{3} + x = 5$

23. $\frac{3}{2} + y = -2$

24. $-\frac{1}{4} + z = \frac{3}{4}$

25. $x - 2.1 = 3.4$

26. $-4.2 = y + 1.7$

27. $-2.6 = z - 3.9$

28 $x + 2\frac{1}{2} = 5$ **29.** $y - 3\frac{2}{3} = 4$ **30.** $1\frac{3}{8} + z = \frac{1}{4}$

In Exercises 31–34, is the statement true *or* false? *If the statement is false, tell why.*

31. -2 is a solution to $x + 2 = 0$.   **32.** $y + 3 = 5$ is a linear equation.

33. Zero is a solution of $x = 2x$.   **34.** $x + 2 = x$ is an identity.

FOR REVIEW

Exercises 35–38 review material from Chapters 1 and 2 to help prepare for the next section. Simplify each expression.

35. $2(5x)$ **36.** $(-1)(-z)$ **37.** $(-4)\left(-\frac{1}{4}y\right)$ **38.** $\left(-\frac{2}{5}\right)\left(-\frac{5}{2}x\right)$

ANSWERS   1–3. Answers will vary.   4. 2   5. 4   6. 0   7. any number is a solution.   8. no solution   9. 0   10. y 11. $y - 1$ 12. no   13. 4   14. 13   15. -16 16. 3   17. -1 18. 2   19. 3   20. -4 21. -5 22. $\frac{14}{3}$ 23. $-\frac{7}{2}$ 24. 1   25. 5.5   26. -5.9 27. 1.3   28. $2\frac{1}{2}$ 29. $7\frac{2}{3}$ 30. $-1\frac{1}{8}$ 31. true   32. true   33. true   34. false (contradiction)   35. $10x$ 36. z 37. y 38. x

3.1 EXERCISES B

Give an example of each of the following.

1. A false equation   **2.** An identity   **3.** An equation that is neither true nor false

Solve the following equations by direct observation or inspection.

4. $x + 5 = 12$ **5.** $3y = 15$ **6.** $4 + z = 4 - z$

7. $1 + x = 1 + x$ **8.** $y + 2 = y + 8$ **9.** $8z = 3z$

Use the equation $x + 3 = 19$ to answer Exercises 10–12.

10. What is the right side of the equation?   **11.** What is the solution?   **12.** Is this a contradiction?

Solve and check.

13. $x + 2 = 12$ **14.** $y - 3 = 10$ **15.** $z + 13 = -2$

16. $x - 11 = -9$ **17.** $y + 17 = 18$ **18.** $z + 9 = 18$

19. $5 = x + 7$ **20.** $-3 = y - 3$ **21.** $-4 = z + 3$

22. $\frac{1}{3} + x = 8$ **23.** $\frac{1}{3} + y = \frac{7}{3}$ **24.** $-\frac{1}{4} + z = 1$

25. $x - 2.3 = 5.7$ **26.** $-5.7 = y - 2.3$ **27.** $-3.8 = z - 1.2$

28. $x + 1\frac{3}{4} = 3$ **29.** $y - 2\frac{1}{5} = 7$ **30.** $3\frac{2}{3} + z = \frac{1}{9}$

In Exercises 31–34, is the statement true *or* false? *If the statement is false, tell why.*

31. -4 is a solution to $4 + x = 0$.

32. $z + 2 = 8$ is a linear equation.

33. A solution of $3y = y$ is 0.

34. $y + 1 = 1 - y$ is a contradiction.

FOR REVIEW

Exercises 35–38 review material from Chapters 1 and 2 to help prepare for the next section. Simplify each expression.

35. $8(4x)$

36. $\dfrac{1}{3}(3y)$

37. $\left(-\dfrac{1}{2}\right)(-2x)$

38. $\left(-\dfrac{2}{3}\right)\left(-\dfrac{3}{2}z\right)$

3.1 EXERCISES C

Solve. A calculator would be helpful for these exercises.

1. $3.75 - x = 7\dfrac{2}{5}$

2. $0.00125 + y = \dfrac{1}{1000}$

3. $125\dfrac{7}{8} = 95.2 - z$

[Answer: -30.675]

3.2 THE MULTIPLICATION-DIVISION RULE

In Section 2.1 we were careful to choose equations for which the coefficient of the variable was 1. When the coefficient of the variable is a number other than 1, we need to carry out an additional step using the following rule.

Multiplication-Division Rule

An equivalent equation is obtained if both sides of an equation are multiplied or divided by the same nonzero quantity.

Suppose a, b, and c are real numbers with $c \neq 0$.

$$\text{If } a = b, \text{ then } ac = bc \text{ and } \frac{a}{c} = \frac{b}{c}.$$

For example, if we start with the true equation $6 = 6$ and then multiply both sides by 2 or divide both sides by 2,

$$6 \cdot 2 = 6 \cdot 2 \quad \text{and} \quad \frac{6}{2} = \frac{6}{2},$$

the resulting equations are also true since $12 = 12$ and $3 = 3$.

EXAMPLE 1 USING THE MULTIPLICATION-DIVISION RULE	PRACTICE EXERCISE 1

Solve. $2x = 10$

$\dfrac{1}{2} \cdot 2x = \dfrac{1}{2} \cdot 10$ Multiply both sides by the reciprocal of 2, which is $\frac{1}{2}$

$1 \cdot x = 5$

$x = 5$

Solve. $9y = -27$

The solution is 5.

Check: $2 \cdot \boxed{5} \stackrel{?}{=} 10$ Replace x with 5 in the original equation

$10 = 10$

Answer: -3

Equivalently, in Example 1 we could divide both sides of the equation by 2.

$$2x = 10$$

$$\frac{2x}{2} = \frac{10}{2}$$ Divide both sides by 2

$$\frac{2}{2} \cdot x = 5$$

$$1 \cdot x = 5$$

$$x = 5$$

| EXAMPLE 2   USING THE MULTIPLICATION-DIVISION RULE | PRACTICE EXERCISE 2 |

Solve. $\dfrac{1}{3}x = 18$

Solve. $\dfrac{1}{8}y = -1$

$3 \cdot \dfrac{1}{3}x = 3 \cdot 18$ Multiply both sides by the reciprocal of $\frac{1}{3}$, which is 3

$1 \cdot x = 54$

$x = 54$

The solution is 54. Check.

Answer: -8

Notice in Examples 1 and 2 that we used the multiplication-division rule to make the coefficient of the variable equal to 1. This is the necessary final step in solving an equation. With practice, many of the steps may be performed mentally.

| EXAMPLE 3   USING THE MULTIPLICATION-DIVISION RULE | PRACTICE EXERCISE 3 |

Solve. $\dfrac{x}{\frac{4}{5}} = 20$

Solve. $\dfrac{z}{\frac{1}{7}} = -35$

In an equation of this type, simplify the left side first.

$$\frac{x}{\frac{4}{5}} = \frac{\frac{x}{1}}{\frac{4}{5}} = \frac{x}{1} \div \frac{4}{5} = \frac{x}{1} \cdot \frac{5}{4} = \frac{5}{4} \cdot x$$

Thus we are actually solving

$$\frac{5}{4} \cdot x = 20$$

$$\frac{4}{5} \cdot \frac{5}{4} \cdot x = \frac{4}{5} \cdot 20$$ Multiply both sides by $\frac{4}{5}$

$$1 \cdot x = 16 \qquad \frac{4}{5} \cdot 20 = \frac{4 \cdot 20}{5} = \frac{4 \cdot 4}{1} = 16$$

$$x = 16$$

The solution is 16.

Check: $\dfrac{16}{\frac{4}{5}} \overset{?}{=} 20$

$\dfrac{5}{4} \cdot 16 \overset{?}{=} 20$

$20 = 20$

Answer: -5

| **EXAMPLE 4** EQUATION WITH DECIMALS | **PRACTICE EXERCISE 4** |

Solve. $-2.7x = 54$

$$\dfrac{-2.7x}{-2.7} = \dfrac{54}{-2.7}$$ Divide both sides by -2.7

$$\dfrac{-2.7}{-2.7} \cdot x = \dfrac{54}{-2.7}$$

$$1 \cdot x = -20$$

$$x = -20$$

The solution is -20.

Check: $-2.7\,(-20) \overset{?}{=} 54$

$54 = 54$

Solve. $0.6y = -7.2$

Answer: -12

We can remember the multiplication-division rule with the statement, "multiply or divide both sides of the equation by the same quantity."

3.2 EXERCISES A

Use the multiplication-division rule to solve. Check all solutions.

1. $5x = 25$

2. $56 = 8y$

3. $-11z = 77$

4. $12x = -84$

5. $-72 = -y$

6. $-2z = 2$

7. $-5x = -50$

8. $-60y = 360$

9. $\dfrac{z}{1} = 10$

10. $\dfrac{2}{3}x = 10$

11. $\dfrac{y}{4} = -3$

12. $-\dfrac{2}{5}z = 8$

13. $\dfrac{x}{\frac{1}{3}} = 27$

14. $\dfrac{y}{\frac{7}{2}} = -6$

15 $\dfrac{z}{-\frac{1}{8}} = 16$

16 $\dfrac{2x}{3} = 4$

17. $\dfrac{3y}{2} = -6$

18. $\dfrac{4z}{-3} = -8$

19. $4x = -24.8$

20. $1.2y = 7.2$

21. $-0.8z = -6.4$

22 $2\dfrac{1}{2}x = 10$

23. $-3\dfrac{1}{3}y = 20$

24. $-4\dfrac{1}{5}z = -63$

FOR REVIEW

Solve.

25. $-8 + x = -12$

26. $-\dfrac{3}{4} + y = -\dfrac{7}{4}$

27. $7.4 + z = -3.2$

28. Is $2 + x = x + 2$ an identity?

29. Is 0 a solution of $x - 3 = -3 - x$?

ANSWERS: 1. 5   2. 7   3. −7   4. −7   5. 72   6. −1   7. 10   8. −6   9. 10   10. 15   11. −12   12. −20   13. 9   14. −21   15. −2   16. 6   17. −4   18. 6   19. −6.2   20. 6   21. 8   22. 4   23. −6   24. 15   25. −4   26. −1   27. −10.6   28. yes   29. yes

3.2 EXERCISES B

Use the multiplication-division rule to solve. Check.

1. $7x = 49$

2. $72 = 6y$

3. $-11z = 44$

4. $12x = -96$

5. $-63 = -y$

6. $-3z = 48$

7. $-20x = -300$

8. $30y = -240$

9. $\dfrac{z}{1} = -30$

10. $\dfrac{2}{3}x = 14$

11. $\dfrac{y}{5} = -3$

12. $-\dfrac{2}{7}z = 8$

13. $\dfrac{x}{\frac{1}{2}} = 10$

14. $\dfrac{y}{\frac{1}{9}} = 27$

15. $\dfrac{z}{-\frac{2}{3}} = 12$

16. $\dfrac{3x}{5} = 6$

17. $\dfrac{4y}{3} = -2$

18. $\dfrac{8z}{-5} = -16$

19. $3x = -25.5$

20. $1.2y = 8.4$

21. $-1.2z = -4.8$

22. $3\frac{2}{3}x = 22$

23. $-4\frac{1}{4}y = 17$

24. $-1\frac{3}{5}z = -32$

FOR REVIEW

Solve.

25. $-21 + x = -25$

26. $-\frac{5}{6} + y = \frac{7}{12}$

27. $1.9 + z = -6.7$

28. Is $x + 1 = x - 1$ a contradiction?

29. Is -6 a solution of $x - 6 = 0$?

3.2 EXERCISES C

Solve. A calculator might be helpful in these exercises.

1. $0.0568x = 0.006816$

2. $9\frac{7}{12}y = -7\frac{3}{16}$

3. $-15\frac{3}{4}z = 0.525$

[Answer: $-0.0\overline{3}$]

3.3 SOLVING EQUATIONS BY COMBINING RULES

=========== STUDENT GUIDEPOSTS ===========

1 Equations with Similar Terms

2 Equations with the Variable on Both Sides

3 Equations with Parentheses

4 Equations with Fractional or Decimal Coefficients

5 Summary of Methods

When we solve an equation, we isolate the variable on one side so we can find the solution by inspection. If we need to use both rules to do this, we generally use the addition-subtraction rule before the multiplication-division rule, as in the following example.

EXAMPLE 1 Using Both Rules	PRACTICE EXERCISE 1

Solve. $3x + 5 = 11$

$3x + 5 - 5 = 11 - 5$ Use addition-subtraction rule first to subtract 5 from both sides

$3x + 0 = 6$

$3x = 6$

$\frac{1}{3} \cdot 3x = \frac{1}{3} \cdot 6$ Now use multiplication-division rule to multiply both sides by $\frac{1}{3}$

$1 \cdot x = 2$

$x = 2$

The solution is 2.

Solve. $4y + 10 = 2$

Check: $3 \cdot 2 + 5 \stackrel{?}{=} 11$

$6 + 5 \stackrel{?}{=} 11$

$11 = 11$

Answer: -2

❶ EQUATIONS WITH SIMILAR TERMS

Sometimes the sides of an equation have similar terms. These terms should be combined before attempting to apply the rules.

EXAMPLE 2 EQUATION WITH SIMILAR TERMS	PRACTICE EXERCISE 2

Solve. $15 = 4y + 3y$

$15 = 7y$ Combine similar terms

$\frac{1}{7} \cdot 15 = \frac{1}{7} \cdot 7y$ Multiply both sides by $\frac{1}{7}$

$\frac{15}{7} = 1 \cdot y$

$\frac{15}{7} = y$

The solution is $\frac{15}{7}$. Check by substituting in the original equation.

Solve. $9z - 3z = 2$

Answer: $\frac{1}{3}$

❷ EQUATIONS WITH THE VARIABLE ON BOTH SIDES

When an equation has terms involving the variable on both sides, we use the addition-subtraction rule to get them together on the same side so they can then be combined.

EXAMPLE 3 VARIABLE ON BOTH SIDES	PRACTICE EXERCISE 3

Solve. $7x - 2 = 2x + 3$

$7x - 2 + 2 = 2x + 3 + 2$ Add 2 to both sides

$7x + 0 = 2x + 5$

$7x = 2x + 5$

$7x - 2x = 2x - 2x + 5$ Subtract $2x$ from both sides

$5x = 0 + 5$

$5x = 5$

$\frac{1}{5} \cdot 5x = \frac{1}{5} \cdot 5$ Multiply both sides by $\frac{1}{5}$

$1 \cdot x = 1$

$x = 1$

The solution is 1. Check.

Solve. $4 + 3y = -8 - y$

Answer: -3

To solve an equation, use the addition-subtraction rule to isolate all terms involving the variable on one side of the equation and then combine them. Next apply, if necessary, the multiplication-division rule. Always remember to combine similar terms before applying the rules.

EXAMPLE 4   VARIABLE IN SEVERAL TERMS

Solve. $2 - x + 10 = 3x + 2x + 24$

$$-x + 12 = 5x + 24 \qquad \text{Combine similar terms}$$
$$-x + 12 - 12 = 5x + 24 - 12 \qquad \text{Subtract 12 from both sides}$$
$$-x = 5x + 12$$
$$-x - 5x = 5x - 5x + 12 \qquad \text{Subtract } 5x \text{ from both sides}$$
$$-6x = 12$$
$$\left(-\frac{1}{6}\right)(-6x) = \left(-\frac{1}{6}\right)12 \qquad \text{Multiply both sides by } -\frac{1}{6}$$
$$1 \cdot x = -2$$
$$x = -2$$

The solution is -2.
Check: $2 - (-2) + 10 \overset{?}{=} 3(-2) + 2(-2) + 24$
$$2 + 2 + 10 \overset{?}{=} -6 - 4 + 24$$
$$14 = 14$$

PRACTICE EXERCISE 4

Solve. $4y + 2 - y = 3 + 5y - 9$

Answer: 4

---

③ EQUATIONS WITH PARENTHESES

When solving an equation containing parentheses, remove the parentheses using the distributive property. The resulting equation can then be solved using previous methods.

EXAMPLE 5   EQUATION WITH PARENTHESES

Solve. $3(x + 6) = 21$

$$3 \cdot x + 3 \cdot 6 = 21 \qquad \text{To remove parentheses, multiply both 6 and } x \text{ by 3}$$
$$3x + 18 = 21$$
$$3x + 18 - 18 = 21 - 18 \qquad \text{Subtract 18}$$
$$3x = 3$$
$$\frac{1}{3} \cdot 3x = \frac{1}{3} \cdot 3 \qquad \text{Multiply by } \frac{1}{3}$$
$$1 \cdot x = 1$$
$$x = 1$$

The solution is 1. Check.

PRACTICE EXERCISE 5

Solve. $5(z + 2) = 10$

Answer: 0

EXAMPLE 6   EQUATION WITH PARENTHESES

Solve $6z - (3z - 4) - 14$

$$6z - 3z + 4 = 14 \qquad \text{Remove parentheses by changing signs}$$

PRACTICE EXERCISE 6

Solve. $4y - (y - 8) = -1$

$$3z + 4 = 14$$
$$3z + 4 - 4 = 14 - 4 \qquad \text{Subtract 4}$$
$$3z = 10$$
$$z = \frac{10}{3}$$

The solution is $\frac{10}{3}$.

Check: $6\left(\dfrac{10}{3}\right) - \left(3\left(\dfrac{10}{3}\right) - 4\right) \overset{?}{=} 14$

$$20 - (10 - 4) \overset{?}{=} 14$$
$$20 - 6 \overset{?}{=} 14$$
$$14 = 14$$

Answer: -3

EXAMPLE 7   EQUATION WITH PARENTHESES

Solve. $2x - 4(2x - 4) = 22 - 3(x - 5)$

$$2x - 8x + 16 = 22 - 3x + 15 \qquad \text{Remove parentheses}$$
$$-6x + 16 = 37 - 3x \qquad \text{Combine like terms}$$
$$-6x + 16 - 16 = 37 - 3x - 16 \qquad \text{Subtract 16}$$
$$-6x = 21 - 3x$$
$$-6x + 3x = 21 - 3x + 3x \qquad \text{Add } 3x$$
$$-3x = 21$$
$$\left(-\frac{1}{3}\right)(-3x) = \left(-\frac{1}{3}\right)21 \qquad \text{Multiply by } -\tfrac{1}{3}$$
$$1 \cdot x = -7$$
$$x = -7$$

The solution is -7. Check.

PRACTICE EXERCISE 7

Solve.

$3z - 2(1 - z) = 7 - 5(z + 3)$

Answer: $-\frac{3}{5}$

---

❹ EQUATIONS WITH FRACTIONAL OR DECIMAL COEFFICIENTS

If an equation has fractional coefficients, it helps to simplify first by multiplying both sides by the least common denominator (LCD) of all fractions. For example, to solve

$$\frac{1}{2}x + \frac{3}{4} = \frac{5}{6},$$

first multiply both sides by the least common denominator 12.

$$12\left(\frac{1}{2}x + \frac{3}{4}\right) = 12 \cdot \frac{5}{6}$$

$$12 \cdot \frac{1}{2}x + 12 \cdot \frac{3}{4} = 12 \cdot \frac{5}{6} \qquad \text{Use distributive law}$$

$$6x + 9 = 10$$

We have less chance of making an error solving this equation than the original. The same remarks apply to equations with decimal coefficients. First eliminate the decimals by multiplying both sides by an appropriate power of 10. For example, to solve

$$1.5y + 3.25 = 6$$

multiply both sides by 100 to clear all decimals.

$$100(1.5y + 3.25) = 100 \cdot 6$$
$$150y + 325 = 600$$
$$150y = 275$$
$$y = \frac{275}{150} = \frac{11}{6}$$

We have combined several steps and performed some of the calculations mentally in the above example. With practice, you should be able to do the same.

5 SUMMARY OF METHODS

We conclude this section by summarizing how to solve linear equations.

To Solve a Linear Equation

1. Simplify both sides by clearing parentheses, fractions, or decimals. Collect like terms.

2. Use the addition-subtraction rule to isolate all variable terms on one side and all constant terms on the other side. Collect like terms when possible.

3. Use the multiplication-division rule to obtain a variable with coefficient of 1.

4. Check the solution by substituting in the original equation.

5. If an identity results, the original equation has every real number as a solution. If a contradiction results, there is no solution.

3.3 EXERCISES A

Solve each equation. Check all solutions.

1. $2x + 1 = 5$

2. $3y - 1 = 8$

3. $-4z + 3 = 5$

4. $\frac{3}{4}x + \frac{1}{4} = 1$

5. $\frac{2}{5} - y = \frac{3}{10}$

6. $12 = -1.2z + 24$

7. $2x + 3x = 10$

8. $3y - y = 7$

9. $\frac{1}{4}z + \frac{1}{2}z = -9$

10. $9.2x - 3.1x = 12.2$

11 $-\frac{1}{7}y + y = 18$

12. $3z = 5 - 7z$

13. $32 - 8x = 3 - 9x$

14 $-7 + 21y + 23 = 16$

15. $z + 3z = 8 - 2z + 10$

16. $x + 9 + 6x = 2 + x + 1$

17 $6y - 4y + 1 = 12 + 2y - 11$

18. $2.1x - 3.2 = -8.4x - 45.2$

19 $3y + \dfrac{5}{2}y + \dfrac{3}{2} = \dfrac{1}{2}y + \dfrac{5}{2}y$

Remove parentheses and solve.

20. $3(2x + 1) = 21$

21. $10 = 5(y - 20)$

22. $2(5z + 1) = 42$

23. $3x - 4(x + 2) = 5$

24. $2y - (13 - 2y) = 59$

25. $3z - (z + 4) = 0$

26. $5(x + 4) - 4(x + 3) = 0$

27. $4(y - 3) - 6(y + 1) = 0$

28. $9z - (3z - 18) = 36$

29 $5(2 - 3x) = 15 - (x + 7)$

30. $8y - (2y + 5) = 0$

31. $9z - (5z - 2) = 8$

32. $7(x - 5) = 10 - (x + 1)$

33. $(y - 9) - (y + 6) = 4y$

34. $8(2z + 1) = 4(7z + 7)$

35. $5 + 3(x + 2) = 3 - (x + 2)$

36 $\dfrac{1}{3}(6x - 9) = \dfrac{1}{2}(8x - 4)$

37 $5(x + 1) - 4x = x - 5$

38 $3(2 - 4x) = 4(2x - 1) - 2(1 + x)$

39. $2 + y = 3 - 2[1 - 2(y + 1)]$

40. $-2(z - 1) - 3(2z + 1) = -9$

41 $-4(-x + 1) + 3(2x + 3) - 7x = -10$

FOR REVIEW

Solve.

42. $\dfrac{x}{-\frac{1}{5}} = 20$

43. $\dfrac{y}{-5} = -2$

44. $\dfrac{3z}{7} = -6$

45. $-2\dfrac{3}{8}x = -19$

46. $y - \dfrac{2}{3} = \dfrac{1}{9}$

47. $3.6 - z = 7.2$

ANSWERS: 1. 2   2. 3   3. $-\frac{1}{2}$ 4. 1   5. $\frac{1}{10}$ 6. 10   7. 2   8. $\frac{7}{2}$ 9. -12 10. 2   11. 21   12. $\frac{1}{2}$ 13. -29 14. 0
15. 3   16. -1 17. every real number (identity)   18. -4 19. $-\frac{3}{5}$ 20. 3   21. 22   22. 4   23. -13 24. 18   25. 2
26. -8 27. -9 28. 3   29. $\frac{1}{7}$ 30. $\frac{5}{6}$ 31. $\frac{3}{2}$ 32. $\frac{11}{2}$ 33. $-\frac{15}{4}$ 34. $-\frac{5}{3}$ 35. $-\frac{5}{2}$ 36. $-\frac{1}{2}$ 37. no solution   38. $\frac{2}{3}$
39. -1 40. 1   41. -5 42. -4 43. 10   44. -14 45. 8   46. $\frac{7}{9}$ 47. -3.6

3.3 EXERCISES B

Solve and check.

1. $3x + 1 = 7$

2. $4y - 2 = 10$

3. $-2z + 5 = 9$

4. $\dfrac{2}{3}x + \dfrac{1}{3} = 1$

5. $\dfrac{3}{5} - y = \dfrac{7}{10}$

6. $10 = -0.5z + 5$

7. $3x + 7x = 40$

8. $4y - y = 27$

9. $\dfrac{1}{8}z + \dfrac{5}{8}z = -6$

10. $1.3x + 2.7x = 4.4$

11. $-\dfrac{1}{5}y + y = 8$

12. $5z = 2 - 9z$

13. $25 - 7x = 4 - 10x$

14. $-2 + 8y + 27 = 1$

15. $z + 2z = 3 - 2z + 17$

16. $x + 3 + 4x = 2 + x + 1$

17. $3y - y + 10 = 14 + 2y - 4$

18. $6.4 - 4.2x = 16.8x + 90.4$

19. $2x - \dfrac{1}{3} + \dfrac{2}{3}x = \dfrac{4}{9} + \dfrac{1}{3}x$

Clear parentheses and solve.

20. $2(3x + 1) = 14$

21. $9 = 3(4y - 1)$

22. $7(2z - 1) = -35$

23. $4x - 2(x + 1) = -1$

24. $7y - (4 - 3y) = 11$

25. $2z - (z + 1) = 0$

26. $6(x + 2) - 2(2x + 1) = 0$

27. $4(2y - 1) - 7(y + 2) = 0$

28. $6z - (2z - 5) = 0$

29. $3(2 - 4x) = 13 - (x + 1)$

30. $7y - (2y + 15) = 0$

31. $8z - (3z - 1) = 0$

32. $4(x - 2) = 12 - (x + 3)$

33. $(y - 2) - (y + 2) = 4y$

34. $9(2z - 1) = 3(z + 2)$

35. $7 + 3(x + 1) = 5 - (x + 1)$

36. $\frac{1}{2}(2y - 4) = \frac{2}{3}(9y + 3)$

37. $6(z - 1) - 3z = 3z + 8$

38. $2(1 - 3x) = 5(2x + 1) - 3(1 + x)$

39. $4 + y = 8 - 3[2 - (y + 2)]$

40. $-3(z - 4) - 2(3z + 1) = -8$

41. $-7(-x + 2) + 5(2x + 1) - 11x = 3$

FOR REVIEW

Solve.

42. $-\frac{3}{5}x = 6$

43. $\dfrac{y}{-\frac{1}{7}} = 14$

44. $\frac{2z}{9} = -12$

45. $-3\frac{1}{7}x = -88$

46. $\frac{3}{4} - y = \frac{1}{2}$

47. $8.4 + z = -9.2$

3.3   EXERCISES C

Solve and check.

1. $0.02x - (0.56x - 7.33) = 2.15 - (0.7x - 0.11)$

2. $3\frac{2}{3}x + 7\frac{1}{9} - \left(\frac{x}{6} - 4\frac{1}{9}\right) = -\left(2\frac{1}{6} - 6\frac{2}{9}x\right)$
$\left[\text{Answer: } \frac{241}{49}\right]$

3. $x(x - 3) - 3x[1 - (x - 1] = 4x(x - 5) + 22$

4. $-3[1 - (x^2 - 2)] - 2x(1 - x) = -5[4 - x(x - 2)]$
$\left[\text{Answer: } -\frac{11}{8}\right]$

3.4   THE LANGUAGE OF PROBLEM SOLVING

STUDENT GUIDEPOSTS

1 Problem Solving

2 Translating Words into Equations

1 PROBLEM SOLVING

To solve a word problem using algebra, we need to do two things. First we need to translate the *words* of the problem into an algebraic equation, and second, we need to solve the equation. We have learned how to solve many simple equations; now we will concentrate on translating word problems into equations. Some common terms and their symbolic translations are given below.

Symbol	+	−	·	÷	=
Terms	sum	minus	times	divided by	equals
	sum of	less	of	quotient of	is
	plus	less than	product	ratio	is equal to
	and	diminished by	product of		is as much as
	added to	difference	multiplied by		is the same as
	increased by	difference between			the result is
	more than	subtracted from			
		decreased by			

❷ TRANSLATING WORDS INTO EQUATIONS

Any letter (we often use x) can be used to stand for the unknown or desired quantity. Some examples of translation follow.

A number increased by 3 is 12.
$$x + 3 = 12$$

Three times a number diminished by 6 is the same as 12.
$$3 \cdot x - 6 = 12$$

My age seven years ago was eleven.
$$x \quad -7 \quad = \quad 11$$

5% of a number is 10.
$$0.05 \cdot x = 10$$

Although we often use x as the variable in a translation, it is sometimes helpful to use another letter more indicative of the quantity it represents. For example, we might use t for time, w for wages, A for area, or M for Mary's age.

EXAMPLE 1 TRANSLATING PHRASES INTO SYMBOLS

Select a variable to represent each quantity and translate the phrase into symbols.

Word phrase	*Symbolic translation*
(a) The time plus 4 hours	$t + 4$
(b) My wages less $100 for taxes	$w - 100$
(c) One half of the area	$\frac{1}{2}A$
(d) Twice Mary's age in 3 years	$2(M + 3)$
(e) $2000 less the cost	$2000 - c$

PRACTICE EXERCISE 1

Select a variable and translate each phrase into symbols.

(a) My age increased by 6 years

(b) The price plus $10 in tax

(c) Twice the area

(d) Five miles less than the distance

(e) Five miles less the distance

Answer: (a) $a + 6$ (b) $p + 10$
(c) $2A$ (d) $d - 5$ (e) $5 - d$

EXAMPLE 2 TRANSLATING SENTENCES INTO SYMBOLS

Use x for the variable and translate each word sentence into symbols.

Word expression	Symbolic translation
(a) Five times a number is 20.	$5x = 20$
(b) The product of a number and 7 is 35.	$7x = 35$
(c) Twice a number, increased by 3, is 11.	$2x + 3 = 11$
(d) Six is 4 less than twice a number.	$6 = 2x - 4$
(e) Six is 4 less twice a number.	$6 = 4 - 2x$
(f) One tenth of a number is 13.	$\dfrac{1}{10}x = 13$
(g) Five times a number, minus twice the number, equals 10.	$5x - 2x = 10$
(h) A number subtracted from 5 is 12.	$5 - x = 12$
(i) 5 subtracted from a number is 12.	$x - 5 = 12$
(j) A number divided by 9 is equivalent to $\dfrac{2}{3}$.	$\dfrac{x}{9} = \dfrac{2}{3}$

PRACTICE EXERCISE 2

Use x for the variable and translate into symbols.

(a) Twice a number is 50.

(b) The product of a number and 8 is 64.

(c) Three times a number, increased by 2, is 25.

(d) Ten is 2 less a number.

(e) Ten is 2 less than a number.

(f) Three fifths of a number is 9.

(g) Twice a number, decreased by four times the number, is the same as the number.

(h) Her age diminished by 3 is 1.

(i) Three diminished by her age is 1.

(j) His weight divided by 3 is the same as 24.

Answers: (a) $2x = 50$ (b) $8x = 64$
(c) $3x + 2 = 25$ (d) $10 = 2 - x$
(e) $10 = x - 2$ (f) $\frac{3}{5}x = 9$
(g) $2x - 4x = x$ (h) $x - 3 = 1$
(i) $3 - x = 1$ (j) $\frac{x}{3} = 24$

3.4 EXERCISES A

Select a variable to represent each quantity and translate the phrase into symbols.

1. The sum of a number and 7

2. A number subtracted from 10

3. The product of a number and 3

4. A number divided by 13

5. The time plus 8 hours

6. The time 6 hours ago

7. His salary plus $200

8. My wages less $50

9. Twice the volume

10. $3000 less the cost

11. A number added to its reciprocal

12. Her score less 20 points

13. 300 more than twice the number of votes

14. 300 more than the number of votes, doubled

15. 4% of the selling price

16. Cost plus 8% of the cost

Using x for the variable, translate the following into symbols.

17. 7 times a number is 25.

18. A number increased by 8 equals 12.

19. A number decreased by 15 is equal to 35.

20. The product of a number and 6 is 42.

21. Twice a number, increased by 3, is 27.

22. Three times a number, diminished by 8, equals 24.

23. A number less 5 is 12.

24. 8 is 4 less than three times a number.

25. A number diminished by 3 is 9.

26. Seven times a number, less 2, is 31.

27. The product of a number and 3, decreased by 5, equals 34.

28. When 10 is added to three times a number the result is the same as twice the number, plus 12.

29. 3% of a number is 12.

30. The sum of two numbers is 24 and one of them is three times the other.

31. Seven times 2 less than a number is 31. (Compare with Exercise 26.)

32. Twice the sum of a number and 5 is three times the number.

33. Seven more than a number is 13.

34. Four times my age in 3 years is 48.

35. When 4 is divided by some number the quotient is the same as $\frac{1}{2}$.

36. One half a number, less twice the reciprocal of the number is $\frac{3}{2}$.

FOR REVIEW

Solve and check.

37. $-\dfrac{1}{4}y + y = 27$

38. $15 - 3z = 5 + 7z$

39. $2(y - 3) = 2(4y + 1) - 3(y + 1)$

40. $2 + x = 4 - 2[1 - (x - 3)]$

ANSWERS: 1. $n + 7$ 2. $10 - n$ 3. $3n$ 4. $\frac{n}{13}$ 5. $t + 8$ 6. $t - 6$ 7. $s + 200$ 8. $w - 50$ 9. $2V$ 10. $3000 - c$ 11. $n + \frac{1}{n}$ 12. $s - 20$ 13. $2v + 300$ 14. $2(v + 300)$ 15. $0.04p$ 16. $c + 0.08c$ 17. $7x = 25$ 18. $x + 8 = 12$ 19. $x - 15 = 35$ 20. $6x = 42$ 21. $2x + 3 = 27$ 22. $3x - 8 = 24$ 23. $x - 5 = 12$ 24. $8 = 3x - 4$ 25. $x - 3 = 9$ 26. $7x - 2 = 31$ 27. $3x - 5 = 34$ 28. $10 + 3x = 2x + 12$ 29. $0.03x = 12$ 30. $x + 3x = 24$ 31. $7(x - 2) = 31$ 32. $2(x + 5) = 3x$ 33. $7 + x = 13$ 34. $4(x + 3) = 48$ 35. $\frac{4}{x} = \frac{1}{2}$ 36. $\frac{1}{2}n - 2\frac{1}{n} = \frac{3}{2}$ or $\frac{n}{2} - \frac{2}{n} = \frac{3}{2}$ 37. 36   38. 1   39. $-\frac{5}{3}$ 40. 6

3.4 EXERCISES B

Select a variable to represent each quantity and translate the phrase into symbols.

 1. The sum of a number and 12

 2. A number subtracted from 33

 3. The product of a number and 5

 4. A number divided by 7

5. The time plus 3 hours

6. The time 14 hours ago

7. Her salary plus $500

8. My wages less $75

9. Three times the volume

10. $5000 less the cost

11. A number less its reciprocal

12. His score less 28 points

13. 100 more than triple the number of votes

14. 100 more than the number of votes, tripled

15. 24% of a number

16. Cost plus 10% of the cost

Using x for the variable, translate the following into symbols.

17. 4 times a number is 32.

18. A number increased by 3 equals 14.

19. A number decreased by 13 is equal to 49.

20. The product of a number and 8 is 96.

21. Twice a number, increased by 4, is 86.

22. Three times a number, diminished by 17, equals 42.

23. A number less 4 is 21.

24. 9 is 3 less than five times a number.

25. A number diminished by 2 is 18.

26. Twelve times a number, less 17, is 39.

27. The product of a number and 7, decreased by 8, equals 29.

28. When 15 is added to three times a number the result is the same as twice the number, plus 40.

29. 9% of a number is 15.

30. The sum of two numbers is 86, and one of them is three times the other.

31. Eleven times 4 less than a number is 25.

32. Twice the sum of a number and 8 is four times the number.

33. Six more than a number is 29.

34. Five times my age in 2 years is 200.

35. When 7 is divided by some number the quotient is the same as $\frac{4}{9}$.

36. One third a number, less three times the reciprocal of the number is $\frac{4}{3}$.

FOR REVIEW

Solve and check.

37. $-\frac{1}{6}y + y = 75$

38. $12 - 4z = 2z + 12$

39. $3(y - 4) = 2(3y + 1) - (y - 3)$

40. $1 + x = 3 - 2[2 - (x - 5)]$

3.4   EXERCISES C

Using x for the variable, translate the following into symbols.

1. If two-thirds of eight more than a number is increased by five times the reciprocal of the number, the result is twice the number, decreased by 8.

2. A number divided by six less than the number is the same as the product of the number and three more than the number.

3.5 BASIC APPLICATIONS AND PERCENT

1 METHOD FOR SOLVING APPLIED PROBLEMS

Success with applied problems (or word problems) comes with practice. Problem solving will be easier if you follow these steps.

To Solve an Applied Problem

1. Read the problem (perhaps several times) and determine what quantity (or quantities) you are asked to find.

2. Represent the unknown quantity (or quantities) using a single letter.

3. Determine which expressions are equal and write an equation.

4. Solve the equation and state the answer to the problem.

5. Check to see if the answer (or answers) satisfy the conditions of the problem.

EXAMPLE 1   NUMBER PROBLEM

Twice a number, increased by 13, is 71. What is the number?
   Let x = the number. The problem may be symbolized as follows.

$$\underset{2\,\cdot}{\text{twice a number}}\;\underset{x}{}\;\underset{+}{\text{increased by}}\;\underset{13}{13}\;\underset{=}{\text{is}}\;\underset{71}{71}$$

$$2 \cdot x + 13 = 71$$
$$2x + 13 - 13 = 71 - 13$$
$$2x = 58$$
$$x = 29$$

The number is 29.
Check: Twice 29 plus 13 is indeed 71.

PRACTICE EXERCISE 1

Three times a number, decreased by 24, is 60. What is the number?

Answer: 28

////////////// **CAUTION** //////////////

Always be neat and complete when working word problems. Time-consuming errors happen when you try to take shortcuts. Writing out what the variable means helps in solving the problem.

//////////

| EXAMPLE 2   APPLICATION TO AGE | PRACTICE EXERCISE 2 |

Bob is three times as old as Dick, and the sum of their ages is 48 years. How old is each?

Let x = Dick's age,
    $3x$ = Bob's age.

$$x + 3x = 48$$
$$4x = 48$$
$$x = 12$$
$$\text{and } 3x = 36$$

Dick is 12 years old and Bob is 36 years old.
Check: $36 = 3 \cdot 12$ and $12 + 36 = 48$.

The sum of two numbers is 77. If one number is three-fourths of the other, find the two numbers

Answer: 33 and 44

② CONSECUTIVE INTEGER APPLICATIONS

Consecutive integers are integers which immediately follow each other in regular counting order. For example, 13, 14, and 15 are consecutive integers. In general, if x is an integer, the next consecutive integer is $x + 1$, and the next after that is $x + 2$. An example of consecutive *even* integers is 2, 4, 6. In this case, if x is the integer, the next consecutive even integer is $x + 2$, and the next after that is $x + 4$. The same is true for consecutive odd integers such as 3, 5, 7. If x is the first, $x + 2$ is the next and $x + 4$ follows.

| EXAMPLE 3   CONSECUTIVE INTEGERS | PRACTICE EXERCISE 3 |

The sum of three consecutive integers is 96. Find the numbers.

Let x = first integer,
    $x + 1$ = next consecutive integer (Why?),
    $x + 2$ = third consecutive integer (Why?).

$$x + (x + 1) + (x + 2) = 96$$
$$3x + 3 = 96 \quad \text{Combine like terms}$$
$$3x = 93 \quad \text{Subtract 3}$$
$$x = 31$$
$$\text{and } x + 1 = 32$$
$$\text{and } x + 2 = 33$$

The numbers are 31, 32, and 33.
Check: $31 + 32 + 33 = 96$ and the three integers are consecutive.

Wendy and Diane were born in consecutive years. If the sum of their ages is 43 and Diane is the older, how old is each?

Answer: Wendy is 21 and Diane is 22.

| EXAMPLE 4   CONSECUTIVE EVEN INTEGERS | PRACTICE EXERCISE 4 |

The sum of two consecutive even integers is 2 less than twice the larger. Find the numbers.

Let x = the first even integer,
    $x + 2$ = the second (larger) even integer.

The sum of three consecutive odd integers is 135. Find the numbers.

$$x + (x + 2) = 2(x + 2) - 2$$
$$2x + 2 = 2x + 4 - 2 \qquad \text{Clear parentheses}$$
$$2x + 2 = 2x + 2 \qquad \text{Combine like terms}$$
$$2 = 2 \qquad \text{Subtract } 2x$$

Since we obtained an identity, any value of x will make the equation true. Therefore, *any* two consecutive even integers solve the problem.

Answer: **43, 45, 47**

③ PERCENT TRANSLATIONS

Many applied problems involve percent and can be restated into one of three basic forms. The following examples illustrate these three types.

What is 25% of 40?
What percent of 60 is 12?
8 is 40% of what number?

Letting x represent the unknown number in each statement and recalling that the word *of* translates to *times,* these three statements can be translated into equations. The first becomes

What is 25% of 40?

$$x = 0.25 \cdot 40.$$

Recall from Chapter 1 that 25% = 0.25

Solving, we have $x = (0.25)(40) = 10$. The second statement translates to

What percent of 60 is 12?

$$x \cdot 60 = 12.$$

Solving,
$$x = \frac{12}{60} = 0.2.$$

In percent notation, x becomes 20%.
    Finally, the third statement becomes

8 is 40% of what number?

$$8 = 0.40 \cdot x.$$

Thus,
$$x = \frac{8}{0.40} = 20.$$

EXAMPLE 5 SOLVING PERCENT PROBLEMS	**PRACTICE EXERCISE 5**

(a) 75% of 128 is what?
The equation to solve is $(0.75)(128) = x$.

$$96 = x$$

(b) 18 is what percent of 48?
The equation to solve is $18 = x \cdot 48$.

$$\frac{18}{48} = x \qquad \text{Divide by 48}$$

$$0.375 = x$$

In percent notation, $x = 37.5\%$.

(a) 36% of 225 is what?

(b) 102 is what percent of 68?

Answers: (a) **81**   (b) **150%**

| EXAMPLE 6 APPLICATION TO SPORTS | PRACTICE EXERCISE 6 |

Manny Black had a batting percentage (percent of hits in times at bat) of 40% in a series. If he had 12 hits, how many times was he at bat?

   In effect, we are asked: 12 is 40% of what?

Let x be the number of times at bat.

$$12 = (0.40)x$$

$$\frac{12}{0.40} = x \qquad \text{Divide by 0.40}$$

$$30 = x$$

Thus, he went to bat 30 times in the series.

For the season, basketball player Andy Hurd shot 90.5% from the free throw line. If he made 181 free throws, how many times did he go to the free throw line?

Answer: 200

❹ COMMISSION RATE AND PERCENT INCREASE

Some salespersons receive all or part of their income as a percent of their sales. This is called a **commission.** The **commission rate** is the percent of the sales that the person receives. That is,

$$(\text{commission}) = (\text{commission rate}) \cdot (\text{total sales}).$$

| EXAMPLE 7 COMMISSION APPLICATION | PRACTICE EXERCISE 7 |

Paul received a commission of $720 on the sale of a machine. If his commission rate is 8% of the selling price, what was the selling price?

   The problem can be written: 8% of what is $720?

Let x be the selling price.

$$(0.08)x = 720$$

$$x = \frac{720}{0.08}$$

$$x = 9000$$

Thus, the machine sold for $9000.

Kerry received a commission of $462 on the sale of a new car. If her commission rate is 3% of the selling price, what was the selling price?

Answer: $15,400

| EXAMPLE 8 PERCENT INCREASE APPLICATION | PRACTICE EXERCISE 8 |

Due to inflation, the price of an item rose 25% over its price the year before. What was the price last year if the price this year is 60¢?

   Problems such as this can be solved by using the equation

$$(\text{previous price}) + (\text{increase in price}) = (\text{new price}).$$

Let x be last year's price.

$$x + (0.25) \cdot x = 60$$

$$1.25x = 60 \qquad \begin{aligned} x + (0.25) \cdot x &= 1 \cdot x + (0.25) \cdot x \\ &= (1 + 0.25) \cdot x = 1.25x \end{aligned}$$

$$x = \frac{60}{1.25} = 48$$

Last year's price was 48¢.

Due to overstocking, a dealer is forced to sell a typewriter at a 20% discount. What was the original price of the typewriter if he sells it for $288?

Answer: $360

⑤ INTEREST APPLICATIONS

Money that we pay to borrow money or the money earned on savings is called **interest.** The money borrowed or saved is called the **principal,** and the interest is calculated as a percent of the principal. This percent is called the **interest rate.** For example, suppose you borrow $500 for one year and the interest rate is 12% per year. The interest charged would be

$$(0.12)(\$500) = \$60.$$

If you needed the money for two years, the interest rate would be

$$2(0.12) = 0.24 = 24\%,$$

with the total interest for two years amounting to

$$(0.24)(\$500) = \$120.$$

This type of interest is called **simple interest** since interest is charged only on the principal and not on the interest itself.

EXAMPLE 9 APPLICATION TO SIMPLE INTEREST	PRACTICE EXERCISE 9

What sum of money invested at 4% simple interest will amount to $988 in one year?

Interest problems like this can be solved by using the equation

$$(\text{amount invested}) + (\text{interest}) = (\text{total amount}).$$

Let x be the amount invested.

$$x + (0.04) \cdot x = 988 \qquad \text{Interest} = (0.04)x$$
$$(1.04) \cdot x = 988$$
$$x = \frac{988}{1.04} = 950$$

The amount invested is $950.

Missie must have $1100 available at the end of one year. What amount must she invest now at 10% simple interest to have the desired amount at the end of the year?

Answer: $1000

3.5 EXERCISES A

Solve. (Some problems have been started.)

1. If 14 is added to four times a number, the result is 38. Find the number.

  Let x = the desired number,
   $4x$ = four times the number.

2. The sum of the two numbers is 180. The larger number is five times the smaller number. Find the two numbers.

  Let x = the smaller number.

3. The sum of two numbers is 98. If one is six times the other, find the two numbers.

4. Two-thirds of a number is 124; what is the number?

  Let x = the desired number,
   $\frac{2}{3}x$ = two thirds the desired number.

5. Two-thirds of the human body is water. If your body contains 124 pounds of water, how much do you weigh? [*Hint:* Compare with Exercise 4.]

6. The first of two numbers is four times the second. If the first is 30,168, what is the second?

7. The area of Lake Superior is four times the area of Lake Ontario. If the area of Lake Superior is 30,168 mi^2, what is the area of Lake Ontario? [*Hint:* Compare with Exercise 6.]

8. The sum of three consecutive integers is 78. Find the three integers.

Let x = the first integer,
$x + 1$ = the second integer,
_____ = the third integer.

9. The sum of two consecutive odd integers is 96. Find the two integers.

Let x = the first odd integer,
$x + 2$ = the next odd integer.

10. The sum of two consecutive even integers is 162. Find the two integers.

11. Maria is 4 years older than Juan. The sum of their ages is 32 years. How old is each?

Let x = Juan's age,
$x + 4$ = Maria's age.

12. George is 5 years older than Sue. If the sum of their ages is 47, how old is each?

13. How old is Marvin if you get 100 years when you multiply his age by 6 and add 16 years?

14. If three-fourths of a number is 27, what is the number?

15. The sum of two numbers is 9, and one is twice the other. What are the numbers?

Let x = the first number,
$2x$ = the second number.

16. A board is 9 ft long. It is to be cut into two pieces in such a way that one piece is twice as long as the other. How long is each piece? [*Hint:* Compare with Exercise 15.]

17 A board is 17 ft long. It is to be cut into two pieces in such a way that one piece is seven ft longer than the other. How long is each piece?

18 Tony is four times as old as Angela and half as old as Theresa. The sum of the three ages is 39 years. How old is each?

19. 45 is 15% of what?

$45 = 0.15 \cdot x$

20. 105 is 20% of what?

21. 25% of 164 is what?

$0.25 \cdot 164 = x$

22. 18% of 90 is what?

23. What is 30% of 420?

24. What percent of 90 is 15?

$x \cdot 90 = 15$

25. What percent of 320 is 20?

26. 40 is what percent of 200?

27. 25 is what percent of 500?

28. 25 is 4% of what?

29. In 150-ml acid solution there are 45 ml of acid. What is the percent of acid in the solution?

30. A basketball player hit 120 free throws in 150 attempts. What percent of her shots did she make?

31. Juan received a commission of $104.50 on the sale of a typewriter. If his commission rate is 11% of the selling price, what was the selling price?

Let x = selling price,
$0.11x$ = his commission on sales.

32 Percy received $31.50 interest on his savings last year. If he was paid 6% simple interest, what amount did he have invested?

33. A baseball player got 42 hits one season. If his batting average was 28%, how many times was he up to bat?

34 A dress is discounted 30% and the sale price is $51.45. What was the original price?

35. If the sales-tax rate is 5% and the marked price plus tax of a mixer is $37.59, what is the marked price?

> Let x = marked price,
> $0.05x$ = tax on the marked price,
> $x + 0.05x$ = price plus tax.

36. The price of a package of gum rose 20% last year. If the present price is 42¢, what was the price last year?

37. If the present retail price of an item is 65¢ and due to increased overhead the price will have to be raised 20%, what will be the new price?

38 What sum invested at $4\frac{1}{2}$% simple interest will amount to $1254 in one year?

39. A baseball player got 21 hits in 70 times at bat. What was his batting average?

40. A family spent $200 a month for food. This was 16% of their monthly income. What was their monthly income?

41. The sales-tax rate in Murphyville is 4%. How much tax would be charged on a purchase of $12.50?

42. What sum at 4% simple interest will amount to $1508 in 1 year?

43. The area of Greenland is 25% of the area of the United States. What is the area of Greenland if the area of the U.S. is 3,615,000 mi^2?

44. The human brain is $2\frac{1}{2}$% of the total body weight. If Nick's brain weighs 4 lb, how much does Nick weigh?

FOR REVIEW

Select a variable to represent each quantity and translate the phrase into symbols.

45. Price plus 10% of the price

46. The reciprocal of a number less twice the number

The following exercises review formulas from geometry that will help you prepare for the next section. These formulas with figures are on the inside cover of this text and were reviewed in the Geometry Appendix. A calculator will be helpful in some exercises.

47. A rectangle with length l and width w has area A given by $A = lw$. Find the area of a rectangle with length 4.5 ft and width 3.2 ft.

48. A triangle with base b and altitude h has area A given by $A = \frac{1}{2}bh$. Find the area of a triangle with base $\frac{5}{4}$ in and altitude $\frac{8}{3}$ in.

49. A circle with radius r has circumference C given by $C = 2\pi r$. Find the circumference of a circle with diameter 9.2 yd, correct to the nearest tenth of a yard.

50. A trapezoid with bases b_1 and b_2 and altitude h has area A given by $A = \frac{1}{2}(b_1 + b_2)h$. Find the area, to the nearest tenth of a square centimeter, of a trapezoid with bases 8.6 cm and 12.3 cm and altitude 4.9 cm.

ANSWERS: 1. 6   2. 150, 30   3. 14, 84   4. 186   5. 186 lb   6. 7542   7. 7542 mi^2 8. 25, 26, 27   9. 47, 49
10. 80, 82   11. Juan is 14, Maria is 18.   12. George is 26 and Sue is 21.   13. 14   14. 36   15. 3, 6   16. 3 ft and
6 ft   17. 5 ft and 12 ft   18. Angela is 3, Tony is 12, Theresa is 24.   19. 300   20. 525   21. 41   22. 16.2   23. 126
24. $16\frac{2}{3}$%   25. $6\frac{1}{4}$%   26. 20%   27. 5%   28. 625   29. 30%   30. 80%   31. $950   32. $525   33. 150   34. $73.50
35. $35.80   36. 35¢   37. 78¢   38. $1200   39. 30%   40. $1250   41. 50¢   42. $1450   43. 903,750 mi^2 44. 160 lb
45. $p + 0.10p$ 46. $\frac{1}{n} - 2n$ 47. 14.4 ft^2 48. $\frac{5}{3}$ in^2 49. 28.9 yd   50. 51.2 cm^2

3.5 EXERCISES B

1. If 9 is added to six times a number, the result is 39. Find the number.

2. The sum of two numbers is 72. The larger number is five times the smaller number. Find the two numbers.

3. The sum of two numbers is 48. If one is five times the other, find the two numbers.

4. Three-fourths of a number is 288; what is the number?

5. Three-fourths of the contents of a filled tank is water. If the tank is holding 288 gallons of water, how much fluid is in it?

6. The first of two numbers is five times the second. If the first is 1250, what is the second?

7. The number of acres in the Henderson farm is five times the number of acres in the Carlson farm. If the Henderson farm has 1250 acres, how many acres are there in the Carlson farm?

8. The sum of three consecutive odd integers is 195. Find the integers.

9. The sum of two consecutive even integers is 90. What are the integers?

10. The sum of two consecutive integers is 203. Find the integers.

11. Barb is 9 years older than Larry. Twice the sum of their ages is 126. How old is each?

12. Hector is 7 years younger than Pedro. If the sum of their ages is 39, how old is each?

13. How old is Burford if you get 99 years when you multiply his age by 10 and subtract 31 years?

14. Two thirds of a number is 62; what is the number?

15. The sum of two numbers is 168, and one is three times the other. What are the numbers?

16. A rope is 168 feet long. It is to be cut into two pieces in such a way that one piece is three times as long as the other. How long is each piece?

17. A steel rod is 22 m long. It is cut into two pieces in such a way that one piece is 6 m longer than the other. How long is each piece?

18. Hortense is five times as old as Wally and half as old as Lew. The sum of their ages is 80 years. How old is each?

19. 48 is 12% of what?

20. 18 is 4% of what?

21. 35% of 180 is what?

22. 50% of 280 is what?

23. What is 3% of 1250?

24. What percent of 85 is 17?

25. What percent of 60 is 210?

26. 15 is what percent of 300?

27. 2.5 is what percent of 7.5?

28. 1230 is 60% of what?

29. In a 250-ml salt solution there are 15 ml of salt. What is the percent of salt in the solution?

30. A basketball player hit 12 shots in 20 attempts during one game. What was his shooting percent?

31. Blue Carpet Realty received a commission of $4950 on the sale of a house. If the commission rate is 6%, what was the selling price of the house?

32. Terry McGinnis received $845 interest on her savings last year. If she is paid 13% simple interest, what amount did she have invested?

33. A quarterback completed 21 passes in one game for a completion percentage of 60%. How many passes did he attempt?

34. A sport coat was discounted 20% and the sale price was $68. What was the original price?

35. Fred now makes $18,920 per year. What was his salary before he received a 10% raise?

36. The price of a pet mouse rose 30% last year. If the present price is $1.56, what was the price one year ago?

37. Due to inflation, the cost of an item rose $7.20. This was a 12% increase. What was the former price? The new price?

38. What sum invested at $8\frac{1}{2}$% simple interest will amount to $1410.50 in one year?

39. A prize fighter won 38 of his 40 professional fights. What was his percent of wins?

40. The Perez family spent $125 last week for food. This was 20% of their weekly income. What was their weekly income?

41. The sales-tax rate in Canyon City is 4%. How much tax would be charged on a purchase of $327?

42. What sum at 11% simple interest will amount to $7215 in 1 year?

43. The area of the Wyler Ranch is 35% of the area of the Rodriguez Ranch. What is the area of the Wyler Ranch if the area of the Rodriguez Ranch is 1200 mi^2?

44. One year the population of the United States was 5% of the total world population. If the U.S. population was 200,000,000, what was the world population?

FOR REVIEW

Select a variable to represent each quantity and translate the phrase into symbols.

45. Cost increased by 15% of the cost

46. A number less twice its reciprocal

The following exercises review formulas from geometry that will help you prepare for the next section. These formulas with figures are on the inside cover of this text and are reviewed in the Geometry Appendix. A calculator will be helpful in some exercises.

47. The perimeter P of a rectangle with length l and width w is given by $P = 2l + 2w$. Find the perimeter of a rectangle with length 8.6 m and width 5.3 m.

48. The area A of a parallelogram with base b and altitude h is given by $A = bh$. Find the area of a parallelogram with base 246 cm and altitude 188 cm.

49. The volume V of a cylinder with radius r and height h is given by $V = \pi r^2 h$. Find the volume of a cylinder, correct to the nearest tenth of a cubic foot, if the radius is 5.4 ft and height 11.2 ft.

50. The volume V of a sphere with radius r is given by $V = \frac{4}{3}\pi r^3$. Find the volume, to the nearest tenth of a cubic inch, of a sphere with radius 3.8 inches.

3.5 EXERCISES C

Solve.

1. A rope 27 m long is to be cut into three pieces in such a way that the second piece is one-fourth the first and the third is 3 m longer than the second. Find the length of each piece.

2. A retailer uses a 30% markup to make a profit on suits. If Chris Katsaropoulos bought a suit for $124.02 and that included a 6% sales tax, what was the original cost of the suit to the dealer? [Answer: $90.00]

3.6 GEOMETRY AND MOTION PROBLEMS

STUDENT GUIDEPOSTS

  1 Geometry Problems   **2** Motion Problems

1 GEOMETRY PROBLEMS

Many geometry problems require knowledge of basic formulas that are summarized inside the back cover and reviewed in the Geometry Appendix. As you read through a problem, try to determine which formula is appropriate. It is always helpful to make a sketch, and label the given parts of the figure.

EXAMPLE 1 APPLICATION OF GEOMETRY	PRACTICE EXERCISE 1

Find the width of a rectangle if its area is 192 ft^2 and its length is 16 ft.

    To work this problem, we need to know that the area of a rectangle is given by $A = lw$. Make a sketch as in Figure 3.1.
Let w = width of rectangle.

$$A = lw$$
$$192 = 16\,w$$
$$12 = w$$

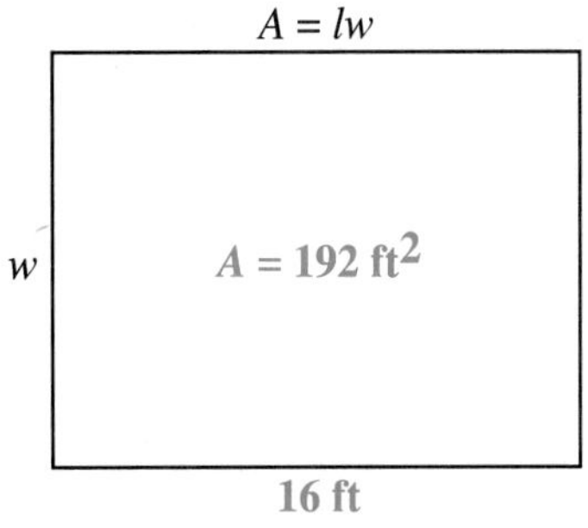

$A = lw$

w $A = 192 \text{ ft}^2$

16 ft

Figure 3.1

The width is 12 ft. Note that it is necessary to include the units in the answer.

The area of a rectangular garden is 19.8 m^2 and its width is 3.6 m. Find the length of the garden.

Answer: 5.5 m

EXAMPLE 2   APPLICATION OF GEOMETRY

If the perimeter of a rectangular room is 32 ft and its length is three times its width, find its dimensions. Make a sketch as in Figure 3.2.

The perimeter of a rectangle is given by $P = 2l + 2w$.

Let w = width of room,
$3w$ = length of room.

$P = 2l + 2w$
$32 = 2(3w) + 2(w)$
$32 = 6w + 2w$
$32 = 8w$
$4 = w$
and $12 = 3w$

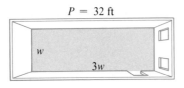

$P = 32$ ft

w

$3w$

Figure 3.2

The length of the room is 12 ft and the width is 4 ft.

PRACTICE EXERCISE 2

A farmer encloses a pasture with 20 miles of fence. If the width of the pasture is one-fourth the length, find the dimensions of the pasture.

Answer: 2 miles by 8 miles

EXAMPLE 3   GEOMETRY PROBLEM

The first angle of a triangle is three times as large as the second. The third angle is 40° larger than the first angle. What is the measure of each angle?

The sum of the measures of the angles of any triangle is 180°. Make a sketch as in Figure 3.3.

Let $3x$ = measure of the first angle,
x = measure of the second angle,
$3x + 40$ = measure of the third angle.

$x + 3x + 3x + 40 = 180$
$7x + 40 = 180$
$7x = 140$
$x = 20$
and $3x = 60$
and $3x + 40 = 100$

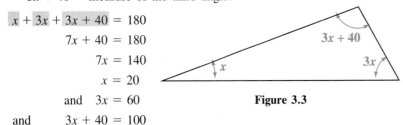

$3x + 40$

x

$3x$

Figure 3.3

The measures are 20°, 60°, and 100°.

PRACTICE EXERCISE 3

The second angle of a triangle is one-half as large as the first, and the third is 45° less than three times the first. Find the measure of each.

Answer: 50°, 25°, 105°

EXAMPLE 4   APPLICATION OF GEOMETRY

The circumference of a circular flower bed is 182 ft. How many feet of pipe do we need to reach from the edge of the bed to a fountain in the center of the bed? The circumference of a circle is given by $C = 2\pi r$. [Use 3.14 for π]. Make a sketch as in Figure 3.4.

$C = 2\pi r$
$182 = 2(3.14)r$
$\dfrac{182}{2(3.14)} = r$

r

$C = 182$ ft

Figure 3.4

$28.98 \approx r$ To the nearest hundredth

We need about 29 ft of pipe.

PRACTICE EXERCISE 4

The circumference of a circle is 46 in. Use 3.14 for π and find the approximate length (to the nearest tenth) of the diameter.

Answer: 14.6 in

In Example 4 we used the symbol ≈ to represent the phrase "is *approximately equal* to." It is important to use ≈ whenever an approximate or rounded number is indicated.

EXAMPLE 5 **APPLICATION OF GEOMETRY**	**PRACTICE EXERCISE 5**

A cylindrical storage bin with radius 10 ft is to be built. How high will it need to be to store 10,200 ft^3 of grain?

The volume of a cylinder is given by $V = \pi r^2 h$. [Use 3.14 for π.]

A cylindrical tank with radius 8 m has surface area 965 m^2. Use 3.14 for π and find the approximate height of the tank. The formula for surface area is $A = 2\pi rh + 2\pi r^2$.

Let h = height of tank.

$V = \pi r^2 h$

Figure 3.5

$$V = \pi r^2 h$$
$$10{,}200 = (3.14)(10)^2 h$$
$$10{,}200 = 314h$$
$$\frac{10{,}200}{314} = h$$
$$32.48 \approx h \qquad \text{To the nearest hundredth}$$

The bin must be about 32.5 ft high.

Answer: 11.2 m

② MOTION PROBLEMS

Problems that involve distances and rates of travel often result in simple linear equations. The distance d that an object travels in a given time t at a fixed rate r is given by the formula

$$\textbf{(distance)} = \textbf{(rate)} \cdot \textbf{(time)} \quad \text{or} \quad d = rt.$$

For example, a boy walking at a rate of 4 mph for 3 hr will travel a distance of

$$d = rt = 4 \cdot 3 = 12 \text{ miles.}$$

Most types of motion problems depend in some way on the formula $d = rt$.

EXAMPLE 6   MOTION PROBLEM

An automobile is driven 336 mi in 7 hr. How fast (at what rate) was the car driven?

Let r = average rate of travel,
336 = distance d driven,
7 = time t of travel.
We need to solve the following equation:

$$336 = r(7)$$
$$\frac{336}{7} = r$$
$$48 = r.$$

Thus, the average rate of travel was 48 mph.

PRACTICE EXERCISE 6

Kevin rides his bike 162 km at an average rate of 36 km/hr. How long did he ride?

Answer: 4.5 hr

EXAMPLE 7   MOTION PROBLEM

Two hikers leave the same camp, one traveling east and the other traveling west. The one going east is hiking at a rate 1 mph faster than the other, and after 5 hr they are 35 mi apart. How fast is each hiking?

Make a sketch like the one in Figure 3.6.

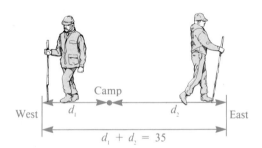

Figure 3.6

Let r = rate of hiker going west,
$r + 1$ = rate of hiker going east,
  $d_1 = 5r$ = distance the westbound hiker travels,
  $d_2 = 5(r + 1)$ = distance the eastbound hiker travels.
The equation we need to solve is

$$d_1 + d_2 = 5r + 5(r + 1) = 35$$
$$5r + 5r + 5 = 35$$
$$10r + 5 = 35$$
$$10r = 30$$
$$r = 3.$$

Thus, the westbound hiker is traveling 3 mph and the eastbound hiker is traveling 4 mph.

PRACTICE EXERCISE 7

Two boats leave an island at the same time, one sailing north and the other south. If one travels 5 mph faster than the other, and they are 140 mi apart in 4 hr, at what rate is each sailing?

Answer: 15 mph, 20 mph

3.6 EXERCISES A

Solve. If necessary, check the inside front cover for a list of formulas. (Some problems have been started.)

1. The perimeter of a rectangle is 60 in. If the length is 4 in more than the width, find the dimensions.

   Let w = width of rectangle,
   $w + 4$ = length of rectangle.

   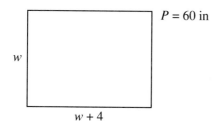
   $P = 60$ in
   w
   $w + 4$

2. If one angle of a triangle is 60° more than the smallest angle and the third angle is six times the smallest angle, find the measure of each angle.

   Let x = smallest angle,
   $x + 60$ = another angle,
   $6x$ = third angle.

3. Find the length of a rectangle if its area is 104 in^2 and its width is 8 in.

4. Find the height of a triangle whose base is 12 cm and whose area is 84 cm^2.

5. A circular patio has a radius of 14 ft. How many feet of edging material will be needed to enclose the patio? [Use $\pi \approx 3.14$.]

6. In a triangle, the longest side is twice the shortest side, and the third side is 5 ft shorter than the longest side. Find the three sides if the perimeter is 45 ft.

7. Find the height of a parallelogram whose base is 19 in and whose area is 133 in^2.

   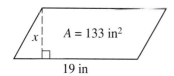
   x $A = 133$ in^2
   19 in

8. If two angles of a triangle are equal and the third angle is equal to the sum of the first two, find the measure of each.

   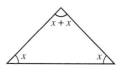
   $x + x$
   x x

9. An isosceles triangle has two equal sides called legs and a third side called the base. If each leg of an isosceles triangle is four times the base and the perimeter is 108 m, find the length of the legs and base.

   $4x$ $4x$
   x

10. The perimeter of a rectangle is 630 in. Find the dimensions if the length is 25 in more than the width.

11. Find the height of a trapezoid if the area is 247 m^2 and the bases are 16 m and 22 m. [*Hint:* $A = \frac{1}{2}(b_1 + b_2)h$]

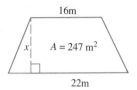

16m

x $A = 247$ m^2

22m

12. Find the base of a parallelogram if the perimeter is 88 cm and a side is 20 cm.

13. Find the area of a circular garden having a diameter of 96 yd.

14 The volume of a grain storage silo in the shape of a rectangular solid is 4725 m^3. If both the width and the height are 15 m, what is the length?

15. Find the height of a cylindrical tank with volume 3200 m^3 and radius 8 m.

h

8 m

16. The perimeter of a square is 48 ft. What is the length of a side?

17 A cube with edge 10 ft is submerged in a rectangular tank containing water. If the tank is 40 ft by 50 ft, how much does the level of water in the tank rise?

18. Find the volume, rounded to the nearest tenth, of a sphere with radius 2 m.

19 The sphere in Exercise 18 is dropped into a rectangular tank that is 10 m long and 8 m wide. To the nearest tenth, how much does this raise the water level?

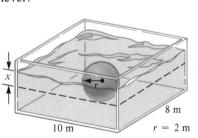

x

8 m

10 m $r = 2$ m

20 A rancher wishes to enclose a pasture that is 3 mi long and 2 mi wide with a fence selling for 35¢ per linear foot. How much will the project cost?

21. Mary Lou Mercer walks for 7 hours at a rate of 5 km/hr. How far does she walk?

22. Bob Packard runs a distance of 20 miles at a rate of 8 mph. How many minutes does he run? [*Hint:* Note the units used.]

23. A hiker crossed a valley by walking 5 mph the first 2 hr and 3 mph the next 4 hr. What was the total distance that she hiked?

24. Two trains leave Omaha, one traveling east and the other traveling west. If one is moving 20 mph faster than the other, and if after 4 hr they are 520 mi apart, how fast is each going?

Omaha

25 Two ships sail north from Bermuda. One is traveling at 22 knots and the other at 17 knots. How many nautical miles will they be from each other after 8 hours?

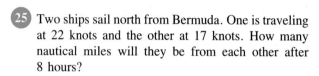

Bermuda

26. Two cars that are 550 mi apart and whose speeds differ by 10 mph are moving towards each other. What is the speed of each if they meet in 5 hr?

FOR REVIEW

27. The sum of three consecutive odd integers is 135. What are the integers?

28. Twice a number, less 7, is the same as three times the number, less 30. What is the number?

29. Gus is 7 years older than Jan. If twice the sum of their ages is 66, how old is each?

30. A wire is 42 in long. It is to be cut into three pieces in such a way that the second is twice the first and the third is equal to the sum of the lengths of the first two. How long is each piece?

31. 54 is 120% of what?

32. 30 is what percent of 150?

33. If the cost of an item rose 8¢, a 16% increase, what was the old price? What is the new price?

34. If the sales-tax rate is 3%, how much tax would be charged on a purchase of $18.00?

35. If a student answered 68 questions correctly on an 80-question exam, what was her percent score?

36. Betsy earned $11,130 one year after she received a 6% raise. What was her former salary?

3.6 EXERCISES B

Solve.

1. The perimeter of a rectangle is 52 ft. If the length is 8 ft more than the width, find the dimensions.

2. If one angle of a triangle is 72° more than the smallest angle, and the third angle is seven times the smallest angle, find the measure of each angle.

3. The area of a rectangle is 32.5 ft^2 and the width is 5 ft. What is the length?

4. If the area of a triangle is 78 cm^2 and its base is 12 cm, find its height.

5. A circular garden is to be enclosed with edging material. How many meters of edging will be needed if the radius of the garden is 5 m? [Use $\pi \approx 3.14$.]

6. In a triangle, the longest side is three times as long as the shortest side, and the third side is 2 ft shorter than the longest side. Find the three sides if the perimeter is 26 ft.

7. Find the height of a parallelogram with a base of 17 cm and an area of 93.5 cm^2.

8. If two angles of a triangle are equal and the third angle is equal to twice their sum, find the measure of each.

9. An isosceles triangle with a base of 13 in has a perimeter of 53 in. Find the length of its legs.

10. The perimeter of a rectangular table is 90 in. Find its dimensions if the length is 15 in more than the width.

11. What is the height of a trapezoid with area 24 yd^2 and bases of length 4.5 yd and 3.5 yd?

12. Find the base of a parallelogram if the perimeter is 100 in and a side is 15 in.

13. What is the area of a circular rug having diameter 82 inches?

14. The volume of a rectangular solid is 26,250 cm^3. If both the width and the height are 25 cm, what is the length?

15. Find the height of a cylindrical storage tank with volume 1105.3 ft^3 and radius 4 ft.

16. The perimeter of a square is 72 yd. What is the length of a side?

17. A cube with edge 8 in is submerged in a rectangular tank containing water. If the tank is 10 in by 20 in, how much does the level of water in the tank rise?

18. Find the volume of a sphere with radius 8 cm.

19. If the sphere in Exercise 18 is submerged in a rectangular tank 20 cm wide by 30 cm long, how much will the level of the water increase?

20. How much will it cost to enclose a garden with a picket fence if the garden is 12 yd long, 8 yd wide, and fencing costs 85¢ per linear foot?

21. A ship travels for 15 hr at 20 knots, where a knot is 1 nautical mile per hour. How many nautical miles has it traveled?

22. Jessie Meyer drove her boat a distance of 45 miles at a rate of 30 mph. How many minutes did she drive?

23. Barry Silverstein traveled by car for 6 hr at a rate of 55 mph and then by boat for 2 hr at 24 mph. What was the total distance traveled?

24. Two campers leave Rocky Mountain National Park, one traveling north and the other traveling south. After 3 hr they are 255 mi apart, and one has been traveling 5 mph faster than the other. How fast is each traveling?

25. Two truckers leave Atlanta at the same time heading west. How far apart will they be in 13 hr if one is traveling at a rate of 60 mph and the other at 56 mph?

26. Two families leave home at 8:00 A.M., planning to meet for a picnic at a point between them. One travels at a rate of 45 mph and the other travels at 55 mph. If they live 500 miles apart, at what time do they meet?

FOR REVIEW

27. The sum of three consecutive even integers is 222. What are the integers?

28. Five times a number, decreased by 8, is the same as four times 1 more than the number. Find the number.

29. Lucy is seven years older than her sister. In 3 years she will be twice as old as her sister is now. How old is each now?

30. A 36-m steel rod is cut into three pieces. The second piece is twice as long as the first and the third is three times as long as the second. How long is each piece?

31. 540 is 150% of what?

32. 98 is what percent of 140?

33. If the cost of a calculator rose $1.92 and this was a 12% increase, what was the old price? What is the new price?

34. If the sales-tax rate is 6%, how much tax will be charged on the purchase of a wallet costing $14.50?

35. During July a car dealer sold 45 cars out of his total inventory of 54 cars. What percent of his inventory was sold?

36. After an 8% salary increase, Henry makes $24,300. What was his salary before the raise?

3.6 EXERCISES C

Solve.

1. The area shown required 124 m of fencing to enclose. Find the dimensions by finding the value of x.

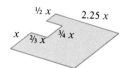

2. Pam drove her car at an average speed of 55 mph for one-half the distance of a trip and then traveled by boat at 20 mph the second half. How many miles did she travel if the total trip took 9 hours?
[Answer: 264 mi]

SUPPLEMENTARY APPLIED PROBLEMS

The following exercises use the techniques presented in the previous three sections.

1. Seven more than twice Herman's age is equal to 73. How old is Herman?

2. The perimeter of Ivan Danielson's yard is 528 yd. If the yard is rectangular in shape with length 24 yd more than the width, what are the dimensions?

3. On a recent field trip, Cindy noticed that the bus was completely full and that there were 14 more girls than boys on the bus. If the bus has a capacity of 54 children, how many boys were on the bus?

4. Bob, Mary, and Sharon were born in consecutive years. If the sum of their ages is 36, how old is each?

5 A pole is standing in a pond. If one-fourth of the height of the pole is in the sand at the bottom of the pond, 9 ft are in the water, and three-eighths of the total height is in the air above the water, what is the length of the pole?

6. In an election with only two candidates, the winner received 50 votes more than twice the total for the loser. If a total of 1310 votes were cast, how many did each candidate receive?

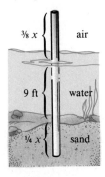

7. Four-fifths of the operating budget of a community college goes for faculty salaries. If the total of all faculty salaries is $4,000,000, what is the total operating budget?

8. If $16.80 is charged as tax on the sale of a washing machine priced at $420, what is the tax rate?

9. A fuel tank contained 840 gallons of fuel. An additional quantity of fuel was added making a total of 1092 gallons. What was the percent increase?

10 The measure of the second angle of a triangle is one-fourth the measure of the third and 30° more than the first. Find the measure of each.

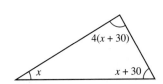

11. International Car Rental will rent a compact car for $12.50 per day plus 10¢ per mile driven. If Charlene rented a compact for two days and was billed a total of $73.00, how many miles did she drive?

12. Tim Chandler bought a pair of running shoes on sale for $33.60. If the sale price was discounted 30% from the original price, what was the original price?

13 A rectangular patio is twice as long as it is wide. If its length were decreased 8 ft and the width were increased 8 ft, it would be square. What are its dimensions?

14. The relationship between °F and °C is given by the formula $F = \frac{9}{5}C + 32$. Determine °C for a temperature of 50°F.

15. How much simple interest will be earned on a deposit of $800 for one year at an interest rate of 12% per year?

16. Barbara purchased a Ford Bronco for $16,500. If the options totaled $5800, what percent of the purchase price was the base price of the vehicle?

17. Leah was able to do 40 more situps than Lynn on a recent fitness test. If the total for both girls was 290, how many situps did Lynn do?

18. The perimeter of a square is eleven cm less than five times the length of a side. Find the length of a side.

19 The population of Deserted, Utah was 740 in 1990. This was a 20% decrease in the population of 1980. What was the population in 1980?

20 Gene won $10,000 in the Illinois Lottery. He invested part in a savings account that earned 14% simple interest, and the rest in a fund that paid 11% simple interest. If at the end of one year he earned an income of $1310 from the two, how much was invested at each rate?

21. The Arnotes plan to sell their home. They must receive $82,000 after deducting the sales commission of 6% on the selling price. Rounded to the nearest dollar, at what selling price should the house be listed?

22. Dr. Cotera wishes to enclose a rectangular plot of land with 140 yards of fencing in such a way that the length is four times the width. What will be the dimensions of the plot?

23. On a recent vacation, Leona traveled one-third the total distance by car, 400 miles by boat, and one-half the total distance by air. What was the total distance traveled on the vacation?

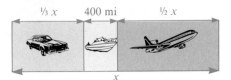

24 Becky must have an average of 90 on four tests in geology to get an A in the course. What is the lowest score she can make on the fourth test if her first three grades are 96, 78, and 91?

25. When ice floats in water, approximately $\frac{8}{9}$ of the height of the ice is below the surface of the water. If the tip of an iceberg is 35 feet above the surface of the water, what is the approximate depth of the iceberg?

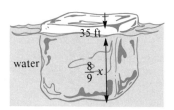

26. A boat traveled from Puerto Vallarta to an island and back again in a total time of 10.5 hours. How far is the island from Puerto Vallarta if the boat averaged 15 mph going to the island and 20 mph returning?

27. In a school election, Morton beats his two opponents by 230 and 175 votes, respectively. If there were a total of 1635 votes cast, how many did Morton receive?

28. A meteorite weighing 4.5 tons was discovered near Meteor Crater, Arizona, in 1918. If the meteorite contains about 87% iron, how many pounds of iron does it contain?

29. After draining their swimming pool to make several minor repairs, the Bonnetts plan to refill it using their garden hose. Suppose the hose lets water in at a rate of 20 gallons per minute and the pool holds 36,000 gallons. Will the pool be filled by noon on Sunday if the water is turned on at 9:00 A.M. Saturday?

30. To estimate the trout population in Lake Louise, 400 banded trout were released on June 1, and the lake was closed to all fishing. On June 10, a sample of 100 trout were caught, 6 of which were banded. As a result, the ranger estimated that 6% of all the trout in Lake Louise were banded. Approximately how many trout were in the lake?

ANSWERS: 1. 33 yr   2. 120 yd, 144 yd   3. 20   4. Bob is 11, Mary is 12, Sharon is 13.   5. 24 ft   6. 420, 890   7. $5,000,000   8. 4%   9. 30%   10. 5°, 35°, 140°   11. 480 mi   12. $48.00   13. 32 ft, 16 ft   14. 10°C   15. $96   16. 64.8%   17. 125   18. 11 cm   19. 925   20. $7000 in savings, $3000 in the fund   21. $87,234   22. 14 yd, 56 yd   23. 2400 mi   24. 95   25. 315 ft   26. 90 mi   27. 680 votes   28. 7830 lb   29. No. It will take 30 hours to fill the pool.   30. 6667 trout

3.7 SOLVING AND GRAPHING LINEAR INEQUALITIES

▬ STUDENT GUIDEPOSTS ▬

① Graphing Equations
② Graphing Inequalities
③ Solving Linear Inequalities
④ Addition-Subtraction Rule
⑤ Multiplication-Division Rule
⑥ Solving by a Combination of Rules
⑦ Method for Solving Linear Inequalities
⑧ Applications Involving Inequalities

Often in algebra we try to picture abstract ideas. This was done in Chapter 2, for instance, using a number line. Recall that a number line is marked off in unit lengths with each point on the line associated with a real number. The origin is identified with the number zero, positive numbers correspond to points to the right of zero, and negative numbers correspond to points to the left of zero. We can identify any real number with exactly one point on a number line, and, conversely, every point on a number line corresponds to exactly one real number.

① GRAPHING EQUATIONS

Linear equations in one variable like those we studied earlier can be graphed on a number line by plotting the points that correspond to solutions of the equation.

EXAMPLE 1 GRAPHING SOLUTIONS TO EQUATIONS	**PRACTICE EXERCISE 1**

Graph $2x + 1 = 3$.

First solve the equation.

$$2x + 1 - 1 = 3 - 1 \quad \text{Subtract 1}$$
$$2x = 2$$
$$x = 1$$

The solution is $x = 1$. Plot the point corresponding to 1 on a number line, as in Figure 3.7.

Graph $3x - 2 = 4$.

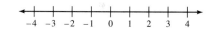

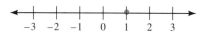

Figure 3.7

Answer: Graph the solution $x = 2$.

To graph $2x + 1 = 1 + 2x$, first solve the equation.

$$2x + 1 - 1 = 1 - 1 + 2x \quad \text{Subtract 1}$$
$$2x = 2x$$
$$2x - 2x = 2x - 2x \quad \text{Subtract } 2x$$
$$0 = 0$$

Since $0 = 0$ is always true, this equation is an identity and every number is a solution. Then every number must be plotted, and the graph of the solution is the entire number line shown in Figure 3.8.

Figure 3.8

Suppose we graph $x + 1 = x$. If we solve the equation, we obtain the following result.

$$x - x + 1 = x - x \qquad \text{Subtract } x$$
$$1 = 0$$

This equation is a contradiction because $1 \neq 0$. There are no solutions to the equation so we have no points to plot.

② GRAPHING INEQUALITIES

Graphing equations is a relatively simple procedure. The procedure for graphing **inequalities,** statements containing the symbols $<$, $\leq$, $>$, or $\geq$, is shown with the following figures.

In Figure 3.9, the graph of $x > 3$ has an open circle at 3. This means that the number 3 is not included among the solutions, while the colored line shows that all points to the right of 3 are included.

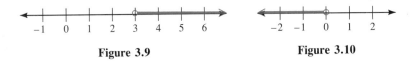

Figure 3.9   **Figure 3.10**

The graph of $x < 0$ is shown in Figure 3.10. In Figure 3.11, the graph of $x \geq -1$ has a solid circle at the point -1, indicating that -1 is included in the graph of $x \geq -1$. Finally, the graph of $x \leq 1$ is given in Figure 3.12.

Figure 3.11   **Figure 3.12**

③ SOLVING LINEAR INEQUALITIES

Now that we know how to graph simple inequalities such as $x > 3$, $x < 0$, $x \geq -1$, and $x \leq 1$, we can start to solve and graph more complex ones. Statements containing the inequality symbols $<$, $\leq$, $>$, and $\geq$, such as

$$x + 1 < 7, \quad 2x - 1 \geq 3, \quad 3x + 1 \leq 4x - 5, \quad \text{and} \quad x + 5 > 2(x - 3)$$

are called **linear inequalities.** A **solution** to a linear inequality is a number that, when substituted for the variable, makes the inequality true. For example, $-6, 0$, and 5 are some of the solutions to $x + 1 < 7$ since

$$-6 + 1 < 7, \quad 0 + 1 < 7, \quad \text{and} \quad 5 + 1 < 7$$

are all true.

Solving linear inequalities is much like solving linear equations since most of the same rules apply. There is one exception that will be discussed shortly. Like equivalent equations, **equivalent inequalities** have exactly the same solutions. Solving an inequality is a matter of transforming it to an equivalent inequality for which the solution is obvious.

④ ADDITION-SUBTRACTION RULE

If we start with the true inequality

$$7 < 15$$

and add 4 to both sides, we obtain another true inequality.

$$7 + 4 < 15 + 4$$
$$11 < 19$$

Likewise, if we subtract 3 from both sides, we again obtain a true inequality.

$$7 - 3 < 15 - 3$$
$$4 < 12$$

These observations lead to the addition-subtraction rule for inequalities.

Addition-Subtraction Rule

An equivalent inequality is obtained if the same quantity is added to or subtracted from both sides of an inequality.
Suppose a, b, and c are real numbers.

If $a < b$ then $a + c < b + c$ and $a - c < b - c$.

We use this rule to solve inequalities in very much the same way that we used the similar addition-subtraction rule to solve equations. The key is to isolate the variable on one side of the inequality.

EXAMPLE 2 ADDITION-SUBTRACTION RULE	PRACTICE EXERCISE 2

Solve and graph.

$$x + 1 < 7$$
$$x + 1 - 1 < 7 - 1 \quad \text{Subtract 1 from both sides}$$
$$x + 0 < 6$$
$$x < 6$$

All numbers less than 6 are solutions, and the graph is given in Figure 3.13. The fact that 6 is not part of the solution is indicated by the open circle.

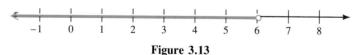

Figure 3.13

Solve and graph.

$$y + 7 > 3$$

$$\overset{\longleftarrow}{\underset{-6\,-5\,-4\,-3\,-2\,-1\ \ 0\ \ 1\ \ 2\ \ 3\ \ 4\ \ 5\ \ 6}{\mid\ \mid\ \mid\ \mid\ \mid\ \mid\ \mid\ \mid\ \mid\ \mid\ \mid\ \mid\ \mid}}\overset{\longrightarrow}{}$$

Answer: Graph $y > -4$.

Most of the equations we have solved have had only one or two solutions. For inequalities, there are many solutions.

| **EXAMPLE 3   ADDITION-SUBTRACTION RULE** | **PRACTICE EXERCISE 3** |

Solve and graph.

$$x - 3 \geq 5$$
$$x - 3 + 3 \geq 5 + 3 \qquad \text{Add 3 to both sides}$$
$$x + 0 \geq 8$$
$$x \geq 8$$

All numbers greater than or equal to 8 are solutions. Generally, we indicate the solution simply by writing $x \geq 8$. The graph is given in Figure 3.14. Notice that 8 is part of the solution, so the circle at the point corresponding to 8 is solid.

Figure 3.14

Solve and graph.

$$y - 15 \leq -12$$

Answer: Graph $y \leq 3$.

---

5 MULTIPLICATION-DIVISION RULE

If we start with the true inequality

$$6 < 14$$

and multiply both sides by 3, we obtain the true inequality

$$18 < 42.$$

However, if we multiply both sides by -2, we obtain the false inequality

$$-12 < -28. \qquad \text{This is false}$$

To obtain a true inequality here, we need to reverse the inequality symbol, that is, change $<$ to $>$. These observations lead to the multiplication-division rule for inequalities.

Multiplication-Division Rule

An equivalent inequality is obtained in the following situations.

1. Each side of the inequality is multiplied or divided by the same *positive* quantity.

$$\text{If } c > 0 \text{ and } a < b \text{ then } ac < bc \text{ and } \frac{a}{c} < \frac{b}{c}.$$

2. Each side of the inequality is multiplied or divided by the same *negative* quantity and the inequality symbol is reversed.

$$\text{If } c < 0 \text{ and } a < b \text{ then } ac > bc \text{ and } \frac{a}{c} > \frac{b}{c}.$$

| EXAMPLE 4 MULTIPLICATION-DIVISION RULE | PRACTICE EXERCISE 4 |

Solve and graph.

$$4x > 16$$

$$\frac{1}{4} \cdot 4x > \frac{1}{4} \cdot 16 \qquad \text{$\frac{1}{4}$ is positive so inequality symbol remains the same}$$

$$1 \cdot x > 4$$

$$x > 4$$

The graph is given in Figure 3.15.

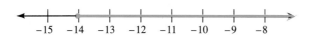

Figure 3.15

PRACTICE EXERCISE 4

Solve and graph.

$$\frac{1}{2}y < 3$$

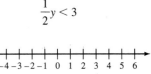

Answer: Graph $y < 6$.

| EXAMPLE 5 MULTIPLICATION-DIVISION RULE | PRACTICE EXERCISE 5 |

Solve and graph.

$$-\frac{1}{2}x \le 7$$

$$\downarrow$$

$$(-2)\left(-\frac{1}{2}x\right) \ge (-2)(7) \qquad \begin{array}{l}\text{Multiply both sides by -2 and \textit{reverse}}\\ \text{the inequality symbol}\end{array}$$

$$1 \cdot x \ge -14$$

$$x \ge -14$$

The graph is given in Figure 3.16.

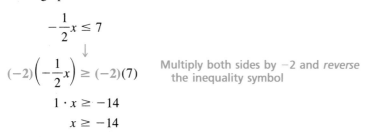

Figure 3.16

PRACTICE EXERCISE 5

Solve and graph.

$$-7y \ge -35$$

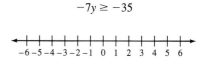

Answer: Graph $y \le 5$.

/////////// **CAUTION** ///////////

To multiply or divide by a negative number, always reverse the inequality symbol. Thus, if $-x < -5$, then $x > 5$ (not $x < 5$). This is the only substantial difference between solving an equation and solving an inequality.

///////////

⑥ SOLVING BY A COMBINATION OF RULES

We now solve inequalities that require both the addition-subtraction rule and the multiplication-division rule. As with solving equations, the basic idea is to isolate the variable on one side of the inequality.

EXAMPLE 6 COMBINATION OF RULES

Solve and graph.

$$2x + 5 < 7$$

$2x + 5 - 5 < 7 - 5$ Subtract 5

$$2x < 2$$

$\dfrac{1}{2} \cdot 2x < \dfrac{1}{2} \cdot 2$ Multiply by $\frac{1}{2}$ or divide by 2; the inequality remains the same

$$1 \cdot x < 1$$

$$x < 1$$

The graph is given in Figure 3.17.

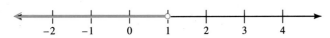

Figure 3.17

PRACTICE EXERCISE 6

Solve and graph.

$$3y - 7 \geq 5$$

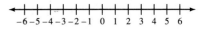

Answer: Graph $y \geq 4$.

EXAMPLE 7 COMBINATION OF RULES

Solve and graph.

$$6 - 4x \geq 2 - 3x$$

$6 - 6 - 4x \geq 2 - 6 - 3x$ Subtract 6

$$-4x \geq -4 - 3x$$

$-4x + 3x \geq -4 - 3x + 3x$ Add $3x$

$$-x \geq -4$$

$(-1)(-x) \leq (-1)(-4)$ Multiply by -1 and *reverse* inequality symbol

$$x \leq 4$$

The graph is given in Figure 3.18.

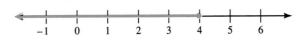

Figure 3.18

PRACTICE EXERCISE 7

Solve and graph.

$$9 - 8y < 6 - 5y$$

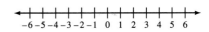

Answer: Graph $y > 1$.

Often an inequality involves parentheses. When this occurs, first remove all parentheses and then proceed as in previous cases.

EXAMPLE 8 INEQUALITY WITH PARENTHESES

Solve. $-6(2 + x) \geq 2(4x - 2)$

$-12 - 6x \geq 8x - 4$ Remove parentheses

$-12 + 12 - 6x \geq 8x - 4 + 12$ Add 12

$-6x \geq 8x + 8$

$-6x - 8x \geq 8x - 8x + 8$ Subtract $8x$

PRACTICE EXERCISE 8

Solve and graph.

$$5(z - 3) > -3(7 - z)$$

$$-14x \geq 8$$
$$\downarrow$$
$$\left(-\frac{1}{14}\right)(-14x) \leq \left(-\frac{1}{14}\right) \cdot 8$$ Multiply by $-\frac{1}{14}$ and *reverse* inequality symbol
$$x \leq -\frac{8}{14}$$
$$x \leq -\frac{4}{7}$$

The graph is given in Figure 3.19.

Figure 3.19

Answer: Graph $z > -3$.

⑦ METHOD FOR SOLVING LINEAR INEQUALITIES

The method for solving a linear inequality is summarized below.

To Solve a Linear Inequality

1. Simplify both sides of the inequality by clearing parentheses, fractions, and decimals.

2. Use the addition-subtraction rule to isolate all variable terms on one side and all constant terms on the other side. Collect like terms when possible.

3. Use the multiplication-division rule to obtain a variable with coefficient of 1. Be sure to reverse the inequality symbol whenever multiplying or dividing both sides by a negative number.

⑧ APPLICATIONS INVOLVING INEQUALITIES

Some applied problems result in inequalities.

EXAMPLE 9   EDUCATION APPLICATION

PRACTICE EXERCISE 9

For Beth to get an A in her Spanish course, she must earn a total of 360 points on four tests each worth 100 points. If she got scores of 87, 96, and 91 on the first three tests, find (using an inequality) the range of scores she could make on the fourth test to get an A.

Let s = score Beth must get on the fourth test. Since the total of her four test scores must be greater than or equal to 360, we must solve

$$87 + 96 + 91 + s \geq 360.$$
$$274 + s \geq 360$$
$$s \geq 86 \quad \text{Subtract 274 from both sides}$$

Beth must make a score of 86 or better on the fourth test to get an A in the course.

If Angie's age is tripled then increased by 5, the result is less than or equal to 25 plus her age. Angie is at most how old?

Answer: 10 yr ($a \leq 10$)

3.7   EXERCISES A

Graph the equation on a number line.

1. $2x - 3 = 1$

2. $x + 5 = x$

3. $x - 1 = x - 1$

4. $\dfrac{y}{\frac{1}{4}} = 20$

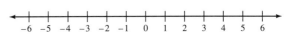

5. $-\dfrac{1}{10} - \dfrac{3}{5}y = \dfrac{1}{5}$

6. $a - 3 + 4a = 2 + a - 5$

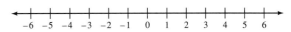

7. $3(z - 1) - 2(z + 3) = -7$

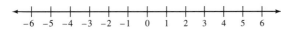

8. $12 - x = 3 + 2[4 - (x - 1)]$

Graph the inequality on a number line.

9. $x > 2$

10. $x \le 2$

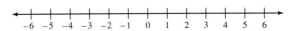

11. $y \le -1$

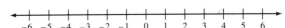

12. $y > -1$

13. $2a > 10$

14. $-2a \le -8$

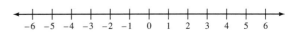

15. $1 - 3x < 8$

16. $2x + 3 > 5x - 3$

17. $-8y - 21 \ge 7 - 15y$

18. $25 \ge 5(3 - z) - 10$

19. $4(a - 1) < 2(3a + 2)$

20. $-2a + 3 + a < 2(a + 3)$

Solve.

21. $x + 9 > 12$

22. $x - 3 < 14$

23. $x + 1 \ge -5$

24. $z + 2.3 > 4.7$

25. $3.9 > z - 5.2$

26. $z - \dfrac{3}{4} \le \dfrac{2}{3}$

27 $2a + 1 > a - 3$

28. $a - 4 \le 2a + 5$

29. $3a - 11 \ge 2a + 9$

30. $3(b + 1) > 2b - 5$

31. $3x > 12$

32. $\frac{1}{3}x < -4$

33 $-\frac{1}{4}y \ge 2$

34. $-3y < 7$

35. $2.1y > 4.2$

36. $-2 < -5z$

37. $-a \le -7$

38. $-3.1z \le 9.3$

39. $-\frac{2}{3}z > \frac{4}{3}$

40. $3b \ge -\frac{1}{2}$

41. $-3b \ge \frac{1}{2}$

42. $1 - 3x \ge -8$

43. $-8y - 7 \ge 21 - 15y$

44. $4(z - 3) \le 3(2z + 4)$

45 $3(2x + 3) - (3x + 2) < 12$

46. $5(x + 3) + 4 \ge x - 1$

47. $y + 3 - 4y > 2(y + 12)$

48. $3 - 2z < 5(z - 7)$

49. $0.05 + 3(z - 1.2) > 2z$

50 $\frac{3}{4}z - \frac{3}{8} < \frac{3}{2} + \frac{1}{8}z$

Are the following statements true *or* false? *If the statement is false, tell why.*

51. If $x < 9$, then $-2x < -18$.

52. If $x > 4$, then $3x > 12$.

53. If $x \le 7$, then $x + 1 \le 8$.

54. If $x < -1$, then $-x > 1$.

55. If $x < 3$, then $x - 7 > -4$.

56. If $-x < -10$, then $1 - x < -9$.

Solve.

57 The product of a number and 3 is greater than or equal to the number less 8. Find the numbers that satisfy this.

58. If twice my age, increased by 7, is greater than or equal to 31 less my age, I am at least how old?

59. Professor Packard will give an F to any student with a point total less than 180 in a course having three 100 point exams. Burford made 38 points on the first exam and 54 on the second. Determine the minimum score he could make on the third exam to avoid failing the course.

60 For Darrell Fosberg to win a trip to Bermuda, his new car sales must average 50 over the three-month period June, July, and August. If he sold 47 cars in June and 62 cars in July, how many cars must he sell in August to qualify for the trip?

FOR REVIEW

61. The length of a rectangular room is 5 yd more than its width. If its perimeter is 54 yd, find its dimensions.

62. Find the volume, rounded to the nearest tenth, of a sphere with radius 3 cm. [Use 3.14 for π.]

63. Roy drove 378 miles in 7 hr. What was his average speed?

64. Two families leave their homes at the same time planning to meet at a point between them. If one travels 55 mph, the other travels 50 mph, and they live 273 miles apart, how long will it take for them to meet?

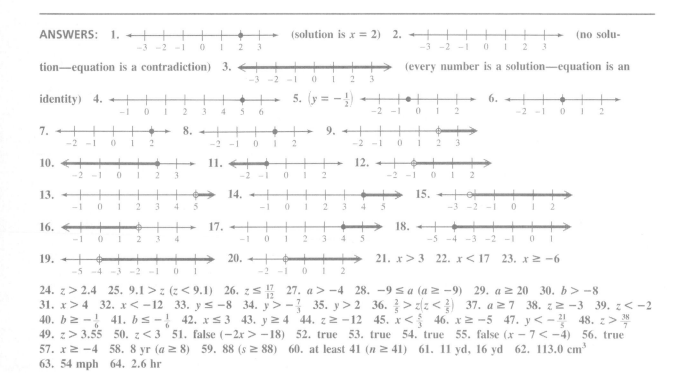

ANSWERS: **1.** (solution is $x = 2$) **2.** (no solution—equation is a contradiction) **3.** (every number is a solution—equation is an identity) **4.** **5.** $\left(y = -\frac{1}{2}\right)$ **6.** **7.** **8.** **9.** **10.** **11.** **12.** **13.** **14.** **15.** **16.** **17.** **18.** **19.** **20.** **21.** $x > 3$ **22.** $x < 17$ **23.** $x \geq -6$

24. $z > 2.4$ **25.** $9.1 > z$ ($z < 9.1$) **26.** $z \leq \frac{17}{12}$ **27.** $a > -4$ **28.** $-9 \leq a$ ($a \geq -9$) **29.** $a \geq 20$ **30.** $b > -8$
31. $x > 4$ **32.** $x < -12$ **33.** $y \leq -8$ **34.** $y > -\frac{7}{3}$ **35.** $y > 2$ **36.** $\frac{2}{5} > z$ ($z < \frac{2}{5}$) **37.** $a \geq 7$ **38.** $z \geq -3$ **39.** $z < -2$
40. $b \geq -\frac{1}{6}$ **41.** $b \leq -\frac{1}{6}$ **42.** $x \leq 3$ **43.** $y \geq 4$ **44.** $z \geq -12$ **45.** $x < \frac{5}{3}$ **46.** $x \geq -5$ **47.** $y < -\frac{21}{5}$ **48.** $z > \frac{38}{7}$
49. $z > 3.55$ **50.** $z < 3$ **51.** false ($-2x > -18$) **52.** true **53.** true **54.** true **55.** false ($x - 7 < -4$) **56.** true
57. $x \geq -4$ **58.** 8 yr ($a \geq 8$) **59.** 88 ($s \geq 88$) **60.** at least 41 ($n \geq 41$) **61.** 11 yd, 16 yd **62.** 113.0 cm^3
63. 54 mph **64.** 2.6 hr

3.7 EXERCISES B

Graph the equation on a number line.

1. $2x - 5 = 1$

2. $4x + 3 = 4x - 3$

3. $x - 5 = x - 5$

4. $\dfrac{y}{\frac{1}{3}} = 9$

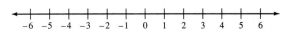

5. $-\dfrac{1}{4} - \dfrac{3}{8}y = \dfrac{1}{2}$

6. $a - 2 + 3a = 4 + a - 5$

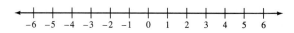

7. $2(z - 2) - 3(z + 1) = -4$

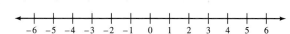

8. $7 - 2x = 3 + [2 - (x - 3)]$

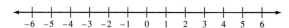

Graph the inequality on a number line.

9. $x > 1$

10. $x \leq 1$

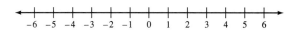

11. $y \leq -3$

12. $y > -3$

13. $3a < 12$

14. $-3a > -9$

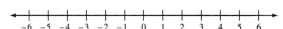

15. $1 - 2x < 4$

16. $3x + 1 < x - 5$

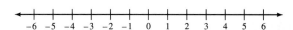

17. $-4y - 4 \geq 3 - 11y$

18. $3(z + 7) - 18 \leq 2z$

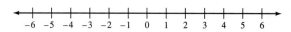

19. $5(a - 2) \geq 2(3a - 4)$

20. $-3a - 4 + 2a \geq 3(a + 2)$

Solve.

21. $x + 3 > 2$

22. $x - 4 \leq 7$

23. $x + 3 \geq -4$

24. $z + 1.9 < 3.9$

25. $4.2 < z - 8.1$

26. $z - \dfrac{2}{3} \leq \dfrac{3}{4}$

27. $a - 5 \leq 2a + 3$

28. $2a + 11 > a - 3$

29. $4a - 12 \geq 3a - 1$

30. $2(b + 5) \geq b + 7$

31. $2x > 14$

32. $\frac{1}{3}x < -8$

33. $-\frac{1}{3}y < 2$

34. $-5y < 11$

35. $2.1y \leq 6.3$

36. $-5a > -30$

37. $-a > -10$

38. $-4.1z > 8.2$

39. $-\frac{1}{3}z \leq \frac{2}{3}$

40. $4b \geq -\frac{1}{5}$

41. $-4b \geq \frac{1}{5}$

42. $1 - 4x \geq -7$

43. $-5y - 2 < 32 + 12y$

44. $5(y - 2) > 4(2y + 1)$

45. $2(3x + 1) - (5x + 4) \geq 2$

46. $3x - 1 \leq x - (3x - 14)$

47. $x - 2 + 2x \leq 2(x + 5)$

48. $3 - 2(z + 1) \geq 5 - z$

49. $0.01 + 2(z + 1.5) > 3z$

50. $\frac{1}{4}z + \frac{3}{8} \geq \frac{1}{2} - \frac{3}{8}z$

Are the following statements true *or* false? *If the statement is false, tell why.*

51. If $x > 7$, then $-2x < -14$.

52. If $x < 5$, then $3x < 15$.

53. If $x \leq -1$, then $x + 1 \leq 0$.

54. If $-x \geq 1$, then $x \geq 1$.

55. If $x < 5$, then $x - 4 < 1$.

56. If $-x \leq -5$, then $2 - x \leq -3$.

Solve.

57. The product of a number and 5 is less than or equal to that number increased by 8. Find the numbers that satisfy this.

58. If Jeff's age is doubled, then diminished by 4, the result is greater than or equal to 5 plus his age. Jeff is at least how old?

59. To get an A in chemistry, Paula must have at least 270 points out of a total of 300 in the course. She has made 87 points on the first test and 98 on the second. At least how many points must she get on the third and final exam to receive an A?

60. Each team of four members in a tug-of-war can have no more than 600 lb total weight to qualify for the finals. Ron's team has three members weighing 120 lb, 140 lb, and 128 lb. What is the greatest amount Ron can weigh in order for his team to qualify?

FOR REVIEW

61. The width of a rectangular table is 22 in less than its length. If its perimeter is 216 in, find its dimensions.

62. A circular cylinder has volume 117.75 cm^3 and radius 5 cm. What is the height of the cylinder? [Use 3.14 for π.]

63. Martha hiked 25 miles at an average rate of 3 mph. What length of time did she hike?

64. Two planes leave Atlanta at the same time, one heading east at 400 mph and the other heading west at 460 mph. How long will it take for them to be 3010 miles apart?

3.7 EXERCISES C

Solve.

1. $\dfrac{x-5}{3} - \dfrac{2x+1}{2} \le x - (2x-1)$

2. The temperature of a hot-spring bath in Mexico is advertised to stay between 35°C and 40°C. What is this range of temperature in degrees Fahrenheit?   [Answer: $95° \le F \le 104°$]

CHAPTER 3 REVIEW

KEY WORDS

3.1 An **equation** is a statement that two quantities are equal.

A **solution** to an equation is a number which makes the equation true.

A **linear equation** can be written in the form $ax + b = 0$.

A **conditional equation** is true for some replacements of the variable and false for others.

An **identity** has the entire set of real numbers for its solution set.

A **contradiction** has no solutions.

Equivalent equations have exactly the same solutions.

3.5 **Consecutive integers** are integers which immediately follow each other in regular counting order.

A **commission** is income as a percent of total sales.

Interest is money we pay to borrow money or money earned on savings.

3.7 A **linear inequality** is a statement involving the inequality symbols $<$, $\le$, $>$, or $\ge$, such as $x + 1 < 7$, $2x - 1 \ge 3$, $3x + 1 \le 4x - 5$, and $x + 5 > 2(x - 3)$.

A **solution** to a linear inequality is a number that, when substituted for the variable, makes the inequality true.

Equivalent inequalities have exactly the same solutions.

KEY CONCEPTS

3.1 When using the addition-subtraction rule to solve an equation, add or subtract the same expression on both sides.

3.2 When using the multiplication-division rule to solve an equation, multiply by a number that makes the coefficient of x equal to 1. For example, to solve

$$\frac{x}{\frac{4}{5}} = 20$$

multiply by $\frac{4}{5}$, *not* by $\frac{5}{4}$.

3.3 1. Use the addition-subtraction rule before the multiplication-division rule when a combination is necessary to solve an equation.

2. When solving an equation involving parentheses, first use the distributive law to clear all parentheses.

3.5 1. Write out complete descriptions of the variables and terms when solving applied problems.

2. Two consecutive even (or odd) integers can be named x and $x + 2$. Two consecutive integers can be named x and $x + 1$.

3. When solving percent-increase problems add the increase to the original amount. For example, an amount of money x plus 5% interest can be represented as $x + 0.05x = (1 + 0.05)x = (1.05)x$.

3.6   **1.** An accurate sketch is often helpful for solving a geometry problem.

      2. The formula $d = rt$, or (distance) = (rate) · (time), is used in many motion problems.

3.7   **1.** The same expression can be added or subtracted on both sides of an inequality.

      2. When you multiply or divide both sides of an inequality by a negative number, *reverse* the symbol of inequality.

For example, $-x < 5$
$$(-1)(-x) > (-1)(5)$$
$$x > -5.$$

3. When graphing $x \geq c$ or $x \leq c$ on a number line, the point corresponding to c is part of the graph, shown with a solid dot at c. When graphing $x > c$ or $x < c$, the point corresponding to c is *not* part of the graph, shown with an ''open'' dot at c.

REVIEW EXERCISES

Part I

3.1   **1.** Use the equation $2(x + 1) = 2x + 2$ to answer Exercises (a)–(f).

      (a) What is the variable? _____   **(d)** What is the solution? _____

      (b) What is the left side? _____   **(e)** Is this an identity? _____

      (c) What is the right side? _____   **(f)** Is this a contradiction? _____

Solve for x.

    2. $x + 9 = 13$ **3.** $\dfrac{2}{3} - x = \dfrac{4}{9}$

3.2   **4.** $-2.3x = 4.6$ **5.** $\dfrac{x}{\frac{1}{3}} = 15$

3.3   **6.** $2x + 3 = x + 3 - 4x$ **7.** $x - \dfrac{1}{3} = -8x$

    8. $3(x - 4) + 5 = 2(x + 1) - 3$ **9.** $5(2x - 2) - 3(x - 4) = 0$

10. $x - (x + 3) = x - 6$ **11.** $20 - 3(x + 5) = 0$

3.4 *Letting x represent the unknown number, translate the following into symbols.*

12. A number increased by 5 is 7.

13. Four times a number is 23.

14. Twice a number, increased by 7, is -12.

15. When 5 is added to six times a number, the result is the same as 3 less than the number.

3.5 **16.** Fred is 12 years older than Bertha. Twice the sum of their ages is 100. How old is each?

17. The sum of three consecutive even integers is 138. Find the integers.

18. If the sales-tax rate is 4%, how much tax would be charged on a purchase of $420?

19. 30% of what is 198?

20. 14% of 50 is what?

21. What percent of 900 is 585?

22. The price of an item rose 15% last year. If the present price is $41.40, what was the price last year?

23. What sum of money invested at 12% simple interest will amount to $716.80 in 1 year?

3.6 **24.** The diameter of a circle is 7.56 cm. What is the radius?

25. The area of a rectangle is 25.2 ft^2 and the length is 6 ft. What is the width?

26. The perimeter of a rectangle is 80 ft. If the length is 10 ft more than the width, what are the dimensions?

27. Two trains leave the same city, one heading north and the other south. If one train is moving 5 mph faster than the other, and if after 4 hr they are 556 mi apart, how fast is each traveling?

Graph the equation or inequality on a number line.

3.7   **28.** $3x + 1 = 10$ **29.** $\dfrac{y}{\frac{1}{8}} = 16$

30. $4 - z = 8 - [1 - (z - 3)]$ **31.** $x > 3$

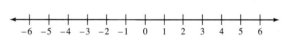

32. $-2a \leq -2$ **33.** $3(y - 1) \leq 2(y - 2)$

Solve.

34. $x - 3 > 4$ **35.** $y + 2 \leq -5$

36. In solving an inequality, when is the symbol of inequality reversed?

Solve.

37. $\dfrac{1}{5}x \leq 35$ **38.** $-4y > 24$ **39.** $6 - 2x > -4 + 3x$

40. $5 + x > 3 - (x + 10)$ **41.** $3(x + 2) \leq 5x - 4$ **42.** $2 - (x - 2) < 7(x - 3)$

Part II

Solve.

43. After receiving a 20% discount on the selling price, Holly paid $6.60 for a record. What was the price of the record before the discount?

44. A sphere with radius 7 cm is submerged in a rectangular tank 20 cm wide and 25 cm long. How much will the level of the water rise in the tank?

45. $-y = -4.7$

46. $1\frac{1}{2} = z + 2$

47. $3(x - 2) = 5 - (x + 1)$

48. $6x + 1 < 2(x - 3)$

49. $-3x \geq \dfrac{6}{5}$

50. $4 - 2(x - 1) \leq 6 - (x - 1)$

Answer true *or* false. *If the statement is false, tell why.*

51. A statement that two quantities are equal is called an inequality.

52. The equation $x + 5 = 5 + x$ is an example of an identity.

53. The equation $x + 5 = x - 5$ is an example of a conditional equation.

54. Two equations which have exactly the same solutions are called equivalent.

Graph each inequality.

55. $x - 2 \leq -5$

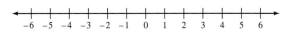

56. $-\dfrac{1}{4}y > 1$

57. $2 - 3(x + 1) < x + 3$

58. $3x - 6 \geq 4 - (x - 6)$

ANSWERS: 1. (a) x (b) $2(x + 1)$ (c) $2x + 2$ (d) every real number (e) yes (f) no 2. 4 3. $\frac{2}{9}$ 4. -2 5. 5 6. 0 7. $\frac{1}{27}$ 8. 6 9. $-\frac{2}{7}$ 10. 3 11. $\frac{5}{3}$ 12. $x + 5 = 7$ 13. $4x = 23$ 14. $2x + 7 = -12$ 15. $6x + 5 = x - 3$ 16. Fred is 31, Bertha is 19. 17. 44, 46, 48 18. $16.80 19. 660 20. 7 21. 65% 22. $36 23. $640 24. 3.78 cm 25. 4.2 ft 26. 15 ft, 25 ft 27. 72 mph, 67 mph 28. ⟵|——|——|——|——|——|——|——|——|——●|——|——|——⟶
   -6 -5 -4 -3 -2 -1 0 1 2 3 4 5 6

29. ⟵|——|——|——|——|——|——|——●|——|——|——|——|——⟶   30. ⟵|——|——|——|——|——|——●|——|——|——|——|——|——⟶   31. ⟵|——|——|——|——|——|——|——|——|——|——|——●═══⟶
      -6 -5 -4 -3 -2 -1 0 1 2 3 4 5 6   -6 -5 -4 -3 -2 -1 0 1 2 3 4 5 6   -6 -5 -4 -3 -2 -1 0 1 2 3 4 5 6

32. ⟵|——|——|——|——|——|——|══●——|——|——|——|——|——⟶   33. ⟵═══●——|——|——|——|——|——|——|——|——|——|——|——⟶   34. $x > 7$ 35. $y \leq -7$ 36. when both sides are multi-
      -6 -5 -4 -3 -2 -1 0 1 2 3 4 5 6   -6 -5 -4 -3 -2 -1 0 1 2 3 4 5 6

plied or divided by a negative number 37. $x \leq 175$ 38. $y < -6$ 39. $x < 2$ 40. $x > -6$ 41. $x \geq 5$ 42. $x > \frac{25}{8}$ 43. $8.25 44. approximately 2.9 cm 45. 4.7 46. $-\frac{1}{2}$ 47. $\frac{5}{2}$ 48. $x < -\frac{7}{4}$ 49. $x \leq -\frac{2}{5}$ 50. $x \geq -1$ 51. false (equality) 52. true 53. false (contradiction) 54. true 55. ⟵═══●——|——|——|——|——|——|——|——|——|——|——|——⟶
                                                                                                                  -6 -5 -4 -3 -2 -1 0 1 2 3 4 5 6

56. ⟵═══●——|——|——|——|——|——|——|——|——|——|——|——⟶   57. ⟵|——|——|——|——|——|——●═══|══|══|══|══|══|══⟶   58. ⟵|——|——|——|——|——|——|——|——|——●═══|══|══|══⟶
      -6 -5 -4 -3 -2 -1 0 1 2 3 4 5 6   -6 -5 -4 -3 -2 -1 0 1 2 3 4 5 6   -6 -5 -4 -3 -2 -1 0 1 2 3 4 5 6

Solve.

1. $x - 5 = 10$

1. _____

2. $4x = 32$

2. _____

3. $x + \dfrac{3}{4} = \dfrac{5}{4}$

3. _____

4. $5.1x = -10.2$

4. _____

5. $\dfrac{1}{4}x = 9$

5. _____

6. $\dfrac{x}{\frac{1}{5}} = 20$

6. _____

7. $3x - 5 = 8x + 10$

7. _____

8. $3(x + 2) - 5(x - 4) = 0$

8. _____

9. $4x - (x + 6) = 3$

9. _____

10. The sum of two consecutive odd integers is 168. What are the integers?

10. _____

11. The price of a dress was $54 but the price was increased by 15%. What is the new price?

11. _____

12. The area of a triangle is 240 cm² and its height is 12 cm. What is the length of the base?

12. _____

13. Two trains that are 840 miles apart and whose speeds differ by 9 mph are traveling towards each other. If they will meet in 8 hours, at what speed is each traveling?

13. _____

Solve the inequalities.

14. $3x - 6 \leq 9x + 12$

14. _____

15. $3 - (2x - 5) > 6x + 4$

15. _____

16. Graph the equation $2x + 5 = 9$ on the number line.

16.

17. Graph the inequality $2(y + 1) \geq 1 - (y + 5)$ on the number line.

17.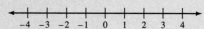

Graphing

4.1 THE RECTANGULAR COORDINATE SYSTEM

In Section 3.7 we graphed linear equations and inequalities in one variable by plotting points on a number line. Now we develop a system in which equations and inequalities in two variables can be graphed. Consider a horizontal number line and a vertical number line as shown in Figure 4.1.

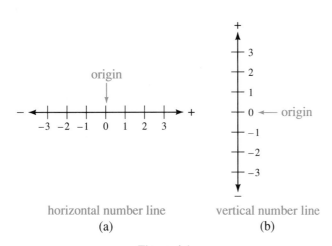

horizontal number line
(a)

vertical number line
(b)

Figure 4.1

1 RECTANGULAR COORDINATE SYSTEM

When the horizontal number line and the vertical number line are placed together so that the two origins coincide and the lines are perpendicular, as in Figure 4.2, the result is called a **rectangular** or **Cartesian coordinate system** (named after French mathematician René Descartes), or a **coordinate plane** (a plane is a flat surface that extends infinitely far in all directions).

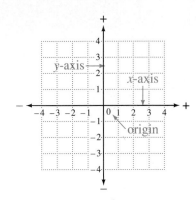

Figure 4.2 Rectangular or Cartesian Coordinate System

The horizontal number line is called the *x*-**axis,** and the vertical number line is called the *y*-**axis.** Collectively the number lines are referred to as **axes,** and the point of intersection of the lines is the **origin.**

❷ ORDERED PAIRS

Just as there is one and only one point on a number line associated with each number, there is one and only one point in a plane associated with each **ordered pair** of numbers. For example, the ordered pair (3, 2) is identified with a point in a coordinate plane as follows:

The first number, 3, the *x*-**coordinate** of the point, is associated with a value on the *x*-axis.

The second number, 2, the *y*-**coordinate,** is associated with a value on the *y*-axis.

The ordered pair (3, 2), is identified with the point where the vertical line through 3 on the *x*-axis intersects the horizontal line through 2 on the *y*-axis. See Figure 4.3. Note that the point (2, 3) is different from (3, 2).

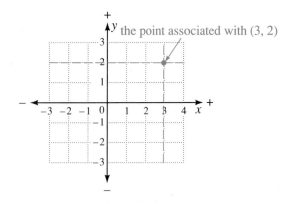

Figure 4.3 Plotting the Point (3, 2)

EXAMPLE 1 PLOTTING POINTS	**PRACTICE EXERCISE 1**

The points associated with (1, 3), (−2, 1), (−3, −2), and (2, −2) are given in the coordinate plane in Figure 4.4. For (1, 3) we go 1 unit right and 3 units up. We find (−2, 1) by going 2 units left and 1 unit up.

Indicate the points associated with (−1, 2), (3, −2), (−1, −1), and (1, 2) in the given coordinate plane.

$(-3, -2)$ is 3 units left and 2 units down. $(2, -2)$ is 2 units right and 2 units down.

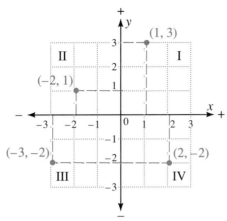

Figure 4.4

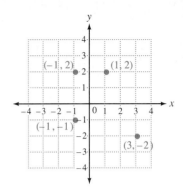

Answer: For $(-1, 2)$ go 1 unit left and 2 units up; for $(3, -2)$ go 3 units right and 2 down; for $(-1, -1)$ go 1 unit left and 1 down; for $(1, 2)$ go 1 unit right and 2 up.

③ QUADRANTS

The axes in a rectangular coordinate system divide the plane into four sections called **quadrants.** The first, second, third, and fourth quadrants are identified by the Roman numerals I, II, III, and IV in Figure 4.4. The x-coordinate (first) and the y-coordinate (second) have the following signs in each quadrant:

$$\text{I: } (+, +), \quad \text{II: } (-, +), \quad \text{III: } (-, -), \quad \text{IV: } (+, -)$$

We often use (x, y) to refer to a general ordered pair of numbers. The point P in the plane associated with the pair (x, y) has x-coordinate x and y-coordinate y. We plot a point P when we identify it with a given pair of numbers in a plane, and we often refer to "the point (x, y)" or write $P(x, y)$.

EXAMPLE 2 DETERMINING ORDERED PAIRS

The points A, B, C, D, E, F, G, and H in Figure 4.5 have coordinates $A(4, 1)$, $B(0, 2)$, $C(0, 0)$, $D(-2, 1)$, $E(-4, -2)$, $F(0, -3)$, $G(1, -2)$, and $H(3, 0)$.

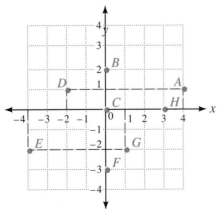

Figure 4.5

PRACTICE EXERCISE 2

Give the coordinates of each point.

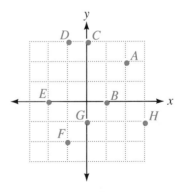

Answer: $A(2, 2)$, $B(1, 0)$, $C(0, 3)$, $D(-1, 3)$, $E(-2, 0)$, $F(-1, -2)$, $G(0, -1)$, $H(3, -1)$

🛡 LINEAR EQUATIONS IN TWO VARIABLES

In Chapter 3 we solved linear equations in one variable. Such equations can always be written in the form

$$ax + b = 0. \qquad a, b \text{ constants and } a \neq 0$$

We now consider linear equations in two variables. A **linear equation in two variables** x and y, is an equation of the form

$$ax + by = c. \qquad a, b, c \text{ constants, } a \text{ and } b \text{ not both zero.}$$

A **solution** to a linear equation in two variables is an ordered pair of numbers which when substituted for the variables results in a true equation. In an *ordered pair* the x-value is always written first and the y-value is always written second.

We can show that the ordered pairs $(1, 3)$ and $(10, 0)$ are two solutions to the linear equation $x + 3y = 10$.

For the ordered pair $(1, 3)$, we must substitute 1 for x and 3 for y in the equation.

$$x + 3y = 10$$
$$1 + 3(3) = 10 \qquad x = 1 \text{ and } y = 3$$
$$1 + 9 = 10$$
$$10 = 10 \qquad \text{True}$$

Thus, $(1, 3)$ is a solution.

For the ordered pair $(10, 0)$, we substitute 10 for x and 0 for y.

$$x + 3y = 10$$
$$10 + 3(0) = 10 \qquad x = 10 \text{ and } y = 0$$
$$10 + 0 = 10$$
$$10 = 10 \qquad \text{True}$$

Thus, $(10, 0)$ is a solution. You might verify that $(4, 2)$ and $\left(0, \frac{10}{3}\right)$ are also solutions to this equation. However, $(9, 0)$ is *not* a solution.

$$x + 3y = 10$$
$$9 + 3(0) \stackrel{?}{=} 10 \qquad x = 9 \text{ and } y = 0$$
$$9 + 0 \stackrel{?}{=} 10$$
$$9 \neq 10 \qquad \text{False}$$

EXAMPLE 3 COMPLETING ORDERED PAIRS	PRACTICE EXERCISE 3

Given the equation $3x + 2y = 6$, complete the ordered pairs so that they are solutions to the equation.

$$(0, \ ), \ (\ ,0), \ (1, \ ), \ (\ ,1), \ (-2, \ )$$

Given the equation $4x - 3y = 12$, complete the ordered pairs so that they are solutions to the equation.

$$(0, \ ), \ (\ , 0), \ (2, \ ), \ (\ , -2), \ (-3, \ )$$

To complete the ordered pair $(0,\ )$, substitute 0 for x in $3x + 2y = 6$ and solve for y.

$$3(0) + 2y = 6$$
$$2y = 6$$
$$y = 3$$

Thus, the completed ordered pair is $(0, 3)$.

To complete the ordered pair $(\ , 0)$, substitute 0 for y in $3x + 2y = 6$ and solve for x.

$$3x + 2(0) = 6$$
$$3x = 6$$
$$x = 2$$

The ordered pair is $(2, 0)$.

To complete the ordered pair $(1,\ )$, substitute 1 for x and solve for y.

$$3(1) + 2y = 6$$
$$3 + 2y = 6$$
$$2y = 3$$
$$y = \frac{3}{2}$$

The completed ordered pair is $\left(1, \frac{3}{2}\right)$.

Similarly, substitute 1 for y and solve for x to complete $(\ , 1)$, obtaining $\left(\frac{4}{3}, 1\right)$, and to complete $(-2,\ )$, substitute -2 for x and solve to obtain $(-2, 6)$.

Answer: $(0, -4)$, $(3, 0)$, $\left(2, -\frac{4}{3}\right)$, $\left(\frac{3}{2}, -2\right)$, $(-3, -8)$

EXAMPLE 4   APPLICATION TO BUSINESS

Mr. Paducci has a small business that manufactures wood-burning stoves. He has found that the cost, y, in dollars of producing a certain number, x, of stoves is given by the equation

$$y = 150x + 80.$$

Find the cost to produce 1, 2, and 5 stoves.

Complete the ordered pairs $(1,\ )$, $(2,\ )$, and $(5,\ )$. Let $x = 1$ and solve for y.

$$y = 150(1) + 80$$
$$y = 230$$

Thus, $(1,\ )$ becomes $(1, 230)$. Similarly, $(2,\ )$ becomes $(2, 380)$ and $(5,\ )$ becomes $(5, 830)$. This means that it costs Mr. Paducci \$230 to produce 1 stove, \$380 to produce 2 stoves, and \$830 to produce 5 stoves.

PRACTICE EXERCISE 4

Laura Hayes manufactures small appliances. She uses the equation $y = 45x + 60$ to find the cost, y, of making x appliances. Find the cost to produce 2 appliances, 7 appliances, and 20 appliances.

Answer: \$150, \$375, \$960

4.1 EXERCISES A

1. Plot the points associated with the given pairs of numbers: $A(2, 4)$, $B(4, -1)$, $C(-3, 4)$, $D(-3, 0)$, $E(2, 0)$, $F(-2, -2)$, $G(1, -4)$, and $H(4, -4)$.

2. Give the coordinates of the points A, B, C, D, E, F, G, and H.

3. Plot the points associated with the given pairs of numbers: $M\left(\frac{1}{2}, 2\right)$, $N\left(-\frac{3}{2}, 3\right)$, $P\left(-2, -\frac{3}{4}\right)$, and $Q\left(3, -\frac{5}{2}\right)$.

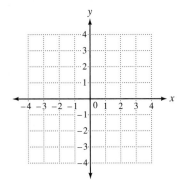

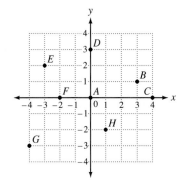

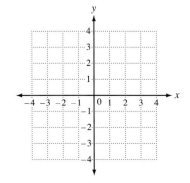

4. In which quadrant are the points M, N, P, and Q of Exercise 3 located?

Give the quadrant in which each of the following is located.

5. $(-8, -8)$

6. $(6, -3)$

7. $(4, 12)$

8. $(-2, 7)$

9. What are the four regions called into which a plane is separated by a Cartesian coordinate system?

10. What is the x-coordinate of the point named by the ordered pair $(-2, 5)$?

11. What is the y-coordinate of the point named by the ordered pair $(3, -8)$?

12. What are the coordinates of the origin?

13. What is another name for the horizontal axis in a Cartesian coordinate system?

14. If the first coordinate of a point is positive and the second is negative, in which quadrant is the point located?

15. If the x-coordinate of a point is negative and the y-coordinate is positive, in which quadrant is the point located?

Using the given equation, complete each ordered pair so that it will be a solution to the equation.

16. $x - y = 2$
    - **(a)** $(0, \ )$ **(b)** $(, 0)$
    - **(c)** $(4, \ )$ **(d)** $(, -3)$

17 $x - 5 = 0$
    - **(a)** $(0, \ )$ **(b)** $(, 0)$
    - **(c)** $(5, \ )$ **(d)** $(, -10)$

18. $x + y = 5$
    (a) $(0, \ \ )$ **(b)** $(\ , 0)$
    (c) $(3, \ \ )$ **(d)** $(\ , -4)$

19. $2x - y = 4$
    (a) $(0, \ \ )$ **(b)** $(\ , 0)$
    (c) $(-2, \ \ )$ **(d)** $(\ , 6)$

20. $x + 5y = 10$
    (a) $(0, \ \ )$ **(b)** $(\ , 0)$
    (c) $(-10, \ \ )$ **(d)** $(\ , 3)$

21. $5x - 2y = 10$
    (a) $(0, \ \ )$ **(b)** $(\ , 0)$
    (c) $(7, \ \ )$ **(d)** $(\ , -4)$

22 The distance, y, traveled by a car averaging 55 mph over x hours is given by the equation

$$y = 55x.$$

Complete the ordered pairs $(1, \ \ )$, $(5, \ \ )$, and $(10, \ \ )$. How far does the car travel **(a)** in 1 hour? **(b)** in 5 hours? **(c)** in 10 hours?

23. Draw the triangle with vertices $(3, 4)$, $(-3, 1)$, and $(1, -2)$.

24. Draw the rectangle with corners $(3, 1)$, $(1, 3)$, $(-3, -1)$, and $(-1, -3)$.

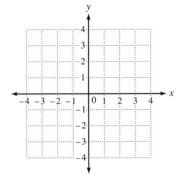

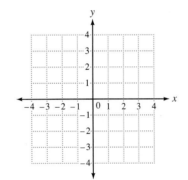

FOR REVIEW

Graph the equation on a number line.

25. $a + 3 = 3(a - 1)$

26. $3x + 8 = 8 + 3x$

Graph the inequality on a number line.

27. $2x \geq 7$

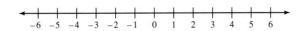

28. $3(y - 1) < 4 - (y - 1)$

ANSWERS:

1.

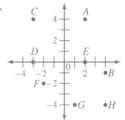

2. $A(0, 0)$,
$B(3, 1)$,
$C(4, 0)$,
$D(0, 3)$,
$E(-3, 2)$,
$F(-2, 0)$,
$G(-4, -3)$,
$H(1, -2)$.

3.

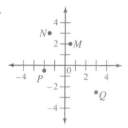

4. M:I, N:II, P:III, Q:IV   5. III   6. IV   7. I
8. II   9. quadrants   10. -2 11. -8
12. $(0, 0)$ 13. x-axis   14. IV   15. II
16. (a) $(0, -2)$ (b) $(2, 0)$ (c) $(4, 2)$
(d) $(-1, -3)$ 17. (a) x cannot be 0 (b) $(5, 0)$
(c) $(5, \text{any number})$ (d) $(5, -10)$
18. (a) $(0, 5)$ (b) $(5, 0)$ (c) $(3, 2)$ (d) $(9, -4)$
19. (a) $(0, -4)$ (b) $(2, 0)$ (c) $(-2, -8)$ (d) $(5, 6)$ 20. (a) $(0, 2)$ (b) $(10, 0)$ (c) $(-10, 4)$ (d) $(-5, 3)$ 21. (a) $(0, -5)$
(b) $(2, 0)$ (c) $\left(7, \frac{25}{2}\right)$ (d) $\left(\frac{2}{5}, -4\right)$ 22. $(1, 55)$, $(5, 275)$, $(10, 550)$ (a) 55 miles (b) 275 miles (c) 550 miles

23.

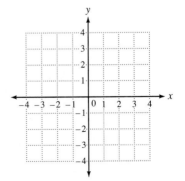

24.

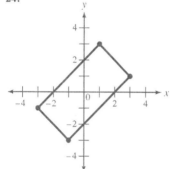

25.

26.

27.

28.

4.1 EXERCISES B

1. Plot the points associated with the given pairs of numbers: $A(1, 3)$, $B(2, -1)$, $C(-3, 4)$, $D(-1, 1)$, $E(-3, -3)$, $F(2, -3)$.

2. Give the coordinates of the points A, B, C, D, E, and F in the figure below.

3. Plot the points associated with the given pairs of numbers: $M\left(-\frac{1}{2}, 1\right)$, $N\left(\frac{3}{2}, 2\right)$, $P\left(2, -\frac{3}{4}\right)$, and $Q\left(-3, -\frac{5}{2}\right)$.

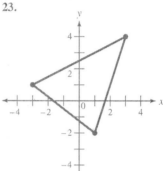

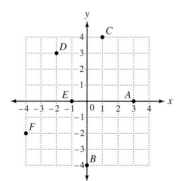

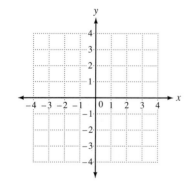

4. In which quadrants are the points *M, N, P,* and *Q* of Exercise 3 located?

Give the quadrant in which each of the following is located.

5. (4, −1)   **6.** (−4, 3)   **7.** (6, 6)   **8.** (−12, −4)

9. What is another name for a Cartesian coordinate system?

10. What is the *y*-coordinate of the point named by the ordered pair (−2, 5)?

11. What is the *x*-coordinate of the point named by the ordered pair (3, −8)?

12. What is the name of the point with coordinates (0, 0)?

13. What is another name for the vertical axis in a Cartesian coordinate system?

14. If the first coordinate of a point is negative and the second is positive, in which quadrant is the point located?

15. If the *x*-coordinate of a point is negative and the *y*-coordinate is also negative, in which quadrant is the point located?

Using the given equation, complete each ordered pair so that it will be a solution to the equation.

16. $x + y = 2$
    (a) (0,   )   **(b)** (, 0)
    (c) (3,   )   **(d)** (, −2)

17. $y + 1 = 0$
    (a) (0,   )   **(b)** (, 0)
    (c) (3,   )   **(d)** (, −1)

18. $x - y = 3$
    (a) (0,   )   **(b)** (, 0)
    (c) (2,   )   **(d)** (, −4)

19. $2x + y = 4$
    (a) (0,   )   **(b)** (, 0)
    (c) (−2,   )   **(d)** (, 6)

20. $x - 5y = 10$
    (a) (0,   )   **(b)** (, 0)
    (c) (−10,   )   **(d)** (, 3)

21. $5x + 2y = 10$
    (a) (0,   )   **(b)** (, 0)
    (c) (7,   )   **(d)** (, −4)

22. The cost, *y*, in dollars of producing a number, *x*, of items has been estimated by the equation

$$y = 300x + 15.$$

Complete the ordered pairs (1,), (3,), and (10,). Give the cost of producing **(a)** 1 item, **(b)** 3 items, **(c)** 10 items.

23. Draw the triangle with vertices (−2, 7), (1, 1), and (−4, −5).

24. Draw the rectangle with corners (−4, 2), (−1, 5), (5, −1), and (2, −4).

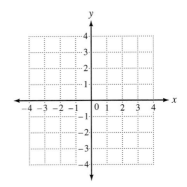

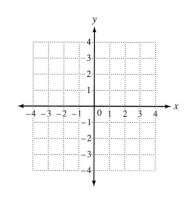

FOR REVIEW

Graph the equation on a number line.

25. $2a + 1 = 3(a - 2)$

26. $3x - 8 = 3x + 8$

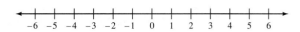

Graph the inequality on a number line.

27. $3x < -5$

28. $2(y - 3) \geq 3 - (y - 6)$

4.1 EXERCISES C

Using the given equation, complete each ordered pair so that it will be a solution to the equation. A calculator might be helpful.

1. $0.012x - 1.565y = 1.878$
   (a) $(0, \quad)$ **(b)** $(\quad, 0)$
   (c) $(234.75, \quad)$ **(d)** $(\quad, 2.4)$
   [Answers: **(a)** $(0, -1.2)$; **(c)** $(234.75, 0.6)$]

2. $-\dfrac{3}{16}x + \dfrac{7}{12}y = 3\dfrac{5}{8}$

   (a) $(0, \quad)$ **(b)** $(\quad, 0)$
   (c) $(-\dfrac{5}{9}, \quad)$ **(d)** $(\quad, 1\dfrac{3}{7})$

4.2 GRAPHING LINEAR EQUATIONS

STUDENT GUIDEPOSTS

1 Graphing Equations
2 General Form of a Linear Equation

3 Graphing Using Intercepts

① GRAPHING EQUATIONS

The **graph** of an equation in two variables x and y is the set of points in a Cartesian coordinate system that corresponds to solutions of the equation. Since there are usually infinitely many solutions, we cannot find and plot each pair. Generally, we plot enough points to see a pattern, and then connect these points with a line or curve to graph the equation.

An excellent way to display a collection of ordered-pair solutions to an equation such as

$$y = 2x + 5$$

is to make a table of values. Choose several values for x and substitute these values into the equation to compute the corresponding value for y. Place each y-value beside the x-value used to calculate it. (Calculations are usually done mentally or as scratch work.)

Substitution	Results in $y = 2x + 5$	x	y
$x = 0$	$y = 2(0) + 5 = 5$	0	5
$x = 1$	$y = 2(1) + 5 = 7$	1	7
$x = -1$	$y = 2(-1) + 5 = 3$	-1	3
$x = 2$	$y = 2(2) + 5 = 9$	2	9
$x = -2$	$y = 2(-2) + 5 = 1$	-2	1
$x = 3$	$y = 2(3) + 5 = 11$	3	11
$x = -3$	$y = 2(-3) + 5 = -1$	-3	-1

This table lists seven (of the infinitely many) solutions to the equation $y = 2x + 5$.

$(0, 5)$, $(1, 7)$, $(-1, 3)$, $(2, 9)$, $(-2, 1)$, $(3, 11)$, $(-3, -1)$

Now plot the points that correspond to these ordered-pair solutions in a rectangular coordinate system, as in Figure 4.6.

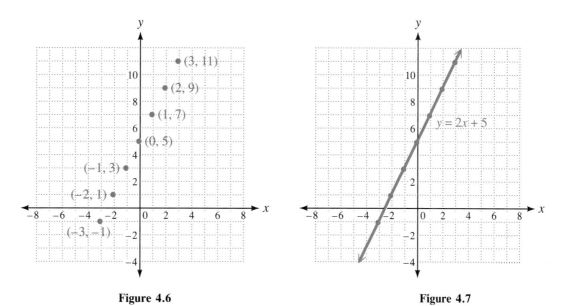

Figure 4.6   **Figure 4.7**

It appears that all seven points lie on a straight line. It is reasonable to assume that the graph of this equation is the straight line passing through these seven points, as in Figure 4.7.

To Graph an Equation in the Two Variables *x* and *y*

1. Make a table of values. These values represent the ordered-pair solutions of the equation.
2. Plot the points that correspond to the ordered-pair solutions in a Cartesian coordinate system.
3. Connect the points with a line or curve.

| EXAMPLE 1   GRAPHING BY PLOTTING POINTS | PRACTICE EXERCISE 1 |

Graph $y + x = 3$.

Before making a table, it is often helpful to solve the equation for one of the variables (usually for y):

$$y = -x + 3.$$

To make a table of values, choose several values for x and substitute each into the equation to compute the corresponding value of y. See the table at the side. Plot the points that correspond to the ordered pairs from the table:

$$(0, 3), \quad (1, 2), \quad (-1, 4), \quad (2, 1), \quad (-2, 5), \quad (3, 0), \quad (-3, 6).$$

Connecting these points gives us a straight line, as in Figure 4.8.

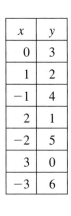

x	y
0	3
1	2
-1	4
2	1
-2	5
3	0
-3	6

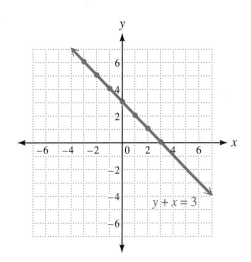

Figure 4.8

Graph $2x + y = -2$.

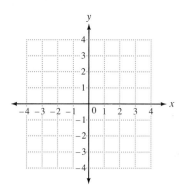

Answer: Straight line passing through $(0, -2)$ and $(-1, 0)$.

In the first part of this chapter, we graphed an equation in one variable on a number line. Usually, for graphing purposes, an equation in one variable such as

$$x = 5 \quad \text{or} \quad y + 2 = 0$$

is thought of as an equation in two variables with the coefficient of the missing variable equal to zero. That is,

$$x = 5 \text{ is the same as } x + 0 \cdot y = 5$$

and

$$y + 2 = 0 \text{ is the same as } 0 \cdot x + y + 2 = 0.$$

With this in mind, such equations can be graphed in a Cartesian coordinate system.

EXAMPLE 2 LINES PARALLEL TO THE AXES

(a) Graph $x = 5$ in a rectangular coordinate system.

Solutions to this equation always have an x-coordinate of 5 and can have any number as y-coordinate. For example,

$$(5, 0), \quad (5, -1), \quad (5, 1), \quad (5, 2)$$

are all solutions, since $x = 5$ is the same as $x + 0 \cdot y = 5$, and we know

$$5 + 0 \cdot (\text{any number}) = 5 + 0 = 5.$$

Plot the points from the table and draw a line through them. The graph of $x = 5$ is a straight line parallel to the y-axis, 5 units to the right of the y-axis. See Figure 4.9.

x	y
5	0
5	1
5	-1
5	2
5	-2
5	3
5	-3

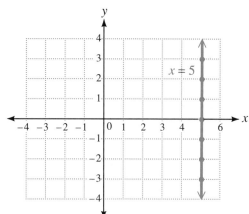

Figure 4.9

(b) Graph $y + 2 = 0$ in a Cartesian coordinate system.

We can write this equation as $y = -2$. Thus, solutions to this equation always have a y-coordinate of -2 and can have any number as x-coordinate. For example,

$$(0, -2), \quad (1, -2), \quad (-1, -2), \quad (2, -2)$$

are all solutions since $y + 2 = 0$ is the same as $0 \cdot x + y + 2 = 0$, and we know that

$$0 \cdot (\text{any number}) + (-2) + 2 = -2 + 2 = 0.$$

x	y
0	-2
1	-2
-1	-2
2	-2
-2	-2
3	-2
-3	-2

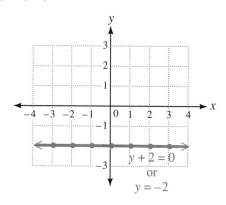

Figure 4.10

Plot the points given in the table and draw a line through them. The graph of $y + 2 = 0$ ($y = -2$) is a straight line parallel to the x-axis, 2 units below the x-axis. (See Figure 4.10).

PRACTICE EXERCISE 2

(a) Graph $x + 3 = 0$.

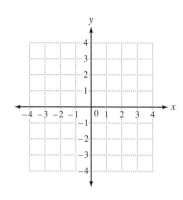

(b) Graph $y = \dfrac{5}{2}$.

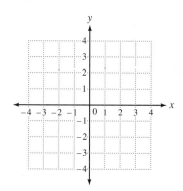

Answers: (a) a straight line parallel to the y-axis, 3 units to the left of the y-axis (b) a straight line parallel to the x-axis, $\frac{5}{2}$ units above the x-axis

If we are asked to graph an equation such as $x = 5$ or $y + 2 = 0$, we must first know whether it is considered an equation in one variable or an equation in two variables having a zero coefficient on the missing variable. In the first case, the graph would be a point on a number line, while in the second, the graph would be a straight line in a rectangular coordinate system. From this point on, we will look only at equations in two variables, and all graphs will be in the coordinate plane.

② GENERAL FORM OF A LINEAR EQUATION

In graphing we can sometimes minimize the number of points that we plot by knowing the nature of the graph. An equation of the form

$$ax + by + c = 0$$

is called a **linear equation in two variables.** The numbers a, b, and c are constant real numbers, a and b not both zero, and this form of the equation is called the **general form.** The graph of a **lin**ear equation is always a straight **line** and can be determined by plotting just two points. Linear equations are also called **first-degree equations** since the variables are raised to the first power only. The following table provides practice at identifying linear equations.

Equation	Nature	General form
$2x + y = 7$	linear	$2x + y - 7 = 0$
$x = -y - 8$	linear	$x + y + 8 = 0$
$2y = 3x - 5$	linear	$3x - 2y - 5 = 0$
$x + 5 = 0$	linear	$x + 0 \cdot y + 5 = 0$
$2y = 3$	linear	$0 \cdot x + 2y - 3 = 0$
$y = x^2 + 3$	not linear (x to second power)	
$x - y^3 + 7 = 0$	not linear (y to third power)	
$y = \frac{5}{x}$	not linear (cannot be put in the form $ax + by + c = 0$)	

③ GRAPHING USING INTERCEPTS

Suppose we are given a linear equation such as

$$2x - 3y = 6.$$

Knowing that the graph of a linear equation is a straight line and that only two points are needed to determine a straight line, our work graphing linear equations can be shortened considerably. Rather than making a table of values that includes many solutions, we need only two solutions. In most instances, the two pairs that are easiest to find are the **intercepts.** The point at which a line crosses the x-axis, where y is zero, is the **x-intercept.** The point at which a line crosses the y-axis, where x is zero, is the **y-intercept.** To find the intercepts, fill in the following table.

x	y
0	
	0

The x-intercept, which is a point on the x-axis, has y-coordinate 0 while the y-intercept, which is a point on the y-axis, has x-coordinate 0. We substitute 0 for x in $2x - 3y = 6$ and solve for y to find the y-intercept.

$$2(0) - 3y = 6$$
$$0 - 3y = 6$$
$$-3y = 6$$
$$y = -2$$

Find the x-intercept by substituting 0 for y in $2x - 3y = 6$ and solving.

$$2x - 3(0) = 6$$
$$2x = 6$$
$$x = 3$$

The completed table

x	y
0	-2
3	0

displays the y-intercept $(0, -2)$ and the x-intercept $(3, 0)$. Plot these two intercepts and draw the straight line through them for the graph of $2x - 3y = 6$ in Figure 4.11. We can check our work by showing that $(-3, -4)$ satisfies the equation and is on the graph in Figure 4.11.

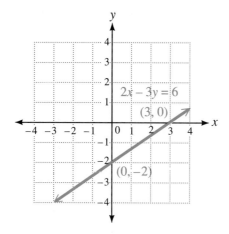

Figure 4.11

To Graph a Linear Equation $ax + by + c = 0$

1. If $a \neq 0$ and $b \neq 0$, find the x- and y-intercepts, plot them, and draw the line through them. If both intercepts are $(0, 0)$, find and plot another point.
2. If $a = 0$, $y = $ a constant and the graph is a line parallel to the x-axis.
3. If $b = 0$, $x = $ a constant and the graph is a line parallel to the y-axis.

| EXAMPLE 3 GRAPHING USING INTERCEPTS | PRACTICE EXERCISE 3 |

(a) Graph $3x + 4y = 12$.

First find the x- and y-intercepts by completing the following table.

x	y
0	
	0

When $x = 0$, $4y = 12$, so that $y = 3$. When $y = 0$, $3x = 12$, so that $x = 4$. The completed table,

x	y
0	3
4	0

shows that the x-intercept is $(4, 0)$ and the y-intercept is $(0, 3)$. Plot $(0, 3)$ and $(4, 0)$ and connect the points to obtain the graph. See Figure 4.12.

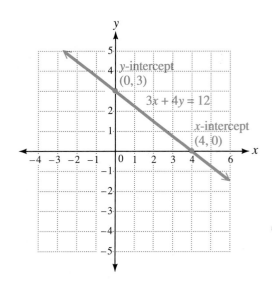

Figure 4.12

(b) Graph $x + 3y = 0$.

First find the x- and y-intercepts by completing the table.

x	y
0	
	0

(a) Graph $-6x + y = 6$.

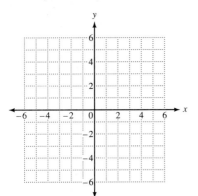

(b) Graph $4x - 3y = 0$.

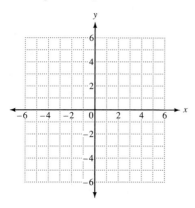

This time both intercepts are (0, 0), so we need to find another point on the line. When $x = 3$, $3 + 3y = 0$, so that $y = -1$. This leads to the table below.

x	y
0	0
3	-1

Plot (0, 0) and (3, -1) and connect the points to obtain the graph in Figure 4.13.

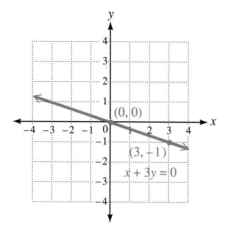

Figure 4.13

Answers: (a) straight line through x-intercept $(-1, 0)$ and y-intercept $(0, 6)$ (b) straight line through x- and y-intercept $(0, 0)$ and the point $(3, 4)$

EXAMPLE 4   VERTICAL AND HORIZONTAL LINES

(a) Graph $x = 2$.

In this case, with no y term, the solution will be $(2, y)$ for any number y. Thus, the graph is the line through the x-intercept $(2, 0)$ parallel to the y-axis, as in Figure 4.14.

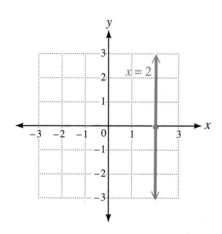

Figure 4.14

PRACTICE EXERCISE 4

(a) Graph $2x = -10$.

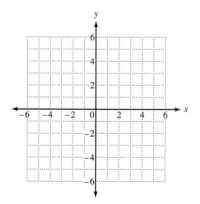

(b) Graph $2y = -2$.

The equation can be simplified to $y = -1$. In this case, with no x term, the solutions will be $(x, -1)$ for any number x. Thus, the graph is the line through the y-intercept $(0, -1)$ parallel to the x-axis, as shown in Figure 4.15.

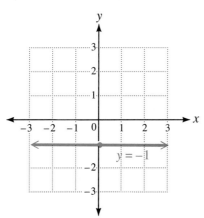

Figure 4.15

(b) Graph $y = -\dfrac{7}{2}$.

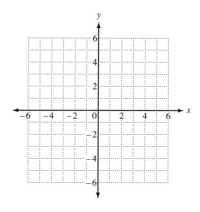

Answers: **(a) line through x-intercept $(-5, 0)$ parallel to the y-axis** **(b) line through y-intercept $\left(0, -\frac{7}{2}\right)$ parallel to the x-axis**

4.2 EXERCISES A

Each of the following is an equation in the two variables x and y. Make a table of values and graph each equation in a rectangular coordinate system.

1. $y = x + 2$

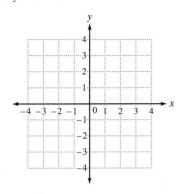

2. $x - y = 1$

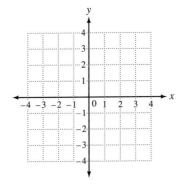

3. $y = 3x + 1$

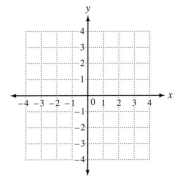

4. $y = \dfrac{1}{2}x - 1$

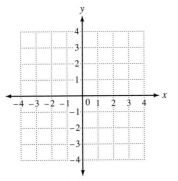

5. $y = 2$

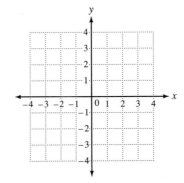

6. $x = -1$

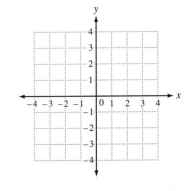

7. Which of the following are linear equations? Explain.
   (a) $2x + y = -4$
   (b) $x - y^3 = 5$
   (c) $2xy = 7$

   (d) $\dfrac{3}{x} + y = 7$
   (e) $x^2 + y^2 = 5$
   (f) $x = y - 8$

8. Why is a linear equation also called a first-degree equation?

9. What is an equation of the form $ax + by + c = 0$ ($a \neq 0$ or $b \neq 0$) called?

10. An equation of the type $x = c$ (c a constant) has as its graph a line parallel to which axis?

11. An equation of the type $y = c$ (c a constant) has as its graph a line parallel to which axis?

12. What is a point where a graph crosses the x-axis called?

13. What is a point where a graph crosses the y-axis called?

14. To graph a general linear equation, only two points are necessary. What are the best points to use?

Find the intercepts and graph the following.

15. $y + 2x - 4 = 0$
16. $3y - 2x = 12$
17. $y + x = 0$

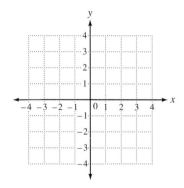
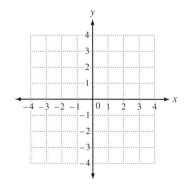
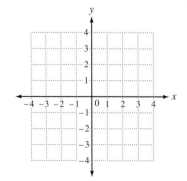

18. $y = 3$ **19.** $2x - 1 = 0$ **20.** $y + 2x = 0$

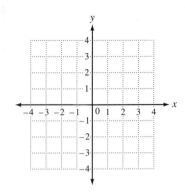 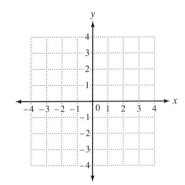 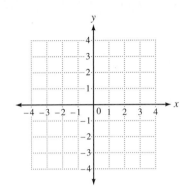

21. What are the intercepts of the line $x = 0$? What is the graph?

FOR REVIEW

Using the given equation, complete each ordered pair so that it will be a solution to the equation.

22. $5x - 3y = 15$
    (a) $(0, \quad)$ **(b)** $(\quad, 0)$
    (c) $(\quad, -1)$ **(d)** $(-1, \quad)$

23. $y = 3$
    (a) $(1, \quad)$ **(b)** $(-1, \quad)$
    (c) $(5, \quad)$ **(d)** $(-3, \quad)$

24. The annual salary, y, of a salesman is given in terms of his total sales, x, by the equation

$$y = \$10,000 + (0.1)x.$$

Complete the ordered pairs $(10,000, \quad)$, $(20,000, \quad)$, and $(50,000, \quad)$. How much does the salesman earn if he sells **(a)** 10,000 items, **(b)** 20,000 items, **(c)** 50,000 items?

25. Find the area of the rectangle with corners $(4, -3)$, $(4, 5)$, $(-3, -3)$, and $(-3, 5)$.

The following exercises will help you prepare for the next section. Starting with the point $A(1, 2)$, give the coordinates of point B.

26. B is 2 units up from A and then 1 unit to the right.

27. B is 2 units down from A and then 3 units to the right.

28. B is 3 units up from A and then 2 units to the left.

29. B is 3 units down from A and then 4 units to the left.

ANSWERS:

1.
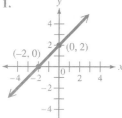

2.
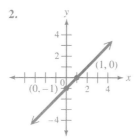

3.
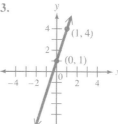

4.
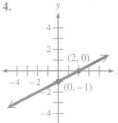

5.
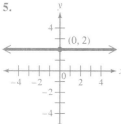

6.
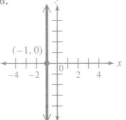

7. (a) and (f) are the only linear equations
8. the variables appear only to the first power
9. the general form of a linear equation
10. y-axis
11. x-axis
12. an x-intercept
13. a y-intercept
14. intercepts

15. x-intercept $(2, 0)$;
    y-intercept $(0, 4)$

16. x-intercept $(-6, 0)$;
    y-intercept $(0, 4)$

17. x-intercept $(0, 0)$;
    y-intercept $(0, 0)$;
    another point is $(1, -1)$

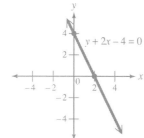

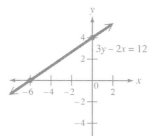

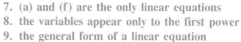

18. no x-intercept;
    y-intercept $(0, 3)$

19. x-intercept $\left(\frac{1}{2}, 0\right)$;
    no y-intercept

20. x-intercept $(0, 0)$;
    y-intercept $(0, 0)$;
    another point is $(1, -2)$

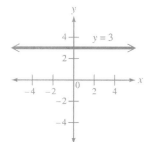

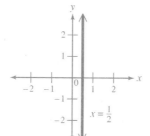

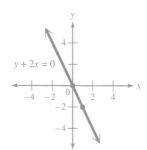

21. x-intercept $(0, 0)$; y-intercept; every point on the y-axis; the graph is the y-axis   22. (a) $(0, -5)$ (b) $(3, 0)$
(c) $\left(\frac{12}{5}, -1\right)$ (d) $\left(-1, \frac{-20}{3}\right)$ 23. (a) $(1, 3)$ (b) $(-1, 3)$ (c) $(5, 3)$ (d) $(-3, 3)$ 24. $(10,000, 11,000)$,
$(20,000, 12,000)$, $(50,000, 15,000)$ (a) \$11,000   (b) \$12,000   (c) \$15,000   25. 56 square units   26. $(2, 4)$ 27. $(4, 0)$
28. $(-1, 5)$ 29. $(-3, -1)$

4.2 EXERCISES B

Each of the following is an equation in the two variables x and y. Make a table of values and graph each equation in a rectangular coordinate system.

1. $y = x + 1$

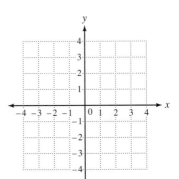

2. $y = 3x - 1$

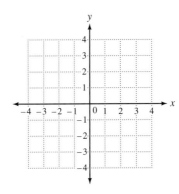

3. $y = \frac{1}{2}x + 1$

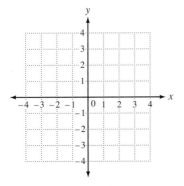

4. $x + y = -1$

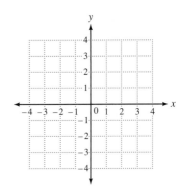

5. $x = 4$

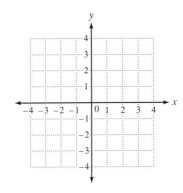

6. $y = -2$

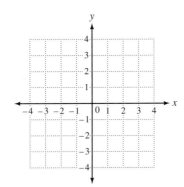

7. Which of the following are linear equations? Explain.

    (a) $3x - y = 8$ **(b)** $x + y^2 = 7$ **(c)** $3xy = -1$

    (d) $\frac{5}{y} + x = 3$ **(e)** $x^2 + y^2 = 16$ **(f)** $x = y - 3$

8. Why is a first-degree equation also called a linear equation?

9. Give the general form of a linear equation.

10. Give the general form of the equation of a line parallel to the *x*-axis.

11. Give the general form of the equation of a line parallel to the *y*-axis.

12. The *x*-intercept of the graph of a line is a point on which axis?

13. The *y*-intercept of the graph of a line is a point on which axis?

14. In general, what are the two best points to use when graphing a linear equation?

Find the intercepts and graph the following.

15. $y - 2x + 4 = 0$ **16.** $3y + 2x = -12$ **17.** $2y + x = 0$

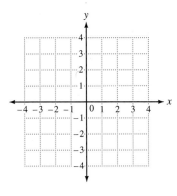   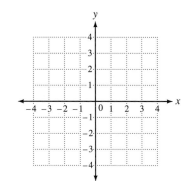   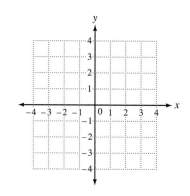

18. $y = -4$ **19.** $2x + 3 = 0$ **20.** $x - 3y = 0$

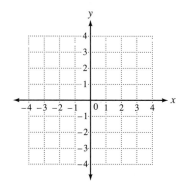   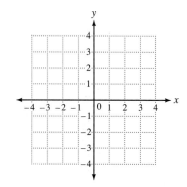   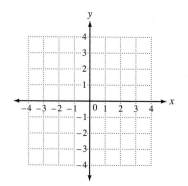

21. What are the intercepts of the line $y = 0$? What is the graph?

Using the given equation, complete each ordered pair so that it will be a solution to the equation.

22. $x - 3y = 6$

    (a) $(0, \quad)$ **(b)** $(\quad, 0)$

    (c) $(\quad, -1)$ **(d)** $(-1, \quad)$

23. $x = -1$

    (a) $(\quad, 1)$ **(b)** $(\quad, -1)$

    (c) $(\quad, 5)$ **(d)** $(\quad, -3)$

24. The number, y, of deer that can live in a forest pre-serve is related to the number of acres, x, in the preserve. If this relationship is given by the equation

$$y = 0.5x + 3$$

complete the ordered pairs $(20, \quad)$, $(100, \quad)$, and $(200, \quad)$. How many deer can the preserve sustain if it contains **(a)** 20 acres? **(b)** 100 acres? **(c)** 200 acres?

25. Find the area of the triangle with vertices $(1, 2)$, $(-3, -2)$, and $(5, -2)$.

The following exercises will help you prepare for the next section. Starting with the point $A\ (-3, 5)$ give the coordinates of point B.

26. B is 4 units up from A and then 3 units to the right.

27. B is 4 units down from A and then 2 units to the right.

28. B is 1 unit up from A and then 4 units to the left.

29. B is 6 units down from A and then 1 unit to the left.

4.2 EXERCISES C

Graph.

1. $0.05x - 0.02y = 0.10$

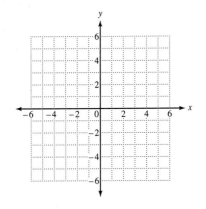

2. $\dfrac{2}{5}x + \dfrac{3}{10}y = \dfrac{6}{5}$

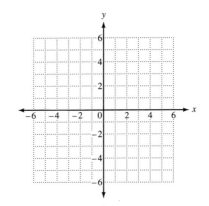

4.3 SLOPE OF A LINE

1 DEFINITION OF SLOPE

The graph of a linear equation may be horizontal (parallel to the x-axis), may be vertical (parallel to the y-axis), may "slope" upward from lower to upper right, or may "slope" downward from upper left to lower right. The way that a line slopes or does not slope can be precisely defined.

The formal definition of slope uses any two points on the line. The coordinates of the points are written (x_1, y_1) and (x_2, y_2) where the subscript distinguishes between the two while identifying the x- and y-coordinates. We read x_1 as "x-sub-one" and y_1 as "y-sub-one," for example. The *slope* of the line passing through (x_1, y_1) and (x_2, y_2) is defined as the ratio of the vertical change, *rise,* to the horizontal change, *run,* as we move from (x_1, y_1) to (x_2, y_2) along the line. See Figure 4.16. This ratio will be the same for any two points on the line.

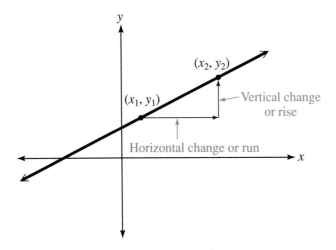

Figure 4.16   Rise and Run of a Line

We can define the rise and run as numbers by using the coordinates of the points. The ratio of these numerical values gives us a precise definition of slope. Refer to Figure 4.17.

The Slope of a Line
Let $P(x_1, y_1)$ and $Q(x_2, y_2)$ be two points on a nonvertical line. The **slope** of the line, denoted m, is given by the equation $$m = \frac{y_2 - y_1}{x_2 - x_1} = \frac{\text{change in } y\text{-coordinates}}{\text{change in } x\text{-coordinates}} = \frac{\text{rise}}{\text{run}}.$$

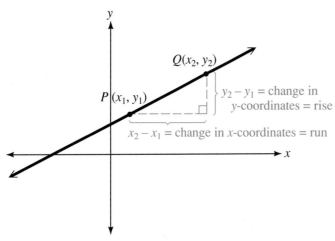

Figure 4.17 Slope $= \dfrac{y_2 - y_1}{x_2 - x_1} = \dfrac{\text{rise}}{\text{run}}$

EXAMPLE 1 SLOPE OF A LINE

Find the slope of the line passing through the given points.

(a) (4, 3) and (1, 2).

See the graph in Figure 4.18. Suppose we identify point $P(x_1, y_1)$ with (1, 2) and point $Q(x_2, y_2)$ with (4, 3). The slope will then be given by

$$m = \frac{y_2 - y_1}{x_2 - x_1} = \frac{3 - 2}{4 - 1} = \frac{1}{3}.$$

What happens if we identify $P(x_1, y_1)$ with (4, 3) and, $Q(x_2, y_2)$ with (1, 2)? In this case we have

$$m = \frac{y_2 - y_1}{x_2 - x_1} = \frac{2 - 3}{1 - 4} = \frac{-1}{-3} = \frac{1}{3}.$$

Thus, we see that *the slope is the same regardless of how the two points are identified.*

PRACTICE EXERCISE 1

Find the slope of the line passing through the given points.

(a) (5, 2) and (3, 7)

(b) (6, −3) and (−1, 8)

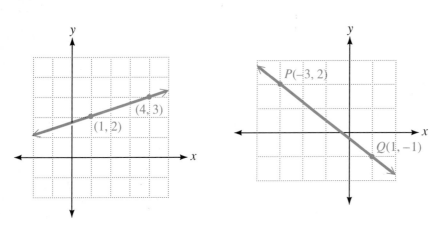

Figure 4.18   **Figure 4.19**

(b) $(-3, 2)$ and $(1, -1)$.

Let us identify $P(x_1, y_1)$ with $(-3, 2)$ and $Q(x_2, y_2)$ with $(1, -1)$ as in Figure 4.19. The slope is given by

$$m = \frac{y_2 - y_1}{x_2 - x_1} = \frac{(-1) - (2)}{(1) - (-3)} \qquad \text{Watch signs}$$

$$= \frac{-3}{1 + 3} = -\frac{3}{4}.$$

Answers: (a) $-\frac{5}{2}$ (b) $-\frac{11}{7}$

///////////// CAUTION //////////////

Make sure that the coordinates are subtracted in the same order. *Do not compute*

$$\frac{y_2 - y_1}{x_1 - x_2}. \qquad \text{This is wrong}$$

///////////

EXAMPLE 2   ZERO AND UNDEFINED SLOPE

Find the slope of the line passing through the given points

(a) $(3, 2)$ and $(-1, 2)$.

Identifying $P(x_1, y_1)$ with $(3, 2)$ and $Q(x_2, y_2)$ with $(-1, 2)$ in Figure 4.20, we obtain

$$m = \frac{y_2 - y_1}{x_2 - x_1} = \frac{2 - 2}{-1 - 3} = \frac{0}{-4} = 0.$$

PRACTICE EXERCISE 2

Find the slope of the line passing through the given points.

(a) $(-3, -4)$ and $(5, -4)$

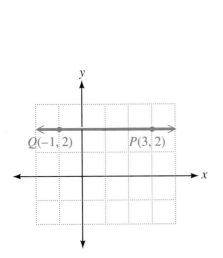

Figure 4.20

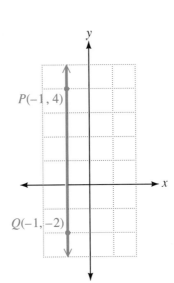

Figure 4.21

(b) $(-1, 4)$ and $(-1, -2)$.

Identifying $P(x_1, y_1)$ with $(-1, 4)$ and $Q(x_2, y_2)$ with $(-1, -2)$ in Figure 4.21, we obtain

$$m = \frac{y_2 - y_1}{x_2 - x_1} = \frac{-2 - 4}{-1 - (-1)} = \frac{-6}{-1 + 1} = \frac{-6}{0}, \text{ which is undefined.}$$

In this case we say that the slope of the line is undefined.

(b) $(3, -3)$ and $(3, 0)$

Answers: (a) 0   (b) undefined slope

② NATURE OF THE SLOPE OF A LINE

The nature of the slope of a line is summarized below.

Summary of the Slope of a Line

1. A line which slopes from lower left to upper right has a **positive slope.**
2. A line which slopes from upper left to lower right has a **negative slope.**
3. A horizontal line (parallel to the x-axis) has **zero slope.**
4. A vertical line (parallel to the y-axis) has **undefined slope.**

The graphs in Figure 4.22 show these four cases.

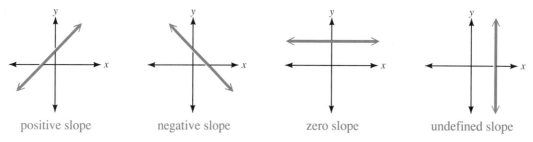

positive slope   negative slope   zero slope   undefined slope

Figure 4.22   Summary of Slopes

③ PARALLEL LINES

Slope can be used to determine when two lines are **parallel** (never intersect).

Parallel Lines

If two nonvertical lines have slopes m_1 and m_2 and $m_1 = m_2$, then the lines are parallel. (Equal slopes determine parallel lines).

EXAMPLE 3   PARALLEL LINES

Verify that the line l_1 through $(1, 8)$ and $(-2, -1)$ and the line l_2 through $(2, 4)$ and $(-1, -5)$ in Figure 4.23 are parallel.

The slope of l_1 is $m_1 = \frac{8 - (-1)}{1 - (-2)} = \frac{8 + 1}{1 + 2} = \frac{9}{3} = 3.$

PRACTICE EXERCISE 3

Is the line l_1 through $(4, -6)$ and $(-3, -1)$ parallel to the line l_2 through $(-2, 5)$ and $(7, -1)$?

The slope of l_2 is $m_2 = \dfrac{4 - (-5)}{2 - (-1)} = \dfrac{4 + 5}{2 + 1} = \dfrac{9}{3} = 3.$

Since $m_1 = m_2$, the lines are parallel.

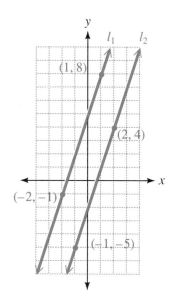

Figure 4.23

Answer: $m_1 = -\frac{5}{7}$, $m_2 = -\frac{2}{3}$, no

④ PERPENDICULAR LINES

Two lines are **perpendicular** if they intersect at right (90°) angles. The slopes of perpendicular lines have the following property.

Perpendicular Lines
If two nonvertical lines have slopes m_1 and m_2 and $m_1 m_2 = -1$, then the lines are perpendicular.

EXAMPLE 4 PERPENDICULAR LINES

Verify that the line l_1 through $(-1, 3)$ and $(2, -1)$ and the line l_2 through $(2, 1)$ and $(-2, -2)$ in Figure 4.24 are perpendicular.

The slope of l_1 is $m_1 = \dfrac{3 - (-1)}{-1 - 2} = \dfrac{3 + 1}{-3} = -\dfrac{4}{3}$

The slope of l_2 is $m_2 = \dfrac{1 - (-2)}{2 - (-2)} = \dfrac{1 + 2}{2 + 2} = \dfrac{3}{4}.$

Since $m_1 m_2 = \left(-\frac{4}{3}\right)\left(\frac{3}{4}\right) = -1$, l_1 and l_2 are perpendicular.

PRACTICE EXERCISE 4

Verify that the line l_1 through $(5, -2)$ and $(6, 7)$ is perpendicular to the line l_2 through $(-1, -3)$ and $(8, -4)$.

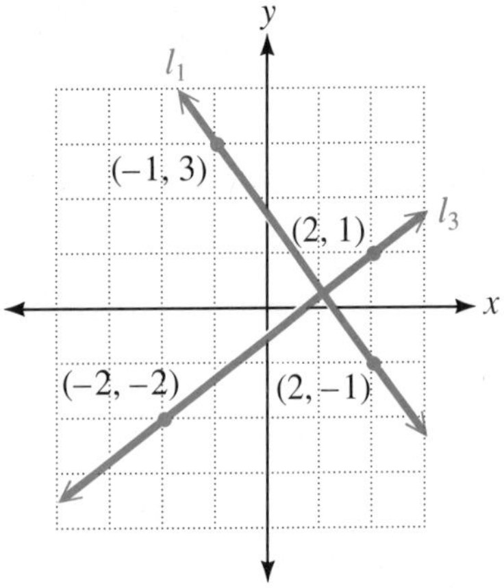

Figure 4.24 Perpendicular Lines

Answer: $m_1 = 9$, $m_2 = -\frac{1}{9}$, $m_1 m_2 = 9\left(-\frac{1}{9}\right) = -1$

4.3 EXERCISES A

Find the slope of the line passing through the given pair of points.

1. $(3, 5)$ and $(1, 3)$

2. $(-2, 3)$ and $(-4, 1)$

3. $(-7, 1)$ and $(3, 9)$

4. $(2, 7)$ and $(2, -3)$

5. $(3, 4)$ and $(-1, 4)$

6. $(0, 2)$ and $(5, 0)$

Answer Exercises 7–10 with one of the following phrases: (a) *positive slope* (b) *negative slope* (c) *zero slope* (d) *undefined slope.*

7. A line parallel to the *y*-axis has ————————————.

8. A line that slopes from lower left to upper right has ————————————.

9. A line perpendicular to the *y*-axis has ————————————.

10. The *x*-axis has ————————————.

In Exercises 11–13, m_1 and m_2 represent the slopes of two lines. State whether the lines are parallel, perpendicular, or neither.

11. $m_1 = -3$ and $m_2 = -3$

12. $m_1 = -2$ and $m_2 = \frac{1}{2}$

13. $m_1 = \frac{1}{5}$ and $m_2 = 5$

14 Do the three points $(2, 3)$, $(0, 2)$, and $(-2, 1)$ all lie on the same straight line? Explain using slopes.

Find the slope of the given line by first finding two points on the line then using the definition of slope.

15. $4x - y + 7 = 0$

16 $5x + 1 = 0$

17. $2 - y = 0$

18. $3x + 5y = 0$ $\qquad\qquad$ **19.** $x + y = 5$ $\qquad\qquad$ **20.** $2x - 3y + 1 = 0$

21. Verify that the line l_1 through $(-2, -1)$ and $(1, 5)$ and the line l_2 through $(4, 3)$ and $(-1, -7)$ are parallel.

22. Verify that the line l_1 through $(-1, 3)$ and $(2, 4)$ and the line l_2 through $(6, -1)$ and $(5, 2)$ are perpendicular.

FOR REVIEW

Give the intercepts and graph the following.

23. $x - 6y = 2$

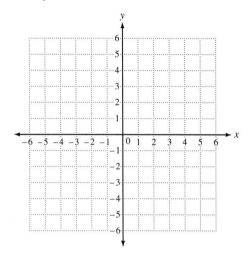

24. $5x + y = 0$

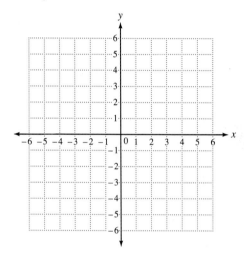

25. $x = -5$

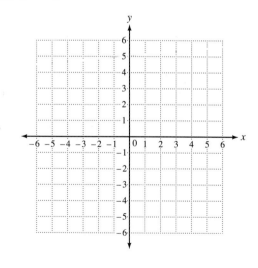

26. $y = 6$

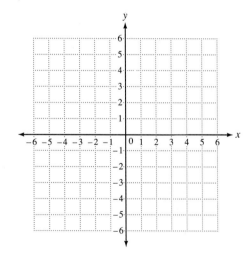

27. Is $\frac{3}{x} + \frac{2}{y} = 5$ a linear equation? Explain.

ANSWERS: 1. 1 2. 1 3. $\frac{4}{5}$ 4. undefined slope 5. 0 6. $-\frac{2}{5}$ 7. undefined slope 8. positive slope 9. zero slope 10. zero slope 11. parallel 12. perpendicular 13. neither 14. Yes. The slope of the line through (2, 3) and (0, 2) is $\frac{1}{2}$, the same as the slope of the line through (0, 2) and (−2, 1). 15. 4 16. undefined slope 17. 0 18. $-\frac{3}{5}$ 19. −1 20. $\frac{2}{3}$ 21. They are parallel since both have slope 2. 22. They are perpendicular since $m_1 = \frac{1}{3}$, $m_2 = -3$, and $m_1 m_2 = -1$.

23. x-intercept (2, 0);
    y-intercept $\left(0, -\frac{1}{3}\right)$

24. x-intercept (0, 0);
    y-intercept (0, 0);
    another point is (1, −5)

25. x-intercept (−5, 0);
    no y-intercept

26. no x-intercept;
    y-intercept (0, 6)

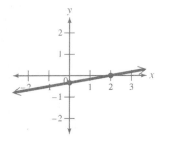
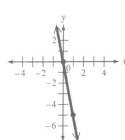
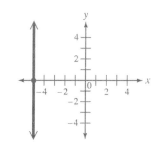
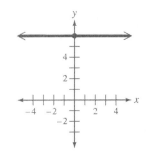

27. No. It cannot be written
    in the form $ax + by + c = 0$.

4.3 EXERCISES B

Find the slope of the line passing through the given pairs of points.

1. (4, 8) and (10, 2)

2. (−2, 1) and (−6, 5)

3. (6, 1) and (6, −5)

4. (4, −1) and (−2, −1)

5. (−3, −1) and (−2, −3)

6. (0, 1) and (−2, 0)

Answer Exercises 7–10 with one of the following phrases: (a) positive slope (b) negative slope (c) zero slope (d) undefined slope.

7. A line parallel to the x-axis has ___?___.

8. A line that slopes from upper left to lower right has ___?___.

9. A line perpendicular to the x-axis has ___?___.

10. The y-axis has ___?___.

In Exercises 11–13, m_1 and m_2 represent the slopes of two lines. State whether the lines are parallel, perpendicular, or neither.

11. $m_1 = \frac{1}{4}$ and $m_2 = \frac{1}{4}$

12. $m_1 = \frac{1}{5}$ and $m_2 = -5$

13. $m_1 = \frac{3}{2}$ and $m_2 = \frac{2}{3}$

14. Do the three points (1, 2), (−1, −1), and (3, 6) all lie on the same straight line? Explain using slopes.

Find the slope of the given line by first finding two points on the line then using the definition of slope.

15. $x + 4y - 9 = 0$ **16.** $3y - 7 = 0$ **17.** $4 + x = 0$

18. $2x - 7y = 0$ **19.** $y - x = 3$ **20.** $3x - 5y + 2 = 0$

21. Verify that the line l_1 through $(4, -2)$ and $(1, 1)$ and the line l_2 through $(3, 3)$ and $(7, -1)$ are parallel.

22. Verify that the line l_1 through $(-2, 4)$ and $(2, 5)$ and the line l_2 through $(3, 2)$ and $(4, -2)$ are perpendicular.

FOR REVIEW

Give the intercepts and graph the following.

23. $6x + y = 3$ **24.** $3x + y = 0$

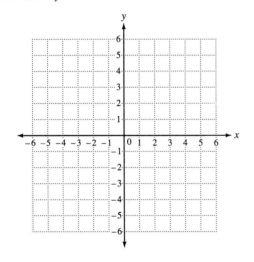   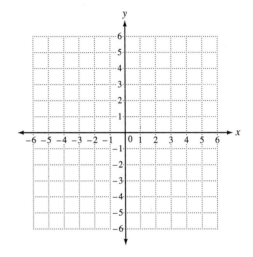

25. $y = 5$ **26.** $x = -3$

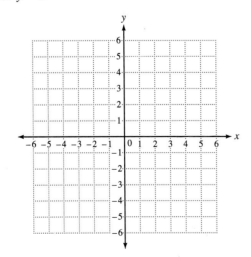   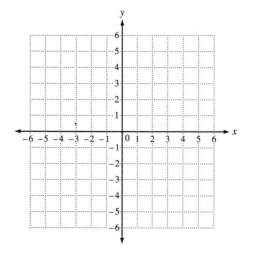

27. Is $3x^2 + 2y^2 = 5$ a linear equation? Explain.

4.3 EXERCISES C

1. Tell whether the lines $ax + c = 0$ and $by + c = 0$ are parallel, perpendicular, or neither.

2. Find the slope of the line with equation $ax + by + c = 0$. [*Hint: a, b,* and *c* are constants.]

4.4 FORMS OF THE EQUATION OF A LINE

STUDENT GUIDEPOSTS

1 General Form
2 Slope-Intercept Form
3 Graphing Using Slope
4 Point-Slope Form

1 GENERAL FORM

The equation of a line can be written in several forms. In Section 4.2 the **general form** was defined to be

$$ax + by + c = 0 \qquad a, b, c \text{ constants}$$

where *a* and *b* are not both zero. For example, consider the equation $x - 3y + 6 = 0$.

2 SLOPE-INTERCEPT FORM

Suppose we solve this equation for *y*. The result is a special form of the equation called the **slope-intercept form.**

$$x - 3y + 6 = 0$$
$$-3y = -x - 6$$
$$y = \frac{1}{3}x + 2 \qquad \text{Divide through by } -3$$

To explain why this is called the slope-intercept form first calculate *y* when $x = 0$.

$$y = \frac{1}{3}(0) + 2 = 2$$

Thus, the *y*-intercept of the line is (0, 2). Also, if *x* increases by 3, say from 0 to 3, *y* increases by 1. That is, *y* was 2 when $x = 0$, and when $x = 3$,

$$y = \frac{1}{3}(3) + 2 = 3.$$

Hence the slope of the line is

$$m = \frac{\text{change in } y}{\text{change in } x} = \frac{\text{rise}}{\text{run}} = \frac{1}{3}.$$

When an equation is in slope-intercept form the coefficient of *x* is the slope, and the constant term is the *y*-coordinate of the *y*-intercept of the line.

> ### Slope-Intercept Form of the Equation of a Line
>
> If the equation of a line is solved for y, the resulting equation is in **slope-intercept form**
>
> $$y = mx + b,$$
>
> where m is the slope of the line and $(0, b)$ is the y-intercept.

EXAMPLE 1   FINDING THE SLOPE AND Y-INTERCEPT

What are the slope and y-intercept of the line with equation $4x + 2y + 1 = 0$?

First solve for y to obtain the slope-intercept form.

$$4x + 2y + 1 = 0$$
$$2y = -4x - 1$$
$$y = -2x - \frac{1}{2} \quad \text{Divide by 2}$$

$$m = -2 = \text{slope} \quad \left(0, -\frac{1}{2}\right) = y\text{-intercept} \quad \left[b = -\frac{1}{2}\right]$$

PRACTICE EXERCISE 1

What are the slope and y-intercept of the line with equation $3x - 6y + 10 = 0$?

Answer: slope is $\frac{1}{2}$, y-intercept is $\left(0, \frac{5}{3}\right)$

Example 1 shows how to find the slope of a nonvertical line when its equation is given in general form. Solve the equation for y to obtain the slope-intercept form. The slope is always the numerical coefficient of x.

EXAMPLE 2   USING THE SLOPE-INTERCEPT FORM

Find the general form of the equation of the line with slope -2 and y-intercept $(0, 5)$.

We first find the slope-intercept form of the equation by substituting -2 for m and 5 for b in

$$y = mx + b.$$
$$y = -2x + 5$$

By writing all terms on the left side of this equation we obtain the general form

$$2x + y - 5 = 0.$$

PRACTICE EXERCISE 2

Find the general form of the equation of the line with slope $\frac{2}{3}$ and y-intercept $(0, -3)$.

Answer: $2x - 3y - 9 = 0$

③ GRAPHING USING SLOPE

In Section 4.2 we learned how to graph a line by finding the intercepts. The graph of a line in slope-intercept form can be obtained in a different way using what we know about slope. This technique is illustrated by graphing the equation $y = \frac{2}{3}x - 1$.

Since the y-intercept is $(0, -1)$, we know the graph passes through this point. But since there are infinitely many lines containing $(0, -1)$, we need to find the one with slope $\frac{2}{3}$. To do this we find a second point on the line that can be obtained from $(0, -1)$ by considering the rise and run specified by a slope of $\frac{2}{3}$. If we start at $(0, -1)$ and move up 2 units (a rise of 2) then move right 3 units (a run of 3), we are at the point $(3, 1)$. The line passes through $(0, -1)$ and $(3, 1)$, as shown in Figure 4.25.

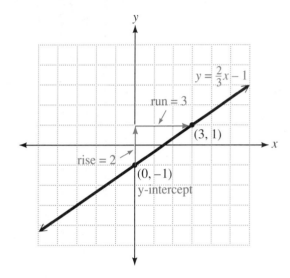

Figure 4.25 Graphing with Slope

4 POINT-SLOPE FORM

If we know that the slope of a line is m and that (x_1, y_1) is a point on the line, we can obtain the equation of the line. Suppose (x, y) is an arbitrary point on the line. Then using the formula for the slope of a line with (x, y) as (x_2, y_2), we have

$$m = \frac{y - y_1}{x - x_1}.$$

Multiplying both sides of this equation by $x - x_1$,

$$m(x - x_1) = y - y_1,$$

gives us another form of the equation of a line.

Point-Slope Form of the Equation of a Line
To find the equation of a line which has slope m and passes through the point (x_1, y_1), substitute these values into the **point-slope form** $$y - y_1 = m(x - x_1).$$

EXAMPLE 3   USING POINT-SLOPE FORM

Find the slope-intercept form of the equation of the line with slope -3 passing through the point $(1, -2)$. Graph the line.

   Use the point-slope form of the equation with $m = -3$ and $(x_1, y_1) = (1, -2)$.

PRACTICE EXERCISE 3

Find the slope-intercept form of the equation of the line with slope -5 passing through the point $(2, -7)$. Graph the line.

$$y - y_1 = m(x - x_1)$$
$$y - (-2) = -3(x - 1) \qquad y_1 = -2,\ x_1 = 1,\ m = -3$$
$$y + 2 = -3x + 3 \qquad \text{Watch the signs}$$
$$y = -3x + 1 \qquad \text{Subtract 2 to obtain slope-intercept form}$$

Since the slope $m = -3 = \frac{-3}{1} = \frac{\text{rise}}{\text{run}}$, the rise is -3 and the run is 1. Start at the y-intercept $(0, 1)$ and move down 3 units (the rise is -3), then right 1 unit (the run is 1). This locates another point on the line, $(1, -2)$. See Figure 4.26.

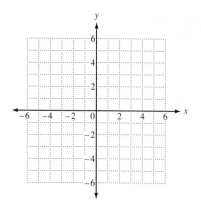

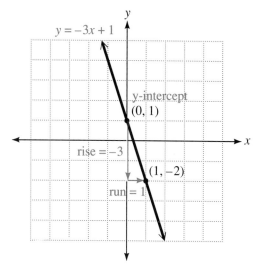

Figure 4.26

Answer: $y = -5x + 3$

To find the equation of a line passing through two points, we use both the slope formula and the point-slope form. This is shown in the next example.

| **EXAMPLE 4** EQUATION OF A LINE THROUGH TWO POINTS | **PRACTICE EXERCISE 4** |

Find the general form of the equation of the line passing through two points $(3, -2)$ and $(-1, 5)$.

    First find the slope of the line using $(3, -2) = (x_1, y_1)$ and $(-1, 5) = (x_2, y_2)$.

$$m = \frac{y_2 - y_1}{x_2 - x_1} = \frac{5 - (-2)}{-1 - 3} = \frac{7}{-4} = -\frac{7}{4}$$

Next, substitute $-\frac{7}{4}$ for m and $(3, -2)$ for (x_1, y_1) in the point-slope form. (We would get the same equation by using $(-1, 5)$ for (x_1, y_1).)

$$y - y_1 = m(x - x_1).$$
$$y - (-2) = -\frac{7}{4}(x - 3)$$
$$4(y - (-2)) = -7(x - 3) \qquad \text{Multiply both sides by 4}$$
$$4y + 8 = -7x + 21$$
$$7x + 4y - 13 = 0$$

This is the general form of the equation of the desired line.

Find the general form of the equation of the line passing through the points $(1, 7)$ and $(-2, 3)$.

Answer: $4x - 3y + 17 = 0$

We conclude this section with a summary of forms of the equation of a line. Note that the names *slope-intercept* and *point-slope form* help you decide which form should be used in a particular problem.

Forms of the Equation of a Line	
General Form	$ax + by + c = 0$
Slope-Intercept Form	
	$y = mx + b$ Slope m, y-intercept $(0, b)$
Point-Slope Form	
	$y - y_1 = m(x - x_1)$ Slope m, point (x_1, y_1) on line

4.4 EXERCISES A

Write each equation in slope-intercept form and give the slope and y-intercept.

1. $5x + y - 12 = 0$

2. $2x - 5y + 10 = 0$

3. $4y + 7 = 0$

4. $5x + 1 = 0$

5. $x + y = 5$

6. $2x - 3y + 1 = 0$

Write each equation in slope-intercept form and use the y-intercept and slope to graph the equation.

7. $x + y + 2 = 0$

8. $2x - 5y + 10 = 0$

9. $4x + 3y = 15$

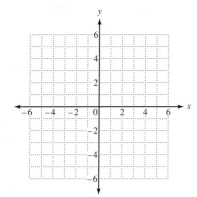 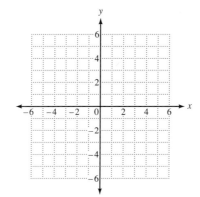 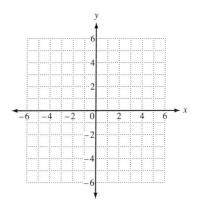

Find the general form of the equation of the line with the given slope and y-intercept.

10. $m = 2$, $(0, 4)$

11. $m = 4$, $(0, -3)$

12. $m = -\dfrac{1}{2}$, $(0, 6)$

Find the general form of the equation of the line with the given slope passing through the given point.

13. $m = 3$, $(2, -4)$

14. $m = -2$, $(1, 5)$

15 $m = -\dfrac{1}{2}$, $(3, -2)$

Find the general form of the equation of the line passing through the given points.

16. $(2, 5)$ and $(4, 7)$

17. $(-1, 3)$ and $(2, 4)$

18 $(3, -2)$ and $(-1, -1)$

Some business problems can be described by a linear equation. For example, the cost y of producing a number of items x is given by $y = mx + b$, where b is the overhead cost (the cost when no items are produced) and m is the variable cost (the cost of producing a single item). Use this information to find the cost equation in 19–20.

19 Overhead cost: $25
Variable cost: $10

20. Overhead cost: $300
Variable cost: $10.50.

21. Which of the two equations, $x = 2$ or $y = -3$, is the equation of the vertical line through $(2, -3)$?

22. Which of the two equations, $x = 2$ or $y = -3$, is the equation of the horizontal line through $(2, -3)$?

23. Find the general form of the equation of the line through $(-1, 5)$ that is parallel to the line $3x + 2y - 4 = 0$.

FOR REVIEW

The following exercises review material from Section 3.7 to help you prepare for the next section. Solve each inequality.

24. $2x + 1 < 5$

25. $-2(x - 1) \geq 8$

26. $3(x - 1) - (2 - x) \geq 5$

ANSWERS: **1.** $y = -5x + 12$; slope is -5, y-intercept is $(0, 12)$ **2.** $y = \frac{2}{5}x + 2$; slope is $\frac{2}{5}$, y-intercept is $(0, 2)$ **3.** $y = -\frac{7}{4} = 0 \cdot x - \frac{7}{4}$; slope is 0, y-intercept is $\left(0, -\frac{7}{4}\right)$ **4.** The equation cannot be solved for y, hence it has no slope-intercept form. Also, it has undefined slope and no y-intercept. **5.** $y = -x + 5$; slope is -1, y-intercept is $(0, 5)$ **6.** $y = \frac{2}{3}x + \frac{1}{3}$; slope is $\frac{2}{3}$, y-intercept is $\left(0, \frac{1}{3}\right)$

7. $y = -x - 2$

8. $y = \frac{2}{5}x + 2$

9. $y = -\frac{4}{3}x + 5$

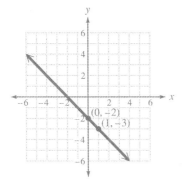

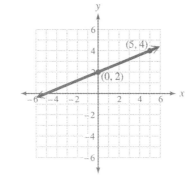

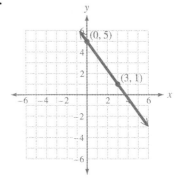

10. $2x - y + 4 = 0$ **11.** $4x - y - 3 = 0$ **12.** $x + 2y - 12 = 0$ **13.** $3x - y - 10 = 0$ **14.** $2x + y - 7 = 0$ **15.** $x + 2y + 1 = 0$ **16.** $x - y + 3 = 0$ **17.** $x - 3y + 10 = 0$ **18.** $x + 4y + 5 = 0$ **19.** $y = 10x + 25$ **20.** $y = 10.50x + 300$ **21.** $x = 2$ **22.** $y = -3$ **23.** $3x + 2y - 7 = 0$ **24.** $x < 2$ **25.** $x \leq -3$ **26.** $x \geq \frac{5}{2}$

4.4 EXERCISES B

Write each equation in slope-intercept form and give the slope and y-intercept.

1. $7x - y + 13 = 0$

2. $5x - 9y + 9 = 0$

3. $2x - 1 = 0$

4. $4y + 12 = 0$

5. $y - x = 3$

6. $3x - 5y + 2 = 0$

Write each equation in slope-intercept form and use the y-intercept and slope to graph the equation.

7. $x + y - 6 = 0$

8. $3x - 4y - 12 = 0$

9. $5x + 2y = 6$

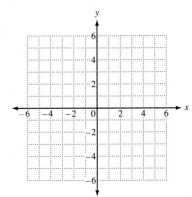
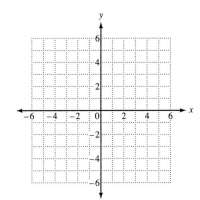
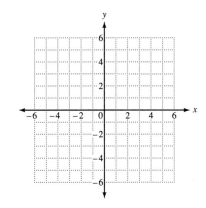

Find the general form of the equation of the line with the given slope and y-intercept.

10. $m = 3$, $(0, 1)$

11. $m = 5$, $(0, -2)$

12. $m = -\dfrac{1}{4}$, $(0, 2)$

Find the general form of the equation of the line with the given slope passing through the given point.

13. $m = 5$, $(-3, 1)$

14. $m = -4$, $(2, 6)$

15. $m = -\dfrac{1}{5}$, $(-1, 4)$

Find the general form of the equation of the line passing through the given points.

16. $(2, 6)$ and $(5, 9)$

17. $(-3, 2)$ and $(2, 5)$

18. $(-4, 6)$ and $(-2, -2)$

Some business problems can be described by a linear equation. For example, the cost y of producing a number of items x is given by $y = mx + b$, where b is the overhead cost (the cost when no items are produced) and m is the variable cost (the cost of producing a single item). Use this information to find the cost equation in 19–20.

19. Overhead cost: $50
   Variable cost: $30

20. Overhead cost: $500
   Variable cost: $8.50.

21. Which of the two equations, $x = -1$ or $y = 5$, is the equation of the vertical line through $(-1, 5)$?

22. Which of the two equations, $x = -1$ or $y = 5$, is the equation of the horizontal line through $(-1, 5)$?

23. Find the general form of the equation of the line through $(-1, 5)$ that is perpendicular to the line $3x + 2y - 4 = 0$.

FOR REVIEW

The following exercises review material from Section 3.7 to help you prepare for the next section. Solve each inequality.

24. $1 - 3x \le -5$ **25.** $4(2 - x) > 7 - 3x$ **26.** $3(x + 2) - (1 - 2x) \ge 10$

4.4 EXERCISES C

1. Find the general form of the equation of the horizontal line through the point $(6, -4)$.

2. Find the general form of the equation of the vertical line through the point $(5, 8)$.

3. The midpoint of the line segment joining (x_1, y_1) and (x_2, y_2) is $\left(\frac{x_1 + x_2}{2}, \frac{y_1 + y_2}{2}\right)$. Find the general form of the perpendicular bisector of the line segment joining $(2, -1)$ and $(-6, 3)$.   [Answer: $2x - y + 5 = 0$]

4.5 GRAPHING LINEAR INEQUALITIES IN TWO VARIABLES

STUDENT GUIDEPOSTS

1 Linear Inequalities in Two Variables

2 Graph of Solution

3 Test Point

4 Method of Graphing

In Section 3.7 we graphed linear inequalities in one variable, such as

$$3x + 2 > 5 \quad \text{and} \quad x - 1 \le 3(x - 5) + 1,$$

on a number line. For example, if we solve

$$3x + 2 > 5$$
$$3x > 3 \qquad \text{Subtract 2}$$
$$x > 1, \qquad \text{Divide by 3}$$

the solution $x > 1$ is graphed in Figure 4.27.

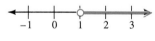

Figure 4.27

1 LINEAR INEQUALITIES IN TWO VARIABLES

We now consider **linear inequalities in two variables** in which the variables are raised only to the first power. For example,

$$2x + y < -1$$

is a linear inequality in the two variables x and y. A **solution** to such an inequality is an ordered pair of numbers which when substituted for x and y yields a true statement. Thus, $(-1, 0)$ is a solution to $2x + y < -1$ since by replacing x with -1 and y with 0 we obtain

$$2x + y < -1$$
$$2(-1) + 0 < -1$$
$$-2 < -1, \quad \text{which is true.}$$

On the other hand, $(4, -3)$ is not a solution since

$$2x + y < -1$$
$$2(4) + (-3) < -1$$
$$8 - 3 < -1$$
$$5 < -1, \quad \text{is false.}$$

EXAMPLE 1 SOLUTIONS TO AN INEQUALITY	PRACTICE EXERCISE 1

Tell whether $(-1, -3)$, $(1, 1)$ and $(-2, 0)$, are solutions to $3x - 2y \geq 1$.

$$3(-1) - 2(-3) \geq 1$$
$$-3 + 6 \geq 1$$
$$3 \geq 1 \qquad \text{True}$$

$(-1, -3)$ is a solution.

$$3(1) - 2(1) \geq 1$$
$$3 - 2 \geq 1$$
$$1 \geq 1 \qquad \text{True}$$

$(1, 1)$ is a solution.

$$3(-2) - 2(0) \geq 1$$
$$-6 - 0 \geq 1$$
$$-6 \geq 1 \qquad \text{False}$$

$(-2, 0)$ is not a solution.

Tell whether $(3, 0)$, $(2, 3)$, and $(1, 1)$ are solutions to $-5x + 4y < -1$.

Answer: $(3, 0)$ is a solution, $(2, 3)$ is not a solution, $(1, 1)$ is not a solution.

② GRAPH OF SOLUTION

The set of all solutions to a linear inequality in two variables can be displayed in a Cartesian coordinate system. Consider

$$2x + y > -3.$$

If we replace the inequality symbol with an equal sign, the resulting equation is

$$2x + y = -3.$$

To graph this equation we first plot the intercepts $\left(-\frac{3}{2}, 0\right)$ and $(0, -3)$ and draw the line as in Figure 4.28. Notice that the graph of the line divides the plane into a region above the line, the line itself, and a region below the line.

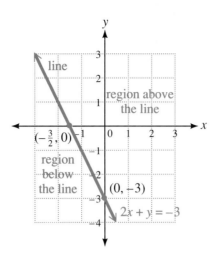

Figure 4.28

➌ TEST POINT

The graph of an inequality such as $2x + y > -3$ consists of all the points on one side of the boundary line $2x + y = -3$. The graph of $2x + y < -3$ is all the points on the other side of the line. To determine the correct side of the line to graph for an inequality such as $2x + y > -3$, we select any point not on the line as a **test point.** If we use $(0, 0)$ in this case the arithmetic is easy:

$$2x + y > -3$$
$$2(0) + 0 > -3$$
$$0 > -3. \qquad \text{This is true}$$

Since this inequality is true, we graph the points on the side of the line containing the test point $(0, 0)$. If the inequality to be graphed had been $2x + y < -3$, a false inequality would have resulted using $(0, 0)$ as a test point:

$$2x + y < -3$$
$$2(0) + 0 < -3$$
$$0 < -3. \qquad \text{This is false}$$

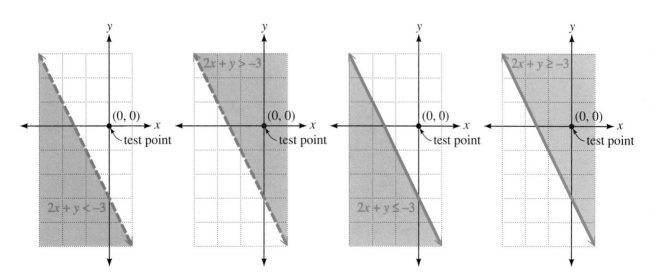

Figure 4.29

In the case when a false inequality is obtained, shade the region that does *not* contain the test point. The graphs of both $2x + y < -3$ and $2x + y > -3$ are shown in Figure 4.29 by shading the region that satisfies the inequality. The figure also has graphs of $2x + y \leq -3$ and $2x + y \geq -3$. We have used a dashed line for the boundary when the inequality is $<$ or $>$ to show that the boundary is *not* part of the graph. For the inequalities $\leq$ and $\geq$ a solid line is used to indicate that the boundary *is* part of the graph.

EXAMPLE 2 GRAPHING INEQUALITIES	PRACTICE EXERCISE 2

(a) Graph $x + 3y > 6$.

Graph the line $x + 3y = 6$ using the intercepts $(0, 2)$ and $(6, 0)$. Since the inequality is $>$, use a dashed line. Select the test point $(0, 0)$. (It is not on the line and the arithmetic is easy with $(0, 0)$.)

$$x + 3y > 6$$
$$0 + 3(0) > 6$$
$$0 > 6 \quad \text{This is false}$$

The inequality is false. Shade the region in Figure 4.30 that does not contain $(0, 0)$.

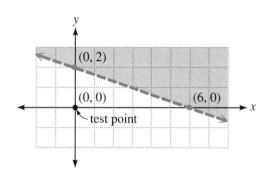

Figure 4.30

(b) Graph $2x - 3y \leq 0$.

Since both intercepts of $2x - 3y = 0$ are $(0, 0)$, we need to find another point on the boundary line. If $x = 3$ then $y = 2$, so $(3, 2)$ is a second point. Draw a solid line through $(0, 0)$ and $(3, 2)$ as in Figure 4.31.

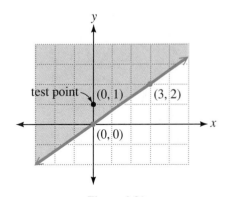

Figure 4.31

This time $(0, 0)$ cannot be used as a test point. Choosing $(0, 1)$ as a test point we obtain a true inequality:

(a) Graph $x + 3y \leq 6$.

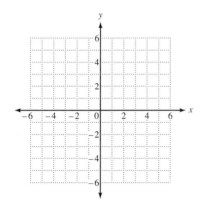

(b) Graph $2x - 3y > 0$.

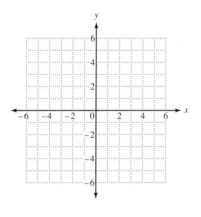

$$2x - 3y \leq 0$$
$$2(0) - 3(1) \leq 0$$
$$-3 \leq 0.$$

Thus, we shade the region containing the test point (0, 1).

Some linear inequalities in two variables have either the *x*-term or the *y*-term missing. The solution techniques are the same for these equations, but sometimes our work can be simplified by first solving the inequality for the remaining variable.

EXAMPLE 3   INEQUALITIES INVOLVING VERTICAL AND HORIZONTAL

PRACTICE EXERCISE 3

(a) Graph $2x - 4 \geq 0$.

$$2x - 4 \geq 0$$
$$2x \geq 4$$
$$x \geq 2$$

Replacing $\geq$ with $=$ we recognize the graph of $x = 2$ as a vertical line with *x*-intercept (2, 0). Since the inequality is $\geq$, we draw the boundary $x = 2$ as a solid line and use (0, 0) as a test point.

$$x \geq 2$$
$$0 \geq 2 \quad \text{This is false}$$

Since the inequality is false, shade the region not containing (0, 0) in Figure 4.32.

(b) Graph $3y + 3 < 0$.

$$3y + 3 < 0$$
$$3y < -3$$
$$y < -1$$

Replacing $<$ with $=$ we recognize the graph of $y = -1$ as a horizontal line with *y*-intercept (0, −1). Since the inequality is $<$, we draw the boundary $y = -1$ as a dashed line and use (0, 0) as a test point.

$$y < -1$$
$$0 < -1 \quad \text{This is false}$$

Since the inequality is false, shade the region not containing (0, 0) in Figure 4.33.

(a) Graph $2x - 4 < 0$.

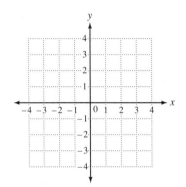

(b) Graph $3y + 3 \geq 0$.

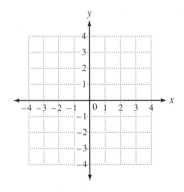

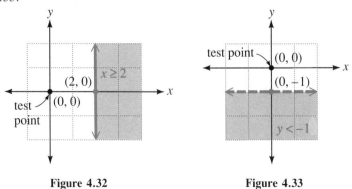

Figure 4.32   Figure 4.33

❹ METHOD OF GRAPHING

We now summarize the techniques that we have learned.

To Graph a Linear Inequality in Two Variables

1. Graph the boundary line using a dashed line if the inequality is $<$ or $>$ and a solid line if it is $\leq$ or $\geq$.

2. Choose a test point that is not on the boundary line and substitute it into the inequality.

3. Shade the region that includes the test point if a true inequality is obtained, and shade the region that does not contain the test point if a false inequality results.

4.5 EXERCISES A

1. In the graph of the inequality $3x - 2y \leq 5$, would the line $3x - 2y = 5$ be solid or dashed?

2. In the graph of the inequality $2x + y > 9$, would the line $2x + y = 9$ be solid or dashed?

In the following exercises the boundary line for the inequality has been given. Complete each graph by shading the appropriate region.

3. $x + y \leq 5$ 4. $2x - y < 6$

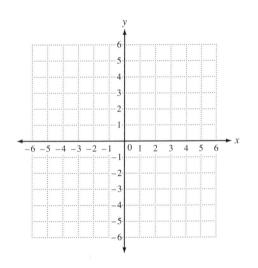   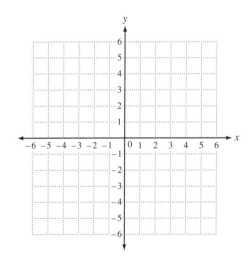

5. $3x + 12 > 0$

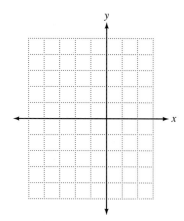

6. $1 - y \geq 0$

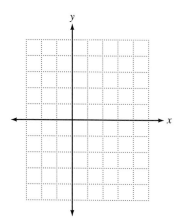

Graph the linear inequalities in two variables.

7. $x + y > 3$

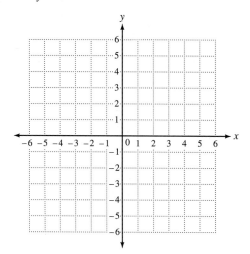

8. $x + y \leq 3$

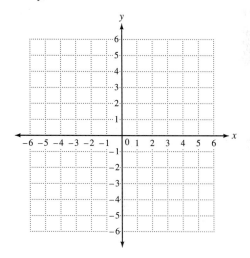

9. $x - y \geq -2$

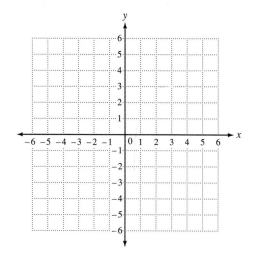

10. $x - y < -2$

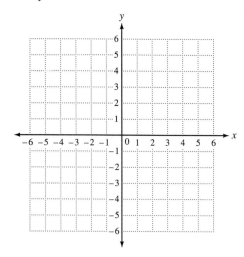

11. $x + 4y < 4$

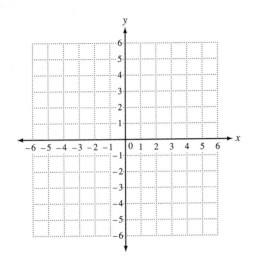

12. $x + 4y \geq 4$

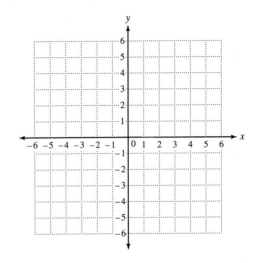

13. $4x + 12 > 0$

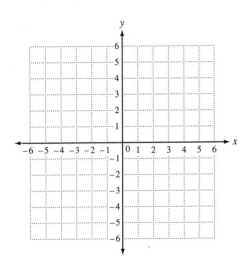

14. $4x + 12 \leq 0$

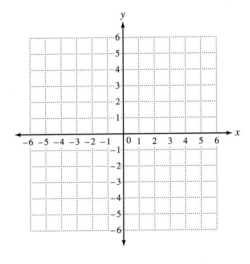

15. $2y - 8 \leq 0$

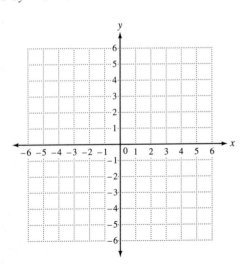

16. $2y - 8 > 0$

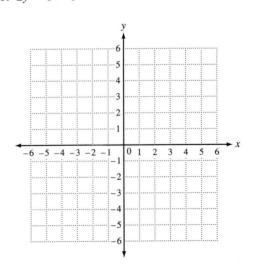

17. $x + y < 0$

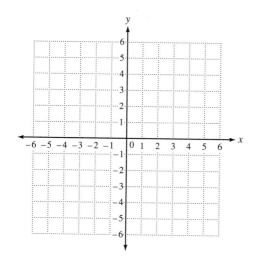

18. $x + y \geq 0$

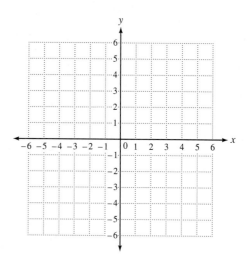

FOR REVIEW

Find the slope of the line passing through the given pair of points.

19. $(-3, 7)$ and $(10, -1)$

20. $(6, 3)$ and $(6, -5)$

21. $(2, -5)$ and $(-8, -5)$

Write each equation in slope-intercept form and give the slope and y-intercept.

22. $12x - 9y + 36 = 0$

23. $5y - 15 = 0$

Find the general form of the equation of the line.

24. $m = \dfrac{1}{5}$, through $(-1, 4)$

25. through $(2, 6)$ and $(-4, -3)$

26. Verify that the line l_1 between $(-9, 2)$ and $(0, 5)$ and the line l_2 between $(1, -4)$ and $(19, 2)$ are parallel.

ANSWERS: 1. solid   2. dashed

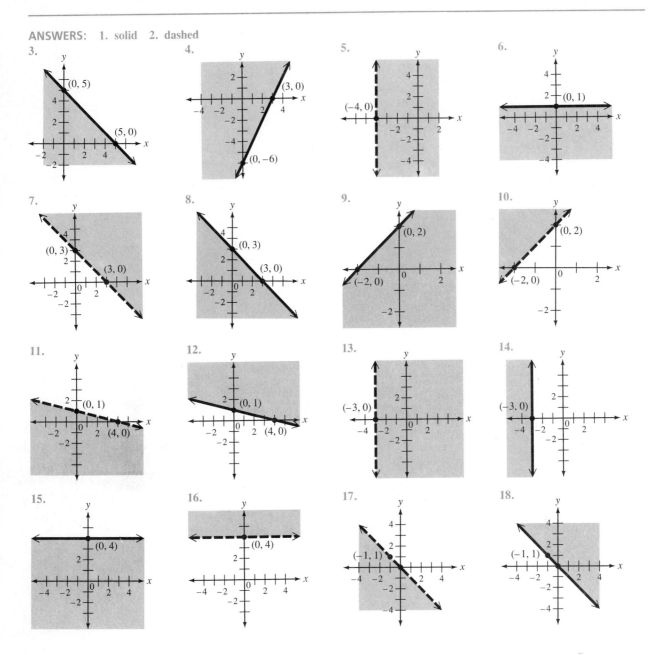

19. $-\frac{8}{13}$ 20. undefined slope   21. 0 (zero slope)   22. $y = \frac{4}{3}x + 4$; slope is $\frac{4}{3}$, y-intercept is (0, 4)   23. $y = 3 = 0 \cdot x + 3$; slope is 0, y-intercept is (0, 3)   24. $x - 5y + 21 = 0$ 25. $3x - 2y + 6 = 0$ 26. The lines are parallel since both have slope $\frac{1}{3}$.

4.5 EXERCISES B

1. In the graph of the inequality $5x + y \geq 1$, would the line $5x + y = 1$ be solid or dashed?

2. In the graph of the inequality $4x - 2y < 5$, would the line $4x - 2y = 5$ be solid or dashed?

In the following exercises the boundary line for the inequality has been given. Complete each graph by shading the appropriate region.

3. $x + y > 5$

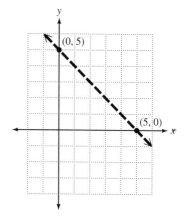

4. $2x - y \geq 6$

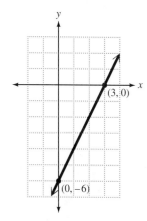

5. $3x + 12 \leq 0$

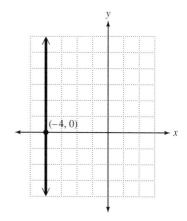

6. $1 - y < 0$

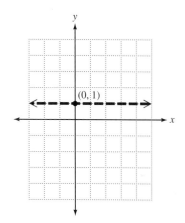

Graph the linear inequalities in two variables.

7. $x + y > 2$

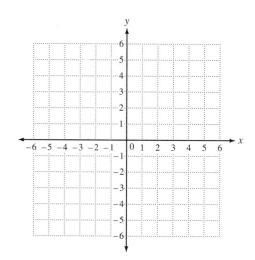

8. $x + y \leq 2$

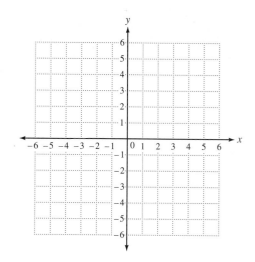

9. $x - y \geq -1$

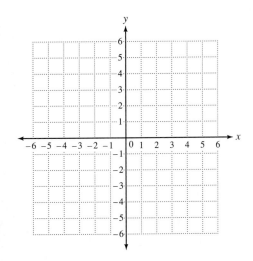

10. $x - y < -1$

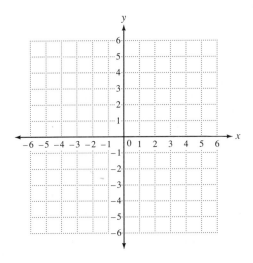

11. $3x + 4y < 12$

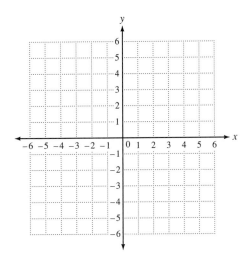

12. $3x + 4y \geq 12$

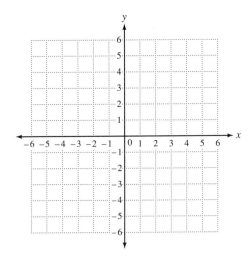

13. $4x + 8 > 0$

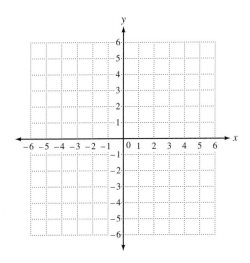

14. $4x + 8 \leq 0$

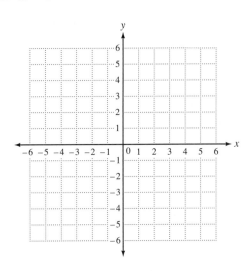

15. $2y - 6 \leq 0$

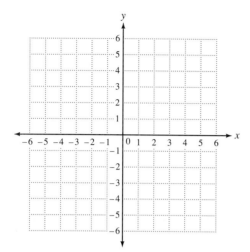

16. $2y - 6 > 0$

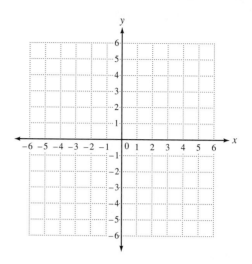

17. $2x + y < 0$

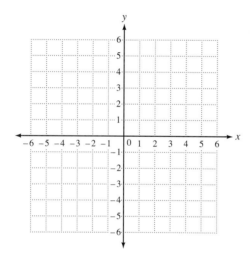

18. $2x + y \geq 0$

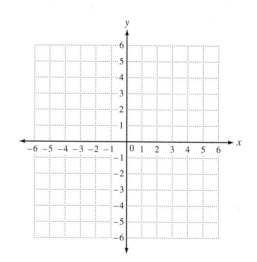

FOR REVIEW

Find the slope of the line passing through the given pair of points.

19. $(-5, 5)$ and $(3, -2)$

20. $(-3, 10)$ and $(2, 10)$

21. $(-9, 0)$ and $(-9, -7)$

Write each equation in slope-intercept form and give the slope and y-intercept.

22. $14x - 7y + 21 = 0$

23. $2y + 1 = 0$

Find the general form of the equation of the line.

24. $m = -3$, through $(-2, 5)$

25. through $(8, -1)$ and $(-4, 2)$

26. Verify that the line l_1 between $(3, 6)$ and $(5, -4)$ and the line l_2 between $(1, -2)$ and $(-1, 8)$ are parallel.

4.5 EXERCISES C

Tell whether the given point satisfies both of the inequalities.

1. $3x - y < 4$
     $y \geq -2$
  (a) $(0, 2)$ **(b)** $(2, 0)$

  (c) $(-1, 5)$ **(d)** $(-3, -4)$
  [Answers: **(c)** yes; **(d)** no]

2. $x + 4y \geq 5$
     $-2x + 3y < 0$
  (a) $(0, -1)$ **(b)** $(-1, 0)$

  (c) $(5, 2)$ **(d)** $(-3, 4)$

CHAPTER 4 REVIEW

KEY WORDS

4.1 A **rectangular** or **Cartesian coordinate system** is formed by using both a horizontal and a vertical number line.

The **x-axis** is the horizontal number line in a rectangular coordinate system.

The **y-axis** is the vertical number line in a rectangular coordinate system.

An **ordered pair** has x-coordinate in the first position and y-coordinate in the second position.

The **quadrants** are the four sections formed by the axes in a rectangular coordinate system.

A **linear equation in two variables** x and y is an equation of the form $ax + by = c$.

A **solution** to a linear equation in two variables is an ordered pair of numbers which when substituted for the variables results in a true equation.

4.2 The **graph** of an equation in two variables x and y is the set of points in a rectangular coordinate system that corresponds to solutions of the equation.

The **general form** of a linear equation is $ax + by + c = 0$.

The **x-intercept** is the point where the line crosses the x-axis.

The **y-intercept** is the point where the line crosses the y-axis.

4.3 The **slope** of a line gives a measure of its direction in a coordinate system.

Parallel lines never intersect.

Perpendicular lines intersect at right angles.

4.5 A **linear inequality in two variables** is an expression using $<$, $\leq$, $>$, or $\geq$ in which the variables are raised to the first power.

A **test point** is a point used to determine which side of the line is the solution of a linear inequality.

KEY CONCEPTS

4.2 **1.** When graphing an equation in a Cartesian coordinate system, construct a table of values.

2. Two convenient points to use for graphing a linear equation are the intercepts (if they exist).

3. An equation of the form $x = c$ is parallel to the y-axis with x-intercept $(c, 0)$.

4. An equation of the form $y = c$ is parallel to the x-axis with y-intercept $(0, c)$.

4.3 **1.** The slope of the line passing through points (x_1, y_1) and (x_2, y_2) is given by

$$m = \frac{y_2 - y_1}{x_2 - x_1} = \frac{\text{change in } y\text{-coordinates}}{\text{change in } x\text{-coordinates}}$$

and can be positive, negative, zero, or undefined.

2. Let m_1 and m_2 be the slopes of two distinct lines. If $m_1 = m_2$, the lines are parallel.

3. Let m_1 and m_2 be the slopes of two distinct lines. If $m_1 m_2 = -1$, the lines are perpendicular.

4.4 **1.** The slope-intercept form of the equation of a line is

$$y = mx + b,$$

where m is the slope and $(0, b)$ is the y-intercept.

2. The best way to find the slope of a line quickly is to solve the equation for y (if possible), putting the equation into slope-intercept form. The coefficient of x is the slope.

3. The point-slope form of the equation of a line is

$$y - y_1 = m(x - x_1),$$

where m is the slope and (x_1, y_1) is any point on the line.

4.5 To graph a linear inequality in two variables, first graph the boundary line using a dashed line (for $<$ or $>$) or a solid line (for $\leq$ or $\geq$). Choose a test point not on the boundary line and substitute it into the inequality. Shade the region containing the test point if a true inequality results, and shade the region not containing the test point if a false inequality results.

REVIEW EXERCISES

Part I

4.1 1. Plot the points associated with the pairs $A(-1, 3)$, $B(0, 0)$, $C(1, -2)$, $D(4, 2)$, and $E(-4, -3)$, and state which quadrant each is in.

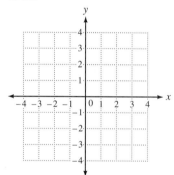

2. Complete the ordered pairs $(0,)$, $(, 0)$, and $(-2,)$ so that they will be solutions to the equation $2x + 3y = 6$.

3. The number, y, of bacteria in a culture is approximated using time x, in days, by the equation $y = 4000x + 5000$.
Complete the ordered pair $(3,)$. After 3 days, what is the bacteria count?

4.2 4. Which of the following are linear equations?
    (a) $2x + 3 = y$ **(b)** $xy = 5$ **(c)** $x + y^2 = 2$
    (d) $x^3 + y = 3$ **(e)** $x = 4$ **(f)** $3y = -1$

Give the intercepts and graph in the Cartesian coordinate system.

5. $3x - 4y = 12$

x-intercept $=$

y-intercept $=$

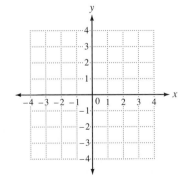

6. $2x + 3 = 0$

x-intercept $=$

y-intercept $=$

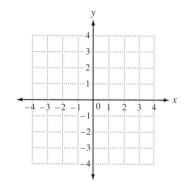

7. $x - 2y = 0$
x-intercept $=$
y-intercept $=$

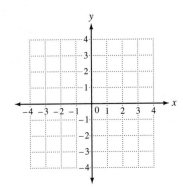

8. $y - 3 = 0$
x-intercept $=$
y-intercept $=$

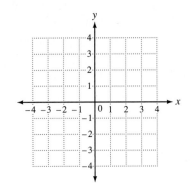

4.3 *Find the slope of the line through the given pair of points.*

9. $(-9, -2)$ and $(3, 6)$ **10.** $(0, 2)$ and $(-3, 0)$ **11.** $(-1, 8)$ and $(-1, 7)$

12. What can be said about two distinct lines that both have slope $\frac{2}{3}$?

13. What can be said about two lines with slopes $\frac{1}{3}$ and -3?

4.4 **14.** Write $8x + y - 3 = 0$ in slope-intercept form and give the slope and y-intercept.

15. Write $3y + 4 = 0$ in slope-intercept form and give the slope and y-intercept.

16. Find the general form of the equation of the line with slope $-\frac{2}{3}$ and y-intercept $(0, 4)$.

17. Find the general form of the equation of the line passing through $(-1, -2)$ and $(5, 3)$.

4.5 *Graph the following in a Cartesian coordinate system.*

18. $x - 2y > -2$

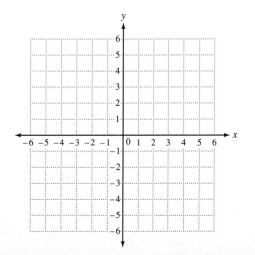

19. $2y + 4 \leq 0$

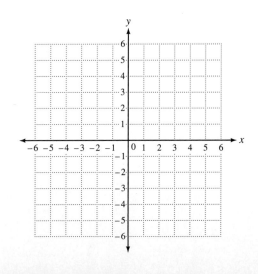

20. $3 - 3x < 0$

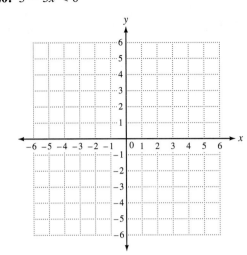

21. $3x + 2y \geq 6$

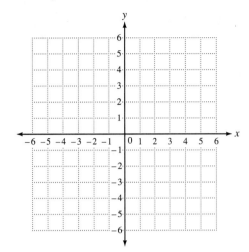

Part II

22. What can be said about two lines both with slope $\frac{1}{5}$ and y-intercept -5?

Find the slope of the line through the given pair of points.

23. $(-1, 2)$ and $(0, 5)$

24. $(-6, -4)$ and $(3, -4)$

Write each equation in slope-intercept form and give its slope and y-intercept.

25. $3x - 2y = 8$

26. $-x + 5y + 4 = 0$

27. Find the general form of the equation of a line through $(6, -1)$ and $(4, 5)$.

28. Find the general form of the equation of a line through $(-2, 7)$ and parallel to a line with slope $-\frac{1}{3}$.

29. Graph $2x + y \leq 4$.

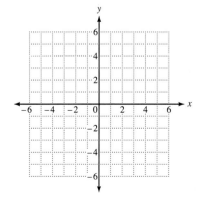

30. Use the y-intercept and slope to graph $3x + y + 2 = 0$.

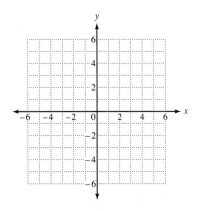

ANSWERS:

1. A is in II:
   B is origin;
   C is in IV;
   D is in I;
   E is in III

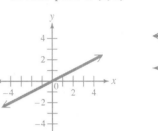

2. $(0, 2)$, $(3, 0)$, $\left(-2, \frac{10}{3}\right)$ 3. $(3, 17{,}000)$; $17{,}000$ 4. (a), (e), and (f) are linear

5. x-intercept $(4, 0)$;
   y-intercept $(0, -3)$

6. x-intercept $\left(-\frac{3}{2}, 0\right)$;
   no y-intercept

7. x-intercept $(0, 0)$;
   y-intercept $(0, 0)$;
   another point is $(2, 1)$

8. no x-intercept;
   y-intercept $(0, 3)$

9. $\frac{2}{3}$ 10. $\frac{2}{3}$ 11. undefined slope   12. They are parallel.   13. They are perpendicular.   14. $y = -8x + 3$; slope is -8; y-intercept is $(0, 3)$ 15. $y = 0 \cdot x - \frac{4}{3}$; slope is 0; y-intercept is $\left(0, -\frac{4}{3}\right)$ 16. $2x + 3y - 12 = 0$ 17. $5x - 6y - 7 = 0$

18.

19.

20.

21.

22. They coincide.   23. 3   24. 0   25. $y = \frac{3}{2}x - 4$; slope is $\frac{3}{2}$; y-intercept is $(0, -4)$ 26. $y = \frac{1}{5}x - \frac{4}{5}$; slope is $\frac{1}{5}$; y-intercept is $\left(0, -\frac{4}{5}\right)$ 27. $3x + y - 17 = 0$ 28. $x + 3y - 19 = 0$

29.

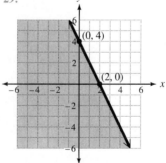

30.

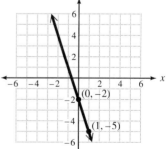

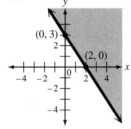

1. The point with coordinates $(2, -3)$ is located in which quadrant?

1. _____

2. Complete the ordered pair $(-3, \quad)$ so that it will be a solution to the equation $x + 5y = -1$.

2. _____

3. The number, y, of items in a store during the week is approximated by using time x, in days, in the equation $y = -200x + 1000$. Complete the ordered pair $(3, \quad)$ to find the number of items in the store after three days.

3. _____

4. True or false: $x - 2y^2 = 3$ is a linear equation.

4. _____

5. Find the slope of the line through the points $(2, -3)$ and $(-4, -5)$.

5. _____

6. Write $2x + 5y = 20$ in slope-intercept form and give the slope and the y-intercept.

6. _____

7. Find the general form of the equation of the line with slope -6 and passing through $(4, -5)$.

7. _____

8. Are lines with slopes $-\frac{1}{7}$ and 7 parallel, perpendicular, or neither?

8. _____

Graph each equation in the given rectangular coordinate system.

9. $y = 2x + 4$

9.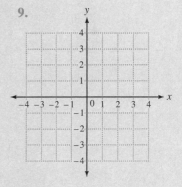

10. $3x + y = 0$

10.

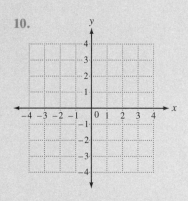

Graph each inequality in the given rectangular coordinate system.

11. $2y + x - 3 > 0$

11.

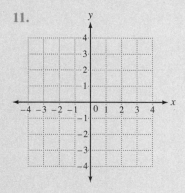

12. $y - 4 \leq 0$

12.

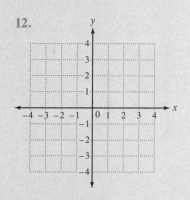

Systems of Linear Equations

5.1 PARALLEL, COINCIDING, AND INTERSECTING LINES

1 SYSTEMS OF LINEAR EQUATIONS

In Chapter 4 we discovered that a solution to a linear equation in two variables such as

$$2x - y + 1 = 0$$

is an ordered pair of numbers that when substituted for x and y makes the equation true. Often in applied situations, two quantities are compared using two different equations. For example, a small businessman may discover that total sales and profit are related in two different ways. If the following pair of linear equations describes the relationships between sales x and profit y, it is called a **system of two linear equations in two variables,** or simply a **system of equations.**

$$3x - 2y + 1 = 0$$
$$3x + y - 5 = 0$$

A **solution** to a system of equations is an ordered pair of numbers (x, y) that is a solution to *both* equations. We can verify that $(1, 2)$ is a solution to the above system by direct substitution.

$$3(1) - 2(2) + 1 \overset{?}{=} 0 \quad x = 1 \text{ and } y = 2 \quad 3(1) + (2) - 5 \overset{?}{=} 0$$
$$3 - 4 + 1 \overset{?}{=} 0 \qquad\qquad\qquad 3 + 2 - 5 \overset{?}{=} 0$$
$$0 = 0 \qquad\qquad\qquad\qquad 0 = 0$$

2 GRAPHING A SYSTEM

Finding a solution to a system of equations is easier to understand if we first examine the graphs of the system. Remember that a linear equation in x and y has a straight line for its graph. When the linear equations in a given system are graphed together in a rectangular coordinate system, one of three possibilities occurs:

1. The lines coincide (the two equations represent the same line).

2. The lines are parallel (do not intersect).

3. The lines intersect in exactly one point.

The following three pairs of linear equations, graphed in Figure 5.1, illustrate these three possible relationships.

(a) $3x - 2y + 1 = 0$ (c) $3x - 2y + 1 = 0$ (e) $3x - 2y + 1 = 0$
(b) $6x - 4y + 2 = 0$ (d) $3x - 2y - 4 = 0$ (f) $3x + y - 5 = 0$

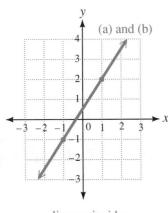
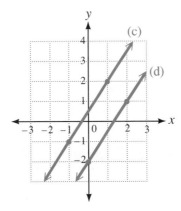
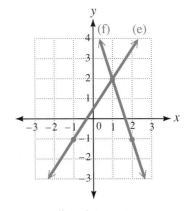

lines coincide   lines are parallel   lines intersect
                                                         in exactly one point

Figure 5.1

In the first case, lines (a) and (b) coincide, in the second case, lines (c) and (d) are parallel, and in the third case, lines (e) and (f) intersect in one point.

❸ RELATION OF THE GRAPHS OF A SYSTEM

We need a way to tell quickly which case we have in a given situation. Let us solve each linear equation above for y, writing it in slope-intercept form.

Lines coincide

(a) $y = \dfrac{3}{2}x + \dfrac{1}{2}$

Slopes are y-intercepts
the same.   are the same.

(b) $y = \dfrac{3}{2}x + \dfrac{1}{2}$

Lines are parallel

(c) $y = \dfrac{3}{2}x + \dfrac{1}{2}$

Slopes are y-intercepts
the same.   are different.

(d) $y = \dfrac{3}{2}x - 2$

Lines intersect in exactly one point

(e) $y = \dfrac{3}{2}x + \dfrac{1}{2}$

    ↑
Slopes are
different.
    ↓

(f) $y = -3\,x + 5$

Notice that all lines with equation $y = b$ are parallel since the slopes are zero. Now we have covered all cases except when one (or both) of the equations in a system is of the form $ax = c$, since such equations cannot be solved for y. However, recalling that the graph of an equation of this form is always a line parallel to the y-axis, parallel or coinciding lines result only if both equations in the system are of this type.

EXAMPLE 1   INTERSECTING LINES

Tell how the graphs of the system are related.

$$2x + 3y + 1 = 0$$
$$x - 2y + 5 = 0$$

Solve for y.

$$3y = -2x - 1 \qquad\qquad -2y = -x - 5$$

$$\boxed{\dfrac{1}{3}} \cdot 3y = \dfrac{1}{3}(-2x - 1) \qquad \boxed{\left(-\dfrac{1}{2}\right)}(-2y) = \boxed{\left(-\dfrac{1}{2}\right)}(-x - 5)$$

$$y = -\dfrac{2}{3}x - \dfrac{1}{3} \qquad\qquad y = \dfrac{1}{2}x + \dfrac{5}{2}$$

Since the coefficients of x, the slopes of the lines, are $-\frac{2}{3}$ and $\frac{1}{2}$ (unequal), the lines are neither parallel nor coinciding, they intersect.

PRACTICE EXERCISE 1

Tell how the graphs of the system are related.

$$2x - 3y + 1 = 0$$
$$x + 5y - 10 = 0$$

Answer: They intersect.

EXAMPLE 2   PARALLEL LINES

Tell how the graphs of the system are related.

$$2x - 3y + 1 = 0$$
$$4x - 6y - 3 = 0$$

Solve for y.

$$-3y = -2x - 1 \qquad\qquad -6y = -4x + 3$$

$$\boxed{\left(-\dfrac{1}{3}\right)}(-3y) = \boxed{\left(-\dfrac{1}{3}\right)}(-2x - 1) \qquad \boxed{\left(-\dfrac{1}{6}\right)}(-6y) = \boxed{\left(-\dfrac{1}{6}\right)}(-4x + 3)$$

$$y = \dfrac{2}{3}x + \dfrac{1}{3} \qquad\qquad y = \dfrac{4}{6}x - \dfrac{3}{6}$$

$$y = \dfrac{2}{3}x - \dfrac{1}{2}$$

Since the coefficients of x, the slopes of the lines, are both $\frac{2}{3}$ (equal), the lines are either parallel or coinciding. Since the constants $\frac{1}{3}$ and $-\frac{1}{2}$ are unequal, the lines have two different y-intercepts, namely $\left(0, \frac{1}{3}\right)$ and $\left(0, -\frac{1}{2}\right)$, and the lines do not coincide. Thus, the lines are parallel.

PRACTICE EXERCISE 2

Tell how the graphs of the system are related.

$$5x - 10y + 20 = 0$$
$$-x + 2y - 4 = 0$$

Answer: They coincide.

EXAMPLE 3 COINCIDING LINES

Tell how the graphs of the system are related.

$$x - 2y + 2 = 0$$
$$-3x + 6y - 6 = 0$$

Solve for y.

$$-2y = -x - 2 \qquad\qquad 6y = 3x + 6$$

$$\left(-\frac{1}{2}\right)(-2y) = \left(-\frac{1}{2}\right)(-x - 2) \qquad \frac{1}{6}(6y) = \frac{1}{6}(3x + 6)$$

$$y = \frac{1}{2}x + 1 \qquad\qquad y = \frac{3}{6}x + \frac{6}{6}$$

$$y = \frac{1}{2}x + 1$$

Since both coefficients of x are $\frac{1}{2}$ and both constants are 1, the lines coincide.

PRACTICE EXERCISE 3

Tell how the graphs of the system are related.

$$-3x + y - 5 = 0$$
$$6x - 2y + 3 = 0$$

Answer: They are parallel.

EXAMPLE 4 ONE LINE VERTICAL

Tell how the graphs are related.

$$2x + 3y - 6 = 0$$
$$2x - 6 = 0$$

Since $2x - 6 = 0$ becomes $2x = 6$ or $x = 3$, its graph is a line parallel to the y-axis, 3 units to the right of the y-axis. Since there is a y term in $2x + 3y - 6 = 0$, the graph of this equation cannot be a line parallel to the y-axis. Thus, the two lines must intersect in one point.

PRACTICE EXERCISE 4

Tell how the graphs of the system are related.

$$7x + 3 = 0$$
$$-2x + 5 = 0$$

Answer: They are parallel.

EXAMPLE 5 ONE LINE HORIZONTAL

Tell how the graphs are related.

$$3y + 5 = 0$$
$$4x - 2y + 3 = 0$$

Solve for y.

$$3y = -5 \qquad\qquad -2y = -4x - 3$$

$$\left(\frac{1}{3}\right)(3y) = \left(\frac{1}{3}\right)(-5) \qquad \left(-\frac{1}{2}\right)(-2y) = \left(-\frac{1}{2}\right)(-4x - 3)$$

$$y = -\frac{5}{3} \qquad\qquad y = \frac{4}{2}x + \frac{3}{2}$$

$$y = 0 \cdot x - \frac{5}{3} \qquad\qquad y = 2x + \frac{3}{2}$$

Since the coefficients of x, 0 and 2, are not equal, the lines intersect in one point.

PRACTICE EXERCISE 5

Tell how the graphs of the system are related.

$$2x - 3y + 6 = 0$$
$$5y - 1 = 0$$

Answer: They intersect.

5.1 EXERCISES A

*Complete Exercises 1–4 with one of the words (**a**) parallel, (**b**) coinciding, or (**c**) intersecting.*

1. When two linear equations are both solved for y and the coefficients of x are unequal, the graphs are _____ lines.

2. When two linear equations are both solved for y and the coefficients of x are equal as are the constants, the graphs are _____ lines.

3. If the y term is missing in both of two linear equations, the graphs are (**a**) _____ lines or lines (**b**) _____ to the y-axis.

4. If the y-term is missing in only one of two linear equations, the graphs are _____ lines.

Tell how the graphs of the system are related.

5. $-2x + 3y + 5 = 0$
   $-2x - 3y + 5 = 0$

6. $x - 5y + 2 = 0$
   $2x - 10y + 4 = 0$

7 $2x - 7y + 1 = 0$
   $-6x + 21y + 3 = 0$

8. $2y + x = 5$
   $2x = 10$

9. $x = 3y + 7$
   $y = 3x + 7$

10 $2y = 8$
   $2x = 8$

11. $2x + 1 = 3$
   $4x - 3 = 1$

12. $x + y = 5$
   $3y = 15$

13. $8y - 5 = -3x$
   $-6x + 10 = 16y$

Tell whether $(-3, 2)$ is a solution to the given system.

14 $2x + 3y = 0$
   $x + 8y = 19$

15. $3x + y + 7 = 0$
   $x + 4y - 5 = 0$

16. $3y = 6$
   $x - 5y = -13$

Tell whether $(4, 0)$ is a solution to the given system.

17. $2x - 3y = 8$
   $x - 4 = 0$

18. $x + y - 2 = 0$
   $2y + 2x = 4$

19. $2y + 1 = x$
   $2x - 8 = y$

FOR REVIEW

To help you prepare for graphing systems in the next section, the following exercises review material from Chapter 4. Find the intercepts of each line.

20. $2x - y = 4$

21. $6x + 2y = 5$

22. $3x - 5y + 2 = 0$

ANSWERS: 1. intersecting **2.** coinciding **3.** (a) coinciding (b) parallel **4.** intersecting **5.** intersecting **6.** coinciding **7.** parallel **8.** intersecting **9.** intersecting **10.** intersecting **11.** coinciding **12.** intersecting **13.** coinciding **14.** no **15.** yes **16.** yes **17.** yes **18.** no **19.** no **20.** x-intercept: $(2, 0)$; y-intercept: $(0, -4)$ **21.** x-intercept: $\left(\frac{5}{6}, 0\right)$; y-intercept: $\left(0, \frac{5}{2}\right)$ **22.** x-intercept: $\left(-\frac{2}{3}, 0\right)$; y-intercept: $\left(0, \frac{2}{5}\right)$

5.1 EXERCISES B

Complete Exercises 1–4 with one of the following words: (a) parallel, (b) coinciding, or (c) intersecting.

1. When the slopes of the lines in a system of equations are equal but the y-intercepts are unequal, the graphs are _____ lines.

2. When the slopes of the lines in a system of equations are unequal, the graphs are _____ lines.

3. If the x term is missing in both equations in a system of equations, the graphs are **(a)**_____ lines or are lines **(b)**_____ to the x-axis.

4. If the x term is missing in only one of the equations in a system of equations, the graphs are _____ lines.

Tell how the graphs of the system are related.

5. $-5x + y - 3 = 0$
   $5x - y - 3 = 0$

6. $x + 2y + 3 = 0$
   $-3x - 6y - 9 = 0$

7. $4x - y + 1 = 0$
   $x - 4y - 1 = 0$

8. $2x + 3y = 1$
   $y = 2$

9. $x = 3y - 1$
   $y = -x - 1$

10. $3x = 1$
   $3y = 1$

11. $3y + 1 = 4$
   $-y + 2 = 1$

12. $y - 3x = 11$
   $2x = 8$

13. $x + 2 = y$
   $y + 1 = x$

Tell whether $(1, 0)$ is a solution to the given system.

14. $2x + 3y = 2$
   $-x - 4y = -1$

15. $3x - y = 1$
   $x + y = 1$

16. $3x = 3$
   $-x + 9y = -1$

Tell whether $(-2, 1)$ is a solution to the given system.

17. $x - y = -3$
   $y - 1 = 0$

18. $2x - 3y + 7 = 0$
   $6y - 4x = 14$

19. $y + 2x = 3$
   $2x + 5 = y$

FOR REVIEW

To help you prepare for graphing systems in the next section, the following exercises review material from Chapter 4. Find the intercepts of each line.

20. $x - 3y = 6$

21. $4x + 5y = 8$

22. $2x - 4y + 7 = 0$

5.1 EXERCISES C

1. Find the value of a so that the line with equation $ax - 3y + 7 = 0$ will be parallel to the line with equation $5x + y - 2 = 0$. [*Hint:* Write each equation in slope-intercept form.]

2. Find the value of c so that the line with equation $2x - 3y + c = 0$ will have the same y-intercept as the line with equation $4x + 2y + 3 = 0$.

5.2 SOLVING SYSTEMS OF EQUATIONS BY GRAPHING

1 THE GRAPHING METHOD

Consider the system of equations

$$x + y = 1$$
$$2x - y = 5.$$

Solve each equation for y.

$$y = -x + 1$$
$$y = 2x - 5$$

The slopes are different, so the lines intersect in exactly one point. Since every point on each line corresponds to an ordered pair of numbers that is a solution to that equation, the point of intersection must correspond to the ordered pair that solves both equations. Hence it is the solution to the system. If we graph both equations in the same rectangular coordinate system, as in Figure 5.2, it appears that the point of intersection has coordinates $(2, -1)$. We can check to see if $(2, -1)$ is indeed a solution by substitution.

$$x + y = 1 \qquad\qquad 2x - y = 5$$
$$2 + (-1) \stackrel{?}{=} 1 \qquad\qquad 2(2) - (-1) \stackrel{?}{=} 5$$
$$1 = 1 \qquad\qquad\qquad 4 + 1 \stackrel{?}{=} 5$$
$$5 = 5$$

This method for solving a system is the **graphing method.**

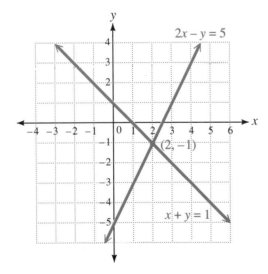

Figure 5.2

2 FINDING THE NUMBER OF SOLUTIONS

If the graphs of the lines in a system are parallel, there can be no point of intersection, so the system has no solution. If the graphs of the lines coincide, then any solution to one equation is also a solution to the other equation, and the system has infinitely many solutions. These observations are summarized in the following statement.

Number of Solutions to a System

A system of equations has:

1. infinitely many solutions if the graphs of the equations coincide;
2. no solution if the graphs of the equations are parallel;
3. exactly one solution if the graphs of the equations intersect in one point.

EXAMPLE 1 INFINITELY MANY SOLUTIONS

Find the number of solutions to the system.

$$3x - 4y + 1 = 0$$
$$8y - 2 = 6x$$

Solve each equation for y.

$$-4y = -3x - 1 \qquad\qquad 8y = 6x + 2$$

$$\left(-\frac{1}{4}\right)(-4y) = \left(-\frac{1}{4}\right)(-3x - 1) \qquad \left(\frac{1}{8}\right)(8y) = \left(\frac{1}{8}\right)(6x + 2)$$

$$y = \frac{3}{4}x + \frac{1}{4} \qquad\qquad y = \frac{6}{8}x + \frac{2}{8}$$

$$y = \frac{3}{4}x + \frac{1}{4}$$

Since the lines coincide (why?), there are infinitely many solutions to the system.

PRACTICE EXERCISE 1

Find the number of solutions to the system.

$$x - 5y + 8 = 0$$
$$3y + 7 = -2x$$

Answer: exactly one

EXAMPLE 2 NO SOLUTION

Find the number of solutions to the system.

$$x - 5y + 1 = 0$$
$$15y - 3x = 1$$

Solve for y.

$$-5y = -x - 1 \qquad\qquad 15y = 3x + 1$$

$$y = \frac{1}{5}x + \frac{1}{5} \qquad\qquad y = \frac{3}{15}x + \frac{1}{15}$$

$$y = \frac{1}{5}x + \frac{1}{15}$$

Since the lines are parallel (why?), there is no solution to the system.

PRACTICE EXERCISE 2

Find the number of solutions to the system.

$$3x + 5y = -7$$
$$6x + 10y - 5 = 0$$

Answer: no solution

EXAMPLE 3 EXACTLY ONE SOLUTION

Find the number of solutions to the system.

$$3x - 5 = 0$$
$$3x + y = -5$$

Since the y term is missing in the first equation, it cannot be solved for y. The graph of $3x - 5 = 0$, however, is a line parallel to the y-axis, so it intersects the line $3x + y = -5$ in only one point (why?) Thus, there is exactly one solution to the system.

PRACTICE EXERCISE 3

Find the number of solutions to the system.

$$4x - 12 = 0$$
$$2x = 6$$

Answer: infinitely many

| EXAMPLE 4   USING THE GRAPHING METHOD | PRACTICE EXERCISE 4 |

Solve the system by graphing.

$$x - 2y = 1$$
$$2x + y = 7$$

We graph each equation by finding its intercepts.

$x - 2y = 1$

x	y
0	$-\dfrac{1}{2}$
1	0

$2x + y = 7$

x	y
0	7
$\dfrac{7}{2}$	0

The point of intersection of the two lines, graphed in Figure 5.3, appears to have coordinates (3, 1). We check by substitution in both equations.

$x - 2y = 1$ $2x + y = 7$

$3 - 2(1) \overset{?}{=} 1$ $2(3) + 1 \overset{?}{=} 7$

$3 - 2 \overset{?}{=} 1$ $6 + 1 \overset{?}{=} 7$

$1 = 1$ $7 = 7$

Thus, the solution to the system is (3, 1).

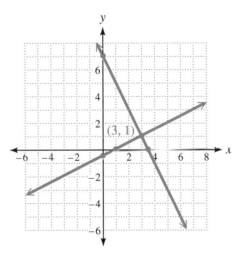

Figure 5.3

Solve the system by graphing.

$$3x - 2y = 12$$
$$2x + 3y = -5$$

Answer: $(2, -3)$

5.2 EXERCISES A

1. How many solutions will a system of equations have if the lines in the system are parallel?

2. How many solutions will a system of equations have if the lines in the system are intersecting?

3. How many solutions will a system of equations have if the lines in the system are coinciding?

Find the number of solutions to the given system. For those that have exactly one solution, find that solution by graphing. Check each solution.

4. $x + 3y = 4$
$-x + y = 0$

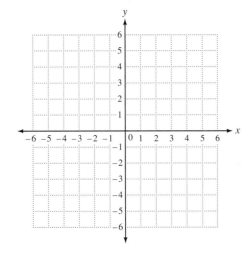

5 $x + 3y = 4$
$-2x - 6y = -8$

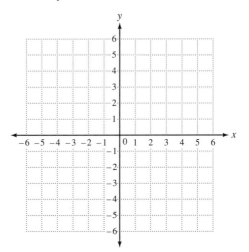

6. $x + 3y = 4$
$-2x - 6y = 0$

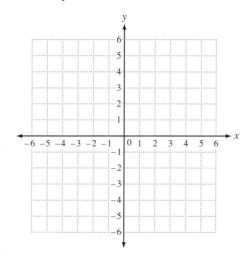

7. $2x + y = 5$
$2x - y = 3$

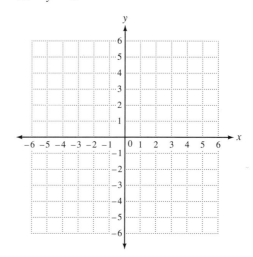

8 $x + y = 3$
$x + 3y = -1$

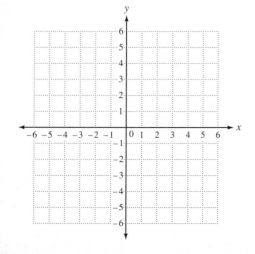

9. $2x + y = 3$
$x + y = 3$

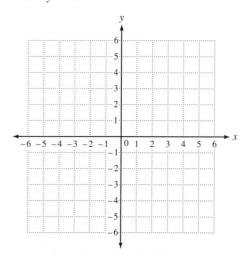

10. $3x - y = 1$
$\quad\ \ 2y = 4$

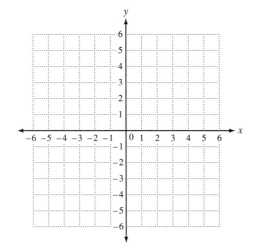

11. $2y = 3$
$\quad\ 3x = 2$

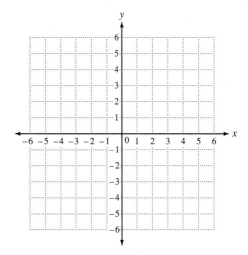

12. $x = y + 1$
$\quad\ \ y = x - 1$

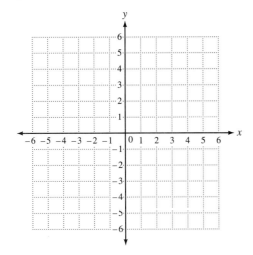

13 $\quad x + 4y = 1$
$\quad -x - 4y = 1$

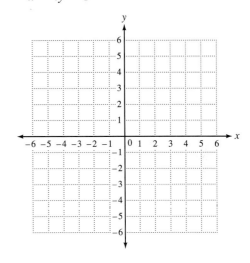

14. $\quad 4x - 3y = 2$
$\quad -2x + \ y = -1$

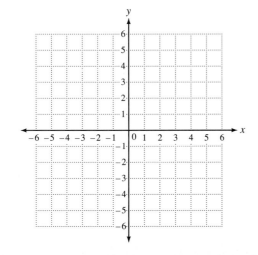

15. $2x + y = 0$
$\quad\ 2x - y = 0$

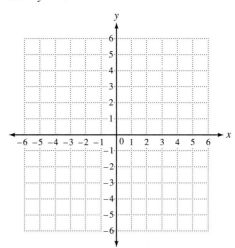

FOR REVIEW

Tell whether (0, 4) is a solution to the given system.

16. $2x + 3y - 12 = 0$
$\quad\;\;\, x - 2y + \;8 = 0$

17. $\;\; x - y = -4$
$\quad\;\; 2x + y = -4$

18. $\qquad\qquad\quad y = 4$
$\qquad 2x - 2y + 8 = 0$

ANSWERS: 1. none   2. one (exactly one)   3. infinitely many   4. one solution: (1, 1)   5. infinitely many solutions
6. no solution   7. one solution: (2, 1)   8. one solution: (5, −2)   9. one solution: (0, 3)   10. one solution: (1, 2)
11. one solution: $\left(\frac{2}{3}, \frac{3}{2}\right)$ 12. infinitely many solutions   13. no solution   14. one solution: $\left(\frac{1}{2}, 0\right)$ 15. one solution: (0, 0)   16. yes   17. no   18. yes

5.2 EXERCISES B

1. If a system of equations has exactly one solution, how are the graphs of the system related?

2. If a system of equations has no solution, how are the graphs of the system related?

3. If a system of equations has infinitely many solutions, how are the graphs of the system related?

Find the number of the solutions to the given system. For those that have exactly one solution, find that solution by graphing. Check each solution.

4. $x + 2y = 1$
$\quad\; x - y = -2$

5. $\;\; x + 2y = 1$
$\quad -x - 2y = -1$

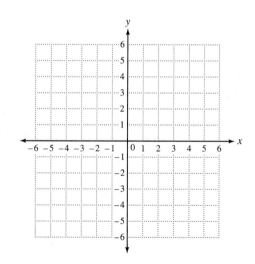

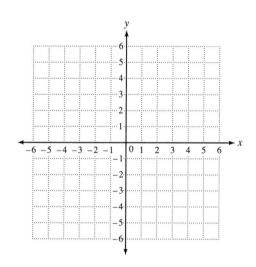

6. $x + 2y = 1$
$-x - 2y = 1$

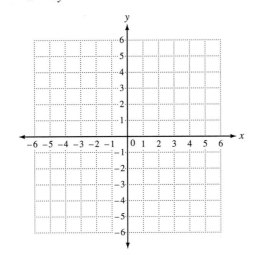

7. $x + 2y = 5$
$x - 2y = -3$

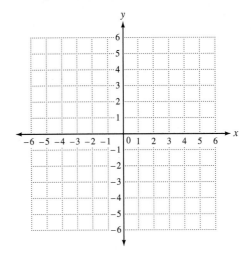

8. $x - y = -4$
$5x + y = 4$

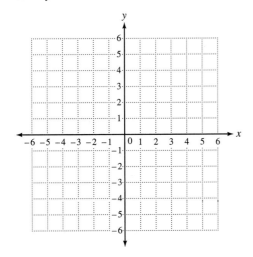

9. $x + y = 2$
$x + 2y = -1$

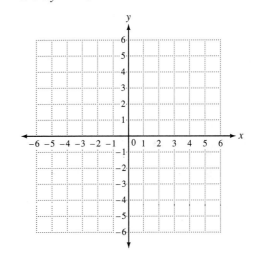

10. $2x + 3y = 6$
$3x = 3$

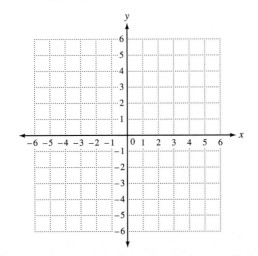

11. $3x = 4$
$4y = 3$

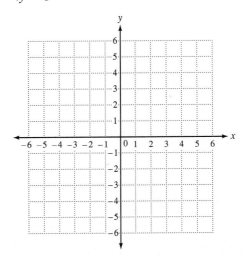

12. $x - 5y = 2$
$-x + 5y = 2$

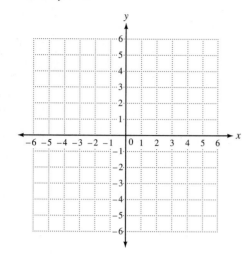

13. $y = x + 2$
$x = y - 2$

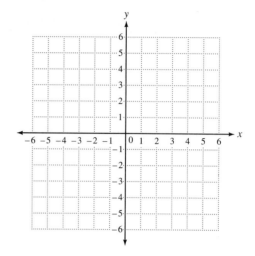

14. $3x - 3y = -1$
$x + 6y = 2$

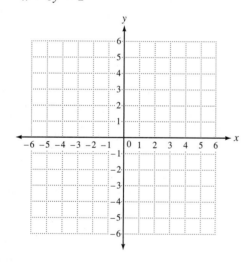

15. $x + 5y = 0$
$x - 5y = 0$

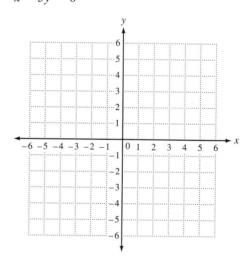

FOR REVIEW

Tell whether $(1, -5)$ *is a solution to the given system.*

16. $2x - y - 7 = 0$
$x + y = -4$

17. $x - 2y = 11$
$2x - 7 = y$

18. $x + 5 = 0$
$y - 1 = 0$

5.2 EXERCISES C

1. Find the value of c so that the system will have infinitely many solutions.

$x - 5y + c = 0$
$-3x + 15y - 7 = 0$ $\left[\text{Answer:} \quad c = \frac{7}{3}\right]$

2. Find the value of a so that the system will have no solution.

$ax - 5y + 8 = 0$
$4x + y - 1 = 0$

5.3 SOLVING SYSTEMS OF EQUATIONS BY SUBSTITUTION

① The Substitution Method   **②** Contradictions and Identities

Solving a system by graphing is time-consuming, and estimating the point of intersection depends on the accuracy of the graph. A better method for solving some systems is the **method of substitution.** Consider the system

$$x - 2y = 1$$
$$2x + y = 7.$$

We will change the system into one equation in one unknown. Solve the first equation for x,

$$x - 2y = 1$$
$$x = 2y + 1,$$

and substitute this value of x into the second equation.

$$2x + y = 7$$
$$2(2y + 1) + y = 7 \qquad x = 2y + 1$$

The result is an equation in the single variable y, which we can solve.

$$4y + 2 + y = 7$$
$$5y + 2 = 7$$
$$5y = 5$$
$$y = 1 \qquad \text{This is the } y\text{-coordinate of the solution}$$

Substitute this value of y into either of the original equations. Using the first, solve for x.

$$x - 2y = 1$$
$$x - 2(1) = 1 \qquad y = 1$$
$$x - 2 = 1$$
$$x = 3 \qquad \text{This is the } x\text{-coordinate of the solution}$$

We check the possible solution pair (3, 1) by substituting it into both original equations.

$$
\begin{array}{ll}
x - 2y = 1 & 2x + y = 7 \\
3 - 2(1) \stackrel{?}{=} 1 & 2(3) + 1 \stackrel{?}{=} 7 \\
3 - 2 \stackrel{?}{=} 1 & 6 + 1 \stackrel{?}{=} 7 \\
1 = 1 & 7 = 7
\end{array}
$$

Thus, the solution to the system is (3, 1), the same result obtained in Example 4 in Section 5.2 using the graphing method.

To Solve a System of Equations Using the Substitution Method

1. Solve one of the equations for one of the variables.
2. Substitute that value of the variable in the *remaining* equation.
3. Solve this new equation and substitute the numerical solution into either of the two *original* equations to find the numerical value of the second variable.
4. Check your solution in both original equations.

EXAMPLE 1 Using Substitution

Solve by the substitution method.

$$5x + 3y = 17$$
$$x + 3y = 1$$

We could solve either equation for either variable. However, since the coefficient of x in the second equation is 1, we can avoid fractions if we solve for x using that equation.

$$x = 1 - 3y$$

Substitute $1 - 3y$ for x in the first equation.

$$5x + 3y = 17$$
$$5(1 - 3y) + 3y = 17$$
$$5 - 15y + 3y = 17$$
$$5 - 12y = 17$$
$$-12y = 12$$
$$y = -1 \quad \textit{y-coordinate of the solution}$$

Substitute -1 for y in the second equation.

$$x + 3y = 1$$
$$x + 3(-1) = 1$$
$$x - 3 = 1$$
$$x = 4 \quad \textit{x-coordinate of the solution}$$

The solution is $(4, -1)$.

Check:
$$5x + 3y = 17 \qquad x + 3y = 1$$
$$5(4) + 3(-1) \overset{?}{=} 17 \qquad (4) + 3(-1) \overset{?}{=} 1$$
$$20 - 3 \overset{?}{=} 17 \qquad 4 - 3 \overset{?}{=} 1$$
$$17 = 17 \qquad 1 = 1$$

PRACTICE EXERCISE 1

Solve by the substitution method.

$$3x + y = 1$$
$$2x + 3y = 10$$

Start by solving the first equation for y and substituting into the second equation.

$$y = -3x + 1$$
$$2x + 3(-3x + 1) = 10$$

Answer: $(-1, 4)$

EXAMPLE 2 System with No Solution

Solve by substitution.

$$4x + 2y = 1$$
$$2x + y = 8$$

If we solve the second equation for y, we obtain

$$y = 8 - 2x.$$

Substitute this value into the first equation.

$$4x + 2y = 1$$
$$4x + 2(8 - 2x) = 1$$
$$4x + 16 - 4x = 1$$
$$16 = 1$$

PRACTICE EXERCISE 2

Solve by substitution.

$$10x + 5y = 2$$
$$-2x - y = 4$$

But $16 \neq 1$. What went wrong? Return to the original system and solve both equations for y.

$$2y = 1 - 4x \quad \text{and} \quad y = 8 - 2x$$
$$y = \frac{1}{2} - \frac{4}{2}x \qquad\qquad y = -2x + 8$$
$$y = -2x + \frac{1}{2}$$

Both coefficients of x are -2 but the constant terms are different. Thus the two lines are parallel, and there is no solution. If there is no solution (the lines are parallel), substitution results in an equation that is a contradiction (such as $16 = 1$).

Answer: no solution

EXAMPLE 3   SYSTEM WITH INFINITELY MANY SOLUTIONS

Solve by substitution.

$$x + 2y = 1$$
$$2y = 1 - x$$

We solve the first equation for x.

$$x = 1 - 2y$$

Substitute this value into the second equation.

$$2y = 1 - (1 - 2y)$$
$$2y = 1 - 1 + 2y \qquad \text{Watch the signs}$$
$$2y = 2y$$

Clearly, $2y = 2y$ no matter what number replaces y so that we end up with an equation that is an identity. Return to the original equations and solve them for y.

$$2y = 1 - x \qquad\qquad 2y = 1 - x$$
$$2y = -x + 1 \qquad\qquad 2y = -x + 1$$
$$y = -\frac{1}{2}x + \frac{1}{2} \qquad\qquad y = -\frac{1}{2}x + \frac{1}{2}$$

The coefficients of x are equal, as are the constants. The two lines coincide so there are infinitely many solutions to this system (any pair of numbers that is a solution to one equation is a solution to both, hence a solution to the system).

PRACTICE EXERCISE 3

Solve by substitution.

$$3x - 6y = 9$$
$$-x + 2y = -3$$

Answer: infinitely many solutions

---

② CONTRADICTIONS AND IDENTITIES

Examples 2 and 3 lead to the following.

> ### When Solving a System of Equations
>
> 1. If you get a **contradiction** (an equation that is never true), the lines are parallel and there is no solution to the system.
>
> 2. If you get an **identity** (an equation that is always true for every value of the variable), the lines coincide and there are infinitely many solutions to the system.

5.3   EXERCISES A

1. Suppose while solving a system of equations using substitution we obtain the equation $3x = 3x$. What would this tell us about the system?

2. Suppose while solving a system using substitution we obtain the equation $5 = 0$. What would this tell us about the system?

Solve the following systems using the substitution method.

3. $y = 3x + 2$
  $5x - 2y = -7$

4. $2x - 3y = 14$
  $x = y + 10$

5. $3x + y = 35$
  $x - 2y = 7$

6. $13x - 4y = -66$
  $5x + 2y = 10$

7 $3x - 5y = 19$
  $2x - 4y = 16$

8. $3x - y = 7$
  $4x - 5y = 2$

9. $x - 3y = 14$
  $x - 2 = 0$
  [*Hint:* Start by solving for x in the second equation.]

10 $3x - 3y = 1$
  $x - y = -1$

11 $2x + 2y = -6$
  $-x - y = 3$

12. $y = 5$
  $4x - y = 19$

13. $2x + 3y = 5$
  $4x + 7y = 11$

14 $3x + 5y = 30$
  $5x + 3y = 34$

15. Which variable in which equation is easiest to solve for in Exercise 14? (The next section explains a method for solving equations that may be better for systems such as this.)

FOR REVIEW

16. If the lines in a system of equations are parallel, how many solutions does the system have?

Solve by the graphing method.

17. $x + y = 5$
  $3x - y = -1$

18. $2x + y = 4$
  $x - 2 = 0$

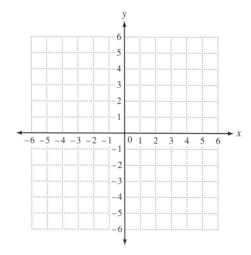

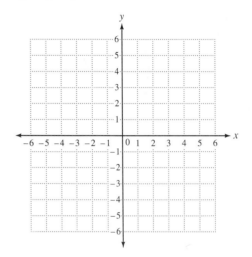

ANSWERS: 1. There are infinitely many solutions; any solution to one equation is a solution to the other and therefore a solution to the system.   2. There are no solutions.   3. (3, 11); first equation is already solved for y 4. (16, 6); second equation is already solved for x 5. (11, 2); better to solve for x in second equation   6. (−2, 10)   7. (−2, −5)   8. (3, 2)   9. (2, −4)   10. no solution   11. infinitely many solutions   12. (6, 5)   13. (1, 1); neither variable is better to solve for than the other   14. (5, 3)   15. Neither variable is better to solve for than the other.   16. no solution   17. (1, 4)   18. (2, 0)

5.3 EXERCISES B

1. Suppose while solving a system of equations using substitution we obtain the equation $-3 = 0$. What would this tell us about the system?

2. Suppose while solving a system using substitution we obtain the equation $x + 1 = x + 1$. What would this tell us about the system?

Solve the following systems using the substitution method.

3. $y = -2x + 3$
  $3x - 4y = -1$

4. $5x - y = -2$
  $x = 3y - 6$

5. $x + y = 0$
  $5x + 2y = -3$

6. $2x - 8y = -10$
  $x - 3y = -2$

7. $2x - y = 3$
  $-4x + 2y = -6$

8. $4x - y = 13$
  $x + 2y = 1$

9. $3x - y = 2$
$-6x + 2y = 2$

10. $2x - y = -2$
$y - 4 = 0$

11. $4x + y = -2$
$-2x - 3y = 1$

12. $3x - y = 14$
$x - 4 = 0$

13. $2x + 3y = -4$
$3x - 2y = 7$

14. $3x + 7y = 32$
$7x + 3y = 8$

15. What is the difficulty in using the substitution method on a system like the one in Exercise 14?

FOR REVIEW

16. If the lines in a system of equations are coinciding, how many solutions does the system have?

Solve by the graphing method.

17. $3x - y = -4$
$-x + y = 0$

18. $x - 3y = 3$
$y + 1 = 0$

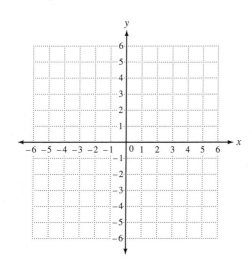

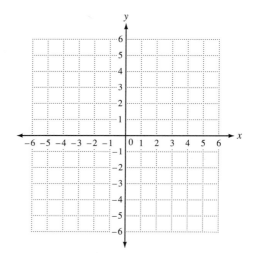

5.3 EXERCISES C

 Solve using the substitution method.

1. $0.03x - 0.01y = 0.12$
$0.04x - 0.05y = 0.20$

2. $\dfrac{x}{10} + \dfrac{2y}{5} = 1$

$\dfrac{2x}{5} + \dfrac{3y}{2} = 2$ [Answer: $(-70, 20)$]

5.4 SOLVING SYSTEMS OF EQUATIONS BY ELIMINATION

1 THE ELIMINATION METHOD

Consider the following system.

$$3x + 5y = 30$$
$$5x + 3y = 34$$

Solving for either variable in either equation will result in fractional coefficients. Another method, better than substitution for systems like this one, is known as the **elimination method.** It is based on the addition-subtraction rule from Chapter 3: if the same quantity is added to or subtracted from both sides of an equation, another equation with the same solutions is obtained. Suppose we illustrate the method using the following system.

$$x + y = 6$$
$$3x - y = 2.$$

In the second equation, $3x - y$ and 2 are both names for the same number, 2. Therefore we can add $3x - y$ to the left side of the first equation, and 2 to the right of it, to obtain another equation with the same solution.

$$(x + y) + (3x - y) = 6 + 2$$
$$4x = 8 \quad y \text{ has been eliminated}$$

Generally, we add the equations vertically.

$$
\begin{array}{rl}
x + y = 6 & \\
\underline{3x - y = 2} & \\
4x \quad\;\; = 8 & \\
x = 2 & \text{Divide by 4 to obtain the } x\text{-coordinate} \\
& \text{of the solution}
\end{array}
$$

By adding the two equations, we obtain one equation in one variable, x. Once we know x is 2, we substitute this value into either one of the original equations (just as in the substitution method) to find the value of y.

$$2 + y = 6 \quad \text{Substitute in the first equation since it is simpler}$$
$$y = 4 \quad \text{The } y\text{-coordinate of the solution}$$

Thus, the solution is (2, 4). Check this by substituting into both equations.

EXAMPLE 1 USING THE ELIMINATION METHOD	PRACTICE EXERCISE 1

Solve by the elimination method.

$$x + y = 5$$
$$-2x + y = -4$$

Solve by the elimination method.

$$4x + 3y = -1$$
$$2x - 3y = 13$$

If we add both equations, the resulting equation still has both variables. (Try it.) So in this case, we subtract instead of adding, to obtain an equation in one variable.

$$
\begin{array}{r}
x + y = 5 \\
+2x - y = 4 \\
\hline
3x = 9 \\
x = 3
\end{array}
$$

Change all signs when subtracting and add

Substitute 3 for x in the first equation.

$$
\begin{aligned}
3 + y &= 5 \\
y &= 2
\end{aligned}
$$

The solution is the ordered pair $(3, 2)$. Check by substitution.

Answer: $(2, -3)$

② MULTIPLICATION RULE

Sometimes addition or subtraction does not yield one equation in only one variable. When this occurs, we may have to multiply one (or both) of the equations by a number so that the coefficients of one variable are negatives of each other. We then add to eliminate that variable.

EXAMPLE 2   USING MULTIPLICATION AND ELIMINATION

Solve by the elimination method.

$$
\begin{aligned}
2x - 5y &= 13 \\
x + 2y &= 11
\end{aligned}
$$

We would not obtain one equation with one variable by simply adding or subtracting in this case. However, if we multiply both sides of the second equation by -2 the resulting equation is

$$-2x - 4y = -22. \quad \text{-2 times second equation}$$

Now add this equation to the first equation.

$$
\begin{array}{r}
2x - 5y = 13 \\
-2x - 4y = -22 \\
\hline
-9y = -9 \\
y = 1
\end{array}
$$

Substitute 1 for y in the second equation.

$$
\begin{aligned}
x + 2(1) &= 11 \\
x + 2 &= 11 \\
x &= 9
\end{aligned}
$$

The solution is $(9, 1)$. Check by substitution.

Sometimes the multiplication rule has to be applied to both equations before adding or subtracting. In Example 3 this technique is used to solve the system at the beginning of this section.

EXAMPLE 3   USING MULTIPLICATION AND ELIMINATION

Solve by the elimination method.

$$3x + 5y = 30$$
$$5x + 3y = 34$$

Multiply the first equation by -5 and the second equation by 3 (making the coefficients of x negatives of each other).

$$-15x - 25y = -150 \qquad \text{-5 times first equation}$$
$$15x + 9y = 102 \qquad \text{3 times second equation}$$

Add to get one equation in one variable.

$$-16y = -48$$
$$y = 3$$

Substitute 3 for y in the first equation.

$$3x + 5(3) = 30$$
$$3x + 15 = 30$$
$$3x = 15$$
$$x = 5$$

The solution is (5, 3). Check by substitution.

PRACTICE EXERCISE 3

Solve by the elimination method.

$$2x + 7y = -4$$
$$5x + 3y = 19$$

Answer: $(5, -2)$

Compare this example with your work on Exercise 14-A in the last section. In this case the elimination method is faster and easier to use than the substitution method.

To Solve a System by the Elimination Method

1. Write both equations in the form $ax + by = c$.

2. If necessary, apply the multiplication rule to each equation so that addition or subtraction will eliminate one of the variables.

3. Solve the resulting single-variable equation. Substitute this value into one of the original equations. The resulting pair of numbers is the ordered-pair solution to the system.

4. Check your answer by substitution in both original equations.

EXAMPLE 4   CONTRADICTIONS AND IDENTITIES

Solve by the elimination method.

(a) $2x = 3y + 5$
$\ -2x + 3y = -1$

Rewrite the first equation as $2x - 3y = 5$. Notice that by adding the equations, we can eliminate x.

$$2x - 3y = 5$$
$$\underline{-2x + 3y = -1}$$
$$0 = 4$$

PRACTICE EXERCISE 4

Solve by the elimination method.

(a) $-3x + 9y = -18$
$\ \ x - 3y = 6$

We actually eliminated both variables and obtained a contradiction. Just as with the substitution method, this means there is no solution to the system. Verify that the lines are parallel by solving each equation for y.

(b) $2x + y = 3$
$\qquad -4x - 2y = -6$

(b) $\qquad 8x = -4y + 12$
$\qquad\quad 4x + 2y = -3$

Multiply the first equation by 2 and add.

$$
\begin{array}{r}
4x + 2y = 6 \\
-4x - 2y = -6 \\
\hline
0 = 0
\end{array}
$$

Again, both variables are eliminated by this process, but this time we obtain an identity. This means there are infinitely many solutions to the system. Verify that the lines coincide by solving each equation for y.

Answers: (a) infinitely many solutions   (b) no solution

---

❸ METHOD TO USE

The question is often asked: "Which method should I use?" If a coefficient of one of the variables is 1 (or -1), it may be better to solve first for that variable using the substitution method. If none of the coefficients is 1 (or -1), it is probably better to use the elimination method. Keep in mind that with either method, the system has no solution if at any step a contradiction is obtained, and the system has infinitely many solutions if at any step an identity is obtained.

5.4   EXERCISES A

1. If we obtain an equation such as $0 = -3$ when using the elimination method, how many solutions does the system have?

2. If we obtain an equation such as $0 = 0$ when using the elimination method, how many solutions does the system have?

Solve the following systems using the elimination method. Check by substitution.

3. $x + y = 6$
$\quad x - y = 4$

4. $3x - 2y = 12$
$\quad 2x + 2y = 13$

5 $6x + 5y = 11$
$\quad 3x - 7y = -4$

6. $x - \quad y = 16$
$\quad x - 3y = 2$

7. $9x - y = 19$
$3x + 7y = -1$

8. $4x + y = -3$
$-8x - 2y = 6$

9. $-2y = 2 - x$
$3x - 6y = -4$

10 $2x + 3y = -4$
$1 + 2y = -3x$

11 $3x + 7y = -21$
$7x + 3y = -9$

12. $5x - 2y = 1$
$3x + 7y = -24$

Decide whether the substitution or the elimination method is more appropriate, then use it to solve the system.

13. $x + 5y = 0$
$x - 5y = 20$

14. $x - 4y = -6$
$3x - 5y = -4$

15. $4x - 5y = 9$
$5x + 4y = 1$

16 $3x + 3y = 3$
$4x + 4y = -3$

17 $2x + 3y = -4$
$5x + 7y = -10$

18. $3x = 9$
$x - y = 7$

19. $y = 3x + 1$
$4x + 5y = 24$

20. $y - 5 = 0$
$x + 2 = 0$

FOR REVIEW

Exercises 21–25 review material from Section 3.4 to help you prepare for the next section. Use x for the variable and translate each sentence into symbols. Do not solve.

21. The sum of a number and 5 is 12.

22. Four times my age in 2 years is 60.

23. The measure of an angle is 10° less than twice its measure.

24. 40% of the cost of an item is $20.

25. $60 is the sale price of an item discounted by 20%.

ANSWERS: 1. no solution   2. infinitely many   3. (5, 1)   4. $\left(5, \frac{3}{2}\right)$ 5. (1, 1)   6. (23, 7)   7. (2, −1)   8. infinitely many solutions   9. no solution   10. (1, −2)   11. (0, −3)   12. (−1, −3)   13. (10, −2)   14. (2, 2)   15. (1, −1)   16. no solution   17. (−2, 0)   18. (3, −4)   19. (1, 4)   20. (−2, 5)   21. $x + 5 = 12$ 22. $4(x + 2) = 60$ 23. $x = 2x - 10$ 24. $0.40x = 20$ 25. $60 = x - 0.20x$

5.4 EXERCISES B

1. When solving a system of equations using the elimination method, if a contradiction results, how many solutions does the system have?

2. When solving a system of equations using the elimination method, if an identity results, how many solutions does the system have?

Solve the following systems using the elimination method. Check by substitution.

3. $x + y = 3$
$x - y = -1$

4. $2x - 3y = -7$
$3x + 3y = 27$

5. $x - y = 0$
$x - 5y = -12$

6. $3x - 2y = 10$
$5x + 3y = 4$

7. $x + 3y = 1$
$3x + y = 11$

8. $2x + 3y = -4$
$3x - 2y = 7$

9. $5x - 2y = 1$
$7 + 10x = 4y$

10. $2x - 2y = 5$
$4y = 3x - 7$

11. $4x - 2y = 2$
$-2x + y = -1$

12. $3x + 2y = -10$
$5x - 3y = 15$

Decide whether the substitution or the elimination method is more appropriate, then use it to solve the system.

13. $x + 3y = 4$
$x + 8y = -6$

14. $x - 7y = 44$
$5x + 3y = -8$

15. $5x + 7y = -2$
$7x + 5y = 2$

16. $3x - 7y = -4$
$7x + 3y = 10$

17. $2x = 8$ **18.** $2x + 5y = -10$ **19.** $x = 4y - 3$ **20.** $y + 7 = 0$
       $x - y = 1$ $3x + 11y = -22$ $2x + 3y = 5$ $x - 3 = 0$

FOR REVIEW

Exercises 21–25 review material from Section 3.4 to help you prepare for the next section. Use x for the variable and translate each sentence into symbols. Do not solve.

21. A number less 2 is 14.

22. Twice a number added to 6 is 28.

23. The measure of an angle is 25° less than three times its measure.

24. 15% of the population is 3000.

25. Her salary plus a 12% increase is $22,400.

5.4 EXERCISES C

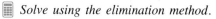 *Solve using the elimination method.*

1. $0.015x - 0.007y = 0.032$
       $-0.009x + 0.014y = -0.001$
       $\left[\text{Answer: } \left(3, \frac{13}{7}\right)\right]$

2. $-\dfrac{3x}{4} + \dfrac{7y}{10} = \dfrac{1}{5}$

   $\dfrac{5x}{2} - \dfrac{4y}{3} = -\dfrac{1}{12}$

5.5 APPLICATIONS USING SYSTEMS

STUDENT GUIDEPOSTS

1 Problem-Solving Techniques   **3** Motion Problems
2 Mixture Problems

1 PROBLEM-SOLVING TECHNIQUES

Many applied problems involving two unknown quantities can be translated into a system of linear equations in two variables. The following steps may be helpful and are illustrated in the examples.

> **To Solve an Applied Problem Using a System of Equations**
>
> 1. Read the problem several times and try to determine what is being asked. Let x represent one of the unknowns and y the other.
> 2. Write two different equations using the variables x and y. Simplify the equations (if necessary) by reducing the size of the numerical coefficients and eliminating fractions or decimals.
> 3. Solve the system of equations.
> 4. Check to see that the answers actually solve the original problem.

| **EXAMPLE 1** NUMBER PROBLEM | **PRACTICE EXERCISE 1** |

The sum of two numbers is 21 and their difference is 9. Find the numbers.

Let x = the larger number,
y = the smaller number.

Then
$x + y = 21$ Represents "their sum is 21"
$x - y = 9$ Represents "their difference is 9"

We should solve the system by the simplest method. In this case, one method is as easy as the other so we add the two equations.

$$2x = 30$$
$$x = 15$$

Now substitute 15 for x in $x + y = 21$.

$$15 + y = 21$$
$$y = 6$$

The numbers are 15 and 6. Check.

If twice one number is added to a second number, the result is 25. If the first number is the larger of the two and their difference is 5, find the numbers.

Let x = the first number,
y = the second number.

Answer: 10, 5

| **EXAMPLE 2** PROBLEM INVOLVING AGES | **PRACTICE EXERCISE 2** |

The sum of Bob's and Joe's ages is 14. In 2 years Bob will be twice as old as Joe. What are their present ages?

Let x = Bob's present age,
y = Joe's present age.

Then both Bob and Joe will age 2 years and

$$x + 2 = \text{Bob's age in 2 years,}$$
$$y + 2 = \text{Joe's age in 2 years.}$$

In two years Bob will be twice as old as Joe

$$(x + 2) = 2 (y + 2)$$

$$x + 2 = 2(y + 2)$$
$$x + 2 = 2y + 4$$
$$-2 = 2y - x \quad \text{or} \quad -x + 2y = -2$$

Since the sum of their ages is 14, $x + y = 14$. This gives us the following system.

$$-x + 2y = -2$$
$$x + y = 14$$

Add these equations. $3y = 12$
$$y = 4$$

Now substitute 4 for y in $x + y = 14$.

$$x + 4 = 14$$
$$x = 10$$

Bob is 10 and Joe is 4. Check.

Annette is $\frac{3}{4}$ as old as Mona. How old is each if the sum of their ages is 84?

Let x = Annette's age,
y = Mona's age,
$\frac{3}{4}y$ = three-fourths Mona's age.

Answer: Annette is 36, Mona is 48.

| EXAMPLE 3 GEOMETRY PROBLEM | PRACTICE EXERCISE 3 |

Two angles are **supplementary** (their sum is 180°). If one is 60° more than twice the other, find the angles.

Two angles are **complementary** (their sum is 90°). If one is 12° less than twice the other, find the angles.

    Let x = one angle,
        y = the other angle.

Then $x + y = 180.$ The angles are supplementary

If x is 60° more than $2y$, then we would add 60° to $2y$ to get x. This gives the following system.

$$x + y = 180$$
$$2y + 60 = x$$

Substitute $2y + 60$ for x in $x + y = 180$.

$$(2y + 60) + y = 180$$
$$3y + 60 = 180$$
$$3y = 120$$
$$y = 40$$

Now substitute 40 for y in $x + y = 180$.

$$x + 40 = 180$$
$$x = 140$$

The angles are 40° and 140°. Check.

Answer: 34°, 56°

❷ MIXTURE PROBLEMS

A useful application of systems is the **mixture problem.** The two equations in the system of the mixture problem are the *quantity equation* and the *value equation.*

    To set up a value equation, we need to look at the idea of *total value*. For a given number of units, each of equal value, the total value can be computed by

        total value = (value per unit)(number of units).

    The following are examples of total value calculations.

Price Given Per Pound
Find the value of 5 lb of candy worth $1.10 per lb.

$$\text{total value} = (\text{value per lb})(\text{number of lb})$$
$$= (\$1.10)(5)$$
$$= \$5.50$$

Value of Coins
Find the value of 30 nickels.

$$\text{total value} = (\text{value per nickel})(\text{number of nickels})$$
$$= (5¢)(30)$$
$$= 150¢ \quad \text{or} \quad \$1.50$$

    The next examples are applications of the total value idea.

Percent Solutions

Find the amount of acid in 16 liters of 20% acid solution.

$$\text{total value} = \text{amount of acid in solution}$$
$$= (\text{percent acid})(\text{liters of solution})$$
$$= (0.20)(16)$$
$$= 3.2 \text{ liters of acid}$$

Interest

Find the annual interest on $8000 invested at 6.5% simple interest.

$$\text{total value} = \text{amount of interest}$$
$$= (\text{percent interest})(\text{amount invested})$$
$$= (0.065)(\$8000)$$
$$= \$520$$

These examples will be helpful in solving the following mixture problems.

EXAMPLE 4   MIXTURE OF CANDY AND NUTS

A man wishes to mix candy selling for 75¢ a pound with nuts selling for 50¢ per pound to obtain a party mix to be sold for 60¢ per pound. How many pounds of each must be used to obtain 50 pounds of the mixture?

Let x = number of pounds of candy,
y = number of pounds of nuts.

The first equation we set up is the quantity equation

$$(\text{no. lb of candy}) + (\text{no. lb of nuts}) = (\text{no. lb of mixture}),$$

which translates to $x + y = 50.$

Next we find the value equation. Since the value of a quantity equals the sum of the values of its parts,

$$(\text{total value of candy}) + (\text{total value of nuts}) = \text{total value of mixture}.$$

$$75 \cdot x = \text{value of candy}$$
$$50 \cdot y = \text{value of nuts}$$
$$60 \cdot 50 = \text{value of mixture}$$

This gives the value equation

$$75x + 50y = 60 \cdot 50.$$

We need to solve the following system of equations.

$$x + \quad y = 50$$
$$75x + 50y = 3000$$

Solve the first equation for x,

$$x = 50 - y,$$

and substitute in the second equation

$$75\,(50 - y) + 50y = 3000$$
$$3750 - 75y + 50y = 3000$$
$$-25y = -750$$
$$y = 30$$

PRACTICE EXERCISE 4

Dr. Lynn Mowen wants to mix a 10% alcohol solution and a 70% alcohol solution to obtain 30 liters of 50% alcohol solution. How much of each must she use to obtain the desired mixture?

Let x = liters of 10% solution to use,
y = liters of 70% solution to use.

$$x + y = \underline{\hspace{2cm}}$$
$$0.10x + 0.70y = 0.50(30)$$

Now substitute 30 for y in $x = 50 - y$.

$$x = 50 - 30 = 20$$

The man must use 20 lb of candy and 30 lb of nuts.

Answer: 10 liters of 10% and 20 liters of 70%

EXAMPLE 5   MIXTURE OF COINS

A boy has 25 nickels and dimes. If the coins' total value is $2.10, find the number of nickels and the number of dimes in the collection.

Let x = number of nickels,
  y = number of dimes.

The quantity equation

  (no. of nickels) + (no. of dimes) = (no. coins in the collection)

translates to $x + y = 25$.

The value equation

  (value of nickels) + (value of dimes) = (value of collection)

becomes $5 \cdot x + 10 \cdot y = 210$.   Convert to cents

Solve the first equation for y,

$$y = 25 - x,$$

and substitute into the second.

$$5x + 10\,(25 - x) = 210$$
$$5x + 250 - 10x = 210$$
$$-5x = -40$$
$$x = 8$$

Now substitute 8 for x in $y = 25 - x$.

$$y = 25 - 8 = 17$$

There are 8 nickels and 17 dimes in the collection.

Jose Garcia invests a total of $8000 in two accounts, one paying 10% simple interest and the other 12% simple interest. How much did he invest in each account if his total interest for the year was $930?

Answer: $1500 at 10% and $6500 at 12%

---

③ MOTION PROBLEMS

Another type of problem that can be solved using systems is the **motion problem.**

EXAMPLE 6   MOTION PROBLEM

A boat travels 60 mi upstream in 4 hr and returns to the starting place in 3 hr. What is the speed of the boat in still water, and what is the speed of the stream?

Let x = speed of the boat in still water,
  y = speed of the stream,
  $x - y$ = speed of the boat going upstream (against current),
  $x + y$ = speed of the boat going downstream (with current).

A boat can travel 30 mi upstream in 3 hr and return in 2 hr. What are the speed of the boat in still water and the speed of the stream?

Since $d = rt$ (distance equals rate times time), we have the following system.

$$60 = (x - y)(4)$$
$$60 = (x + y)(3)$$

Here we can simplify the system by dividing the first equation by 4 and the second equation by 3.

$$
\begin{aligned}
x - y &= 15 \\
\underline{x + y} &= \underline{20} \\
2x &= 35 \qquad \text{Add the equations} \\
x &= \frac{35}{2} \\
x &= 17.5
\end{aligned}
$$

Substituting 17.5 into $x + y = 20$ gives

$$17.5 + y = 20$$
$$y = 2.5.$$

Thus, the speed of the boat is 17.5 mph, and the speed of the stream is 2.5 mph.

Answer: speed of boat is 12.5 mph, speed of stream is 2.5 mph

5.5   EXERCISES A

Solve. (Some problems have been started.)

1. The sum of two numbers is 39 and their difference is 13. Find the numbers.

Let x = the first number,
    y = the other number.

$x + y =$
$x - y =$

2. Mike is 2 years younger than Burford. Four years from now the sum of their ages will be 36. How old is each now?

Let x = Burford's present age,
    y = Mike's present age,

$x + 4$ = Burford's age in 4 years,
$y + 4$ = Mike's age in 4 years.

3. Two angles are supplementary (their sum is 180°) and one is 5° more than six times the other. Find the angles.

4. If three times the smaller of two numbers is increased by five times the larger, the result is 195. Find the two numbers if their sum is 47.

Let x = the smaller number,
    y = the larger number

$3x + 5y =$

5. Mr. Smith has two sons. If the sum of their ages is 19 and the difference between their ages is 3, how old are his sons?

6. Bill is three times as old as Jim. Find the age of each if the sum of their ages is 24.

7. One-third of one number is the same as twice another. The sum of the first and five times the second is 33. Find the numbers.

8. Two angles are *complementary* (their sum is 90°) and their difference is 66°. Find the angles.

9. A 30-ft rope is cut into two pieces. One piece is 8 ft longer than the other. How long is each?

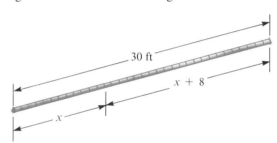

10. The starting players on a basketball team scored 40 points more than the reserves. If the team scored a total of 90 points, how many points did the starters score?

11. A candy mix sells for $1.10 per lb. If it is composed of two kinds of candy, one worth 90¢ per lb and the other worth $1.50 per lb, how many pounds of each would be in 30 lb of the mixture?

12. A collection of nickels and dimes is worth $2.85. If there are 34 coins in the collection, how many of each are there?

Let x = number of nickels,
  y = number of dimes,
   $5x$ = value of nickels (in cents),
   $10y$ = value of dimes (in cents).

$5x + 10y =$
$x + y =$

13. A collection of 13 coins consists of dimes and quarters. If the value of the collection is $2.80, how many of each are there?

14. A man wishes to mix two grades of nuts selling for 48¢ and 60¢ per lb. How many pounds of each must he use to get a 20-lb mixture that sells for 54¢ per lb? (Does your answer seem reasonable?)

15. There were 450 people at a play. If the admission price was $2.00 for adults and 75¢ for children and the receipts were $600.00, how many adults and how many children were in attendance?

16. A painter needs to have 24 gallons of paint that contains 30% solvent. How many gallons of paint containing 20% solvent and how many gallons containing 60% solvent should be mixed to obtain the desired paint?

17 When riding his bike against the wind, Terry Larsen can go 40 mi in 4 hr. He takes 2 hr to return riding with the wind. What are Terry's average speed in still air and the average wind speed?

18. Victoria Harris can ride her bike 120 mi with the wind in 5 hr and return against the wind in 10 hr. What are her speed in still air and the speed of the wind?

19 Vince McKee has $10,000 to invest, part at 10% simple interest, and the rest at 12% simple interest. If he wants to have $1160 in interest at the end of the year, how much should he invest at each rate?

20. Wendy Young has $20,000 invested. Part of the money earns 9% simple interest and the rest earns 11% simple interest. If her interest income for one year was $2060, how much did she have invested at each rate?

21 International Car Rentals charges a daily fee plus a mileage fee. Jack Pritchard was charged $82.00 for 3 days and 400 mi, while Bill Poole was charged $120.00 for 5 days and 500 mi. What are the daily rate and the mileage rate?

22. Harder-Than-Ever Car Rentals charged Terry McGinnis $76.00 for 2 days and 300 mi. For the same type of car, Linda Youngman was charged $132.00 for 6 days and 100 mi. What are the daily fee and the mileage fee?

FOR REVIEW

Solve each system using either the elimination or the substitution method, whichever seems most appropriate.

23. $2x - 3y = 20$
$7x + 5y = -54$

24. $x - 11y = -5$
$7x + 13y = -35$

ANSWERS: 1. 13, 26 2. Burford is 15, Mike is 13. 3. 25°, 155° 4. 20, 27 5. 11 yr, 8 yr 6. 6 yr, 18 yr
7. 18, 3 8. 12°, 78° 9. 11 ft, 19 ft 10. 65 11. 10 lb ($1.50), 20 lb (90¢) 12. 11 nickels, 23 dimes 13. 3 dimes,
10 quarters 14. 10 lb (48¢), 10 lb (60¢) 15. 210 adults, 240 children 16. 18 gal of 20%, 6 gal of 60% 17. Terry:
15 mph, wind: 5 mph 18. Victoria: 18 mph, wind: 6 mph 19. $2000 at 10%, $8000 at 12% 20. $7000 at 9%,
$13,000 at 11% 21. $14 per day, 10¢ per mile 22. $20 per day, 12¢ per mile 23. (−2, −8) 24. (−5, 0)

5.5 EXERCISES B

Solve.

1. If three times the smaller of two numbers is increased by two times the larger, the result is 16. Find the two numbers if their sum is 7.

2. Julie is 4 years older than Randy. Five years from now the sum of their ages will be 24. How old is each now?

3. Two angles are supplementary (their sum is 180°) and one is 15° more than twice the other. Find the angles.

4. If four times the larger of two numbers is decreased by six times the smaller, the result is 4. Find the two numbers if they sum up to 16.

5. Tim Chandler has two sons. If the sum of their ages is 30 and the difference between their ages is 4, how old are his sons?

6. Sherry Maducci is five times as old as Mac Davis. Find the age of each if the sum of their ages is 42.

7. One half of one number is the same as three times a second. The sum of twice the second and the first is 80. Find the numbers.

8. Two angles are complementary (their sum is 90°) and one is 15° more than four times the other. Find each angle.

9. A 45-inch wire is cut into two pieces. One piece is 11 inches longer than the other. How long is each?

10. The starting five on the Memphis State basketball team scored 35 more points than the reserves. If the team scored a total of 85 points, how many points did the starters score?

11. A collection of nickels and dimes is worth $2.00. If the number of dimes minus the number of nickels is 5, how many of each are there?

12. A man wishes to make a party mix consisting of nuts worth 50¢ per lb and candy worth 80¢ per lb. How many pounds of each must he use to make 100 lb of the mixture that will sell for 68¢ per lb?

13. Mark Bonnett has 40 coins in his piggy bank, consisting of dimes and quarters. If the value of the collection is $5.80, how many of each are in the bank?

14. A grocer mixes two types of candy selling for 50¢ and 70¢ per lb. How many pounds of each must he use to obtain a 30-lb mixture that sells for 60¢ per lb?

15. There were 12,000 people at a rock concert at Colorado State University. If the admission price was $7.00 for a student with an I.D. and $10.00 for a nonstudent, how many of each were in attendance if the total receipts amounted to $94,500?

16. Dr. Karen Winter needs 20 liters of 70% alcohol solution in her clinic. How many liters of 40% solution and how many liters of 90% solution must be mixed to obtain the desired solution?

17. While driving his boat upstream, Charlie Moore can go 75 mi in 5 hr. It takes him 3 hr to make the return trip downstream. What are the average speed of the boat in still water and the average stream speed?

18. Maria Lopez traveled 60 mi downstream in 2 hr but needed 5 hr to go back upstream to her starting point. What are the average speed of her boat in still water and the average speed of the stream?

19. Grady and Barbara wish to invest $5000, part at 8% simple interest, and the rest at 14% simple interest. If they must earn $610 by the end of one year, how much should they invest at each rate?

20. A student has $1200 invested. Part of this money earns 12% simple interest, and the rest earns 10% simple interest. How much is invested at each rate if the interest income for the year will be $140?

21. Horn Rentals charges a daily fee and a mileage fee to rent a truck. Syd was charged $375 for 5 days and 1500 mi while Barry was charged $165 for 3 days and 600 mi. What are the daily rate and the mileage rate?

22. Better-Than-Ever Car Rentals charges a daily fee and a mileage fee. Renee paid $110 for 10 days and 200 mi, while Pamela rented the same type of car and paid $168 for 6 days and 800 mi. What are the daily fee and the mileage fee?

FOR REVIEW

Solve the system using either the elimination or the substitution method, whichever seems most appropriate.

23. $3x + 2y = 2$
$5x - 2y = 30$

24. $4x - y = 6$
$3x + 5y = -30$

5.5   EXERCISES C

Solve. A calculator would be helpful in these exercises.

1. On a Pacific island Angie is charged $157.92 after driving a rented car for 420 km in 5 days. Andy paid $217.56 for 7 days and 560 km. If each bill includes a 5% pleasure tax, what was the daily fee and what was the fee charged per km? [*Hint:* Find the fees before tax first.]

2. John and Pat Woods invested $22,000, part at 14% simple interest and the rest at 9% simple interest. How much was invested at 14% if the total interest received in three years was $8040?
[Answer: $14,000]

CHAPTER 5 REVIEW

KEY WORDS

5.1 A **system of equations** is a pair of linear equations.

A **solution** to a system of equations is an ordered pair of numbers (x, y) that is a solution to both equations.

Coinciding lines are the graphs of equations that represent the same line.

Parallel lines are lines that do not intersect.

Intersecting lines have exactly one point in common.

5.5 A **mixture problem** involves two quantities mixed together and is solved by finding a quantity equation and a value equation.

A **motion problem** makes use of the equation $d = rt$, distance equals rate times time.

KEY CONCEPTS

5.1, A system of equations has:
5.2 **1.** infinitely many solutions if the lines coincide,

2. no solution if the lines are parallel,

3. exactly one solution if the lines intersect.

5.2 The method of graphing is not usually used to solve a system since it is time-consuming and depends on our ability to graph accurately.

5.3, 5.4

1. To solve a system of equations, use either the substitution method or the elimination method, whichever seems more appropriate.

2. If you reach an identity (such as $0 = 0$) when solving a system, the system has infinitely many solutions.

3. If you reach a contradiction (such as $3 = 0$) when solving a system, the system has no solution.

4. When solving a system, be sure to find the value of both variables, not just one of them. For example, when solving

$$
\begin{array}{rl}
x + y = 3 & \\
\underline{x - y = 1} & \\
2x = 4 & \text{Adding the two} \\
x = 2 &
\end{array}
$$

do not stop here. Solve for y also.

5.5

1. The total value of a given number of units each of equal value is found with the following equation.

(total value) = (value per unit)(no. of units)

2. The value of a quantity is equal to the sum of the values of its parts.

REVIEW EXERCISES

Part I

5.1 *Tell how the graphs of the system are related.*

1. $2x + y = -3$
$-4x - 2y = 6$

2. $x - 3y = 7$
$4x - 12y = -3$

3. $x + y = 3$
$2x + y = -4$

5.2, 5.3, 5.4 *Solve by whichever method seems more appropriate.*

4. $3x - y = 9$
$2x + 5y = -11$

5. $2x - 3y = -14$
$5x + 2y = 3$

6. $8x - 2y = 4$
$-4x + y = -2$

7. $x + 3 = 0$
$3x + 2y = -7$

8. $3x + y = -2$
$-6x - 2y = -2$

9. $x - 8 = 0$
$2y + 1 = 0$

5.5 *Solve.*

10. The sum of two numbers is 37 and twice one minus the other is 14. Find the numbers.

11. The sum of Barb's and Cindy's ages is 17. In 4 years Barb will be twice as old as Cindy is now. What is the present age of each?

12. A collection of nickels and dimes is worth $2.00. If the number of dimes minus the number of nickels is 5, how many of each are there?

13. A dealer wishes to mix tea worth 80¢ per lb with tea worth $1.00 per lb to obtain a 50-lb mixture worth 94¢ per lb. How many pounds of each must he use?

14. A boat travels 96 mi downstream in 4 hr and returns to the starting point in 6 hr. What is the speed of the boat in still water?

15. Tina invested $7000, part at 8% simple interest and the rest at 12% simple interest. How much did she have invested at 12% if her total interest income for the year was $640?

Part II

Solve.

16. A chemist has one solution that is 50% acid and a second that is 25% acid. If she wishes to obtain 10 liters of a 40% acid solution, how many liters of each should be combined to obtain the mixture?

17. Mike Brown flew his plane 600 mi with the wind in 5 hr and returned against the wind in 6 hr. What was the speed of the wind?

18. $2x + y = 0$
$x - y = 6$

19. $4x - y = -3$
$6x - 2y = -7$

20. $3x + 2 = 0$
$2y - 3 = 0$

21. Two angles are complementary (their sum is 90°) and one is 6° more than five times the other. Find each angle.

22. Mrs. Carlson has two sons. If the sum of their ages is 17 and the difference between their ages is 7, how old are her sons?

ANSWERS: 1. coinciding   2. parallel   3. intersecting   4. (2, −3)   5. (−1, 4)   6. infinitely many solutions
7. (−3, 1)   8. no solution   9. $\left(8, -\frac{1}{2}\right)$ 10. 17, 20   11. Barb is 10, Cindy is 7.   12. 10 nickels, 15 dimes
13. 15 pounds (80¢), 35 pounds ($1.00)   14. 20 mph   15. $2000   16. 6 liters of 50%, 4 liters of 25%
17. 10 mph   18. (2, −4)   19. $\left(\frac{1}{2}, 5\right)$ 20. $\left(-\frac{2}{3}, \frac{3}{2}\right)$ 21. 14°, 76°   22. 12 years, 5 years

1. Without graphing, tell how the graphs of the pair of linear equations are related (intersecting, parallel, or coinciding).

   $x + 3y = 1$
   $2x - y = 5$

   1. _____

2. Tell whether $(4, -2)$ is a solution to the following system.

   $3x + 4y = 4$
   $-2x - y = -10$

   2. _____

Solve the system of equations by whichever method seems more appropriate.

3. $2x - 5y = 11$
   $x + 3y = 0$

   3. _____

4. $3x - 2y = 1$
   $-6x + 4y = -1$

   4. _____

5. In a system of equations, if the lines are coinciding, how many solutions will the system have?

   5. _____

6. The sum of two numbers is 18 and their difference is 12. Find the two numbers.

6. _____

7. Two angles are supplementary and one is 12° more than six times the other. Find each angle.

7. _____

8. A collection of nickels and dimes has a value of $3.70. If there are 52 coins in the collection, how many of each are there?

8. _____

9. An airplane can fly 1000 mi with the wind in 2 hr and return against the wind in 2.5 hr. What is the speed of the plane in still air?

9. _____

10. A specialty store wishes to sell a party mix of candy and nuts for $1.80 per lb. If candy is $1.20 per lb and nuts cost $2.20 per lb, how many pounds of each should be used to make 100 lb of the mixture?

10. _____

Exponents and Polynomials

6.1 INTEGER EXPONENTS

1 EXPONENTIAL NOTATION

In Section 1.5 we introduced exponents and exponential notation.

Exponential Notation

If a is any number and n is a positive integer,

$$a^n = \underbrace{a \cdot a \cdot a \cdots a,}_{n \text{ factors}}$$

where a is the base, n the exponent, and a^n the exponential expression.

EXAMPLE 1 USING EXPONENTIAL NOTATION	PRACTICE EXERCISE 1

Write without using exponents.

(a) $4^5 = \underbrace{4 \cdot 4 \cdot 4 \cdot 4 \cdot 4}_{5 \text{ factors}}$ The product is 1024

(b) $3y^2 = \underbrace{3yy}_{2 \text{ ys as factors}}$ 3 *is not* squared

(c) $(3y)^2 = \underbrace{(3y)(3y)}_{2 \text{ (3y)s as factors}}$ 3 *is* squared

(d) $2^2 + 3^2 = \underbrace{2 \cdot 2}_{2 \text{ 2s}} + \underbrace{3 \cdot 3}_{2 \text{ 3s}}$ This simplifies to 13, *not* $(2 + 3)^2 = 5^2 = 25$

Write without using exponents.

(a) 9^7

(b) $6w^3$

(c) $(6w)^3$

(d) $4^2 + 8^2$

(e) 1^{27}

(f) $(-1)^5$

(g) $(-1)^{14}$

(e) $1^{32} = \underbrace{1 \cdot 1 \cdot 1 \cdots 1}_{32 \text{ factors}} = 1$ 1 to any power is always 1

(f) $(-1)^3 = (-1)(-1)(-1)$
$= (+1)(-1) = -1$ −1 to an *odd* power is −1

(g) $(-1)^4 = (-1)(-1)(-1)(-1)$
$= (-1)(-1) = 1$ −1 to an *even* power is +1 or 1

Answers: (a) $9 \cdot 9 \cdot 9 \cdot 9 \cdot 9 \cdot 9 \cdot 9$
(b) $6 \cdot w \cdot w \cdot w$ (c) $(6w)(6w)(6w)$
(d) $4 \cdot 4 + 8 \cdot 8$ (e) 1   (f) −1
(g) 1

CAUTION

Two of the most common errors made with exponents have been shown in the above examples:

$$3y^2 \neq (3y)^2 \quad \text{and} \quad 2^2 + 3^2 \neq (2+3)^2.$$

Parts **(f)** and **(g)** of Example 1 illustrate the following general rule: An odd power of a negative number is negative, and an even power of a negative number is positive.

❷ PRODUCT RULE

When we combine terms containing exponential expressions by multiplication, division, or taking powers, our work can be simplified by using the basic properties of exponents. For example,

$$a^3 \cdot a^2 = \underbrace{(a \cdot a \cdot a)}_{3 \text{ factors}}\underbrace{(a \cdot a)}_{2 \text{ factors}} = \underbrace{a \cdot a \cdot a \cdot a \cdot a}_{5 \text{ factors}} = a^5.$$

When two exponential expressions *with the same base* are multiplied, the product is that base raised to the sum of the exponents on the original expressions.

Product Rule for Exponents

If a is any number, and m and n are positive integers.

$$a^m a^n = a^{m+n}.$$

(To multiply powers with the same base, add exponents.)

EXAMPLE 2   USING THE PRODUCT RULE

Find the product.

(a) $a^3 \cdot a^4 = a^{3+4} = a^7$

(b) $5^3 \cdot 5^7 = 5^{3+7} = 5^{10}$ *Not* 25^{10}

(c) $2^3 \cdot 2^2 \cdot 2^6 = 2^{3+2+6} = 2^{11}$ The rule applies to more than two factors

(d) $3x^3 \cdot x^2 = 3x^{3+2} = 3x^5$ The 3 is not raised to the powers

PRACTICE EXERCISE 2

Find the product.

(a) $w^5 \cdot w^8$

(b) $8^2 \cdot 8^8$

(c) $3^2 \cdot 3^3 \cdot 3^5 \cdot 3^{11}$

(d) $7a^5 \cdot a^{12}$

Answers: (a) w^{13} (b) 8^{10}
(c) 3^{21} (d) $7a^{17}$

③ QUOTIENT RULE

When two powers with the same base are divided, for example,

$$\frac{a^5}{a^2} = \frac{\overbrace{a \cdot a \cdot a \cdot a \cdot a}^{5 \text{ factors}}}{\underbrace{a \cdot a}_{2 \text{ factors}}} = \frac{a \cdot a \cdot a \cdot \cancel{a} \cdot \cancel{a}}{\cancel{a} \cdot \cancel{a}} = \underbrace{a \cdot a \cdot a}_{3 \text{ factors}} = a^3,$$

the quotient can be found by raising the base to the difference of the exponents $(5 - 2 = 3)$.

Quotient Rule for Exponents

If a is any number except zero, and m, n, and $m - n$ are positive integers, then

$$\frac{a^m}{a^n} = a^{m-n}.$$

(To divide powers with the same base, subtract exponents.)

EXAMPLE 3   USING THE QUOTIENT RULE

Find the quotient.

(a) $\dfrac{a^7}{a^3} = a^{7-3} = a^4$

(b) $\dfrac{5^8}{5^5} = 5^{8-5} = 5^3$

(c) $\dfrac{2^3}{3^4}$ Cannot be simplified using the rule of exponents since the bases are different

(d) $\dfrac{3x^3}{x^2} = 3x^{3-2} = 3x^1 = 3x$ $\quad x^1 = x$

(e) $\dfrac{2yy^3}{y^2} = \dfrac{2y^1y^3}{y^2} = \dfrac{2y^{1+3}}{y^2}$

$\qquad = \dfrac{2y^4}{y^2} = 2y^{4-2} = 2y^2$

---

④ POWER RULE

When a power is raised to a power, for example,

$$(a^2)^3 = \underbrace{(a^2)(a^2)(a^2)}_{3 \text{ factors}} = (a \cdot a)(a \cdot a)(a \cdot a)$$
$$= \underbrace{a \cdot a \cdot a \cdot a \cdot a \cdot a}_{6 \text{ factors}} = a^6,$$

the resulting exponential expression can be found by raising the base to the product of the exponents $(2 \cdot 3 = 6)$.

Power Rule

If a is any number, and m and n are positive integers,

$$(a^m)^n = a^{mn}.$$

(To raise a power to a power, multiply exponents.)

/////////////// **C A U T I O N** ///////////////

Do not confuse the power rule with the product rule. For example,

$$(a^5)^2 = a^{5 \cdot 2} = a^{10},$$

but

$$a^5 a^2 = a^{5+2} = a^7.$$

///////////

EXAMPLE 4   USING THE POWER RULE

Find the powers.

(a) $(a^3)^8 = a^{3 \cdot 8} = a^{24}$

(b) $2(x^3)^2 = 2x^{3 \cdot 2} = 2x^6$

PRACTICE EXERCISE 4

Find the powers.

(a) $(w^5)^7$

(b) $8(a^4)^3$

Answers: **(a)** w^{35} **(b)** $8a^{12}$

---

⑤ POWERS OF PRODUCTS AND QUOTIENTS

A product or quotient of expressions is often raised to a power. For example,

$$(3x^2)^4 = \underbrace{(3x^2) \cdot (3x^2) \cdot (3x^2) \cdot (3x^2)}_{4 \text{ factors}}$$

$$= \underbrace{3 \cdot 3 \cdot 3 \cdot 3}_{4 \text{ factors}} \cdot \underbrace{x^2 \cdot x^2 \cdot x^2 \cdot x^2}_{4 \text{ factors}} = 3^4(x^2)^4,$$

and

$$\left(\frac{2}{y^2}\right)^3 = \underbrace{\frac{2}{y^2} \cdot \frac{2}{y^2} \cdot \frac{2}{y^2}}_{3 \text{ factors}} = \frac{\overbrace{2 \cdot 2 \cdot 2}^{3 \text{ factors}}}{\underbrace{y^2 \cdot y^2 \cdot y^2}_{3 \text{ factors}}} = \frac{2^3}{(y^2)^3}.$$

These illustrate the next rule.

Powers of Products and Quotients

If a and b are any numbers, and n is a positive integer, then

$$(a \cdot b)^n = a^n \cdot b^n \quad \text{and} \quad \left(\frac{a}{b}\right)^n = \frac{a^n}{b^n} \quad (b \text{ not zero}).$$

| EXAMPLE 5 POWERS OF PRODUCTS AND QUOTIENTS | PRACTICE EXERCISE 5 |

Simplify.

(a) $(2y)^5 = 2^5 \cdot y^5 = 2^5 y^5 = 32y^5$

(b) $(3a^2b^3)^4 = 3^4 \cdot (a^2)^4 \cdot (b^3)^4$ Raise each factor to the fourth power

$\qquad = 3^4 a^8 b^{12}$

$\qquad = 81a^8 b^{12}$

(c) $\left(\dfrac{2}{x^3}\right)^4 = \dfrac{2^4}{(x^3)^4}$ $\left(\dfrac{a}{b}\right)^n = \dfrac{a^n}{b^n}$

$\qquad = \dfrac{16}{x^{3\cdot4}}$ $(a^m)^n = a^{mn}$

$\qquad = \dfrac{16}{x^{12}}$

(d) $\left(\dfrac{3a^2}{b}\right)^3 = \dfrac{(3a^2)^3}{b^3}$ $\left(\dfrac{a}{b}\right)^n = \dfrac{a^n}{b^n}$

$\qquad = \dfrac{3^3(a^2)^3}{b^3}$ $(a \cdot b)^n = a^n \cdot b^n$

$\qquad = \dfrac{27a^6}{b^3}$

Practice Exercise 5:

Simplify.

(a) $(5a)^3$

(b) $(2u^3 w^5)^2$

(c) $\left(\dfrac{3}{z^2}\right)^3$

(d) $\left(\dfrac{4x}{y^3}\right)^2$

Answers: (a) $125a^3$ (b) $4u^6 w^{10}$ (c) $\frac{27}{z^6}$ (d) $\frac{16x^2}{y^6}$

CAUTION

Rules similar to the product and quotient rules above do not exist for sums and differences. For example,

$$(2^2 + 3^2)^3 \quad is\ not \quad (2^2)^3 + (3^2)^3,$$

and

$$(1^3 - 4^3)^2 \quad is\ not \quad (1^3)^2 - (4^3)^2.$$

Also, exponential expressions with different bases cannot be combined by adding exponents. For example, in general,

$$a^2 \cdot b^5 \quad is\ not \quad (ab)^7.$$

6 ZERO EXPONENTS

We know that if a is not zero,

$$\frac{a^m}{a^n} = a^{m-n}.$$

Suppose we extend the quotient rule to include $m = n$. Then

$$\frac{a^m}{a^m} = a^{m-m} = a^0 \quad \text{and also} \quad \frac{a^m}{a^m} = 1.$$

(Any number divided by itself is 1.) This suggests the following definition.

Zero Exponent

If a is any number except zero,

$$a^0 = 1.$$

EXAMPLE 6 USING ZERO EXPONENTS	PRACTICE EXERCISE 6

Simplify.

(a) $5^0 = 1$

(b) $(2a^2b^3)^0 = 1$ (assuming $a \neq 0$ and $b \neq 0$)

Simplify.

(a) $21^0 = $ _____.

(b) If $x \neq 0$ and $y \neq 0$,
$(7xy^5)^0 = $ _____.

Answers: (a) 1 (b) 1

7 NEGATIVE EXPONENTS

Again, consider

$$\frac{a^m}{a^n} = a^{m-n} \ (a \neq 0).$$

What happens when $n > m$? For example, if we let $n = 5$ and $m = 2$ and we extend the quotient rule to include $m - n < 0$, we have

$$\frac{a^m}{a^n} = \frac{a^2}{a^5} = a^{2-5} = a^{-3}.$$

If we look at the problem another way, we have

$$\frac{a^2}{a^5} = \frac{\cancel{a} \cdot \cancel{a}}{\cancel{a} \cdot \cancel{a} \cdot a \cdot a \cdot a} = \frac{1}{a \cdot a \cdot a} = \frac{1}{a^3}.$$

Thus, we conclude that $a^{-3} = \frac{1}{a^3}$. This suggests a way to define exponential expressions with negative integer exponents.

Negative Exponents
If $a \neq 0$ and n is a positive integer ($-n$ is a negative integer), then $$a^{-n} = \frac{1}{a^n}.$$

EXAMPLE 7 USING NEGATIVE EXPONENTS	PRACTICE EXERCISE 7

Simplify and write without negative exponents.

(a) $5^{-3} = \dfrac{1}{5^3} = \dfrac{1}{125}$ 5^{-3} is *not* -5^3 nor $(-3)(5)$

(b) $4^{-2} = \dfrac{1}{4^2} = \dfrac{1}{16}$

(c) $\dfrac{1}{3^{-2}} = \dfrac{1}{\frac{1}{3^2}} = \dfrac{1}{\frac{1}{9}} = 1 \cdot \dfrac{9}{1} = 9 = 3^2$

(d) $(-2)^{-5} = \dfrac{1}{(-2)^5} = \dfrac{1}{-32} = -\dfrac{1}{32}$

Simplify and write without negative exponents.

(a) 7^{-2}

(b) 2^{-5}

(c) $\dfrac{1}{5^{-1}}$

(d) $(-6)^{-2}$

Answers: (a) $\frac{1}{49}$ (b) $\frac{1}{32}$ (c) 5

(d) $\frac{1}{36}$

Example 7 shows that we can "remove" negative exponents simply by moving an exponential expression with a negative exponent from numerator to denominator (or denominator to numerator) and changing the sign of the exponent.

CAUTION

a^{-n} is not equal to $-a^n$ nor to $(-n)a$. As shown in Example 7 (a), 5^{-3} is $\frac{1}{125}$ and not -5^3, which is -125, nor $(-3)(5)$, which is -15.

---

⑧ SUMMARY OF RULES FOR EXPONENTS

All the rules of exponents stated for positive integer exponents apply to all integer exponents: positive, negative, and zero. These are summarized as follows.

Rules for Exponents

Let a and b be any two numbers, m and n any two integers.

1. $a^m \cdot a^n = a^{m+n}$

2. $\dfrac{a^m}{a^n} = a^{m-n} \quad (a \neq 0)$

3. $(a^m)^n = a^{mn}$

4. $(a \cdot b)^n = a^n b^n$

5. $\left(\dfrac{a}{b}\right)^n = \dfrac{a^n}{b^n} \quad (b \neq 0)$

6. $a^0 = 1 \quad (a \neq 0)$

7. $a^{-n} = \dfrac{1}{a^n} \quad (a \neq 0)$

8. $\dfrac{1}{a^{-n}} = a^n \quad (a \neq 0)$

EXAMPLE 8 USING ALL RULES

Simplify and write without negative exponents.

(a) $(2a)^{-1} = \dfrac{1}{(2a)^1} = \dfrac{1}{2a}$ $(2a)^{-1}$ is *not* $-2a$

(b) $2a^{-1} = 2\dfrac{1}{a} = \dfrac{2}{a}$ The exponent -1 is only on a, not on 2

(c) $y^3 y^{-5} = y^{3+(-5)} = y^{-2} = \dfrac{1}{y^2}$

(d) $\dfrac{a^2 b^{-3}}{a^{-1} b^5} = a^{2-(-1)} b^{-3-5} = a^3 b^{-8} = a^3 \cdot \dfrac{1}{b^8} = \dfrac{a^3}{b^8}$

(e) $\dfrac{1}{x^{-3}} = \dfrac{1}{\frac{1}{x^3}} = 1 \cdot \dfrac{x^3}{1} = x^3$

(f) $(2a^{-2}b)^{-3} = 2^{-3}(a^{-2})^{-3} b^{-3} = \dfrac{1}{2^3} a^{(-2)(-3)} \dfrac{1}{b^3}$

$= \dfrac{1}{8} \cdot a^6 \dfrac{1}{b^3} = \dfrac{a^6}{8b^3}$

PRACTICE EXERCISE 8

Simplify and write without negative exponents.

(a) $(5w)^{-2}$

(b) $5w^{-2}$

(c) $u^{-7} u^5$

(d) $\dfrac{x^4 y^{-3}}{x^{-1} y^2}$

(e) $\dfrac{1}{m^{-7}}$

(f) $(6y^{-3}z^{-1})^{-2}$

(g) $\left(\dfrac{3w^{-1}}{x^2}\right)^{-3}$

(g) $\left(\dfrac{a^2}{2y^{-3}}\right)^{-2} = \dfrac{(a^2)^{-2}}{2^{-2}(y^{-3})^{-2}} = \dfrac{a^{-4}}{\frac{1}{2^2}y^6} = \dfrac{\frac{1}{a^4}}{\frac{y^6}{2^2}}$

$$= \dfrac{1}{a^4} \cdot \dfrac{2^2}{y^6} = \dfrac{4}{a^4 y^6}$$

Answers: (a) $\frac{1}{25w^2}$ (b) $\frac{5}{w^2}$

(c) $\frac{1}{u^2}$ (d) $\frac{x^5}{y^5}$ (e) m^7 (f) $\frac{y^6 z^2}{36}$

(g) $\frac{w^3 x^6}{27}$

EXAMPLE 9 EVALUATION OF EXPONENTIAL EXPRESSIONS	**PRACTICE EXERCISE 9**

Evaluate the following when $a = -2$ and $b = 3$.

(a) $a^{-1} = (-2)^{-1} = \dfrac{1}{-2} = -\dfrac{1}{2}$ $(-2)^{-1}$ is *not* $+2$

(b) $\dfrac{a^{-2}}{b} = \dfrac{(-2)^{-2}}{3} = \dfrac{\frac{1}{(-2)^2}}{3} = \dfrac{\frac{1}{4}}{3} = \dfrac{1}{4} \cdot \dfrac{1}{3} = \dfrac{1}{12}$

(c) $(a + b)^{-1} = (-2 + 3)^{-1} = 1^{-1} = \dfrac{1}{1} = 1$

(d) $a^{-1} + b^{-1} = (-2)^{-1} + (3)^{-1}$

$$= \dfrac{1}{-2} + \dfrac{1}{3} = -\dfrac{3}{6} + \dfrac{2}{6} = -\dfrac{1}{6}$$

Evaluate the following when $x = -3$ and $y = 5$.

(a) x^{-1}

(b) $\dfrac{x^2}{y^{-1}}$

(c) $(y - x)^{-1}$

(d) $y^{-1} - x^{-1}$

Answers: (a) $\frac{1}{-3}$ (b) 45   (c) $\frac{1}{8}$

(d) $\frac{8}{15}$

6.1 EXERCISES A

Write in exponential notation.

1. $8 \cdot 8 \cdot 8 \cdot 8$

2. $2 \cdot 2 \cdot y \cdot y \cdot y$

3. $(2x)(2x)(2x)(2x)$

Write without using exponents.

4. $2y^4$

5. $(2y)^4$

6. $a^2 + b^2$

Simplify and write without negative exponents.

7. $x^2 \cdot x^5$

8. $a^3 \cdot a^2 \cdot a^4$

9. $2y^2 \cdot y^8$

10. $\dfrac{a^4}{a^3}$

11. $\dfrac{2y^5}{y^3}$

12. $(a^3)^4$

13. $(2x^3)^4$

14. $2(x^3)^4$

15. $\left(\dfrac{2}{x^3}\right)^4$

16. $\dfrac{a^3}{b^5}$

17. 5^0

18. 0^0

19. $(2a)^0 (a \neq 0)$

20. $(2x)^{-1}$

21. $2x^{-1}$

22. $\dfrac{2x^7}{x^9}$

23. $3y^4 y^{-7}$

24. $(3y)^{-2}$

25. $3y^{-2}$

26. $(3y^{-2})^3$

27 $\left(\dfrac{2y}{x^3}\right)^{-2}$

28. $\dfrac{b^{-2}}{a^{-4}}$

29. $\dfrac{a^{-4}b^2}{b^{-2}}$

30 $\dfrac{a^{-2}b^2}{a^4 b^{-3}}$

31. $(3xy^2)(4x^2 y^3)$

32. $(2x^2 y^2)^2 (3x^2 y)^3$

33. $(-xy)(2x^3 y)(4xy^3)$

34. $\dfrac{x^3 y^{-5}}{x^4 y^{-6}}$

35. $\dfrac{3^0 a^3 b^{-8}}{ab^4}$

36. $\dfrac{3^{-1} x^{-1} y^{-5}}{x^{-6} y^2}$

37. $(x^{-2} y^{-1})^{-2}$

38. $(x^2 y^{-3})^{-4}$

39. $(3x^{-1} y)^{-2}$

40. $\left(\dfrac{a^{-5}}{b^{-1}}\right)^{-1}$

41. $\left(\dfrac{2a^{-3}}{b^3}\right)^{-2}$

42. $\left(\dfrac{2a^3 b^{-2}}{a^{-5} b}\right)^{-3}$

Evaluate when $a = -2$ and $b = 3$.

43. $3a^2$

44. $(3a)^2$

45. $-3a^2$

46. $(-3a)^2$

47. $-b^2$

48. $(-b)^2$

49. $a^2 - b^2$

50. $(a - b)^2$

51 a^{-2}

52 $-2a$

53 $-a^2$

54 $a^{-2} + b^{-2}$

55 $(a + b)^{-2}$

56. $\dfrac{a^{-1}}{b^{-2}}$

57. a^{-3}

58. $(-a)^{-3}$

59 $a^{-1} b^{-1}$

60 $(ab)^{-1}$

ANSWERS: 1. 8^4 2. $2^2 y^3$ 3. $(2x)^4$ 4. $2yyyy$ 5. $(2y)(2y)(2y)(2y)$ 6. $aa + bb$ 7. x^7 8. a^9 9. $2y^{10}$ 10. $a^1 = a$ 11. $2y^2$ 12. a^{12} 13. $16x^{12}$ 14. $2x^{12}$ 15. $\frac{16}{x^{12}}$ 16. cannot be simplified further 17. 1 18. undefined 19. 1 20. $\frac{1}{2x}$ 21. $\frac{2}{x}$ 22. $\frac{2}{x^2}$ 23. $\frac{3}{y^3}$ 24. $\frac{1}{9y^2}$ 25. $\frac{3}{y^2}$ 26. $\frac{27}{y^6}$ 27. $\frac{x^6}{4y^2}$ 28. $\frac{a^4}{b^2}$ 29. $\frac{b^4}{a^4}$ 30. $\frac{b^5}{a^6}$ 31. $12x^3 y^5$ 32. $108x^{10} y^7$ 33. $-8x^5 y^5$ 34. $\frac{y}{x}$ 35. $\frac{a^2}{b^{12}}$ 36. $\frac{x^5}{3y^7}$ 37. $x^4 y^2$ 38. $\frac{y^{12}}{x^8}$ 39. $\frac{x^2}{9y^2}$ 40. $\frac{a^5}{b}$ 41. $\frac{a^6 b^6}{4}$ 42. $\frac{b^9}{8a^{24}}$ 43. 12 44. 36 45. -12 46. 36 47. -9 48. 9 49. -5 50. 25 51. $\frac{1}{4}$ 52. 4 53. -4 54. $\frac{13}{36}$ 55. 1 56. $-\frac{9}{2}$ 57. $-\frac{1}{8}$ 58. $\frac{1}{8}$ 59. $-\frac{1}{6}$ 60. $-\frac{1}{6}$

6.1 EXERCISES B

Write in exponential notation.

1. $4 \cdot 4 \cdot z \cdot z \cdot z$

2. $(3w)(3w)(3w)$

3. $3www$

Write without using exponents.

4. $4y^3$

5. $(4y)^3$

6. $x^2 + y^2$

Simplify and write without negative exponents.

7. $y^2 \cdot y^7$

8. $x^3 \cdot x^2 \cdot x^6$

9. $2z^2 \cdot z^5$

10. $\dfrac{b^5}{b^2}$

11. $\dfrac{2y^7}{y^4}$

12. $(w^3)^5$

13. $(2c^2)^4$

14. $2(y^3)^5$

15. $\left(\dfrac{2}{a^2}\right)^3$

16. $\dfrac{x^3}{y^4}$

17. 7^0

18. -0^0

19. $(4x)^0 (x \neq 0)$

20. $(5y)^{-1}$

21. $5y^{-1}$

22. $\dfrac{3a^3}{a^7}$

23. $4z^4z^{-9}$

24. $(5w)^{-2}$

25. $5w^{-2}$

26. $(2a^{-2})^3$

27. $\left(\dfrac{2y}{x^3}\right)^2$

28. $\dfrac{b^{-5}}{a^{-3}}$

29. $\dfrac{a^{-3}b^{-3}}{b^{-4}}$

30. $\dfrac{x^{-2}y^4}{x^3y^{-3}}$

31. $(-5x^2y^2)(3x^3y)$

32. $(x^3y^2)^3(2x^3y)^2$

33. $(-2x)(5x^5y)(-3xy^2)$

34. $\dfrac{a^4b^{-3}}{a^{-2}b^2}$

35. $\dfrac{5^0a^{-5}b^{-2}}{a^4b^{-1}}$

36. $\dfrac{2^{-2}x^3y^{-4}}{x^{-2}y^{-2}}$

37. $(a^{-1}b^{-5})^{-1}$

38. $(a^{-6}b^3)^{-3}$

39. $(4a^{-1}b^{-1})^{-3}$

40. $\left(\dfrac{x^{-4}}{y^{-2}}\right)^{-2}$

41. $\left(\dfrac{3x^4}{y^3}\right)^{-1}$

42. $\left(\dfrac{3x^{-2}y^{-1}}{x^3y^{-5}}\right)^{-4}$

Evaluate when $x = -3$ and $y = 2$.

43. $4x^2$

44. $(4x)^2$

45. $-4x^2$

46. $(-4x)^2$

47. $-x^2$

48. $(-x)^2$

49. $y^2 - x^2$

50. $(y - x)^2$

51. x^{-2}

52. $-3y$

53. $-y^3$

54. $x^{-2} + y^{-2}$

55. $(x + y)^{-2}$ **56.** $\dfrac{x^{-1}}{y^{-2}}$ **57.** x^{-3}

58. $(-x)^{-3}$ **59.** $x^{-1}y^{-1}$ **60.** $(xy)^{-1}$

6.1 EXERCISES C

Simplify and write without negative exponents.

1. $\dfrac{a^{-2}b^3c^2}{a^{-3}b^{-2}c^{-2}}$

2. $\dfrac{3^0x^{-6}(y^{-1})^{-2}}{x^{-2}y^{-3}}$

3. $\dfrac{2^{-2}(x^2)^{-3}y^3z^{-2}}{3^{-1}x^{-1}(yz)^{-1}}$

$\left[\text{Answer:}\quad \dfrac{3y^4}{4x^5z}\right]$

4. $\left(\dfrac{5^0a^{-6}b^2c^3}{a^2b^{-1}c^{-2}}\right)^{-1}$

5. $\left(\dfrac{2^{-1}x^{-5}y^{-8}}{4^{-1}x^3y^{-4}}\right)^{-2}$

6. $\left[\left(\dfrac{a^2b^{-2}}{a^{-5}b^{-1}}\right)^{-1}\right]^{-2}$

$\left[\text{Answer:}\quad \dfrac{a^{14}}{b^2}\right]$

6.2 SCIENTIFIC NOTATION

STUDENT GUIDEPOSTS

1 Scientific Notation   **2** Calculations Using Scientific Notation

When we compute with very large or very small numbers on a calculator, problems can arise because the display is limited (usually 8 digits). For example, if we use a calculator to multiply

$$(290{,}000)(15{,}000),$$

the product might be given as in Figure 6.1.

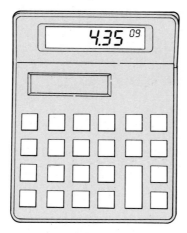

Figure 6.1

The number 4.35×10^9 is *scientific notation* for 4,350,000,000, which has ten digits, too many for the display. This shorthand notation depends on the use of integer exponents.

❶ SCIENTIFIC NOTATION

A scientist might use the number
$$235,000,000,000,000,000,000$$

but, instead of writing out all the zeros, he or she would write
$$2.35 \times 10^{20}.$$

This short form is easier to use in computations. Likewise, the number
$$0.000000000057$$

could be written as 5.7×10^{-11}.

A number is written in **scientific notation** if it is the product of a power of 10 and a number that is greater than or equal to 1 and less than 10.

To Write a Number in Scientific Notation

1. Move the decimal point to the position immediately to the right of the first nonzero digit.
2. Multiply by a power of ten that is equal to the number of decimal places moved. The exponent on 10 is positive if the original number is greater than 10 and negative if the number is less than 1.

EXAMPLE 1 CONVERTING TO SCIENTIFIC NOTATION

Write in scientific notation.

first nonzero digit
↓
(a) $2,500,000 = 2.5 \times 10^6$ Count 6 decimal places
6 places

first nonzero digit
↓
(b) $0.0000025 = 2.5 \times 10^{-6}$ Count 6 decimal places
6 places

(c) $4,321,000,000 = 4.321 \times 10^9$
9 places

(d) $0.00000000001 = 1 \times 10^{-11}$
11 places

(e) $0.1 = 1 \times 10^{-1}$
1 place

(f) $4.8 = 4.8 \times 10^0$

PRACTICE EXERCISE 1

Write in scientific notation.

(a) 18,300,000

(b) 0.000087

(c) 65,240,000,000,000

(d) 0.0000000001

(e) 0.5

(f) 9.7

Answers: (a) 1.83×10^7
(b) 8.7×10^{-5} (c) 6.524×10^{13}
(d) 1×10^{-10} (e) 5×10^{-1}
(f) 9.7×10^0

EXAMPLE 2 CONVERTING TO STANDARD NOTATION

Write in standard notation.
(a) $5.4 \times 10^5 = 540,000$ Count 5 decimal places

(b) $5.4 \times 10^{-5} = 0.000054$ Count 5 decimal places

PRACTICE EXERCISE 2

Write in standard notation.
(a) 6.1×10^7

(b) 6.1×10^{-7}

(c) $8.94 \times 10^{13} = 89,400,000,000,000$

(d) $2.113 \times 10^{-8} = 0.00000002113$

(c) 5.35×10^{12}

(d) 9.03×10^{-6}

Answers: (a) 61,000,000
(b) 0.00000061
(c) 5,350,000,000,000
(d) 0.00000903

② CALCULATIONS USING SCIENTIFIC NOTATION

Scientific notation not only shortens the notation for many numbers, but also helps in calculations.

EXAMPLE 3 CALCULATING WITH SCIENTIFIC NOTATION

Perform the indicated operations using scientific notation.

(a) $(30,000)(2,000,000) = (3 \times 10^4)(2 \times 10^6)$
$\qquad = (3 \cdot 2) \times (10^4 \times 10^6)$ Change order
$\qquad = 6 \times 10^{10}$ Add exponents

(b) $(2.4 \times 10^{-12})(4.0 \times 10^{11}) = (2.4)(4.0) \times (10^{-12} \times 10^{11})$
$\qquad\qquad\qquad = 9.6 \times 10^{-1}$

(c) $\dfrac{3.2 \times 10^{-1}}{1.6 \times 10^5} = \dfrac{3.2}{1.6} \times \dfrac{10^{-1}}{10^5} = 2 \times 10^{-6}$

PRACTICE EXERCISE 3

Perform the indicated operations using scientific notation.

(a) $(5,000,000)(200,000)$

(b) $(6.4 \times 10^{-10})(0.8 \times 10^9)$

(c) $\dfrac{8.1 \times 10^{-2}}{2.7 \times 10^{-6}}$

Answers: (a) 1×10^{12}
(b) 5.12×10^{-1} (c) 3×10^4

6.2 EXERCISES A

Write in scientific notation.

1. 370

2. 0.0037

3. 98,000

4. 0.00012

5. 8360

6. 0.00279

7. 0.0000000000756

8. 2,650,000,000

9. 0.01

Write in standard notation.

10. 2.3×10^2

11. 2.3×10^{-2}

12. 8.7×10^{-5}

13. 4.58×10^6

14. 7.51×10^{-8}

15. 6.64×10^{10}

Perform the indicated operations using scientific notation.

16. $(4 \times 10^5)(1 \times 10^6)$

17. $(40,000,000)(20,000)$

18. $(1 \times 10^{-10})(5 \times 10^7)$

19 $(0.0000022)(300)$

20. $\dfrac{3.3 \times 10^{12}}{1.1 \times 10^{-2}}$

21. $\dfrac{0.0000006}{0.03}$

22. The measure of one calorie is equal to 0.000000278 kilowatt-hours. Write this number in scientific notation.

23. The distance that light will travel in 1 year is approximately 5,870,000,000,000 miles. Write the number in scientific notation.

FOR REVIEW

Evaluate when $a = -2$ and $b = 5$.

24. a^{-1}

25. a^{-2}

26. $-2a$

27. b^{-2}

28. $-2b$

29. $a^{-1} + b^{-1}$

30. $(a + b)^{-1}$

31. $(a - b)^{-2}$

32. $a^{-2} - b^{-2}$

ANSWERS: 1. 3.7×10^2 2. 3.7×10^{-3} 3. 9.8×10^4 4. 1.2×10^{-4} 5. 8.36×10^3 6. 2.79×10^{-3} 7. 7.56×10^{-11} 8. 2.65×10^9 9. 1×10^{-2} 10. 230 11. 0.023 12. 0.000087 13. 4,580,000 14. 0.0000000751 15. 66,400,000,000 16. 4×10^{11} 17. 8×10^{11} 18. 5×10^{-3} 19. 6.6×10^{-4} 20. 3×10^{14} 21. 2×10^{-5} 22. 2.78×10^{-7} 23. 5.87×10^{12} mi 24. $-\frac{1}{2}$ 25. $\frac{1}{4}$ 26. 4 27. $\frac{1}{25}$ 28. -10 29. $-\frac{3}{10}$ 30. $\frac{1}{3}$ 31. $\frac{1}{49}$ 32. $\frac{21}{100}$

6.2 EXERCISES B

Write in scientific notation.

1. 5400

2. 0.0054

3. 386

4. 0.000028

5. 33,000

6. 0.1

7. 0.000000000254

8. 4,620,000,000

9. 0.001

Write in standard notation.

10. 8.4×10^3

11. 8.4×10^{-3}

12. 5.42×10^{-7}

13. 7.25×10^7

14. 2.06×10^{-6}

15. 4.99×10^{12}

Perform the indicated operations using scientific notation.

16. $(5 \times 10^3)(1 \times 10^4)$

17. $(60,000,000)(50,000)$

18. $(1 \times 10^{-12})(3 \times 10^8)$

19. $(0.00000044)(500)$

20. $\dfrac{4.4 \times 10^{15}}{1.1 \times 10^{-3}}$

21. $\dfrac{0.00000008}{0.04}$

22. The earth is approximately 93,000,000 miles from the sun. Write this number in scientific notation.

23. Chemists use 602,000,000,000,000,000,000,000, known as Avogadro's number, in calculations. Write this number in scientific notation.

FOR REVIEW

Evaluate when $x = -3$ and $y = 2$.

24. x^{-1}

25. x^{-2}

26. $-2x$

27. y^{-2}

28. $-2y$

29. $x^{-1} + y^{-1}$

30. $(x + y)^{-1}$

31. $(x - y)^{-2}$

32. $x^{-2} - y^{-2}$

6.2 EXERCISES C

Perform the indicated operations using scientific notation.

1. $\dfrac{(2.5 \times 10^{-3})(4.2 \times 10^{-8})}{(5.0 \times 10^7)(8.4 \times 10^{-10})}$

2. $\dfrac{(0.000036)(0.0001)^{-1}}{(1,200,000)(1 \times 10^7)^{-2}}$ [Answer: 3×10^7]

6.3 BASIC CONCEPTS OF POLYNOMIALS

STUDENT GUIDEPOSTS

1 Polynomials

2 Monomials, Binomials, and Trinomials

3 Degree of a Polynomial

4 Ascending and Descending Order

5 Like Terms

6 Evaluating Polynomials

In Chapter 1 we defined a **variable** as a letter that represents a number and said that an **algebraic expression** involves sums, differences, products, quotients, or powers of numbers and variables. The **terms** in an algebraic expression are the parts that are separated by plus and minus signs. For example,

$$3x^2 - y + 2uv^3 - 9$$

is an algebraic expression with four terms, $3x^2$, $-y$, $2uv^3$, and -9. Notice that the minus sign goes with the terms $-y$ and -9.

① POLYNOMIALS

A **polynomial** is an algebraic expression whose terms are products of numbers and variables with whole-number exponents.

THESE ARE POLYNOMIALS:

$$3x^2 + x - 5, \quad 2m + 1, \quad 3x, \quad \frac{1}{2}z^2 - 5z^3, \quad \text{and} \quad -2x^2y + xy^2 + 3$$

The polynomial $3x^2 + x - 5$ has three terms, $3x^2$, x (or $1 \cdot x$), and -5. The **numerical coefficients** (usually just called the coefficients) of these terms are 3, 1, and -5. Similarly, the polynomial $\frac{1}{2}z^2 - 5z^3$ has terms $\frac{1}{2}z^2$ and $-5z^3$ with coefficients $\frac{1}{2}$ and -5.

THESE ARE NOT POLYNOMIALS:

$$3\sqrt{x} + 5, \quad \frac{x + 2}{x - y}, \quad x^{-2}$$

The examples above are algebraic expressions that are not polynomials. In a polynomial, a variable cannot appear under a radical, in a denominator, or with a negative exponent.

② MONOMIALS, BINOMIALS, AND TRINOMIALS

A polynomial with one term is a **monomial,** a **binomial** has two terms, and a **trinomial** has three terms. When a polynomial has more than three terms, no special name is used and we simply call it a polynomial. Study the polynomials in the following table.

Polynomial	Type	Terms	Coefficients
$5x - 2$	binomial	$5x, -2$	$5, -2$
6	monomial	6	6
$\frac{1}{3}y^2 + 2y - 3$	trinomial	$\frac{1}{3}y^2, 2y, -3$	$\frac{1}{3}, 2, -3$
$7a^3 + 2a^2 - 6a + 5$	polynomial	$7a^3, 2a^2, -6a, 5$	$7, 2, -6, 5$
$0.5x^3 + 0.1$	binomial	$0.5x^3, 0.1$	$0.5, 0.1$
$-4b^2a^3$	monomial	$-4b^2a^3$	-4

A **polynomial in one variable,** such as $3x^3 + 5x^2 - x + 7$, has the same variable in each of its terms. A **polynomial in several variables,** such as $-2a^2b - 3ab - b$, has two or more variables. Our primary interest is in polynomials in one variable, but some properties of polynomials with several variables will also be discussed.

③ DEGREE OF A POLYNOMIAL

For polynomials in one variable, the **degree of a term** is the exponent on the variable. The **degree of a polynomial** is the degree of the term of highest degree.

	EXAMPLE 1 DEGREE OF A TERM

For the polynomial $-3y^2 + 9y^4 - 6y^7 + y + 8$, list the terms and their degree. Also give the degree of the polynomial.

Term	Degree of the term	Reason
$-3y^2$	2	Exponent on y is 2
$9y^4$	4	Exponent on y is 4
$-6y^7$	7	Exponent on y is 7
y	1	Exponent on y is 1, since $y = y^1$
8	0	Exponent on y is 0, since $8 = 8 \cdot 1 = 8y^0$ (Remember that $y^0 = 1$)

Since $-6y^7$ has the highest degree, 7, the degree of the polynomial is 7.

PRACTICE EXERCISE 1

For the polynomial $-2x^3 + 9x^5 + 11$, list the terms and their degree. Also give the degree of the polynomial.

Answer: The term $-2x^3$ has degree 3, the term $9x^5$ has degree 5, and the term 11 has degree 0. The degree of the polynomial is 5.

④ ASCENDING AND DESCENDING ORDER

If the polynomial in Example 1 is written in the order

$$-6y^7 + 9y^4 - 3y^2 + y + 8,$$

we say that it is in **descending order.** Note that the highest degree term is first, followed by the next highest, and so forth. The constant term is last since 8 can be written as $8y^0$. Written in **ascending order** the same polynomial would appear as

$$8 + y - 3y^2 + 9y^4 - 6y^7.$$

Most of the time we will use descending order.

	EXAMPLE 2 ASCENDING AND DESCENDING ORDER

Write $-3x + 5 - x^6 + x^3$ in both descending and ascending order.

$-x^6 + x^3 - 3x + 5$	Descending order
$5 - 3x + x^3 - x^6$	Ascending order

PRACTICE EXERCISE 2

Write $y^3 - 3y^5 + 12 - y$ in both descending and ascending order.

Answer: Descending: $-3y^5 + y^3 - y + 12$; ascending: $12 - y + y^3 - 3y^5$

⑤ LIKE TERMS

Terms that have the variable raised to the same power are called **like terms.** The like terms for the polynomial

$$7a^2 - 3a^3 + 2a - 5a^2 + 1 - 2a - 5,$$

for example, are

$$7a^2 \text{ and } -5a^2, \quad 2a \text{ and } -2a, \quad \text{and} \quad 1 \text{ and } -5.$$

A polynomial like this one can be simplified by **collecting** or **combining like terms** using the distributive law. Both words, collecting and combining, are used to describe this operation which is illustrated in the next example.

EXAMPLE 3 COLLECTING AND COMBINING LIKE TERMS

Collect like terms and write each polynomial in descending order.

(a) $7a^2 - 3a^3 + 2a - 5a^2 + 1 - 2a - 5$

$= \boxed{7a^2 - 5a^2} - 3a^3 + \boxed{2a - 2a} + \boxed{1 - 5}$ Commute to collect like terms

$= (7 - 5)a^2 - 3a^3 + (2 - 2)a + (1 - 5)$ Use distributive law to combine like terms

$= 2a^2 - 3a^3 + 0 \cdot a - 4$

$= 2a^2 - 3a^3 - 4$

$= -3a^3 + 2a^2 - 4$ Descending order

(b) $-8x^3 + x^3 - 3x + 2 + 5x - 7$

$= -8x^3 + x^3 - 3x + 5x + 2 - 7$ Commute

$= (-8 + 1)x^3 + (-3 + 5)x + (2 - 7)$ Distributive law

$= -7x^3 + 2x - 5$

(c) $-3x^3 + x^7 - 7x - 4x^7 + x^3 + 1$

$= x^7 - 4x^7 - 3x^3 + x^3 - 7x + 1$

$= (1 - 4)x^7 + (-3 + 1)x^3 - 7x + 1$ This step may be omitted

$= -3x^7 - 2x^3 - 7x + 1$

PRACTICE EXERCISE 3

Collect like terms and write each polynomial in descending order.

(a) $3z^4 + z - 2z^4 - 5z + 1$

(b) $-9w^6 + 2w + 3w^6 - 5 - w^6 + 11$

(c) $y^3 - 3y + 6y^3 + 1 + 3y - 1 - 5y^3$

Answers: (a) $z^4 - 4z + 1$ (b) $-7w^6 + 2w + 6$ (c) $2y^3$

Like terms in polynomials with several variables must contain exactly the same variables raised to the same power. The following table lists several terms along with like and unlike terms.

Term	Like term	Unlike term
$-2xy$	$5xy$	$6y$
$8a^2b^3$	$-2a^2b^3$	$3a^3b^2$
$9xy^4z$	$4xy^4z$	$-7x^2y^4z$
2	-10	$5a^2b^2$

EXAMPLE 4 COLLECTING AND COMBINING LIKE TERMS

Collect like terms.

$3x^2y - 5xy - 2x - 4x^2y + 4x + 2xy^2$

$= 3x^2y - 4x^2y - 5xy - 2x + 4x + 2xy^2$

$= (3 - 4)x^2y - 5xy + (-2 + 4)x + 2xy^2$ This step may be omitted

$= -x^2y - 5xy + 2x + 2xy^2$

PRACTICE EXERCISE 4

Collect like terms.

$6u^2v - 3uv + 5u + 9uv - 5u^2v + uv^2$

Answer: $u^2v + uv^2 + 6uv + 5u$

6 EVALUATING POLYNOMIALS

Since a variable represents a real number, a polynomial in that variable also represents a real number. We can **evaluate** a polynomial when specific values for the variable(s) are given. The value of a polynomial will usually be different when different values for the variable(s) are used.

EXAMPLE 5 EVALUATING POLYNOMIALS

Evaluate the polynomials for the given values of the variables.

(a) $5x^2 - 3x + 1$ for $x = -2$

$$5x^2 - 3x + 1 = 5(-2)^2 - 3(-2) + 1 \quad \text{Substitute } -2 \text{ for } x$$
$$= 5(4) + 6 + 1 \quad \text{Square first, then multiply}$$
$$= 20 + 6 + 1 = 27 \quad \text{Add}$$

(b) $5x^2 - 3x + 1$ for $x = 2$

$$5x^2 - 3x + 1 = 5(2)^2 - 3(2) + 1 \quad \text{Substitute 2 for } x$$
$$= 5(4) - 6 + 1$$
$$= 20 - 6 + 1 = 15$$

Notice from (a) and (b) that the values of the polynomial are different for the different values of x.

(c) $3ab + 2a - 5b$ for $a = -1$ and $b = 0$

$$3ab + 2a - 5b = 3(-1)(0) + 2(-1) - 5(0) \quad \text{Substitute } -1 \text{ for } a$$
$$= 0 - 2 - 0 = -2 \quad \text{and 0 for } b$$

PRACTICE EXERCISE 5

Evaluate the polynomials for the given values of the variables.

(a) $6y^3 - y + 5$ for $y = -1$

(b) $6y^3 - y + 5$ for $y = 1$

(c) $-2uv + v - 7u + 1$ for $u = -2$ and $v = 0$

Answers: (a) 0   (b) 10   (c) 15

EXAMPLE 6 MANUFACTURING APPLICATION

The profit made by a machine manufacturer who sells x machines per week is given by the expression $100x^3 - 2500$. This means that fixed costs are \$2500 and the profit increases with the sale of each machine.

(a) Find the profit when 5 machines are sold.

$$100x^3 - 2500 = 100(5)^3 - 2500 \quad \text{Substitute 5 for } x$$
$$= 100(125) - 2500$$
$$= 12{,}500 - 2500$$
$$= 10{,}000$$

The profit for the week was \$10,000.

(b) Find the profit when 2 machines are sold.

$$100x^3 - 2500 = 100(2)^3 - 2500$$
$$= 100(8) - 2500$$
$$= 800 - 2500$$
$$= -1700$$

Since the profit is negative, the company lost \$1700 that week.

PRACTICE EXERCISE 6

The cost in dollars of manufacturing a particular type of circuit board is given by $0.45n^2 + 180$, where n is the number of boards produced.

(a) Find the cost when 10 boards are made.

(b) Find the overhead cost that results when no boards are made ($n = 0$).

Answers: (a) \$225   (b) \$180

6.3 EXERCISES A

Fill in the following table.

Polynomial	Type	Terms	Coefficients
1. $-3x^2 + 2$	(a) _____	(b) _____	(c) _____
2. $2a$	(a) _____	(b) _____	(c) _____
3. $-4y^2 + 2y^3 - 7y + 1$	(a) _____	(b) _____	(c) _____
4. $a^2 - 2a + \dfrac{1}{3}$	(a) _____	(b) _____	(c) _____
5. -6	(a) _____	(b) _____	(c) _____
6. $3x - 4y$	(a) _____	(b) _____	(c) _____
7. $7a^2b^3$	(a) _____	(b) _____	(c) _____
8. $x^2 - 2xy + y^2$	(a) _____	(b) _____	(c) _____

Give the degree of each of the following terms.

9. $3x^2$ **10.** $9b$ **11.** -5 **12.** $-7b^{14}$

Give the degree of each of the following polynomials.

13. $x^2 - 6x^3 + x^4 - 9x$ **14.** $-3y^2 + 9y^5 + 2y - y^7$ **15.** $a - 2$

16. 6 **17.** $2x^3 - x$ **18.** $-y^2 - 8y + y^4 - 3y^3 + 5$

Collect like terms and write in descending order.

19. $8x - 3x$ **20.** $-5y + 2y$ **21.** $10a^2 - 7a + 5a^2 + 3a$

22. $-4b^2 - 9 + 12 - 6b^2$ **23.** $2x^4 - 2x^3 + x^3 - 3x^4 + 7 - 5$ **24.** $3x^3 - 7 + 4x^2 - 3x^3 + 6 + 2x^3$

25. $7y^3 + 3y^2 - 7y^3 - 3y^2$ **26.** $4y^4 - 2y^3 + 2y^3 - 4y^4 + 8$

27. $6a^2 + 7a + 8a^2 - 3 - 7a - 9a^2 + 1$ **28** $-8x^{10} + x^5 - 2x^{10} - 7x^5 + 1 - x^{10} + 3$

29. $\dfrac{3}{4}y^2 - \dfrac{1}{8}y^2$ **30.** $-0.5b^3 + 0.77b^3$

Collect like terms.

31. $2xy - 5xy$ **32.** $5x^2 + 2y^2$ **33.** $2xy - 3y^2 + 5xy + y^2$

34. $a^2b^2 - ab + a^2b^2 + ab$ **35** $-4x^2y + 2xy^2 + x^2 - 3x^2y$ **36.** $5ab^2 - 3ab + 3ab - 5ab^2$

Evaluate the polynomials for the given value of the variables.

37. $3x + 2$ for $x = 5$ **38** $7y^2 - 2y - 5$ for $y = -2$

39. $8a^3 - 5$ for $a = -3$ **40.** $2a^2 + 5b^2$ for $a = 1$ and $b = -1$

41 The cost in dollars of manufacturing x bolts is given by the expression $0.05x + 15.5$. Find the cost when 440 bolts are made.

42. The cost of typing a manuscript is given as two times the number of pages plus ten. Use x as the number of pages to be typed and write a polynomial to describe this cost. Find the cost of typing a one-hundred-page manuscript.

FOR REVIEW

Write in scientific notation.

43. 265,000   **44.** 0.000000902

Write in standard notation.

45. 6.75×10^{-4} **46.** 1.06×10^8

Exercises 47–50 review material from Chapter 2 to help you prepare for the next section. Simplify each expression by removing the parentheses.

47. $-(-x)$ **48.** $-(x - 1)$ **49.** $-(-x - 1 + a)$ **50.** $-(a + x - 2)$

ANSWERS: 1. (a) binomial   (b) $-3x^2, 2$ (c) $-3, 2$ 2. (a) monomial   (b) $2a$ (c) 2   3. (a) polynomial   (b) $-4y^2$, $2y^3, -7y, 1$ (c) $-4, 2, -7, 1$ 4. (a) trinomial   (b) $a^2, -2a, \frac{1}{3}$ (c) $1, -2, \frac{1}{3}$ 5. (a) monomial   (b) -6 (c) -6 6. (a) binomial   (b) $3x, -4y$ (c) $3, -4$ 7. (a) monomial   (b) $7a^2b^3$ (c) 7   8. (a) trinomial   (b) $x^2, -2xy$, y^2 (c) $1, -2, 1$ 9. 2   10. 1   11. 0   12. 14   13. 4   14. 7   15. 1   16. 0   17. 3   18. 4   19. $5x$ 20. $-3y$ 21. $15a^2 - 4a$ 22. $-10b^2 + 3$ 23. $-x^4 - x^3 + 2$ 24. $2x^3 + 4x^2 - 1$ 25. 0   26. 8   27. $5a^2 - 2$ 28. $-11x^{10} - 6x^5 + 4$ 29. $\frac{5}{8}y^2$ 30. $0.27b^3$ 31. $-3xy$ 32. $5x^2 + 2y^2$ (no like terms)   33. $7xy - 2y^2$ 34. $2a^2b^2$ 35. $-7x^2y + 2xy^2 + x^2$ 36. 0   37. 17   38. 27   39. -221 40. 7   41. \$37.50   42. $2x + 10$, \$210   43. 2.65×10^5 44. 9.02×10^{-7} 45. 0.000675   46. 106,000,000   47. x 48. $-x + 1$ 49. $x + 1 - a$ 50. $-a - x + 2$

6.3 EXERCISES B

Give the type and list the terms and coefficients of the following polynomials.

1. $7x - 5$ **2.** $3y^2 - 2y + 3$ **3.** $-4a^3 + 6a + a^4 - 7$ **4.** 22

5. $-12b^{15}$ **6.** $-8a + b$ **7.** $4x^2 + 4xy - y^2$ **8.** $9x^4y^6$

Give the degree of each of the following terms.

9. $5y^7$ **10.** $-2x^{32}$ **11.** $4a^0$ **12.** $44b^{12}$

Give the degree of each of the following polynomials.

13. $3 + 4x^{10}$

14. $6y^4 + 12y^2 - 8y + 7y^5$

15. $-2 - 3a$

16. -9

17. $-x^4 - x^2 + 9x$

18. $5y^{10} + 14y^{20} - 2y^{15} + 3$

Collect like terms and write in descending order.

19. $-2x + 10x$

20. $y - 9y$

21. $a^3 - 3a^2 + 4a^3 - a^2$

22. $5 - b^4 + 3b^4 - 8$

23. $3x - 8x^2 + 5x^3 - 2x^3 + 4x^2 - 2x$ **24.** $3y + 2 - 7y^2 - 2y - 5$

25. $11a^3 - 17a^4 + 8a^2 - 5 + a^3 - 6 + a^4$

26. $-3b + 21b^7 - 5b^3 + 12b - b^7$

27. $10 - 3y + 7y^2 - 8y^3$

28. $-22x^3 + 17x^5 - 4x^3 + 2$

29. $0.2b - 0.8b^2 + 0.7b^2$

30. $\dfrac{3}{2}a^3 - \dfrac{1}{3}a^4 + \dfrac{1}{4}a^3 + \dfrac{2}{9}a^4$

Collect like terms.

31. $-9xy + 7xy$

32. $5x^2y^2 - 5$

33. $x^2 - 3xy + 8xy - 4x^2$

34. $2a^2b - a^2b^2 - 2a^2b + a^2b^2$

35. $2x^2y^2 - 3xy^2 + 5x^2y^2 + xy^2$

36. $5ab - 2a^2 - 2a^2 + 5ab + 6$

Evaluate the polynomials for the given values of the variables.

37. $-8x + 5$ for $x = 4$

38. $-2y^2 + 3y - 2$ for $y = -3$

39. $2a^3 + a^2$ for $a = 5$

40. $5a^2b^2 - 2$ for $a = 9$ and $b = 0$

41. The profit in dollars when x pairs of shoes are sold is given by the expression $2x^2 - 120$. Find the profit when 10 pairs are sold.

42. The cost of making dresses is described as 10 times the number of dresses plus 8. Use x as the number of dresses and write a polynomial to describe this cost. Find the cost of making 20 dresses.

FOR REVIEW

Write in scientific notation.

43. 0.0000205

44. 8,600,000,000

Write in standard notation.

45. 8.11×10^5

46. 6.25×10^{-2}

Exercises 47–50 review material from Chapter 2 to help you prepare for the next section. Simplify each expression by removing the parentheses.

47. $-(x + 1)$

48. $-(-x + 1)$

49. $-(a - x + 2)$

50. $-(-a - x + 2)$

6.3 EXERCISES C

The degree of a term of polynomial in several variables is found by adding the exponents on the variables. Find the degree of each term.

1. $3x^2y^3$ **2.** $-8xy^6$ **3.** $6xyz$ [Answer: 3] **4.** $-9a^4b^2c^5$

Write each polynomial in descending powers of x.

5. $7xy^3 - 4x^2y + 8x^4y^4$ **6.** $-3x^2yz^4 + 4yx - 8xyz$
   [Answer: $8x^4y^4 - 4x^2y + 7xy^3$]

6.4 ADDITION AND SUBTRACTION OF POLYNOMIALS

STUDENT GUIDEPOSTS

1 Adding Polynomials   2 Subtracting Polynomials

1 ADDING POLYNOMIALS

Adding polynomials is simply a matter of collecting like terms. We work as follows.

To Add Polynomials

1. Indicate the addition with a plus sign.
2. Remove parentheses and collect like terms.

EXAMPLE 1   ADDING POLYNOMIALS

Add $2x^2 - 3x + 5$ and $-x^2 + 6x + 2$.

$(2x^2 - 3x + 5) + (-x^2 + 6x + 2)$

$= 2x^2 - 3x + 5 - x^2 + 6x + 2$ Remove parentheses

$= (2 - 1)x^2 + (-3 + 6)x + (5 + 2)$ Collect like terms using the distributive law

$= x^2 + 3x + 7$

PRACTICE EXERCISE 1

Add $3y^3 - y + 5$ and $-2y^3 + 6y - 3$.

Answer: $y^3 + 5y + 2$

When adding polynomials in one variable, we usually arrange the terms in descending order. This aids in collecting like terms and helps avoid forgetting a term.

| **EXAMPLE 2** ARRANGING IN DESCENDING ORDER AND ADDING | **PRACTICE EXERCISE 2** |

Add $-2x^3 - 4x^4 + 36 - 3x$ and $-14x^2 + 3x^3 - 6 + 5x$.

$(-4x^4 - 2x^3 - 3x + 36) + (3x^3 - 14x^2 + 5x - 6)$ Arrange in descending order and indicate addition

$\quad = -4x^4 - 2x^3 - 3x + 36 + 3x^3 - 14x^2 + 5x - 6$ Remove parentheses

$\quad = -4x^4 + (3 - 2)x^3 - 14x^2 + (5 - 3)x + (36 - 6)$ Collect like terms using the distributive law

$\quad = -4x^4 + x^3 - 14x^2 + 2x + 30$

Add $2z - 5z^6 + z^3 - 5$ and $4 + z^6 - 8z + 3z^2$.

Answer: $-4z^6 + z^3 + 3z^2 - 6z - 1$

Another way to add polynomials is to arrange like terms in vertical columns as illustrated in Example 3. This technique will also be used later when we multiply polynomials.

| **EXAMPLE 3** ADDING IN COLUMNS | **PRACTICE EXERCISE 3** |

Add $2 - 3x^2$, $-x^4 + 7x - 3x^3 - 5$, and $-8x^3 + 4x^2 - 7$ using vertical columns.

$$
\begin{array}{l}
\, - 3x^2 + 2 \qquad \text{Note the spaces where terms are missing} \\
-x^4 -\, 3x^3 + 7x - 5 \\
\, -\, 8x^3 + 4x^2 - 7x \\
\hline
-x^4 - 11x^3 + x^2 + 0x - 3 = -x^4 - 11x^3 + x^2 - 3
\end{array}
$$

Add $y^5 + 1 - 3y$, $y - 2y^4$, and $2y - y^3 + 3y^5 - 5$ using the column method.

Answer: $4y^5 - 2y^4 - y^3 - 4$

② SUBTRACTING POLYNOMIALS

Since addition is commutative, the arrangement of the polynomials for addition does not matter. However, when subtracting we must be sure to write the problem in the right order. For example, to subtract

$$2x + 5 \quad \text{from} \quad x^2 - 3x + 1,$$

we need to write $\qquad (x^2 - 3x + 1) - (2x + 5).$

However, if we are given, for example,

$$(y^2 + 3) - (-2y + 1),$$

the problem is already set up for us.

Recall the definition of subtraction given in Chapter 1.

$$a - b = a + (-b)$$

That is, to subtract one expression from another, we add the negative of the second expression to the first. If a and b represent polynomials, $-b$ is found by changing all the signs in b. We can proceed as follows.

To Subtract Polynomials

1. Indicate the subtraction by putting a minus sign before the polynomial to be subtracted.
2. Remove parentheses, changing *all* signs in the polynomial being subtracted.
3. Collect like terms as in addition and simplify.

It helps to arrange terms in descending order when subtracting polynomials in one variable.

EXAMPLE 4 SUBTRACTING POLYNOMIALS

Subtract $7x - 3x^2$ from $4 - 2x - 4x^2$.

$(-4x^2 - 2x + 4) - (-3x^2 + 7x)$ Arrange in descending order and indicate the subtraction

$= -4x^2 - 2x + 4 + 3x^2 - 7x$ Remove parentheses, changing signs

$= -x^2 - 9x + 4$ Combine like terms

PRACTICE EXERCISE 4

Subtract $5z^2 + 2$ from $1 - z^2 + 8z$.

Answer: $-6z^2 + 8z - 1$

CAUTION

Be sure to change *all* signs in the polynomial being subtracted. This is shown in Example 4.

To do the same subtraction vertically, change the signs in $7x - 3x^2$. It becomes $-7x + 3x^2$. Now arrange the terms in vertical columns.

$$\begin{array}{ll} -4x^2 - 2x + 4 & \\ +3x^2 - 7x & \text{Change signs} \\ \hline -x^2 - 9x + 4 & \text{Add} \end{array}$$

EXAMPLE 5 SUBTRACTING POLYNOMIALS

Subtract $-3x + 7x^4 + 5x^5 - 2$ from $2x^4 - 8x^5 - 4x + 12x^2 - x^3$.

$(-8x^5 + 2x^4 - x^3 + 12x^2 - 4x) - (5x^5 + 7x^4 - 3x - 2)$

$= -8x^5 + 2x^4 - x^3 + 12x^2 - 4x - 5x^5 - 7x^4 + 3x + 2$ Change signs

$= -8x^5 - 5x^5 + 2x^4 - 7x^4 - x^3 + 12x^2 - 4x + 3x + 2$ Commute

$= -13x^5 - 5x^4 - x^3 + 12x^2 - x + 2$

Subtracting vertically gives the same result.

$$\begin{array}{ll} -8x^5 + 2x^4 - x^3 + 12x^2 - 4x & \text{Arrange in descending order} \\ -5x^5 - 7x^4 \qquad\qquad\quad + 3x + 2 & \text{Change signs} \\ \hline -13x^5 - 5x^4 - x^3 + 12x^2 - x + 2. & \text{Add} \end{array}$$

PRACTICE EXERCISE 5

Subtract $1 - 6w^6 + 2w - 5w^5$ from $4w + 2w^6 - 7 - 8w^5 + w^3$.

Answer: $8w^6 - 3w^5 + w^3 + 2w - 8$

EXAMPLE 6   ADDING AND SUBTRACTING

Perform the indicated operations.

$(5x^3 - 6 + x^2) + (7x^2 - 3) - (-4x^3 + 7x - 5)$

$= (5x^3 + x^2 - 6) + (7x^2 - 3) - (-4x^3 + 7x - 5)$

$= 5x^3 + x^2 - 6 + 7x^2 - 3 + 4x^3 - 7x + 5$ Change signs on
polynomial being
subtracted

$= (5 + 4)x^3 + (1 + 7)x^2 - 7x + (5 - 6 - 3)$

$= 9x^3 + 8x^2 - 7x - 4$

We add and subtract vertically as follows.

$$\begin{aligned}
5x^3 + x^2 \quad\quad - 6 & \\
7x^2 \quad\quad - 3 & \\
+ 4x^3 \quad\quad\quad - 7x + 5 & \quad\text{Change all signs}\\
\hline
9x^3 + 8x^2 - 7x - 4. & \quad\text{Add}
\end{aligned}$$

PRACTICE EXERCISE 6

Perform the indicated operations.

$(6y^4 + 5 - y^3) - (4 - y^3)$
$\quad\quad + (y^2 - 5y^4 + 2)$

Answer: $y^4 + y^2 + 3$

For polynomials in several variables, the procedures are exactly the same. Be sure that only like terms are combined.

EXAMPLE 7   ADDING IN SEVERAL VARIABLES

Add $8x^2 + 2y^2 - 3xy$ and $6x^2 - 7y^2 - 9xy + 3$.

$(8x^2 + 2y^2 - 3xy) + (6x^2 - 7y^2 - 9xy + 3)$

$= 8x^2 + 2y^2 - 3xy + 6x^2 - 7y^2 - 9xy + 3$

$= (8 + 6)x^2 + (2 - 7)y^2 + (-3 - 9)xy + 3$

$= 14x^2 - 5y^2 - 12xy + 3$

PRACTICE EXERCISE 7

Add $6a^3 + b^2 - 4ab$ and
$2b^2 - 3a^3 + 2ab - 5$.

Answer: $3a^3 + 3b^2 - 2ab - 5$

EXAMPLE 8   SUBTRACTING IN SEVERAL VARIABLES

Subtract $3ab - 5a + 9$ from $-6ab + 2a + b$.

$(-6ab + 2a + b) - (3ab - 5a + 9)$

$= -6ab + 2a + b - 3ab + 5a - 9$ Change signs

$= (-6 - 3)ab + (2 + 5)a + b - 9$ Collect like terms

$= -9ab + 7a + b - 9$

PRACTICE EXERCISE 8

Subtract $6uv - 3v^2 + 12$ from
$v^2 - 2uv + 5$.

Answer: $4v^2 - 8uv - 7$

EXAMPLE 9   SUBTRACTING SEVERAL TIMES

Perform the indicated operations.

$(7x^2y^2 - 2xy) - (5x^2y + 9xy) - (-x^2y^2 + 3xy^2 - 4xy)$

$= 7x^2y^2 - 2xy - 5x^2y - 9xy + x^2y^2 - 3xy^2 + 4xy$ Change signs

$= (7 + 1)x^2y^2 + (-2 - 9 + 4)xy - 5x^2y - 3xy^2$ Collect like terms

$= 8x^2y^2 - 7xy - 5x^2y - 3xy^2$

Note that $-5x^2y$ and $-3xy^2$ are not like terms.

PRACTICE EXERCISE 9

Perform the indicated operations.

$(9a^3b^3 + ab) - (3ab - a^3b^3)$
$\quad\quad + (8a - 5a^3b^3 + 2ab)$

Answer: $5a^3b^3 + 8a$

6.4 EXERCISES A

Add.

1. $3x - 5$ and $-8x + 4$

2. $7x^2 + 6$ and $x^2 - 2$

3. $y^2 + 3y - 5$ and $-8y^2 - 5y + 9$

4. $2y - 3$ and $-4y^2 + 2$

5. $3x^5 - 2x^3 + 5x^2$ and $-5x^5 + 8x^4 - 7x^2$

6. $-x^6 + 2x^4$ and $2x^7 - x^6 + 5x^2 + 2$

7. $8y - 6y^2 + 2$ and $-5 + 2y + 8y^2$

8. $6y^5 - 2y + y^4$ and $3 - 5y^2 + 21y + 2y^5$

9. $2a^2 + 3a$, $-5a^2 + 6$, and $-9a + 2$

10 $-4a^3 + 7a^4 + 3a + 2$, $5 - 3a + 7a^3$, and $17a^4 - 5 + 12a^3$

11. $2 - x^2 + 7x$, $9x - 7 + x^3$, $-6x$, and $12 - 3x^3$

12. $25x^4 - 7x^5$, $12x - 17x^2 + 40x^3 - 18x^4$, and $x - 21 + 18x^5$

13. $\begin{array}{r} 3y^2 - 2y + 1 \\ -2y^2 + 2y + 8 \\ \hline \end{array}$

14. $\begin{array}{r} 5x^5 \quad\ + \ x^3 \qquad - 2x \\ 2x^5 - 8x^4 - 7x^3 + 6x^2 \\ \hline \end{array}$

15. $\begin{array}{r} 5x^3 + 6x^2 \qquad - 7 \\ -4x^4 + 3x^3 - 3x^2 - 8x \\ 3x^4 - 2x^3 + 4x^2 \qquad + 1 \\ \hline \end{array}$

16 $\begin{array}{r} 0.03y^3 - 0.75y^2 - 3y + 2 \\ -0.15y^3 \qquad\quad + 5y - 0.3 \\ 0.21y^3 - 0.13y^2 \qquad + 0.6 \\ \hline \end{array}$

Subtract.

17. $2x + 5$ from $3x - 6$

18. $6x^2 - 2$ from $4x^2 + 5$

19. $-3y^2 + 2y - 7$ from $-8y^2 + y - 9$

20. $-3y + 2$ from $y^2 - 4y - 5$

21. $a^4 + 5a^3 - 3a^2$ from $7a^5 - 2a^4 - a^2 + 3a$

22. $3a - 2a^2 + 7a^3$ from $a^4 - 2a^2 + a^3 - 2$

23 $3x^4 - 7x^5 + 15x - 32$ from $56x - 93x^3 + 21x^4 + 32x^5$

24. $7x^{10} - 4x^5 + 1$ from $3x^5 + 1 - x^6 - 3x^2$

Perform the indicated operations.

25. $(-8y^2 + 4) - (7y^2 - 3)$

26. $(-8a^2 - 2a + 5) - (-8a^2 + a + 1)$

27. $(6y^7 - y + y^5 + 2) - (2y + 3y^2 - 4y^5 - 2)$

28 $(8y^{10} - y^8) - (3y^{12} + 2y^{10} - y^8)$

29. $(3y^2 + 2) + (-5y - 5) - (y^2 - 2y + 10)$

30. $(-2a^2 + 3a) - (a^2 + a + 1) - (a^2 - 2a - 5)$

31 $(9x^4 + 3x^3 + 8x) + (3x^4 + x^3 - 7x^2) - (12x^4 - 3x^2 + x)$

32. $(9y^4 + 3y^3 + 8y) - (3y^4 + y^3 - 7y^2) - (12y^4 - 3y^2 + y)$

33. $(4a^2b^2 - 2ab) + (5a^2b^2 + 9ab)$

34. $(3a^2 + 2b^2 + 3) + (-6a^2 + 2b^2 - 2)$

35. $(4x^2y^2 - 2xy) - (5x^2y^2 + 9xy)$

36. $(3x^2 + 2y^2 + 3) - (-6x^2 + 2y^2 - 2)$

37. $(-2a^2b + ab - 4ab^2) + (6a^2b + 4ab^2)$

38 $(-2a^2b + ab - 4ab^2) - (6a^2b + 4ab^2)$

39. $(6x^2y - xy) + (3x^2y - 7xy^2) - (4xy - 5xy^2)$

40 $(6x^2y - xy) - (3x^2y - 7xy^2) - (4xy - 5xy^2)$

FOR REVIEW

Give the degree of each polynomial.

41. $-6y + 8y^5 + y^4 - 2$

42. 14

43. $x^{10} - 6x^{20} + x^{30}$

Evaluate the polynomial for the given value of the variable.

44. $-7a + 3$ for $a = -2$

45. $3y^3 + 2y^2$ for $y = -1$

46. The profit in dollars when x suits are sold is given by the expression $7x - 50$. Find the profit when 60 suits are sold.

Exercises 47–50 review material from Chapter 2 to help you prepare for the next section. Find the product.

47. $(3x)(-2x)$ **48.** $(-3x)(-2x)$ **49.** $(x^2)(4x)$ **50.** $(2x^2)(-5x^2)$

ANSWERS: 1. $-5x - 1$ 2. $8x^2 + 4$ 3. $-7y^2 - 2y + 4$ 4. $-4y^2 + 2y - 1$ 5. $-2x^5 + 8x^4 - 2x^3 - 2x^2$ 6. $2x^7 - 2x^6 + 2x^4 + 5x^2 + 2$ 7. $2y^2 + 10y - 3$ 8. $8y^5 + y^4 - 5y^2 + 19y + 3$ 9. $-3a^2 - 6a + 8$ 10. $24a^4 + 15a^3 + 2$ 11. $-2x^3 - x^2 + 10x + 7$ 12. $11x^5 + 7x^4 + 40x^3 - 17x^2 + 13x - 21$ 13. $y^2 + 9$ 14. $7x^5 - 8x^4 - 6x^3 + 6x^2 - 2x$ 15. $-x^4 + 6x^3 + 7x^2 - 8x - 6$ 16. $0.09y^3 - 0.88y^2 + 2y + 2.3$ 17. $x - 11$ 18. $-2x^2 + 7$ 19. $-5y^2 - y - 2$ 20. $y^2 - y - 7$ 21. $7a^5 - 3a^4 - 5a^3 + 2a^2 + 3a$ 22. $a^4 - 6a^3 - 3a - 2$ 23. $39x^5 + 18x^4 - 93x^3 + 41x + 32$ 24. $-7x^{10} - x^6 + 7x^5 - 3x^2$ 25. $-15y^2 + 7$ 26. $-3a + 4$ 27. $6y^7 + 5y^5 - 3y^2 - 3y + 4$ 28. $-3y^{12} + 6y^{10}$ 29. $2y^2 - 3y - 13$ 30. $-4a^2 + 4a + 4$ 31. $4x^3 - 4x^2 + 7x$ 32. $-6y^4 + 2y^3 + 10y^2 + 7y$ 33. $9a^2b^2 + 7ab$ 34. $-3a^2 + 4b^2 + 1$ 35. $-x^2y^2 - 11xy$ 36. $9x^2 + 5$ 37. $4a^2b + ab$ 38. $-8a^2b + ab - 8ab^2$ 39. $9x^2y - 5xy - 2xy^2$ 40. $3x^2y - 5xy + 12xy^2$ 41. 5 42. 0 43. 30 44. 17 45. -1 46. $\$370$ 47. $-6x^2$ 48. $6x^2$ 49. $4x^3$ 50. $-10x^4$

6.4   EXERCISES B

Add.

1. $9x + 3$ and $-4x - 2$

2. $-8x^2 + 5$ and $3x^2 + 2$

3. $8a^2 + a - 7$ and $-a + 3$

4. $2a^4 - 5a^2 + 6$ and $3a^2$

5. $6x^4 + 2x^3 - 7x^2$ and $-9x^5 + 3x^4 - x^3 + 7$

6. $6x^9 - 3x^6 + 2x^3$ and $2x^8 - 4x^6 - 9x^3$

7. $-8y^3 + 5 - 6y^2$ and $3 - 2y^2 + y^3$

8. $3y - 5y^3 - 2y^2 + 2$ and $9 + 2y^2 - y^3$

9. $-12a^2 - 6$, $2a^2 + 5a$, and $a - 3$

10. $-9a^3 + a^4 - 6$, $2a - 5a^4 + a^3$, and $8a^4 - 2a^3 + 5$

11. $x^2 - x^3 + x$, $5x - 6x^3 + 3$, 9, and $3 + 2x^3$

12. $-3x^5 + 9x^2$, $6 - 12x^5 + 6x^2$, and $3 - 6x + 5x^5 + x^2$

13. $\begin{aligned} -9y^2 + 2y - 6 \\ \underline{6y^2 + 3y + 9} \end{aligned}$

14. $\begin{aligned} -8x^4 \qquad\quad - 2x^2 + 6x - 8 \\ \underline{7x^4 + 12x^3 + 2x^2 \qquad + 4} \end{aligned}$

15. $\begin{aligned} - 2x^3 + 8x^2 - 9x + 9 \\ 7x^4 \qquad\quad + 2x^2 + 9x - 8 \\ \underline{-6x^4 + 6x^3 - 4x^2 \qquad + 7} \end{aligned}$

16. $\begin{aligned} 0.23y^3 + 0.98y^2 - 0.6y + 0.8 \\ 0.54y^3 - 0.82y^2 + 0.1y \\ \underline{-0.77y^3 \qquad\qquad + 0.9y - 0.3} \end{aligned}$

Subtract.

17. $3x - 5$ from $2x + 8$

18. $2x^2 - 9$ from $-5x^2 + 7$

19. $6y^2 - 2y + 8$ from $-4y^2 - 2y + 1$

20. $-3y^2 + 9$ from $2y^2 - 9y + 2$

21. $2a^4 - 3a^3 + 5a^2$ from $-9a^5 + 2a^4 - 3a^2 + 2a$

22. $5a - 3a^2 + 5a^3$ from $a^4 + 5a^2 + 5a^3 + 3$

23. $5x^4 + 7x^5 - 8x + 20$ from $19x + 20x^3 + 72x^4 - 10x^5$ **24.** $10x^8 + 10x^6 - 13$ from $2x^4 - 13 + 8x^8 - x^2$

Perform the indicated operations.

25. $(7y^2 - 8) - (-2y^2 + 7)$

26. $(9a^2 + 7a - 4) - (-2a^2 + 7a - 7)$

27. $(-6y + 4y - y^5 + 8) - (5y - 4y^2 + 7y^5 - 2)$

28. $(3y^{14} - 2y^{10}) - (-y^{14} + 2y^{10} + y^6)$

29. $(5y + 6) + (-2y^2 + 5) - (2y^2 - 3y - 8)$

30. $(-8a^2 - 2a) - (2a^2 + 3a - 1) - (a^2 - 9a + 6)$

31. $(3x^4 - 8x^3 - 4x) + (2x^4 - 8x^3 + 5x^2) - (5x^4 + 5x^2 - 6x)$

32. $(3y^4 - 8y^3 - 4y) - (2y^4 - 8y^3 + 5y^2) - (5y^4 + 5y^2 - 6y)$

33. $(2a^2b^2 + 5ab) + (9a^2b^2 - 2ab)$

34. $(4a^2 - 3b^2 + 9) + (-4a^2 - 5b^2 + 3)$

35. $(8x^2y^2 - 9xy) - (-4x^2y^2 + 7xy)$

36. $(5x^2 + 3y^2 - 7) - (-2x^2 + 3y^2 + 7)$

37. $(5a^2b - 2ab - 2ab^2) + (2a^2b - 6ab^2)$

38. $(5a^2b - 2ab - 2ab^2) - (2a^2b - 6ab^2)$

39. $(11x^2y - 2xy) + (4x^2y - 9xy^2) - (8xy - 10xy^2)$

40. $(11x^2y - 2xy) - (4x^2y - 9xy^2) - (8xy - 10xy^2)$

FOR REVIEW

Give the degree of each polynomial.

41. $8y^2 - 6y^{10} - y^{14}$

42. $22a$

43. $x^{20} + 2x^{40}$

Evaluate the polynomial for the given value of the variable.

44. $6a^2 - 2a$ for $a = 3$

45. $y^4 - 6y^2 + 8$ for $y = -2$

46. The cost in dollars when x suits are made is given by the expression $4x^2 + 5x - 100$. Find the cost when 15 suits are made.

Exercises 47–50 review material from Chapter 2 to help you prepare for the next section. Find the product.

47. $(-7y)(4y)$

48. $(-7y)(-4y)$

49. $(-3x^3)(6x)$

50. $(6x^3)(5x^3)$

6.4 EXERCISES C

Perform the indicated operations.

1. $\left(\dfrac{1}{3}x^4 - \dfrac{1}{2}x^3 + \dfrac{7}{8}\right) + \left(-\dfrac{1}{9}x^4 + \dfrac{3}{4}x^3 - \dfrac{2}{3}x^2 + x\right) + \left(\dfrac{2}{9}x^4 - \dfrac{3}{2}x^3 + \dfrac{4}{3}x^2 + \dfrac{1}{3}x + \dfrac{3}{8}\right)$

$\left[\text{Answer: } \dfrac{4}{9}x^4 - \dfrac{5}{4}x^3 + \dfrac{2}{3}x^2 + \dfrac{4}{3}x + \dfrac{5}{4}\right]$

2. $(100a^3 - 97a^2 + 21a - 105) + (16a^3 + 45a^2 - 115a) - (88a^3 - 19a^2 - 47)$

3. $(-2.13a^2 + 3.25a + 7.98) - (0.32a^3 - 2.10a + 1.92)$ [Answer: $-0.32a^3 - 2.13a^2 + 5.35a + 6.06$]

4. $\left(-\dfrac{1}{9}x^5 + \dfrac{1}{3}x^4 - \dfrac{2}{3}x + \dfrac{1}{3}\right) - \left(-\dfrac{5}{9}x^5 - \dfrac{2}{9}x^4 + \dfrac{8}{9}x + \dfrac{7}{9}\right)$

6.5 MULTIPLICATION OF POLYNOMIALS

―――――――――――――― STUDENT GUIDEPOSTS ――――――――――――――

1 Multiplying Monomials
2 Multiplying a Binomial by a Monomial
3 Multiplying Polynomials
4 The FOIL Method

1 MULTIPLYING MONOMIALS

In Section 6.1 we multiplied powers of a variable by adding their exponents using the product rule

$$x^m x^n = x^{m+n}.$$

This rule is used repeatedly when we multiply polynomials. We begin by reviewing multiplication of monomials.

EXAMPLE 1 MONOMIAL TIMES A MONOMIAL	PRACTICE EXERCISE 1

Multiply.

(a) $(-3x^2)(7x^5) = (-3)(7)x\ x$ The order of the factors can be changed
$$= -21x \qquad \text{Use } x^m x^n = x^{m+n}$$
$$= -21x^7$$

(b) $(5x^4)(-x) = (5)(-1)x^4 x$ $-x = (-1) \cdot x$
$$= -5x^{4+1} \qquad x = x^1$$
$$= -5x^5$$

Multiply.

(a) $(4y^3)(-5y^7)$

(b) $(-8y^2)(-y)$

Answers: (a) $-20y^{10}$ (b) $8y^3$

2 MULTIPLYING A BINOMIAL BY A MONOMIAL

In Section 2.4 we introduced the distributive law. Examples in that section showed multiplication of a binomial by a monomial. Example 2 reviews this idea.

EXAMPLE 2 MONOMIAL TIMES A BINOMIAL	PRACTICE EXERCISE 2

Multiply.

(a) $a\,(3a^2 + 2) = (a)\,(3a^2) + (a)\,(2)$ Distributive law
$$= 3aa^2 + 2a$$
$$= 3a^3 + 2a$$

(b) $-3a^2\,(8a^3 - 5a) = (-3a^2)\,(8a^3) + (-3a^2)\,(-5a)$ Distributive law
$$= (-3)(8)a^2 a^3 + (-3)(-5)a^2 a$$
$$= -24a^{2+3} + 15a^{2+1}$$
$$= -24a^5 + 15a^3$$

Multiply.

(a) $x(5 + 3x^3)$

(b) $-4x^3(2x^2 - 7x)$

Answers: (a) $5x + 3x^4$
(b) $-8x^5 + 28x^4$

3 MULTIPLYING POLYNOMIALS

Polynomials can be multiplied by repeated use of the distributive law. This is illustrated in Example 3.

EXAMPLE 3 USING THE DISTRIBUTIVE LAW	**PRACTICE EXERCISE 3**

Multiply.

(a) $(3x + 2)(x^2 - 5x)$

$= (3x + 2)\,x^2 + (3x + 2)(-5x)$ Distributive law

$= (3x)(x^2) + (2)(x^2) + (3x)(-5x) + (2)(-5x)$ Distributive law

$= 3x^3 + 2x^2 - 15x^2 - 10x$

$= 3x^3 - 13x^2 - 10x$

(b) $(4x + 1)(3x^2 - x + 5)$

$= (4x + 1)\,3x^2 + (4x + 1)(-x) + (4x + 1)\,5$

$= (4x)(3x^2) + (1)(3x^2) + (4x)(-x) + (1)(-x) + (4x)(5) + (1)(5)$

$= 12x^3 + 3x^2 - 4x^2 - x + 20x + 5$

$= 12x^3 - x^2 + 19x + 5$

Multiply.

(a) $(2z + 1)(z^3 - 3z)$

(b) $(5z - 3)(2z^2 + z - 7)$

Answers: (a) $2z^4 + z^3 - 6z^2 - 3z$
(b) $10z^3 - z^2 - 38z + 21$

Notice that the second step in each part of Example 3 contains all possible products of terms in the first polynomial with terms in the second. This leads to the following rule.

To Multiply Two Polynomials, Neither of Which Is a Monomial

1. Multiply each term in one by each term in the other.
2. Collect and combine like terms.

④ THE FOIL METHOD

The distributive law guarantees that when two polynomials are multiplied, all products of all terms are found. Now suppose we use the following method to find the product in Example 3(a).

$(3x + 2)(x^2 - 5x) = (3x)(x^2) + (3x)(-5x) + (2)(x^2) + (2)(-5x)$

$= 3x^3 - 15x^2 + 2x^2 - 10x$

$= 3x^3 - 13x^2 - 10x$

Notice that the letters F, O, I, L spell "FOIL" and stand for the **F**irst terms, **O**utside terms, **I**nside terms, and **L**ast terms. Remember this word, and you will not omit terms when multiplying a binomial by a binomial.

FOIL Method

$(a + b) \cdot (c + d) = a \cdot c + a \cdot d + b \cdot c + b \cdot d$

EXAMPLE 4   MULTIPLYING BINOMIALS USING FOIL

Use the FOIL method to multiply the binomials.

(a) $(x - 3) \cdot (x + 4) = x^2 + 4x - 3x - 12$
$= x^2 + x - 12$

(b) $(x + 7)(x + 5) = x^2 + 5x + 7x + 35$
$= x^2 + 12x + 35$

PRACTICE EXERCISE 4

Use the FOIL method to multiply the binomials.

(a) $(a - 2)(a + 8)$

(b) $(a + 2)(a + 9)$

Answers: (a) $a^2 + 6a - 16$
(b) $a^2 + 11a + 18$

When we study factoring in Chapter 7, we will start with trinomials such as $x^2 + x - 12$ in Example 4(a) and be expected to find the factors $(x - 3)$ and $(x + 4)$. Practicing the FOIL method now will help us understand factoring later.

EXAMPLE 5   USING THE FOIL METHOD

Use the FOIL method to multiply the binomials.

(a) $(2x - 3)(3x + 7) = 6x^2 + 14x - 9x - 21$
$= 6x^2 + 5x - 21$

(b) $(2a - 5)(2a + 5) = 4a^2 + 10a - 10a - 25$
$= 4a^2 + 0 \cdot a - 25$
$= 4a^2 - 25$

(c) $(5x + 3)(x + 8) = 5x^2 + 40x + 3x + 24$
$= 5x^2 + 43x + 24$

(d) $(7y - 2)(3y - 8) = 21y^2 - 56y - 6y + 16$
$= 21y^2 - 62y + 16$

(e) $(10a + 5)(3a - 5) = 30a^2 - 50a + 15a - 25$
$= 30a^2 - 35a - 25$

PRACTICE EXERCISE 5

Use the FOIL method to multiply the binomials.

(a) $(3y - 1)(2y + 4)$

(b) $(6x - 7)(6x + 7)$

(c) $(8z + 3)(z - 1)$

(d) $(4a - 5)(2a - 3)$

(e) $(12x + 1)(5x - 2)$

Answers: (a) $6y^2 + 10y - 4$
(b) $36x^2 - 49$ (c) $8z^2 - 5z - 3$
(d) $8a^2 - 22a + 15$ (e) $60x^2 - 19x - 2$

A third way to multiply is to write one polynomial above the other and then arrange like terms in vertical columns.

$$
\begin{array}{r}
x^2 - 5x \\
3x + 2 \\
\hline
3x^3 - 15x^2 \\
2x^2 - 10x \\
\hline
3x^3 - 13x^2 - 10x
\end{array}
$$

3x multiplied by each term of the top polynomial
2 multiplied by each term of the top polynomial
Sum

The third method is especially good when trinomials or larger polynomials are part of the multiplication.

EXAMPLE 6 USING VERTICAL COLUMNS

Multiply.

(a)
$$5x^3 - 2x + 3$$
$$\underline{-7x^2 + 4x}$$
$$-35x^5 \qquad\quad + 14x^3 - 21x^2 \qquad\qquad -7x^2 \text{ multiplied by top}$$
$$\text{polynomial}$$
$$\underline{\qquad 20x^4 \qquad\quad - 8x^2 + 12x} \quad 4x \text{ multiplied by top polynomial}$$
$$-35x^5 + 20x^4 + 14x^3 - 29x^2 + 12x \quad \text{Sum}$$

We leave spaces for missing terms so we can add like terms in columns.

(b)
$$a^2 + 2a + 4$$
$$\underline{a - 2}$$
$$a^3 + 2a^2 + 4a \qquad\qquad\qquad a \text{ multiplied by top polynomial}$$
$$\underline{\quad - 2a^2 - 4a - 8} \qquad -2 \text{ multiplied by top polynomial}$$
$$a^3 \qquad\qquad - 8 = a^3 - 8 \quad \text{Sum}$$

(c)
$$5x^2 - 3x + 7$$
$$\underline{-4x^2 + x + 8}$$
$$-20x^4 + 12x^3 - 28x^2 \qquad\qquad -4x^2 \text{ multiplied by top polynomial}$$
$$5x^3 - 3x^2 + 7x \qquad\qquad x \text{ multiplied by top polynomial}$$
$$\underline{\qquad\qquad\quad 40x^2 - 24x + 56} \quad 8 \text{ multiplied by top polynomial}$$
$$-20x^4 + 17x^3 + 9x^2 - 17x + 56 \quad \text{Sum}$$

All the procedures used above can be applied to the multiplication of polynomials in several variables.

EXAMPLE 7 POLYNOMIALS IN TWO VARIABLES

Use the FOIL method to multiply the binomials in two variables.

(a) $(2x + y)(x - 3y) = 2x^2 - 6xy + xy - 3y^2$
$$= 2x^2 - 5xy - 3y^2$$

(b) $(2a - 5b)(2a + 5b) = 4a^2 + 10ab - 10ab - 25b^2$
$$= 4a^2 + 0 \cdot ab - 25b^2$$
$$= 4a^2 - 25b^2$$

(c) $(5x + 3y)(x + 8y) = 5x^2 + 40xy + 3xy + 24y^2$
$$= 5x^2 + 43xy + 24y^2$$

EXAMPLE 8 USING VERTICAL COLUMNS

Use vertical columns to multiply.

$$3x^2 - 4xy + 5y^2$$
$$\underline{6x - 7y}$$
$$18x^3 - 24x^2y + 30xy^2 \qquad\qquad 6x \text{ multiplied by top polynomial}$$
$$\underline{\quad - 21x^2y + 28xy^2 - 35y^3} \quad -7y \text{ multiplied by top polynomial}$$
$$18x^3 - 45x^2y + 58xy^2 - 35y^3 \quad \text{Sum}$$

We conclude this section by solving an applied problem.

PRACTICE EXERCISE 6

Multiply.

(a) $3y^2 - y + 2$
$$\underline{y^2 + 2y}$$

(b) $4x^2 + 6x + 9$
$$\underline{2x - 3}$$

(c) $2a^2 + 5a - 6$
$$\underline{-3a^2 + a + 5}$$

Answers: (a) $3y^4 + 5y^3 + 4y$
(b) $8x^3 - 27$ (c) $-6a^4 - 13a^3 + 33a^2 + 19a - 30$

PRACTICE EXERCISE 7

Use the FOIL method to multiply the binomials in two variables.

(a) $(5a + 2b)(a - b)$

(b) $(3x + 7y)(3x - 7y)$

(c) $(6z - 4w)(5z - 3w)$

Answers: (a) $5a^2 - 3ab - 2b^2$
(b) $9x^2 - 49y^2$ (c) $30z^2 - 38zw + 12w^2$

PRACTICE EXERCISE 8

Use vertical columns to multiply.

$$4a^2 + 3ab + b^2$$
$$\underline{5a - 2b}$$

Answer: $20a^3 + 7a^2b - ab^2 - 2b^3$

| EXAMPLE 9   BUSINESS APPLICATION | PRACTICE EXERCISE 9 |

The Valley Video Center sells blank video cassettes. From past experience it has been shown that the number n of cassettes that can be sold each week depends on the price p in dollars charged for each according to the equation $n = 1000 - 100p$. Find a formula for the weekly revenue R in terms of p. Use this formula to find the revenue when the price of a cassette is $5, $8, and $10.

Find a formula for weekly revenue from a small item if $n = 300 - 20p$. Use the formula to find R when the price is $1, $2, and $3.

The **revenue** R produced by selling n items for p dollars per item is given by $R = np$. In our case, the number of cassettes sold per week is related to the price of a cassette by $n = 1000 - 100p$. Notice that this relationship tells us that when p is $9, $n = 1000 - 100(9) = 100$, and when p is reduced to $8, $n = 1000 - 100(8) = 200$. As a result, we see that more cassettes can be sold when the price is lowered. The formula for the revenue produced weekly is

$$R = np = (1000 - 100p)p$$
$$= 1000p - 100p^2.$$

When p is $5,

$$R = 1000(5) - 100(5)^2$$
$$= 5000 - 2500 = 2500,$$

so the revenue is $2500. When p is $8,

$$R = 1000(8) - 100(8)^2$$
$$= 8000 - 6400 = 1600,$$

so the revenue is $1600. Finally, when p is $10,

$$R = 1000(10) - 100(10)^2$$
$$= 10{,}000 - 10{,}000 = 0,$$

so the revenue drops off to $0. In other words, the consumer views $10 as too much and refuses to purchase cassettes at this price.

Answer: $R = 300p - 20p^2$; $280, $520, $720

6.5 EXERCISES A

Multiply the following monomials.

1. $(2)(6x)$

2. $(-8y)(4)$

3. $(-7a^2)(-4a)$

4. $(4y^2)(2y^2)$

5. $(-3x^4)(2x)$

6. $(8y^3)(-9y^2)$

7. $(-13x^7)(10x^3)$

8. $(-8y^6)(-7y^5)$

Multiply using the distributive law.

9. $2(x + 3)$

10. $3x(x - 4)$

11. $-4y(y + 8)$

12. $-6a^2(4a + 5)$

13. $3(2x^2 - 3x + 1)$

14. $4y(y^2 - 8y - 5)$

15. $2y^3(-3y^2 - 2y + 5)$

16 $-10x^2(x^5 - 6x^3 + 7x^2)$

17. $-9a^3(4a^4 - 3a^2 + 2)$

18. $(x + 8)(x + 4)$

19. $(2a - 3)(3a - 2)$

20. $(x - 5)(x^2 - 3x + 2)$

21. $(2y + 7)(3y^2 + y - 1)$

22. $(a^2 + 1)(a^4 - a^2 + 1)$

23 $(x^2 + 4x - 2)(2x^2 - x + 3)$

Use the FOIL method to multiply the following binomials.

24. $(x + 10)(x + 2)$

25. $(a - 6)(a + 6)$

26. $(y + 5)(y - 3)$

27. $(x - 7)(x - 3)$

28. $(a + 8)(a + 8)$

29. $(x - 12)(x + 4)$

30. $(a - 8)(a - 8)$

31. $(3x + 2)(x + 5)$

32. $(5a - 3)(a - 7)$

33. $(2y - 7)(y + 10)$

34. $(7x + 3)(x - 5)$

35. $(2a + 3)(3a + 5)$

36. $(4y - 3)(2y - 5)$

37 $(5x - 2)(3x + 4)$

38. $(9a + 2)(2a - 7)$

39. $(10y - 1)(4y + 5)$

40. $(7x + 5)(3x - 8)$

41 $(2z^2 + 1)(z^2 - 2)$

Multiply.

42. $3a^2 + 5$
  $\underline{a - 4}$

43. $2a^3 - 3a$
  $\underline{7a^2 - 5}$

44. $3y^3 - 2$
  $\underline{5y^2 + 4y}$

45. $3a^2 - 5$
  $\underline{3a^2 - 5}$

46. $7y^3 - 2$
  $\underline{7y^3 + 2}$

47. $5x^4 + 3x^2$
  $\underline{8x^2 - 7}$

48. $3x^2 - 5x + 2$
  $\underline{x + 4}$

49. $-7y^2 - 3y + 10$
  $\underline{2y - 1}$

50. $12a^3 - 3a + 8$
  $\underline{4a^2 + 7a}$

51. $12a^3 - 6a + 4$
$7a^2 - 3$

52 $0.3x^2 + 0.2$
$0.5x - 0.7$

53. $\dfrac{2}{3}y^2 - \dfrac{3}{4}$
$\dfrac{1}{2}y + 5$

54. $a^2 - 2a + 2$
$a^2 + 2a - 2$

55 $3y^2 + 5y - 6$
$y^2 - 3y + 2$

56. $-4x^3 + 2x - 5$
$6x^2 - x + 1$

Multiply the polynomials in two variables.

57. $2xy(x^2 + 2xy)$

58. $3a^2b(a^2b + 2ab - ab^2)$

59 $(2a - 3b)(5a + b)$

60. $(x + 3y)(x + 2y)$

61. $(4x - y)(4x + y)$

62. $(7a - 2b)(2a - 5b)$

63. $3x^2 - 2xy + 4y^2$
$2x + y$

64. $a^2b^2 + ab - 3$
$a^2 - b^2$

Perform the following operations.

65 $(x - 1)(x + 1)(x + 2)$

66. $(y + 3)(y - 2)(y + 2)$

67 The length of a rectangle measures $(2x + 1)$ feet and the width is $(3x - 2)$ feet. Find the area of the rectangle in terms of x.

68. The dimensions of a box, given in centimeters, are y, $y + 1$, and $2y + 3$. Find the volume of the box in terms of y.

69. A wholesale distributor who sells clocks knows that the number n of clocks she can sell each month is related to the price in dollars p of each clock by the equation $n = 5000 - 100p$. Find an equation for the monthly revenue R in terms of p and use it to find R when p is \$10, \$30, and \$45.

70. The length l of a rectangular pasture is to be 300 m longer than the width w. Find an equation that gives the area in terms of the width and use it to give the area when the width is 40 m, 100 m, and 500 m.

FOR REVIEW

Perform the indicated operations.

71. $(3x^2 + 2x - 1) + (2x + 1) - (x^2 - 2)$

72. $(-4y^2 + 2y) + (y^3 - 2y) - (y^2 + 3y - 5)$

73. $(3a^2 - 5a + 5) - (-a^5 + 4a^2 - 2) - (6a - 7)$

74. $(x^2 - y^2) - (2x^2 - 3y^2) - (5x^2 - y^2)$

In Exercises 75–80, find the special products using the FOIL method. This will help you prepare for the next section.

75. $(x + 5)(x + 5)$

76. $(x - 5)(x - 5)$

77. $(x + 5)(x - 5)$

78. $(3x + 2y)(3x - 2y)$

79. $(3x - 2y)(3x - 2y)$

80. $(3x + 2y)(3x + 2y)$

ANSWERS: 1. $12x$ 2. $-32y$ 3. $28a^3$ 4. $8y^4$ 5. $-6x^5$ 6. $-72y^5$ 7. $-130x^{10}$ 8. $56y^{11}$ 9. $2x + 6$
10. $3x^2 - 12x$ 11. $-4y^2 - 32y$ 12. $-24a^3 - 30a^2$ 13. $6x^2 - 9x + 3$ 14. $4y^3 - 32y^2 - 20y$ 15. $-6y^5 - 4y^4 + 10y^3$
16. $-10x^7 + 60x^5 - 70x^4$ 17. $-36a^7 + 27a^5 - 18a^3$ 18. $x^2 + 12x + 32$ 19. $6a^2 - 13a + 6$ 20. $x^3 - 8x^2 + 17x - 10$
21. $6y^3 + 23y^2 + 5y - 7$ 22. $a^6 + 1$ 23. $2x^4 + 7x^3 - 5x^2 + 14x - 6$ 24. $x^2 + 12x + 20$ 25. $a^2 - 36$
26. $y^2 + 2y - 15$ 27. $x^2 - 10x + 21$ 28. $a^2 + 16a + 64$ 29. $x^2 - 8x - 48$ 30. $a^2 - 16a + 64$ 31. $3x^2 + 17x + 10$
32. $5a^2 - 38a + 21$ 33. $2y^2 + 13y - 70$ 34. $7x^2 - 32x - 15$ 35. $6a^2 + 19a + 15$ 36. $8y^2 - 26y + 15$
37. $15x^2 + 14x - 8$ 38. $18a^2 - 59a - 14$ 39. $40y^2 + 46y - 5$ 40. $21x^2 - 41x - 40$ 41. $2z^4 - 3z^2 - 2$
42. $3a^3 - 12a^2 + 5a - 20$ 43. $14a^5 - 31a^3 + 15a$ 44. $15y^5 + 12y^4 - 10y^2 - 8y$ 45. $9a^4 - 30a^2 + 25$ 46. $49y^6 - 4$
47. $40x^6 - 11x^4 - 21x^2$ 48. $3x^3 + 7x^2 - 18x + 8$ 49. $-14y^3 + y^2 + 23y - 10$ 50. $48a^5 + 84a^4 - 12a^3 + 11a^2 + 56a$
51. $84a^5 - 78a^3 + 28a^2 + 18a - 12$ 52. $0.15x^3 - 0.21x^2 + 0.10x - 0.14$ 53. $\frac{1}{3}y^3 + \frac{10}{3}y^2 - \frac{3}{8}y - \frac{15}{4}$
54. $a^4 - 4a^2 + 8a - 4$ 55. $3y^4 - 4y^3 - 15y^2 + 28y - 12$ 56. $-24x^5 + 4x^4 + 8x^3 - 32x^2 + 7x - 5$ 57. $2x^3y + 4x^2y^2$
58. $3a^4b^2 + 6a^3b^2 - 3a^3b^3$ 59. $10a^2 - 13ab - 3b^2$ 60. $x^2 + 5xy + 6y^2$ 61. $16x^2 - y^2$ 62. $14a^2 - 39ab + 10b^2$
63. $6x^3 - x^2y + 6xy^2 + 4y^3$ 64. $a^4b^2 + a^3b - 3a^2 - a^2b^4 - ab^3 + 3b^3$ 65. $x^3 + 2x^2 - x - 2$ 66. $y^3 + 3y^2 - 4y - 12$
67. $(6x^2 - x - 2)$ ft^2 68. $(2y^3 + 5y^2 + 3y)$ cm^3 69. $R = 5000p - 100p^2$; \$40,000; \$60,000; \$22,500
70. $A = w^2 + 300w$; 13,600 m^2; 40,000 m^2; 400,000 m^2 71. $2x^2 + 4x + 2$ 72. $y^3 - 5y^2 - 3y + 5$
73. $a^5 - a^2 - 11a + 14$ 74. $-6x^2 + 3y^2$ 75. $x^2 + 10x + 25$ 76. $x^2 - 10x + 25$ 77. $x^2 - 25$ 78. $9x^2 - 4y^2$
79. $9x^2 - 12xy + 4y^2$ 80. $9x^2 + 12xy + 4y^2$

6.5 EXERCISES B

Multiply the following monomials.

1. $(3)(5x)$

2. $(-4y)(8)$

3. $(9a)(2a^3)$

4. $(-3a^3)(5a^2)$

5. $(-10y)(-8y^5)$

6. $(a^7)(a^7)$

7. $(18x^8)(-2x^3)$

8. $(-10y^7)(-20y^8)$

Multiply using the distributive law.

9. $5x(x - 6)$

10. $2x(-3x + 5)$

11. $-8x(3x - 4)$

12. $-9a(2a^2 - 3)$

13. $5(3x^2 - 2x - 3)$

14. $3y(y^2 - 5y + 4)$

15. $6y^2(-8y^2 + 5y - 3)$

16. $6x^3(2x^7 - 3x^6 + 5x^4)$

17. $-10a^5(-2a^5 + 2a^4 - 8a^2)$

18. $(x + 5)(x + 10)$

19. $(5a - 2)(2a - 7)$

20. $(x - 3)(x^2 + 5x - 4)$

21. $(3y + 8)(2y^2 - 2y + 3)$

22. $(a^2 - 1)(a^4 + a^2 + 1)$

23. $(2x^2 + 3x - 2)(x^2 - 5x + 1)$

Use the FOIL method to multiply the following binomials.

24. $(x + 3)(x - 9)$ **25.** $(a + 8)(a - 8)$ **26.** $(x - 4)(x - 5)$

27. $(y - 12)(y + 3)$ **28.** $(x + 2)(x - 8)$ **29.** $(a - 9)(a - 9)$

30. $(y - 9)(y + 4)$ **31.** $(5x + 3)(x + 2)$ **32.** $(4a - 5)(a - 6)$

33. $(7y - 2)(y + 5)$ **34.** $(8x + 1)(x - 10)$ **35.** $(4a + 5)(5a + 4)$

36. $(3y - 5)(4y - 7)$ **37.** $(2x - 9)(5x + 3)$ **38.** $(10a + 3)(5a - 9)$

39. $(7y - 3)(8y + 3)$ **40.** $(9x + 7)(2x - 1)$ **41.** $(3z^2 - 1)(z^2 + 1)$

Multiply.

42. $3x + 4$
$5x - 8$

43. $3a^3 - 5a$
$4a^2 - 7$

44. $2y^2 - 9$
$3y^2 + 5y$

45. $4x^2 + 9$
$4x^2 + 9$

46. $8y^3 + 5$
$8y^3 - 5$

47. $10x^4 + 7x$
$5x^2 - 2x$

48. $2x^2 - 8x + 3$
$x + 3$

49. $-6y^2 + 2y - 5$
$3y - 2$

50. $8a^4 - 5a + 1$
$3a^2 + 5$

51. $15a^3 + 5a^2 - 3$
$6a^2 - 7$

52. $1.2x^2 + 4.5$
$0.1x - 1.2$

53. $\dfrac{1}{2}y^2 - \dfrac{5}{4}$
$\dfrac{4}{3}y + 7$

54. $a^2 + 3a - 1$
$a^2 - 3a + 1$

55. $4y^2 + 3y - 5$
$y^2 - 6y + 4$

56. $-2x^4 + 5x^2 - 3$
$8x^2 - 3x + 2$

Multiply the polynomials in two variables.

57. $5xy(3xy + y^2)$ **58.** $6a^3b^2(2a^4b - 3a^2b^2 + ab)$ **59.** $(x + 4y)(x + 6y)$

60. $(3a - 5b)(7a + 8b)$ **61.** $(6x - y)(6x + y)$ **62.** $(3a - 8b)(10a - 3b)$

63. $5x^2 - 5xy + 6y^2$
$4x + 3y$

64. $2a^2b^2 - 3ab + 9$
$ab - 2$

Perform the following operations.

65. $(y - 3)(y + 3)(y - 2)$ **66.** $(x + 4)(x + 2)(x - 4)$

67. The base of a triangle measures $(3y + 2)$ inches and the height is $(2y - 1)$ inches. Find the area of the triangle in terms of y.

68. The dimensions of a box, given in feet, are x, $x + 2$, and $3x + 5$. Find the volume of the box in terms of x.

69. The number n of new magazine subscriptions that can be sold during each month is related to the price in dollars p of a subscription by the equation $n = 200 - 10p$. Find an equation that gives the monthly revenue R in terms of p and use it to find R when p is \$15, \$10, and \$8.

70. A rectangular metal plate is to be constructed in such a way that the length is 10 cm more than twice the width. Find an equation that gives the area of the top of the plate and use it to find the area when the width is 30 cm, 70 cm, and 120 cm.

FOR REVIEW

Perform the indicated operations.

71. $(-4x^2 + 5x - 2) + (5x - 2) - (3x^2 - 2x)$

72. $(9y^2 - 6) + (y^4 - 5y^2) - (2y^2 - 5y + 7)$

73. $(7a^2 - 4a + 10) - (a^3 - 3a^2 + 5) - (-a + 4)$

74. $(x^2y^2 - xy + 2) - (5x^2y^2 - 3) - (4xy - 11)$

In Exercises 75–80, find the special products using the FOIL method. This will help you prepare for the next section.

75. $(y + 9)(y + 9)$

76. $(y - 9)(y - 9)$

77. $(y - 9)(y + 9)$

78. $(5x - 4y)(5x + 4y)$

79. $(5x - 4y)(5x - 4y)$

80. $(5x + 4y)(5x + 4y)$

6.5 EXERCISES C

Perform the indicated operations.

1. $(x + y)(x^2 - xy + y^2)$ [Answer: $x^3 + y^3$]

2. $(x - y)(x^2 + xy + y^2)$

3. $(a - 5b)(2a + b)(2a - 3b)$

4. $(a - b)[3a^2 - (a + b)(a - 2b)]$
[Answer: $2a^3 - a^2b + ab^2 - 2b^3$]

6.6 SPECIAL PRODUCTS

=== STUDENT GUIDEPOSTS ===

1 Difference of Squares

2 Perfect Square Trinomials

1 DIFFERENCE OF SQUARES

In Section 6.5 we learned how to multiply binomials using the FOIL method. Certain products of binomials occur often enough to merit special consideration. For example, suppose we use the FOIL method to multiply the two binomials $x + 6$ and $x - 6$.

$$(x + 6)(x - 6) = x^2 - 6x + 6x - 6^2$$
$$= x^2 - 36$$

Notice that the sum of the middle terms is zero. This always happens when two binomials of the form $a + b$ and $a - b$ are multiplied.

Difference of Two Squares

The product of a sum and a difference is the square of the first minus the square of the second.

$$(a + b)(a - b) = a^2 - b^2$$

Since the order of multiplication can be changed, we also know that $(a - b)(a + b) = a^2 - b^2$.

EXAMPLE 1 DIFFERENCE OF TWO SQUARES	**PRACTICE EXERCISE 1**

Find each product using both FOIL and the difference of squares formula.

(a) $(x - 7)(x + 7)$.

FOIL method	Using $(a - b)(a + b) = a^2 - b^2$
$(x - 7)(x + 7) = x^2 + 7x - 7x - 49$	$(x - 7)(x + 7) = \boxed{x}^2 - \boxed{7}^2$
$\qquad\qquad = x^2 - 49$	$\qquad\qquad = x^2 - 49$

(b) $(3z + 4)(3z - 4)$.

FOIL method	Using $(a + b)(a - b) = a^2 - b^2$
$(3z + 4)(3z - 4)$	$(3z + 4)(3z - 4) = \boxed{(3z)}^2 - \boxed{4}^2$
$\quad = 9z^2 - 12z + 12z - 16$	$\qquad\qquad = 9z^2 - 16$
$\quad = 9z^2 - 16$	

(c) $(7x^2 - 2x)(7x^2 + 2x)$.

FOIL method	Using $(a - b)(a + b) = a^2 - b^2$
$(7x^2 - 2x)(7x^2 + 2x)$	$(7x^2 - 2x)(7x^2 + 2x) = \boxed{(7x^2)}^2 - \boxed{(2x)}^2$
$\quad = 49x^4 + 14x^3 - 14x^3 - 4x^2$	$\qquad\qquad = 49x^4 - 4x^2$
$\quad = 49x^4 - 4x^2$	

Find each product using both FOIL and the difference of squares formula.

(a) $(a - 5)(a + 5)$

(b) $(4x + 7)(4x - 7)$

(c) $(2y^2 - y)(2y^2 + y)$

Answers: (a) $a^2 - 25$
(b) $16x^2 - 49$ (c) $4y^4 - y^2$

---

❷ PERFECT SQUARE TRINOMIALS

Squaring a binomial also takes a special form. For example, suppose we square $x + 6$.

$$(x + 6)^2 = (x + 6)(x + 6) = x^2 + 6x + 6x + 36 \qquad \text{FOIL method}$$
$$= x^2 + 2(6x) + 36$$
$$= x^2 + 12x + 36$$

This illustrates one of the perfect square formulas.

Perfect Square Trinomial

The square of a binomial of the form $a + b$ is the square of the first plus twice the product of first and second plus the square of the second.

$$(a + b)^2 = (a + b)(a + b) = a^2 + 2ab + b^2$$

EXAMPLE 2   SQUARING $a + b$

Find each product first using both FOIL, and the perfect square formula.

(a) $(y + 3)(y + 3)$.

FOIL method	Using $(a + b)^2 = a^2 + 2ab + b^2$
$(y + 3)(y + 3) = y^2 + 3y + 3y + 9$	$(y + 3)^2 = y^2 + 2\,(y)(3) + 3^2$
$\qquad\qquad = y^2 + 6y + 9$	$\qquad\quad = y^2 + 6y + 9$

(b) $(5x + 4)(5x + 4)$.

FOIL method	Using $(a + b)^2 = a^2 + 2ab + b^2$
$(5x + 4)(5x + 4)$	$(5x + 4)^2 = (5x)^2 + 2\,(5x)(4) + 4^2$
$= 25x^2 + 20x + 20x + 16$	$\qquad\quad = 25x^2 + 40x + 16$
$= 25x^2 + 40x + 16$	

PRACTICE EXERCISE 2

Find each product using both FOIL and the perfect square formula.

(a) $(x + 8)(x + 8)$

(b) $(7y + 2)(7y + 2)$

Answers: (a) $x^2 + 16x + 64$
(b) $49y^2 + 28y + 4$

To illustrate the second perfect square formula, look at the square of the binomial $x - 6$.

$$(x - 6)^2 = (x - 6)(x - 6) = x^2 - 6x - 6x + 36 \qquad \text{FOIL method}$$
$$= x^2 - 2(6x) + 36 = x^2 - 12x + 36$$

Perfect Square Trinomial

The square of a binomial of the form $a - b$ is the square of the first minus twice the product of first and second plus the square of the second.

$$(a - b)^2 = (a - b)(a - b) = a^2 - 2ab + b^2$$

EXAMPLE 3   SQUARING $a - b$

Find each product using both FOIL and the perfect square formula.

(a) $(y - 2)(y - 2)$

FOIL method	Using $(a - b)^2 = a^2 - 2ab + b^2$
$(y - 2)(y - 2)$	$(y - 2)^2 = y^2 - 2(y)(2) + 2^2$
$= y^2 - 2y - 2y + 4$	$\qquad\quad = y^2 - 4y + 4$
$= y^2 - 4y + 4$	

(b) $(5x - 4)(5x - 4)$.

FOIL method	Using $(a - b)^2 = a^2 - 2ab + b^2$
$(5x - 4)(5x - 4)$	$(5x - 4)^2 = (5x)^2 - 2\,(5x)(4) + 4^2$
$= 25x^2 - 20x - 20x + 16$	$\qquad\quad = 25x^2 - 40x + 16$
$= 25x^2 - 40x + 16$	

PRACTICE EXERCISE 3

Find each product using both FOIL and the perfect square formula.

(a) $(x - 9)(x - 9)$

(b) $(7a - 1)(7a - 1)$

Answers: (a) $x^2 - 18x + 81$
(b) $49a^2 - 14a + 1$

///////////// **CAUTION** ///////////

A common mistake students make when squaring a binomial is to write

$$(a - b)^2 = a^2 - b^2 \qquad \text{THIS IS WRONG}$$

or $\qquad (a + b)^2 = a^2 + b^2. \qquad \text{THIS IS WRONG}$

To see that $(a + b)^2 \neq a^2 + b^2$, substitute 1 for a and 2 for b. Then,

$$(a + b)^2 = (1 + 2)^2 = 3^2 = 9.$$

However, $\qquad a^2 + b^2 = (1)^2 + (2)^2 = 1 + 4 = 5.$

Don't forget the middle term $2ab$ when finding $(a + b)^2$, or $-2ab$ when finding $(a - b)^2$.

//////////

EXAMPLE 4 PRODUCTS WITH TWO VARIABLES	**PRACTICE EXERCISE 4**

Use the special formulas in this section to find the following products involving two variables.

(a) $(2x - 3y)(2x + 3y) = (2x)^2 - (3y)^2 \qquad (a - b)(a + b) = a^2 - b^2$
$$= 4x^2 - 9y^2$$

(b) $(2x + 3y)^2 = (2x)^2 + 2(2x)(3y) + (3y)^2 \quad (a + b)^2 = a^2 + 2ab + b^2$
$$= 4x^2 + 12xy + 9y^2$$

(c) $(2x - 3y)^2 = (2x)^2 - 2(2x)(3y) + (3y)^2 \quad (a - b)^2 = a^2 - 2ab + b^2$
$$= 4x^2 - 12xy + 9y^2$$

(d) $\left(\dfrac{1}{3}x - \dfrac{1}{2}y\right)^2 = \left(\dfrac{1}{3}x\right)^2 - 2\left(\dfrac{1}{3}x\right)\left(\dfrac{1}{2}y\right) + \left(\dfrac{1}{2}y\right)^2 \quad \begin{array}{l}(a - b)^2 = \\ a^2 - 2ab + b^2\end{array}$

$$= \dfrac{1}{9}x^2 - \dfrac{1}{3}xy + \dfrac{1}{4}y^2$$

Use the special formulas to find each product.

(a) $(5a - b)(5a + b)$

(b) $(9a + 2b)^2$

(c) $(9a - 2b)^2$

(d) $\left(\dfrac{1}{3}x + \dfrac{1}{2}y\right)^2$

Answers: (a) $25a^2 - b^2$
(b) $81a^2 + 36ab + 4b^2$ (c) $81a^2 - 36ab + 4b^2$ (d) $\frac{1}{9}x^2 + \frac{1}{3}xy + \frac{1}{4}y^2$

6.6 EXERCISES A

Multiply using special formulas.

1. $(x - 3)(x + 3)$

2. $(y + 7)(y - 7)$

3. $(x + 5)^2$

4. $(x - 5)^2$

5. $(y - 8)(y + 8)$

6. $(a + 10)(a - 10)$

7. $(x - 12)^2$

8. $(x + 12)^2$

9. $(2y + 5)(2y - 5)$

10. $(7a - 1)(7a + 1)$

11. $(2x + 7)^2$

12. $(2x - 7)^2$

13 $(4y - 9)(4y + 9)$

14. $(5a + 7)(5a - 7)$

15. $(3a + 1)^2$

16. $(3a - 1)^2$

17. $(5x - 3)(5x + 3)$

18. $(5y + 1)(5y - 1)$

19. $(6a + 7)^2$

20. $(6a - 7)^2$

21. $(4x + 8)^2$

22. $(3y - 9)^2$

23. $(5a - 10)(5a + 10)$

24. $(3x + 12)(3x - 12)$

25. $(5y^2 + 1)^2$

26. $(5y^2 - 1)^2$

27. $(5a^2 - 7)(5a^2 + 7)$

28. $(2x^2 - 3x)(2x^2 + 3x)$

29. $(2y^2 + 3y)^2$

30. $(2y^2 - 3y)^2$

31 $(0.7y - 3)^2$

32. $\left(\dfrac{1}{2}a - \dfrac{1}{3}\right)\left(\dfrac{1}{2}a + \dfrac{1}{3}\right)$

33. $(5x - 2y)(5x + 2y)$

34. $(5x + 2y)^2$

35. $(5x - 2y)^2$

36. $(x^2 + y^2)(x^2 - y^2)$

37 $(2x^2 - y)^2$

38. $(8x + 3y)^2$

39 $(a^2 + 2b)^2$

Perform the following operations.

40 $(x + 1)^2 - (x - 1)^2$

41. $(a - 2b)^2 + (a + 2b)^2$

42. $(z - 4)(z + 4) - (z + 4)^2$

43. $(w + 2)^2 - (w + 2)(w - 2)$

44 A field that is rectangular in shape is $(2a + 7)$ miles long and $(a - 1)$ miles wide. Find the area of the field in terms of a.

To multiply the polynomials in Exercises 45–50, use the formulas

$$(a + b)(a^2 - ab + b^2) = a^3 + b^3$$
$$(a - b)(a^2 + ab + b^2) = a^3 - b^3.$$

45. $(x + 2)(x^2 - 2x + 4)$

46. $(x - 2)(x^2 + 2x + 4)$

47. $(y - 5)(y^2 + 5y + 25)$

48. $(y + 5)(y^2 - 5y + 25)$

49. $(3x + y)(9x^2 - 3xy + y^2)$

50. $(3x - y)(9x^2 + 3xy + y^2)$

FOR REVIEW

Multiply.

51. $-3a^2(ab^2 - ab + 7)$

52. $(5a + 7)(3a - 8)$

53. $4x^2 - 9$
$\underline{2x\ + 3}$

54. $2x^2 - xy + y^2$
$\underline{\quad x\ - y}$

Exercises 55–57 review material from Section 6.1 to help you prepare for the next section. Simplify each expression using the quotient rule for exponents.

55. $\dfrac{15x^2}{3x}$

56. $\dfrac{26x^2y}{13xy}$

57. $\dfrac{-33x^3y^5}{33x^4y^2}$

ANSWERS: 1. $x^2 - 9$ 2. $y^2 - 49$ 3. $x^2 + 10x + 25$ 4. $x^2 - 10x + 25$ 5. $y^2 - 64$ 6. $a^2 - 100$ 7. $x^2 - 24x + 144$
8. $x^2 + 24x + 144$ 9. $4y^2 - 25$ 10. $49a^2 - 1$ 11. $4x^2 + 28x + 49$ 12. $4x^2 - 28x + 49$ 13. $16y^2 - 81$
14. $25a^2 - 49$ 15. $9a^2 + 6a + 1$ 16. $9a^2 - 6a + 1$ 17. $25x^2 - 9$ 18. $25y^2 - 1$ 19. $36a^2 + 84a + 49$
20. $36a^2 - 84a + 49$ 21. $16x^2 + 64x + 64$ 22. $9y^2 - 54y + 81$ 23. $25a^2 - 100$ 24. $9x^2 - 144$ 25. $25y^4 + 10y^2 + 1$
26. $25y^4 - 10y^2 + 1$ 27. $25a^4 - 49$ 28. $4x^4 - 9x^2$ 29. $4y^4 + 12y^3 + 9y^2$ 30. $4y^4 - 12y^3 + 9y^2$
31. $0.49y^2 - 4.2y + 9$ 32. $\frac{1}{4}a^2 - \frac{1}{9}$ 33. $25x^2 - 4y^2$ 34. $25x^2 + 20xy + 4y^2$ 35. $25x^2 - 20xy + 4y^2$ 36. $x^4 - y^4$
37. $4x^4 - 4x^2y + y^2$ 38. $64x^2 + 48xy + 9y^2$ 39. $a^4 + 4a^2b + 4b^2$ 40. $4x$ 41. $2a^2 + 8b^2$ 42. $-8z - 32$
43. $4w + 8$ 44. $(2a^2 + 5a - 7)$ mi² 45. $x^3 + 8$ 46. $x^3 - 8$ 47. $y^3 - 125$ 48. $y^3 + 125$ 49. $27x^3 + y^3$
50. $27x^3 - y^3$ 51. $-3a^3b^2 + 3a^3b - 21a^2$ 52. $15a^2 - 19a - 56$ 53. $8x^3 + 12x^2 - 18x - 27$
54. $2x^3 - 3x^2y + 2xy^2 - y^3$ 55. $5x$ 56. $2x$ 57. $-\frac{y^3}{x}$

6.6 EXERCISES B

Multiply using special formulas.

1. $(x - 4)(x + 4)$ **2.** $(y + 5)(y - 5)$ **3.** $(x + 10)^2$ **4.** $(x - 10)^2$

5. $(y - 9)(y + 9)$ **6.** $(a + 12)(a - 12)$ **7.** $(x - 4)^2$ **8.** $(x + 4)^2$

9. $(3y - 7)(3y + 7)$ **10.** $(5a - 2)(5a + 2)$ **11.** $(3x + 4)^2$ **12.** $(3x - 4)^2$

13. $(8y - 1)(8y + 1)$ **14.** $(6a - 5)(6a + 5)$ **15.** $(9a + 1)^2$ **16.** $(9a - 1)^2$

17. $(10x - 3)(10x + 3)$ **18.** $(3y + 5)(3y - 5)$ **19.** $(4a + 5)^2$ **20.** $(4a - 5)^2$

21. $(2x + 10)^2$ **22.** $(6y - 3)^2$ **23.** $(2a - 10)(2a + 10)$ **24.** $(6x + 3)(6x - 3)$

25. $(3y^2 + 2)^2$ **26.** $(3y^2 - 2)^2$ **27.** $(2a^2 - 5)(2a^2 + 5)$ **28.** $(3x^2 + 4x)(3x^2 - 4x)$

29. $(3y^2 + 4y)^2$ **30.** $(3y^2 - 4y)^2$ **31.** $(0.4y - 5)^2$ **32.** $\left(\dfrac{2}{3}a - \dfrac{1}{4}\right)\left(\dfrac{2}{3}a + \dfrac{1}{4}\right)$

33. $(7x - 3y)(7x + 3y)$ **34.** $(7x + 3y)^2$ **35.** $(7x - 3y)^2$ **36.** $(x^3 + y^3)(x^3 - y^3)$

37. $(3x - y^2)^2$ **38.** $(8x - 3y)^2$ **39.** $(a^2 - 2b)^2$

Perform the following operations.

40. $(x - 1)^2 - (x + 1)^2$ **41.** $(a + 3b)^2 + (a - 3b)^2$

42. $(y - 2)(y + 2) - (y - 2)^2$ **43.** $(z + 3)^2 - (z + 3)(z - 3)$

44. A picture frame is in the shape of a square with sides $(3z + 2)$ feet in length. Find the area of the frame in terms of z.

Use the formulas

$$(a + b)(a^2 - ab + b^2) = a^3 + b^3$$
$$(a - b)(a^2 + ab + b^2) = a^3 - b^3$$

to multiply the polynomials in Exercises 45–50.

45. $(x + 3)(x^2 - 3x + 9)$ **46.** $(x - 3)(x^2 + 3x + 9)$ **47.** $(y - 4)(y^2 + 4y + 16)$

48. $(y + 4)(y^2 - 4y + 16)$ **49.** $(2x + 3y)(4x^2 - 6xy + 9y^2)$ **50.** $(2x - 3y)(4x^2 + 6xy + 9y^2)$

FOR REVIEW

Multiply.

51. $4ab(-2a^2b - 6a + 9b)$ **52.** $(2a - 5)(4a - 3)$

53. $3x^2 + 11$
$\underline{3x - 2}$

54. $5x^2y^2 - 2xy + 3$
$\underline{2xy \quad - 1}$

Exercises 55–57 review material from Section 6.1 to help you prepare for the next section. Simplify each expression using the quotient rule for exponents.

55. $\dfrac{-24a^3}{8a^2}$ **56.** $\dfrac{64a^3b^2}{16ab^5}$ **57.** $\dfrac{12a^7b}{24a^3b}$

6.6 EXERCISES C

Perform the indicated operations. Assume all exponents are positive integers.

1. $[2x^2 - (x - 2y)][2x^2 + (x - 2y)]$
[*Hint:* Multiply before removing parentheses.]

2. $(2x + y - z)^2$

3. $(a^n + b^n)(a^n - b^n)$

4. $(a^n - b^n)^2$ [Answer: $a^{2n} - 2a^n b^n + b^{2n}$]

6.7 DIVISION OF POLYNOMIALS

═══════════════ STUDENT GUIDEPOSTS ═══════════════

❶ Dividing a Polynomial by a Monomial

❷ Dividing a Polynomial by a Binomial

① DIVIDING A POLYNOMIAL BY A MONOMIAL

To illustrate the rule for dividing a polynomial by a monomial, look at the following example.

$$\frac{3x^3 - x^2}{x} = \frac{1}{x}[3x^3 - x^2] \qquad \text{Dividing by } x \text{ is the same as multiplying by } \tfrac{1}{x}$$

$$= \frac{1}{x}(3x^3) - \frac{1}{x}(x^2) \qquad \text{Distributive law}$$

$$= \frac{3x^3}{x} - \frac{x^2}{x} \qquad \text{Multiplication by } \tfrac{1}{x} \text{ is the same as dividing by } x$$

$$= 3x^{3-1} - x^{2-1} \qquad \frac{a^m}{a^n} = a^{m-n}$$

$$= 3x^2 - x$$

In general, we omit the first two steps, write

$$\frac{3x^3 - x^2}{x} = \frac{3x^3}{x} - \frac{x^2}{x},$$

and use the quotient rule for exponents.

> **To Divide a Polynomial by a Monomial**
>
> 1. Divide each term of the polynomial by the monomial.
> 2. Use the rule
>
> $$\frac{x^m}{x^n} = x^{m-n}$$
>
> to divide the variables.

EXAMPLE 1   DIVIDING BY A MONOMIAL

Divide.

(a) $\dfrac{14x^3 - 7x^2 + 28x - 7}{7} = \dfrac{14x^3}{7} - \dfrac{7x^2}{7} + \dfrac{28x}{7} - \dfrac{7}{7}$

$$= 2x^3 - x^2 + 4x - 1$$

(b) $\dfrac{8x^4 - 6x^2 + 12x}{-2x} = \dfrac{8x^4}{-2x} - \dfrac{6x^2}{-2x} + \dfrac{12x}{-2x}$

$$= -4x^3 + 3x - 6 \qquad \tfrac{x^4}{x} = x^{4-1} = x^3, \tfrac{x^2}{x} = x^{2-1}$$
$$= x, \tfrac{x}{x} = x^{1-1} = x^0 = 1$$

(c) $\dfrac{7x^5 - 35x^4 + 40x}{5x^2} = \dfrac{7x^5}{5x^2} - \dfrac{35x^4}{5x^2} + \dfrac{40x}{5x^2}$

$$= \frac{7}{5}x^3 - 7x^2 + \frac{8}{x} \qquad \tfrac{x}{x^2} = \tfrac{1}{x}$$

PRACTICE EXERCISE 1

Divide.

(a) $\dfrac{6a^4 - 9a^3 + 12a^2 + 3a - 15}{3}$

(b) $\dfrac{25y^5 - 15y^3 + 30y}{-5y}$

(c) $\dfrac{18w^6 - 27w^4 - 9w^2}{9w^2}$

Answers: (a) $2a^4 - 3a^3 + 4a^2 + a - 5$ (b) $-5y^4 + 3y^2 - 6$
(c) $2w^4 - 3w^2 - 1$

//////////////// **C A U T I O N** ////////////////

Is $\frac{x+5}{5}$ equal to $x + 1$? The answer is no. Every number in the numerator must be divided by 5. The correct quotient is

$$\frac{x+5}{5} = \frac{x}{5} + \frac{5}{5} = \frac{x}{5} + 1.$$

//////////

If a division problem is expressed using the sign $\div$, change to the notation used in the examples above. To find $(3a^3b^3 - 9a^2b + 27ab) \div 9ab$ we write

$$\frac{3a^3b^3 - 9a^2b + 27ab}{9ab} = \frac{3a^3b^3}{9ab} - \frac{9a^2b}{9ab} + \frac{27ab}{9ab}$$

$$= \frac{1}{3}a^2b^2 - a + 3.$$

② DIVIDING A POLYNOMIAL BY A BINOMIAL

The procedure for dividing a polynomial by a binomial closely follows long division of one whole number by another. It might help to refer to the following numerical problem as you read through the rules given below.

```
        43 ÷ 37
        │ 61 ÷ 37
        │ │ 247 ÷ 37
        │ │ │ 252 ÷ 37
        ↓ ↓ ↓ ↓
        1166       Remainder 30
    37)43172
        37 ←──── First digit of quotient times divisor 37
        61       Subtract and bring down next digit of dividend, 43172
        37 ←──── Second digit of quotient times 37
        247      Subtract and bring down next digit of dividend
        222 ←──── Third digit of quotient times 37
        252      Subtract and bring down next digit
        222 ←──── Fourth digit of quotient times 37
        30       Subtract; no more digits in dividend
```

The answer is 1166 with a remainder of 30, or $1166 + \frac{30}{37}$.

To Divide a Polynomial by a Binomial

1. Arrange the terms of both polynomial and binomial in descending order and set up as in long division.

2. Divide the first term of the polynomial (the dividend) by the first term of the binomial (the divisor) to obtain the first term of the quotient.

3. Multiply the first term of the quotient by the binomial and subtract the result from the dividend. Bring down the next terms to obtain a new polynomial which becomes the new dividend.

4. Divide the new dividend polynomial by the binomial. Continue the process until the variable in the first term of the remainder dividend is raised to a lower power than the variable in the first term of the divisor.

| EXAMPLE 2 DIVIDING BY A BINOMIAL | PRACTICE EXERCISE 2 |

Divide $6 - 5x + x^2$ by $x - 2$.

1. Arrange terms in descending order.

$$x - 2\overline{)x^2 - 5x + 6}$$

2. Divide the first term of the polynomial by the first term of the binomial.

$$\text{equals}\quad \overset{x}{x - 2\overline{)x^2} - 5x + 6} \qquad x^2 \div x = x$$
$$\text{divided by}$$

3. Multiply the first term of the quotient by the binomial and subtract the results from the dividend. Bring down the next terms to obtain a new dividend.

$$\begin{array}{r} \overset{\text{times}}{x} \\ x - 2\overline{)x^2 - 5x + 6} \\ \underset{\text{equals}}{x^2 - 2x} \\ \hline -3x + 6 \end{array}$$

$x(x - 2) = x^2 - 2x$

Subtract $x^2 - 2x$ from $x^2 - 5x$ and bring down $+ 6$

4. Divide the new dividend polynomial by the binomial, using Steps 2 and 3.

$$\begin{array}{r} x - 3 \\ x - 2\overline{)x^2 - 5x + 6} \\ \underline{x^2 - 2x} \\ -3x + 6 \\ \underline{-3x + 6} \\ 0 \end{array}$$

Divide $-3x + 6$ by $x - 2$

Multiply $x - 2$ by -3

No variable in the new dividend; the process terminates

Divide $8x - 9 + x^2$ by $x - 1$.

$$\text{equals}\quad \overset{x}{(x) - 1\overline{)(x^2) + 8x - 9}}\leftarrow\text{Descending}$$
$$\text{order}$$
$$\text{divided by}\qquad \underline{x^2 - \ x}\leftarrow x(x - 1)$$
$$9x \leftarrow\text{Subtract like}$$
$$\text{terms:}$$
$$x^2 - x^2 = 0$$
$$\text{and } 8x - (-x) =$$
$$8x + x = 9x$$

Bring down the -9 and continue by dividing x into $9x$.

Answer: $x + 9$

| EXAMPLE 3 DIVIDING BY A BINOMIAL | PRACTICE EXERCISE 3 |

Divide $a^3 + 27$ by $a + 3$.

The quotient of a^3 and a

$$\begin{array}{r} a^2 - 3a + 9\leftarrow \\ a + 3\overline{)a^3 \qquad\quad + 27} \\ \underline{a^3 + 3a^2}\leftarrow \\ -3a^2 \qquad + 27 \\ \underline{-3a^2 - 9a}\leftarrow \\ 9a + 27 \\ \underline{9a + 27}\leftarrow \\ 0 \end{array}$$

The quotient of $-3a^2$ and a
The quotient of $9a$ and a
Leave space for missing terms
The quotient a^2 times the divisor $a + 3$
Subtract $a^3 + 3a^2$ from a^3 and bring down 27
The quotient $-3a$ times $a + 3$
Subtract $-3a^2 - 9a$ from $3a^2 + 27$
The quotient 9 times $a + 3$
Subtract; no variable in new dividend

The answer is $a^2 - 3a + 9$ with a remainder of 0.

Divide $x^3 - 8$ by $x - 2$.

Answer: $x^2 + 2x + 4$

Note that we can check our work in any division problem by multiplying the divisor by the quotient to obtain the dividend. We can check the work in Example 3 as follows.

$$(a + 3)(a^2 - 3a + 9) = a^3 - 3a^2 + 9a + 3a^2 - 9a + 27$$
$$= a^3 + 27$$

EXAMPLE 4 DIVIDING BY A BINOMIAL

Divide $3x^4 - 19x^3 + 27x^2 - 41x + 32$ by $x - 5$.

$$
\begin{array}{r}
3x^3 - 4x^2 + 7x - 6 \\
x - 5\overline{)3x^4 - 19x^3 + 27x^2 - 41x + 32} \\
\underline{3x^4 - 15x^3} \\
-4x^3 + 27x^2 - 41x + 32 \\
\underline{-4x^3 + 20x^2} \\
7x^2 - 41x + 32 \\
\underline{7x^2 - 35x} \\
-6x + 32 \\
\underline{-6x + 30} \\
2
\end{array}
$$

Divide $3x^4$ by x

No variable in new dividend; process ends

The answer is $3x^3 - 4x^2 + 7x - 6$ with remainder 2, or $3x^3 - 4x^2 + 7x - 6 + \frac{2}{x-5}$.

PRACTICE EXERCISE 4

Divide $2y^4 + 9y^3 - 3y^2 - 25y + 10$ by $y + 4$.

Answer: $2y^3 + y^2 - 7y + 3 - \frac{2}{y+4}$

EXAMPLE 5 DIVIDING BY A BINOMIAL

Divide $20x^3 + 18x^2 + 21x + 40$ by $5x + 7$.

$$
\begin{array}{r}
4x^2 - 2x + 7 \\
5x + 7\overline{)20x^3 + 18x^2 + 21x + 40} \\
\underline{20x^3 + 28x^2} \\
-10x^2 + 21x + 40 \\
\underline{-10x^2 - 14x} \\
35x + 40 \\
\underline{35x + 49} \\
-9
\end{array}
$$

No variable in new dividend

The answer is $4x^2 - 2x + 7$ with remainder -9, or $4x^2 - 2x + 7 - \frac{9}{5x+7}$.

PRACTICE EXERCISE 5

Divide $16a^3 - 8a^2 + 5a - 3$ by $4a - 3$.

Answer: $4a^2 + a + 2 + \frac{3}{4a-3}$

EXAMPLE 6 DIVIDING BY A BINOMIAL

Divide $4x^4 - 8x^3 - 12x^2 + 44x - 15$ by $4x^2 - 8$.

$$
\begin{array}{r}
x^2 - 2x - 1 \\
4x^2 - 8\overline{)4x^4 - 8x^3 - 12x^2 + 44x - 15} \\
\underline{4x^4 \qquad - 8x^2} \\
-8x^3 - 4x^2 + 44x - 15 \\
\underline{-8x^3 \qquad + 16x} \\
-4x^2 + 28x - 15 \\
\underline{-4x^2 \qquad + 8} \\
28x - 23
\end{array}
$$

Keep like terms in same column

x is raised to first power; process ends

The answer is $x^2 - 2x - 1 + \frac{28x - 23}{4x^2 - 8}$.

PRACTICE EXERCISE 6

Divide $9x^4 + 6x^3 + 3x^2 + 7x - 1$ by $3x^2 - 1$.

Answer: $3x^2 + 2x + 2 + \frac{9x + 1}{3x^2 - 1}$

6.7 EXERCISES A

Divide.

1. $\dfrac{3x^3 - 9x^2 + 27x + 3}{3}$

2. $\dfrac{14x^4 - 7x^2 + 28}{7}$

3. $(25a^5 - 20a^4 + 15a^3) \div (5a^3)$

4. $(y^8 - y^6 + y^5) \div (y^3)$

5. $\dfrac{7x^5 - 35x^4 + 14x^2}{14x^2}$

6. $\dfrac{8x^3 - 64x^2}{8x^2}$

7 $(3a^{12} - 9a^6 + 27a^5 + 81a^4) \div (9a^2)$

8. $(y^6 - 3y^2) \div (2y^2)$

9. $\dfrac{-8x^3 + 6x^2 - 4x}{-2x}$

10 $\dfrac{-8x^3 + 6x^2 - 4x}{0.2x}$

11. $\dfrac{y^3 + 2y^2 + y}{y^2}$

12. $\dfrac{-8y^4 + 16y^3 - 4y}{2y^2}$

13 $\dfrac{-5a^2b + 3ab - 2a}{-a}$

14. $\dfrac{-6x^2y^3 + 9x^2y}{3xy}$

15. $\dfrac{27x^2y^4 - 18xy^6 + 36xy^3}{9xy^3}$

16. $\dfrac{12x^4y^4 - 42x^2y^2}{6x^3y^3}$

17. $(x^2 - 5x + 6) \div (x - 3)$

18. $(a^2 + 5a + 6) \div (a + 3)$

19. $(y^2 - 2y - 3) \div (y + 1)$

20. $(x^2 - 9x - 5) \div (x - 7)$

21. $(3y + 2 + y^2) \div (y + 2)$
[*Hint:* Remember descending order.]

22 $(6 + 8y - y^2) \div (4 - y)$

23 $(3a^2 - 23a + 40) \div (a - 5)$

24. $(14y - 6y^2 + 80) \div (3y + 8)$

25. $4x + 6\overline{)28x^3 + 26x^2 - 44x - 36}$

26. $2x - 3\overline{)10x^4 - 15x^3 \quad - 8x + 12}$

27. $3a - 12\overline{)6a^3 - 18a^2 + 33a - 80}$

28 $y + 2\overline{)y^5 \quad + 32}$

29. $3x^2 + 2\overline{)9x^3 + 30x^2 + x - 10}$

30 $x^2 - 1\overline{)x^4 \quad - 1}$

FOR REVIEW

Multiply using special formulas.

31. $(4x - 1)(4x + 1)$

32. $(6x - 5)^2$

33. $(2a + 13)^2$

34. $(a^2 - 5)(a^2 + 5)$

35. $(y^3 - 2)^2$

36. $(2y^2 + 7)^2$

To prepare for factoring polynomials in the next chapter, factor each integer into a product of primes. (See Section 1.1.)

37. 45

38. 210

39. 748

40. 1625

ANSWERS: 1. $x^3 - 3x^2 + 9x + 1$ 2. $2x^4 - x^2 + 4$ 3. $5a^2 - 4a + 3$ 4. $y^5 - y^3 + y^2$ 5. $\frac{1}{2}x^3 - \frac{5}{2}x^2 + 1$
6. $x - 8$ 7. $\frac{1}{3}a^{10} - a^4 + 3a^3 + 9a^2$ 8. $\frac{1}{2}y^4 - \frac{3}{2}$ 9. $4x^2 - 3x + 2$ 10. $-40x^2 + 30x - 20$ 11. $y + 2 + \frac{1}{y}$
12. $-4y^2 + 8y - \frac{2}{y}$ 13. $5ab - 3b + 2$ 14. $-2xy^2 + 3x$ 15. $3xy - 2y^3 + 4$ 16. $2xy - \frac{7}{xy}$ 17. $x - 2$ 18. $a + 2$
19. $y - 3$ 20. $x - 2 - \frac{19}{x - 7}$ 21. $y + 1$ 22. $y - 4 - \frac{22}{y - 4}$ 23. $3a - 8$ 24. $-2y + 10$ 25. $7x^2 - 4x - 5 - \frac{6}{4x + 6}$
26. $5x^3 - 4$ 27. $2a^2 + 2a + 19 + \frac{148}{3a - 12}$ 28. $y^4 - 2y^3 + 4y^2 - 8y + 16$ 29. $3x + 10 - \frac{5x + 30}{3x^2 + 2}$ 30. $x^2 + 1$
31. $16x^2 - 1$ 32. $36x^2 - 60x + 25$ 33. $4a^2 + 52a + 169$ 34. $a^4 - 25$ 35. $y^6 - 4y^3 + 4$ 36. $4y^4 + 28y^2 + 49$
37. $3 \cdot 3 \cdot 5$ 38. $2 \cdot 3 \cdot 5 \cdot 7$ 39. $2 \cdot 2 \cdot 11 \cdot 17$ 40. $5 \cdot 5 \cdot 5 \cdot 13$

6.7 EXERCISES B

Divide.

1. $\dfrac{5x^4 - 25x^2 + 45x - 20}{5}$

2. $\dfrac{26x^3 - 16x^2 + 10}{-2}$

3. $(55a^6 - 22a^4 + 33a^2) \div (11a^2)$

4. $(-y^{12} + y^7 - y^5) \div (y^4)$

5. $\dfrac{-100x^7 + 50x^5 - 20x^3}{10x^2}$

6. $\dfrac{15x^3 - 30x^2 + 5x}{30x}$

7. $(35a^9 - 49a^7 - 14a^6 + 56a^5) \div (7a^4)$

8. $(-8y^4 - 6y^3 - 12y^2) \div (-3y^2)$

9. $\dfrac{54x^6 + 81x^4 - 36x^3}{9x^3}$

10. $\dfrac{54x^6 + 81x^4 - 36x^3}{0.9x^3}$

11. $\dfrac{y^4 - 6y^3 + y^2}{y^3}$

12. $\dfrac{-3y^6 + 12y^4 - 6y^2}{3y^4}$

13. $\dfrac{4a^2b^2 - 5a^2b + 7ab^2}{ab}$

14. $\dfrac{18x^3y^4 - 12x^4y^3}{3x^3y^3}$

15. $\dfrac{44x^6y^6 + 20x^4y^2 - 28x^2y^4}{4x^2y}$

16. $\dfrac{-5x^6y^4 + 10x^4y^6}{5x^6y^6}$

17. $(x^2 - 6x + 8) \div (x - 4)$

18. $(a^2 + 6a + 8) \div (a + 2)$

19. $(y^2 - 5y - 6) \div (y + 1)$

20. $(x^2 + 7x - 8) \div (x + 5)$

21. $(7y + 12 + y^2) \div (y + 4)$

22. $(3 - 5y - y^2) \div (3 - y)$

23. $(3a^2 + 13a - 30) \div (a + 6)$

24. $(7a - 12 + 12a^2) \div (4a - 3)$

25. $(2y^3 + y^2 - y + 1) \div (2y + 3)$

26. $(6x^3 + x^2 - 7x + 2) \div (3x - 1)$

27. $(a^4 - 1) \div (a^2 + 1)$

28. $(x^5 - 32) \div (x - 2)$

29. $(21x^4 - 7x^3 - 6x + 2) \div (3x - 1)$

30. $(2x^5 - x^3 + 16x^2 - 8) \div (2x^2 - 1)$

FOR REVIEW

Multiply using special formulas.

31. $(3x - 5)(3x + 5)$

32. $(3x + 7)^2$

33. $(2x - 11)^2$

34. $(a^2 + 7)(a^2 - 7)$ **35.** $(y^3 + 5)^2$ **36.** $(3y^2 - 4)^2$

To prepare for factoring polynomials in the next chapter, factor each integer into a product of primes. (See Section 1.1.)

37. 42 **38.** 462 **39.** 910 **40.** 2805

6.7 EXERCISES C

Divide.

1. $(2x^6 - 7x^3 - 30) \div (2x^3 + 5)$

2. $(6x^6 + x^3 - 12) \div (3x^3 - 4)$ [Answer: $2x^3 + 3$]

3. $(5x^5 + 21 - 26x + x^2 + 10x^4 - 11x^3) \div (x^2 + 2x - 3)$

4. $(-34x^4 - 5x^3 + 3x^5 + 6x^6 + 54x^2 - 56 + 7x) \div (2x^2 + x - 8)$ [Answer: $3x^4 - 5x^2 + 7$]

CHAPTER 6 REVIEW

KEY WORDS

6.1 An **exponential expression** is of the form a^n where a is the **base** and n is the **exponent.**

6.2 A number is written in **scientific notation** if it is the product of a power of 10 and a number that is greater than or equal to 1 and less than 10.

6.3 A **polynomial** is an algebraic expression having as terms products of numbers and variables with whole-number exponents.

The **numerical coefficient** is the numerical multiplier of the term.

A **monomial** is a polynomial with one term.

A **binomial** is a polynomial with two terms.

A **trinomial** is a polynomial with three terms.

For a polynomial in one variable, the **degree of a term** is the exponent on the variable.

The **degree of a polynomial** is the degree of the term of highest degree.

KEY CONCEPTS

6.1 **1.** If a and b are any numbers, and m and n are natural numbers, the following rules for exponents hold.

(a) $a^m \cdot a^n = a^{m+n}$

(b) $\dfrac{a^m}{a^n} = a^{m-n}$, if $a \neq 0$

(c) $(a^m)^n = a^{mn}$

(d) $(a \cdot b)^n = a^n \cdot b^n$

(e) $\left(\dfrac{a}{b}\right)^n = \dfrac{a^n}{b^n}$, if $b \neq 0$

(f) $a^0 = 1$, if $a \neq 0$ and 0^0 is undefined.

2. In an expression such as $2y^3$, only y is cubed, not $2y$. That is, $2y^3$ is not the same as $(2y)^3$, which is equal to $8y^3$.

6.3 Only like terms can be collected. For example, $5x + 3 \neq 8x$ since $5x$ and 3 are *not* like terms. Similarly, $x^2 - 2x + 1$ has no like terms to combine.

6.4 Change all signs when removing parentheses preceded by a minus sign. For example, $3x - (2x^2 - 4x - 2) = 3x - 2x^2 + 4x + 2$.

6.5 Use the FOIL method to multiply binomials, and write out all details.

6.6 **1.** $(a + b)(a - b) = a^2 - b^2$
2. $(a + b)^2 = a^2 + 2ab + b^2$ (not $a^2 + b^2$)
3. $(a - b)^2 = a^2 - 2ab + b^2$ (not $a^2 - b^2$)

6.7 Write polynomials in descending order before dividing.

REVIEW EXERCISES

Part I

6.1 *Write in exponential notation.*

1. *aaaaa*   **2.** $(3z)(3z)(3z)$ **3.** $(a + b)(a + b)$

Write without using exponents.

4. b^7 **5.** $-2x^3$ **6.** $(-2x)^3$

Simplify and write without negative exponents.

7. $2y^7y^2$ **8.** $(3x^2)^3$ **9.** $(x^2y^3)^5$

10. $2a^0 \quad (a \neq 0)$ **11.** $(2a)^0 \quad (a \neq 0)$ **12.** $\dfrac{2b^2b^{-5}}{b^7}$

13. $5a^{-1}$ **14.** $(5a)^{-1}$ **15.** $\dfrac{a^{-1}}{b^{-1}}$

16. $(a^2b^{-3})^{-2}$ **17.** $\dfrac{3^0a^{-5}b^7}{a^4b^{-1}}$ **18.** $\left(\dfrac{2^{-1}x^2y^{-4}}{x^{-2}y^{-2}}\right)^{-1}$

6.2 *Write in scientific notation.*

19. 0.000000411 **20.** $549{,}000{,}000{,}000$

Write without using scientific notation.

21. 6.15×10^{12} **22.** 4×10^{-8}

6.3 *Tell whether the following are monomials, binomials, or trinomials.*

23. $4x + 1$ **24.** $3x^2$ **25.** $5y^4 + y^2$ **26.** $-7y^3 + y - 5$

Give the degree of each polynomial.

27. $6x^4$ **28.** $15x^2 - 14x + 6x^7 - 4x^3$

29. $-x^3 + 14x^4 - x^{10} + 8x^2$ **30.** $2x^8 - 4x^4 - 16$

Collect and combine like terms and write in descending order.

31. $5x + 2 - 6x$ **32.** $6y^2 - 2y + 8y^2 + 5 - 3y^2 + y$

33. $4a^3 - 6a + 7a^2 - 4a^3 + a - a^2$ **34.** $-7x^2 - 30x^4 + x^5 - 3x + 5x^5 - 4x + 22x^4 - 5$

Collect and combine like terms.

35. $8xy + 5 - 4xy - 8$ **36.** $-a^2b^3 - a^2b^2 - 5a^2b^2 - ab^2 + 5a^2b^3$

Solve.

37. The profit in dollars when x refrigerators are sold is given by the expression $4x^3 - 140$. Find the profit when 5 refrigerators are sold.

6.4   **38.** Add $5x^2 - 4x + 5$ and $-7x^2 + 4x - 8$.   **39.** Subtract $-2x^2 - 1$ from $x^3 - 5x^2 + 14$.

Perform the indicated operations.

40. $(4x^3 - x^4 + 6x - 2) + (3x - x^5 + 2x^3 - 5x^4 + 2)$

41. $(16x - 2x^4 + 3x^2 - 5) - (-6 + x^4 - 3x^3 - 3x^2 + 2x)$

42. $(6x^2y^2 - 5xy) + (-4xy + 3) - (3x^2y^2 - 8xy + 5)$

Add.

43.
$$
\begin{array}{r}
x^4 - 2x^3 + 7x^2 - 20 \\
-8x^4 - 5x^2 + 3x + 7 \\
\hline
36x^4 - 18x^3 + 5x - 31
\end{array}
$$

44.
$$
\begin{array}{r}
6.1x^3 - 0.2x^2 + 1.1x + 5 \\
-5.7x^3 + 2.3x - 7 \\
\hline
1.1x^2 - 2.3x + 8
\end{array}
$$

6.5   *Multiply.*
6.6

45. $(x + 8)(x - 10)$ **46.** $(2x + 1)(3x - 4)$ **47.** $(4a - 3)(4a + 3)$

48. $(4a - 3)^2$ **49.** $(4a + 3)^2$ **50.** $(5y - 6)(4y - 7)$

51. $\left(\dfrac{3}{4} + 2y^3\right)\left(\dfrac{3}{4} - 2y^3\right)$ **52.** $(4x - 9)(x + 2)$ **53.** $(x + 4)(2x^2 + x - 3)$

54.
$$
\begin{array}{r}
3x^2 - 2x + 1 \\
x^2 + x - 2 \\
\hline
\end{array}
$$

55.
$$
\begin{array}{r}
2x^2 + xy - 3y^2 \\
x + 4y \\
\hline
\end{array}
$$

56.
$$
\begin{array}{r}
6x^2y^2 - xy + 5 \\
2xy - 3 \\
\hline
\end{array}
$$

57. $(7a - b)(4a + 3b)$ **58.** $(5a - 7b)^2$

59. $(4x - y)(4x + y)$ **60.** $(8x + y)^2$

6.7   *Divide.*

61. $\dfrac{25x^4 - 50x^3 + 10x^2}{5x^2}$ **62.** $\dfrac{x^{10} - x^8 + x^6}{-x^4}$

63. $\dfrac{6x^4y^4 - 2x^3y^3 + 10x^2y^2}{2x^2y^2}$

64. $\dfrac{-14a^4b^2 + 21a^2b^4 - 35ab^2}{-7a^2b^2}$

65. $(x^2 + 3x - 28) \div (x - 4)$

66. $(8x^2 + 10x + 3) \div (2x + 1)$

67. $2x + 3\overline{)14x^4 + 27x^3 + 5x^2 - 16x - 15}$

68. $x - 5\overline{)x^3 \qquad\qquad -100}$

Part II

69. The number n of records that can be sold during each week is related to the price in dollars p of each record by the equation $n = 1200 - 100p$. Find an equation that gives the weekly revenue R in terms of p and use it to find R when p is \$10.

70. Light travels at the rate of 186,000 mi per sec. Write this number in scientific notation.

Perform the indicated operation.

71. $(2x + 1)(5x - 2)$

72. $(6x^2y^2 - 3xy) - (-2x^2y^2 + 1)$

73. $(x^2 + 10x + 21) \div (x + 7)$

74. $(6x^5 - 4x^3) + (3x^5 + x^2) - (-x^3 + x^2)$

75. $(7x + 2y)(7x - 2y)$

76. $(6x^2y^3 - 4x^2y - 2xy^2) \div (-2xy)$

77. $(5a - b)^2$

78. $(8x + 3)^2$

79. Subtract $-x^3 - 2$ from $x^3 - 2$.

80. $3a^2b(4a^3b^4 - 5a^2b - 2ab)$

Simplify and write without negative exponents.

81. $(-2x^2)^3$

82. $-2(x^2)^3$

83. $\left(\dfrac{x^3}{y^2}\right)^{-2}$

84. $(a^2b^{-4})^{-2}$

85. $\dfrac{2^0x^{-4}y^3}{x^2y^{-3}}$

86. $\left(\dfrac{ab^{-2}}{a^{-3}b^{-3}}\right)^{-1}$

Give the degree of each polynomial.

87. $6x^3 - 4x^2 + 2x^5 - 1$

88. $-7y + 6y^3 - 2 + y^2$

Tell whether the following are monomials, binomials, or trinomials.

89. $3x + 2$

90. $x^2 - 6x + 7$

ANSWERS: 1. a^5 2. $(3z)^3$ 3. $(a+b)^2$ 4. $bbbbbb$ 5. $-2xxx$ 6. $(-2x)(-2x)(-2x)$ 7. $2y^9$ 8. $27x^6$ 9. $x^{10}y^{15}$
10. 2 11. 1 12. $\frac{2}{b^{10}}$ 13. $\frac{5}{a}$ 14. $\frac{1}{5a}$ 15. $\frac{b}{a}$ 16. $\frac{b^6}{a^4}$ 17. $\frac{b^8}{a^9}$ 18. $\frac{2y^2}{x^4}$ 19. 4.11×10^{-7} 20. 5.49×10^{11}
21. 6,150,000,000,000 22. 0.00000004 23. binomial 24. monomial 25. binomial 26. trinomial 27. 4 28. 7
29. 10 30. 8 31. $-x + 2$ 32. $11y^2 - y + 5$ 33. $6a^2 - 5a$ 34. $6x^5 - 8x^4 - 7x^2 - 7x - 5$ 35. $4xy - 3$
36. $4a^2b^3 - 6a^2b^2 - ab^2$ 37. \$360 38. $-2x^2 - 3$ 39. $x^3 - 3x^2 + 15$ 40. $-x^5 - 6x^4 + 6x^3 + 9x$
41. $-3x^4 + 3x^3 + 6x^2 + 14x + 1$ 42. $3x^2y^2 - xy - 2$ 43. $29x^4 - 20x^3 + 2x^2 + 8x - 44$ 44. $0.4x^3 + 0.9x^2 + 1.1x + 6$
45. $x^2 - 2x - 80$ 46. $6x^2 - 5x - 4$ 47. $16a^2 - 9$ 48. $16a^2 - 24a + 9$ 49. $16a^2 + 24a + 9$ 50. $20y^2 - 59y + 42$
51. $\frac{9}{16} - 4y^6$ 52. $4x^2 - x - 18$ 53. $2x^3 + 9x^2 + x - 12$ 54. $3x^4 + x^3 - 7x^2 + 5x - 2$ 55. $2x^3 + 9x^2y + xy^2 - 12y^3$
56. $12x^3y^3 - 20x^2y^2 + 13xy - 15$ 57. $28a^2 + 17ab - 3b^2$ 58. $25a^2 - 70ab + 49b^2$ 59. $16x^2 - y^2$
60. $64x^2 + 16xy + y^2$ 61. $5x^2 - 10x + 2$ 62. $-x^6 + x^4 - x^2$ 63. $3x^2y^2 - xy + 5$ 64. $2a^2 - 3b^2 + \frac{5}{a}$ 65. $x + 7$
66. $4x + 3$ 67. $7x^3 + 3x^2 - 2x - 5$ 68. $x^2 + 5x + 25 + \frac{25}{x-5}$ 69. $R = 1200p - 100p^2$; \$2000
70. 1.86×10^5 mi per sec. 71. $10x^2 + x - 2$ 72. $8x^2y^2 - 3xy - 1$ 73. $x + 3$ 74. $9x^5 - 3x^3$ 75. $49x^2 - 4y^2$
76. $-3xy^2 + 2x + 2y$ 77. $25a^2 - 10ab + b^2$ 78. $64x^2 + 48x + 9$ 79. $2x^3$ 80. $12a^5b^5 - 15a^4b^2 - 6a^3b^2$
81. $-8x^6$ 82. $-2x^6$ 83. $\frac{y^4}{x^6}$ 84. $\frac{b^8}{a^4}$ 85. $\frac{y^6}{x^6}$ 86. $\frac{1}{a^4b}$ 87. 5 88. 3 89. binomial 90. trinomial

1. Write in exponential notation. $5 \cdot 5 \cdot a \cdot a \cdot a$

1. _____

Simplify and write without negative exponents.

2. $2y^5y^2$

2. _____

3. $\dfrac{a^7}{a^3}$

3. _____

4. $(-2x^2)^3$

4. _____

5. z^{-3}

5. _____

6. $\dfrac{x^{-6}y^3}{x^2y^{-4}}$

6. _____

7. Evaluate y^{-3} when $y = -4$.

7. _____

8. Evaluate. $\dfrac{3.9 \times 10^{-4}}{1.3 \times 10^{-2}}$

8. _____

9. Is $3x^2 + 5$ a monomial, binomial, or trinomial?

9. _____

10. Give the degree of $x^3 - 7x^2 + 10x^4 - 8$.

10. _____

11. Collect and combine like terms and write in descending order.

$3y^3 - 7y^5 + y^3 - 4y + 6 + y^5 - 2y$

11. _____

12. Collect and combine like terms.

$3x^2y^2 - 4x^2 + 3x^2y - 7x^2y^2 + 2x^2y$

12. _____

13. The cost in dollars of making x pairs of shoes is given by the expression $20x + 45$. Find the cost when 12 pairs of shoes are made.

13. _____

Add.

14. $(3x + 2) + (-5x + 8)$

14. _____

15. $(6x^3 + 5x^2 - 3x^4 + 2) + (-8x^2 + 4x^3 + 6x^4 + 5)$

15. _____

16. $(3a^2b^2 - 2ab + 5) + (4a^2b^2 + 8ab - 3)$

16. _____

Subtract.

17. $(-5x + 4) - (2x - 3)$ 17. _____

18. $(4x^2 - 6x^4 + 2) - (-2x^2 + x^3 - 2x^4 + 1)$ 18. _____

19. $(8a^2b - 3ab^2 - 2ab) - (4a^2b + ab^2 - ab)$ 19. _____

Multiply.

20. $-4x^2(3x^2 - 4x + 1)$ 20. _____

21. $(x + 5)(x - 8)$ 21. _____

22. $(4a - 3)(2a + 7)$ 22. _____

23. $(y + 8)(3y^2 - 2y + 1)$ 23. _____

24. $(2x + y)(2x - y)$ 24. _____

25. $(3x - 4y)^2$ 25. _____

26. $(2x + 5)^2$ 26. _____

27. $(3x + 2)(2x^2 - 3x + 4)$ 27. _____

28. $(x^2 - 3)(x^2 + 3)$ 28. _____

Divide.

29. $\dfrac{25x^4 - 10x^3 - 15x^2}{5x^2}$ 29. _____

30. $(3x^3 - 11x^2 + 10x - 12) \div (x - 3)$ 30. _____

Factoring Polynomials

7.1 COMMON FACTORS AND GROUPING

=== STUDENT GUIDEPOSTS ===

1 Greatest Common Factor
2 Factoring by Removing the GCF
3 Factoring by Grouping

Factoring is the reverse of multiplying. In Chapter 1 we factored integers such as 28 by writing

$$28 = 4 \cdot 7 = 2 \cdot 2 \cdot 7 = 2^2 \cdot 7.$$

In this form, we see that 4 and 7 are two **factors** of 28, (as are 1, 2, 14, and 28), while 2 and 7 are also **prime factors** (2 and 7 are prime numbers). Algebraic expressions can also be factored into prime factors. For example,

$$28x^3 = 2 \cdot 2 \cdot 7 \cdot x \cdot x \cdot x = 2^2 \cdot 7 \cdot x^3,$$

has 2, 7, and x as its prime factors. A **common factor** of two expressions is a factor that occurs in each of them. For instance, a common factor of 14 and 28 is 7 since

$$14 = 2 \cdot 7 \quad \text{and} \quad 28 = 2^2 \cdot 7.$$

Other common factors of 14 and 28 are 2 and 14.

1 GREATEST COMMON FACTOR

The **greatest common factor (GCF)** of two integers is the largest factor common to both integers. Thus, 14 is the greatest common factor of 14 and 28. As shown in the next example, it is easier to find the greatest common factor if integers are expressed as a product of primes.

EXAMPLE 1 GREATEST COMMON FACTOR (GCF) OF INTEGERS

Find the greatest common factor of the integers.

(a) 35 and 75

$$35 = 5 \cdot 7 \quad \text{and} \quad 75 = 3 \cdot 5 \cdot 5$$

GCF: 5

PRACTICE EXERCISE 1

Find the greatest common factor of the integers.

(a) 55 and 33

(b) 54 and 90

$$54 = 2 \cdot 3 \cdot 3 \cdot 3 \quad \text{and} \quad 90 = 2 \cdot 3 \cdot 3 \cdot 5$$

GCF: $2 \cdot 3 \cdot 3 = 18$

(b) 140 and 84

Answers: (a) **11**   (b) **28**

We can also find common factors of monomials such as $14x^3$ and $28x^2$. First we write the monomials in factored form.

$$14x^3 = 2 \cdot 7 \cdot x^3 = \boxed{2 \cdot 7 \cdot x^2} \cdot x$$
$$28x^2 = 2^2 \cdot 7 \cdot x^2 = 2 \cdot \boxed{2 \cdot 7 \cdot x^2}$$

Factors common to both

Each shaded factor is a common factor of $14x^3$ and $28x^2$, and the shaded product, $2 \cdot 7 \cdot x^2 = 14x^2$, is the greatest common factor of the two monomials. Usually we are interested in finding the greatest common factor of the terms of a polynomial. For instance, $14x^2$ is the greatest common factor of the terms of the polynomial $14x^3 + 28x^2$. The following rule summarizes the technique.

To Find the Greatest Common Factor

1. Write each term as a product of prime factors, expressing repeated factors as powers.

2. Select the factors that are common to all terms and raise each to the *lowest* power that it occurs in any one term.

3. The product of these factors is the greatest common factor (GCF).

EXAMPLE 2   GCF OF A POLYNOMIAL

Find the greatest common factor of the terms of each polynomial.

(a) $12x - 9 = 2^2 \cdot 3 \cdot x - 3 \cdot 3$
Common factors: 3
Lowest power of each: 3^1
GCF: **3**

(b) $18x^2 + 30x = 2 \cdot 3 \cdot 3 \cdot x \cdot x + 2 \cdot 3 \cdot 5 \cdot x$
Common factors: 2, 3, x
Lowest power of each: $2^1, 3^1, x^1$
GCF: $2 \cdot 3 \cdot x = \mathbf{6x}$

(c) $3x^4 + 12x^2 = 3 \cdot x^2 \cdot x^2 + 2^2 \cdot 3 \cdot x^2$
Common factors: 3, x
Lowest power of each: $3^1, x^2$
GCF: $3 \cdot x^2 = 3x^2$

(d) $7y + 3 = 1 \cdot 7 \cdot y + 1 \cdot 3$
Common factors: 1
Lowest power of each: 1^1
GCF: 1

(e) $14x^4 - 28x^3 + 21x^2 = 2 \cdot 7 \cdot x^2 \cdot x^2 - 2^2 \cdot 7 \cdot x^2 \cdot x + 3 \cdot 7 \cdot x^2$
Common factors: 7, x
Lowest power of each: $7^1, x^2$
GCF: $7 \cdot x^2 = 7x^2$

PRACTICE EXERCISE 2

Find the greatest common factor of the terms of each polynomial.

(a) $15a - 25$

(b) $24y^2 + 20y$

(c) $7w^5 + 21w^3$

(d) $9a + 8$

(e) $18y^5 - 27y^4 + 45y^3$

(f) $22u^3v^3 - 44u^3v + 77u^2v^4$

(f) $5x^2y^3 - 15xy^2 = 5 \cdot x \cdot x \cdot y^2 \cdot y - 3 \cdot 5 \cdot x \cdot y^2$
Common factors: $5, x, y$
Lowest power of each: $5^1, x^1, y^2$
GCF: $5 \cdot x \cdot y^2 = 5xy^2$

Answers: (a) 5 (b) $4y$ (c) $7w^3$
(d) 1 (e) $9y^3$ (f) $11u^2v$

② FACTORING BY REMOVING THE GCF

A polynomial is said to be **factored by removing common factors** when the greatest common factor is removed using the distributive law. For example, since

$$14x^3 + 28x^2 = 14x^2(x + 2)$$

we have factored $14x^3 + 28x^2$ using the distributive law by removing the greatest common factor $14x^2$. Example 3 illustrates the technique summarized below.

> ### To Factor by Removing the Greatest Common Factor
>
> **1.** Write each term as a product of primes.
>
> **2.** Find the greatest common factor.
>
> **3.** Write each term in the form
>
> > **GCF · remaining factors.**
>
> **4.** Use the distributive property to remove the **GCF**. The factors left in each term are the **remaining factors.**
>
> **5.** To check, multiply the GCF by the polynomial factor.

EXAMPLE 3 REMOVING THE GCF

Factor the polynomials given in Example 2.

(a) $12x - 9 = 2^2 \cdot 3 \cdot x - 3 \cdot 3$ GCF is 3

$= 3(2^2 \cdot x - 3)$ Remove the GCF using the distributive property

$= 3(4x - 3)$ Keep the GCF as a multiplier

(b) $18x^2 + 30x = 2 \cdot 3 \cdot 3 \cdot x \cdot x + 2 \cdot 3 \cdot 5 \cdot x$ GCF is $6x$

$= 2 \cdot 3 \cdot x(3x + 5)$ Distributive property

$= 6x(3x + 5)$ Keep the GCF, $6x$, as a multiplier

(c) $3x^4 + 12x^2 = 3x^2 \cdot x^2 + 3x^2 \cdot 4$ GCF is $3x^2$

$= 3x^2(x^2 + 4)$ Distributive property

(d) $7y + 3 = 1 \cdot 7y + 1 \cdot 3$ GCF is 1

$= 7y + 3$ Cannot be factored

(e) $14x^4 - 28x^3 + 21x^2 = 7x^2 \cdot 2x^2 - 7x^2 \cdot 4x + 7x^2 \cdot 3$ GCF is $7x^2$

$= 7x^2(2x^2 - 4x + 3)$

(f) $5x^2y^3 - 15xy^2 = 5 \cdot x^2 \cdot y^3 - 3 \cdot 5 \cdot x \cdot y^2$

$= 5xy^2 \cdot xy - 5xy^2 \cdot 3$ GCF is $5xy^2$

$= 5xy^2(xy - 3)$

PRACTICE EXERCISE 3

Factor by removing the greatest common factor.

(a) $15a - 25$

(b) $24y^2 + 20y$

(c) $7w^5 + 21w^3$

(d) $9a + 8$

(e) $18y^5 - 27y^4 + 45y^3$

(f) $22u^3v^3 - 44u^3v + 77u^2v^4$

Answers: (a) $5(3a - 5)$
(b) $4y(6y + 5)$ (c) $7w^3(w^2 + 3)$
(d) $9a + 8$ is a prime polynomial
(e) $9y^3(2y^2 - 3y + 5)$
(f) $11u^2v(2uv^2 - 4u + 7v^3)$

When the greatest common factor is 1, we say the polynomial cannot be factored by removing the common factor. Polynomials like $7y + 3$ in Example 3(d), which cannot be factored, are called **prime polynomials.**

////////////// **CAUTION** //////////////

When told to factor a polynomial such as $18x^2 + 30x$, a common error is to give

$$2 \cdot 3 \cdot 3 \cdot x \cdot x + 2 \cdot 3 \cdot 5 \cdot x$$

for the answer. Here the terms of the polynomial have been factored, but the polynomial itself has not. The correct factorization is given in Example 3(b).

//////////

Often when a polynomial consists of many negative terms, or when its leading coefficient is negative, we factor out the negative of the greatest common factor of its terms, as shown in the next example.

| EXAMPLE 4   FACTORING OUT A NEGATIVE FACTOR | PRACTICE EXERCISE 4 |

Factor $-6x^5 - 12x^4 - 15x$.

$$-6x^5 - 12x^4 - 15x = (-3x)2x^4 + (-3x)4x^3 + (-3x)5$$
$$= -3x(2x^4 + 4x^3 + 5)$$

In this example the greatest common factor is $3x$; thus, another way to factor is $3x(-2x^4 - 4x^3 - 5)$. Either answer is considered correct, but the first is sometimes preferred.

Factor $-10x^4 - 35x^3 - 40x^2$.

Answer: $-5x^2(2x^2 + 7x + 8)$

| EXAMPLE 5   FACTORING WITH 1 LEFT IN TERM | PRACTICE EXERCISE 5 |

Factor $3x^2y^3 - xy$.

$$3x^2y^3 - xy = xy \cdot 3xy^2 - xy \cdot 1 \qquad \text{GCF is } xy$$
$$= xy(3xy^2 - 1) \qquad \text{Distributive property}$$

Factor $6u^4v^4 + u^2v^3$.

Answer: $u^2v^3(6u^2v + 1)$

////////////// **CAUTION** //////////////

When we factored xy out of $xy \cdot 1$ in Example 5, we were left with a 1 in the term. Do not leave this term out and write $xy(3xy^2)$ as the factors of $3x^2y^3 - xy$. If we multiply $xy(3xy^2)$ we get $3x^2y^3$ which is just the first term of the given polynomial.

//////////

Sometimes a polynomial contains grouping symbols. For example, the polynomial

$$3x(x + 2) + 5(x + 2)$$

is really a binomial made of the two terms $3x(x + 2)$ and $5(x + 2)$. These terms have a greatest common factor $(x + 2)$.

$$3x(x + 2) + 5(x + 2) = 3x \cdot (x + 2) + 5 \cdot (x + 2) \qquad \text{GCF is } (x + 2)$$
$$= (3x + 5)(x + 2) \qquad \text{Distributive property}$$

Here the $(x + 2)$ is a single factor of each term and is removed using the distributive property.

EXAMPLE 6 POLYNOMIALS CONTAINING GROUPING SYMBOLS

Factor $3x(a + b) + y(a + b)$.

$$3x(a + b) + y(a + b) = 3x \cdot (a + b) + y \cdot (a + b) \qquad \text{GCF is } (a + b)$$
$$= (3x + y)(a + b) \qquad \text{Distributive property}$$

PRACTICE EXERCISE 6

Factor $2y(a - b) - x(a - b)$.

Answer: $(2y - x)(a - b)$

---

③ FACTORING BY GROUPING

Notice that if we clear the parentheses in the polynomial in Example 6, we obtain

$$3x(a + b) + y(a + b) = 3xa + 3xb + ya + yb.$$

Suppose we are given the polynomial on the right to factor. We may do so by grouping the first two terms and the last two terms. Then we factor as in Example 6. This process is called **factoring by grouping** and is illustrated in the next example.

EXAMPLE 7 FACTORING BY GROUPING

Factor each polynomial by grouping.

(a) $3xa + 3xb + ya + yb$

$= (3xa + 3xb) + (ya + yb)$ Group the first two terms and the last two terms

$= 3x(a + b) + y(a + b)$ Use the distributive law on each group

$= (3x + y)(a + b)$ Use the distributive law to factor out $(a + b)$

(b) $5ay - 5ax + 3by - 3bx = (5ay - 5ax) + (3by - 3bx)$

$$= 5a(y - x) + 3b(y - x)$$
$$= (5a + 3b)(y - x)$$

PRACTICE EXERCISE 7

Factor each polynomial by grouping.

(a) $2ay + by + 2ax + bx$

(b) $6xu - 5yu + 6xv - 5yv$

Answers: (a) $(2a + b)(y + x)$
(b) $(6x - 5y)(u + v)$

Factoring by grouping should be tried when the polynomial has four terms. We conclude this section with an application of factoring.

EXAMPLE 8 APPLICATION OF FACTORING

If a rocket is fired straight up with a velocity of 128 feet per second, the equation $h = -16t^2 + 128t$ gives the height h of the rocket at any time t. Factor the right side of the equation and use this factored form to find the height of the rocket after 6 seconds.

Factoring, we obtain

$$h = -16t^2 + 128t = -16t^2 + 16 \cdot 8t$$
$$= -16t \, (t - 8). \qquad \text{Notice the minus sign}$$

PRACTICE EXERCISE 8

Given a rocket as in Example 8 with $h = -16t^2 + 64t$, factor and find the height after 4 seconds.

Substitute 6 for t to find h.

$$h = -16t(t - 8)$$
$$= -16(6)(6 - 8)$$
$$= -16(6)(-2) = 192$$

The rocket will be 192 ft high after 6 sec.

Answers: $h = -16t(t - 4)$; 0

You never need to make a mistake in a factoring problem since you can always check your work by multiplying the factors and comparing the product with the given polynomial.

7.1 EXERCISES A

Find the greatest common factor.

1. 10, 15

2. 28, 42

3. 17, 23

4. $10x^2$, $15x$

5. $28y^3$, $42y$

6. $17y^2$, $23y^4$

7. 10, 15, 35

8. 28, 42, 12

9. 17, 23, 16

10. $10x^2$, $15x$, $35x^3$

11. $28y^3$, $42y^2$, $12y^5$

12. $17y^2$, $23y^4$, 16

13. $6x^3$, $12x^2$, $18x^2$

14 $36y^4$, $6y^3$, $42y^5$

15. $8y^2$, $9x$, $12y^5$

16. $5x^2y$, $40xy^2$

17. $16x^2y^2$, $12x^3y^4$, $8xy^5$

18. x^4y^3, x^5y^2, x^3

Find the greatest common factor of the terms of the polynomials.

19. $4x + 8$

20. $8y^2 - 16y$

21. $3a^3 - 9a^2$

22. $15x^3 - 25x^2$

23. $9y^4 + 18y^2$

24. $50a^{10} + 75a^8$

25. $4x^3 + 2x^2 - 6x$

26. $18y^5 - 24y^3 + 36y$

27 $3a(a + 2) + 5(a + 2)$

28. $6x^2y^2 - 4x^2y$

29 $5a^3b^3 - 15a^2b^2 + 10ab^2$

30. $x(a + b) + y(a + b)$

Factor.

31. $3x + 9$

32. $21x - 14$

33. $24y - 6$

34. $4a^2 + 2a$

35. $18y^2 + 11y$

36. $23a^2 - 5$

37. $x^{10} - x^8 + x^6$

38. $6y^4 - 24y^2 + 12y$

39. $16a^5 + 48a^3 - 24$

40. $60x^3 + 50x^2 - 25x$

41 $-6y^{10} - 8y^8 - 4y^5$

42. $a^2 + 2a + 2$

43. $x^2y^2 + xy$

44. $6a^2b - 2ab^2$

45. $27x^3y^2 + 45xy^2$

46. $15x^3y^3 + 5x^2y^2 + 10xy$

47 $a^2(a + 2) + 3(a + 2)$

48. $x^2(a + b) + y^2(a + b)$

Factor by grouping.

49. $a^3 + 2a^2 + 3a + 6$

50 $x^2a + x^2b + y^2a + y^2b$

51. $a^2b - a^2 + 5b - 5$

52. $x^3 + 2x^2 - 7x - 14$

53. $a^3b^2 + a^3 + 2b^2 + 2$

54. $x^2y - 3x^2 + y - 3$

55. $a^4b + a^3 - ab^3 - b^2$

56 $-x^2y - x^2 - 3y - 3$

57. If a principal of P dollars is invested in an account with an annual interest rate of 9% compounded annually, the amount A in the account after one year is given by the equation $A = P + 0.09P$. Factor the right side of the equation and use this factored form to find the amount in the account in one year when P is   **(a)** $100,   **(b)** $1000, and   **(c)** $4537.35.

58. When asked to factor $4x^3 + 8x^2$, Burford gave $2x(2x^2 + 4x)$ for the answer. What is wrong with Burford's work? What is the correct answer?

59. What is wrong with the following factoring problem?

$$-3x^2y^2 + 6x^2y = -3x^2y(y + 2)$$

60. What is the GCF of the terms of the binomial $2xy + 4x^2$? What is the GCF of the terms of the binomial $5xy - 10y$? Find the product of $2xy + 4x^2$ and $5xy - 10y$ and give the GCF of the terms of this polynomial. What can you conclude?

FOR REVIEW

Use the FOIL method to multiply.

61. $(x + 3)(x + 7)$

62. $(x + 3)(x - 7)$

63. $(x - 3)(x + 7)$

64. $(x - 3)(x - 7)$

65. $(x + 4y)(x + 3y)$

66. $(x + 4y)(x - 3y)$

ANSWERS: 1. 5 2. 14 3. 1 4. $5x$ 5. $14y$ 6. y^2 7. 5 8. 2 9. 1 10. $5x$ 11. $2y^2$ 12. 1 13. $6x^2$ 14. $6y^3$ 15. 1 16. $5xy$ 17. $4xy^2$ 18. x^3 19. 4 20. $8y$ 21. $3a^2$ 22. $5x^2$ 23. $9y^2$ 24. $25a^8$ 25. $2x$ 26. $6y$ 27. $a + 2$ 28. $2x^2y$ 29. $5ab^2$ 30. $a + b$ 31. $3(x + 3)$ 32. $7(3x - 2)$ 33. $6(4y - 1)$ 34. $2a(2a + 1)$ 35. $y(18y + 11)$ 36. cannot be factored 37. $x^6(x^4 - x^2 + 1)$ 38. $6y(y^3 - 4y + 2)$ 39. $8(2a^5 + 6a^3 - 3)$ 40. $5x(12x^2 + 10x - 5)$ 41. $-2y^5(3y^5 + 4y^3 + 2)$ 42. cannot be factored 43. $xy(xy + 1)$ 44. $2ab(3a - b)$ 45. $9xy^2(3x^2 + 5)$ 46. $5xy(3x^2y^2 + xy + 2)$ 47. $(a + 2)(a^2 + 3)$ 48. $(a + b)(x^2 + y^2)$ 49. $(a + 2)(a^2 + 3)$ 50. $(a + b)(x^2 + y^2)$ 51. $(b - 1)(a^2 + 5)$ 52. $(x + 2)(x^2 - 7)$ 53. $(b^2 + 1)(a^3 + 2)$ 54. $(y - 3)(x^2 + 1)$ 55. $(ab + 1)(a^3 - b^2)$ 56. $-(x^2 + 3)(y + 1)$ 57. $A = P(1 + 0.09) = P(1.09)$; (a) $109 (b) $1090 (c) $4945.71 58. He did not factor out the GCF. The answer should be $4x^2(x + 2)$ 59. The answer should be $-3x^2y(y - 2)$. 60. $2x$; $5y$; $10xy$; the GCF of the product is the product of GCFs. 61. $x^2 + 10x + 21$ 62. $x^2 - 4x - 21$ 63. $x^2 + 4x - 21$ 64. $x^2 - 10x + 21$ 65. $x^2 + 7xy + 12y^2$ 66. $x^2 + xy - 12y^2$

7.1 EXERCISES B

Find the greatest common factor.

1. 12, 18

2. 32, 48

3. 11, 29

4. $12x, 18x^3$

5. $32y^4, 48y^2$

6. $11y^3, 29y^2$

7. 12, 18, 9

8. 32, 48, 64

9. 11, 29, 30

10. $12x^4, 18x^2, 9x^3$

11. $32y^3, 48y^5, 64y^4$

12. $11y^2, 29y^3, 30$

13. $15x^2, 20x^3, 5x^4$

14. $22y^5, 33y^4, 44y^3$

15. $36y^4, 15x^2, 5xy$

16. $6x^2y^3, 24x^2y$

17. $30x^5y^5, 2x^2y^4, 12x^4y^3$

18. x^8y^7, x^4y^6, y^5

Find the greatest common factor of the terms of the polynomials.

19. $9x + 3$

20. $22y^2 - 11y$

21. $15a^3 - 10a^2$

22. $16x^4 - 24x^2$

23. $14y^5 + 35y^4$

24. $100a^{16} + 200a^{14}$

25. $5x^3 + 15x^2 - 15x$

26. $48y^6 - 24y^5 + 12y^2$

27. $7a(a - 5) + 6(a - 5)$

28. $9x^3y^3 - 12xy^4$

29. $4a^3b^3 + 6a^2b^2 - 6ab^2$

30. $2a(x + y) + b(x + y)$

Factor.

31. $5y - 20$

32. $15x - 35$

33. $54y + 12$

34. $22y - 11$

35. $25x^4 - 15x^3$

36. $16a^2 - 7$

37. $x^{12} - x^6 + x^4$

38. $26y^5 - 13y^3 + 39y^2$

39. $8a^6 - 18a^3 + 12$

40. $90x^4 - 45x^3 + 180x^2$

41. $-5y^8 - 10y^6 - 15y^4$

42. $a^2 + a + 5$

43. $x^4y^4 - x^2y$

44. $11a^3b^3 - 22a^2b^2$

45. $24x^4y^3 + 36x^3y^4$

46. $28x^4y^4 + 14x^3y^3 + 35x^2y^2$

47. $2a^3(a - 5) + 5(a - 5)$

48. $a^2(x + y) + 2b(x + y)$

Factor by grouping.

49. $a^3 + 7a^2 + 2a + 14$

50. $a^2x + a^2y + b^2x + b^2y$

51. $a^3b - 2a^3 + 4b - 8$

52. $x^3 + 5x^2 - 2x - 10$

53. $a^5b^3 + 2a^5 + 6b^3 + 12$

54. $xy - 2y + x - 2$

55. $ab^2 + ab - 5b - 5$

56. $-x^3y^2 - 3x^3y - 7y - 21$

57. If a bullet is fired straight up with a velocity of 256 feet per second, the equation $h = -16t^2 + 256t$ gives the height h of the bullet at any time t. Factor the right side of the equation and use this factored form to find the height of the bullet after **(a)** 6 sec, **(b)** 8 sec, **(c)** 10 sec, and **(d)** 16 sec.

58. When Burford was told to factor $12x^2 - 8x$ he gave $2 \cdot 2 \cdot 3 \cdot x \cdot x - 2 \cdot 2 \cdot 2 \cdot x$ for the answer. What is wrong with Burford's work? What is the correct answer?

59. What is wrong with the following factoring problem?

$$4x^2y^3 - 2xy = 2xy(2xy^2)$$

60. What is the GCF of the terms of the binomial $2x - 4$? What is the GCF of the terms of the binomial $3x + 6$? Find the product of $2x - 4$ and $3x + 6$ and give the GCF of the terms of this trinomial. What can you conclude?

FOR REVIEW

Use the FOIL method to multiply.

61. $(x + 2)(x + 3)$

62. $(x + 2)(x - 3)$

63. $(x - 2)(x + 3)$

64. $(x - 2)(x - 3)$

65. $(x - 2y)(x - 2y)$

66. $(x + 2y)(x + 2y)$

7.1 EXERCISES C

Factor by grouping. Do not collect like terms first.

1. $6x^2 + 21x - 10x - 35$

2. $10x^2 - 15xy - 2xy + 3y^2$

3. $a^2b^2c^2 - 5abc + 3abc - 15$

4. $6a^2b^2c - 15ab - 4abc^2 + 10c$
[Answer: $(3ab - 2c)(2abc - 5)$]

7.2 FACTORING TRINOMIALS OF THE FORM $x^2 + bx + c$

STUDENT GUIDEPOSTS

1 Review of FOIL

3 Factoring $x^2 + bxy + cy^2$

2 Factoring $x^2 + bx + c$

1 REVIEW OF FOIL

To factor a trinomial into the product of two binomials, we need to recognize the pattern for multiplying binomials in reverse. Consider the following multiplication.

$$(x + 3)(x + 7) = x \cdot x + x \cdot 7 + 3 \cdot x + 3 \cdot 7$$
$$= x^2 + 7x + 3x + 21$$
$$= x^2 + 10x + 21$$

Notice that x^2 is the product of the first terms Ⓕ, 21 is the product of the last terms Ⓛ, and $10x$ is the sum of the product Ⓞ and Ⓘ.

If we are given the trinomial $x^2 + 10x + 21$ to factor, we need to reverse the multiplication steps above to find $(x + 3)(x + 7)$. That is, we must fill in the blanks in

$$x^2 + 10x + 21 = (x + \underline{\quad})(x + \underline{\quad}).$$

The numbers in the blanks must have a product of 21 and a sum of 10.

$$x^2 + 10x + 21 = (x + \underline{3})(x + \underline{7})$$

2 FACTORING $x^2 + bx + c$

In general, to factor $x^2 + bx + c$ we look for a pair of integers whose product is c and whose sum is b.

$$x^2 + bx + c = (x + \underline{\quad})(x + \underline{\quad})$$

To Factor $x^2 + bx + c$
1. Write $x^2 + bx + c = (x + \underline{\hspace{1em}})(x + \underline{\hspace{1em}})$.
2. List all pairs of integers whose product is c.
3. Find the pair from this list whose sum is b (if there is one).
4. Fill in the blanks with this pair.

EXAMPLE 1 FACTORING TRINOMIALS

Factor $x^2 + 6x + 8$.   ($b = 6$ and $c = 8$)

$$x^2 + 6x + 8 = (x + \underline{\hspace{1em}})(x + \underline{\hspace{1em}})$$

Factors of $c = 8$	Sum of factors
1, 8	$1 + 8 = 9$
2, 4	$2 + 4 = 6$
$-1, -8$	$-1 + (-8) = -9$
$-2, -4$	$-2 + (-4) = -6$

Since the product of 2 and 4 is 8, which equals c, and the sum of 2 and 4 is 6, which equals b, we write

$$x^2 + 6x + 8 = (x + \underline{2})(x + \underline{4}).$$

That is, the blanks are filled in with the shaded pair above. To check, we multiply using the FOIL method.

$$(x + 2)(x + 4) = x^2 + 4x + 2x + 8 = x^2 + 6x + 8$$

PRACTICE EXERCISE 1

Factor $x^2 + 9x + 18$.

Factors of $c = 18$	Sum of factors
1, 18	$1 + 18 = 19$
2, 9	$2 + 9 = 11$
3, 6	$3 + 6 = 9$

Answer: $(x + 3)(x + 6)$

EXAMPLE 2 FACTORING TRINOMIALS IN ONE VARIABLE

Factor each trinomial.

(a) $x^2 + 5x + 6$ ($b = 5$ and $c = 6$)

$$x^2 + 5x + 6 = (x + \underline{\hspace{1em}})(x + \underline{\hspace{1em}})$$

Factors of $c = 6$	Sum of factors
6, 1	$6 + 1 = 7$
2, 3	$2 + 3 = 5$
$-6, -1$	$-6 + (-1) = -7$
$-2, -3$	$-2 + (-3) = -5$

We fill in the blanks with the shaded pair obtaining

$$x^2 + 5x + 6 = (x + \underline{2})(x + \underline{3}).   2 \cdot 3 = 6 \text{ and } 2 + 3 = 5$$

Note in the table that since $c = 6 > 0$, the factors in each pair must have the same sign. Also, since $b = 5 > 0$, we know that the negative factors will not work.

PRACTICE EXERCISE 2

Factor each trinomial.

(a) $x^2 + 10x + 21$

(b) $x^2 - 10x + 21$

(b) $x^2 - 5x + 6$ $(b = -5$ and $c = 6)$

$$x^2 - 5x + 6 = (x + \_\_)(x + \_\_)$$

Factors of $c = 6$	Sum of factors
6, 1	$6 + 1 = 7$
2, 3	$2 + 3 = 5$
$-6, -1$	$-6 + (-1) = -7$
$-2, -3$	$-2 + (-3) = -5$

$x^2 - 5x + 6 = (x - 2)(x - 3)$ $(-2)(-3) = 6$ and $-2 + (-3) = -5$

With $c > 0$ the factors in each pair must have the same sign. Since $b = -5$, positive factors will not work.

(c) $x^2 + x - 6$ $(b = 1$ and $c = -6)$

$$x^2 + x - 6 = (x + \_\_)(x + \_\_)$$

Factors of $c = -6$	Sum of factors
6, -1	$6 + (-1) = 5$
-6, 1	$-6 + 1 = -5$
2, -3	$2 + (-3) = -1$
-2, 3	$-2 + 3 = 1$

$x^2 + x - 6 = (x - 2)(x + 3)$ $-2 \cdot 3 = -6$ and $-2 + 3 = 1$

Since $c = -6 < 0$, the factors of c must have opposite signs.

(d) $x^2 - x - 6$ $(b = -1$ and $c = -6)$

$$x^2 - x - 6 = (x + \_\_)(x + \_\_)$$

Factors of $c = -6$	Sum of factors
6, -1	$6 + (-1) = 5$
-6, 1	$-6 + 1 = -5$
2, -3	$2 + (-3) = -1$
-2, 3	$-2 + 3 = 1$

$x^2 - x - 6 = (x + 2)(x - 3)$ $2 \cdot (-3) = -6$ and $2 + (-3) = -1$

(c) $x^2 + 4x - 21$

(d) $x^2 - 4x - 21$

Answers: (a) $(x + 3)(x + 7)$
(b) $(x - 3)(x - 7)$
(c) $(x - 3)(x + 7)$
(d) $(x + 3)(x - 7)$

3 FACTORING $x^2 + bxy + cy^2$

To factor trinomials in two variables such as $x^2 + bxy + cy^2$, we use the same procedures as above. But we now fill in the blanks in

$$x^2 + 5xy + 6y^2 = (x + \_\_y)(x + \_\_y).$$

The following table compares factoring trinomials in two variables with factoring similar trinomials in one variable.

One variable	Two variables
$x^2 + 5x + 6 = (x + 2)(x + 3)$	$x^2 + 5xy + 6y^2 = (x + 2y)(x + 3y)$
$x^2 - 5x + 6 = (x - 2)(x - 3)$	$x^2 - 5xy + 6y^2 = (x - 2y)(x - 3y)$
$x^2 + x - 6 = (x - 2)(x + 3)$	$x^2 + xy - 6y^2 = (x - 2y)(x + 3y)$
$x^2 - x - 6 = (x + 2)(x - 3)$	$x^2 - xy - 6y^2 = (x + 2y)(x - 3y)$

EXAMPLE 3 FACTORING TRINOMIALS IN TWO VARIABLES

PRACTICE EXERCISE 3

Factor each trinomial.

(a) $x^2 + 7xy + 10y^2$ ($b = 7$ and $c = 10$)

$$x^2 + 7xy + 10y^2 = (x + \underline{}y)(x + \underline{}y)$$

Factors of $c = 10$	Sum of factors
10, 1	$10 + 1 = 11$
5, 2	$5 + 2 = 7$
$-10, -1$	$-10 + (-1) = -11$
$-5, -2$	$-5 + (-2) = -7$

$$x^2 + 7xy + 10y^2 = (x + 5y)(x + 2y) \quad 5 \cdot 2 = 10 \text{ and } 5 + 2 = 7$$

(b) $x^2 - xy - 12y^2$ ($b = -1$ and $c = -12$)

$$x^2 - xy - 12y^2 = (x + \underline{}y)(x + \underline{}y)$$

Factors of $c = -12$	Sum of factors
12, -1	$12 + (-1) = 11$
$-12, 1$	$-12 + 1 = -11$
6, -2	$6 + (-2) = 4$
$-6, 2$	$-6 + 2 = -4$
4, -3	$4 + (-3) = 1$
$-4, 3$	$-4 + 3 = -1$

$$x^2 - xy - 12y^2 = (x - 4y)(x + 3y)   \quad (-4) \cdot 3 = -12 \text{ and } -4 + 3 = -1$$

PRACTICE EXERCISE 3

Factor each trinomial.

(a) $x^2 + 7xy + 12y^2$

(b) $x^2 + xy - 12y^2$

Answers: (a) $(x + 4y)(x + 3y)$
(b) $(x + 4y)(x - 3y)$

With practice you will be able to find the factors of c that are more likely to sum to b without constructing the complete table.

Some trinomials are not in the form $x^2 + bxy + cy^2$ but can be put into this form by removing a common factor.

EXAMPLE 4 TRINOMIALS WITH A COMMON FACTOR

PRACTICE EXERCISE 4

Factor $3x^2 - 12xy + 12y^2$.

$$3x^2 - 12xy + 12y^2 = 3 \cdot x^2 - 3 \cdot 4xy + 3 \cdot 4y^2$$
$$= 3(x^2 - 4xy + 4y^2)$$

PRACTICE EXERCISE 4

Factor $7x^2 + 21xy - 70y^2$.

We can now factor $x^2 - 4xy + 4y^2$ where $b = -4$ and $c = 4$.

$$x^2 - 4xy + 4y^2 = (x + \underline{}y)(x + \underline{}y)$$

Factors of $c = 4$	Sum of factors
4, 1	$4 + 1 = 5$
2, 2	$2 + 2 = 4$
$-4, -1$	$-4 + (-1) = -5$
$-2, -2$	$-2 + (-2) = -4$

$x^2 - 4xy + 4y^2 = (x - 2y)(x - 2y)$ $(-2)(-2) = 4$ and $-2 + (-2) = -4$

Thus, $3x^2 - 12xy + 12y^2 = 3(x - 2y)(x - 2y)$.

Answer: $7(x + 5y)(x - 2y)$

Always include the common factor as part of the answer. In Example 4 we included the factor 3 in the final product.

Not all trinomials can be factored. One example is $x^2 + 5x + 2$. The only factors of $c = 2$ are 2 and 1, but $2 + 1$ is 3 and not $b = 5$.

Factoring, just like any other skill, is learned by practice. The more time spent and the more problems worked, the more proficient we become.

EXAMPLE 5   APPLICATION OF FACTORING

A small company manufactures wood-burning stoves. The total cost c of producing n stoves is given by the equation $c = -200n^2 + 1200n + 1400$. Factor the right side of this equation and use this factored form to find the cost of producing 4 stoves.

Factoring gives the following.

$$c = -200n^2 + 1200n + 1400$$
$$= -200(n^2 - 6n - 7) \quad \text{The GCF is } -200$$
$$= -200(n - 7)(n + 1)$$

Substitute 4 for n to find c.

$$c = -200(4 - 7)(4 + 1)$$
$$= -200(-3)(5) = 3000$$

It would cost \$3000 to produce 4 stoves.

PRACTICE EXERCISE 5

If the profit on stoves is $p = -10n^2 + 200n - 960$, factor the right side of the equation and find the profit on 10 stoves.

Answers: $p = -10(n - 12)(n - 8)$; \$40

7.2 EXERCISES A

Factor and check by multiplying.

1. $x^2 + 4x + 3$

2. $x^2 + 2x - 3$

3. $x^2 - 2x - 3$

4. $x^2 - 4x + 3$

5. $u^2 - 12u + 35$

6. $u^2 - 2u - 35$

7. $u^2 + 12u + 35$

8. $u^2 + 2u - 35$

9. $y^2 + 10y + 21$

10 $y^2 + 5y - 24$

11. $x^2 - 12x + 27$

12. $x^2 + 4x - 45$

13. $y^2 - y - 56$

14 $x^2 - 2x - 63$

15. $x^2 - 2x - 120$

16. $x^2 + 4x - 77$

17. $x^2 + 4xy + 3y^2$

18. $x^2 + 2xy - 3y^2$

19. $x^2 - 2xy - 3y^2$

20 $u^2 - 9uv + 20v^2$

21. $x^2 - 4xy + 3y^2$

22. $u^2 + uv - 20v^2$

23. $u^2 + 9uv + 20v^2$

24. $u^2 - uv - 20v^2$

25. $x^2 + 13xy - 30y^2$

26. $x^2 + 11xy + 24y^2$

27. $u^2 - 8uv + 15v^2$

28. $u^2 - 3uv - 40v^2$

29. $x^2 - 10xy + 24y^2$

30. $x^2 + 11xy + 30y^2$

31. $u^2 + 12uv + 36v^2$

32. $u^2 - uv - 42v^2$

33. $x^2 + xy - 90y^2$

34. $x^2 - 4xy - 32y^2$

35 $u^2 - 22uv + 121v^2$

36. $u^2 + 18uv + 77v^2$

37. A child standing on a bridge 48 ft above the surface of the water throws a rock upward with a velocity of 32 ft per second. The equation that gives the height h in feet of the rock above the water in terms of time t in seconds is $h = -16t^2 + 32t + 48$. Factor the right side of this equation and use this factored form to find the height of the rock after **(a)** 1 sec, **(b)** 2 sec, and **(c)** 3 sec.

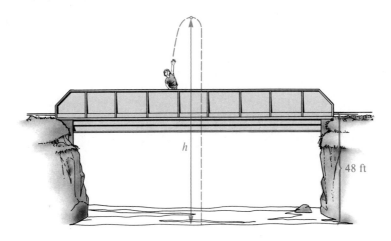

38. What is wrong with the work in the following factoring problem?

$$2x^2 + 10x - 48 = 2(x^2 + 5x - 24)$$
$$= (x - 3)(x + 8)$$

FOR REVIEW

Find the GCF and factor.

39. $35x - 70$

40. $88y^3 - 33y^2$

41. $9a^3 + 24a^2 - 15a$

42. $6x^3y^2 - 12x^2y^3$

43. $28a^4b^4 - 14a^3b^3 - 21a^2b^2$

44. $2a(x + y) - 3b(x + y)$

Factor by grouping.

45. $x^3 - 6x^2 + 5x - 30$

46. $5ax^2 + 2bx^2 + 5ay^2 + 2by^2$

The following review material from Chapter 6 will help you prepare for the next section. Use the FOIL method to multiply.

47. $(2x + 1)(x + 3)$

48. $(3x + 2)(x + 1)$

49. $(3x - 5)(2x - 1)$

50. $(5x - 2)(2x + 1)$

51. $(2x + 5)(x - 2)$

52. $(2x + 1)(x - 3)$

7.2 EXERCISES B

Factor and check by multiplying.

1. $x^2 + 6x + 5$ 2. $x^2 + 4x - 5$ 3. $x^2 - 4x - 5$ 4. $x^2 - 6x + 5$

5. $u^2 - 8u + 15$ 6. $u^2 - 2u - 15$ 7. $u^2 + 8u + 15$ 8. $u^2 + 2u - 15$

9. $y^2 - 10y + 21$ 10. $x^2 - 5x - 24$ 11. $u^2 + 12u + 27$ 12. $y^2 + y - 20$

13. $x^2 - 4x - 32$ 14. $u^2 - 2u - 48$ 15. $y^2 + 16y + 60$ 16. $x^2 + 8x - 33$

17. $x^2 + 6xy + 5y^2$ 18. $x^2 + 4xy - 5y^2$ 19. $x^2 - 4xy - 5y^2$ 20. $x^2 - 6xy + 5y^2$

21. $u^2 - 9uv + 18v^2$ 22. $u^2 + 3uv - 18v^2$ 23. $u^2 + 9uv + 18v^2$ 24. $u^2 - 3uv - 18v^2$

25. $x^2 + 6xy - 27y^2$ 26. $x^2 + 12xy + 20y^2$ 27. $u^2 - 11uv + 28v^2$ 28. $u^2 + uv - 30v^2$

29. $x^2 - 13xy + 40y^2$ 30. $x^2 + 16xy + 63y^2$ 31. $u^2 + 8uv + 16v^2$ 32. $u^2 - 3uv - 40v^2$

33. $x^2 + 6xy - 72y^2$ 34. $x^2 - 5xy - 50y^2$ 35. $u^2 - 26uv + 169v^2$ 36. $u^2 + 19uv + 88v^2$

37. A company can produce n items at a total cost of c dollars where c is given by $c = -100n^2 + 600n + 1600$. Factor the right side of this equation and use this factored form to find the cost of producing   **(a)** 0 items,   **(b)** 2 items,   **(c)** 3 items, and   **(d)** 6 items.

38. What is wrong with the work in the following factoring problem?

$$-4x^2 + 12x + 40 = -4(x^2 + 3x - 10)$$
$$= -4(x + 5)(x - 2)$$

FOR REVIEW

Find the GCF and factor.

39. $18x + 36$ 40. $35y^4 - 21y^2$ 41. $16a^5 - 24a^4 - 32a^3$

42. $25x^4y^5 + 45x^3y^6$ 43. $8a^5b^5 + 20a^3b^4 - 32a^2b^5$ 44. $6u(x^2 + y^2) + 7(x^2 + y^2)$

Factor by grouping.

45. $x^3 + 3x^2 - 7x - 21$

46. $2a^2x - 2a^2y + b^2x - b^2y$

The following review material from Chapter 6 will help you prepare for the next section. Use the FOIL method to multiply.

47. $(3x + 1)(x + 7)$

48. $(5x + 2)(x + 3)$

49. $(3x + 5)(2x - 1)$

50. $(2x - 1)(x + 10)$

51. $(5x - 3)(x + 6)$

52. $(2x - 5)(2x + 7)$

7.2 EXERCISES C

Factor.

1. $x^2 + \dfrac{1}{3}x - \dfrac{2}{9}$

2. $x^2 - 0.04x + 0.0003$

3. $(x + 2)^2 - 10(x + 2) + 16$
[Answer: $x(x - 6)$]

4. $x^4 + 5x^2 + 6$

7.3 FACTORING TRINOMIALS OF THE FORM $ax^2 + bx + c$

STUDENT GUIDEPOSTS

1. Factoring $ax^2 + bx + c$
2. Factoring $ax^2 + bxy + cy^2$
3. Factoring $ax^2 + bx + c$ by Grouping

1 FACTORING $ax^2 + bx + c$

In Section 7.2 we concentrated on factoring trinomials for which the coefficient of the squared term is 1. Now we consider trinomials of the form $ax^2 + bx + c$, with $a \neq 1$. As before, it helps to look at the pattern in multiplication.

$$(5x + 2)(2x + 3) = 5x \cdot 2x + 5x \cdot 3 + 2 \cdot 2x + 2 \cdot 3$$
$$= 10x^2 + 15x + 4x + 6$$
$$= 10x^2 + 19x + 6$$

The term $10x^2$ is the product of the first terms Ⓕ, the 6 is the product of the last terms Ⓛ, and $19x$ is the sum of the products Ⓞ and Ⓘ. With $a \neq 1$ we can use the following trial and error method to factor. (Another method is presented later.)

To Factor $ax^2 + bx + c$
1. Write $ax^2 + bx + c = (\_\_x + \_\_)(\_\_x + \_\_)$.
2. List all pairs of integers whose product is a and try each of these in the first blanks in each binomial factor.
3. List all pairs of integers whose product is c and try each of these in the second blanks.
4. Use trial and error to determine which pair (if one exists) gives $\textcircled{0} + \textcircled{I} = b$.

EXAMPLE 1 FACTORING $ax^2 + bx + c$

Factor $2x^2 + 7x + 3$. $(a = 2,\ b = 7,\ c = 3)$

$$2x^2 + 7x + 3 = (\_\_x + \_\_)(\_\_x + \_\_)$$

Factors of 3 · Factors of 2

Since all terms of the trinomial are positive, we need only list the positive factors of a and c. We list the factors of c in both orders as a reminder to try all possibilities.

Factors of $a = 2$	Factors of $c = 3$
2, 1	3, 1
	1, 3

$$2x^2 + 7x + 3 = (\_\_x + \_\_)(\_\_x + \_\_)$$
$$= (2x + \_\_)(x + \_\_) \quad \text{The only factors of } a \text{ are 2 and 1}$$
$$\overset{?}{=} (2x + 3)(x + 1) \quad \text{Does not work because } 2x + 3x = 5x \ne 7x$$
$$\overset{?}{=} (2x + 1)(x + 3) \quad \text{This works because } 6x + x = 7x$$
$$2x^2 + 7x + 3 = (2x + 1)(x + 3)$$

To check we multiply.

$$(2x + 1)(x + 3) = 2x^2 + 6x + x + 3 = 2x^2 + 7x + 3$$

In Example 1 we listed only the factors of c (1 and 3) in both orders and not the factors of a (1 and 2). It is not necessary to list both pairs in both orders. For example, $(2x + 1)(x + 3)$ and $(x + 3)(2x + 1)$ are two factorizations that are the same by the commutative law of multiplication.

EXAMPLE 2 FACTORING $ax^2 + bx + c$

Factor $6x^2 - 13x + 5$. $(a = 6,\ b = -13,\ c = 5)$

$$6x^2 - 13x + 5 = (\_\_x + \_\_)(\_\_x + \_\_)$$

Factors of 5 · Factors of 6

PRACTICE EXERCISE 1

Factor $3x^2 + 5x + 2$.

Factors of $a = 3$	Factors of $c = 2$
3, 1	2, 1
	1, 2

$$(\_\_x + \_\_)(\_\_x + \_\_)$$

Factors of 2 · Factors of 3

Answer: $(3x + 2)(x + 1)$

PRACTICE EXERCISE 2

Factor $10x^2 + x - 2$.

Here the factors of $c = 5 > 0$ must both be negative to obtain the term $-13x$. As before we list them in both orders.

Factors of $a = 6$	Factors of $c = 5$	
6, 1	$-5, -1$	Try 6, 1 with both $-5, -1$ and $-1, -5$
3, 2	$-1, -5$	Try 3, 2 with both $-5, -1$ and $-1, -5$

$6x^2 - 13x + 5 = (\_\_x + \_\_)(\_\_x + \_\_)$

$\overset{?}{=} (6x - 5)(x - 1)$ Does not work because $-6x - 5x = -11x \neq -13x$

$\overset{?}{=} (6x - 1)(x - 5)$ Does not work because $-30x - x = -31x \neq -13x$

$\overset{?}{=} (3x - 5)(2x - 1)$ This works because $-3x - 10x = -13x$

$6x^2 - 13x + 5 = (3x - 5)(2x - 1)$

To check we multiply.

$(3x - 5)(2x - 1) = 6x^2 - 3x - 10x + 5 = 6x^2 - 13x + 5$

Answer: $(5x - 2)(2x + 1)$

EXAMPLE 3 FACTORING $ax^2 + bx + c$

PRACTICE EXERCISE 3

Factor $8y^2 - 10y - 7$. $(a = 8, b = -10, c = -7)$
The factors of $c = -7$ will have opposite signs.

Factor $8y^2 + 10y - 7$.

Factors of $a = 8$	Factors of $c = -7$
8, 1	7, -1
4, 2	$-7, 1$
	1, -7
	$-1, 7$

With this many cases to test, we try to minimize the number of trials. If 8 and 1 are used, the middle term will be either too big, because $8 \cdot 7y - 1 \cdot 1y = 55y$, or too small, because $8 \cdot 1y - 7 \cdot 1y = y$. Thus, we try 4 and 2 as factors of 8.

$8y^2 - 10y - 7 = (\_\_y + \_\_)(\_\_y + \_\_)$

$= (4y + \_\_)(2y + \_\_)$ Try 4, 2

$\overset{?}{=} (4y + 7)(2y - 1)$ Does not work

$\overset{?}{=} (4y - 7)(2y + 1)$ This works

$8y^2 - 10y - 7 = (4y - 7)(2y + 1)$

Check this by multiplying.

Answer: $(4y + 7)(2y - 1)$

EXAMPLE 4 FACTORING OUT A COMMON FACTOR

PRACTICE EXERCISE 4

Factor $-4x^2 - 2x + 20$.
Unlike Examples 1, 2, and 3 this trinomial has a common factor. First we factor out the common factor including -1 to make a positive. If a is positive we only have to consider positive factors of a.

$-4x^2 - 2x + 20 = \quad (2x^2 + x - 10)$

Factor $-6x^2 + 15x + 9$.

Now we factor $2x^2 + x - 10$ where $a = 2$, $b = 1$, and $c = -10$.

Factors of $a = 2$	Factors of $c = -10$
2, 1	10, -1 and -10, 1
	-1, 10 and 1, -10
	5, -2 and -5, 2
	-2, 5 and 2, -5

$-4x^2 - 2x + 20 = -2(2x^2 + x - 10)$

$= -2(2x + \underline{\ \ })(x + \underline{\ \ })$

$\overset{?}{=} -2(2x + 10)(x - 1)$ Factors of 10 give too large a middle term

$\overset{?}{=} -2(2x + 5)(x - 2)$ This works

$-4x^2 - 2x + 20 = -2(2x + 5)(x - 2)$

Answer: $-3(2x + 1)(x - 3)$

⚠ CAUTION

Do not forget to include the common factor in the answer. The common factor -2 was included in Example 4.

---

❷ FACTORING $ax^2 + bxy + cy^2$

Trinomials of the form $ax^2 + bxy + cy^2$ can be factored by filling in the blanks as indicated below.

$$ax^2 + bxy + cy^2 = (\underline{\ \ }x + \underline{\ \ }y)(\underline{\ \ }x + \underline{\ \ }y)$$

EXAMPLE 5   FACTORING $ax^2 + bxy + cy^2$

PRACTICE EXERCISE 5

Factor $3x^2 - 14xy + 8y^2$.   ($a = 3$, $b = -14$, $c = 8$)
Both factors of $c = 8$ must be negative since $b = -14$.

Factor $3a^2 + ab - 10b^2$.

Factors of $a = 3$	Factors of $c = 8$
3, 1	-8, -1
	-1, -8
	-4, -2
	-2, -4

$3x^2 - 14xy + 8y^2 = (3x + \underline{\ \ }y)(x + \underline{\ \ }y)$

$\overset{?}{=} (3x - 8y)(x - y)$ Does not work

$\overset{?}{=} (3x - y)(x - 8y)$ Does not work

$\overset{?}{=} (3x - 4y)(x - 2y)$ Does not work

$\overset{?}{=} (3x - 2y)(x - 4y)$ This works

$3x^2 - 14xy + 8y^2 = (3x - 2y)(x - 4y)$

Answer: $(3a - 5b)(a + 2b)$

| EXAMPLE 6 FACTORING $ax^2 + bxy + cy^2$ | PRACTICE EXERCISE 6 |

Factor $10u^2 + 7uv - 3v^2$. $(a = 10, b = 7, c = -3)$

Factor $10x^2 - 13xy + 3y^2$.

Factors of $a = 10$	Factors of $c = -3$
10, 1	3, −1
5, 2	−3, 1
	1, −3
	−1, 3

$10u^2 + 7uv - 3v^2 = (\_\_u + \_\_v)(\_\_u + \_\_v)$

$\overset{?}{=} (10u + 3v)(u - v)$ Does not work since $-10uv + 3uv = -7uv$

$\overset{?}{=} (10u - 3v)(u + v)$ We know this works from the first trial

$10u^2 + 7uv - 3v^2 = (10u - 3v)(u + v)$

Answer: $(10x - 3y)(x - y)$

❸ FACTORING $ax^2 + bx + c$ BY GROUPING

There is another method for factoring trinomials. Consider the following procedure.

$$2x^2 + 13x + 15 = 2x^2 + 10x + 3x + 15 \quad \text{13x = 10x + 3x}$$
$$= (2x^2 + 10x) + (3x + 15) \quad \text{Group terms}$$
$$= 2x(x + 5) + 3(x + 5) \quad \text{Factor the groups}$$
$$= (2x + 3)(x + 5) \quad \text{Factor out the common factor x + 5}$$

We have factored $2x^2 + 13x + 15$ by grouping the appropriate terms. But how did we decide to write $13x = 10x + 3x$? Notice that 10 and 3 are factors of $2 \cdot 15 = 30$. That is, 10 and 3 are factors of the product ac in $ax^2 + bx + c$.

To Factor $ax^2 + bx + c$ by Grouping

1. Find the product ac.
2. List the factors of ac until a pair is found with sum b.
3. Write bx as a sum using these factors as coefficients of x.
4. Factor the result by grouping.

| EXAMPLE 7 FACTORING BY GROUPING | PRACTICE EXERCISE 7 |

Factor $3x^2 - 10x + 7$ by grouping. The product $ac = 3 \cdot 7 = 21$.

Factor $3x^2 + 10x + 7$ by grouping.

Factors of $ac = 21$	Sum of factors
21, 1	$21 + 1 = 22$
−21, −1	$-21 + (-1) = -22$
7, 3	$7 + 3 = 10$
−7, −3	$-7 + (-3) = -10$

The factors -7 and -3 will work so we write $-10x = -7x - 3x$.

$3x^2 - 10x + 7 = 3x^2 - 7x - 3x + 7$

$\qquad = x(3x - 7) - 1 \cdot (3x - 7)$ Factor out x and -1

$\qquad = (x - 1)(3x - 7)$ The common factor is   Answer: $(x + 1)(3x + 7)$

| **EXAMPLE 8**   FACTORING $ax^2 + bxy + cy^2$ BY GROUPING | **PRACTICE EXERCISE 8** |

Factor $5x^2 + 6xy - 8y^2$ by grouping.

For two variables the procedure is the same. $ac = 5(-8) = -40$. We try factors that seem most likely to have a sum of 6.

Factor $5x^2 - 6xy - 8y^2$ by grouping.

Factors of $ac = -40$	Sum of factors
8, -5	$8 + (-5) = 3$
-8, 5	$-8 + 5 = -3$
-10, 4	$-10 + 4 = -6$
10, -4	$10 + (-4) = 6$

Write $6xy$ as $10xy - 4xy$.

$5x^2 + 6xy - 8y^2 = 5x^2 + 10xy - 4xy - 8y^2$

$\qquad = 5x(x + 2y) - 4y(x + 2y)$ Factor groups

$\qquad = (5x - 4y)(x + 2y)$ Common factor is $x + 2y$ Answer: $(5x + 4y)(x - 2y)$

7.3   EXERCISES A

Factor and check by multiplying.

1. $2x^2 + 7x + 5$

2. $2x^2 - 5x + 3$

3. $2u^2 - 5u - 3$

4. $2u^2 + 13u - 7$

5. $2y^2 + 13y - 24$

6. $2y^2 + 19y + 24$

7. $2y^2 - 19y + 24$

8. $2y^2 - 13y - 24$

9. $3z^2 - 14z - 5$

10 $6z^2 - 13z - 28$

11. $5x^2 - 33x + 40$

12. $-6x^2 - 39x - 54$
[*Hint:* Don't forget the common factor.]

13. $7u^2 - 14u + 7$

14. $16u^2 - 16u + 4$

15. $y^2 - 9$ [*Hint:* Note that $b = 0$, $a = 1$, and $c = -9$.]

16. $3y^3 + 11y^2 + 10y$

(17) $-45x^2 + 150x - 125$

18. $6x^2 - 3x + 21$

19. $6u^2 - 23u + 20$

20. $5u^2 + 7u - 24$

21. $2x^2 - 5xy + 3y^2$

(22) $2x^2 + 7xy + 5y^2$

23. $2u^2 - 5uv - 3v^2$

24. $2u^2 + 13uv - 7v^2$

25. $3x^2 - 10xy + 3y^2$

26. $3x^2 + 13xy + 14y^2$

27. $5u^2 + 21uv + 4v^2$

28. $6u^2 + uv - v^2$

29. $-3x^2 + 18xy - 24y^2$

30. $x^2 + xy + y^2$

(31) $4u^2 - v^2$

32. $3u^3v + 14u^2v^2 - 5uv^3$

33. $6x^2 - 35xy - 6y^2$

34. $6x^2 + 29xy + 30y^2$

35. $14x^2 + 30xy + 16y^2$

36. $3x^3y - 2x^2y^2 - xy^3$

FOR REVIEW

Factor.

37. $x^2 + 17x + 72$

38. $x^2 - 17x + 72$

39. $y^2 - 2y - 80$

40. $x^2 + 6xy - 27y^2$

41. $x^2 - 18xy + 81y^2$

42. $-3x^2 + 9x + 120$

Exercises 43–48 review material from Section 6.6 to help you prepare for the next section. Find the following special products.

43. $(x + 3)(x - 3)$

44. $(x + 3)^2$

45. $(x - 3)^2$

46. $(4u + 5)^2$

47. $(4u - 5)^2$

48. $(4u + 5)(4u - 5)$

ANSWERS: 1. $(2x + 5)(x + 1)$ 2. $(2x - 3)(x - 1)$ 3. $(2u + 1)(u - 3)$ 4. $(2u - 1)(u + 7)$ 5. $(2y - 3)(y + 8)$ 6. $(2y + 3)(y + 8)$ 7. $(2y - 3)(y - 8)$ 8. $(2y + 3)(y - 8)$ 9. $(3z + 1)(z - 5)$ 10. $(2z - 7)(3z + 4)$ 11. $(5x - 8)(x - 5)$ 12. $(-3)(2x + 9)(x + 2)$ 13. $7(u - 1)(u - 1)$ 14. $4(2u - 1)(2u - 1)$ 15. $(y + 3)(y - 3)$ 16. $y(3y + 5)(y + 2)$ 17. $(-5)(3x - 5)(3x - 5)$ 18. $3(2x^2 - x + 7)$; the trinomial cannot be factored   19. $(2u - 5)(3u - 4)$ 20. $(5u - 8)(u + 3)$ 21. $(2x - 3y)(x - y)$ 22. $(2x + 5y)(x + y)$ 23. $(2u + v)(u - 3v)$ 24. $(2u - v)(u + 7v)$ 25. $(3x - y)(x - 3y)$ 26. $(3x + 7y)(x + 2y)$ 27. $(5u + v)(u + 4v)$ 28. $(3u - v)(2u + v)$ 29. $(-3)(x - 4y)(x - 2y)$ 30. cannot be factored   31. $(2u + v)(2u - v)$ 32. $uv(3u - v)(u + 5v)$ 33. $(6x + y)(x - 6y)$ 34. $(3x + 10y)(2x + 3y)$ 35. $2(7x + 8y)(x + y)$ 36. $xy(3x + y)(x - y)$ 37. $(x + 8)(x + 9)$ 38. $(x - 8)(x - 9)$ 39. $(y + 8)(y - 10)$ 40. $(x - 3y)(x + 9y)$ 41. $(x - 9y)(x - 9y)$ 42. $-3(x + 5)(x - 8)$ 43. $x^2 - 9$ 44. $x^2 + 6x + 9$ 45. $x^2 - 6x + 9$ 46. $16u^2 + 40u + 25$ 47. $16u^2 - 40u + 25$ 48. $16u^2 - 25$

7.3   EXERCISES B

Factor and check by multiplying.

1. $2x^2 + 5x + 3$

2. $2x^2 - 7x + 5$

3. $2u^2 + 3u - 14$

4. $2u^2 - 9u - 5$

5. $2u^2 + 9u - 5$

6. $2y^2 + 17y + 30$

7. $2x^2 - 15x + 25$

8. $2u^2 - 19u - 10$

9. $3y^2 + 10y - 8$

10. $6x^2 + x - 15$

11. $5u^2 - 33u + 18$

12. $-4y^2 - 22y - 28$

13. $5x^2 - 20x + 20$

14. $27u^2 + 18u + 3$

15. $y^2 - 25$

16. $2x^4 + 7x^3 + 3x^2$

17. $-16u^2 + 80u - 100$

18. $6y^2 + 36y + 36$

19. $20x^2 - 13x + 2$

20. $4u^2 - 4u - 35$

21. $2x^2 + 5xy + 3y^2$

22. $2x^2 - 7xy + 5y^2$

23. $2u^2 + 3uv - 14v^2$

24. $2u^2 - 9uv - 5v^2$

25. $3x^2 - 7xy + 2y^2$

26. $3x^2 + 22xy + 7y^2$

27. $5u^2 + 7uv + 2v^2$

28. $6u^2 - uv - 2v^2$

29. $-5x^2 + 25xy - 30y^2$

30. $x^2 - 2xy + 2y^2$

31. $9u^2 - v^2$

32. $3u^4 - 4u^3v - 4u^2v^2$

33. $8x^2 - 7xy - y^2$

34. $6x^2 + 25xy + 11y^2$

35. $14x^2 - 30xy + 16y^2$

36. $3x^3y + 2x^2y^2 - xy^3$

FOR REVIEW

Factor.

37. $x^2 + 17x + 70$

38. $x^2 - 17x + 70$

39. $y^2 - 2y - 80$

40. $x^2 - 5xy - 66y^2$

41. $x^2 + 24xy + 144y^2$

42. $-5x^2 - 20x + 105$

Exercises 43–48 review material from Section 6.6 to help you prepare for the next section. Find the following special products.

43. $(y + 4)(y - 4)$

44. $(y + 4)^2$

45. $(y - 4)^2$

46. $(3u + 7)^2$

47. $(3u - 7)^2$

48. $(3u + 7)(3u - 7)$

7.3 EXERCISES C

Factor.

1. $2x^2 - \dfrac{1}{5}x - \dfrac{1}{25}$

2. $0.02x^2 - 0.9x + 4$

3. $2(3x - 1)^2 - (3x - 1) - 1$
   [*Hint:* Substitute u for $3x - 1$.]

4. $3x^4 + 10x^2 - 8$

7.4 FACTORING PERFECT SQUARE TRINOMIALS AND DIFFERENCE OF SQUARES

═══════════════ STUDENT GUIDEPOSTS ═══════════════

 Perfect Square Trinomials   ❷ Difference of Two Squares

❶ PERFECT SQUARE TRINOMIALS

We could factor trinomials such as

$$x^2 + 6x + 9 = (x + 3)(x + 3)$$
$$\text{and} \quad 4y^2 - 4y + 1 = (2y - 1)(2y - 1)$$

with the methods of the previous two sections. However, by recognizing these as *perfect square trinomials,* we can save some of the time used in trial-and-error methods. In Section 6.6 we used the formulas

$$(a + b)^2 = a^2 + 2ab + b^2$$
$$(a - b)^2 = a^2 - 2ab + b^2$$

to square binomials. Now we consider the reverse operation, factoring trinomials like those on the right into the square of a binomial. To factor $x^2 + 6x + 9$, we first observe that it fits the appropriate formula.

$$
\begin{aligned}
x^2 + 6x + 9 &= x^2 + 6x + 3^2 &&\quad x^2 \text{ and } 3^2 \text{ are perfect squares}\\
&= x^2 + 2 \cdot x \cdot 3 + 3^2 &&\quad x = a \text{ and } 3 = b \text{ in } a^2 + 2ab + b^2\\
&= (x + 3)^2 &&\quad a^2 + 2ab + b^2 = (a + b)^2
\end{aligned}
$$

From the formula and this example, it is easy to see that a **perfect square trinomial** must contain two terms that are perfect squares (x^2 and 9 in the example). The remaining term must be plus or minus twice the product of the numbers whose squares form the other terms ($2 \cdot x \cdot 3 = 6x$ in the example). Notice that $x^2 + 10x + 9 = (x + 1)(x + 9)$ is *not* a perfect square trinomial because of its middle term, even though both x^2 and 9 are perfect squares.

EXAMPLE 1 FACTORING PERFECT SQUARES	PRACTICE EXERCISE 1

Factor the trinomials.

(a) $x^2 + 8x + 16 = x^2 + 8x + 4^2$ x^2 and 16 are perfect squares

$\qquad\qquad\qquad = x^2 + 2 \cdot x \cdot 4 + 4^2$ $x = a$ and $4 = b$ in $a^2 + 2ab + b^2$

$\qquad\qquad\qquad = (x + 4)^2$ $a^2 + 2ab + b^2 = (a + b)^2$

Notice that we have $(x + 4)^2$ instead of $(x - 4)^2$ since $8x$ is positive.

(b) $x^2 - 8x + 16 = x^2 - 2 \cdot x \cdot 4 + 4^2$ $x = a$ and $4 = b$

$\qquad\qquad\qquad = (x - 4)^2$ Note the minus sign in this case

(c) $16u^2 + 40u + 25 = (4u)^2 + 40u + 5^2$ $4u = a$ and $5 = b$

$\qquad\qquad\qquad = (4u)^2 + 2(4u)(5) + 5^2$ $2ab = 2(4u)(5) = 40u$

$\qquad\qquad\qquad = (4u + 5)^2$

(d) $16u^2 - 40u + 25 = (4u)^2 - 2(4u)(5) + 5^2$

$\qquad\qquad\qquad = (4u - 5)^2$

Factor the trinomials.

(a) $y^2 + 22y + 121$

(b) $y^2 - 22y + 121$

(c) $9u^2 + 42u + 49$

(d) $9u^2 - 42u + 49$

Answers: **(a)** $(y + 11)^2$
(b) $(y - 11)^2$ **(c)** $(3u + 7)^2$
(d) $(3u - 7)^2$

To Factor Perfect Square Trinomials

1. Determine if two terms are perfect squares and the remaining one is twice the product of the numbers whose squares are the other terms.

2. Use the formulas

$$a^2 + 2ab + b^2 = (a + b)^2$$
$$a^2 - 2ab + b^2 = (a - b)^2.$$

EXAMPLE 2 FACTORING PERFECT SQUARES	PRACTICE EXERCISE 2

Factor the trinomials in two variables.

(a) $x^2 + 18xy + 81y^2 = x^2 + 18xy + (9y)^2$ x^2 and $(9y)^2$ are perfect squares

$\qquad\qquad\qquad = x^2 + 2 \cdot x \cdot 9y + (9y)^2$ $x = a$ and $9y = b$

$\qquad\qquad\qquad = (x + 9y)^2$ Use the formula

(b) $2u^2 - 40uv + 200v^2 = 2(u^2 - 20uv + 100v^2)$ Factor out common factor

$\qquad\qquad\qquad = 2(u^2 - 2 \cdot u \cdot 10v + (10v)^2)$ $u = a$ and $10v = b$

$\qquad\qquad\qquad = 2(u - 10v)^2$

Factor the trinomials in two variables.

(a) $x^2 - 18xy + 81y^2$

(b) $3u^2 + 60uv + 300v^2$

Answers: **(a)** $(x - 9y)^2$
(b) $3(u + 10v)^2$

❷ DIFFERENCE OF TWO SQUARES

Another special formula can be used whenever we factor the difference of two squares.

$$a^2 - b^2 = (a + b)(a - b)$$

Observe that we are factoring a binomial in this case.

EXAMPLE 3 FACTORING A DIFFERENCE OF SQUARES	PRACTICE EXERCISE 3

Factor the binomials.

(a) $x^2 - 9 = x^2 - 3^2$ x^2 and 3^2 are perfect squares

$= (x + 3)(x - 3)$ $x = a$ and $3 = b$

(b) $y^2 - 121 = y^2 - 11^2$ y^2 and 11^2 are perfect squares

$= (y + 11)(y - 11)$ $y = a$ and $11 = b$

(c) $16u^2 - 25 = (4u)^2 - 5^2$ $(4u)^2$ and 5^2 are perfect squares

$= (4u + 5)(4u - 5)$ $4u = a$ and $5 = b$

(d) $5x^3 - 500x = 5x(x^2 - 100)$ Factor out common factor

$= 5x(x^2 - 10^2)$

$= 5x(x + 10)(x - 10)$

(e) $y^6 - 49 = (y^3)^2 - 7^2$ $(y^3)^2$ and 7^2 are perfect squares

$= (y^3 + 7)(y^3 - 7)$ $y^3 = a$ and $7 = b$

(f) $x^2 + 9 = x^2 + 3^2$ Cannot be factored

Factor the binomials.

(a) $x^2 - 1$

(b) $y^2 - 169$

(c) $9u^2 - 49$

(d) $2y^2 - 288$

(e) $z^4 - 4$

(f) $w^2 + 25$

Answers: **(a)** $(x + 1)(x - 1)$
(b) $(y + 13)(y - 13)$
(c) $(3u + 7)(3u - 7)$
(d) $2(z + 12)(z - 12)$
(e) $(z^2 + 2)(z^2 - 2)$
(f) **cannot be factored**

⟋⟋⟋⟋⟋⟋⟋⟋⟋ CAUTION ⟋⟋⟋⟋⟋⟋⟋⟋⟋

Notice in Example 3(f) that the sum of two squares cannot be factored. Do not make the mistake of giving $(a + b)^2$ as the factors of $a^2 + b^2$.

$$(a + b)^2 = a^2 + 2ab + b^2 \neq a^2 + b^2$$

Also, $(a - b)^2 \neq a^2 - b^2.$

To Factor a Difference of Two Squares

1. Determine if the binomial is the difference of two perfect squares.
2. Use the formula

$$a^2 - b^2 = (a + b)(a - b).$$

EXAMPLE 4 Difference of Squares in Two Variables	**Practice Exercise 4**

Factor the binomials in two variables.

(a) $25x^2 - 9y^2 = (5x)^2 - (3y)^2$ $\quad$ $(5x)^2$ and $(3y)^2$ are perfect squares

$\qquad\qquad\quad = (5x + 3y)(5x - 3y)$ $\quad$ $5x = a$ and $3y = b$

(b) $27u^3 - 48uv^2 = 3u(9u^2 - 16v^2)$ $\qquad$ Factor out common factor

$\qquad\qquad\qquad = 3u[(3u)^2 - (4v)^2]$ $\qquad$ $3u = a$ and $4v = b$

$\qquad\qquad\qquad = 3u(3u + 4v)(3u - 4v)$

(c) $x^4 - y^4 = (x^2)^2 - (y^2)^2$ $\qquad$ $x^2 = a$ and $y^2 = b$

$\qquad\quad = (x^2 + y^2)(x^2 - y^2)$ $\qquad$ Now $x = a$ and $y = b$ in the difference of two squares $x^2 - y^2$

$\qquad\quad = (x^2 + y^2)(x + y)(x - y)$ $\quad$ Factor again

Factor the binomials in two variables.

(a) $100a^2 - b^2$

(b) $50x^3 - 162x$

(c) $x^4 + y^4$

Answers: (a) $(10a + b)(10a - b)$
(b) $2x(5x + 9)(5x - 9)$ (c) **cannot be factored**

7.4 EXERCISES A

Factor using the special formulas.

1. $x^2 + 10x + 25$

2. $x^2 - 10x + 25$

3. $x^2 - 25$

4. $x^2 + 25$

5. $u^2 - 14u + 49$

6. $u^2 + 14u + 49$

7. $u^2 - 49$

8. $u^2 + 49$

9. $x^2 - 24x + 144$

10. $x^2 - 121$

11 $9u^2 + 6u + 1$

12. $9u^2 - 6u + 1$

13. $4y^2 - 12y + 9$

14. $4y^2 - 9$

15. $25x^2 + 20x + 4$

16. $4x^2 + 20x + 25$

17. $9u^2 - 25$

18. $9u^2 + 25$

19 $-12y^2 + 60y - 75$

20. $4y^2 - 36y + 81$

21. $9x^2 + 30x + 25$

22. $x^2 + x + 4$

23. $8u^3 - 8u$

24. $u^6 - 16$

25. $5y^3 - 100y^2 + 500y$

26. $7y^4 - 63$

27. $x^2 + 6xy + 9y^2$

28 $25x^2 - 10xy + y^2$

29. $64u^2 - 9v^2$

30 $u^4 - v^4$

Before we complete this exercise set we summarize the factoring techniques we have learned.

To Factor a Polynomial

1. Factor out any common factor, including -1 if a is negative in $ax^2 + bx + c$ or $ax^2 + bxy + cy^2$.

2. To factor the difference of two squares use the formula
$$a^2 - b^2 = (a + b)(a - b).$$

3. To factor a trinomial use the techniques of Section 7.2 or Section 7.3, but also consider perfect squares using the formulas
$$a^2 + 2ab + b^2 = (a + b)^2 \quad \text{and} \quad a^2 - 2ab + b^2 = (a - b)^2.$$

4. To factor a polynomial of more than three terms, try to factor by grouping.

Factor.

31. $x^2 + 9x + 18$

32. $x^2 - 12x + 36$

33. $u^2 - 14u + 48$

34. $u^2 + 10u + 16$

35. $4y^2 - 49$

36. $4y^2 - 28y + 49$

37. $-6x^2 + 486$

38. $2x^2 + 24x + 64$

39. $25u^2 - 20u + 4$

40. $16u^2 - 9$

41. $4y^2 - 16y + 15$

42. $40y^2 + 11y - 2$

43. $3x^9 - 147x^3$

44. $98x^6 - 18$

45. $x^2 - 9y^2$

46. $3x^2 - 19xy - 14y^2$

47. $36u^2 - 60uv + 25v^2$

48. $18u^2 - 98v^2$

49. $100x^5 - 60x^4y + 9x^3y^2$

50. $(x + y)^2 - 25$ [*Hint:* $x + y = a$, $5 = b$]

FOR REVIEW

Exercises 51–56 review material from Chapter 3 to help you prepare for the next section. Solve the following equations.

51. $x - 3 = 0$

52. $3x + 1 = 0$

53. $2x + 5 = 0$

54. $6x - 1 = 0$

55. $4x = 0$

56. $3x + 9 = 0$

ANSWERS: 1. $(x + 5)^2$ 2. $(x - 5)^2$ 3. $(x + 5)(x - 5)$ 4. cannot be factored 5. $(u - 7)^2$ 6. $(u + 7)^2$
7. $(u + 7)(u - 7)$ 8. cannot be factored 9. $(x - 12)^2$ 10. $(x + 11)(x - 11)$ 11. $(3u + 1)^2$ 12. $(3u - 1)^2$
13. $(2y - 3)^2$ 14. $(2y + 3)(2y - 3)$ 15. $(5x + 2)^2$ 16. $(2x + 5)^2$ 17. $(3u + 5)(3u - 5)$ 18. cannot be factored
19. $(-3)(2y - 5)^2$ 20. $(2y - 9)^2$ 21. $(3x + 5)^2$ 22. cannot be factored 23. $8u(u + 1)(u - 1)$ 24. $(u^3 + 4)(u^3 - 4)$
25. $5y(y - 10)^2$ 26. $7(y^2 + 3)(y^2 - 3)$ 27. $(x + 3y)^2$ 28. $(5x - y)^2$ 29. $(8u + 3v)(8u - 3v)$
30. $(u^2 + v^2)(u + v)(u - v)$ 31. $(x + 3)(x + 6)$ 32. $(x - 6)^2$ 33. $(u - 6)(u - 8)$ 34. $(u + 2)(u + 8)$
35. $(2y + 7)(2y - 7)$ 36. $(2y - 7)^2$ 37. $-6(x + 9)(x - 9)$ 38. $2(x + 4)(x + 8)$ 39. $(5u - 2)^2$
40. $(4u + 3)(4u - 3)$ 41. $(2y - 3)(2y - 5)$ 42. $(8y - 1)(5y + 2)$ 43. $3x^3(x^3 + 7)(x^3 - 7)$ 44. $2(7x^3 + 3)(7x^3 - 3)$
45. $(x + 3y)(x - 3y)$ 46. $(3x + 2y)(x - 7y)$ 47. $(6u - 5v)^2$ 48. $2(3u + 7v)(3u - 7v)$ 49. $x^3(10x - 3y)^2$
50. $(x + y + 5)(x + y - 5)$ 51. 3 52. $-\frac{1}{3}$ 53. $-\frac{5}{2}$ 54. $\frac{1}{6}$ 55. 0 56. -3

7.4 EXERCISES B

Factor using the special formulas.

1. $x^2 + 12x + 36$

2. $x^2 - 12x + 36$

3. $x^2 - 36$

4. $x^2 + 36$

5. $u^2 - 16u + 64$

6. $u^2 + 16u + 64$

7. $u^2 - 64$

8. $u^2 + 64$

9. $x^2 - 22x + 121$

10. $x^2 - 169$

11. $4u^2 + 4u + 1$

12. $4u^2 - 4u + 1$

13. $9y^2 - 12y + 4$

14. $9y^2 - 4$

15. $36x^2 + 12x + 1$

16. $9x^2 + 30x + 25$

17. $4u^2 - 25$ **18.** $4u^2 + 25$ **19.** $-32y^2 + 48y - 18$ **20.** $9y^2 - 42y + 49$

21. $25x^2 + 30x + 9$ **22.** $x^2 + 2x + 4$ **23.** $4u^3 - 16u$ **24.** $u^6 - 81$

25. $-4y^2 - 64y - 256$ **26.** $y^4 - 81$ **27.** $x^2 + 8xy + 16y^2$ **28.** $36x^2 - 12xy + y^2$

29. $49u^2 - 16v^2$ **30.** $81u^4 - v^4$

Factor.

31. $x^2 + 12x + 35$ **32.** $x^2 + 20x + 100$ **33.** $u^2 - 22u + 121$ **34.** $u^2 - 14u + 45$

35. $9y^2 - 64$ **36.** $-4y^2 + 400$ **37.** $9x^2 - 60x + 100$ **38.** $28x^2 - 41x + 15$

39. $-10u^2 - 21u + 10$ **40.** $18u^2 - 98$ **41.** $42y^2 - 3y - 9$ **42.** $16y^2 + 6y - 27$

43. $25x^2 - 90x + 81$ **44.** $x^2 - 3x - 180$ **45.** $x^2 - 25y^2$ **46.** $2x^2 - 52xy + 338y^2$

47. $49u^2 + 42uv + 9v^2$ **48.** $200u^2 - 242v^2$ **49.** $9x^4y^2 - 60x^3y^3 + 100x^2y^4$

50. $(x - y)^2 - 16$

FOR REVIEW

Exercises 51–56 review material from Chapter 3 to help you prepare for the next section. Solve the following equations.

51. $x - 9 = 0$ **52.** $7x + 4 = 0$ **53.** $5x + 6 = 0$

54. $4x - 2 = 0$ **55.** $\frac{1}{2}x = 0$ **56.** $6x + 12 = 0$

7.4 EXERCISES C

To factor the binomials in Exercises 1–6, use the formulas

$$x^3 + y^3 = (x + y)(x^2 - xy + y^2)$$
$$\text{and} \quad x^3 - y^3 = (x - y)(x^2 + xy + y^2).$$

1. $a^3 + 8$
[*Hint:* $a^3 + 8 = a^3 + 2^3$]

2. $a^3 - 8$

3. $27b^3 - 8$

4. $27b^3 + 8$

5. $8a^3 - 125b^3$ [Answer:
$(2a - 5b)(4a^2 + 10ab + 25b^2)$]

6. $8a^3 + 125b^3$

7.5 FACTORING TO SOLVE EQUATIONS

1 ZERO-PRODUCT RULE

A property of the number zero is used to solve certain types of equations. We know that any number times zero yields a zero product. For example,

$$4 \cdot 0 = 0 \quad \text{and} \quad 0 \cdot \frac{2}{3} = 0.$$

Moreover, if a product of two or more numbers is zero, then at least one of the factors must be zero. This property, called the zero-product rule, gives us a useful method for solving equations.

> ### Solving Equations Using the Zero-Product Rule
>
> If an equation has a product of two or more factors on one side of the equation and 0 on the other, the solutions to the equation are found by setting each factor equal to 0.
>
> If $a \cdot b = 0$, then $a = 0$ or $b = 0$.

The zero-product rule can be used to solve for x in equations of the form

$$(x - 3)(x + 5) = 0.$$

That is, when $x - 3$ and $x + 5$ are multiplied to give zero, the rule says either $x - 3 = 0$ or $x + 5 = 0$. Thus, one solution of

$$(x - 3)(x + 5) = 0$$

is found by solving

$$x - 3 = 0$$
$$x = 3. \quad \text{One solution is 3}$$

Another solution comes by solving

$$x + 5 = 0$$
$$x = -5. \quad \text{Another solution is } -5$$

There are two solutions to the equation, 3 and -5. To check the solutions we substitute back into the original equations.

Check of 3	*Check of* -5
$(x - 3)(x + 5) = 0$	$(x - 3)(x + 5) = 0$
$(3 - 3)(3 + 5) \overset{?}{=} 0$	$(-5 - 3)(-5 + 5) \overset{?}{=} 0$
$0 \cdot 8 = 0$	$(-8) \cdot 0 = 0$
So 3 checks.	And -5 checks.

| EXAMPLE 1 USING THE ZERO-PRODUCT RULE | PRACTICE EXERCISE 1 |

Solve $(2x + 1)(x - 1) = 0$.

$$2x + 1 = 0 \qquad \text{or} \qquad x - 1 = 0$$
$$2x + 1 - 1 = 0 - 1 \qquad x - 1 + 1 = 0 + 1$$
$$2x = -1 \qquad\qquad x = 1$$
$$\frac{1}{2} \cdot 2x = \frac{1}{2} \cdot (-1)$$
$$x = -\frac{1}{2}$$

The solutions are $-\frac{1}{2}$ and 1.

Solve $(3x + 2)(x - 9) = 0$.

Answer: $-\frac{2}{3}$ and 9

| EXAMPLE 2 USING THE ZERO-PRODUCT RULE | PRACTICE EXERCISE 2 |

Solve $x(3x + 1) = 0$.

$$x = 0 \quad \text{or} \qquad 3x + 1 = 0$$
$$3x + 1 - 1 = 0 - 1$$
$$3x = -1$$
$$\frac{1}{3} \cdot 3x = \frac{1}{3}(-1)$$
$$x = -\frac{1}{3}$$

The solutions are 0 and $-\frac{1}{3}$.

Solve $y(y + 5) = 0$.

Answer: 0 and -5

When one of the factors has only one term, say x, as in Example 2, one of the solutions will always be zero. Be sure that you give this solution, and do not divide both sides of the equation by x.

② SOLVING QUADRATIC EQUATIONS

Suppose we solve $(x - 2)(x + 3) = 0$.

$$x - 2 = 0 \quad \text{or} \quad x + 3 = 0$$
$$x = 2 \qquad\qquad x = -3$$

The solutions are 2 and -3. The equation $(x - 2)(x + 3) = 0$ is the factored form of $x^2 + x - 6 = 0$, which is a *quadratic equation*. A **quadratic equation** has the form

$$ax^2 + bx + c = 0.$$

Thus, the zero-product rule can sometimes be used to solve quadratic equations, a subject that we will study in more detail in Chapter 10.

To Solve an Equation by Factoring

1. Write all terms on the left side of the equation leaving zero on the right side.
2. Factor the left side.
3. Use the zero-product rule, set each factor equal to zero, and solve the resulting equations.
4. Check possible solutions in the original equation.

EXAMPLE 3 SOLVING A QUADRATIC EQUATION	PRACTICE EXERCISE 3

Solve $x^2 + 4x = 21$.

$$x^2 + 4x - 21 = 0 \qquad \text{Rewrite with all terms on the left}$$

$$(x + 7)(x - 3) = 0 \qquad \text{Factor the trinomial}$$

$$x + 7 = 0 \quad \text{or} \quad x - 3 = 0 \qquad \text{Use zero-product rule}$$

$$x = -7 \qquad\qquad x = 3$$

$$\begin{array}{cc} \textit{Check of } -7 & \textit{Check of } 3 \\ (-7)^2 + 4(-7) \overset{?}{=} 21 & (3)^2 + 4(3) \overset{?}{=} 21 \\ 49 - 28 \overset{?}{=} 21 & 9 + 12 \overset{?}{=} 21 \\ 21 = 21 & 21 = 21 \end{array}$$

The solutions are -7 and 3.

Solve $2x^2 - 11x = 40$

Answer: 8 and $-\frac{5}{2}$

CAUTION

Using the zero-product rule, we can only solve equations in the form $a \cdot b = 0$. If $a \cdot b = 5$, we cannot conclude that $a = 5$ or $b = 5$. Nor can we apply this rule to $a + b = 0$ or $a - b = 0$.

EXAMPLE 4 SOLVING A QUADRATIC EQUATION	PRACTICE EXERCISE 4

Solve $x(x - 4) = 5$.

The zero-product rule does not apply immediately because this product does *not* equal zero. When this happens, try to rewrite it as a product that does equal zero. Begin by clearing the parentheses.

$$x^2 - 4x = 5$$

$$x^2 - 4x - 5 = 0 \qquad \text{Write all terms on the left}$$

$$(x - 5)(x + 1) = 0 \qquad \text{Factor}$$

$$x - 5 = 0 \quad \text{or} \quad x + 1 = 0 \qquad \text{Zero-product rule}$$

$$x = 5 \qquad\qquad x = -1$$

The solutions are 5 and -1.

Solve $x(x + 5) = 14$.

Answer: 2 and -7

EXAMPLE 5 LINEAR EQUATION	**PRACTICE EXERCISE 5**

Solve $(2x + 1) - (x + 5) = 0$.

Solve $(x + 2) - (3x + 8) = 0$.

   The zero-product rule does not apply because the left side of the equation is a difference, *not* a product. To solve, we remove parentheses and then combine like terms.

$$2x + 1 - x - 5 = 0$$
$$x - 4 = 0$$
$$x = 4$$

The solution is 4. Notice that this time after clearing parentheses, the result is a linear equation, not a quadratic equation which would require the zero-product rule. Also, it cannot be solved by setting $2x + 1$ and $x + 5$ equal to zero.

Answer: -3

   The zero-product rule can be extended to include products of three or more factors. That is, if $a \cdot b \cdot c = 0$ then $a = 0$ or $b = 0$ or $c = 0$. We use this fact in the next example.

EXAMPLE 6 PRODUCT OF THREE FACTORS	**PRACTICE EXERCISE 6**

Solve $x^3 + 2x^2 = 15x$.

Solve $x^3 = 5x^2 - 6x$.

$$x^3 + 2x^2 - 15x = 0 \qquad \text{Subtract } 15x \text{ from both sides}$$
$$x(x^2 + 2x - 15) = 0 \qquad \text{Factor out the GCF } x$$
$$x(x - 3)(x + 5) = 0 \qquad \text{Factor the trinomial}$$
$$x = 0 \quad \text{or} \quad x - 3 = 0 \quad \text{or} \quad x + 5 = 0 \qquad \text{Set each factor equal to 0}$$
$$x = 3 \qquad\qquad x = -5$$

The solutions are 0, 3, and -5. Check these in the original equation.

Answer: 0, 2, and 3

7.5 EXERCISES A

Solve.

1. $(x - 5)(x + 7) = 0$

2. $(x + 6)(x + 1) = 0$

3. $(y - 8)(y - 9) = 0$

4. $(y + 10)(y - 20) = 0$

5. $(u + 2)(u - 2) = 0$

6. $(u - 0.5)(u + 0.2) = 0$

7. $\left(x - \dfrac{1}{2}\right)\left(x + \dfrac{2}{5}\right) = 0$

8. $(x - 8)^2 = 0$

9. $(3y + 2)(y - 6) = 0$

10. $(y + 8)(y - 8) = 0$

11. $(5u + 1)(2u - 3) = 0$

12. $(8u - 5)(3u - 8) = 0$

13 $x(3x + 7) = 0$

14. $(2x - 5)(x + 12) = 0$

15. $y^2 - 5y + 6 = 0$

16. $y^2 - y - 6 = 0$

17. $u^2 - 8u + 7 = 0$

18. $u^2 + u - 20 = 0$

19. $x^2 + 10x + 25 = 0$

20. $x^2 - 13x + 40 = 0$

21. $y^2 - y - 42 = 0$

22. $y^2 - 16 = 0$

23 $4u^2 - 8u = 0$

24. $4u^2 + 12u + 9 = 0$

25. $9x^2 - 49 = 0$

26 $(3x - 5) - (x + 7) = 0$

27 $y(y + 3) = 10$

28. $y^2 - 2y = 35$

29. $u^2 = -2u + 35$

30. $16u^2 - 42u + 5 = 0$

31. $(x - 3)(x - 2)(x + 1) = 0$

32. $(2y)(y + 1)(y + 5) = 0$

33. $2x^3 - 4x^2 - 6x = 0$

FOR REVIEW

Factor.

34. $x^2 - 4x - 21$

35. $x^2 + 14x + 49$

36. $25y^2 - 16$

37. $7y^2 - 70y + 175$

38. $2x^2 + 9xy - 5y^2$

39. $45x^2 - 80y^2$

Exercise 40 reviews material from Chapter 3 to help you prepare for the next section. Solve the applied problem.

40. After receiving a 9% raise Lynn Harris now makes $34,880 per year. What was her salary before the raise?

ANSWERS: 1. 5, −7   2. −6, −1   3. 8, 9   4. −10, 20   5. −2, 2   6. 0.5, −0.2   7. $\frac{1}{2}$, −$\frac{2}{5}$ 8. 8   9. −$\frac{2}{3}$, 6
10. −8, 8   11. −$\frac{1}{5}$, $\frac{3}{2}$ 12. $\frac{5}{8}$, $\frac{8}{3}$ 13. 0, −$\frac{7}{3}$ 14. $\frac{5}{2}$, −12   15. 2, 3   16. −2, 3   17. 1, 7   18. −5, 4   19. −5
20. 5, 8   21. −6, 7   22. −4, 4   23. 0, 2   24. −$\frac{3}{2}$ 25. $\frac{7}{3}$, −$\frac{7}{3}$ 26. 6   27. 2, −5   28. −5, 7   29. 5, −7   30. $\frac{1}{3}$, $\frac{5}{2}$
31. 3, 2, −1   32. 0, −1, −5   33. 0, −1, 3   34. $(x + 3)(x − 7)$ 35. $(x + 7)^2$ 36. $(5y + 4)(5y − 4)$ 37. $7(y − 5)^2$
38. $(2x − y)(x + 5y)$ 39. $5(3x + 4y)(3x − 4y)$ 40. $32,000

7.5 EXERCISES B

Solve.

1. $(x + 2)(x − 6) = 0$

2. $(x + 8)(x + 3) = 0$

3. $(y − 5)(y − 7) = 0$

4. $(y + 8)(y + 12) = 0$

5. $(u + 6)(u − 11) = 0$

6. $(u + 4)(u − 4) = 0$

7. $(x − 0.7)(x + 0.3) = 0$

8. $\left(x + \dfrac{1}{3}\right)\left(x + \dfrac{2}{3}\right) = 0$

9. $(2y − 3)(y + 8) = 0$

10. $(y − 5)^2 = 0$

11. $(5u − 6)(2u + 1) = 0$

12. $(3u + 14)(2u − 5) = 0$

13. $x(4x + 3) = 0$

14. $(2x − 7)(2x + 9) = 0$

15. $y^2 + 8y + 15 = 0$

16. $y^2 − 2y − 15 = 0$

17. $u^2 − 11u + 18 = 0$

18. $u^2 − 3u − 40 = 0$

19. $x^2 − 18x + 81 = 0$

20. $x^2 + 14x + 40 = 0$

21. $2y^2 − 13y + 6 = 0$

22. $49y^2 − 1 = 0$

23. $3u^2 − 27u = 0$

24. $9u^2 − 18u + 9 = 0$

25. $25y^2 − 16 = 0$

26. $(5x + 1) − (x − 7) = 0$

27. $y(y − 8) = −16$

28. $−63 = y^2 − 16y$

29. $63 − u^2 = 2u$

30. $6u^2 − 29u − 5 = 0$

31. $(x − 2)(x + 3)(x − 1) = 0$

32. $4y(y − 1)(y + 3) = 0$

33. $2x^3 − 4x^2 − 6x = 0$

FOR REVIEW

Factor.

34. $x^2 + 4x - 32$

35. $4x^2 - 12x + 9$

36. $81y^2 - 49$

37. $-6y^2 - 72y - 216$

38. $15x^2 + xy - 2y^2$

39. $40x^2 - 1210y^2$

Exercise 40 reviews material from Chapter 3 to help you prepare for the next section. Solve the applied problem.

40. The perimeter of a rectangle is 38 m. Find the dimensions if the width is 3 m less than the length.

7.5 EXERCISES C

Solve.

1. $(x + 3)(2x - 5)(x - 4) = 0$

2. $(x - 3)^2 - 3(x - 3) = 4$

3. $x^4 - 16 = 0$

4. $9(x + 2)^2 - 6(x + 2) = -1$
$\left[\text{Answer:} \quad -\frac{5}{3}\right]$

5. $16x^4 - 8x^2 + 1 = 0$

6. $5(x + 4)^2 = -9(x + 4) + 2$
[*Hint:* Substitute u for $x + 4$.]

7.6 APPLICATIONS OF FACTORING

STUDENT GUIDEPOSTS

1 Factoring to Solve Applied Problems   3 Business Applications
2 Geometry Applications

1 FACTORING TO SOLVE APPLIED PROBLEMS

Sometimes an applied problem can be translated into an equation that is solvable using the technique in Section 7.5.

| **EXAMPLE 1** NUMBER PROBLEM | **PRACTICE EXERCISE 1** |

If two times the square of a number minus three times the number is 27, find the number.

Let x = the number.

$$\underset{\substack{\downarrow \\ 2x^2}}{\text{two times the square}\atop\text{of the number}} \quad \underset{\substack{\downarrow \\ -}}{\text{minus}} \quad \underset{\substack{\downarrow \\ 3x}}{\text{three times}\atop\text{the number}} \quad \underset{\substack{\downarrow \\ =}}{\text{is}} \quad \underset{\substack{\downarrow \\ 27}}{27}$$

$2x^2 - 3x = 27$

$2x^2 - 3x - 27 = 0$ Subtract 27 from both sides

$(2x - 9)(x + 3) = 0$ Factor

$2x - 9 = 0$ or $x + 3 = 0$ Zero-product rule

$2x = 9$ $x = -3$

$x = \dfrac{9}{2}$

The solutions are $\frac{9}{2}$ and -3.

The product of 1 less than a number and 5 more than the same number is 16. Find the number.

Let x = the number,
$x - 1 = 1$ less than the number,
$x + 5 = 5$ more than the number.
Solve the equation
$(x - 1)(x + 5) = $ ___.

Answer: 3 or -7

| **EXAMPLE 2** CONSECUTIVE INTEGERS | **PRACTICE EXERCISE 2** |

The product of two consecutive even integers is 440. Find the integers.

Let n = first integer,
$n + 2$ = next consecutive even integer,
$n(n + 2)$ = product of the two consecutive even integers.

$n(n + 2) = 440$ Product is 440

$n^2 + 2n = 440$ Remove parentheses

$n^2 + 2n - 440 = 0$ Subtract 440 from both sides

$(n + 22)(n - 20) = 0$ Factor

$n + 22 = 0$ or $n - 20 = 0$

$n = -22$ $n = 20$

$n + 2 = -20$ $n + 2 = 22$

Thus, one solution is -22 and -20 while the other is 20 and 22.

The product of two consecutive positive integers is 132. Find the integers.

Let n = first integer,
$n + 1$ = next consecutive integer.

Answer: 11 and 1

---

② GEOMETRY APPLICATIONS

Some geometry problems may also require factoring. It always helps to make a sketch for a geometry problem. Also, be sure that you answer the stated question and that your answer is reasonable. Geometry formulas are found on the inside covers and the Geometry Appendix gives a discussion of geometric figures.

| **EXAMPLE 3** GEOMETRY APPLICATION | **PRACTICE EXERCISE 3** |

If the area of a rectangle is 48 m^2 and the length is three times the width, find the dimensions of the rectangle.

Make a sketch, as in Figure 7.1.

A triangle has area 60 ft^2, and its base is 2 ft longer than its height. Find the length of each.

Figure 7.1

Let x = width of rectangle in meters,
  $3x$ = length of rectangle in meters.
For a rectangle.

$$\underset{\downarrow}{\text{width}} \cdot \underset{\downarrow}{\text{length}} = \underset{\downarrow}{\text{area.}}$$

$$x \cdot 3x = 48$$
$$3x^2 = 48$$
$$3x^2 - 48 = 0$$
$$3(x^2 - 16) = 0$$
$$x^2 - 16 = 0 \qquad \text{Divide both sides by 3. } \tfrac{0}{3} = 0$$
$$(x - 4)(x + 4) = 0$$
$$x - 4 = 0 \quad \text{or} \quad x + 4 = 0$$
$$x = 4 \qquad\qquad x = -4 \qquad \text{Rule out } x = -4 \text{ since we want the width of a rectangle}$$

Thus, since $x = 4$ and $3x = 12$, the rectangle is 4 m by 12 m.

Answer: base: 12 ft; height: 10 ft

---

❸ BUSINESS APPLICATIONS

There are numerous applications to business.

| **EXAMPLE 4   RETAIL APPLICATION** | **PRACTICE EXERCISE 4** |

A clothing store owner finds that her daily profit on jeans is given by $P = n^2 - 6n - 20$, where n is the number of jeans sold. How many jeans must be sold for a profit of $260?
  Since $P = \$260$, we need to solve the following equation.

$$P = n^2 - 6n - 20 = 260 \qquad \text{Profit is to be \$260}$$
$$n^2 - 6n - 280 = 0 \qquad \text{Set equal to zero}$$
$$(n + 14)(n - 20) = 0 \qquad \text{Factor}$$
$$n + 14 = 0 \quad \text{or} \quad n - 20 = 0$$
$$n = -14 \qquad\qquad n = 20$$

Since n is the number of jeans sold, the -14 answer is discarded. Thus, 20 jeans must be sold to make $260. To check we evaluate P for $n = 20$.

$$P = n^2 - 6n - 20$$
$$= (20)^2 - 6(20) - 20 \qquad \text{Substitute 20 for } n$$
$$= 400 - 120 - 20$$
$$= 260 \qquad \text{This checks}$$

A chemical reaction is described by the equation $C = 2n^2 - 7n + 1$, where n is always positive. Find n when C is 16.

Answer: 5

EXAMPLE 5 PRODUCTION APPLICATION

The IBX Corporation, producer of components for personal computers, has discovered that the number n of a particular type of component that it can sell each week is related to the price p of the component by the equation $n = 1100 - 100p$. Find the price that IBX should set on each component to produce a weekly revenue of $2800.

The revenue equation is $R = np$ where n is the number of components sold and p is the price of each component. Since $n = 1100 - 100p$, substituting we have

$$R = np$$
$$= (1100 - 100p)p$$
$$= 1100p - 100p^2.$$

We want to find p when R is 2800.

$$2800 = 1100p - 100p^2$$

Add $100p^2$ and subtract $1100p$ from both sides.

$$100p^2 - 1100p + 2800 = 0$$
$$100(p^2 - 11p + 28) = 0 \quad \text{Factor out 100}$$
$$100(p - 4)(p - 7) = 0 \quad \text{Factor}$$
$$p - 4 = 0 \quad \text{or} \quad p - 7 = 0 \quad \text{Zero-product rule}$$
$$p = 4 \qquad\qquad p = 7$$

If components are sold for $4 or for $7, the weekly revenue will be $2800.

PRACTICE EXERCISE 5

Suppose in Example 5 $n = 500 - 50p$. Find the price if the weekly revenue is $1250.

Answer: $5

7.6 EXERCISES A

Solve.

1. The product of 5 more than a number and 3 less than the number is zero. Find the number.

   Let x = the number,
        $x + 5$ = 5 more than the number,
        $x - 3$ = 3 less than the number.

2. If the square of a number plus seven times the number is 44, what is the number?

3. The product of a number and 4 less than the number is 21. Find the number.

4. The length of a rectangle is 5 cm more than the width. If the area is 24 cm^2, find the dimensions.

5 The product of two consecutive integers is 240. Find the integers.

6. If the square of a number is increased by 3 times the number, the result is 70. Find the number.

7. The product of two consecutive odd integers is 143. Find the integers.

Let n = first integer,
$n + 2$ = next odd integer.

8 The sum of the squares of two consecutive even positive integers is 100. Find the integers.

9. The area of a rectangle is 84 ft^2 and the length is 5 feet more than the width. Find the dimensions of the rectangle.

```
┌─────────────────────┐
│                     │
│   A = 84 ft²        │ x
│                     │
└─────────────────────┘
        x + 5
```

10 The area of a square is numerically 4 less than the perimeter. Find the length of the side in cm.

```
        x
   ┌─────────┐
   │          │
 x │ A = x²   │ x
   │          │
   │ P = 4x   │
   └─────────┘
        x
```

11. The area of a triangle is 98 cm^2. If the base is four times the height, find the base and height.

12. The area of a triangle is 56 in^2. If the base is 6 in less than the height, find the base and height.

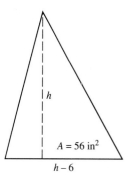

h

$A = 56$ in^2

$h - 6$

13. A toy box is 20 in high. The length is 3 in longer than the width. If the volume is 80 in^3, find the length and width.

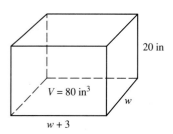

20 in

$V = 80$ in^3

w

$w + 3$

14. The width of a box is 9 cm. The height is 6 cm less than the length. Find the height and length if the volume is 360 cm^3.

The profit on a small appliance is given by the equation $P = n^2 - 3n - 60$ where n is the number of appliances sold per day. Use this equation in Exercises 15–18.

15 What is the profit when 30 appliances are sold?

16. What is the profit when 5 appliances are sold?

17 How many appliances were sold on a day when the profit was $120?

18. How many appliances were sold on a day when there was a $20 loss?

The number of ways of choosing two people from an organization to serve on a committee is given by $N = \frac{1}{2}n(n - 1)$, where n is the number of people in the organization. Use this equation to do Exercises 19–22.

19 How many committees of two can be formed from a group with 7 members?

20. How many committees of two can be formed from a club with 10 members?

21 If 10 different committees of two can be formed from the members of a club, how many members are in the club?

22. If 28 different committees of two can be formed from the members of a sorority, how many women are in the sorority?

23. A chemical reaction is described by the equation $C = 2n^2 - 7n + 1$. Find n when $C = 16$ if n is positive.

24. The relationship between the number n of radios a company can sell per month and the price of each radio p is given by the equation $n = 1700 - 100p$. Find the price at which a radio should be sold to produce a monthly revenue of $7000.

FOR REVIEW

Solve.

25. $(2x - 5)(x + 10) = 0$

26. $9x^2 + 6x + 1 = 0$

27. $50y^2 - 32 = 0$

28. $3y^2 = 13y + 10$

The following exercises review material that we covered in Chapter 1. They will help you prepare for the next section. Perform the indicated operations.

29. $\dfrac{0}{4}$

30. $\dfrac{4}{0}$

31. $\dfrac{4}{4}$

32. $\dfrac{-4}{4}$

Supply the missing numerator.

33. $\dfrac{1}{2} = \dfrac{?}{6}$ **34.** $\dfrac{2}{3} = \dfrac{?}{15}$ **35.** $\dfrac{-3}{7} = \dfrac{?}{42}$ **36.** $-\dfrac{10}{21} = \dfrac{?}{105}$

Reduce each fraction to lowest terms.

37. $\dfrac{20}{50}$ **38.** $\dfrac{72}{96}$ **39.** $\dfrac{30}{180}$ **40.** $\dfrac{144}{252}$

ANSWERS: 1. −5 or 3 2. −11 or 4 3. −3 or 7 4. 3 cm, 8 cm 5. −16, −15 or 15, 16 6. 7 or −10
7. −13, −11; 11, 13 8. 6, 8 9. 7 ft by 12 ft 10. 2 cm 11. 28 cm, 7 cm 12. 8 in, 14 in 13. 4 in, 1 in
14. 4 in, 10 in 15. \$750 16. −\$50 (loss of \$50) 17. 15 18. 8 19. 21 20. 45 21. 5 22. 8 23. 5
24. \$7 or \$10 25. $\frac{5}{2}$, −10 26. $-\frac{1}{3}$ 27. $-\frac{4}{5}$, $\frac{4}{5}$ 28. $-\frac{2}{3}$, 5 29. 0 30. undefined 31. 1 32. −1 33. 3 34. 10
35. −18 36. −50 37. $\frac{2}{5}$ 38. $\frac{3}{4}$ 39. $\frac{1}{6}$ 40. $\frac{4}{7}$

7.6 EXERCISES B

Solve.

1. The product of 4 more than a number and 6 less than the number is 0. Find the number.

2. The square of a number, plus 15, is the same as 8 times the number. Find the number.

3. The product of 4 more than a number and 6 less than the number is 24. Find the number.

4. The length of a rectangle is 10 m more than the width. If the area is 144 m², find the dimensions.

5. If the length of the side of a square is increased by 3 inches, the area is 49 in². Find the length of the side of the original square.

6. The product of two consecutive integers is 132. Find the integers.

7. The product of two consecutive even integers is 224. Find the integers.

8. The sum of the squares of two consecutive odd positive integers is 202. Find the integers.

9. The area of a rectangle is 40 cm² and the length is 3 cm more than the width. Find the dimensions of the rectangle.

10. The area of a square is numerically 12 more than the perimeter. Find the length of a side in ft.

11. The area of a triangle is 150 cm². If the base is three times the height, find the base and height.

12. The area of a triangle is 30 in². If the base is 4 in more than the height, find the base and height.

13. A shipping box is 6 in high. The length must be 2 in more than the width and the volume is 210 in³. Find the length and width.

14. The length of a box is 12 cm and the height is 4 cm less than the width. If the volume of the box is 384 cm³, find the height and width.

The profit on one type of shoe is given by the equation $P = n^2 - 5n - 200$, where n is the number of shoes sold per day.
Use this equation to do Exercises 15–18.

15. What is the profit if 20 pairs of shoes are sold?

16. What is the profit if 10 pairs of shoes are sold?

17. How many pairs of shoes were sold on a day when the profit was \$550?

18. How many pairs of shoes were sold on a day when there was a loss of \$50?

In a basketball league containing n teams, the number of different ways that the league champion and second place team can be chosen is given by $N = n(n - 1)$. Use this equation to do Exercises 19–22.

19. How many different first and second place finishers are possible in a league with 8 members?

20. How many different first and second place finishers are possible in the NBA, which has 26 teams?

21. If it is known that there are 240 different ways for two teams to finish first and second in a league, how many teams are in the league?

22. If it has been determined that there are 90 different ways for two teams to finish first and second in a conference, how many teams are in the conference?

23. An engineer found that $E = 3u^2 - 16u + 6$ where u is positive. Find u when $E = 18$.

24. The relationship between the number n of tables a furniture store can sell per month and the price of each table p is given by the equation $n = 3000 - 200p$. At what price should the store sell each table to produce a monthly revenue of $10,800?

FOR REVIEW

Solve.

25. $(3x + 2)(x - 8) = 0$

26. $x^2 - 14x + 49 = 0$

27. $-45y^2 + 245 = 0$

28. $5y^2 = 18y + 8$

The following exercises review material that we covered in Chapter 1. They will help you prepare for the next section. Perform the indicated operations.

29. $\dfrac{0}{9}$

30. $\dfrac{9}{0}$

31. $\dfrac{9}{9}$

32. $\dfrac{9}{-9}$

Supply the missing numerator.

33. $\dfrac{1}{3} = \dfrac{?}{9}$

34. $\dfrac{4}{7} = \dfrac{?}{21}$

35. $\dfrac{6}{-11} = \dfrac{?}{55}$

36. $\dfrac{8}{25} = -\dfrac{?}{150}$

Reduce each fraction to lowest terms.

37. $\dfrac{55}{22}$

38. $\dfrac{125}{175}$

39. $\dfrac{450}{750}$

40. $\dfrac{169}{39}$

7.6 EXERCISES C

Solve.

1. A chemical reaction is described by the equation $C = 6n^2 - 17n - 4$. Find n when $C = 10$ if n is a positive integer. [Answer: no solution]

2. A banker uses the equation $M = (n + 2)(n - 7)$. Find n when $M = -20$. [*Hint:* Do not set each of the given factors equal to -20.]

CHAPTER 7 REVIEW

KEY WORDS

7.1   A **common factor** of two expressions is a factor that occurs in both of them.

The **greatest common factor (GCF)** of two terms is the largest factor common to both terms.

A **prime polynomial** cannot be factored.

7.4   A **perfect square trinomial** is one that can be factored into the form $(a + b)^2$ or $(a - b)^2$.

KEY CONCEPTS

7.1   Always factor completely. For example, factor $20x + 30$ as $10(2x + 3)$, not $5(4x + 6)$.

7.2   To factor $x^2 + bx + c = (x + \underline{\hspace{1em}})(x + \underline{\hspace{1em}})$, list all pairs of integers whose product is c and fill the blanks with the pair from this list whose sum is b.

7.3   To factor $ax^2 + bx + c = (\underline{\hspace{1em}}x + \underline{\hspace{1em}})(\underline{\hspace{1em}}x + \underline{\hspace{1em}})$, list all pairs of integers whose product is a for the first blanks in each binomial factor. Do the same for c using these in the second blanks. By trial and error, select the correct pairs that give the middle term bx. Factoring by grouping may also be used.

7.4   **(1)** $a^2 - b^2 = (a + b)(a - b)$, not $(a - b)^2$

(2) $a^2 + 2ab + b^2 = (a + b)^2$

(3) $a^2 - 2ab + b^2 = (a - b)^2$

7.5   **1.** The zero-product rule,

$$\text{if } a \cdot b = 0, \quad \text{then } a = 0 \text{ or } b = 0,$$

applies only when one side of the equation is zero. For example, if $a \cdot b = 7$, we *cannot* conclude that $a = 7$ or $b = 7$.

2. Do not use the zero-product rule on a zero-sum or zero-difference equation. For example, $(2x + 1) - (x + 5) = 0$ is *not* a zero-product equation. To solve it, clear parentheses and combine terms. *Do not* set $2x + 1$ and $x + 5$ equal to zero.

REVIEW EXERCISES

Part I

7.1   *Factor by removing the greatest common factor.*

1. $8x + 2$

2. $3y^4 - 9y^2$

3. $6a^2 - 1$

4. $50x^3 - 40x^2 + 20x$

5. $6x^4y^2 + 12x^2y^4$

6. $a^2(a + b) + 5(a + b)$

Factor by grouping.

7. $x^3 - 3x^2 + 2x - 6$

8. $x^2y^2 + x^2 + 6y^2 + 6$

7.2,
7.3, *Factor.*
7.4

9. $x^2 + 3x - 10$

10. $9y^2 + 6y + 1$

11. $x^2 + 7x + 10$

12. $y^2 - 8y + 7$

13. $4x^2 - 9$

14. $4y^2 - 12y + 9$

15. $x^2 - 12x - 45$

16. $y^2 + 12y - 45$

17. $x^2 + 10x + 16$

18. $2y^2 - 19y + 42$

19. $5x^2 + 12x - 9$

20. $2y^2 - 50$

21. $x^4 - 25$

22. $y^4 - 81$

23. $6x^2 + 13x - 5$

24. $6y^2 - 13y - 63$

25. $x^2 - 10xy + 25y^2$

26. $x^2 - 13xy + 42y^2$

27. $81x^2 - 64y^2$

28. $81x^2 + 64y^2$

29. $5x^2 - 90xy + 405y^2$

30. $6x^2 + 13xy - 28y^2$

7.5 *Solve.*

31. $(2x - 1)(12x + 5) = 0$

32. $81y^2 - 9 = 0$

33. $4x^2 + 5x + 1 = 0$

34. $y^2 - 4y + 4 = 0$

35. $12x^2 + 6x = 0$

36. $y(y - 5) = 84$

7.6 **37.** The product of a number and 5 less than the number is -6. Find the number.

38. The sum of the squares of two consecutive even integers is 452. Find the integers.

39. The length of a rectangle is 5 times the width. If the area is 180 cm^2, find the dimensions.

40. The profit on the sale of sport coats is given by the equation $P = n^2 - 6n - 27$, where n is the number of coats sold per day.

(a) What is the profit when 12 coats are sold?

(b) How many coats were sold on a day when the profit was $160?

Part II

Solve.

41. $x(x - 3)(x + 8) = 0$

42. $4x^2 + 4x - 15 = 0$

43. $3x^2 = 11x + 4$

44. The equation $n = 2400 - 120p$ gives the relationship between the number n of balls sold and the price p of each ball. What price should be set to have a monthly revenue of $12,000?

Factor.

45. $9x^2 + 30x + 25$

46. $(x + 1)^2 - 25$

47. $x^2 + 4$

48. $2x^2 + ax - 10x - 5a$

49. $8y^4 - 2$

50. $4x^2 - 29x + 7$

ANSWERS: 1. $2(4x + 1)$ 2. $3y^2(y^2 - 3)$ 3. cannot be factored 4. $10x(5x^2 - 4x + 2)$ 5. $6x^2y^2(x^2 + 2y^2)$
6. $(a + b)(a^2 + 5)$ 7. $(x - 3)(x^2 + 2)$ 8. $(y^2 + 1)(x^2 + 6)$ 9. $(x + 5)(x - 2)$ 10. $(3y + 1)^2$ 11. $(x + 5)(x + 2)$
12. $(y - 7)(y - 1)$ 13. $(2x + 3)(2x - 3)$ 14. $(2y - 3)^2$ 15. $(x + 3)(x - 15)$ 16. $(y - 3)(y + 15)$ 17. $(x + 2)(x + 8)$
18. $(2y - 7)(y - 6)$ 19. $(5x - 3)(x + 3)$ 20. $2(y + 5)(y - 5)$ 21. $(x^2 + 5)(x^2 - 5)$ 22. $(y^2 + 9)(y + 3)(y - 3)$
23. $(3x - 1)(2x + 5)$ 24. $(3y + 7)(2y - 9)$ 25. $(x - 5y)^2$ 26. $(x - 6y)(x - 7y)$ 27. $(9x + 8y)(9x - 8y)$ 28. cannot be
factored 29. $5(x - 9y)^2$ 30. $(3x - 4y)(2x + 7y)$ 31. $\frac{1}{2}, -\frac{5}{12}$ 32. $-\frac{1}{3}, \frac{1}{3}$ 33. $-\frac{1}{4}, -1$ 34. 2 35. $0, -\frac{1}{2}$
36. $-7, 12$ 37. 2, 3 38. $-16, -14; 14, 16$ 39. 6 cm by 30 cm 40. (a) $45 (b) 17 41. $0, 3, -8$ 42. $-\frac{5}{2}, \frac{3}{2}$
43. $-\frac{1}{3}, 4$ 44. $10 45. $(3x + 5)^2$ 46. $(x + 6)(x - 4)$ 47. cannot be factored 48. $(2x + a)(x - 5)$
49. $2(2y^2 + 1)(2y^2 - 1)$ 50. $(4x - 1)(x - 7)$

Factor by removing the greatest common factor.

1. $20x + 12$

1. _____

2. $35y - 7$

2. _____

3. $24x^4 - 12x^3 + 18x^2$

3. _____

4. $5x^3y^3 + 40x^3y^2$

4. _____

5. Factor $7x^2y^2 - y^2 + 14x^2 - 2$ by grouping.

5. _____

Factor.

6. $x^2 + 15x + 56$

6. _____

7. $x^2 - 16x + 64$

7. _____

8. $3y^2 - 75$

8. _____

9. $x^2 - 8x - 20$

9. _____

CONTINUED

10. $2x^2 + 13x - 24$

10. _____

11. $3x^2 + 16xy + 5y^2$

11. _____

Solve.

12. $(4y - 5)(2y + 3) = 0$

12. _____

13. $x^2 - x - 30 = 0$

13. _____

14. The product of a number and 3 more than the number is 4. Find the number.

14. _____

15. The area of a triangle is 12 cm^2. If the base is 2 cm less than the height, find the base and height.

15. _____

16. The profit on the sale of one type of dress is $P = n^2 - 3n - 8$, where n is the number of dresses sold per day.
(a) What is the profit when 20 dresses are sold?

16. (a) _____

(b) How many dresses were sold on a day when the profit was $100?

16. (b) _____

CHAPTER 8

Rational Expressions

BASIC CONCEPTS OF ALGEBRAIC FRACTIONS

STUDENT GUIDEPOSTS

1. Algebraic Fractions and Rational Expressions
2. Excluding Values that Make the Denominator Zero
3. Equivalent Fractions
4. Reducing to Lowest Terms
5. Fractions Equivalent to −1

① ALGEBRAIC FRACTIONS AND RATIONAL EXPRESSIONS

In Chapter 1 we reviewed some of the properties of numerical fractions. Now we define an **algebraic fraction** as a fraction that contains a variable in the numerator or denominator or both. The following are algebraic fractions.

$$\frac{xy}{x+1}, \qquad \frac{y^2+1}{y-1}, \qquad \frac{\sqrt{a}+5}{2}, \qquad \frac{7}{(x-2)(x+1)}, \qquad \frac{x+y}{x-y}$$

If an algebraic fraction is the quotient of two polynomials it is called a **rational expression.** Of the algebraic fractions listed above only $\frac{\sqrt{a}+5}{2}$ is *not* a rational expression ($\sqrt{a}+5$ is not a polynomial because a is not raised to a whole number power). The algebraic fractions that we study in this chapter will all be rational expressions.

② EXCLUDING VALUES THAT MAKE THE DENOMINATOR ZERO

With rational expressions, we need to avoid division by zero, which is undefined. Any value of the variable that makes the denominator zero must be excluded from consideration. The expression is defined for all other values of the variable.

To Find the Values Which Must Be Excluded in a Rational Expression

1. Set the denominator equal to zero.
2. Solve this equation; any solution must be excluded.

EXAMPLE 1 FINDING EXCLUDED VALUES

Find the value of the variable that must be excluded in each rational expression.

(a) $\dfrac{2x}{x + 7}$

Set the denominator equal to zero and solve.

$$x + 7 = 0$$
$$x = -7$$

Check: $\dfrac{2(-7)}{-7 + 7} = \dfrac{-14}{0}$

Since division by zero is not defined, -7 must be excluded. The expression is defined for all x except -7.

(b) $\dfrac{(a - 1)(a + 1)}{2}$

Set the denominator equal to zero and solve.

$$2 = 0$$

Since there is no solution to this equation, there are no values to exclude. The expression is defined for all values of a.

(c) $\dfrac{3y}{(y - 1)(y + 4)}$

We need to solve $(y - 1)(y + 4) = 0$, so we use the zero-product rule.

$$y - 1 = 0 \quad \text{or} \quad y + 4 = 0$$
$$y = 1 \qquad\qquad y = -4$$

The values to exclude are 1 and -4. The expression is defined for all values of y except 1 and -4.

(d) $\dfrac{x^2 + 1}{x^2 + 5x + 6}$

We need to solve $x^2 + 5x + 6 = 0$. First factor and then use the zero-product rule.

$$x^2 + 5x + 6 = 0$$
$$(x + 2)(x + 3) = 0$$
$$x + 2 = 0 \quad \text{or} \quad x + 3 = 0$$
$$x = -2 \qquad\qquad x = -3$$

The values to exclude are -2 and -3. The expression is defined for all values of x except -2 and -3.

PRACTICE EXERCISE 1

Find the values of the variable that must be excluded.

(a) $\dfrac{a}{a - 2}$

(b) $\dfrac{x^2 + x - 3}{7}$

(c) $\dfrac{4w}{(w + 1)(w - 7)}$

(d) $\dfrac{y^2}{3y^2 - 11y - 4}$

Answers: (a) 2 (b) There are none. (c) -1 and 7 (d) 4 and $-\frac{1}{3}$

③ EQUIVALENT FRACTIONS

In Chapter 1 we saw that multiplying or dividing the numerator and denominator of a fraction by the same nonzero number gives us an equivalent fraction.

$$\frac{4}{6} = \frac{4 \cdot 3}{6 \cdot 3} = \frac{12}{18} \qquad \frac{4}{6} \text{ is equivalent to } \frac{12}{18}$$

$$\frac{4}{6} = \frac{4 \div 2}{6 \div 2} = \frac{2}{3} \qquad \frac{4}{6} \text{ is equivalent to } \frac{2}{3}$$

We can generalize the idea of equivalent fractions to include algebraic fractions.

Fundamental Principle of Fractions

If the numerator and denominator of an algebraic fraction are multiplied or divided by the same nonzero expression, the resulting fraction is **equivalent** to the original. That is, if $\frac{P}{Q}$ is an algebraic fraction, $Q \neq 0$, and R is a nonzero expression, then

$$\frac{P}{Q} \text{ is equivalent to } \frac{P \cdot R}{Q \cdot R} \text{ and to } \frac{P \div R}{Q \div R}.$$

EXAMPLE 2 **EQUIVALENT FRACTIONS**	**PRACTICE EXERCISE 2**

Are the given fractions equivalent?

Are the given fractions equivalent?

(a) $\dfrac{3}{x + 2}$ and $\dfrac{3(x + 1)}{(x + 2)(x + 1)}$

(a) $\dfrac{a(a + 2)}{(a - 3)(a + 2)}$ and $\dfrac{a}{a - 3}$

$$\frac{3 \cdot (x + 1)}{(x + 2) \cdot (x + 1)} = \frac{3(x + 1)}{(x + 2)(x + 1)} \qquad \text{Numerator and denominator are both multiplied by } x + 1$$

The fractions are equivalent since the first becomes the second when both numerator and denominator are multiplied by $x + 1$.

We may also divide numerator and denominator of the second fraction by $x + 1$ to obtain the first.

$$\frac{3(x + 1) \div (x + 1)}{(x + 2)(x + 1) \div (x + 1)} = \frac{\dfrac{3(x + 1)}{(x + 1)}}{\dfrac{(x + 2)(x + 1)}{(x + 1)}} = \frac{3}{x + 2}$$

This division can be abbreviated by showing the expression is equivalent to $\frac{3}{x+2}$ times 1.

(b) $\dfrac{5}{y + 1}$ and $\dfrac{5}{y - 1}$

$$\frac{3(x + 1)}{(x + 2)(x + 1)} = \frac{3}{x + 2} \cdot \frac{x + 1}{x + 1}$$

$$= \frac{3}{x + 2} \qquad 1 = \frac{3}{x + 2}$$

We usually accomplish this by dividing out the factor common to both numerator and denominator.

$$\frac{3(x + 1)}{(x + 2)(x + 1)} = \frac{3\cancel{(x + 1)}}{(x + 2)\cancel{(x + 1)}} = \frac{3}{x + 2}$$

The word "cancel" is often used to describe this process, but we will use "divide out."

(b) $\dfrac{3}{x+2}$ and $\dfrac{7}{x+1}$

These are not equivalent since there is no expression that one can be multiplied by, in both numerator and denominator, to give the other.

Answers: (a) yes (b) no

④ REDUCING TO LOWEST TERMS

A fraction is **reduced to lowest terms** when 1 (or -1) is the only number or expression that divides both numerator and denominator. For example, we reduce $\frac{30}{12}$ to lowest terms as follows.

$$\frac{30}{12} = \frac{\cancel{2}\cdot\cancel{3}\cdot 5}{\cancel{2}\cdot 2\cdot\cancel{3}} = \frac{5}{2}$$

Factor completely and divide out or cancel common factors

CAUTION

Divide out or cancel *factors* only, never cancel *terms*. The only expressions that can be divided out are those that are multiplied (never added or subtracted) by every other expression in the numerator or denominator. For example,

$$\frac{5\cdot 6}{5} = \frac{\cancel{5}\cdot 6}{\cancel{5}} \quad \text{but} \quad \frac{5+6}{5} \neq \frac{\cancel{5}+6}{\cancel{5}}.$$

We can generalize this procedure to reduce rational expressions.

To Reduce a Rational Expression to Lowest Terms

1. Factor numerator and denominator completely.
2. Divide out all common factors.
3. Multiply the remaining factors in the numerator and multiply the remaining factors in the denominator.

EXAMPLE 3 REDUCING FRACTIONS

Reduce each fraction to lowest terms.

(a) $\dfrac{12x^2}{18x^3} = \dfrac{2\cdot 2\cdot 3\cdot\cancel{x}\cdot\cancel{x}}{2\cdot 3\cdot 3\cdot\cancel{x}\cdot\cancel{x}\cdot x} = \dfrac{2}{3x}$

Factor completely and divide out common factors

With practice you can avoid factoring completely to prime factors. For example, the above might take the form

$$\frac{12x^2}{18x^3} = \frac{\cancel{6}\cdot 2\cdot\cancel{x^2}}{\cancel{6}\cdot 3\cdot\cancel{x^2}\cdot x} = \frac{2}{3x}.$$

PRACTICE EXERCISE 3

Reduce each fraction to lowest terms.

(a) $\dfrac{15y^6}{35y^2}$

(b) $\dfrac{x}{2x^2} = \dfrac{\cancel{x}}{2 \cdot \cancel{x} \cdot x} = \dfrac{1}{2x}$

Notice that when the factors x are divided out, 1 is left in the numerator, not 0.

(c) $\dfrac{2x^2 + x}{3x^3 + 3x} = \dfrac{\cancel{x}(2x + 1)}{3\cancel{x}(x^2 + 1)} = \dfrac{2x + 1}{3(x^2 + 1)}$

(d) $\dfrac{7a + 14}{7} = \dfrac{\cancel{7}(a + 2)}{\cancel{7}} = \dfrac{a + 2}{1} = a + 2$

(e) $\dfrac{2 + 2y}{2y} = \dfrac{\cancel{2}(1 + y)}{\cancel{2}y} = \dfrac{1 + y}{y}$

We cannot divide out the ys. The y in the numerator is a term, not a factor. If we replace y with the number 2, it is easy to see that

$$\frac{1 + y}{y} \text{ is not equal to } \frac{1 + \cancel{y}}{\cancel{y}}.$$

(f) $\dfrac{x^2 + 3xy + 2y^2}{x^2 - 4y^2} = \dfrac{(x + y)(x + 2y)}{(x - 2y)(x + 2y)} = \dfrac{x + y}{x - 2y}$

Why can't we divide out the x and the y in the answer to obtain $-\frac{1}{2}$?

(b) $\dfrac{a - 3}{5(a - 3)}$

(c) $\dfrac{a^2 - 3a}{5a^3 + 10a}$

(d) $\dfrac{3}{3y - 6}$

(e) $\dfrac{-5x}{x - 5}$

(f) $\dfrac{y^2 + 2y - 15}{y^2 - 2y - 35}$

Answers: (a) $\frac{3y^4}{7}$ (b) $\frac{1}{5}$
(c) $\frac{a - 3}{5(a^2 + 2)}$ (d) $\frac{1}{y - 2}$ (e) cannot
be reduced (f) $\frac{y - 3}{y - 7}$

⑤ FRACTIONS EQUIVALENT TO −1

Consider the fraction $\frac{5 - x}{x - 5}$. Our first impression might be that it is reduced to lowest terms. However, if we factor -1 from the numerator, we obtain

$$\frac{5 - x}{x - 5} = \frac{(-1)(-5 + x)}{x - 5}$$

$$= \frac{(-1)\cancel{(x - 5)}}{\cancel{(x - 5)}} \qquad -5 + x = x - 5$$

$$= -1.$$

In general, we can show that if $a \neq b$,

$$\frac{a - b}{b - a} = -1.$$

For example, if $a = 5$ and $b = 2$,

$$\frac{a - b}{b - a} = \frac{5 - 2}{2 - 5} = \frac{3}{-3} = -1.$$

Whenever a fraction of this type results, simply replace it with -1.

CAUTION

Remember that $a + b = b + a$ so that $\frac{a + b}{b + a} = 1$, not -1.

| **EXAMPLE 4** $\text{REDUCING } \frac{a-b}{b-a} \text{ TO } -1$ | **PRACTICE EXERCISE 4** |

Reduce $\dfrac{x^2 - xy}{3y - 3x}$ to lowest terms.

$$\frac{x^2 - xy}{3y - 3x} = \frac{x(x - y)}{3(y - x)} = \frac{x}{3} \cdot \frac{x - y}{y - x}$$

$$= \frac{x}{3} \cdot (-1) \qquad \tfrac{x-y}{y-x} = -1$$

$$= -\frac{x}{3}$$

Reduce $\dfrac{a^2b - a^3}{5a - 5b}$ to lowest terms.

Answer: $-\dfrac{a^2}{5}$

8.1   EXERCISES A

Find the values of the variable that must be excluded.

1. $\dfrac{3}{x + 1}$

2. $\dfrac{a}{a - 7}$

3. $\dfrac{y + 1}{y(y + 3)}$

4. $\dfrac{z + 2}{2(z + 4)}$

5. $\dfrac{x + 2}{(x - 1)(x + 4)}$

6. $\dfrac{a - 3}{(a - 1)(a + 1)}$

7 $\dfrac{2x + 7}{x^2 + 2x + 1}$

8. $\dfrac{3y - 4}{y^2 - 9}$

9. $\dfrac{(x - 1)(x + 1)}{7}$

Are the given fractions equivalent? Explain.

10. $\dfrac{2}{7}, \dfrac{6}{21}$

11. $\dfrac{3}{y}, \dfrac{-3}{-y}$

12. $\dfrac{2x}{4}, \dfrac{x}{4}$

13. $\dfrac{a^2}{3}, \dfrac{-2a^2}{-6}$

14. $\dfrac{x + 1}{2}, \dfrac{3(x + 1)}{6}$

15. $\dfrac{x + 1}{9}, \dfrac{(x + 1)(x - 1)}{9(x - 1)}$

16 $\dfrac{2}{x^2}, \dfrac{2 + x}{x^2 + x}$

17 $\dfrac{x - 1}{3x - 1}, \dfrac{x}{3x}$

18. $\dfrac{z - 2}{z^2 - 4}, \dfrac{1}{z + 2}$

Reduce to lowest terms.

19. $\dfrac{21}{30}$

20. $\dfrac{2y^2}{10y^3}$

21. $\dfrac{4x^3y}{2xy^2}$

22. $\dfrac{22ab^5}{11ab^2}$

23. $\dfrac{3x^2z}{21x^6z^2}$

24. $\dfrac{a(a+1)}{5(a+1)}$

25. $\dfrac{(x+2)(2x-3)}{(x+2)(2x+3)}$

26. $\dfrac{z+9}{z^2+9}$

27. $\dfrac{a+3}{a^2-9}$

28. $\dfrac{w^2+7w}{w^2-49}$

29 $\dfrac{x^2-3x-10}{x^2-6x+5}$

30. $\dfrac{2a(a+5)}{10a+2a^2}$

31. $\dfrac{x-7}{7-x}$

32 $\dfrac{x^2-9}{3-x}$

33. $\dfrac{y+4}{y-4}$

34. $\dfrac{x+y}{x^2-y^2}$

35. $\dfrac{a^2+2ab+b^2}{a+b}$

36 $\dfrac{x^2-7xy+6y^2}{x^2-4xy-12y^2}$

FOR REVIEW

The following exercises review material from Section 1.1 to help you prepare for the next section. Perform the indicated operations.

37. $\dfrac{5}{8} \cdot \dfrac{2}{15}$

38. $\dfrac{7}{5} \cdot 20$

39. $\dfrac{2}{3} \div \dfrac{4}{9}$

40. $\dfrac{4}{5} \div 2$

41. Factor 300 into a product of primes.

ANSWERS: 1. -1 2. 7 3. 0, -3 4. -4 5. 1, -4 6. 1, -1 7. -1 8. 3, -3 9. none 10. yes 11. yes 12. no 13. yes 14. yes 15. yes 16. no 17. no 18. yes 19. $\frac{7}{10}$ 20. $\frac{1}{5y}$ 21. $\frac{2x^2}{y}$ 22. $2b^3$ 23. $\frac{1}{7x^4z}$ 24. $\frac{a}{5}$ 25. $\frac{2x-3}{2x+3}$ 26. $\frac{z+9}{z^2+9}$ 27. $\frac{1}{a-3}$ 28. $\frac{w}{w-7}$ 29. $\frac{x+2}{x-1}$ 30. 1 31. -1 32. $-(x+3)$ 33. $\frac{y+4}{y-4}$ 34. $\frac{1}{x-y}$ 35. $a+b$ 36. $\frac{x-y}{x+2y}$ 37. $\frac{1}{12}$ 38. 28 39. $\frac{3}{2}$ 40. $\frac{2}{5}$ 41. $2 \cdot 2 \cdot 3 \cdot 5 \cdot 5$

8.1   EXERCISES B

Find the values of the variable that must be excluded.

1. $\dfrac{5}{x-6}$

2. $\dfrac{x^2+8}{x+5}$

3. $\dfrac{y-3}{y(y-4)}$

4. $\dfrac{z-8}{2(z+10)}$

5. $\dfrac{x+1}{(x-2)(x+6)}$

6. $\dfrac{a+5}{(a-3)(a+3)}$

7. $\dfrac{3x-1}{x^2-2x+1}$

8. $\dfrac{3y+7}{y^2-16}$

9. $\dfrac{z+5}{z^2+9}$

Are the given fractions equivalent? Explain.

10. $\dfrac{3}{5},\dfrac{6}{10}$

11. $\dfrac{5}{y},\dfrac{-5}{-y}$

12. $\dfrac{3x}{6},\dfrac{x}{6}$

13. $\dfrac{a^3}{4},\dfrac{-2a^3}{-8}$

14. $\dfrac{x+5}{5},\dfrac{2(x+5)}{10}$

15. $\dfrac{x+3}{8},\dfrac{(x+3)(x-4)}{8(x-4)}$

16. $\dfrac{3}{x^3},\dfrac{3-x}{x^3-x}$

17. $\dfrac{z+3}{z^2-9},\dfrac{1}{z-3}$

18. $\dfrac{x^2+2}{3x^2+2},\dfrac{x^2}{3x^2}$

Reduce to lowest terms.

19. $\dfrac{22}{55}$

20. $\dfrac{45x^6}{15x^2}$

21. $\dfrac{9a^4b}{3ab^3}$

22. $\dfrac{25xy^3}{5xy^4}$

23. $\dfrac{6ay}{18a^2y^3}$

24. $\dfrac{a^2(a-5)}{3(a-5)}$

25. $\dfrac{(2x+1)(x+5)}{(2x+1)(x-5)}$

26. $\dfrac{z+16}{z^2+16}$

27. $\dfrac{a+4}{a^2-16}$

28. $\dfrac{x^3+5x^2}{x(x^2-25)}$

29. $\dfrac{x^2-3x-10}{x^2-8x+15}$

30. $\dfrac{3a^2(a+4)}{9a+6a^2}$

31. $\dfrac{8-x}{x-8}$

32. $\dfrac{16-x^2}{x-4}$

33. $\dfrac{2y+3}{2y-3}$

34. $\dfrac{x-y}{x^2-y^2}$

35. $\dfrac{a^2+7ab+10b^2}{a+5b}$

36. $\dfrac{x^2-10xy+25y^2}{x^2-4xy-5y^2}$

FOR REVIEW

The following exercises review material from Section 1.1 to help you prepare for the next section. Perform the indicated operations.

37. $\dfrac{7}{10}\cdot\dfrac{5}{14}$

38. $\dfrac{8}{3}\cdot 27$

39. $\dfrac{3}{4}\div\dfrac{9}{16}$

40. $7\div\dfrac{14}{3}$

41. Factor 420 into a product of primes.

8.1 EXERCISES C

Reduce to lowest terms. The factoring formulas in Exercises C of Section 7.4 are used in these exercises.

1. $\dfrac{5x^2 - 25xy + 30y^2}{10x^2 - 40y^2}$

2. $\dfrac{27x^3 - y^3}{18x^2 + 6xy + 2y^2}$

$\left[\text{Answer: } \dfrac{3x - y}{2}\right]$

3. $\dfrac{x^4 + 8xy^3}{x^4 - 16y^4}$

8.2 MULTIPLICATION AND DIVISION OF FRACTIONS

STUDENT GUIDEPOSTS

1 Multiplying Fractions   **2** Dividing Fractions

1 MULTIPLYING FRACTIONS

In Chapter 1 we saw that the product of two or more fractions is equal to the product of all numerators divided by the product of all denominators. The resulting fraction should be reduced to lowest terms. For example,

$$\frac{3}{4} \cdot \frac{2}{9} = \frac{3 \cdot 2}{4 \cdot 9} = \frac{6}{36}.$$

To reduce $\frac{6}{36}$ to lowest terms, factor the numerator and denominator.

$$\frac{6}{36} = \frac{2 \cdot 3}{2 \cdot 3 \cdot 2 \cdot 3} = \frac{1}{6}$$

To accomplish multiplication and reduction in one process we factor numerators and denominators first, divide out common factors, and then multiply. The resulting fraction is the product.

$$\frac{3}{4} \cdot \frac{2}{9} = \frac{3}{2 \cdot 2} \cdot \frac{2}{3 \cdot 3} = \frac{3 \cdot 2}{2 \cdot 2 \cdot 3 \cdot 3} = \frac{1}{2 \cdot 3} = \frac{1}{6}$$

We can generalize this procedure to multiply algebraic fractions.

To Multiply Algebraic Fractions

1. Factor all numerators and denominators completely.
2. Place the factored product of all numerators over the factored product of all denominators.
3. Divide out common factors before multiplying the remaining numerator factors and multiplying the remaining denominator factors.

EXAMPLE 1 MULTIPLYING FRACTIONS

Multiply.

$\dfrac{x}{6} \cdot \dfrac{3}{2x^2} = \dfrac{x}{2 \cdot 3} \cdot \dfrac{3}{2 \cdot x \cdot x}$ Factor

$= \dfrac{\cancel{x} \cdot \cancel{3}}{2 \cdot \cancel{3} \cdot 2 \cdot \cancel{x} \cdot x}$ Indicate product and divide out common factors

$= \dfrac{1}{2 \cdot 2 \cdot x}$ The numerator is 1, *not* 0

$= \dfrac{1}{4x}$ Multiply

PRACTICE EXERCISE 1

Multiply. $\dfrac{y^2}{15} \cdot \dfrac{5}{y}$

Answer: $\frac{y}{3}$

EXAMPLE 2 MULTIPLYING FRACTIONS

Multiply.

$\dfrac{a^2 - 9}{a^2 - 4a + 4} \cdot \dfrac{a - 2}{a + 3}$

$= \dfrac{(a - 3)(a + 3)}{(a - 2)(a - 2)} \cdot \dfrac{(a - 2)}{(a + 3)}$ Factor

$= \dfrac{(a - 3)\cancel{(a + 3)}\cancel{(a - 2)}}{(a - 2)\cancel{(a - 2)}\cancel{(a + 3)}}$ Divide out common factors

$= \dfrac{a - 3}{a - 2}$ *Do not* divide out the term *a*; it is not a factor

PRACTICE EXERCISE 2

Multiply. $\dfrac{x^2 - 2x}{x^2 + 11x + 28} \cdot \dfrac{x + 7}{x - 2}$

Answer: $\frac{x}{x + 4}$

EXAMPLE 3 MULTIPLYING FRACTIONS

Multiply.

$\dfrac{x^2 - y^2}{xy} \cdot \dfrac{xy - x^2}{x^2 - 2xy + y^2}$

$= \dfrac{(x - y)(x + y)}{xy} \cdot \dfrac{x(y - x)}{(x - y)(x - y)}$ Factor

$= \dfrac{\cancel{(x - y)}(x + y) \cdot \cancel{x} \cdot (y - x)}{\cancel{x} \cdot y \cdot \cancel{(x - y)}(x - y)}$ Divide out common factors

$= \dfrac{x + y}{y} \cdot \dfrac{(y - x)}{(x - y)}$

$= \dfrac{x + y}{y} \cdot (-1)$ $\frac{y - x}{x - y} = -1$

$= -\dfrac{x + y}{y}$

PRACTICE EXERCISE 3

Multiply.

$\dfrac{a^2 - ab}{2b^2 - ab - a^2} \cdot \dfrac{2a^2 + 3ab + b^2}{2a^2 + ab}$

Answer: $-\frac{a + b}{2b + a}$

② DIVIDING FRACTIONS

In Chapter 1 we divided one numerical fraction by another. In

$$\frac{2}{3} \div \frac{3}{5},$$

we divide the dividend $\frac{2}{3}$ by the divisor $\frac{3}{5}$. To find the quotient, we multiply the dividend by the **reciprocal** of the divisor. That is,

$$\frac{2}{3} \div \frac{3}{5} = \frac{2}{3} \cdot \frac{5}{3} = \frac{2 \cdot 5}{3 \cdot 3} = \frac{10}{9}. \qquad \text{$\frac{5}{3}$ is the reciprocal of $\frac{3}{5}$}$$

We use exactly the same process when dividing algebraic fractions.

To Divide Algebraic Fractions

Multiply the dividend by the reciprocal of the divisor.

$$\frac{P}{Q} \div \frac{R}{S} = \frac{P}{Q} \cdot \frac{S}{R}$$

The following table lists several algebraic fractions and their reciprocals.

Fraction	Reciprocal	
$\dfrac{2}{x}$	$\dfrac{x}{2}$	
$\dfrac{x+1}{(x-5)^2}$	$\dfrac{(x-5)^2}{x+1}$	
5	$\dfrac{1}{5}$	5 can be thought of as $\frac{5}{1}$
x	$\dfrac{1}{x}$	x is $\frac{x}{1}$
$\dfrac{1}{3}$	3	3 is $\frac{3}{1}$
$x+y$	$\dfrac{1}{x+y}$	Not $\frac{1}{x} + \frac{1}{y}$. For example, $\frac{1}{2+3} = \frac{1}{5} \neq \frac{1}{2} + \frac{1}{3} = \frac{5}{6}$

EXAMPLE 4 DIVIDING FRACTIONS	PRACTICE EXERCISE 4

Divide.

(a) $\dfrac{x}{x+1} \div \dfrac{x}{2} = \dfrac{x}{x+1} \cdot \dfrac{2}{x}$ $\frac{2}{x}$ is the reciprocal of $\frac{x}{2}$

$\qquad = \dfrac{\cancel{x} \cdot 2}{(x+1) \cdot \cancel{x}}$ Divide out common factor

$\qquad = \dfrac{2}{x+1}$

Divide.

(a) $\dfrac{y-5}{y} \div \dfrac{5}{y}$

(b) $\dfrac{y+3}{y} \div y = \dfrac{y+3}{y} \cdot \dfrac{1}{y}$ $\frac{1}{y}$ is the reciprocal of y

$= \dfrac{y+3}{y^2}$

(b) $\dfrac{b}{a+b} \div b$

(c) $\dfrac{a^2-4a+4}{a^2-9}$

$\dfrac{a^2-4a+4}{a^2-9} \cdot \dfrac{a+3}{a-2}$

$= \dfrac{(a-2)(a-2)}{(a-3)(a+3)} \cdot \dfrac{(a+3)}{(a-2)}$

$= \dfrac{(a-2)(a-2)(a+3)}{(a-3)(a+3)(a-2)}$

$= \dfrac{a-2}{a-3}$ Do not divide out a

(c) $\dfrac{y^2+2y-15}{y^2-25} \div \dfrac{y-3}{y-5}$

Answers: (a) $\frac{y-5}{5}$ (b) $\frac{1}{a+b}$
(c) 1

EXAMPLE 5 DIVIDING FRACTIONS

Divide.

$\dfrac{x^2-3xy+2y^2}{x^2-4y^2} \div (x^2+xy-2y^2)$

$= \dfrac{x^2-3xy+2y^2}{x^2-4y^2} \cdot \dfrac{1}{x^2+xy-2y^2}$ Find reciprocal and multiply

$= \dfrac{(x-y)(x-2y)}{(x+2y)(x-2y)} \cdot \dfrac{1}{(x-y)(x+2y)}$ Factor

$= \dfrac{(x-y)(x-2y)}{(x+2y)(x-2y)(x-y)(x+2y)}$ Divide out common factors

$= \dfrac{1}{(x+2y)^2}$

PRACTICE EXERCISE 5

Divide.

$\dfrac{a^2+8a-33}{a^2-2a-3} \div (a^2+12a+11)$

Answer: $\frac{1}{(a+1)^2}$

8.2 EXERCISES A

Multiply.

1. $\dfrac{1}{x^2} \cdot \dfrac{x}{3}$

2. $\dfrac{3a}{4} \cdot \dfrac{6}{9a^2}$

3. $\dfrac{2y}{5} \cdot \dfrac{25}{4y^3}$

4. $\dfrac{2x}{(x+1)^2} \cdot \dfrac{x+1}{4x^2}$

5. $\dfrac{x^2-36}{x^2-6x} \cdot \dfrac{x}{x+6}$

6. $\dfrac{z^2}{z+2} \cdot \dfrac{z^2-4}{z^2-2z}$

7. $\dfrac{5x + 5}{x - 2} \cdot \dfrac{x^2 - 4x + 4}{x^2 - 1}$

8. $\dfrac{y^2 - 3y - 10}{y^2 - 4y + 4} \cdot \dfrac{y - 2}{y - 5}$

9. $\dfrac{x^2 + 2x + 1}{9x^2} \cdot \dfrac{3x^3}{x^2 - 1}$

10 $\dfrac{z^2 - z - 20}{z^2 + 7z + 12} \cdot \dfrac{z + 3}{z^2 - 25}$

11. $\dfrac{y^2 + 6y + 5}{7y^2 - 63} \cdot \dfrac{7y + 21}{(y + 5)^2}$

12. $\dfrac{16 - x^2}{5x - 1} \cdot \dfrac{5x^2 - x}{16 - 8x + x^2}$

13. $\dfrac{a^2 - 4}{a^2 - 4a + 4} \cdot \dfrac{a^2 - 9a + 14}{a^3 + 2a^2}$

14 $\dfrac{2x^2 - 5xy + 3y^2}{x^2 - y^2} \cdot (x^2 + 2xy + y^2)$

15. $\dfrac{x^2 - y^2}{(x + y)^2} \cdot \dfrac{x + y}{x - y}$

16. $\dfrac{x^2 - 7xy + 6y^2}{x^2 - xy - 2y^2} \cdot \dfrac{x + y}{x - 6y}$

Find the reciprocal.

17. $\dfrac{3}{7}$

18. $\dfrac{1}{2}$

19. 5

20. $\dfrac{1}{x + 1}$

21. $\dfrac{2x - 3}{x + 5}$

22 $z + 2$

Divide.

23. $\dfrac{x}{x + 2} \div \dfrac{x}{3}$

24. $\dfrac{2a}{a + 1} \div \dfrac{5a}{a + 1}$

25. $\dfrac{2(a + 3)}{7} \div \dfrac{4(a + 3)}{21}$

26. $\dfrac{3(x^2 + 5)}{4(x^2 - 3)} \div \dfrac{x^2 + 5}{4(x^2 - 3)}$

27. $\dfrac{5y^4}{y^2 - 1} \div \dfrac{5y^3}{y^2 + 2y + 1}$

28. $\dfrac{6a^2}{a^2 - 25} \div \dfrac{3a^3}{5a^2 - 25a}$

29 $(x + 6) \div \dfrac{x^2 - 36}{x^2 - 6x}$

30. $\dfrac{z^2 - 4z + 4}{z^2 - 1} \div \dfrac{z - 2}{5z + 5}$

31. $\dfrac{y^2 - y - 20}{y^2 + 7y + 12} \div \dfrac{y^2 - 25}{y + 3}$

32. $\dfrac{a^2 + 10a + 21}{a^2 - 2a - 15} \div \dfrac{a + 7}{a^2 - 25}$

33. $\dfrac{y^3 - 64y}{2y^2 + 16y} \div \dfrac{y^2 - 9y + 8}{y^2 + 4y - 5}$

34 $\dfrac{x^2 + 16x + 64}{2x^2 - 128} \div \dfrac{3x^2 + 30x + 48}{x^2 - 6x - 16}$

35 $\dfrac{x^2 - 5xy + 6y^2}{x^2 - 4y^2} \div (x^2 - 2xy - 3y^2)$

36. $\dfrac{x + 2y}{x^2 - y^2} \div \dfrac{x + 2y}{x + y}$

37. $\dfrac{x^2 - 81y^2}{x^2 + 18xy + 81y^2} \div \dfrac{x - 9y}{x + 9y}$

38 $\dfrac{a^2 - y^2}{a^2 - ay} \cdot \dfrac{2a^2 + ay}{a^2 - 4y^2} \div \dfrac{a + y}{a + 2y}$

FOR REVIEW

Find the values of the variable that must be excluded.

39. $\dfrac{y-1}{y^2+2y}$

40. $\dfrac{a}{a^2+5}$

41. $\dfrac{(x+1)(x-5)}{3x}$

Are the given fractions equivalent?

42. $\dfrac{x+5}{2x^2+5}, \dfrac{x}{2x^2}$

43. $\dfrac{x}{3}, \dfrac{x^2+2x}{3(x+2)}$

44. $\dfrac{1}{x+1}, \dfrac{x+1}{1}$

45. Change $\frac{x}{2}$ to an equivalent fraction with the given denominator.

(a) 10

(b) $2x$

(c) $2x^2$

(d) $2(x+1)$

(e) $2(2x-1)$

(f) $2(x^2-1)$

46. Burford reduced the fraction $\frac{1+2x}{1+5x}$ in the following way:

$$\frac{1+2x}{1+5x} = \frac{\cancel{1}+2\cancel{x}}{\cancel{1}+5\cancel{x}} = \frac{2}{5}$$

What is wrong with Burford's work?

The following exercises review material from Section 1.2 to help you prepare for the next two sections. Perform the indicated operations.

47. $\dfrac{7}{5}+\dfrac{3}{5}$

48. $\dfrac{3}{4}-\dfrac{1}{4}$

49. $\dfrac{3}{10}+\dfrac{2}{15}$

50. $\dfrac{5}{6}-\dfrac{4}{21}$

51. $5+\dfrac{2}{3}$

52. $\dfrac{13}{5}-2$

ANSWERS: 1. $\frac{1}{3x}$ 2. $\frac{1}{2a}$ 3. $\frac{5}{2y^2}$ 4. $\frac{1}{2x(x+1)}$ 5. 1 6. z 7. $\frac{5(x-2)}{x-1}$ 8. $\frac{y+2}{y-2}$ 9. $\frac{x(x+1)}{3(x-1)}$ 10. $\frac{1}{z+5}$ 11. $\frac{y+1}{(y-3)(y+5)}$ 12. $\frac{x(4+x)}{4-x}$ 13. $\frac{a-7}{a^2}$ 14. $(2x-3y)(x+y)$ 15. 1 16. $\frac{x-y}{x-2y}$ 17. $\frac{7}{3}$ 18. 2 19. $\frac{1}{5}$ 20. $x+1$ 21. $\frac{x+5}{2x-3}$ 22. $\frac{1}{z+2}$ 23. $\frac{3}{x+2}$ 24. $\frac{2}{5}$ 25. $\frac{3}{2}$ 26. 3 27. $\frac{y(y+1)}{y-1}$ 28. $\frac{10}{a+5}$ 29. x 30. $\frac{5(z-2)}{z-1}$ 31. $\frac{1}{y+5}$ 32. $a+5$ 33. $\frac{y+5}{2}$ 34. $\frac{1}{6}$ 35. $\frac{1}{(x+2y)(x+y)}$ 36. $\frac{1}{x-y}$ 37. 1 38. $\frac{2a+y}{a-2y}$ 39. $0, -2$ 40. none 41. 0 42. no 43. yes 44. no 45. (a) $\frac{5x}{10}$ (b) $\frac{x^2}{2x}$ (c) $\frac{x^3}{2x^2}$ (d) $\frac{x(x+1)}{2(x+1)}$ (e) $\frac{x(2x-1)}{2(2x-1)}$ (f) $\frac{x(x^2-1)}{2(x^2-1)}$ 46. Burford is dividing out terms not factors. It is clear that the two fractions are not equal when x is replaced with a number. 47. 2 48. $\frac{1}{2}$ 49. $\frac{13}{30}$ 50. $\frac{9}{14}$ 51. $\frac{17}{3}$ 52. $\frac{3}{5}$

8.2 EXERCISES B

Multiply.

1. $\dfrac{2}{x^3} \cdot \dfrac{x^2}{6}$

2. $\dfrac{5a^2}{8} \cdot \dfrac{4a}{10a^3}$

3. $\dfrac{7y^3}{3} \cdot \dfrac{15}{5y}$

4. $\dfrac{3x^2}{x+5} \cdot \dfrac{(x+5)^2}{9x}$

5. $\dfrac{x^2-9}{x^2+3x} \cdot \dfrac{x}{x+9}$

6. $\dfrac{z^3}{z+3} \cdot \dfrac{z^2-9}{z^2-3z}$

7. $\dfrac{4x+4}{4x-4} \cdot \dfrac{x^2-2x+1}{x^2-1}$

8. $\dfrac{y^2-5y+6}{y^2-9} \cdot \dfrac{y+2}{y-2}$

9. $\dfrac{12x^4}{x^2+8x+16} \cdot \dfrac{x^2-16}{3x^2}$

10. $\dfrac{z^2+z-20}{z^2-7x+12} \cdot \dfrac{z-3}{z^2-25}$

11. $\dfrac{y^2+8y+7}{5y^2-125} \cdot \dfrac{5y-25}{(y+1)^2}$

12. $\dfrac{9-x^2}{3x-2} \cdot \dfrac{3x^2-2x}{9-6x+x^2}$

13. $\dfrac{a^2-25}{a^2-10a+25} \cdot \dfrac{a^2-8a+15}{a^3-3a^2}$

14. $\dfrac{2x^2+xy-10y^2}{x^2-4y^2} \cdot (x^2+4xy+4y^2)$

15. $\dfrac{x^2-2xy+y^2}{x+y} \cdot \dfrac{(x+y)^2}{x-y}$

16. $\dfrac{x^2-7xy+10y^2}{x^2-3xy-10y^2} \cdot \dfrac{x+2y}{x-2y}$

Find the reciprocal.

17. 6

18. $\dfrac{6}{y}$

19. $7a$

20. $\dfrac{x}{x+6}$

21. $\dfrac{3x-2}{x-2}$

22. $5a-1$

Divide.

23. $\dfrac{x+3}{x} \div \dfrac{4}{x}$

24. $\dfrac{a+5}{3a} \div \dfrac{a+5}{9a}$

25. $\dfrac{6(a+4)}{5} \div \dfrac{4(a+4)}{35}$

26. $\dfrac{8(x^2+x+1)}{6(x^2-2)} \div \dfrac{x^2+x+1}{3(x^2-2)}$

27. $\dfrac{y^2+4y+4}{8y^3} \div \dfrac{y^2-4}{2y^2}$

28. $\dfrac{a^2-16}{5a^3} \div \dfrac{3a^2-12a}{20a}$

29. $\dfrac{z^2+4z+4}{z^2-4} \div \dfrac{z+2}{3z-6}$

30. $\dfrac{x^2-49}{x^2+7x} \div (x-7)$

31. $\dfrac{y^2+y-20}{y^2-7y+12} \div \dfrac{y^2-25}{y-3}$

32. $\dfrac{a^2-10a+21}{a^2+2a-15} \div (5a-35)$

33. $\dfrac{x^2-16x+64}{3x^2-192} \div \dfrac{3x^2-30x+48}{x^2+6x-16}$

34. $\dfrac{5y^2+10y}{2y^3-8y} \div \dfrac{y^2+5y-6}{y^2-3y+2}$

35. $\dfrac{x^2 - y^2}{x - 3y} \div \dfrac{x - y}{x - 3y}$

36. $\dfrac{x^2 + 8xy + 15y^2}{x^2 - 25y^2} \div (x^2 + 2xy - 3y^2)$

37. $\dfrac{9x^2 - y^2}{9x^2 - 6xy + y^2} \div \dfrac{3x + y}{3x - y}$

38. $\dfrac{u^2 - 4v^2}{u^2 + uv - 2v^2} \cdot \dfrac{4u^2 - 4uv - 3v^2}{2u^2 - 3uv - 2v^2} \div \dfrac{3u + v}{u - v}$

FOR REVIEW

Find the values of the variable that must be excluded.

39. $\dfrac{x - 3}{x^2 - 8x}$

40. $\dfrac{a^2 + 1}{a^2 - 9}$

41. $\dfrac{y^2 + 3y}{6}$

Are the given fractions equivalent?

42. $\dfrac{6x^3 - 7}{7x - 7}, \dfrac{6x^3}{7x}$

43. $\dfrac{x^2 - 5x}{3x - 15}, \dfrac{x^2}{3x}$

44. $\dfrac{2x + y}{1}, \dfrac{1}{2x + y}$

45. Change $\frac{x^2}{5}$ to an equivalent fraction with the given denominator.

(a) 35

(b) $5x^3$

(c) $5(x - 2)$

(d) $5(x + 3)$

(e) $5(5x + 1)$

(f) $5(x^3 + 2x)$

46. What is wrong with the following? $\dfrac{3 - 2y}{3 + 5y} = \dfrac{3 - 2\cancel{y}}{3 + 5\cancel{y}} = -\dfrac{2}{5}$

The following exercises review material from Section 1.2 to help you prepare for the next two sections. Perform the indicated operations.

47. $\dfrac{6}{7} + \dfrac{8}{7}$

48. $\dfrac{7}{9} - \dfrac{4}{9}$

49. $\dfrac{5}{6} + \dfrac{5}{18}$

50. $\dfrac{7}{15} - \dfrac{17}{20}$

51. $\dfrac{7}{5} + 6$

52. $3 - \dfrac{15}{2}$

8.2 EXERCISES C

Perform the indicated operations.

1. $\dfrac{x^2 - 5x - 6}{x + 1} \div \dfrac{x^2 - 12x + 36}{x^2 - 1} \cdot \dfrac{x}{x - 1}$

$\left[\text{Answer: } \dfrac{x(x + 1)}{x - 6}\right]$

2. $\dfrac{y^2 - 3y - 4}{y^2 - 1} \div \dfrac{y + 3}{y^2 - 9} \cdot \dfrac{y - 1}{y^2 - 6y + 9}$

3. $\dfrac{a^2 - b^2}{2a + b} \div \left[\dfrac{a - b}{2a^2 + 3ab + b^2} \cdot \dfrac{a - b}{a + b}\right]$

$\left[\text{Answer: } \dfrac{(a + b)^3}{a - b}\right]$

4. $\dfrac{a + 2b}{a^2 - 4b^2} \div \left[\dfrac{a - b}{a^2 - 3ab + 2b^2} \cdot \dfrac{a + 2b}{a - 2b}\right]$

8.3 ADDITION AND SUBTRACTION OF LIKE FRACTIONS

① ADDING AND SUBTRACTING LIKE FRACTIONS

Fractions that have the same denominators are called **like fractions.** For example,

$$\frac{3}{5x} \quad \text{and} \quad \frac{x+2}{5x}$$

are like fractions, while

$$\frac{1}{2x+1} \quad \text{and} \quad \frac{1}{2x}$$

are called **unlike fractions.**

To Add or Subtract Like Fractions

1. Add or subtract the numerators to find the numerator of the answer.

2. Use the common denominator as the denominator of the answer.

3. Reduce the sum or difference to lowest terms.

Symbolically, if P, Q, and R are fractions, with $R \neq 0$, then

$$\frac{P}{R} + \frac{Q}{R} = \frac{P+Q}{R} \quad \text{and} \quad \frac{P}{R} - \frac{Q}{R} = \frac{P-Q}{R}.$$

EXAMPLE 1 OPERATIONS ON LIKE FRACTIONS

Perform the indicated operation.

(a) $\dfrac{3}{7} + \dfrac{2}{7} = \dfrac{3+2}{7}$ Add numerators and place sum over the common denominator 7

$$= \frac{5}{7}$$

(b) $\dfrac{2+x}{x} + \dfrac{x^2+1}{x} = \dfrac{(2+x)+(x^2+1)}{x}$ Add numerators and place sum over common denominator x

$$= \frac{x^2+x+3}{x}$$

(c) $\dfrac{1}{4} - \dfrac{3}{4} = \dfrac{1-3}{4} = \dfrac{-2}{4} = \dfrac{-2}{2 \cdot 2} = -\dfrac{1}{2}$

Always reduce the answer to lowest terms.

(d) $\dfrac{2x+1}{x-5} - \dfrac{x-3}{x-5} = \dfrac{(2x-1)-(x-3)}{x-5}$ Use parentheses in subtraction to avoid making a sign error

$$= \frac{2x+1-x+3}{x-5}$$ Be sure to change the sign to +3

$$= \frac{x+4}{x-5}$$

PRACTICE EXERCISE 1

Perform the indicated operation.

(a) $\dfrac{1}{11} + \dfrac{5}{11}$

(b) $\dfrac{y-1}{y} + \dfrac{3+y^2}{y}$

(c) $\dfrac{5}{12} - \dfrac{1}{12}$

(d) $\dfrac{x^2}{x+y} - \dfrac{y^2}{x+y}$

Answers: (a) $\frac{6}{11}$ (b) $\frac{y^2+y+2}{y}$ (c) $\frac{1}{3}$ (d) $x-y$

///////////// **CAUTION** ///////////////

When subtracting rational expressions, if you enclose the numerators in parentheses, you will eliminate sign errors when the distributive property is used. Notice how this happens in Example 1(d) when $-(x-3)$ becomes $-x+3$.

///////////////

❷ DENOMINATORS DIFFERING IN SIGN ONLY

Some fractions have denominators that are negatives of each other. These can be made into like fractions by changing the sign of both the numerator and the denominator of *one* of the fractions. To do this, multiply both numerator and denominator by -1. Since you are multiplying the fraction by $\frac{-1}{-1}=1$, the result is equivalent to the original fraction. This is illustrated in the next example.

EXAMPLE 2 DENOMINATORS DIFFERING IN SIGN	**PRACTICE EXERCISE 2**

Perform the indicated operations.

(a) $\dfrac{x}{5}+\dfrac{2x+1}{-5}=\dfrac{x}{5}+\dfrac{(-1)(2x+1)}{(-1)(-5)}$ Multiply numerator and denominator by -1

$=\dfrac{x}{5}+\dfrac{(-2x-1)}{5}$ The fractions are now like fractions

$=\dfrac{x+(-2x-1)}{5}$ Add numerators

$=\dfrac{-x-1}{5}$

(b) $\dfrac{2x}{x-3}-\dfrac{x+3}{3-x}=\dfrac{2x}{x-3}-\dfrac{(-1)(x+3)}{(-1)(3-x)}$ $x-3$ and $3-x$ are negatives of each other

$=\dfrac{2x}{x-3}-\dfrac{(-x-3)}{-3+x}$

$=\dfrac{2x}{x-3}-\dfrac{(-x-3)}{x-3}$

$=\dfrac{2x-(-x-3)}{x-3}$

$=\dfrac{2x+x+3}{x-3}$ Watch all signs

$=\dfrac{3x+3}{x-3}$

Perform the indicated operations.

(a) $\dfrac{a+3}{2}+\dfrac{3a-1}{-2}$

(b) $\dfrac{-3u}{u-v}-\dfrac{3u+1}{v-u}$

Answers: (a) $2-a$ (b) $\frac{1}{u-v}$

8.3 EXERCISES A

Perform the indicated operations.

1. $\dfrac{3}{5}+\dfrac{7}{5}$

2. $\dfrac{x}{6}-\dfrac{3}{6}$

3. $\dfrac{2}{x}+\dfrac{7}{x}$

4. $\dfrac{2x + 1}{x - 7} + \dfrac{-1 - 2x}{x - 7}$

5 $\dfrac{2y}{y + 2} - \dfrac{y + 1}{y + 2}$

6. $\dfrac{z - 1}{z + 1} - \dfrac{2z - 1}{z + 1}$

7. $\dfrac{a + 1}{a + 1} + \dfrac{a^2 + 1}{a + 1}$

8. $\dfrac{7x^2}{3x^2 - 1} - \dfrac{4x^2}{3x^2 - 1}$

9. $\dfrac{a}{(a + 1)^2} + \dfrac{1}{(a + 1)^2}$

10 $\dfrac{z}{2} + \dfrac{3z - 1}{-2}$

11. $\dfrac{4}{a} + \dfrac{2a - 3}{-a}$

12. $\dfrac{2x}{-3} + \dfrac{3x + 7}{3}$

13. $\dfrac{7}{2z} - \dfrac{3z + 1}{-2z}$

14 $\dfrac{2a}{a - 1} + \dfrac{3a}{1 - a}$

15. $\dfrac{3x}{1 - x} - \dfrac{x + 1}{x - 1}$

16. $\dfrac{z}{(z - 1)(z + 1)} - \dfrac{1}{(z - 1)(z + 1)}$

17. $\dfrac{3}{a^2 + 2a + 1} - \dfrac{2 - a}{a^2 + 2a + 1}$

18. $\dfrac{x}{6x - 12} - \dfrac{4}{3(4 - 2x)}$

19. $\dfrac{2z - 3}{z^2 + 3z - 4} - \dfrac{z - 7}{z^2 + 3z - 4}$

20 $\dfrac{2x}{x^2 + x - 6} + \dfrac{x - 3}{6 - x - x^2}$

21. $\dfrac{x^2 - 15}{x^2 + 4x + 3} - \dfrac{2x}{x^2 + 4x + 3}$

22. $\dfrac{y}{x} + \dfrac{2y - 1}{x}$

23. $\dfrac{3a}{a+b} - \dfrac{2a-b}{a+b}$

24 $\dfrac{x+y}{x-y} - \dfrac{x+y}{y-x}$

25 $\dfrac{5x}{x+1} + \dfrac{2x-1}{x+1} - \dfrac{3x}{x+1}$

26. $\dfrac{x+2y}{2x-y} - \dfrac{3x-2y}{2x-y} - \dfrac{4y}{2x-y}$

FOR REVIEW

Perform the indicated operations.

27. $\dfrac{3x-6}{2x} \cdot \dfrac{24x^2}{3(x^2-4x+4)}$

28. $\dfrac{a^2-9}{4a+12} \div \dfrac{a-3}{6}$

29. $\dfrac{y^2+10y+21}{y^2-2y-15} \div (y^2+2y-35)$

30. $\dfrac{x^2-5xy+6y^2}{x^2+3xy-10y^2} \cdot \dfrac{x^2+5xy}{x^2-3xy}$

ANSWERS: 1. 2 2. $\frac{x-3}{6}$ 3. $\frac{9}{x}$ 4. 0 5. $\frac{y-1}{y+2}$ 6. $\frac{-z}{z+1}$ 7. $\frac{a^2+a+2}{a+1}$ 8. $\frac{3x^2}{3x^2-1}$ 9. $\frac{1}{a+1}$ 10. $\frac{-2z+1}{2}$ 11. $\frac{7-2a}{a}$ 12. $\frac{x+7}{3}$ 13. $\frac{3z+8}{2z}$ 14. $\frac{-a}{a-1}$ 15. $\frac{4x+1}{1-x}$ 16. $\frac{1}{z+1}$ 17. $\frac{1}{a+1}$ 18. $\frac{x+4}{6(x-2)}$ 19. $\frac{1}{z-1}$ 20. $\frac{1}{x-2}$ 21. $\frac{x-5}{x+1}$ 22. $\frac{3y-1}{x}$ 23. 1 24. $\frac{2(x+y)}{x-y}$ 25. $\frac{4x-1}{x+1}$ 26. $\frac{-2x}{2x-y}$ 27. $\frac{12x}{x-2}$ 28. $\frac{3}{2}$ 29. $\frac{1}{(y-5)^2}$ 30. 1

8.3 EXERCISES B

Perform the indicated operations.

1. $\dfrac{x}{3} + \dfrac{7}{3}$

2. $\dfrac{4}{5} - \dfrac{9}{5}$

3. $\dfrac{5}{3x} - \dfrac{8}{3x}$

4. $\dfrac{6}{x+2} + \dfrac{10}{x+2}$

5. $\dfrac{3x-2}{x-4} + \dfrac{2-3x}{x-4}$

6. $\dfrac{8y}{y-5} - \dfrac{4y+3}{y-5}$

7. $\dfrac{a+2}{a+2} + \dfrac{3a-2}{a+2}$

8. $\dfrac{6x^2}{5x^2+2} - \dfrac{9x^2}{5x^2+2}$

9. $\dfrac{a^2}{a^2+1} - \dfrac{-1}{a^2+1}$

10. $\dfrac{z}{3} + \dfrac{2z-3}{-3}$

11. $\dfrac{9}{a} + \dfrac{5a-2}{-a}$

12. $\dfrac{3x}{-5} + \dfrac{4x-7}{5}$

13. $\dfrac{10}{5z} - \dfrac{5z+4}{-5z}$

14. $\dfrac{5a}{a-4} + \dfrac{7a}{4-a}$

15. $\dfrac{4x}{7-x} - \dfrac{2x+5}{x-7}$

16. $\dfrac{2z}{(2z+1)(2z-1)} - \dfrac{1}{(2z+1)(2z-1)}$

17. $\dfrac{5}{a^2+4a+4} - \dfrac{3-a}{a^2+4a+4}$

18. $\dfrac{2x}{5(x-2)} - \dfrac{3}{10-5x}$

19. $\dfrac{3z+2}{z^2+4z-12} - \dfrac{2z-4}{z^2+4z-12}$

20. $\dfrac{2x}{x^2-2x-15} + \dfrac{x+5}{15+2x-x^2}$

21. $\dfrac{x^2-8}{x^2-4x+3} - \dfrac{-7x}{x^2-4x+3}$

22. $\dfrac{x+1}{y} - \dfrac{x-2}{y}$

23. $\dfrac{a-2b}{a-b} - \dfrac{a+2b}{a-b}$

24. $\dfrac{x-y}{2x-y} - \dfrac{x-y}{y-2x}$

25. $\dfrac{3x-1}{x+2} + \dfrac{4x-2}{x+2} - \dfrac{5x-3}{x+2}$

26. $\dfrac{x+3y}{x-2y} - \dfrac{4x+y}{x-2y} - \dfrac{2y}{x-2y}$

FOR REVIEW

Perform the indicated operations.

27. $\dfrac{5x+5}{x-2} \cdot \dfrac{x^2-4x+4}{x^2-1}$

28. $\dfrac{a^2-16}{5a-20} \div \dfrac{a+4}{10}$

29. $\dfrac{y^2-11y+28}{y^2-2y-35} \div (y^2+y-20)$

30. $\dfrac{2x^2+9xy+4y^2}{x^2+xy-12y^2} \cdot \dfrac{3xy-x^2}{2x^2+xy}$

8.3 EXERCISES C

Perform the indicated operations.

1. $\dfrac{2x}{x-5} - \dfrac{2}{5-x} + \dfrac{3x}{5-x}$ $\left[\text{Answer: } \dfrac{2-x}{x-5}\right]$

2. $\dfrac{5y}{x^2-y^2} + \dfrac{5y}{y^2-x^2} - \dfrac{x-y}{y^2-x^2}$

3. $\dfrac{a+5}{(a-2)(a-3)} - \dfrac{2a-1}{(2-a)(3-a)} + \dfrac{3a-2}{(a-2)(3-a)}$

4. $\dfrac{a+b}{2a-b} - \dfrac{2b-a}{b-2a} - \dfrac{a-b}{2a-b} + \dfrac{3b-a}{b-2a}$
$\left[\text{Answer: } \dfrac{b}{2a-b}\right]$

8.4 ADDITION AND SUBTRACTION OF UNLIKE FRACTIONS

STUDENT GUIDEPOSTS

❶ Least Common Denominator (LCD) **❷** Adding and Subtracting Fractions

Before we can add or subtract unlike fractions, we need to convert them to equivalent like fractions. Recall from Chapter 1 that to add $\frac{2}{15}$ and $\frac{1}{6}$ we must first find a common denominator. One such denominator is $15 \cdot 6 = 90$. Since

$$\frac{2}{15} = \frac{2 \cdot 6}{15 \cdot 6} = \frac{12}{90} \quad \text{and} \quad \frac{1}{6} = \frac{1 \cdot 15}{6 \cdot 15} = \frac{15}{90}$$

we could add as follows.

$$\frac{2}{15} + \frac{1}{6} = \frac{12}{90} + \frac{15}{90}$$

$$= \frac{12 + 15}{90}$$

$$= \frac{27}{90} = \frac{\cancel{9} \cdot 3}{\cancel{9} \cdot 10} = \frac{3}{10}$$

❶ LEAST COMMON DENOMINATOR (LCD)

Often it is wiser to try to find a common denominator smaller than the product of denominators. If we use the *least common denominator*, we shorten the computation needed to reduce the final sum or difference to lowest terms. In the example above, we could have used the least common denominator, 30.

$$\frac{2}{15} + \frac{1}{6} = \frac{2 \cdot 2}{15 \cdot 2} + \frac{1 \cdot 5}{6 \cdot 5} = \frac{4}{30} + \frac{5}{30} = \frac{9}{30} = \frac{3}{10}$$

In the same way, with algebraic fractions we will be concerned with finding the **least common denominator (LCD)** of all fractions.

To Find the LCD of Two or More Fractions

1. Factor the denominators completely.
2. Put each factor in the LCD as many times as it appears in the denominator where it is found the greatest number of times.

EXAMPLE 1 FINDING THE LCD

Find the LCD of the given fractions.

(a) $\dfrac{7}{90}$ and $\dfrac{5}{24}$

$$\overset{\text{one 2 two 3s one 5}}{90 = 2 \cdot 3 \cdot 3 \cdot 5} \quad \text{and} \quad \overset{\text{three 2s one 3}}{24 = 2 \cdot 2 \cdot 2 \cdot 3.}$$

The LCD must consist of three 2s, two 3s, and one 5.

$$\text{LCD} = 2 \cdot 2 \cdot 2 \cdot 3 \cdot 3 \cdot 5 = 360$$

(b) $\dfrac{3}{x}$ and $\dfrac{2}{x+1}$

Since x and $x + 1$ are already completely factored, and since there are no common factors in the two denominators, the LCD $= x(x + 1)$.

(c) $\dfrac{3}{2y}$ and $\dfrac{y+1}{y^2}$

The denominators are essentially factored already. We see that the LCD has one 2 and two ys. Thus, the LCD is $2y^2$.

(d) $\dfrac{2x-3}{x^2-25}$ and $\dfrac{7x}{2x-10}$

$$x^2 - 25 = (x-5)(x+5) \quad \text{and} \quad 2x - 10 = 2(x-5)$$

We need one $(x - 5)$, one $(x + 5)$, and one 2 for the LCD. That is, the LCD $= 2(x-5)(x+5)$.

(e) $\dfrac{y}{y^2-4}$ and $\dfrac{4}{y^2-4y+4}$

$$y^2 - 4 = (y-2)(y+2) \quad \text{and} \quad y^2 - 4y + 4 = (y-2)(y-2)$$

The LCD must consist of two $(y - 2)$s and one $(y + 2)$. Thus, the LCD $= (y-2)^2(y+2)$.

PRACTICE EXERCISE 1

Find the LCD of the given fractions.

(a) $\dfrac{5}{12}$ and $\dfrac{11}{126}$

(b) $\dfrac{5}{a+1}$ and $\dfrac{1}{a-1}$

(c) $\dfrac{7}{x^2}$ and $\dfrac{x+11}{5x}$

(d) $\dfrac{1-y}{7y+21}$ and $\dfrac{4y}{y^2-9}$

(e) $\dfrac{w}{w^2+2w+1}$ and $\dfrac{3}{w^2-5w-6}$

Answers: (a) 252
(b) $(a+1)(a-1)$ (c) $5x^2$
(d) $7(y+3)(y-3)$
(e) $(w+1)^2(w-6)$

② ADDING AND SUBTRACTING FRACTIONS

To add or subtract unlike fractions, we first find their LCD and then transform each fraction into an equivalent fraction having the LCD as its denominator. We illustrate this method in the following numerical example.

$$\frac{2}{15} + \frac{3}{35}$$

$$= \frac{2}{3 \cdot 5} + \frac{3}{5 \cdot 7} \qquad \text{Since } 15 = 3 \cdot 5 \text{ and } 35 = 5 \cdot 7, \text{ the LCD is } 3 \cdot 5 \cdot 7 = 105$$

$$= \frac{2 \cdot 7}{3 \cdot 5 \cdot 7} + \frac{3 \cdot 3}{5 \cdot 7 \cdot 3} \qquad \text{Multiply numerator and denominator of } \frac{2}{15} \text{ by 7 and of } \frac{3}{35} \text{ by 3 so the denominators are the same}$$

$$= \frac{2 \cdot 7 + 3 \cdot 3}{3 \cdot 5 \cdot 7} \qquad \text{Since denominators are now the same, add numerators and place the sum over the LCD}$$

$$= \frac{14 + 9}{3 \cdot 5 \cdot 7} \qquad \text{Leave the denominator in factored form and simplify the numerator}$$

$$= \frac{23}{3 \cdot 5 \cdot 7} = \frac{23}{105} \qquad \text{Since 23 has no factor of 3, 5, or 7, the resulting fraction is in lowest terms}$$

We use exactly the same procedure for rational expressions except we now have polynomials for numerators and denominators.

To Add or Subtract Rational Expressions

1. Rewrite the indicated sum or difference with all denominators expressed in factored form.
2. Find the LCD.
3. Multiply the numerator and denominator of each fraction by all factors present in the LCD but missing in the denominator of the particular fraction.
4. Write out the sum or difference of all numerators, using parentheses if needed, and place the result over the LCD.
5. Simplify and factor (if possible) the resulting numerator and divide any factors common to the LCD.

EXAMPLE 2 ADDING FRACTIONS

Add.

$$\frac{5}{6x} + \frac{3}{10x^2}$$

$$= \frac{5}{2 \cdot 3 \cdot x} + \frac{3}{2 \cdot 5 \cdot x \cdot x} \qquad \text{Factor denominators; LCD} = 2 \cdot 3 \cdot 5 \cdot x \cdot x$$

$$= \frac{5 \cdot 5 \cdot x}{2 \cdot 3 \cdot x \cdot 5 \cdot x} + \frac{3 \cdot 3}{2 \cdot 5 \cdot x \cdot x \cdot 3} \qquad \text{Supply missing factors}$$

$$= \frac{5 \cdot 5 \cdot x + 3 \cdot 3}{2 \cdot 3 \cdot 5 \cdot x \cdot x} \qquad \text{Add numerators over LCD}$$

$$= \frac{25x + 9}{30x^2} \qquad \text{No common factors exist, so this is in lowest terms}$$

PRACTICE EXERCISE 2

Add.

$$\frac{1}{12y^3} + \frac{7}{4y}$$

Answer: $\frac{1 + 21y^2}{12y^3}$

EXAMPLE 3   SUBTRACTING FRACTIONS

Subtract.

$$\frac{2}{y+2} - \frac{2}{y+3}$$ Denominators already factored; LCD $= (y+2)(y+3)$

$$= \frac{2(y+3)}{(y+2)(y+3)} - \frac{2(y+2)}{(y+2)(y+3)}$$ Supply missing factors

$$= \frac{2(y+3) - 2(y+2)}{(y+2)(y+3)}$$ Subtract numerators over LCD

$$= \frac{2y+6-2y-4}{(y+2)(y+3)}$$ Watch signs when using the distributive law

$$= \frac{2}{(y+2)(y+3)}$$ No common factors exist, so this is in lowest terms

PRACTICE EXERCISE 3

Subtract.

$$\frac{3}{a-1} - \frac{1}{a+3}$$

Answer: $\frac{2(a+5)}{(a-1)(a+3)}$

EXAMPLE 4   ADDING FRACTIONS

Add.

(a) $\dfrac{5}{x^2-4} + \dfrac{7}{x^2+2x}$

$$= \frac{5}{(x-2)(x+2)} + \frac{7}{x(x+2)}$$ Factor denominators; LCD $= x(x-2)(x+2)$

$$= \frac{5(x)}{(x-2)(x+2)(x)} + \frac{7(x-2)}{x(x+2)(x-2)}$$ Supply missing factors

$$= \frac{5x + 7(x-2)}{x(x+2)(x-2)}$$ Add numerators

$$= \frac{5x + 7x - 14}{x(x+2)(x-2)}$$ Clear parentheses using the distributive law

$$= \frac{12x - 14}{x(x+2)(x-2)}$$

$$= \frac{2(6x - 7)}{x(x+2)(x-2)}$$ No common factors exist, so this is in lowest terms

(b) $\dfrac{5}{x^2+x-6} + \dfrac{3x}{x^2-4x+4}$

$$= \frac{5}{(x-2)(x+3)} + \frac{3x}{(x-2)(x-2)}$$ LCD $= (x+3)(x-2)^2$

$$= \frac{5(x-2)}{(x-2)(x+3)(x-2)} + \frac{3x(x+3)}{(x-2)(x-2)(x+3)}$$

$$= \frac{5(x-2) + 3x(x+3)}{(x+3)(x-2)^2}$$

$$= \frac{5x - 10 + 3x^2 + 9x}{(x+3)(x-2)^2}$$

$$= \frac{3x^2 + 14x - 10}{(x+3)(x-2)^2}$$ This is in lowest terms

PRACTICE EXERCISE 4

Add.

(a) $\dfrac{3}{y^2-5y} + \dfrac{-2}{y^2-3y-10}$

(b) $\dfrac{y}{y^2+2y-15} + \dfrac{1-y}{y^2-6y+9}$

Answers: (a) $\frac{y+6}{y(y-5)(y+2)}$
(b) $\frac{5-7y}{(y-3)^2(y+5)}$

We use the same procedures even if there are two variables and more than two terms.

| **EXAMPLE 5   OPERATIONS WITH TWO VARIABLES** | **PRACTICE EXERCISE 5** |

Perform the indicated operations.

$$\frac{2x}{x+y} - \frac{y}{x-y} - \frac{-4xy}{x^2-y^2}$$

$$= \frac{2x}{x+y} - \frac{y}{x-y} - \frac{-4xy}{(x+y)(x-y)} \qquad \text{LCD} = (x+y)(x-y)$$

$$= \frac{2x(x-y)}{(x+y)(x-y)} - \frac{y(x+y)}{(x-y)(x+y)} - \frac{-4xy}{(x+y)(x-y)} \qquad \begin{array}{c}\text{Supply missing} \\ \text{factors}\end{array}$$

$$= \frac{2x(x-y) - y(x+y) + 4xy}{(x+y)(x-y)}$$

$$= \frac{2x^2 - 2xy - xy - y^2 + 4xy}{(x+y)(x-y)}$$

$$= \frac{2x^2 + xy - y^2}{(x+y)(x-y)} \qquad \text{This will simplify}$$

$$= \frac{(2x-y)(x+y)}{(x+y)(x-y)} \qquad \text{Divide out } (x+y)$$

$$= \frac{2x-y}{x-y}$$

Perform the indicated operations.

$$\frac{2a}{a+2b} - \frac{b}{a-2b} + \frac{2b^2+ab}{a^2-4b^2}$$

Answer: $\frac{2a}{a+2b}$

8.4   EXERCISES A

Find the LCD of the given fractions.

1. $\frac{1}{20}$ and $\frac{7}{30}$

2. $\frac{2}{39}$ and $\frac{4}{35}$

3. $\frac{y+1}{y^2}$ and $\frac{1}{3y}$

4. $\frac{a+1}{9a^2}$ and $\frac{5}{12a^3}$

5. $\frac{2}{3x}$ and $\frac{x+2}{3x+3}$

6. $\frac{7}{y+1}$ and $\frac{y}{y-1}$

7. $\frac{3}{2z+4}$ and $\frac{-5}{3z+6}$

8. $\frac{2a+1}{a^2-25}$ and $\frac{3}{a+5}$

9. $\frac{3x+2}{3x+6}$ and $\frac{x}{x^2-4}$

10. $\frac{1}{x^2+2x+1}$ and $\frac{2}{x^2-1}$

11. $\frac{3}{z^2-9}$ and $\frac{2}{z^2+z-6}$

12. $\frac{2}{a^2-9}$ and $\frac{2}{a^2+2a-3}$

13. $\dfrac{1}{x^2 - y^2}$ and $\dfrac{3xy}{x - y}$

14. $\dfrac{5x}{x + y}$, $\dfrac{7y}{2x + 2y}$, and $\dfrac{2xy}{3x + 3y}$

15. $\dfrac{16}{5x}$, $\dfrac{5}{3y}$, and $\dfrac{xy}{2x - y}$

Perform the indicated operations.

16. $\dfrac{4}{21} + \dfrac{5}{14}$

17. $\dfrac{5}{12} - \dfrac{11}{30}$

18. $\dfrac{2}{9x} + \dfrac{4}{15x^2}$

19. $\dfrac{a - 5}{a} - \dfrac{3a - 1}{4a}$

20. $\dfrac{5}{y + 5} - \dfrac{3}{y - 5}$

21. $\dfrac{3a}{a + 2} + \dfrac{a}{a - 2}$

22. $\dfrac{2}{3x + 21} - \dfrac{3}{5x + 35}$

23. $\dfrac{4y}{y^2 - 36} - \dfrac{4}{y + 6}$

24. $\dfrac{8}{7 - a} - \dfrac{8a}{49 - a^2}$

25. $\dfrac{3}{2x^2 - 2x} - \dfrac{5}{2x - 2}$

26. $\dfrac{-8}{y^2 - 4} - \dfrac{4}{y + 2}$

27. $\dfrac{8}{a^2 - 16} + \dfrac{1}{a + 4}$

28. $\dfrac{3z + 2}{3z + 6} + \dfrac{z - 2}{z^2 - 4}$

29. $\dfrac{2}{y^2 - 1} + \dfrac{1}{y^2 + 2y - 3}$

30. $\dfrac{2}{z^2 - 9} - \dfrac{2}{z^2 + 2z - 3}$

31. $\dfrac{3}{a^2 - a - 12} - \dfrac{2}{a^2 - 9}$

32. $\dfrac{5}{x^2 - 4x + 3} + \dfrac{7}{x^2 + x - 2}$

33. $\dfrac{y + 1}{y + 4} + \dfrac{4 - y^2}{y^2 - 16}$

34. $\dfrac{2}{a - 1} + \dfrac{a}{a^2 - 1}$

35. $\dfrac{3}{x + 1} + \dfrac{5}{x - 1} - \dfrac{10}{x^2 - 1}$

36. $\dfrac{5x}{x + y} + \dfrac{6y - 2x}{x - y}$

37 $\dfrac{2y}{y+5} - \dfrac{3y}{y-2} - \dfrac{2y^2}{y^2+3y-10}$

38. $\dfrac{1}{x(x-y)} - \dfrac{2}{y(x+y)}$

FOR REVIEW

Perform the indicated operations.

39. $\dfrac{3}{z^2+2z+1} - \dfrac{2-z}{z^2+2z+1}$

40. $\dfrac{2}{2x-1} - \dfrac{3}{1-2x}$

41. $\dfrac{2a-1}{a-5} - \dfrac{6-a}{5-a}$

42. $\dfrac{3-y}{y-7} + \dfrac{2y-5}{7-y}$

43. $\dfrac{2x}{3x-2} - \dfrac{5x}{3x-2} - \dfrac{-2}{3x-2}$

44. $\dfrac{x^2}{x-y} + \dfrac{x^2}{x-y} - \dfrac{y^2}{y-x}$

The following exercises review material that we covered in Chapters 3 and 7. They will help you prepare for the next section. Solve.

45. $2 - 3(x+1) = 5$

46. $y - 4(2y+1) = y + 4$

47. $x^2 - x - 6 = 0$

48. $x^2 + 5x = 0$

49. Eric Wade receives 11% simple interest on his savings. How much interest will he earn at the end of one year if he deposits $875?

50. Two families leave their homes at the same time planning to meet for a reunion at a point between them. If one travels at a rate of 55 mph, the other travels at 50 mph, and they live 273 miles apart, how long will it take for them to meet?

ANSWERS: 1. 60 2. 1365 3. $3y^2$ 4. $36a^3$ 5. $3x(x+1)$ 6. $(y+1)(y-1)$ 7. $6(z+2)$ 8. $(a+5)(a-5)$
9. $3(x+2)(x-2)$ 10. $(x-1)(x+1)^2$ 11. $(z-3)(z+3)(z-2)$ 12. $(a-3)(a+3)(a-1)$ 13. $(x+y)(x-y)$
14. $6(x+y)$ 15. $15xy(2x-y)$ 16. $\frac{23}{42}$ 17. $\frac{1}{20}$ 18. $\frac{2(5x+6)}{45x^2}$ 19. $\frac{a-19}{4a}$ 20. $\frac{2(y-20)}{(y+5)(y-5)}$ 21. $\frac{4a(a-1)}{(a+2)(a-2)}$ 22. $\frac{1}{15(x+7)}$
23. $\frac{24}{(y-6)(y+6)}$ 24. $\frac{56}{(7-a)(7+a)}$ 25. $\frac{3-5x}{2x(x-1)}$ 26. $\frac{-4y}{(y-2)(y+2)}$ 27. $\frac{1}{a-4}$ 28. $\frac{3z+5}{3(z+2)}$ 29. $\frac{3y+7}{(y-1)(y+1)(y+3)}$
30. $\frac{4}{(z-3)(z+3)(z-1)}$ 31. $\frac{a-1}{(a-4)(a+3)(a-3)}$ 32. $\frac{12x-11}{(x-3)(x-1)(x+2)}$ 33. $\frac{-3y}{(y-4)(y+4)}$ 34. $\frac{3a+2}{(a-1)(a+1)}$ 35. $\frac{8}{x+1}$ 36. $\frac{3x^2-xy+6y^2}{(x+y)(x-y)}$
37. $\frac{-y(3y+19)}{(y+5)(y-2)}$ 38. $\frac{y^2+3xy-2x^2}{xy(x+y)(x-y)}$ 39. $\frac{1}{z+1}$ 40. $\frac{5}{2x-1}$ 41. $\frac{a+5}{a-5}$ 42. $\frac{8-3y}{y-7}$ 43. -1 44. $\frac{2x^2+y^2}{x-y}$ 45. -2
46. -1 47. 3 or -2 48. 0 or -5 49. \$96.25 50. 2.6 hr

8.4 EXERCISES B

Find the LCD of the given fractions.

1. $\dfrac{1}{40}$ and $\dfrac{1}{30}$

2. $\dfrac{4}{39}$ and $\dfrac{2}{15}$

3. $\dfrac{3}{x}$ and $\dfrac{8}{7x}$

4. $\dfrac{a+1}{6a^2}$ and $\dfrac{7}{9a^3}$

5. $\dfrac{3}{5x}$ and $\dfrac{x+5}{5x+5}$

6. $\dfrac{3}{x+1}$ and $\dfrac{x}{x-1}$

7. $\dfrac{-8}{3z-6}$ and $\dfrac{2}{7z-14}$

8. $\dfrac{2z+3}{z^2-9}$ and $\dfrac{12}{z+3}$

9. $\dfrac{4x-1}{2x+6}$ and $\dfrac{5x+1}{x^2-9}$

10. $\dfrac{4}{y^2-1}$ and $\dfrac{y}{y^2-2y+1}$

11. $\dfrac{a}{a^2-4}$ and $\dfrac{2a+5}{a^2-a-6}$

12. $\dfrac{6y^2}{y^2-25}$ and $\dfrac{3y}{y^2+y-20}$

13. $\dfrac{1}{4x^2-y^2}$ and $\dfrac{x^2}{2x+y}$

14. $\dfrac{2}{4x-4y}$, $\dfrac{3x}{2x-2y}$, and $\dfrac{3xy}{5x-5y}$

15. $\dfrac{21y}{8x}$, $\dfrac{15x}{7y}$, and $\dfrac{x^2y^2}{x+3y}$

Perform the indicated operations.

16. $\dfrac{2}{15} + \dfrac{3}{25}$

17. $\dfrac{11}{20} - \dfrac{8}{35}$

18. $\dfrac{3}{4y} + \dfrac{5}{8y^2}$

19. $\dfrac{a+8}{2a} - \dfrac{5a-3}{3a}$

20. $\dfrac{6}{x-6} - \dfrac{4}{x+6}$

21. $\dfrac{5y}{y+4} + \dfrac{2y}{y-4}$

22. $\dfrac{7}{4x+8} - \dfrac{5}{7x+14}$

23. $\dfrac{4}{z+2} + \dfrac{8}{z^2-4}$

24. $\dfrac{14}{49-a^2} - \dfrac{7}{7-a}$

25. $\dfrac{5}{3x^2-3x} - \dfrac{7}{3x-3}$

26. $\dfrac{8}{y^2-16} + \dfrac{1}{y+4}$

27. $\dfrac{10}{x^2-25} - \dfrac{3}{x-5}$

28. $\dfrac{2z - 3}{2z - 4} + \dfrac{z + 2}{z^2 - 4}$

29. $\dfrac{2}{a^2 - a - 2} + \dfrac{a}{a^2 - 1}$

30. $\dfrac{y}{y^2 - 1} - \dfrac{y + 2}{y^2 + y - 2}$

31. $\dfrac{4}{x^2 - 9} - \dfrac{4}{x^2 - 2x - 3}$

32. $\dfrac{8}{x^2 + 4x + 3} + \dfrac{3}{x^2 - x - 2}$

33. $\dfrac{y - 3}{y + 5} + \dfrac{9 - y^2}{y^2 - 25}$

34. $\dfrac{4}{a - 2} + \dfrac{2a}{a^2 - 4}$

35. $\dfrac{2}{x + 5} + \dfrac{3}{x - 5} - \dfrac{7}{x^2 - 25}$

36. $\dfrac{2y}{x - y} + \dfrac{3x - y}{x + y}$

37. $\dfrac{3y}{y + 4} - \dfrac{2y}{y - 3} - \dfrac{y^2}{y^2 + y - 12}$

38. $\dfrac{3}{x(x + y)} - \dfrac{2}{y(x - y)}$

FOR REVIEW

Perform the indicated operations.

39. $\dfrac{z}{z^2 + 2z + 1} + \dfrac{2 + z}{z^2 + 2z + 1}$

40. $\dfrac{3x}{x - 1} - \dfrac{-2x}{1 - x}$

41. $\dfrac{y + 2}{y - 2} - \dfrac{y^2 - 2}{2 - y}$

42. $\dfrac{2x^2}{2x^2 - 1} + \dfrac{x^2}{1 - 2x^2}$

43. $\dfrac{6x}{2x - 3} - \dfrac{8x}{2x - 3} - \dfrac{5x}{2x - 3}$

44. $\dfrac{2x^2}{2x - y} + \dfrac{3y^2}{2x - y} - \dfrac{x^2 + y^2}{y - 2x}$

The following exercises review material that we covered in Chapters 3 and 7. They will help you prepare for the next section. Solve.

45. $3 - 4(-x + 2) = 7$

46. $2y - 3(3y - 2) = -y - 6$

47. $x^2 - 5x + 6 = 0$

48. $2x^2 - 6x = x^2$

49. If \$16.80 is charged as tax on the sale of a washing machine priced at \$420, what is the tax rate?

50. Clem hiked to a waterfall at a rate of 3 mph and returned at a rate of 4 mph. If the total time of the trip was 7 hours, how long did it take to reach the waterfall? What was the total distance hiked?

8.4 EXERCISES C

Perform the indicated operations.

1. $\dfrac{2}{x^2 - y^2} - \dfrac{3}{x^2 + 2xy + y^2} - \dfrac{3}{x^2 - 2xy + y^2}$
$\left[\text{Answer:} \quad \dfrac{-4(x^2 + 2y^2)}{(x + y)^2(x - y)^2} \right]$

2. $\dfrac{3}{x^2 - 7xy + 10y^2} + \dfrac{2}{x^2 + 2xy - 8y^2} -$
$\dfrac{2}{x^2 - xy - 20y^2}$

3. $\dfrac{1}{x + 5y} \div \dfrac{1}{x - 5y} - \dfrac{1}{x + 5y} \cdot \dfrac{-6xy}{x - y}$

4. $\dfrac{2x - y}{x^2 - y^2} \div \dfrac{2x - y}{x + y} - \dfrac{x + y}{(x - y)^2} \div \dfrac{x + y}{2y}$
$\left[\text{Answer:} \quad \dfrac{x - 3y}{(x - y)^2} \right]$

8.5 SOLVING FRACTIONAL EQUATIONS

STUDENT GUIDEPOSTS

1 Fractional Equations

3 Work Problems

2 Applications of Fractional Equations

1 FRACTIONAL EQUATIONS

An equation that has one or more algebraic fractions or rational expressions is called a **fractional equation.** Once all the fractions have been eliminated (called **clearing the fractions**) by multiplying both sides by the LCD of the denominators in the equation, the resulting equation can often be solved by the techniques of Chapter 3 or Section 7.5. Every solution to this new equation must be checked in the original equation to be sure that it does not make one of the denominators equal to zero. If this occurs it must be discarded.

To Solve a Fractional Equation
Find the LCD of all fractions in the equation.
Multiply both sides of the equation by the LCD, making sure that *all* terms are multiplied. Once simplified, the resulting equation will be free of fractions.
Solve this equation.
Check your solutions in the original equation to be certain your answers do not make one of the denominators zero.

EXAMPLE 1 SOLVING A FRACTIONAL EQUATION

Solve $\dfrac{2}{5} + \dfrac{x}{15} = \dfrac{1}{3}$.

$15\left(\dfrac{2}{5} + \dfrac{x}{15}\right) = 15 \cdot \dfrac{1}{3}$ Multiply both sides by the LCD, 15

$15 \cdot \dfrac{2}{5} + 15 \cdot \dfrac{x}{15} = 5$ Clear the parentheses using the distributive law

$\qquad\qquad 6 + x = 5$

$\qquad 6 - 6 + x = 5 - 6$ Subtract 6 from both sides

$\qquad\qquad\quad x = -1$

Check: $\dfrac{2}{5} + \dfrac{-1}{15} \stackrel{?}{=} \dfrac{1}{3}$

$\qquad\quad \dfrac{2 \cdot 3}{5 \cdot 3} + \dfrac{-1}{15} \stackrel{?}{=} \dfrac{1 \cdot 5}{3 \cdot 5}$

$\qquad\qquad \dfrac{6}{15} + \dfrac{-1}{15} \stackrel{?}{=} \dfrac{5}{15}$

$\qquad\qquad\qquad \dfrac{5}{15} = \dfrac{5}{15}$

The solution is -1.

PRACTICE EXERCISE 1

Solve $\dfrac{x}{24} + \dfrac{3}{8} = \dfrac{2}{3}$.

Answer: 7

Notice that we are using the LCD in this section to clear the fractions and not to write them as equivalent fractions with the same denominator.

EXAMPLE 2 SOLVING A FRACTIONAL EQUATION

Solve $\dfrac{1}{y} = \dfrac{1}{8 - y}$.

$y(8 - y)\left(\dfrac{1}{y}\right) = y(8 - y)\left(\dfrac{1}{8 - y}\right)$ Multiply both sides by the LCD, $y(8 - y)$

$\qquad\qquad 8 - y = y$

$\qquad\qquad\quad 8 = 2y$ Add y to both sides

$\qquad\qquad\quad 4 = y$

Check: $\dfrac{1}{4} \stackrel{?}{=} \dfrac{1}{8 - 4}$

$\qquad\quad \dfrac{1}{4} = \dfrac{1}{4}$

The solution is 4.

PRACTICE EXERCISE 2

Solve $\dfrac{2}{x + 3} = \dfrac{5}{x - 1}$.

Answer: $-\dfrac{17}{3}$

EXAMPLE 3 SOLVING A FRACTIONAL EQUATION

Solve $\dfrac{3x}{x-3} - 3 = \dfrac{1}{x}$.

$$x(x-3)\left[\dfrac{3x}{x-3} - 3\right] = x(x-3)\left[\dfrac{1}{x}\right]$$ The LCD $= x(x-3)$

$$x(3x) - 3\,x(x-3) = x - 3$$ Be sure to multiply 3 by LCD

$$3x^2 - 3x^2 + 9x = x - 3$$

$$8x = -3$$

$$x = -\dfrac{3}{8}$$

Substituting $-\frac{3}{8}$ into the original equation to check would require much calculation. Instead, we might note that the only two numbers that make a denominator zero are 0 and 3, neither of which is our answer, $-\frac{3}{8}$. After checking our work for mistakes, we conclude that the solution is $-\frac{3}{8}$.

PRACTICE EXERCISE 3

Solve $\dfrac{2x}{x-4} - 2 = \dfrac{-24}{x}$.

Answer: 3

EXAMPLE 4 EQUATION WITH NO SOLUTION

Solve $\dfrac{x}{x+4} - \dfrac{4}{x-4} = \dfrac{x^2+16}{x^2-16}$.

The LCD $= (x-4)(x+4)$.

$$(x-4)(x+4)\left(\dfrac{x}{x+4} - \dfrac{4}{x-4}\right) = (x-4)(x+4)\left(\dfrac{x^2+16}{x^2-16}\right)$$

$$x(x-4) - 4(x+4) = x^2 + 16$$

$$x^2 - 4x - 4x - 16 = x^2 + 16$$

$$-8x - 16 = 16$$

$$-8x = 32$$

$$x = -4$$

Check: $\dfrac{-4}{-4+4} - \dfrac{4}{-4-4} \overset{?}{=} \dfrac{(-4)^2+16}{(-4)^2-16}$.

But $\frac{-4}{-4+4} = \frac{-4}{0}$, which is undefined. Thus, -4 *does not check*. There are no solutions.

PRACTICE EXERCISE 4

Solve $\dfrac{2}{y-3} - \dfrac{12}{y^2-9} = \dfrac{3}{y+3}$.

Answer: no solution (notice that 3 makes the denominators of the fractions on the left side equal to zero)

CAUTION

Be sure to check your answers in the *original* equation. Example 4 shows that sometimes a possible solution may have to be excluded because it makes one or more denominators equal to zero.

In some fractional equations, when the fractions are cleared the resulting equation must be solved by the factoring method that we studied in Chapter 7.

EXAMPLE 5   USING THE ZERO-PRODUCT RULE

Solve $\dfrac{2}{x+2} + \dfrac{3}{x-2} = 1$.

$$(x+2)(x-2)\left[\dfrac{2}{x+2} + \dfrac{3}{x-2}\right] = (x+2)(x-2)(1) \qquad \text{LCD} = (x+2)(x-2)$$

$$2(x-2) + 3(x+2) = x^2 - 4$$

$$2x - 4 + 3x + 6 = x^2 - 4$$

$$5x + 2 = x^2 - 4$$

$$-x^2 + 5x + 6 = 0$$

$$x^2 - 5x - 6 = 0 \qquad \text{Multiply by } -1$$

$$(x-6)(x+1) = 0 \qquad \text{Factor}$$

$$x - 6 = 0 \quad \text{or} \quad x + 1 = 0 \qquad \text{Zero-product Rule}$$

$$x = 6 \qquad\qquad x = -1$$

Check: $\dfrac{2}{6+2} + \dfrac{3}{6-2} \stackrel{?}{=} 1 \qquad \dfrac{2}{-1+2} + \dfrac{3}{-1-2} \stackrel{?}{=} 1$

$$\dfrac{2}{8} + \dfrac{3}{4} \stackrel{?}{=} 1 \qquad\qquad \dfrac{2}{1} + \dfrac{3}{-3} \stackrel{?}{=} 1$$

$$\dfrac{1}{4} + \dfrac{3}{4} = 1 \qquad\qquad 2 - 1 = 1$$

The solutions are 6 and -1.

PRACTICE EXERCISE 5

Solve $\dfrac{y}{y-1} + \dfrac{2}{y+1} = \dfrac{8}{y^2 - 1}$.

Answer: 2 and -5

---

② APPLICATIONS OF FRACTIONAL EQUATIONS

In Chapter 3 we looked at applications that resulted in simple equations. We now consider similar problems that result in fractional equations.

EXAMPLE 6   AGE PROBLEM

The sum of Bill's and Bob's ages is 45 years, and Bill's age divided by Bob's age is $\frac{5}{4}$. How old is each?

Let x = Bill's age,
$45 - x$ = Bob's age.   The sum of x and $45 - x$ is 45

We need to solve the following equation.

$$\dfrac{x}{45 - x} = \dfrac{5}{4} \qquad \text{The LCD} = 4(45 - x)$$

$$4(45 - x)\left(\dfrac{x}{45 - x}\right) = 4(45 - x)\left(\dfrac{5}{4}\right) \qquad \text{Multiply by the LCD}$$

$$4x = (45 - x)5$$

$$4x = 225 - 5x$$

$$9x = 225$$

$$x = 25 \qquad \text{Bill's age is 25}$$

Since $45 - x = 45 - 25 = 20$, Bill is 25 years old and Bob is 20.

PRACTICE EXERCISE 6

Terry's age divided by Pam's age is $\frac{7}{5}$, and the sum of their ages is 24. How old is each?

Let x = Terry's age,
$24 - x$ = Pam's age.

Answer: Terry is 14 and Pam is 10.

| EXAMPLE 7   NUMBER PROBLEM | PRACTICE EXERCISE 7 |

The reciprocal of 2 more than a number is three times the reciprocal of the number. Find the number.

Let x = the number,

$x + 2$ = 2 more than the number,

$\dfrac{1}{x + 2}$ = the reciprocal of 2 more than the number,

$\dfrac{1}{x}$ = the reciprocal of the number,

$3 \cdot \dfrac{1}{x} = \dfrac{3}{x}$ = three times the reciprocal of the number.

The equation we need to solve is $\dfrac{1}{x + 2} = \dfrac{3}{x}$.

$$x(x + 2)\left(\dfrac{1}{x + 2}\right) = x(x + 2)\left(\dfrac{3}{x}\right) \qquad \text{The LCD} = x(x + 2)$$

$$x = (x + 2)3$$
$$x = 3x + 6$$
$$-2x = 6$$
$$x = -3$$

The number is -3.

The numerator of a fraction is 3 more than the denominator, and the value of the fraction is $\frac{4}{3}$. Find the fraction.

Let x = denominator of fraction,

$x + 3$ = numerator of fraction,

$\dfrac{x + 3}{x}$ = the fraction.

The equation to solve is

$$\dfrac{x + 3}{x} = \underline{\qquad}.$$

Answer: $\frac{12}{9}$

③ WORK PROBLEMS

Consider the following problem: Bob can do a job in 2 hours and Pete can do the same job in 4 hours. How long would it take them to do the job if they worked together?

To solve this type of problem, usually called a *work problem,* three principles must be kept in mind:

1. The time it takes to do a job when the individuals work together must be less than the time it takes for the faster worker to complete the job alone. Thus, the time together is *not* the average of the two times (which would be 3 hours in this case). Since Bob can do the job alone in 2 hr, with help the time must clearly be less than 2 hr.

2. If a job can be done in t hr, in 1 hour $\frac{1}{t}$ of the job would be completed. For example, since Bob can do the job in 2 hours, in 1 hour he would do $\frac{1}{2}$ the job. Similarly, Pete would do $\frac{1}{4}$ the job in 1 hr since he does it all in 4 hr.

3. The amount of work done by Bob in 1 hr added to the amount of work done by Pete in 1 hr equals the amount of work done together in 1 hr. So if t is the time it takes to do the job working together,

(amount Bob does in 1 hr) + (amount Pete does in 1 hr)
= (amount done together in 1 hr)

translates to the equation

$$\dfrac{1}{2} + \dfrac{1}{4} = \dfrac{1}{t}.$$

We see that a work problem translates to a fractional equation that may result in either a first-degree equation or a second-degree (quadratic) equation. In our example, we obtain a first-degree equation after we multiply both sides by the LCD, $4t$. We will work only with first-degree equations in this section.

$$4t\left(\frac{1}{2} + \frac{1}{4}\right) = 4t\left(\frac{1}{t}\right)$$

$$4t \cdot \frac{1}{2} + 4t \cdot \frac{1}{4} = 4$$

$$2t + t = 4$$

$$3t = 4$$

$$t = \frac{4}{3}$$

Thus, it would take $\frac{4}{3}$ hr (1 hr 20 min) for Bob and Pete to do the job working together.

EXAMPLE 8 WORK PROBLEM	PRACTICE EXERCISE 8

Jan can wash the dishes in 20 minutes and her mother can wash them in worked together?

    20 = number of minutes for Jan to do the job,

$\frac{1}{20}$ = amount done by Jan in 1 minute,

    15 = number of minutes for Jan's mother to do the job,

$\frac{1}{15}$ = amount done by Jan's mother in 1 minute.

    Let t = time it takes to do the job if Jan and her mother work together.

Then $\frac{1}{t}$ = amount done in 1 minute when they work together.

Thus, we have the following equation.

    (amount by Jan) + (amount by mother) = (amount together)

$$\frac{1}{20} + \frac{1}{15} = \frac{1}{t}$$

$$60t\left(\frac{1}{20} + \frac{1}{15}\right) = 60t\left(\frac{1}{t}\right) \qquad \text{Multiply by the LCD, } 60t$$

$$60t \cdot \frac{1}{20} + 60t \cdot \frac{1}{15} = 60t \cdot \frac{1}{t}$$

$$3t + 4t = 60$$

$$7t = 60$$

$$t = \frac{60}{7} = 8\frac{4}{7}$$

Working together, it would take Jan and her mother $8\frac{4}{7}$ minutes to wash the dishes. Is this reasonable in view of principle (1)?

Barbara can do a job in 3 days and Paula can do the same job in 8 days. How long will it take them to do the job if they work together?

Let t = number of days to do the job together

$\frac{1}{t}$ = amount both do in 1 day

$\frac{1}{3}$ = amount Barbara does in 1 day

$\frac{1}{8}$ = _____

The equation to solve is

$$\frac{1}{3} + \frac{1}{8} = \underline{\hspace{3cm}}.$$

Answer: $\frac{24}{11}$ days

The next example is a variation of a work problem. It asks for the time it takes for one working unit to do a job when the time for the other and the time together are given.

| EXAMPLE 9   FINDING THE TIME OF ONE WORKING UNIT |

PRACTICE EXERCISE 9

When the valves on a 6-inch pipe and a 2-inch pipe are opened together, it takes 5 hr to fill an oil storage tank. If the 6-inch pipe can fill the tank in 7 hr, how long would it take the 2-inch pipe to fill the tank by itself?

$5 =$ number of hr to fill the tank together,

$\frac{1}{5} =$ amount filled together in 1 hr,

$7 =$ number of hr for 6-inch pipe to fill the tank,

$\frac{1}{7} =$ amount filled by the 6-inch pipe in 1 hr.

Let $t =$ number of hr for 2-inch pipe to fill the tank,

$\frac{1}{t} =$ amount filled by the 2-inch pipe in 1 hr.

We have the following equation.

(amount by 6-inch pipe) + (amount by 2-inch pipe) = (amount together)

$$\frac{1}{7} + \frac{1}{t} = \frac{1}{5}$$

$$35t\left(\frac{1}{7} + \frac{1}{t}\right) = 35t\left(\frac{1}{5}\right) \quad \text{The LCD is } 35t$$

$$35t\left(\frac{1}{7}\right) + 35t\left(\frac{1}{t}\right) = 35t \cdot \frac{1}{5}$$

$$5t + 35 = 7t$$

$$35 = 2t$$

$$\frac{35}{2} = t$$

The 2-inch pipe would take $\frac{35}{2}$ hours or $17\frac{1}{2}$ hours to fill the tank alone. (Does this make sense in view of principle (1) of work problems?) Notice that the 6 inches and 2 inches only describe the size of the pipes and do not enter into the calculations.

Working together, Phil Mortensen and his son Brad can build a barn in 45 days. Working by himself, Phil can build the barn in 60 days. How long would it take Brad working alone?

Answer: 180 days

8.5 EXERCISES A

Solve.

1. $\dfrac{3x+5}{6} = \dfrac{4+3x}{5}$

2. $\dfrac{1}{x} - \dfrac{2}{x} = 6$

3. $\dfrac{2}{y+1} = \dfrac{3}{y}$

4 $\dfrac{1}{z-3} + \dfrac{3}{z-5} = 0$

5. $\dfrac{1}{2x-3} - \dfrac{3}{4x-15} = 0$

6. $\dfrac{a-1}{a+1} = \dfrac{a-3}{a+2}$

7. $\dfrac{2y+1}{3y-2} = \dfrac{2y-3}{3y+2}$

8 $\dfrac{z}{z+1} - 1 = \dfrac{1}{z}$

9. $\dfrac{2x}{x+2} - 2 = \dfrac{1}{x}$

10. $\dfrac{1}{y+2} + \dfrac{1}{y-2} = \dfrac{1}{y^2-4}$

11 $\dfrac{4}{z-3} + \dfrac{2z}{z^2-9} = \dfrac{1}{z+3}$

12. $\dfrac{3}{x+3} + \dfrac{1}{x-3} = \dfrac{10}{x^2-9}$

13. $\dfrac{y}{y+3} - \dfrac{3}{y-3} = \dfrac{y^2+9}{y^2-9}$

14. $\dfrac{10}{x^2-25} + 1 = \dfrac{x}{x+5}$

15. $\dfrac{y}{y-4} - 1 = \dfrac{8}{y^2-16}$

16 $\dfrac{3}{x-3} + \dfrac{2}{x+3} = -1$

17. $\dfrac{4}{y-4} + \dfrac{6}{y+4} = -1$

18. $\dfrac{2z}{z-2} - 1 = \dfrac{1}{z^2-4}$

Solve. (Some problems have been started.)

19. Find two numbers whose sum is 42 and whose quotient is $\frac{3}{4}$.

> Let x = one number,
> $42 - x$ = second number. [The sum of the two is
> $x + (42 - x) = 42$.]

20. Two-thirds of a certain number is one more than five-eighths of the number. Find the number.

21. The denominator of a certain fraction is 2 greater than the numerator, and the value of the fraction is $\frac{4}{5}$. Find the fraction.

22. If Joe is three-fourths as old as Jeff and the difference in their ages is 6 years, how old is each?

Let x = Jeff's age,
$\frac{3}{4}x$ = Joe's age.

23 Find the number which when added to both the numerator and denominator of $\frac{5}{7}$ will produce a fraction equivalent to $\frac{4}{5}$.

24. The reciprocal of 3 less than a number is four times the reciprocal of the number. Find the number.

Let x = the desired number,
$x - 3$ = 3 less than the number,
$\dfrac{1}{x - 3}$ = the reciprocal of 3 less than the number.

25 The numerator of a fraction is 4 less than its denominator. If both the numerator and denominator are increased by 1, the value of the fraction is $\frac{2}{3}$. Find the fraction.

26. Mary's age is 4 years more than Ruth's age, and the quotient of their ages is $\frac{7}{5}$. How old is each?

27. The denominator of a fraction exceeds the numerator by 6, and the value of the fraction is $\frac{5}{6}$. Find the fraction.

Let x = numerator of fraction,
$x + 6$ = denominator of fraction.

28 The reciprocal of 4 less than a number is three times the reciprocal of the number. Find the number.

29. If Burford can mow a lawn in 7 hours and Hilda can mow the same lawn in 3 hours, how long would it take them to mow it if they worked together?

7 = number of hr for Burford to do the job
$\frac{1}{7}$ = amount done by Burford in 1 hr
3 = number of hr for Hilda to do the job
$\frac{1}{3}$ = amount done by Hilda in 1 hr
Let t = time to do the job working together
$\frac{1}{t}$ = amount done in 1 hr working together
$\frac{1}{7} + \frac{1}{3} = \frac{1}{t}$

30 Bob can paint a house alone in 12 days. When he and his father work together it takes only 9 days. How long would it take his father to paint the house if he worked alone?

31. A 6-inch pipe can fill a reservoir in 3 weeks. When this pipe and a second pipe are both turned on, it takes only 2 weeks to fill the reservoir. How long would it take the second pipe to fill it by itself?

32. If Irv takes 4 times longer to repair a car than Max, and together they can repair it in 8 hours, how long does it take each working alone?

33. If a 1-inch pipe takes 40 minutes to drain a pond and a 2-inch pipe takes 10 minutes to drain the same pond, how long would it take them to drain it together?

34. If Art can frame a shed in 4 days and together Art and Clyde can frame the same shed in 3 days, how long would it take Clyde to frame it working by himself?

35. An older machine requires twice as much time to do a job as a modern one. Together they accomplish the job in 8 minutes. How long would it take each to do the job alone?

36. Susan can make 100 widgets in 3 hours. It takes Harry 5 hours to produce 100 widgets. How long would it take them to produce 100 widgets if they worked together?

FOR REVIEW

Perform the indicated operations.

37. $\dfrac{3}{5 - x} - \dfrac{2 - x}{x - 5}$

38. $\dfrac{2}{x - 3} + \dfrac{5}{x + 2}$

39. $\dfrac{-5y}{y^2 - 25} - \dfrac{y}{y + 5}$

40. $\dfrac{5}{y^2 - 25} + \dfrac{5}{y^2 - 3y - 10}$

41. $\dfrac{3}{x^2 - x - 2} - \dfrac{2}{x^2 + 6x + 5}$

42. $\dfrac{x + y}{x - y} - \dfrac{x - y}{x + y}$

ANSWERS: 1. $\frac{1}{3}$ 2. $-\frac{1}{6}$ 3. -3 4. $\frac{7}{2}$ 5. -3 6. $-\frac{1}{3}$ 7. $\frac{1}{5}$ 8. $-\frac{1}{2}$ 9. $-\frac{2}{5}$ 10. $\frac{1}{2}$ 11. no solution 12. 4
13. no solution 14. 3 15. -2 16. 1, -6 17. 2, -12 18. -1, -3 19. 18, 24 20. 24 21. $\frac{8}{10}$ 22. Joe is 18,
Jeff is 24 23. 3 24. 4 25. $\frac{7}{11}$ 26. Mary is 14, Ruth is 10 27. $\frac{30}{36}$ 28. 6 29. $\frac{21}{10}$ hr 30. 36 days 31. 6 weeks
32. 10 hr, 40 hr 33. 8 min 34. 12 days 35. 12 min, 24 min 36. $\frac{15}{8}$ hr 37. 1 38. $\frac{7x - 11}{(x - 3)(x + 2)}$ 39. $\frac{-y^2}{y^2 - 25}$
40. $\frac{5(2y + 7)}{(y + 5)(y - 5)(y + 2)}$ 41. $\frac{x + 19}{(x - 2)(x + 1)(x + 5)}$ 42. $\frac{4xy}{(x + y)(x - y)}$

8.5 EXERCISES B

Solve.

1. $\dfrac{5x - 2}{3} = \dfrac{2 + 7x}{4}$

2. $\dfrac{3}{x} - \dfrac{4}{x} = 3$

3. $\dfrac{8}{a + 4} = \dfrac{7}{a - 5}$

4. $\dfrac{2}{z - 4} + \dfrac{3}{z - 6} = 0$

5. $\dfrac{2}{3x - 1} - \dfrac{1}{4x - 5} = 0$

6. $\dfrac{x - 3}{x - 2} = \dfrac{x - 2}{x + 3}$

7. $\dfrac{2y - 3}{4y - 1} = \dfrac{y - 6}{2y - 1}$

8. $\dfrac{2z}{z + 3} - 2 = \dfrac{3}{z}$

9. $\dfrac{x + 1}{x} = 1 + \dfrac{1}{x + 1}$

10. $\dfrac{1}{y + 3} + \dfrac{2}{y - 3} = \dfrac{3}{y^2 - 9}$

11. $\dfrac{5}{z + 2} + \dfrac{3z}{z^2 - 4} = \dfrac{2}{z - 2}$

12. $\dfrac{-2}{x + 5} + \dfrac{1}{x - 5} = \dfrac{4}{x^2 - 25}$

13. $\dfrac{y}{y + 5} - \dfrac{5}{y - 5} = \dfrac{y^2 + 25}{y^2 - 25}$

14. $\dfrac{64}{x^2 - 16} + 2 = \dfrac{2x}{x - 4}$

15. $\dfrac{y}{y + 6} - 1 = \dfrac{6}{y^2 - 36}$

16. $\dfrac{6}{x + 5} + \dfrac{1}{x - 5} = 1$

17. $\dfrac{1}{y - 6} - \dfrac{6}{y + 6} = -1$

18. $\dfrac{3z}{z - 3} - 2 = \dfrac{10}{z^2 - 9}$

Solve.

19. Find two numbers whose sum is 32 and whose quotient is $\frac{3}{5}$.

20. Three-fourths of a number is three more than three-eighths of the number. Find the number.

21. The denominator of a fraction is 6 greater than the numerator, and the value of the fraction is $\frac{7}{9}$. Find the fraction.

22. Art is five-eighths as old as Toni and the difference in their ages is 21 years. How old is each?

23. Find the number which when added to both the numerator and denominator of $\frac{3}{11}$ will produce a fraction equivalent to $\frac{1}{2}$.

24. The reciprocal of 6 less than a number is four times the reciprocal of the number. Find the number.

25. The numerator of a fraction is 5 less than the denominator. If both the numerator and the denominator are decreased by 2, the value of the fraction is $\frac{1}{6}$. Find the fraction.

26. Cindy is 4 years younger than Becky and the quotient of their ages is $\frac{7}{9}$. How old is each?

27. The numerator of a fraction exceeds the denominator by 10, and the value of the fraction is $\frac{7}{5}$. Find the fraction.

28. The reciprocal of 6 more than a number is four times the reciprocal of the number. Find the number.

29. Alfonse can paint a garage in 20 hours and Heidi can do the same job in 10 hours. How long would it take them if they worked together?

30. Walter can dig a well in 8 days. When he and Sean work together they can dig the same size well in 6 days. How long would it take Sean to do the job working alone?

31. If Andy takes 3 times as long to put new brakes on a car as Hortense does, and together they can do the job in 4 hours, how long does it take each working alone?

32. There are two drains to a water tank. If one is used, the tank is drained in 24 minutes. The other requires 32 minutes. How long will it take if both are opened?

33. If Nancy can process an order of 1000 appliances in 20 minutes and together she and Hans can process the order in 12 minutes, how long will it take Hans to process the order by himself?

34. Using old equipment it requires 5 times as long to make 100 sportcoats as with new equipment. If together the equipment can accomplish the job in $7\frac{1}{2}$ hours, how long would it take each working alone?

35. Charles can make 200 dolls in 3 weeks and Wilma can make 200 dolls in $2\frac{1}{2}$ weeks. How long will it take to make 200 dolls if they work together?

36. Bertha can clean the stable in 4 hours. If Bertha and Jose work together they can do the same job in 1 hour. How long would Jose take to do the job working alone?

FOR REVIEW

Perform the indicated operations.

37. $\dfrac{2}{9-x} - \dfrac{3-x}{x-9}$

38. $\dfrac{8}{z-7} + \dfrac{2}{z+1}$

39. $\dfrac{-9y}{y^2-16} - \dfrac{5y}{y-4}$

40. $\dfrac{3}{y^2-6y+8} + \dfrac{7}{y^2-16}$

41. $\dfrac{5}{x^2+x-12} - \dfrac{3}{x^2+10x+24}$

42. $\dfrac{x-y}{x+y} - \dfrac{x+y}{x-y}$

8.5 EXERCISES C

Solve.

1. $\dfrac{3}{x^2-6x+5} - \dfrac{2}{x^2-25} = \dfrac{4}{x^2+4x-5}$

2. $\dfrac{2x}{x^2-3x-4} - \dfrac{x-1}{x^2+2x+1} = \dfrac{x+4}{x^2+5x+4}$
[Answer: 0]

3. A man walks a distance of 12 mi at a rate 8 mph slower than the rate he rides a bicycle for a distance of 24 mi. If the total time of the trip is 5 hr, how fast does he walk? How fast does he ride?

4. A swimming pool can be filled by an inlet pipe in 6 hr. It can be drained by an outlet in 9 hr. If the inlet and outlet are both opened, how long will it take to fill the empty pool?   [Answer: 18 hr]

8.6 RATIO, PROPORTION, AND VARIATION

════════════════ STUDENT GUIDEPOSTS ════════════════

❶ Ratios	❹ Applying Proportions
❷ Proportions	❺ Direct Variation
❸ Means-Extremes Property	❻ Inverse Variation

❶ RATIOS

The **ratio** of one number x to another number y is the quotient

$$x \div y \quad \text{or} \quad \frac{x}{y}.$$

We sometimes express the ratio $\frac{x}{y}$ with the notation $x:y$, which is read "the ratio of x to y." Ratios occur in applications such as percent, rate of speed, gas mileage, unit cost, and number comparisons. Consider the following examples.

Applications	*Ratio*
5% sales tax	$\dfrac{\$5}{\$100} = \$5$ per $\$100$
200 miles in 4 hours	$\dfrac{200 \text{ mi}}{4 \text{ hr}} = 50\dfrac{\text{mi}}{\text{hr}} = 50$ mi per hr
200 miles on 10 gallons of gas	$\dfrac{200 \text{ mi}}{10 \text{ gal}} = 20\dfrac{\text{mi}}{\text{gal}} = 20$ mi per gal
$12 for 2 kg of meat	$\dfrac{\$12}{2 \text{ kg}} = \$6$ per kg
30 children and 6 adults	$\dfrac{30 \text{ children}}{6 \text{ adults}} = 5$ children per adult

② PROPORTIONS

Ratios are used to define *proportions* which can describe many practical problems.

Proportions

An equation which states that two ratios are equal is called a **proportion.** That is, if $\frac{a}{b}$ and $\frac{c}{d}$ are two ratios, the equation

$$\frac{a}{b} = \frac{c}{d}$$

is a proportion. It can be read "a is to b as c is to d."

③ MEANS-EXTREMES PROPERTY

For example,

$$\frac{2}{3} = \frac{4}{6}$$

is a proportion. The numbers 2 and 6 are the **extremes** of the proportion, and the numbers 3 and 4 are the **means.**

2 is to 3 as 4 is to 6
means
extremes

Consider the proportion

$$\frac{a}{b} = \frac{c}{d},$$

with extremes a and d and means b and c. If we multiply both sides by the LCD of the two fractions, bd, we obtain

$$bd\frac{a}{b} = \frac{c}{d}bd$$

$$\text{or} \quad ad = bc.$$

We have just verified the following property of proportions.

Means-Extremes Property of Proportions

In any proportion, the product of the extremes is equal to the product of the means. That is, if

$$\frac{a}{b} = \frac{c}{d} \quad \text{then} \quad ad = bc.$$

This property of proportions allows us to solve for an unknown term. We call this **solving the proportion.** For example, suppose we solve the proportion

$$\frac{15}{25} = \frac{x}{5}.$$

$15 \cdot 5 = 25 \cdot x$ The product of the extremes equals the product of the means

$75 = 25x$ Simplify

$\dfrac{75}{25} = x$ Divide both sides by 25

$3 = x$ Reduce the fraction

④ APPLYING PROPORTIONS

We now show how proportions can be used to solve applied problems. Consider the following: If it takes 3 hours to travel 150 miles, how many hours would it take to travel 250 miles? The simple proportion

$$\text{first time} \rightarrow \frac{3}{150} = \frac{x}{250} \leftarrow \text{second time}$$
$$\text{first distance} \rightarrow \qquad\qquad \leftarrow \text{second distance}$$

where x is the number of hours it takes to travel 250 miles, completely describes this problem. That is, the ratio of the first time to the first distance must equal the ratio of the second time to the second distance. Setting the product of the extremes equal to the product of the means, we have

$$\left(\begin{matrix}\text{first} \\ \text{time}\end{matrix}\right) \cdot \left(\begin{matrix}\text{second} \\ \text{distance}\end{matrix}\right) = \left(\begin{matrix}\text{first} \\ \text{distance}\end{matrix}\right) \cdot \left(\begin{matrix}\text{second} \\ \text{time}\end{matrix}\right)$$

$$3 \quad \cdot \quad 250 \quad = \quad 150 \quad \cdot \quad x$$

or

$$5 = x.$$

It would take 5 hours to go 250 miles.

EXAMPLE 1   AN APPLICATION OF PROPORTION

If a family drinks 5 gallons of milk in two weeks, how many gallons of milk will be consumed in a year (52 weeks)?

    Let x = number of gallons of milk consumed in 52 weeks. The proportion is

$$\begin{matrix}\text{gallons consumed} \rightarrow \\ \text{in 2 weeks}\end{matrix} \quad \frac{5}{2} = \frac{x}{52}. \quad \begin{matrix}\leftarrow \text{gallons consumed} \\ \text{in 52 weeks}\end{matrix}$$

$5 \cdot 52 = 2x$ Product of the extremes equals product of the means

$130 = x$

The family drinks 130 gallons of milk in a year.

PRACTICE EXERCISE 1

A sample of 235 television sets contained 4 that were defective. How many defective sets would you expect to find in a sample of 1645?

Answer: 28

| EXAMPLE 2   AN APPLICATION OF PROPORTION | PRACTICE EXERCISE 2 |

If a secretary spends 37 minutes typing 3 pages of a report, how long will it take her to type the remaining 10 pages?

Let x be the number of minutes it takes to complete the typing. We set up the following proportion.

minutes to type 3 pages → $\dfrac{3}{37} = \dfrac{10}{x}$. minutes to ← type 10 pages

$3x = 370$ Product of the extremes equals product of the means

$x = 123\dfrac{1}{3}.$

It will take another $123\frac{1}{3}$ minutes or 2 hours, 3 minutes, and 20 seconds.

If it costs $18.00 to operate a freezer for 3 months, how much would it cost to operate the freezer for one year?

Answer: $72.00

⑤ DIRECT VARIATION

Another way of looking at proportion problems is in terms of **variation.** For example, the distance that a car travels varies (or changes) as we vary or change the speed. Also, the total cost of a roast will vary as the price per pound varies. The equation in each of these examples can be written as a ratio equal to a constant. This type of variation is called **direct variation.**

Direct Variation

If two variables have a constant ratio k,

$$\frac{y}{x} = k \quad \text{or} \quad y = kx,$$

we say that y **varies directly** as x and call k the **constant of variation.**

One common example of this is the fact that the ratio of the circumference to the diameter of any circle is π.

$$\frac{c}{d} = \pi \quad \text{or} \quad c = \pi d \quad k = \pi$$

In a variation problem if we know the value of y for one value of x, we can find k. With k known we can find y for any x.

| EXAMPLE 3   USING DIRECT VARIATION | PRACTICE EXERCISE 3 |

Find the equation of variation if y varies directly as x and we know that $y = 30$ when $x = 6$. Also, find y when $x = 10$.

First, write an equation stating that y *varies directly as x.* Then substitute $y = 30$ and $x = 6$ to find k.

$y = kx$ y varies directly as x

$30 = k \cdot 6$ $y = 30$ and $x = 6$

$\dfrac{30}{6} = k$ k is a ratio

$5 = k$

Find the equation of variation if u varies directly as v and we know that $u = 4$ when $v = 3$. Use this equation to find u when $v = 45$.

The equation of variation is

$$y = 5x.$$

To find y when $x = 10$, substitute 10 for x in the equation of variation.

$y = 5x$	Equation of variation
$y = 5(10)$	Substitute 10 for x
$y = 50$	

Thus y is 50 when x is 10.

Answer: $u = \frac{4}{3}v$; 60

EXAMPLE 4   APPLYING DIRECT VARIATION

The cost c of steak varies directly as the weight w of the steak purchased. If 8 kg of steak cost \$36, how much will 30 kg of steak cost?

First write the equation for c varies directly as w. Then substitute $c =$ \$36 and $w = 8$ kg to find k.

$c = kw$	c varies directly as w
$36 = k \cdot 8$	$c = \$36$ and $w = 8$ kg
$4.5 = k$	$\frac{36}{8} = 4.5$

The equation of variation that we can use for any weight of meat is

$$c = 4.5\,w.$$

To find the cost of 30 kg of steak substitute 30 for w.

$$c = 4.5w$$
$$c = 4.5(30)$$
$$= 135$$

Thus 30 kg (about 66 lb) of steak would cost \$135.

PRACTICE EXERCISE 4

The weight m of an object on the moon varies directly as its weight e on earth. If a man weighing 150 lb on earth weighs 24 lb on the moon, how much will a module that weighs 9500 lb on earth weigh on the moon?

Answer: 1520 lb

---

⑥ INVERSE VARIATION

In many applications the product of the variables is a constant. This is called **inverse variation.**

Inverse Variation

If two variables have a constant product k,

$$xy = k \quad \text{or} \quad y = \frac{k}{x}$$

where k is the **constant of variation,** we say that y **varies inversely as** x.

| EXAMPLE 5   APPLYING INVERSE VARIATION | PRACTICE EXERCISE 5 |

The time t it takes to drive a fixed distance d varies inversely as the rate r (or speed) of the car. If it takes 3 hours when the speed is 40 mph, how long will it take when the speed is 50 mph?

    The equation of variation in this case is

$$t = \frac{d}{r}.$$

Since the distance is fixed, d is the constant of variation. We are given that $t = 3$ hr when $r = 40$ mph.

$$t = \frac{d}{r}$$

$$3 = \frac{d}{40}$$

$$120 = d$$

Thus

$$t = \frac{120}{r} \quad \text{Equation of variation}$$

$$t = \frac{120}{50} \quad r = 50 \text{ mph}$$

$$t = 2.4.$$

It will take 2.4 hours to go the distance at 50 mph.

The rate r of a car when traveling a fixed distance d varies inversely as the time t of travel. If the rate is 55 mph when the time is 4 hr, what is the rate when the time is 5.5 hr?

Answer: 40 mph

8.6 EXERCISES A

Indicate as a ratio and simplify.

1. 320 mi in 8 hr

2. $10 for 4 lb of meat

3. 20 children for 2 adults

4. 420 mi on 12 gal of gas

5. 200 trees on 4 acres

6. 45°F in 30 min

Solve the proportions.

7. $\dfrac{x}{2} = \dfrac{9}{27}$

8. $\dfrac{y+8}{y-2} = \dfrac{7}{3}$

9. $\dfrac{w-5}{6} = \dfrac{w+6}{2}$

10. $\dfrac{5}{x-7} = \dfrac{4}{x+4}$

11. $\dfrac{y-3}{y+2} = \dfrac{y-2}{y+3}$

12. $\dfrac{z+5}{z-2} = \dfrac{z-7}{z-3}$

Solve. (Some problems have been started.)

13. In an election the winning candidate won by a 5 to 3 margin. If she received 1025 votes, how many votes did the losing candidate receive?

Let x = no. of votes for losing candidate

$\dfrac{1025}{x} = \dfrac{5}{3}$

14. If a boat uses 7 gal of gas to go 51 mi, how many gallons would be needed to go 255 mi?

Let x = no. of gallons needed to go 255 mi

$\dfrac{7}{51} = \dfrac{x}{255}$

15. If 50 feet of wire weigh 15 lb, what will 80 feet of the same wire weigh?

Let x = weight of 80 feet of wire

16. If $\frac{3}{4}$ inch on a map represents 10 mi, how many miles will be represented by 9 in?

Let x = no. of miles represented by 9 in

17. A sample of 184 tires contained 6 that were defective. How many defective tires would you expect in a sample of 1288?

18. If 2 bricks weigh 9 pounds, how much will 558 bricks weigh?

19. In an election, the successful candidate won by a 3 to 2 margin. If he received 324 votes, how many did the losing candidate receive?

20. If a car uses 8 gallons of gas for a trip of 132 miles, how much gas will be used for a trip of 550 miles?

21. The reciprocal of 4 more than a number is equal to five times the reciprocal of the number. Find the number.

22. If Sal is 7 years older than Ev and the ratio of their ages is equal to $\frac{2}{3}$, how old is each?

23. On a map 2 cm represent 14 km. How many centimeters are used to represent 35 km?

24. In a sample of 212 light bulbs, 3 were defective. How many defective bulbs would you expect in a shipment of 3240?

25. Pat was charged a commission of $65.00 on the purchase of 400 shares of stock. What commission would be charged on 700 shares of the same stock?

26 To determine the approximate number of fish in a lake, the State Game and Fish Department caught 100 fish, tagged their fins, and returned them to the water. After a period of time they caught 70 fish and discovered that 14 were tagged. Approximately how many fish are in the lake?

Solve using direct variation.

27. If y varies directly as x, and $y = 10$ when $x = 15$, find y when $x = 42$.

28. If u varies directly as w, and $u = 24$ when $w = 6$, find u when $w = 20$.

29. If y varies directly as x, and $y = 12$ when $x = 9$, find y when $x = 6$.

30 The cost c of peaches varies directly as the total weight w of the peaches. If 12 pounds of peaches cost $4.80, what would 66 pounds cost?

31. The number d of defective tires in a shipment varies directly as the number n of tires in the shipment. If there were 10 defective tires in a shipment of 2200, how many defective tires would you expect in a shipment of 8800 tires?

32. The number g of gallons of gas needed varies directly as the number n of miles traveled. If 25 gallons are needed to travel 650 miles, how many gallons will be needed to travel 3900 miles?

33. The size of Maria's paycheck varies directly as the number of hours she works. If she is paid $334.00 for working 40 hours each week, how much would she be paid for working 2080 hours (1 year)?

34 The amount of money a man spends on recreation varies directly as his total income. If he spends $1920 on recreation when earning $16,000, how much would he spend if his income is increased by $5,000?

Solve using inverse variation.

35. If y varies inversely as x, and $y = 12$ when $x = 9$, find y when $x = 6$.

36. If p varies inversely as q, and $p = 520$ when $q = 80$, find p when $q = 200$.

37 The time t needed to travel a fixed distance varies inversely as the rate r of travel. If 6 hr are needed when the rate is 500 mph, how long would it take at 1200 mph?

38. The rate r of travel for a fixed distance varies inversely as the time t. If the rate is 120 km per hour when the time is 4 hr, find the rate when the time is 6 hr.

39. The volume V of a given amount of gas varies inversely as the pressure P. If the volume is 24 in^3 when the pressure is 8 lb/in^2, what is the volume when the pressure is 6 lb/in^2?

40. The current I in an electrical circuit with constant voltage varies inversely as the resistance R of the circuit. If the current is 3 amps when the resistance is 4 ohms, find I when $R = 6$ ohms.

FOR REVIEW

Solve.

41. $\dfrac{3}{x+1} - \dfrac{2}{x+1} = 5$

42. $\dfrac{3}{2y+1} - \dfrac{1}{2y-1} = 0$

43. $\dfrac{x+1}{x-1} - \dfrac{x-1}{x+1} = 0$

44. $\dfrac{3x}{x-3} - 3 = \dfrac{1}{x}$

45. $\dfrac{1}{y-5} + \dfrac{1}{y+5} = \dfrac{1}{y^2-25}$

46. $\dfrac{2x}{x-3} - 1 = \dfrac{16}{x^2-9}$

47. Phillip can paint a wall in 3 hr and Elizabeth can paint the same size wall in 2 hr. How long would it take to paint the wall if they worked together?

48. Steve can type a manuscript 3 times as fast as Randy. Working together they can do the job in 5 hr. How long would it take each working alone?

The following exercises review material from Chapter 1 to help you prepare for the next section. Find the reciprocal of each fraction.

49. $\dfrac{7}{11}$

50. 4

Divide the fractions.

51. $\dfrac{4}{9} \div \dfrac{8}{27}$

52. $8 \div \dfrac{10}{9}$

8.6 EXERCISES B

Give as a ratio and simplify.

1. 480 mi in 6 hr

2. $6 for 15 lb of apples

3. 60 children for 4 adults

4. 2232 mi on 72 gal of gas

5. 600 bushels from 12 acres

6. 320 gal in 10 min

Solve the proportions.

7. $\dfrac{x}{18} = \dfrac{12}{20}$

8. $\dfrac{y-3}{y-7} = \dfrac{5}{8}$

9. $\dfrac{w+8}{5} = \dfrac{w-4}{3}$

10. $\dfrac{6}{x+10} = \dfrac{5}{x-10}$

11. $\dfrac{y+8}{y-4} = \dfrac{y+1}{y-6}$

12. $\dfrac{z-1}{z-2} = \dfrac{z-3}{z-4}$

Solve.

13. In an election the winning candidate won by a 6 to 5 margin. If the loser received 1520 votes, how many did the winner receive?

14. If a boat uses 12 gallons of gas to go 82 miles, how many gallons would be needed to go 123 miles?

15. If 20 m of hose weighs 6 kg, what will 320 m of the same hose weigh?

16. If 3 cm on a map represent 14 km, how many km are represented by 36 cm?

17. A sample of 350 bolts contained 5 that were defective. How many defective bolts should be expected in a shipment of 9870 bolts?

18. If 5 blocks weigh 14 pounds, how much will 255 blocks weigh?

19. In an election the losing candidate lost by a 3 to 4 margin. If he received 5280 votes, how many votes did the winner receive?

20. A boat uses 12 gal of gas in 4 hr. Under the same conditions how many gallons will be used in 10 hr?

21. The reciprocal of 5 less than a number is equal to the reciprocal of twice the number. Find the number.

22. If Walter is 3 years older than Molly and the ratio of their ages is equal to $\frac{5}{6}$, how old is each?

23. On a map 5 inches represent 24 mi. How many miles are represented by 2 in?

24. In a sample of 280 toasters, 8 were defective. How many defective toasters should be expected in a warehouse containing 42,700 toasters?

25. Marvin received a commission of $1200 on the sale of $150,000 worth of life insurance. If the same person has bought $250,000 worth of insurance, what commission would Marvin have earned?

26. To determine the number of antelope on a game preserve, a ranger caught 50 antelope and tagged their ears. Some time later he captured 18 antelope and discovered that 5 were tagged. Approximately how many antelope are on the preserve?

Solve using direct variation.

27. If y varies directly as x, and $y = 5$ when $x = 35$, find y when $x = 10$.

28. If u varies directly as w, and $u = 52$ when $w = 13$, find u when $w = 22$.

29. If y varies directly as x, and $y = 7$ when $x = 3$, find y when $x = 39$.

30. The cost c of carrots varies directly as the total weight w. If 5 kg of carrots cost \$3.20, what would 80 kg of carrots cost?

31. The number d of defective bolts in a shipment varies directly as the number n of bolts in the shipment. If there were 16 defective bolts in a shipment of 4800, how many would be expected in a shipment of 8400?

32. The number g of gallons of gas needed varies directly as the number n of miles traveled. If 16 gallons are needed to travel 840 miles, how many gallons will be needed to travel 2800 miles?

33. The amount of tax paid on real estate varies directly as its assessed value. If the Austins paid \$320.00 in taxes on a home valued at \$90,000, how much tax will the Wilsons pay on their home valued at \$65,000?

34. The amount of money a family spends on food varies directly as its income. If the Carpenters spend \$3780 on food and have an income of \$18,000, how much will the Woods spend if their income is \$31,000?

Solve using inverse variation.

35. If y varies inversely as x, and $y = 7$ when $x = 3$, find y when $x = 39$.

36. If p varies inversely as q, and $p = 580$ when $q = 20$, find p when $q = 50$.

37. The time t it takes to travel a fixed distance varies inversely as the rate r of travel. If it takes 12 hr when the rate is 24 mph, how long would it take at 36 mph?

38. The rate r of travel for a fixed distance varies inversely as the time t. If the rate is 580 km per hour when the time is 8 hr, find the rate when the time is 20 hr.

39. The volume V of a given amount of gas varies inversely as the pressure P. If the volume is 120 ft^3 when the pressure is 15 lb per sq ft, what is the volume when the pressure is 50 lb per sq ft?

40. The resistance R of an electrical circuit with constant voltage varies inversely as the current I of the circuit. If the resistance is 10 ohms when the current is 5 amps, find the resistance when the current is 25 amps.

FOR REVIEW

Solve.

41. $\dfrac{5}{x-2} - \dfrac{1}{x-2} = 8$

42. $\dfrac{10}{3y-2} + \dfrac{3}{3y+1} = 0$

43. $\dfrac{x-2}{x+2} - \dfrac{x+2}{x-2} = 0$

44. $\dfrac{2x}{x+5} - 2 = \dfrac{1}{x}$

45. $\dfrac{1}{y+6} + \dfrac{1}{y-6} = \dfrac{4}{y^2-36}$

46. $\dfrac{3x}{x+4} - 2 = \dfrac{12}{x^2-16}$

47. Grady can build a log cabin in 5 weeks and Smitty can build the same type of cabin in 4 weeks. How long would it take to build a cabin if they worked together?

48. Lynda can process an order twice as fast as Maria, and working together they can do the job in 3 hours. How long would it take each working alone?

The following exercises review material from Chapter 1 to help you prepare for the next section. Find the reciprocal of each fraction.

49. $-\dfrac{17}{8}$

50. $\dfrac{1}{99}$

Divide the fractions.

51. $-\dfrac{24}{25} \div \dfrac{16}{15}$

52. $\dfrac{49}{36} \div 14$

8.6 EXERCISES C

Solve.

1. In a sample of 120 tires, 4 were defective. Because of manufacturing problems it is estimated that the defective rate will increase by 50% on the next shipment of 480 tires. How many of the new shipment would you predict to be defective?   [Answer: 24]

2. A light-bulb manufacturer detected 12 defective bulbs in a sample of 840. After buying new equipment he was able to decrease the number of defective bulbs by 40 percent. How many defective bulbs should now be expected in a shipment of 1400?

3. If z varies directly as x and inversely as y, show that x varies directly as y and z.

4. The volume V of a gas varies directly as the temperature T and inversely as the pressure P. If the volume is 50 ft^3 when the temperature is 300° and the pressure is 30 lb per sq ft, find V when $T = 400°$ and $P = 20$ lb/ft^2.   [Answer: 100 ft^3]

8.7 SIMPLIFYING COMPLEX FRACTIONS

STUDENT GUIDEPOSTS

① Complex Fractions

② Simplifying Complex Fractions

① COMPLEX FRACTIONS

A **complex fraction** is a fraction that contains at least one other fraction within it. The following are complex fractions.

$$\frac{\frac{2}{3}}{5}, \quad \frac{\frac{1}{2}}{\frac{4}{5}}, \quad \frac{\frac{1}{x}}{3}, \quad \frac{1 + \frac{1}{x}}{2}, \quad \frac{1 + \frac{1}{a}}{1 - \frac{1}{a}}$$

② SIMPLIFYING COMPLEX FRACTIONS

A complex fraction is **simplified** when all its component fractions have been eliminated. There are two basic methods for simplifying a complex fraction. We will present both methods since with some complex fractions one method might be more easily applied than the other. The first method involves writing the numerator and denominator as single fractions and then dividing.

To Simplify a Complex Fraction (Method 1)
1. Change the numerator and denominator to single fractions.
2. Divide the two fractions.
3. Reduce the result to lowest terms.

EXAMPLE 1   SIMPLIFYING COMPLEX FRACTIONS (METHOD 1)

Simplify the complex fractions.

(a) $\dfrac{\dfrac{x}{4}}{\dfrac{x^2}{2}}$

Since the numerator and denominator are already single fractions, all we need to do is divide.

$$\frac{\dfrac{x}{4}}{\dfrac{x^2}{2}} = \frac{x}{4} \div \frac{x^2}{2} = \frac{x}{4} \cdot \frac{2}{x^2} = \frac{\cancel{x} \cdot \cancel{2}}{\cancel{2} \cdot 2 \cdot \cancel{x} \cdot x} = \frac{1}{2x}$$

(b) $\dfrac{1 + \dfrac{1}{2}}{1 - \dfrac{1}{3}}$

$$1 + \frac{1}{2} = \frac{2}{2} + \frac{1}{2} = \frac{3}{2} \qquad \text{Add the numerator fractions}$$

$$1 - \frac{1}{3} = \frac{3}{3} - \frac{1}{3} = \frac{2}{3} \qquad \text{Subtract the denominator fractions}$$

Thus, $\dfrac{1 + \dfrac{1}{2}}{1 - \dfrac{1}{3}} = \dfrac{\dfrac{3}{2}}{\dfrac{2}{3}} = \dfrac{3}{2} \div \dfrac{2}{3} = \dfrac{3}{2} \cdot \dfrac{3}{2} = \dfrac{9}{4}.$

(c) $\dfrac{1 + \dfrac{1}{x}}{2}$

$$1 + \frac{1}{x} = \frac{x}{x} + \frac{1}{x} = \frac{x + 1}{x} \qquad \text{Add the numerator fractions}$$

Thus, $\dfrac{1 + \dfrac{1}{x}}{2} = \dfrac{\dfrac{x + 1}{x}}{2} = \dfrac{x + 1}{x} \div 2 = \dfrac{x + 1}{x} \cdot \dfrac{1}{2} = \dfrac{x + 1}{2x}.$

PRACTICE EXERCISE 1

Simplify the complex fractions.

(a) $\dfrac{\dfrac{y^2}{10}}{\dfrac{y^4}{15}}$

(b) $\dfrac{2 + \dfrac{3}{4}}{3 - \dfrac{1}{6}}$

(c) $\dfrac{2 + \dfrac{2}{y}}{2y}$

Answers: (a) $\frac{3}{2y^2}$ (b) $\frac{33}{34}$
(c) $\frac{y + 1}{y^2}$

The second method involves clearing all fractions within the complex fraction by multiplying numerator and denominator by the LCD of all the internal fractions.

To Simplify a Complex Fraction (Method 2)
1. Find the LCD of all fractions within the complex fraction.
2. Multiply numerator and denominator of the complex fraction by the LCD to obtain an equivalent fraction.
3. Reduce the result to lowest terms.

EXAMPLE 2   SIMPLIFYING A COMPLEX FRACTION (METHOD 2)

Simplify $\dfrac{1 + \dfrac{2}{a}}{1 - \dfrac{4}{a^2}}$

The denominator in $1 + \frac{2}{a}$ is a, and in $1 - \frac{4}{a^2}$ is a^2. Thus, the LCD $= a^2$.

$$\frac{1 + \dfrac{2}{a}}{1 - \dfrac{4}{a^2}} = \frac{\left(1 + \dfrac{2}{a}\right) \cdot a^2}{\left(1 - \dfrac{4}{a^2}\right) \cdot a^2} \qquad \text{Multiply by LCD} = a^2$$

$$= \frac{1 \cdot a^2 + \dfrac{2}{a} \cdot a^2}{1 \cdot a^2 - \dfrac{4}{a^2} \cdot a^2} \qquad \text{Multiply all terms by } a^2$$

$$= \frac{a^2 + 2a}{a^2 - 4}$$

$$= \frac{a(a + 2)}{(a + 2)(a - 2)} \qquad \text{Factor and reduce to lowest terms}$$

$$= \frac{a}{a - 2}$$

PRACTICE EXERCISE 2

Simplify $\dfrac{\frac{1}{x} + 3}{\frac{1}{x^2} - 9}$.

Answer: $\frac{x}{1 - 3x}$

EXAMPLE 3   SIMPLIFYING A COMPLEX FRACTION (METHOD 2)

Simplify $\dfrac{1 + \dfrac{2}{x - 2}}{\dfrac{2}{x + 2} - 1}$.

The denominator in $1 + \frac{2}{x-2}$ is $x - 2$, and in $\frac{2}{x+2} - 1$ is $x + 2$. Thus, the LCD $= (x - 2)(x + 2)$.

$$\frac{1 + \dfrac{2}{x - 2}}{\dfrac{2}{x + 2} - 1} = \frac{\left(1 + \dfrac{2}{x - 2}\right)(x - 2)(x + 2)}{\left(\dfrac{2}{x + 2} - 1\right)(x - 2)(x + 2)} \qquad \text{Multiply by LCD}$$

$$= \frac{1 \cdot (x - 2)(x + 2) + \dfrac{2}{x - 2}(x - 2)(x + 2)}{\dfrac{2}{x + 2}(x - 2)(x + 2) - 1 \cdot (x - 2)(x + 2)}$$

$$= \frac{(x^2 - 4) + 2(x + 2)}{2(x - 2) - (x^2 - 4)}$$

$$= \frac{x^2 - 4 + 2x + 4}{2x - 4 - x^2 + 4} \qquad \begin{array}{l}\text{Watch signs when clearing} \\ \text{parentheses using the distributive law}\end{array}$$

$$= \frac{x^2 + 2x}{-x^2 + 2x}$$

$$= \frac{x(x + 2)}{x(-x + 2)} = \frac{x + 2}{2 - x} \qquad \text{Factor and reduce to lowest terms}$$

PRACTICE EXERCISE 3

Simplify $\dfrac{\frac{1}{y + 2} - 4}{4 + \frac{15}{y - 2}}$.

Answer: $\frac{2 - y}{y + 2}$

8.7 EXERCISES A

Simplify.

1. $\dfrac{\frac{2}{3}}{\frac{1}{3}}$

2. $\dfrac{3 + \frac{2}{3}}{1 - \frac{1}{3}}$

3. $\dfrac{\frac{1}{3} + \frac{1}{5}}{\frac{2}{3} - \frac{3}{5}}$

4. $\dfrac{\frac{2}{x}}{\frac{3}{x}}$

5. $\dfrac{6}{1 + \frac{1}{y}}$

6. $\dfrac{1 - \frac{1}{a}}{3}$

7. $\dfrac{\frac{1}{x} + 3}{\frac{1}{x} - 3}$

8. $\dfrac{y - 1}{y - \frac{1}{y}}$

9. $\dfrac{1 - \frac{4}{y}}{\frac{4 - y}{y}}$

10. $\dfrac{\frac{2}{a} + a}{\frac{a}{2} + a}$

11. $\dfrac{\frac{1}{2x} - 1}{\frac{1}{x} - 2}$

12. $\dfrac{\frac{2}{y} + 3}{2 - \frac{3}{y}}$

13. $\dfrac{\frac{1}{x} + 1}{\frac{1}{x} - 1}$

14. $\dfrac{4 - \frac{1}{a^2}}{2 - \frac{1}{a}}$

15. $\dfrac{z - 2 - \frac{3}{z}}{1 + \frac{1}{z}}$

16. $\dfrac{1 - \frac{2}{y} - \frac{3}{y^2}}{1 + \frac{1}{y}}$

17. $\dfrac{x - 3 + \frac{2}{x}}{x - 4 + \frac{3}{x}}$

18. $\dfrac{\frac{3}{a - 3} + 1}{\frac{3}{a + 3} - 1}$

19 An important formula in physics gives V in terms of S_1 and S_2 by the formula

$$V = \frac{3}{\dfrac{1}{S_1} + \dfrac{1}{S_2}}.$$

Express V as a simple fraction and use the result to find V when S_1 is 2 and S_2 is 5.

20. A transportation engineer uses the formula

$$R = \frac{2}{\dfrac{1}{g} + \dfrac{1}{r}}$$

to calculate average rate of speed on a round trip. Simplify this complex fraction.

FOR REVIEW

Solve.

21. $\dfrac{x + 10}{x - 8} = \dfrac{2}{3}$

22. $\dfrac{y + 3}{5} = \dfrac{y - 7}{3}$

23. $\dfrac{z + 1}{z - 5} = \dfrac{z + 2}{z - 6}$

24. A boat can go 38 mi on 6 gal of gas. How far can it go on 21 gal of gas?

25. If y varies directly as x, and $y = 42$ when $x = 18$, find y when $x = 3$.

26. The cost c of fish varies directly as the total weight w of the fish. If 8 lb of fish cost $18.60, what would 60 lb of fish cost?

27. A recipe calls for 2.5 cups of flour to feed 6 people. How many cups would be needed to feed 18 people?

The following exercises review material from Chapter 1 to help you prepare for the first section of the next chapter. Simplify each expression.

28. 3^2

29. $(-3)^2$

30. -3^2

31. $2^2 \cdot 5^2$

32. $(2 \cdot 5)^2$

33. $|4|$

34. $|-4|$

35. $-(4 - 5)^2$

8.7 EXERCISES B

Simplify.

1. $\dfrac{\dfrac{3}{5}}{\dfrac{2}{5}}$

2. $\dfrac{7+\dfrac{1}{2}}{2-\dfrac{1}{2}}$

3. $\dfrac{\dfrac{1}{6}+\dfrac{1}{8}}{\dfrac{5}{6}-\dfrac{1}{8}}$

4. $\dfrac{\dfrac{4}{x}}{\dfrac{2}{x}}$

5. $\dfrac{10}{1-\dfrac{1}{y}}$

6. $\dfrac{2+\dfrac{1}{a}}{7}$

7. $\dfrac{\dfrac{1}{x}+1}{\dfrac{1}{x}-1}$

8. $\dfrac{2-\dfrac{1}{y}}{\dfrac{2}{y}}$

9. $\dfrac{a-\dfrac{1}{a}}{1-\dfrac{1}{a}}$

10. $\dfrac{\dfrac{3}{a}+a}{\dfrac{a}{3}+a}$

11. $\dfrac{\dfrac{1}{3y}-2}{\dfrac{1}{y}+2}$

12. $\dfrac{\dfrac{1}{a}+\dfrac{2}{a}}{\dfrac{3}{a}+\dfrac{4}{a}}$

13. $\dfrac{\dfrac{2}{x-2}+1}{\dfrac{2}{x+2}-1}$

14. $\dfrac{\dfrac{1}{y^2}-4}{\dfrac{1}{y}-2}$

15. $\dfrac{\dfrac{3}{z}+z+4}{1+\dfrac{3}{z}}$

16. $\dfrac{1-\dfrac{2}{x}-\dfrac{3}{x^2}}{\dfrac{1}{x}+1}$

17. $\dfrac{\dfrac{5}{y+5}-1}{\dfrac{5}{y-5}+1}$

18. $\dfrac{z-5+\dfrac{6}{z}}{z-1-\dfrac{2}{z}}$

19. An environmental chemist might use the formula
$$w=\frac{\dfrac{1}{c}}{1+\dfrac{1}{d}}.$$
Express w as a simple fraction and use the result to find w when c is 3 and d is 14.

20. A navigation expert uses the formula
$$V=\frac{v_1+v_2}{1+\dfrac{v_1v_2}{c^2}}$$
to calculate the velocity of an airplane. Simplify this complex fraction.

FOR REVIEW

Solve.

21. $\dfrac{x-3}{x+4}=\dfrac{7}{8}$

22. $\dfrac{2y+8}{6}=\dfrac{y-9}{6}$

23. $\dfrac{z-3}{z-4}=\dfrac{z-6}{z-5}$

24. In a shipment of 200 tires, 6 were defective. How many defective tires would you expect in a shipment of 9000 tires?

25. If y varies directly as x, and $y = 14$ when $x = 5$, find y when $x = 20$.

26. The distance d traveled varies directly as the rate r of travel. If 280 miles are traveled at a rate of 40 mph, what distance could be traveled at 60 mph?

27. If 1 quart contains 0.95 liters, how many quarts are in 190 liters?

The following exercises review material from Chapter 1 to help you prepare for the first section of the next chapter. Simplify each expression.

28. 5^2

29. $(-5)^2$

30. -5^2

31. $4^2 \cdot 3^2$

32. $(4 \cdot 3)^2$

33. $|9|$

34. $|-9|$

35. $-(6 - 8)^2$

8.7 EXERCISES C

Simplify.

1. $\dfrac{\dfrac{a+b}{a-b} - \dfrac{a-b}{a+b}}{\dfrac{a}{a-b} + \dfrac{b}{a+b}}$

2. $\dfrac{\dfrac{1}{xy} + \dfrac{1}{yz} + \dfrac{1}{xz}}{\dfrac{x+y+z}{xyz}}$

[Answer: 1]

3. $a - \dfrac{a}{1 - \dfrac{a}{1-a}}$

$\left[\text{Answer: } \dfrac{-a^2}{1-2a}\right]$

CHAPTER 8 REVIEW

KEY WORDS

8.1 An **algebraic fraction** is a fraction that contains a variable in the numerator or denominator or both.

A **rational expression** is the quotient of two polynomials.

A fraction is **reduced to lowest terms** when 1 (or -1) is the only number or expression that divides both numerator and denominator.

8.2 The **reciprocal** of the fraction $\frac{a}{b}$ is $\frac{b}{a}$.

8.3 **Like fractions** have the same denominator.
Unlike fractions have different denominators.

8.4 The **least common denominator (LCD)** of two or more fractions is the smallest of all common denominators.

8.5 A **fractional equation** is one that contains one or more algebraic fractions.

8.6 The **ratio** of one number x to another number y is the quotient $x \div y$.

A **proportion** is an equation which states that two ratios are equal.

The **extremes** of the proportion $\frac{a}{b} = \frac{c}{d}$ are a and d.

The **means** of the proportion $\frac{a}{b} = \frac{c}{d}$ are b and c.

If $y = kx$ (**direct variation**) or $y = \frac{k}{x}$ (**inverse variation**), k is called the **constant of variation**.

8.7 A **complex fraction** is a fraction that contains at least one other fraction within it.

KEY CONCEPTS

8.1 1. Always be aware of values that must be excluded when working with algebraic fractions. For example, 4 cannot replace x in $\frac{3}{x-4}$.

2. When reducing fractions, divide out factors only, never terms.

3. When reducing fractions by dividing out common factors, do not make the numerator 0 when it is actually 1. For example,

$$\frac{\cancel{(x+1)}}{3\cancel{(x+1)}} = \frac{1}{3}, \quad not \quad \frac{0}{3}.$$

8.2 When multiplying or dividing fractions, factor numerators and denominators and divide out common factors. Do not find the LCD.

8.3 1. Find the LCD of all fractions when adding
8.4 or subtracting.

2. When subtracting fractions, use parentheses to avoid sign errors. For example,

$$\frac{2x+1}{x-5} - \frac{x-3}{x-5}$$
$$= \frac{2x+1-(x-3)}{x-5}$$
$$= \frac{2x+1-x+3}{x-5} = \frac{x+4}{x-5}.$$

3. Do not divide out terms when adding or subtracting algebraic fractions. For example,

$$\frac{3(x-2)}{(x+2)(x-2)} - \frac{7(x+2)}{(x-2)(x+2)}$$
$$= \frac{3(x-2)-7(x+2)}{(x+2)(x-2)}$$
$$= \frac{3x-6-7x-14}{(x+2)(x-2)},$$

$$not \quad \frac{3\cancel{(x-2)}-7\cancel{(x+2)}}{\cancel{(x+2)}\cancel{(x-2)}}$$

8.5 1. To solve a fractional equation, multiply both sides by the LCD of all fractions.

2. Check all answers in the *original* fractional equation. Watch for division by zero.

8.6 1. If y varies directly as x, then

$$y = kx$$

where k is the constant of variation.

2. If y varies inversely as x, then

$$y = \frac{k}{x}$$

where k is the constant of variation.

8.7 One way to simplify a complex fraction is to write the numerator as a single fraction, the denominator as a single fraction, and divide. Another way is to multiply each term of a complex fraction by the LCD of all fractions. For example,

$$\frac{1+\frac{1}{x}}{1-\frac{1}{x}} = \frac{\left(1+\frac{1}{x}\right)\cdot x}{\left(1-\frac{1}{x}\right)\cdot x}$$
$$= \frac{1\cdot x + \frac{1}{x}\cdot x}{1\cdot x - \frac{1}{x}\cdot x} = \frac{x+1}{x-1}.$$

REVIEW EXERCISES

Part I

8.1 *Find the values of the variable that must be excluded.*

1. $\frac{x+2}{x(x-1)}$

2. $\frac{a}{a^2+1}$

3. $\frac{y^2}{(y+1)(y-5)}$

Are the given fractions equivalent?

4. $\frac{x-2}{5x-2}, \frac{x}{5x}$

5. $\frac{a-3}{a^2-9}, \frac{1}{a+3}$

6. $\frac{x+y}{x}, \frac{x}{x+y}$

Reduce to lowest terms.

7. $\dfrac{x^2 + 2x}{x}$

8. $\dfrac{x^2 - 9}{x + 3}$

9. $\dfrac{x^2 + 2xy + y^2}{x + y}$

8.2 *Perform the indicated operations.*

10. $\dfrac{x}{x + 6} \cdot \dfrac{x^2 - 36}{x^2 - 6x}$

11. $\dfrac{y^2 - y - 2}{y^2 - 2y - 3} \cdot \dfrac{y^2 - 3y}{y + 2}$

12. $\dfrac{x^2 - 4y^2}{x + 2y} \cdot \dfrac{x + y}{x - 2y}$

13. $\dfrac{a + 7}{a - 7} \div \dfrac{a^2 + 7}{a^2 - 49}$

14. $\dfrac{4y^4}{y^2 - 1} \div \dfrac{2y^3}{y^2 - 2y + 1}$

15. $\dfrac{9x^2 - y^2}{x + y} \div \dfrac{3x + y}{x^2 - y^2}$

8.3
8.4 *Perform the indicated operations.*

16. $\dfrac{3x}{x - 2} + \dfrac{2x - 1}{2 - x}$

17. $\dfrac{2x}{x^2 - x} + \dfrac{x}{x^2 - 1}$

18. $\dfrac{2}{x^2 - 4x + 3} + \dfrac{3}{x^2 + x - 2}$

19. $\dfrac{x + 1}{x \quad 3} - \dfrac{x - 1}{3 - x}$

20. $\dfrac{2}{y^2 - y - 12} - \dfrac{1}{y^2 - 9}$

21. $\dfrac{a}{a^2 - 1} - \dfrac{a + 2}{a^2 + a - 2}$

22. $\dfrac{3}{x - 2} + \dfrac{x + 2}{x - 2} - \dfrac{2x}{2 - x}$

23. $\dfrac{6x}{x^2 - 9} - \dfrac{2}{x - 3} - \dfrac{5}{x + 3}$

24. $\dfrac{x}{x + y} + \dfrac{y}{x - y}$

8.5　*Solve.*

25. $\dfrac{2}{z+5} = \dfrac{3}{z}$

26. $\dfrac{3}{x+2} + \dfrac{2}{x-2} = \dfrac{3}{x^2-4}$

27. $\dfrac{x}{x-1} + 1 = \dfrac{x^2+1}{x^2-1}$

28. The denominator of a fraction is 2 less than the numerator. If 2 is added to both the numerator and denominator, the result has value $\frac{4}{3}$. Find the fraction.

29. Al is 9 years older than Jim and the quotient of their ages is $\frac{5}{6}$. How old is each?

30. If Tim can do a job in 3 days and Bob can do the same job in 11 days, how long would it take them to do the job if they worked together?

31. If it takes pipe A three times longer to fill a tank than pipe B and together they can fill it in 6 hours, how long would it take pipe B to fill the tank?

8.6　*Give as a ratio and simplify.*

32. 450 mi on 15 gal of gas

33. 1200 trees on 6 acres

34. $12 for 8 lb of meat

Solve.

35. $\dfrac{x-2}{x+2} = \dfrac{3}{2}$

36. $\dfrac{y+5}{y-4} = \dfrac{y+4}{y-5}$

37. $\dfrac{z-8}{z+3} = \dfrac{z+9}{z-2}$

38. In an election, the winner won by a 6 to 5 margin. If the winner had 270 votes, how many votes did the loser receive?

39. If 70 ft of wire weighs 15 lb, how much will 210 ft of the same wire weigh?

40. If y varies directly as x, and $y = 20$ when $x = 4$, find y when $x = 14$.

41. The cost c of candy varies directly as the total weight w of the candy. If 4 lb of candy cost $10.40, what would 6 lb cost?

8.7 *Simplify.*

42. $\dfrac{2 - \dfrac{1}{8}}{3 + \dfrac{3}{4}}$

43. $\dfrac{4x - \dfrac{1}{x}}{2 - \dfrac{1}{x}}$

44. $\dfrac{y - 2 + \dfrac{1}{y}}{1 - \dfrac{1}{y}}$

Part II

Solve.

45. If Sandy can frame a shed in 4 days and together Sandy and Clyde can frame the same shed in 3 days, how long would it take Clyde to frame it working by himself?

46. If y varies inversely as x, and $y = 45$ when $x = 5$, find y when $x = 15$.

47. $\dfrac{3}{x - 4} + \dfrac{4}{x + 4} = \dfrac{3}{x^2 - 16}$

48. $\dfrac{x + 1}{x} = \dfrac{5}{7}$

49. $x + \dfrac{6}{x} = 5$

50. $\dfrac{x}{x - 2} - 1 = \dfrac{3}{x^2 - 4}$

Find the values of the variable that must be excluded.

51. $\dfrac{x + 5}{(x - 4)(x + 1)}$

52. $\dfrac{y^2}{x^2 - 9}$

Perform the indicated operation and simplify.

53. $\dfrac{2}{x-5} + \dfrac{x}{x+5}$

54. $\dfrac{x+1}{x-2} - \dfrac{x-1}{x+3}$

55. $\dfrac{x^2-4x+3}{x^2+7x+10} \cdot \dfrac{x^2-25}{x^2-9}$

56. $\dfrac{y^2-3y}{y^2-y-6} \div \dfrac{y}{y+2}$

57. $\dfrac{1 + \dfrac{3}{x} - \dfrac{10}{x^2}}{1 - \dfrac{5}{x} + \dfrac{6}{x^2}}$

58. $\dfrac{5}{y-1} - \dfrac{6}{y+1} - \dfrac{1}{2}$

ANSWERS: 1. 0, 1   2. none   3. 5, −1   4. no   5. yes   6. no   7. $x+2$ 8. $x-3$ 9. $x+y$ 10. 1   11. $\frac{y(y-2)}{y+2}$
12. $x+y$ 13. $\frac{(a+7)^2}{a^2+7}$ 14. $\frac{2y(y-1)}{y+1}$ 15. $(3x-y)(x-y)$ 16. $\frac{x+1}{x-2}$ 17. $\frac{3x+2}{(x-1)(x+1)}$ 18. $\frac{5}{(x-3)(x+2)}$ 19. $\frac{2x}{x-3}$
20. $\frac{y-2}{(y-4)(y+3)(y-3)}$ 21. $\frac{-1}{(a+1)(a-1)}$ 22. $\frac{3x+5}{x-2}$ 23. $\frac{9-x}{(x+3)(x-3)}$ 24. $\frac{x^2+y^2}{x^2-y^2}$ 25. −15   26. 1   27. −2   28. $\frac{6}{4}$ 29. Al is
54, Jim is 45   30. $\frac{33}{14}$ days   31. 8 hr   32. 30 mpg   33. 200 trees per acre   34. $1.50 per lb   35. −10   36. no solution   37. $-\frac{1}{2}$ 38. 225 votes   39. 45 lb   40. 70   41. $15.60   42. $\frac{1}{2}$ 43. $2x+1$ 44. $y-1$ 45. 12 days   46. 15
47. 1   48. $-\frac{7}{2}$ 49. 2 or 3   50. $-\frac{1}{2}$ 51. 4, −1   52. 3, −3   53. $\frac{x^2-3x+10}{(x-5)(x+5)}$ 54. $\frac{7x+1}{(x-2)(x+3)}$ 55. $\frac{(x-1)(x-5)}{(x+2)(x+3)}$ 56. 1
57. $\frac{x+5}{x-3}$ 58. $\frac{-y^2-2y+23}{2(y-1)(y+1)}$

1. What are the values of the variable that must be excluded in

   $\dfrac{5}{x(x+5)}$?

   1. _____

2. Are the fractions $\dfrac{y}{5}$ and $\dfrac{6y}{30}$ equivalent?

   2. _____

3. Reduce to lowest terms. $\dfrac{a+2}{a^2+2a}$

   3. _____

4. Multiply. $\dfrac{x^2-16}{7x^2} \cdot \dfrac{35x}{x-4}$

   4. _____

5. Divide. $\dfrac{y}{y^2+y-12} \div \dfrac{y^2+3y}{y^2+7y+12}$

   5. _____

6. Add. $\dfrac{3}{a-b} + \dfrac{2}{b-a}$

   6. _____

7. Add. $\dfrac{x^2-6x}{x^2-4} + \dfrac{x}{x-2}$

   7. _____

8. Subtract. $\dfrac{4}{y^2-1} - \dfrac{3}{y^2-y-2}$

   8. _____

9. Solve. $\dfrac{x}{x+1} = \dfrac{x^2+1}{x^2-1} + \dfrac{3}{x-1}$

9. _____

10. The denominator of a fraction is 3 more than the numerator. If 1 is subtracted from both the numerator and denominator, the result has value $\frac{3}{4}$. Find the fraction.

10. _____

11. A car needed 16 gal of gas to go 352 mi. How many gallons would be needed to go 814 mi?

11. _____

12. The cost c of ground beef varies directly as the total weight w of the meat purchased. If 5 lb of ground beef cost $5.75, what would 12 lb cost?

12. _____

13. Ralph can do a job in 3 days and Harry can do the same job in 5 days. How long would it take them to do the same job if they work together?

13. _____

14. Simplify. $\dfrac{\dfrac{1}{y} - 1}{\dfrac{1}{y} - y}$

14. _____

15. If y varies inversely as x, and $y = 20$ when $x = 4$, find y when $x = 16$.

15. _____

Radicals

9.1 ROOTS AND RADICALS

1 PERFECT SQUARES AND SQUARE ROOTS

To introduce our study of radicals we review material from Section 2.5. Consider the following.

$$2^2 = 2 \cdot 2 = 4$$
$$(-2)^2 = (-2)(-2) = 4$$

When an integer is squared, the result is a **perfect square.** Thus 4 is a perfect square, since $4 = 2^2$. Either of the identical factors of a perfect square is a *square root* of the number. From above, 2 is a square root of 4 and since $(-2)^2 = 4$, -2 is also a square root of 4.

In addition to whole number perfect squares, there are also fractional perfect squares. For example,

$$\frac{4}{9} \text{ is a perfect square since } \frac{4}{9} = \left(\frac{2}{3}\right)^2,$$

$$\text{and } \frac{2}{3} \text{ is a square root of } \frac{4}{9}.$$

There are fractional perfect squares that do not have perfect square numerators and perfect square denominators. For example, $\frac{8}{18}$ is a perfect square since

$$\frac{8}{18} = \frac{4 \cdot \cancel{2}}{9 \cdot \cancel{2}} = \frac{4}{9},$$

which is a perfect square.

We saw in Chapter 2 that some square roots are irrational numbers. As we will see later, these numbers are included in the following definition. If a and x are real numbers such that

$$a^2 = x$$

then a is a **square root** of x. In general, every *positive* real number has two square roots. If a is a square root of x, then both a and $-a$ are square roots of x since

$$a^2 = x \quad \text{and} \quad (-a)^2 = x.$$

The nonnegative square root of a positive real number is called its **principal square root,** and is denoted by a **radical,** $\sqrt{}$. For example, if we write $\sqrt{4}$, we call 4 the **radicand** and 2 its principal square root. That is,

$$\sqrt{4} = 2. \qquad \text{2 is the principal (nonnegative) square root of 4}$$

To indicate the other square root of 4, we place a minus sign in front of the radical. Thus,

$$-\sqrt{4} = -2.$$

The positive and negative square roots of a positive real number can be represented together with the symbol $\pm$. For example,

$$\pm\sqrt{4} = \pm 2$$

represents both 2 and -2, and is read "plus or minus 2." Of course 0 has only one square root, namely 0, since $-0 = 0$.

We have seen that every positive real number has two square roots (one positive and one negative), that 0 has only one square root (namely 0 itself), but what about negative numbers? Does -9 have a square root, for example? The obvious choices would be -3 or 3, but

$$(-3)^2 = 9 \quad \text{and} \quad 3^2 = 9.$$

In fact, since $a^2 \geq 0$ for any real number a, a^2 could never be equal to the *negative* number -9. Thus, since no real number squared can be negative, negative numbers have no real square roots. We say that $\sqrt{-9}$, for example, is not a real number.

EXAMPLE 1 EVALUATING SQUARE ROOTS

Evaluate the following radicals.

(a) $\sqrt{100} = 10 \qquad 10^2 = 100$

(b) $-\sqrt{100} = -10$

(c) $\pm\sqrt{100} = \pm 10$ (two numbers, 10 and -10)

(d) $\sqrt{-100}$ is not a real number

(e) $\sqrt{\dfrac{25}{81}} = \dfrac{5}{9}$

(f) $\sqrt{\dfrac{12}{75}} = \sqrt{\dfrac{4 \cdot 3}{25 \cdot 3}} = \sqrt{\dfrac{4}{25}} = \dfrac{2}{5}$

(g) $\sqrt{1} = 1$

(h) $\sqrt{5^2} = \sqrt{25} = 5$

(i) $\sqrt{(-5)^2} = \sqrt{25} = 5$

(j) $\sqrt{-5^2} = \sqrt{-25}$ is not a real number.
Note the difference between $(-5)^2$ and -5^2.

PRACTICE EXERCISE 1

Evaluate the following radicals.

(a) $\sqrt{169}$

(b) $-\sqrt{169}$

(c) $\pm\sqrt{169}$

(d) $\sqrt{-169}$

(e) $\sqrt{\dfrac{144}{25}}$

(f) $\sqrt{\dfrac{50}{32}}$

(g) $-\sqrt{1}$

(h) $\sqrt{9^2}$

(i) $\sqrt{(-9)^2}$

(j) $\sqrt{-9^2}$

Answers: (a) **13** (b) -13
(c) ± 13 (d) **not a real number**
(e) $\frac{12}{5}$ (f) $\frac{5}{4}$ (g) -1 (h) 9
(i) 9 (j) **not a real number**

---

❷ EVALUATING $\sqrt{x^2}$

In Example 1 we saw that

$$\sqrt{5^2} = 5 \quad \text{and} \quad \sqrt{(-5)^2} = 5.$$

This is also true in general. Thus, if x is a nonnegative number, $\sqrt{x^2} = x$. For example, $\sqrt{5^2} = 5$. However, if x is negative then $\sqrt{x^2} = -x$. For example, $\sqrt{(-5)^2} = -(-5) = 5$. To make it clear that the principal square root of a number is positive, we write

$$\sqrt{x^2} = |x|,$$

where $|x|$ is the absolute value of x. To avoid problems of this nature, we will assume that *all variables represent nonnegative real values*. Then for $x \geq 0$,

$$\sqrt{x^2} = x. \qquad \text{Since } |x| = x \text{ in this case}$$

We will also assume that *all algebraic expressions under radicals are nonnegative real numbers*. Thus, for example,

$$\sqrt{(x - 7)^2} = x - 7,$$

since we are assuming that $x - 7 \geq 0$.

EXAMPLE 2 EVALUATING RADICALS	PRACTICE EXERCISE 2

Evaluate the following radicals assuming all variables and algebraic expressions are nonnegative.

(a) $\sqrt{49} = \sqrt{7^2} = 7$

(b) $\sqrt{y^2} = y$

(c) $\sqrt{25x^2} = \sqrt{5^2 x^2} = \sqrt{(5x)^2} = 5x \qquad a^n b^n = (ab)^n$

(d) $\sqrt{x^2 + 6x + 9} = \sqrt{(x + 3)^2} = x + 3$

(e) $-\sqrt{81} = -\sqrt{9^2} = -9$

(f) $-\sqrt{u^2} = -u$

Evaluate the following radicals assuming all variables and algebraic expressions are nonnegative.

(a) $\sqrt{36}$

(b) $\sqrt{a^2}$

(c) $\sqrt{36a^2}$

(d) $\sqrt{x^2 - 6x + 9}$

(e) $-\sqrt{196}$

(f) $-\sqrt{w^2}$

Answers: (a) 6   (b) a (c) $6a$
(d) $x - 3$ (e) -14 (f) $-w$

❸ USING A CALCULATOR

The square roots we have looked at so far have all resulted in rational numbers. For example, $\sqrt{25}$ is the rational number 5. Many square roots, such as $\sqrt{5}$, do not have this property. Since 5 is not a perfect square, that is, since there is no integer a such that $a^2 = 5$, $\sqrt{5}$ cannot be simplified further. In fact, $\sqrt{5}$ is an irrational number for which an approximate rational number must be used in a practical situation. There are two ways to find such an approximation, with a table of square roots (like the one at the back of this text), or with a calculator. Whenever we need to find the approximate value of a square root, we will use a calculator with a $\boxed{\sqrt{}}$ key.

To find a rational-number approximation for $\sqrt{5}$, use the following steps on your calculator.

$$5 \; \boxed{\sqrt{}} \; \rightarrow \; \boxed{2.2360680}$$

To see that 2.2360680 is indeed an approximation of $\sqrt{5}$, enter 2.2360680 on your calculator and square it using the $\boxed{x^2}$ key.

$$2.2360680 \; \boxed{x^2} \; \rightarrow \; \boxed{5.0000001}$$

4 APPLICATIONS OF RADICALS

In applied problems such as the next example, the approximate value of a square root is needed.

EXAMPLE 3 APPLYING RADICALS	PRACTICE EXERCISE 3

If P dollars are invested in an account for two years and at the end of this period the value of the account is A dollars, then the annual rate of return r on the investment can be found by the formula

$$r = \sqrt{\frac{A}{P}} - 1.$$

Suppose Diane purchased a rare book for $1000 and two years later sold it for $1500. What was the annual rate of return on Diane's investment?

Since $A = 1500$ and $P = 1000$, we substitute into the formula to find r.

$$r = \sqrt{\frac{A}{P}} - 1 = \sqrt{\frac{1500}{1000}} - 1$$

$$\approx 0.2247449$$

The calculator steps used to find r are:

1500 $\boxed{\div}$ 1000 $\boxed{=}$ $\boxed{\sqrt{}}$ $\boxed{-}$ 1 $\boxed{=}$ → $\boxed{0.2247449}$[1]

Rounded to three decimal places, $r \approx 0.225$, which translates to 22.5%. Thus, Diane realized an approximate annual return of 22.5% on her investment.

If Pamela invested $5000 in bonds and two years later sold the bonds for $6200, what was the annual rate of return?

Answer: 11.4%

///////////// **CAUTION** /////////////

Do not get in the habit of reaching immediately for your calculator when given a radical expression. Using a calculator, for example, to evaluate $\sqrt{4}$ would be as ridiculous as using it to find $2 + 3$. Throughout this chapter you should use a calculator only for approximate values of radicals whose square roots are not integers.

///////////

9.1 EXERCISES A

Tell whether each number has 0, 1, or 2 square roots.

1. 25　　　　　**2.** -25　　　　　**3.** 0　　　　　**4.** 175

Evaluate each square root assuming that all variables and algebraic expressions under the radical are nonnegative.

5. $\sqrt{6^2}$　　　　**6.** $\sqrt{(-6)^2}$　　　　**7.** $\sqrt{-6^2}$　　　　**8.** $\sqrt{121}$

---

[1]Calculator steps are given using algebraic logic. If you have a calculator that uses Reverse Polish Notation (RPN), consult your operator's manual.

9. $-\sqrt{121}$ **10.** $\pm\sqrt{121}$ **11.** $\sqrt{\dfrac{9}{4}}$ **12.** $\pm\sqrt{\dfrac{9}{4}}$

13. $\sqrt{\dfrac{144}{25}}$ **14** $\sqrt{\dfrac{1000}{10}}$ **15.** $\sqrt{\dfrac{50}{2}}$ **16.** $\sqrt{-\dfrac{49}{9}}$

17. $\sqrt{(x+1)^2}$ **18.** $\sqrt{(x-1)^2}$ **19** $\sqrt{49a^2}$ **20.** $\sqrt{x^2y^2}$

21. $\sqrt{-x^2y^2}$ **22.** $-\sqrt{x^2y^2}$ **23** $\sqrt{x^2+10x+25}$ **24.** $\sqrt{y^2-8y+16}$

In a particular manufacturing plant, the productivity, p, is related to the work force, w, by the equation $p = \sqrt{w}$.

25 Find p when $w = 25$.   **26.** Find p when $w = 8^2$.   **27.** Find p when $w = 0$.

28 Find w when $p = 4$.   **29.** Find w when $p = 12$.   **30.** Find p when $w = -4$.

In Exercises 31–38, use a calculator to find each square root correct to three decimal places.

31. $\sqrt{3}$ **32.** $\sqrt{8}$ **33.** $\sqrt{41}$ **34.** $\sqrt{73}$

35. $\sqrt{26.5}$ **36.** $\sqrt{34.8}$ **37.** $\sqrt{627.84}$ **38.** $\sqrt{543.09}$

Use the formula given in Example 3 to solve each investment problem given in Exercises 39–40.

39. Suppose Janet Condon invested $300 in the stock market and two years later the stocks were worth $525. Find the approximate annual rate of return on Janet's investment.

40. If Peter Horn purchased an antique car for $4500 and sold it two years later for $6700, what was the approximate annual rate of return on his investment?

FOR REVIEW

Exercises 41–44 review material from Section 6.1 to help you prepare for the next section. Simplify each exponential expression.

41. $9x^5x^2$ **42.** $(2x^3)(18x^7)$ **43.** $\dfrac{8x^6}{2x^2}$ **44.** $\dfrac{50x^2y^3}{2x^6y}$

ANSWERS: 1. 2 2. 0 3. 1 4. 2 5. 6 6. 6 7. meaningless 8. 11 9. −11 10. ±11 (11 and −11) 11. $\frac{3}{2}$ 12. $\pm\frac{3}{2}$ ($\frac{3}{2}$ and $-\frac{3}{2}$) 13. $\frac{12}{5}$ 14. 10 15. 5 16. meaningless 17. $x+1$ 18. $x-1$ 19. $7a$ 20. xy 21. meaningless 22. $-xy$ 23. $x+5$ 24. $y-4$ 25. 5 26. 8 27. 0 28. 16 29. 144 30. no value for p when w is negative 31. 1.732 32. 2.828 33. 6.403 34. 8.544 35. 5.148 36. 5.899 37. 25.057 38. 23.304 39. 32.3% 40. 22.0% 41. $9x^7$ 42. $36x^{10}$ 43. $4x^4$ 44. $\frac{25y^2}{x^4}$

9.1 EXERCISES B

Tell whether each number has 0, 1, or 2 square roots.

1. 81   **2.** −81   **3.** 19   **4.** −0

Evaluate each square root assuming that all variables and algebraic expressions under the radical are nonnegative.

5. $\sqrt{49}$ **6.** $\sqrt{(-7)^2}$ **7.** $\sqrt{-7^2}$ **8.** $\sqrt{144}$

9. $-\sqrt{144}$ **10.** $\pm\sqrt{144}$ **11.** $\pm\sqrt{\dfrac{16}{9}}$ **12.** $\sqrt{\dfrac{121}{16}}$

13. $\sqrt{\dfrac{75}{3}}$ **14.** $\pm\sqrt{\dfrac{64}{25}}$ **15.** $\sqrt{-0}$ **16.** $\sqrt{-\dfrac{9}{16}}$

17. $\sqrt{(y+2)^2}$ **18.** $\sqrt{(y-2)^2}$ **19.** $\sqrt{36x^2}$ **20.** $\sqrt{u^2v^2}$

21. $\sqrt{-u^2v^2}$ **22.** $-\sqrt{u^2v^2}$ **23.** $\sqrt{x^2+12x+36}$ **24.** $\sqrt{y^2-14y+49}$

In a particular retail operation, the cost of production c, is related to the number of items sold, n, by the equation $c = \sqrt{n}$.

25. Find c when $n = 49$.   **26.** Find c when $n = 0$.   **27.** Find c when $n = 3^2$.

28. Find n when $c = 6$.   **29.** Find n when $c = 11$.   **30.** Find c when $n = -1$.

In Exercises 31–38, use a calculator to find each square root correct to three decimal places.

31. $\sqrt{7}$ **32.** $\sqrt{11}$ **33.** $\sqrt{57}$ **34.** $\sqrt{91}$

35. $\sqrt{14.8}$ **36.** $\sqrt{63.7}$ **37.** $\sqrt{340.22}$ **38.** $\sqrt{992.13}$

Use the formula given in Example 3 to solve each investment problem given in Exercises 39–40.

39. Robert Weaver invested $2150 in savings. At the end of two years the account was worth $2490. Find the approximate annual rate of return.

40. Michele Furr bought a painting for $3200. Two years later it was valued at $4100. What was the approximate annual rate of return?

FOR REVIEW

Exercises 41–44 review material from Section 6.1 to help you prepare for the next section. Simplify each exponential expression.

41. $21y^8y^4$ **42.** $(3y^5)(15y^3)$ **43.** $\dfrac{30y^6}{10y^5}$ **44.** $\dfrac{99x^3y^7}{11x^7y}$

9.1 EXERCISES C

Use the definition of cube root, $\sqrt[3]{a^3} = a$, to do the following exercises.

1. $\sqrt[3]{5^3}$ **2.** $\sqrt[3]{27}$ **3.** $\sqrt[3]{8x^3}$ [Answer: $2x$]   **4.** $\sqrt[3]{(2x+1)^6}$

9.2 SIMPLIFYING RADICALS

═══════════════ STUDENT GUIDEPOSTS ═══════════════

1 Square Roots of Products   **3** Rationalizing a Denominator
2 Square Roots of Quotients   **4** Simplified Radicals

1 SQUARE ROOTS OF PRODUCTS

In the last section we saw that $\sqrt{9} = \sqrt{3^2} = 3$, but what about square roots such as $\sqrt{18}$? We might use a calculator but this would result in only an approximation of $\sqrt{18}$. However, since $18 = 9 \cdot 2$, we can write

$$\sqrt{18} = \sqrt{9 \cdot 2} = \sqrt{9} \cdot \sqrt{2}$$
$$= 3\sqrt{2}.$$

This is an example of the following rule.

Simplifying Rule for Products

If $a \geq 0$ and $b \geq 0$, then

$$\sqrt{ab} = \sqrt{a} \sqrt{b}.$$

The square root of a product is equal to the product of square roots.

What do we look for to take advantage of this rule? In the example

$$\sqrt{18} = \sqrt{9 \cdot 2} = \sqrt{9} \sqrt{2} = 3\sqrt{2},$$

9 is a perfect square. Thus we look for perfect square factors of the radicand (18 in this case). Notice that when there are no perfect square factors the rule does not help us. For example, it does no good in trying to simplify $\sqrt{6} = \sqrt{2 \cdot 3}$ since neither 2 nor 3 is a perfect square. In fact $\sqrt{6}$ is in its simplest form.

EXAMPLE 1 SIMPLIFYING SQUARE ROOTS OF PRODUCTS

Simplify the radicals.

(a) $\sqrt{27} = \sqrt{9 \cdot 3}$ 9 is a perfect square

$\quad\quad = \sqrt{9} \sqrt{3}$ $\sqrt{ab} = \sqrt{a} \sqrt{b}$

$\quad\quad = 3\sqrt{3}$ $\sqrt{9} = \sqrt{3^2} = 3$

(b) $\sqrt{52} = \sqrt{4 \cdot 13}$ 4 is a perfect square

$\quad\quad = \sqrt{4} \sqrt{13}$ $\sqrt{ab} = \sqrt{a} \sqrt{b}$

$\quad\quad = 2\sqrt{13}$ $\sqrt{4} = \sqrt{2^2} = 2$

(c) $\sqrt{15} = \sqrt{3 \cdot 5}$ Cannot be simplified since neither 3 nor 5 is a perfect square

$\quad\quad = \sqrt{15}$

(d) $\sqrt{3^4} = \sqrt{(3^2)^2}$ $3^4 = 3^{2 \cdot 2} = (3^2)^2$, which is a perfect square

$\quad\quad = 3^2 = 9$ $\sqrt{(3^2)^2} = 3^2$

PRACTICE EXERCISE 1

Simplify the radicals.

(a) $\sqrt{45}$

(b) $\sqrt{44}$

(c) $\sqrt{55}$

(d) $\sqrt{2^4}$

(e) $2\sqrt{3^6} = 2\sqrt{(3^3)^2}$ $3^6 = 3^{3\cdot2} = (3^3)^2$

$\qquad = 2\cdot3^3 = 2\cdot27 = 54$

(f) $\sqrt{3^7} = \sqrt{3^6\cdot3}$ $3^7 = 3^{6+1} = 3^6\cdot3^1 = 3^6\cdot3$

$\qquad = \sqrt{(3^3)^2}\,\sqrt{3}$

$\qquad = 3^3\sqrt{3} = 27\sqrt{3}$

(e) $3\sqrt{2^6}$

(f) $\sqrt{2^9}$

Answers: (a) $3\sqrt{5}$ (b) $2\sqrt{11}$
(c) cannot be simplified   (d) 4
(e) 24   (f) $16\sqrt{2}$

As Example 1 illustrates, to simplify radicals we need to recall the rules of exponents such as

$$a^{m+n} = a^m a^n \quad \text{and} \quad a^{m\cdot n} = (a^m)^n.$$

Notice in Example 1(d) and (e) when the exponent under the radical is even, the radicand is a perfect square. When the exponent under the radical is odd, we rewrite the radicand as a product with a perfect square factor. For example, in Example 1(f), we rewrite $3^7 = 3^6\cdot3$ and then simplify further. This process also works with variables. For example,

$\sqrt{x^7} = \sqrt{x^6\cdot x}$ x^6 is a perfect square

$\qquad = \sqrt{x^6}\,\sqrt{x}$ $\sqrt{ab} = \sqrt{a}\,\sqrt{b}$

$\qquad = \sqrt{(x^3)^2}\,\sqrt{x}$ $x^6 = x^{3\cdot2} = (x^3)^2$

$\qquad = x^3\sqrt{x}.$

Remember that all our variables are positive.

EXAMPLE 2   SQUARE ROOTS INVOLVING VARIABLES

Simplify the radicals.

(a) $\sqrt{x^4} = \sqrt{(x^2)^2}$ $(x^2)^2$ is a perfect square

$\qquad = x^2$

(b) $\sqrt{x^5} = \sqrt{x^4\cdot x}$ $x^5 = x^{4+1} = x^4\cdot x$

$\qquad = \sqrt{x^4}\,\sqrt{x}$ $\sqrt{ab} = \sqrt{a}\,\sqrt{b}$

$\qquad = \sqrt{(x^2)^2}\,\sqrt{x}$

$\qquad = x^2\sqrt{x}$

(c) $\sqrt{27a^3} = \sqrt{9\cdot3\cdot a^2\cdot a}$ 9 and a^2 are perfect squares

$\qquad = \sqrt{9\cdot a^2\cdot 3a}$ $3a$ will be left under the radical

$\qquad = \sqrt{9}\cdot\sqrt{a^2}\cdot\sqrt{3a}$ The simplifying rule expanded to three factors, $\sqrt{abc} = \sqrt{a}\,\sqrt{b}\,\sqrt{c}$

$\qquad = 3\cdot a\cdot\sqrt{3a}$

$\qquad = 3a\sqrt{3a}$

(d) $\sqrt{x^4y^7} = \sqrt{x^4y^6\cdot y}$ x^4 and y^6 are perfect squares

$\qquad = \sqrt{x^4}\,\sqrt{y^6}\,\sqrt{y}$ $\sqrt{abc} = \sqrt{a}\,\sqrt{b}\,\sqrt{c}$

$\qquad = \sqrt{(x^2)^2}\,\sqrt{(y^3)^2}\,\sqrt{y}$

$\qquad = x^2y^3\sqrt{y}$

PRACTICE EXERCISE 2

Simplify the radicals.

(a) $\sqrt{y^6}$

(b) $\sqrt{y^7}$

(c) $\sqrt{45x^2}$

(d) $\sqrt{a^5b^8}$

(e) $\sqrt{288ab^2} = \sqrt{2 \cdot 144 \cdot a \cdot b^2}$ 144 and b^2 are perfect squares

$= \sqrt{144 \cdot b^2 \cdot 2a}$

$= \sqrt{144}\ \sqrt{b^2}\ \sqrt{2a}$

$= 12b\sqrt{2a}$

(e) $\sqrt{162x^3b^4}$

Answers: (a) y^3 (b) $y^3\sqrt{y}$
(c) $3x\sqrt{5}$ (d) $a^2b^4\sqrt{a}$
(e) $9xb^2\sqrt{2x}$

② SQUARE ROOTS OF QUOTIENTS

We can use the simplifying rule above to simplify

$$\sqrt{\frac{9}{4}} = \sqrt{\left(\frac{3}{2}\right)^2} = \frac{3}{2}.$$

This can also be simplified as

$$\sqrt{\frac{9}{4}} = \frac{\sqrt{9}}{\sqrt{4}} = \frac{3}{2}.$$

This is an example of the following rule which is used for radical expressions involving fractions.

Simplifying Rule for Quotients

If $a \ge 0$ and $b > 0$, then

$$\sqrt{\frac{a}{b}} = \frac{\sqrt{a}}{\sqrt{b}}.$$

The square root of a quotient is equal to the quotient of the square roots.

EXAMPLE 3   SIMPLIFYING SQUARE ROOTS OF QUOTIENTS

Simplify the radicals.

(a) $\sqrt{\frac{27}{4}} = \frac{\sqrt{27}}{\sqrt{4}}$ $\sqrt{\frac{a}{b}} = \frac{\sqrt{a}}{\sqrt{b}}$

$= \frac{\sqrt{9 \cdot 3}}{2}$ $\sqrt{4} = 2$

$= \frac{\sqrt{9}\ \sqrt{3}}{2} = \frac{3\sqrt{3}}{2}$ $\sqrt{ab} = \sqrt{a}\ \sqrt{b}$

(b) $\sqrt{\frac{4x^3}{y^2}} = \frac{\sqrt{4x^3}}{\sqrt{y^2}}$ $\sqrt{\frac{a}{b}} = \frac{\sqrt{a}}{\sqrt{b}}$

$= \frac{\sqrt{4 \cdot x^2 \cdot x}}{y}$ $\sqrt{y^2} = y$

$= \frac{\sqrt{4}\ \sqrt{x^2}\ \sqrt{x}}{y}$ $\sqrt{abc} = \sqrt{a}\ \sqrt{b}\ \sqrt{c}$

$= \frac{2x\sqrt{x}}{y}$

PRACTICE EXERCISE 3

Simplify the radicals.

(a) $\sqrt{\frac{8}{9}}$

(b) $\sqrt{\frac{25a^5}{w^2}}$

(c) $\sqrt{\dfrac{48x^3y^2}{3xy}} = \sqrt{\dfrac{3 \cdot 16 \cdot x^3 \cdot y^2}{3xy}}$ Simplify under the radical first

$\qquad = \sqrt{16x^2y}$ $\dfrac{x^3}{x} = x^{3-1} = x^2, \ \dfrac{y^2}{y} = y^{2-1} = y$

$\qquad = \sqrt{16}\ \sqrt{x^2}\ \sqrt{y}$

$\qquad = 4x\sqrt{y}$

(c) $\sqrt{\dfrac{32a^4b^3}{2a^2b^2}}$

Answers: (a) $\dfrac{2\sqrt{2}}{3}$ (b) $\dfrac{5a^2\sqrt{a}}{w}$

(c) $4a\sqrt{b}$

Notice that in Example 3(c) we used

$$\frac{a^m}{a^n} = a^{m-n}.$$

❸ RATIONALIZING A DENOMINATOR

Consider .he following problem.

$$\sqrt{\frac{9}{7}} = \frac{\sqrt{9}}{\sqrt{7}} = \frac{3}{\sqrt{7}}$$ $\sqrt{7}$ is in the denominator

When a square root is left in the denominator, the radical expression is not considered completely simplified. The process of removing radicals from the denominator (making the denominator a rational number) is called **rationalizing the denominator.** We do this by making the expression under the radical in the denominator a perfect square.

To rationalize denominators we need the following property. Since $\sqrt{ab} = \sqrt{a}\ \sqrt{b}$, if we let $a = b$, then

$$\sqrt{aa} = \sqrt{a}\ \sqrt{a}$$
$$\sqrt{a^2} = \sqrt{a}\ \sqrt{a}$$
$$a = \sqrt{a}\ \sqrt{a}.$$

Thus, $\sqrt{7}\ \sqrt{7} = 7$ and $\sqrt{x^3}\ \sqrt{x^3} = x^3$.

EXAMPLE 4   RATIONALIZING DENOMINATORS

Rationalize the denominators.

(a) $\dfrac{3}{\sqrt{7}} = \dfrac{3\sqrt{7}}{\sqrt{7}\ \sqrt{7}}$ Multiply numerator and denominator by $\sqrt{7}$

$\qquad = \dfrac{3\sqrt{7}}{7}$ Since $\sqrt{7}\ \sqrt{7} = 7$, the denominator is now rational

(b) $\sqrt{\dfrac{25x^2}{y}} = \dfrac{\sqrt{25x^2}}{\sqrt{y}}$ $\sqrt{\dfrac{a}{b}} = \dfrac{\sqrt{a}}{\sqrt{b}}$

$\qquad = \dfrac{\sqrt{25}\ \sqrt{x^2}}{\sqrt{y}}$

$\qquad = \dfrac{5x}{\sqrt{y}}$

$\qquad = \dfrac{5x\sqrt{y}}{\sqrt{y}\ \sqrt{y}} = \dfrac{5x\sqrt{y}}{y}$ The denominator is rationalized

PRACTICE EXERCISE 4

Rationalize the denominators.

(a) $\dfrac{5}{\sqrt{11}}$

(b) $\sqrt{\dfrac{49u^4}{w}}$

(c) $\sqrt{\dfrac{25x^2y}{yz^2}} = \sqrt{\dfrac{25x^2\cancel{y}}{\cancel{y}z^2}}$ Simplify under the radical first

$= \dfrac{\sqrt{25x^2}}{\sqrt{z^2}}$ $\sqrt{\dfrac{a}{b}} = \dfrac{\sqrt{a}}{\sqrt{b}}$

$= \dfrac{5x}{z}$ Do not need to rationalize

(c) $\sqrt{\dfrac{32x^7y^9}{3y^4}}$

Answers: (a) $\dfrac{5\sqrt{11}}{11}$ (b) $\dfrac{7u^2\sqrt{w}}{w}$

(c) $\dfrac{4x^3y^2\sqrt{6xy}}{3}$

④ SIMPLIFIED RADICALS

A Radical Is Considered Simplified

1. When there are no perfect square factors of the radicand.
2. When there are no fractions under the radical.
3. When there are no radicals in the denominator.

EXAMPLE 5   APPLICATION TO PHYSICS

If an object is dropped from a height of h feet, the time t in seconds it takes the object to reach the ground is given by the formula

$$t = \sqrt{\dfrac{h}{16}}.$$

Simplify this radical expression and use it to find the time it would take for a rock to reach the ground if dropped from a hot air balloon 144 feet above the ground. See Figure 9.1.

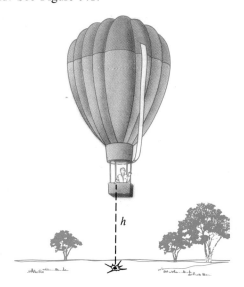

Figure 9.1

Simplify t.

$$t = \sqrt{\dfrac{h}{16}} = \dfrac{\sqrt{h}}{\sqrt{16}} = \dfrac{\sqrt{h}}{4}$$

PRACTICE EXERCISE 5

Simplify the expression

$$t = \sqrt{\dfrac{25h}{169}}$$

and use it to find t when $h = 49$.

Substitute 144 for h.

$$t = \frac{\sqrt{h}}{4} = \frac{\sqrt{144}}{4} = \frac{12}{4} = 3$$

It would take 3 seconds for the rock to reach the ground.

Answer: $\frac{35}{13}$

9.2   EXERCISES A

Simplify each of the radicals. Assume that all variables are positive.

1. $\sqrt{75}$

2. $\sqrt{10}$

3. $\sqrt{32}$

4. $3\sqrt{200}$

5. $\sqrt{147}$

6. $5\sqrt{36}$

7. $\sqrt{5^4}$

8. $\sqrt{625}$

9. $\sqrt{7^3}$

10. $\sqrt{5^5}$

11. $\sqrt{75y^2}$

12. $\sqrt{25x^3}$

13. $\sqrt{75y^3}$

14. $\sqrt{3^2x^3}$

15. $\sqrt{3^3y^3}$

16. $\sqrt{147x^5}$

17. $\sqrt{9x^2y^2}$

18. $\sqrt{27x^3y^2}$

19. $\sqrt{48x^2y^3}$

20. $\sqrt{75x^9y^5}$

21. $\sqrt{\dfrac{75}{49}}$

22. $\sqrt{\dfrac{8}{121}}$

23. $\sqrt{\dfrac{50}{32}}$

24. $\sqrt{\dfrac{25x^2}{y^2}}$

25. $\sqrt{\dfrac{16x^4}{y^4}}$

26. $\sqrt{\dfrac{75x^2}{y^2}}$

27. $\sqrt{\dfrac{75x^2y^4}{3}}$

28 $\sqrt{\dfrac{x^3y^3}{49}}$

29. $\dfrac{5}{\sqrt{3}}$

30. $\dfrac{8}{\sqrt{7}}$

31 $\sqrt{\dfrac{9}{5}}$

32. $\sqrt{\dfrac{45}{5}}$

33. $\sqrt{\dfrac{75x^2y^3}{yz^2}}$

34 $\sqrt{\dfrac{36y^2}{x}}$

35 $\sqrt{\dfrac{25x^3}{y^3z^4}}$

36. $\sqrt{\dfrac{48x^2y^2}{3x^4y^4}}$

In a wholesale operation, it has been estimated that the cost, c, is related to the number of items sold, n, by the equation $c = \sqrt{n}$.

37. Find c if $n = 81$.　　　**38.** Find c if $n = 8$.　　　**39.** Find c if $n = 0$.

40. Find c if $n = 75$.　　　**41.** Find c if $n = 2^4$.　　　**42.** Find c if $n = 2^5$.

A child drops a coin from the top of a building. Use the formula given in Example 5 to find the time it will take for the coin to reach the ground for each height h given in Exercises 43–46. Use a calculator to give the answer correct to the nearest tenth of a second if appropriate.

43. 256 ft　　　　　**44.** 100 ft　　　　　**45.** 150 ft　　　　　**46.** 740 ft

FOR REVIEW

Evaluate.

47. $-\sqrt{121}$　　　　　**48.** $\sqrt{-121}$　　　　　**49.** $\sqrt{-0}$

50. $\sqrt{\dfrac{49}{25}}$　　　　　**51.** $\sqrt{(x-7)^2}$　　　　　**52.** $\sqrt{x^2 - 10x + 25}$

ANSWERS: 1. $5\sqrt{3}$　2. $\sqrt{10}$　3. $4\sqrt{2}$　4. $30\sqrt{2}$　5. $7\sqrt{3}$　6. 30　7. 25　8. 25　9. $7\sqrt{7}$　10. $25\sqrt{5}$　11. $5y\sqrt{3}$
12. $5x\sqrt{x}$　13. $5y\sqrt{3y}$　14. $3x\sqrt{x}$　15. $3y\sqrt{3y}$　16. $7x^2\sqrt{3x}$　17. $3xy$　18. $3xy\sqrt{3x}$　19. $4xy\sqrt{3y}$　20. $5x^4y^2\sqrt{3xy}$
21. $\dfrac{5\sqrt{3}}{7}$　22. $\dfrac{2\sqrt{2}}{11}$　23. $\dfrac{5}{4}$　24. $\dfrac{5x}{y}$　25. $\dfrac{4x^2}{y^2}$　26. $\dfrac{5x\sqrt{3}}{y}$　27. $5xy^2$　28. $\dfrac{xy\sqrt{xy}}{7}$　29. $\dfrac{5\sqrt{3}}{3}$　30. $\dfrac{8\sqrt{7}}{7}$　31. $\dfrac{3\sqrt{5}}{5}$
32. 3　33. $\dfrac{5xy\sqrt{3}}{z}$　34. $\dfrac{6y\sqrt{x}}{x}$　35. $\dfrac{5x\sqrt{xy}}{y^2z^2}$　36. $\dfrac{4}{xy}$　37. 9　38. $2\sqrt{2}$　39. 0　40. $5\sqrt{3}$　41. 4　42. $4\sqrt{2}$
43. 4 sec　44. 2.5 sec　45. 3.1 sec　46. 6.8 sec　47. -11　48. meaningless　49. 0　50. $\dfrac{7}{5}$　51. $x-7$　52. $x-5$

9.2 EXERCISES B

Simplify each of the radicals. Assume that all variables are positive.

1. $\sqrt{98}$　　　　**2.** $\sqrt{35}$　　　　**3.** $\sqrt{50}$　　　　**4.** $2\sqrt{300}$

5. $\sqrt{108}$　　　　**6.** $4\sqrt{64}$　　　　**7.** $\sqrt{3^4}$　　　　**8.** $\sqrt{81}$

9. $\sqrt{5^3}$　　　　**10.** $\sqrt{7^5}$　　　　**11.** $\sqrt{98y^2}$　　　　**12.** $\sqrt{49x^3}$

13. $\sqrt{98y^3}$　　　　**14.** $\sqrt{5^2x^3}$　　　　**15.** $\sqrt{5^3x^3}$　　　　**16.** $\sqrt{125x^5}$

17. $\sqrt{16x^2y^2}$　　　　**18.** $\sqrt{32x^3y^2}$　　　　**19.** $\sqrt{27x^2y^3}$　　　　**20.** $\sqrt{12x^7y^9}$

21. $\sqrt{\dfrac{32}{81}}$　　　　**22.** $\sqrt{\dfrac{75}{144}}$　　　　**23.** $\sqrt{\dfrac{45}{20}}$　　　　**24.** $\sqrt{\dfrac{49x^2}{y^2}}$

25. $\sqrt{\dfrac{81x^4}{y^4}}$ 　　　 **26.** $\sqrt{\dfrac{147x^2}{y^2}}$ 　　　 **27.** $\sqrt{\dfrac{x^5y^5}{25}}$ 　　　 **28.** $\sqrt{\dfrac{147x^4y^7}{3}}$

29. $\dfrac{3}{\sqrt{5}}$ 　　　 **30.** $\dfrac{12}{\sqrt{11}}$ 　　　 **31.** $\sqrt{\dfrac{16}{3}}$ 　　　 **32.** $\sqrt{\dfrac{63}{7}}$

33. $\sqrt{\dfrac{4x^4}{y}}$ 　　　 **34.** $\sqrt{\dfrac{147x^3y^2}{xz^2}}$ 　　　 **35.** $\sqrt{\dfrac{50x^3y^3}{2x^5y^5}}$ 　　　 **36.** $\sqrt{\dfrac{16x^5}{y^5z^2}}$

In a manufacturing process, two quantities p and r are related by the equation $p = \sqrt{r}$.

37. Find p when $r = 64$. 　　　 **38.** Find p when $r = 27$. 　　　 **39.** Find p when $r = 0$.

40. Find p when $r = 32$. 　　　 **41.** Find p when $r = 3^4$. 　　　 **42.** Find p when $r = 3^5$.

A child drops a coin from the top of a building. Use the formula given in Example 5 to find the time it will take for the coin to reach the ground for each height h given in Exercises 43–46. Use a calculator to give the answer correct to the nearest tenth of a second if appropriate.

43. 64 ft 　　　 **44.** 196 ft 　　　 **45.** 270 ft 　　　 **46.** 580 ft

FOR REVIEW

Evaluate.

47. $-\sqrt{144}$ 　　　 **48.** $\sqrt{-144}$ 　　　 **49.** $\sqrt{0}$

50. $\sqrt{\dfrac{81}{4}}$ 　　　 **51.** $\sqrt{(x-3)^2}$ 　　　 **52.** $\sqrt{x^2 + 10x + 25}$

9.2 EXERCISES C

Simplify each of the radicals. Assume all variables are positive.

1. $\sqrt{\dfrac{27x^9y^7}{2x^6}}$ 　　　 **2.** $\sqrt{\dfrac{243x^5y^4}{8xyz^3}}$ 　　　 **3.** $\sqrt{\dfrac{75xy}{9x^5y^6z^7}}$

$\left[\text{Answer: } \dfrac{5\sqrt{3yz}}{3x^2y^3z^4}\right]$

Use $\sqrt[3]{a^3} = a$ in the following exercises.

4. $\sqrt[3]{y^5}$ 　　　 **5.** $\sqrt[3]{\dfrac{81x^5y^4}{8z^3}}$ 　　　 **6.** $\sqrt[3]{\dfrac{24x^4}{y}}$

$\left[\text{Answer: } \dfrac{2x\sqrt[3]{3xy^2}}{y}\right]$

9.3 MULTIPLICATION AND DIVISION OF RADICALS

1 PRODUCT OF SQUARE ROOTS

The simplifying rule for products, introduced in Section 9.2, allows us to simplify radicals like the following.

$$\sqrt{4 \cdot 9} = \sqrt{4}\,\sqrt{9} = 2 \cdot 3 = 6$$

The rule can also be used in the reverse order to multiply radicals. For example,

$$
\begin{aligned}
\sqrt{2}\,\sqrt{18} &= \sqrt{2 \cdot 18} && \text{Reverse of the simplifying rule} \\
&= \sqrt{2 \cdot 2 \cdot 9} && \text{Factor 18 into } 2 \cdot 9 \\
&= \sqrt{4 \cdot 9} && \text{4 and 9 are perfect squares} \\
&= \sqrt{4}\,\sqrt{9} = 2 \cdot 3 = 6.
\end{aligned}
$$

Here we used the simplifying rule for products from Section 9.2 in reverse order. This gives us the rule for multiplying radicals.

Multiplication Rule

If $a \geq 0$ and $b \geq 0$, then

$$\sqrt{a}\,\sqrt{b} = \sqrt{ab}.$$

The product of square roots is equal to the square root of the product.

EXAMPLE 1   MULTIPLYING RADICALS

Multiply and then simplify.

(a) $\sqrt{7}\,\sqrt{7} = \sqrt{7 \cdot 7}$ $\sqrt{a}\,\sqrt{b} = \sqrt{ab}$
$\phantom{\sqrt{7}\,\sqrt{7}} = \sqrt{7^2} = 7$ We did this in Section 9.2

(b) $\sqrt{5}\,\sqrt{20} = \sqrt{5 \cdot 20}$ $\sqrt{a}\,\sqrt{b} = \sqrt{ab}$
$\phantom{\sqrt{5}\,\sqrt{20}} = \sqrt{5 \cdot 5 \cdot 4}$
$\phantom{\sqrt{5}\,\sqrt{20}} = \sqrt{25 \cdot 4}$ 25 and 4 are perfect squares
$\phantom{\sqrt{5}\,\sqrt{20}} = \sqrt{25}\,\sqrt{4}$ $\sqrt{ab} = \sqrt{a}\,\sqrt{b}$
$\phantom{\sqrt{5}\,\sqrt{20}} = 5 \cdot 2 = 10$

(c) $\sqrt{6}\,\sqrt{75} = \sqrt{6 \cdot 75}$ $\sqrt{a}\,\sqrt{b} = \sqrt{ab}$
$\phantom{\sqrt{6}\,\sqrt{75}} = \sqrt{2 \cdot 3 \cdot 3 \cdot 25}$ Factor to find perfect squares
$\phantom{\sqrt{6}\,\sqrt{75}} = \sqrt{9 \cdot 25 \cdot 2}$ 2 will be left under the radical
$\phantom{\sqrt{6}\,\sqrt{75}} = \sqrt{9}\,\sqrt{25}\,\sqrt{2}$
$\phantom{\sqrt{6}\,\sqrt{75}} = 3 \cdot 5 \cdot \sqrt{2} = 15\sqrt{2}$

(d) $\sqrt{5}\,\sqrt{7y} = \sqrt{5 \cdot 7 \cdot y}$ $\sqrt{a}\,\sqrt{b} = \sqrt{ab}$
$\phantom{\sqrt{5}\,\sqrt{7y}} = \sqrt{35y}$ This is considered a simpler form than $\sqrt{5}\,\sqrt{7y}$

PRACTICE EXERCISE 1

Multiply and then simplify.

(a) $\sqrt{11}\,\sqrt{11}$

(b) $\sqrt{2}\,\sqrt{18}$

(c) $\sqrt{48}\,\sqrt{6}$

(d) $\sqrt{2x}\,\sqrt{13}$

(e) $\sqrt{3x} \sqrt{75x} = \sqrt{3x \cdot 75x}$ $\sqrt{a} \sqrt{b} = \sqrt{ab}$

$= \sqrt{3 \cdot 3 \cdot 25 \cdot x^2}$ Look for perfect squares

$= \sqrt{9} \sqrt{25} \sqrt{x^2}$ 9, 25, and x^2 are perfect squares

$= 3 \cdot 5 \cdot x = 15x$

(e) $\sqrt{24a} \sqrt{6a}$

Answers: (a) 11   (b) 6
(c) $12\sqrt{2}$ (d) $\sqrt{26x}$ (e) $12a$

EXAMPLE 2   MULTIPLYING RADICALS

Multiply and then simplify.

(a) $\sqrt{5x^3} \sqrt{3x^3} \sqrt{5x} = \sqrt{5x^3 \cdot 3x^3 \cdot 5x}$ $\sqrt{a} \sqrt{b} \sqrt{c} = \sqrt{abc}$

$= \sqrt{5 \cdot 5 \cdot x^3 \cdot x^3 \cdot 3x}$ Look for perfect squares

$= \sqrt{5^2 \cdot (x^3)^2 \cdot 3x}$ 5^2 and $(x^3)^2$ are perfect squares

$= \sqrt{5^2} \sqrt{(x^3)^2} \sqrt{3x}$

$= 5x^3 \sqrt{3x}$

(b) $\sqrt{2xy} \sqrt{2x^3y^3} = \sqrt{2xy \cdot 2x^3y^3}$ $\sqrt{a} \sqrt{b} = \sqrt{ab}$

$= \sqrt{2^2x^4y^4}$

$= \sqrt{2^2} \sqrt{(x^2)^2} \sqrt{(y^2)^2}$ $x^4 = (x^2)^2$ and $y^4 = (y^2)^2$

$= 2x^2y^2$

(c) $7\sqrt{6ab} \sqrt{3a} = 7\sqrt{6ab \cdot 3a}$ Multiplication rule

$= 7\sqrt{2 \cdot 3 \cdot 3 \cdot a^2 \cdot b}$ 3^2 and a^2 are perfect squares

$= 7\sqrt{3^2a^2 \cdot 2b}$

$= 7\sqrt{3^2} \sqrt{a^2} \sqrt{2b}$

$= 7 \cdot 3 \cdot a \cdot \sqrt{2b}$

$= 21a\sqrt{2b}$

PRACTICE EXERCISE 2

Multiply and then simplify.

(a) $\sqrt{7w^3} \sqrt{2w} \sqrt{7w^5}$

(b) $\sqrt{3ab^3} \sqrt{3ab}$

(c) $5\sqrt{14xy} \sqrt{7y}$

Answers: (a) $7w^4\sqrt{2w}$ (b) $3ab^2$
(c) $35y\sqrt{2x}$

---

2 QUOTIENT OF SQUARE ROOTS

The simplifying rule for quotients, presented in Section 9.2, allows us to simplify problems such as

$$\sqrt{\frac{9}{4}} = \frac{\sqrt{9}}{\sqrt{4}} = \frac{3}{2}.$$

Again, we can reverse this rule to simplify quotients of radicals. For example,

$$\frac{\sqrt{45}}{\sqrt{20}} = \sqrt{\frac{45}{20}} = \sqrt{\frac{5 \cdot 9}{5 \cdot 4}} = \sqrt{\frac{9}{4}} = \frac{\sqrt{9}}{\sqrt{4}} = \frac{3}{2}.$$

We see that a quotient of radicals may be simplified by first considering the radical of the quotient, then reducing the quotient. This technique is summarized in the next rule.

Division Rule

If $a \geq 0$ and $b > 0$, then

$$\frac{\sqrt{a}}{\sqrt{b}} = \sqrt{\frac{a}{b}}.$$

The quotient of square roots is equal to the square root of the quotient.

EXAMPLE 3 Dividing Radicals

Divide and then simplify.

(a) $\dfrac{\sqrt{75}}{\sqrt{12}} = \sqrt{\dfrac{75}{12}}$ $\quad \dfrac{\sqrt{a}}{\sqrt{b}} = \sqrt{\dfrac{a}{b}}$

$\quad = \sqrt{\dfrac{3 \cdot 25}{3 \cdot 4}}$ $\quad$ Look for perfect squares and common factors

$\quad = \sqrt{\dfrac{25}{4}}$ $\quad$ 25 and 4 are perfect squares

$\quad = \dfrac{\sqrt{25}}{\sqrt{4}} = \dfrac{5}{2}$ $\quad \sqrt{\dfrac{a}{b}} = \dfrac{\sqrt{a}}{\sqrt{b}}$

(b) $\dfrac{\sqrt{15x^3}}{\sqrt{3x}} = \sqrt{\dfrac{15x^3}{3x}}$ $\quad \dfrac{\sqrt{a}}{\sqrt{b}} = \sqrt{\dfrac{a}{b}}$

$\quad = \sqrt{\dfrac{3 \cdot 5 \cdot x \cdot x^2}{3x}}$ $\quad$ 3 and x are common factors and x^2 is a perfect square

$\quad = \sqrt{5x^2} = x\sqrt{5}$

(c) $\dfrac{\sqrt{12y^5}}{\sqrt{75y^9}} = \sqrt{\dfrac{12y^5}{75y^9}}$ $\quad$ Division rule

$\quad = \sqrt{\dfrac{3 \cdot 4 \cdot y^5}{3 \cdot 25 \cdot y^5 \cdot y^4}}$ $\quad$ 3 and y^5 are common factors since $y^9 = y^{5+4} = y^5 \cdot y^4$

$\quad = \sqrt{\dfrac{4}{25y^4}}$

$\quad = \dfrac{\sqrt{4}}{\sqrt{25}\ \sqrt{y^4}} = \dfrac{2}{5y^2}$

PRACTICE EXERCISE 3

Divide and then simplify.

(a) $\dfrac{\sqrt{72}}{\sqrt{50}}$

(b) $\dfrac{\sqrt{21a^5}}{\sqrt{7a}}$

(c) $\dfrac{\sqrt{20a}}{\sqrt{405a^3}}$

Answers: (a) $\frac{6}{5}$ (b) $a^2\sqrt{3}$ (c) $\frac{2}{9a}$

EXAMPLE 4 Dividing Radicals

Divide and then simplify.

(a) $\dfrac{\sqrt{3xy^3}}{\sqrt{4x^3y}} = \sqrt{\dfrac{3xy^3}{4x^3y}}$ $\quad$ Division rule

$\quad = \sqrt{\dfrac{3 \cdot x \cdot y^2 \cdot y}{4 \cdot x \cdot x^2 \cdot y}}$ $\quad$ x and y are common factors

$\quad = \sqrt{\dfrac{3y^2}{4x^2}}$

$\quad = \dfrac{\sqrt{y^2}\ \sqrt{3}}{\sqrt{4}\ \sqrt{x^2}} = \dfrac{y\sqrt{3}}{2x}$

PRACTICE EXERCISE 4

Divide and then simplify.

(a) $\dfrac{\sqrt{7ab^3}}{\sqrt{9a^5b}}$

(b) $\dfrac{2\sqrt{3x}}{\sqrt{12x^3}} = 2\sqrt{\dfrac{3x}{12x^3}}$ Division rule

$\quad\quad = 2\sqrt{\dfrac{3 \cdot x}{3 \cdot 4 \cdot x^2 \cdot x}}$ 3 and x are common factors

$\quad\quad = 2\sqrt{\dfrac{1}{4x^2}}$ 1 is left in the numerator

$\quad\quad = \dfrac{2\sqrt{1}}{\sqrt{4}\,\sqrt{x^2}}$

$\quad\quad = \dfrac{2 \cdot 1}{2x} = \dfrac{1}{x}$ 2 is a common factor and $\sqrt{1} = 1$

(c) $\dfrac{\sqrt{2x^2}}{\sqrt{6y}} = \sqrt{\dfrac{2x^2}{6y}}$ Division rule

$\quad\quad = \sqrt{\dfrac{x^2}{3y}}$ 2 is a common factor

$\quad\quad = \dfrac{\sqrt{x^2}}{\sqrt{3y}}$

$\quad\quad = \dfrac{x}{\sqrt{3y}}$

$\quad\quad = \dfrac{x\sqrt{3y}}{\sqrt{3y}\,\sqrt{3y}}$ Rationalize the denominator

$\quad\quad = \dfrac{x\sqrt{3y}}{3y}$

(b) $\dfrac{5\sqrt{2w^4}}{\sqrt{50w^6}}$

(c) $\dfrac{\sqrt{3u^4}}{\sqrt{15v}}$

Answers: (a) $\dfrac{b\sqrt{7}}{3a^2}$ (b) $\dfrac{1}{w}$

(c) $\dfrac{u^2\sqrt{5v}}{5v}$

9.3 EXERCISES A

Multiply and simplify.

1. $\sqrt{2}\,\sqrt{50}$ 　　**2.** $\sqrt{3}\,\sqrt{75}$ 　　**3.** $\sqrt{15}\,\sqrt{5}$ 　　**4.** $\sqrt{18}\,\sqrt{98}$

5. $\sqrt{6}\,\sqrt{7}$ 　　**6.** $\sqrt{6}\,\sqrt{30}$ 　　**7.** $\sqrt{20}\,\sqrt{48}$ 　　**8.** $2\sqrt{24}\,\sqrt{42}$

9. $\sqrt{98}\,\sqrt{75}$ 　　**10.** $\sqrt{2}\,\sqrt{8x}$ 　　**11.** $\sqrt{27x}\,\sqrt{3x}$ 　　**12.** $\sqrt{15y}\,\sqrt{5}$

13. $\sqrt{15y}\ \sqrt{20y}$ **14** $\sqrt{3a^2}\ \sqrt{12a}$ **15.** $\sqrt{3x^3}\ \sqrt{3x}$ **16.** $\sqrt{2xy}\ \sqrt{8xy}$

17 $\sqrt{6x^2y}\ \sqrt{2xy^2}$ **18.** $\sqrt{7x^3y^3}\ \sqrt{42xy^3}$ **19** $\sqrt{5x}\ \sqrt{15xy}\ \sqrt{3y}$ **20.** $\sqrt{2xy}\ \sqrt{6xy}\ \sqrt{9xy}$

Divide and simplify.

21. $\dfrac{\sqrt{50}}{\sqrt{2}}$ **22.** $\dfrac{\sqrt{18}}{\sqrt{8}}$ **23** $\dfrac{\sqrt{98}}{\sqrt{18}}$ **24.** $\dfrac{\sqrt{6}}{\sqrt{50}}$

25. $\dfrac{\sqrt{54}}{\sqrt{50}}$ **26** $\dfrac{5\sqrt{15}}{\sqrt{75}}$ **27.** $\dfrac{\sqrt{4}}{\sqrt{3}}$ **28.** $\dfrac{\sqrt{20}}{\sqrt{35}}$

29. $\dfrac{\sqrt{10}}{\sqrt{12}}$ **30.** $\dfrac{\sqrt{50x^2}}{\sqrt{2x^2}}$ **31** $\dfrac{\sqrt{3y}}{\sqrt{4y}}$ **32** $\dfrac{\sqrt{27a^3}}{\sqrt{3}}$

33. $\dfrac{\sqrt{27x^2}}{\sqrt{3}}$ **34.** $\dfrac{\sqrt{4y}}{\sqrt{y^3}}$ **35.** $\dfrac{\sqrt{9x^2y}}{\sqrt{4y}}$ **36.** $\dfrac{\sqrt{50x^3y^3}}{\sqrt{2xy}}$

37 $\dfrac{5\sqrt{2xy}}{\sqrt{25x}}$ **38** $\dfrac{\sqrt{18x}}{\sqrt{2y}}$ **39.** $\dfrac{\sqrt{25}}{\sqrt{x}}$ **40.** $\dfrac{\sqrt{2y^2}}{\sqrt{5x}}$

In an environmental study, three quantities m, p, and q have been found to be related by the equation $m = \sqrt{p}\,\sqrt{q}$.

41. Find m if $p = 4$ and $q = 9$.   **42.** Find m if $p = 25$ and $q = 8$.   **43.** Find m if $p = 8$ and $q = 18$.

44. Find m if $p = 0$ and $q = 5$.   **45.** Find m if $p = 4$ and $q = -4$.   **46.** Find m if $p = 6$ and $q = 15$.

FOR REVIEW

Simplify.

47. $\sqrt{288x^2}$ **48.** $\sqrt{147x^2y}$ **49.** $\sqrt{\dfrac{98x^2}{25}}$ **50.** $\sqrt{\dfrac{36x^2}{5y}}$

Exercises 51–56 review material from Section 2.4 to prepare for the next section. Use the distributive law to collect like terms.

51. $2x + 5x$ **52.** $8y - 3y$ **53.** $7x + 9y$ **54.** $4x - x + 6x$

55. $\dfrac{1}{2}y - \dfrac{3}{4}y + \dfrac{1}{8}y$ **56.** $4x + 3y - 2x - 9y$

ANSWERS: 1. 10 2. 15 3. $5\sqrt{3}$ 4. 42 5. $\sqrt{42}$ 6. $6\sqrt{5}$ 7. $8\sqrt{15}$ 8. $24\sqrt{7}$ 9. $35\sqrt{6}$ 10. $4\sqrt{x}$ 11. $9x$ 12. $5\sqrt{3y}$ 13. $10y\sqrt{3}$ 14. $6a\sqrt{a}$ 15. $3x^2$ 16. $4xy$ 17. $2xy\sqrt{3xy}$ 18. $7x^2y^3\sqrt{6}$ 19. $15xy$ 20. $6xy\sqrt{3xy}$ 21. 5 22. $\dfrac{3}{2}$ 23. $\dfrac{7}{3}$ 24. $\dfrac{\sqrt{3}}{5}$ 25. $\dfrac{3\sqrt{3}}{5}$ 26. $\sqrt{5}$ 27. $\dfrac{2\sqrt{3}}{3}$ 28. $\dfrac{2\sqrt{7}}{7}$ 29. $\dfrac{\sqrt{30}}{6}$ 30. 5 31. $\dfrac{\sqrt{3}}{2}$ 32. $3a\sqrt{a}$ 33. $3x$ 34. $\dfrac{2}{y}$ 35. $\dfrac{3x}{2}$ 36. $5xy$ 37. $\sqrt{2y}$ 38. $\dfrac{3\sqrt{xy}}{y}$ 39. $\dfrac{5\sqrt{x}}{x}$ 40. $\dfrac{y\sqrt{10x}}{5x}$ 41. 6 42. $10\sqrt{2}$ 43. 12 44. 0 45. no value for m if q is negative 46. $3\sqrt{10}$ 47. $12x\sqrt{2}$ 48. $7x\sqrt{3y}$ 49. $\dfrac{7x\sqrt{2}}{5}$ 50. $\dfrac{6x\sqrt{5y}}{5y}$ 51. $7x$ 52. $5y$ 53. no like terms to collect 54. $9x$ 55. $-\dfrac{1}{8}y$ 56. $2x - 6y$

9.3 EXERCISES B

Multiply and simplify.

1. $\sqrt{5}\,\sqrt{45}$ **2.** $\sqrt{7}\,\sqrt{28}$ **3.** $\sqrt{6}\,\sqrt{2}$ **4.** $\sqrt{20}\,\sqrt{45}$

5. $\sqrt{7}\,\sqrt{11}$ **6.** $\sqrt{10}\,\sqrt{18}$ **7.** $\sqrt{18}\,\sqrt{28}$ **8.** $3\sqrt{35}\,\sqrt{45}$

9. $\sqrt{72}\,\sqrt{48}$ **10.** $\sqrt{5}\,\sqrt{20x}$ **11.** $\sqrt{8x}\,\sqrt{2x}$ **12.** $\sqrt{12y}\,\sqrt{4}$

13. $\sqrt{14y}\,\sqrt{50y}$ **14.** $\sqrt{5a^2}\,\sqrt{20a}$ **15.** $\sqrt{5x}\,\sqrt{5x^3}$ **16.** $\sqrt{3xy}\,\sqrt{27xy}$

17. $\sqrt{10x^2y}\,\sqrt{2xy^2}$ **18.** $\sqrt{5x^3y^3}\,\sqrt{35xy^3}$ **19.** $\sqrt{2x}\,\sqrt{14xy}\,\sqrt{7y}$ **20.** $\sqrt{3xy}\,\sqrt{15xy}\,\sqrt{xy}$

Divide and simplify.

21. $\dfrac{\sqrt{3}}{\sqrt{48}}$

22. $\dfrac{\sqrt{27}}{\sqrt{12}}$

23. $\dfrac{\sqrt{125}}{\sqrt{45}}$

24. $\dfrac{\sqrt{14}}{\sqrt{72}}$

25. $\dfrac{\sqrt{150}}{\sqrt{8}}$

26. $\dfrac{6\sqrt{10}}{\sqrt{72}}$

27. $\dfrac{\sqrt{9}}{\sqrt{2}}$

28. $\dfrac{\sqrt{50}}{\sqrt{14}}$

29. $\dfrac{\sqrt{5}}{\sqrt{28}}$

30. $\dfrac{\sqrt{98x^2}}{\sqrt{2x^2}}$

31. $\dfrac{\sqrt{5y}}{\sqrt{9y}}$

32. $\dfrac{\sqrt{45x^2}}{\sqrt{5}}$

33. $\dfrac{\sqrt{8a^3}}{\sqrt{2}}$

34. $\dfrac{\sqrt{9y}}{\sqrt{y^3}}$

35. $\dfrac{\sqrt{25xy^2}}{\sqrt{9x}}$

36. $\dfrac{\sqrt{98x^3y^3}}{\sqrt{2xy}}$

37. $\dfrac{7\sqrt{3xy}}{\sqrt{49y}}$

38. $\dfrac{\sqrt{16}}{\sqrt{x}}$

39. $\dfrac{\sqrt{75y}}{\sqrt{3x}}$

40. $\dfrac{\sqrt{3x^2}}{\sqrt{7y}}$

During a scientific experiment, it was found that three quantities v, w, and d are related by the equation $v = \dfrac{\sqrt{w}}{\sqrt{d}}$.

41. Find v if $w = 4$ and $d = 25$.

42. Find v if $w = 8$ and $d = 16$.

43. Find v if $w = 8$ and $d = 50$.

44. Find v if $w = 0$ and $d = 5$.

45. Find v if $w = -9$ and $d = 7$.

46. Find v if $w = 21$ and $d = 14$.

FOR REVIEW

Simplify.

47. $\sqrt{242x^2}$

48. $\sqrt{75xy^2}$

49. $\sqrt{\dfrac{147x^2}{16}}$

50. $\sqrt{\dfrac{25x^2}{6y}}$

Exercises 51–56 review material from Section 2.4 to prepare for the next section. Use the distributive law to collect like terms.

51. $4x + 9x$

52. $7y - 2y$

53. $-2x + 3y$

54. $3x + 7x - 6x$

55. $\dfrac{2}{3}y - \dfrac{1}{9}y + \dfrac{1}{18}y$

56. $-5x + 4y + 2x - 6y$

9.3 EXERCISES C

Multiply or divide and simplify.

1. $\sqrt{3x^3y^{-3}}\,\sqrt{6xy^{-1}}$

2. $\dfrac{\sqrt{125x^{-6}y^{-3}}}{\sqrt{45x^4y^{-6}}}$

3. $\dfrac{\sqrt{16x^2y^5z^{-1}}}{\sqrt{6x^3y^{-4}z^{-3}}}$

$\left[\text{Answer: } \dfrac{2y^4z\sqrt{6xy}}{3x}\right]$

Use $\sqrt[3]{a^3} = a$ in the following exercises.

4. $\sqrt[3]{81x^4y^6} \; \sqrt[3]{18x^5y^5}$

5. $\dfrac{\sqrt[3]{9x^4y^7}}{\sqrt[3]{24xy^2}}$

6. $\dfrac{\sqrt[3]{40x^6y}}{\sqrt[3]{25xy^5}}$

$$\left[\text{Answer:} \quad \frac{2x\sqrt[3]{25x^2y^2}}{5y^2} \right]$$

9.4 ADDITION AND SUBTRACTION OF RADICALS

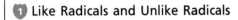

STUDENT GUIDEPOSTS

1 Like Radicals and Unlike Radicals   **2** Adding and Subtracting Radicals

The rules for multiplication and division of radicals,

$$\sqrt{a}\,\sqrt{b} = \sqrt{ab} \quad \text{and} \quad \frac{\sqrt{a}}{\sqrt{b}} = \sqrt{\frac{a}{b}},$$

do not have counterparts relative to addition and subtraction. For example,

$$5 = \sqrt{25} = \sqrt{9+16} \neq \sqrt{9} + \sqrt{16} = 3 + 4 = 7, \qquad 5 \neq 7$$

so that in general,

$$\sqrt{a+b} \neq \sqrt{a} + \sqrt{b}.$$

Also, since

$$4 = \sqrt{16} = \sqrt{25-9} \neq \sqrt{25} - \sqrt{9} = 5 - 3 = 2, \qquad 4 \neq 2$$

in general,

$$\sqrt{a-b} \neq \sqrt{a} - \sqrt{b}.$$

1 LIKE RADICALS AND UNLIKE RADICALS

We may, however, use the distributive law to add or subtract *like radicals*. **Like radicals** have the same radicand. Thus, the terms

$$\sqrt{11}, \quad 3\sqrt{11}, \quad -5\sqrt{11}$$

have like radicals. Also, terms such as

$$\sqrt{x}, \quad -7\sqrt{x}, \quad 100\sqrt{x}$$

contain like radicals. However,

$$\sqrt{11}, \quad \sqrt{x}, \quad 3\sqrt{y}$$

do not have like radicals; they are **unlike radicals.**

2 ADDING AND SUBTRACTING RADICALS

We add or subtract like radicals just as we collect like terms of polynomials. Remember,

$$3x + 5x = (3 + 5)x = 8x.$$

We collect like radicals in the same way.

$$3\sqrt{x} + 5\sqrt{x} = (3 + 5)\sqrt{x} = 8\sqrt{x}$$

EXAMPLE 1 ADDING AND SUBTRACTING LIKE RADICALS

Add or subtract.

(a) $9\sqrt{5} + 4\sqrt{5} = (9+4)\sqrt{5}$ Distributive law, $ac + bc = (a+b)c$
$\qquad = 13\sqrt{5}$

(b) $9\sqrt{5} - 4\sqrt{5} = (9-4)\sqrt{5}$ Distributive law, $ac - bc = (a-b)c$
$\qquad = 5\sqrt{5}$

(c) $3\sqrt{xy} - 10\sqrt{xy} = (3-10)\sqrt{xy}$ Distributive law
$\qquad = -7\sqrt{xy}$

(d) $7\sqrt{x} + 5\sqrt{y}$ cannot be simplified since $\sqrt{x}$ and $\sqrt{y}$ are not like radicals.

PRACTICE EXERCISE 1

Add or subtract.

(a) $2\sqrt{11} + 5\sqrt{11}$

(b) $2\sqrt{11} - 5\sqrt{11}$

(c) $6\sqrt{u} - 5\sqrt{u}$

(d) $8\sqrt{a} - 2\sqrt{w}$

Answers: (a) $7\sqrt{11}$ (b) $-3\sqrt{11}$
(c) $\sqrt{u}$ (d) cannot be simplified
since $\sqrt{a}$ and $\sqrt{w}$ are not like radicals

Some unlike radicals can be changed into like radicals if we simplify. For example, $\sqrt{2}$ and $\sqrt{8}$ are not like radicals since they have different radicands. However,

$$\sqrt{8} = \sqrt{4 \cdot 2} = \sqrt{4}\,\sqrt{2} = 2\sqrt{2}.$$

Now the terms $\sqrt{2}$ and $2\sqrt{2}$ do contain like radicals.

To Add or Subtract Radicals

1. Simplify each radical as much as possible.
2. Use the distributive law to collect any like radicals.

EXAMPLE 2 SIMPLIFYING BEFORE ADDING OR SUBTRACTING

Add or subtract.

(a) $3\sqrt{2} + 5\sqrt{8} = 3\sqrt{2} + 5\sqrt{4 \cdot 2}$ 4 is a perfect square
$\qquad = 3\sqrt{2} + 5\sqrt{4}\,\sqrt{2}$ $\sqrt{ab} = \sqrt{a}\,\sqrt{b}$
$\qquad = 3\sqrt{2} + 5 \cdot 2\sqrt{2}$ $\sqrt{4} = 2$
$\qquad = 3\sqrt{2} + 10\sqrt{2}$ Terms now contain like radicals
$\qquad = (3 + 10)\sqrt{2} = 13\sqrt{2}$ Distributive law

(b) $\sqrt{12} + \sqrt{75} = \sqrt{4 \cdot 3} + \sqrt{25 \cdot 3}$ 4 and 25 are perfect squares
$\qquad = \sqrt{4}\,\sqrt{3} + \sqrt{25}\,\sqrt{3}$ $\sqrt{ab} = \sqrt{a}\,\sqrt{b}$
$\qquad = 2\sqrt{3} + 5\sqrt{3}$ $\sqrt{4} = 2$ and $\sqrt{25} = 5$
$\qquad = (2 + 5)\sqrt{3} = 7\sqrt{3}$ Distributive law

(c) $5\sqrt{12} - 7\sqrt{27} = 5\sqrt{4 \cdot 3} - 7\sqrt{9 \cdot 3}$ 4 and 9 are perfect squares
$\qquad = 5\sqrt{4}\,\sqrt{3} - 7\sqrt{9}\,\sqrt{3}$ $\sqrt{ab} = \sqrt{a}\,\sqrt{b}$
$\qquad = 5 \cdot 2 \cdot \sqrt{3} - 7 \cdot 3 \cdot \sqrt{3}$
$\qquad = 10\sqrt{3} - 21\sqrt{3}$ Terms contain like radicals
$\qquad = (10 - 21)\sqrt{3} = -11\sqrt{3}$ Distributive law

PRACTICE EXERCISE 2

Add or subtract.

(a) $5\sqrt{5} + 3\sqrt{20}$

(b) $7\sqrt{32} + 3\sqrt{18}$

(c) $6\sqrt{8} - \sqrt{50}$

(d) $2\sqrt{25y} - 4\sqrt{36y} = 2\sqrt{25}\ \sqrt{y} - 4\sqrt{36}\ \sqrt{y}$ 25 and 36 are
                          perfect squares

$$= 2 \cdot 5\sqrt{y} - 4 \cdot 6\sqrt{y}$$

$$= 10\sqrt{y} - 24\sqrt{y}$$

$$= (10 - 24)\sqrt{y} = -14\sqrt{y} \quad \text{Distributive law}$$

(d) $8\sqrt{4x} - 3\sqrt{49x}$

Answers: (a) $11\sqrt{5}$ (b) $37\sqrt{2}$
(c) $7\sqrt{2}$ (d) $-5\sqrt{x}$

| **EXAMPLE 3**   ADDING SEVERAL RADICALS | **PRACTICE EXERCISE 3** |

Perform the indicated operations.

(a) $4\sqrt{18} + 3\sqrt{27} - 6\sqrt{12}$

$$= 4\sqrt{9 \cdot 2} + 3\sqrt{9 \cdot 3} - 6\sqrt{4 \cdot 3}$$

$$= 4\sqrt{9}\ \sqrt{2} + 3\sqrt{9}\ \sqrt{3} - 6\sqrt{4}\ \sqrt{3}$$

$$= 4 \cdot 3\sqrt{2} + 3 \cdot 3\sqrt{3} - 6 \cdot 2\sqrt{3}$$

$$= 12\sqrt{2} + 9\boxed{\sqrt{3}} - 12\boxed{\sqrt{3}} \quad \text{Only the last two terms are like terms}$$

$$= 12\sqrt{2} + (9 - 12)\boxed{\sqrt{3}} \quad \text{Distributive law}$$

$$= 12\sqrt{2} - 3\sqrt{3} \quad \text{As simple as possible}$$

(b) $3\sqrt{x} - 2\sqrt{4x} - 5\sqrt{9x} = 3\sqrt{x} - 2\sqrt{4}\ \sqrt{x} - 5\sqrt{9}\ \sqrt{x}$

$$= 3\sqrt{x} - 2 \cdot 2\sqrt{x} - 5 \cdot 3\sqrt{x}$$

$$= 3\boxed{\sqrt{x}} - 4\boxed{\sqrt{x}} - 15\boxed{\sqrt{x}}$$

$$= (3 - 4 - 15)\boxed{\sqrt{x}} = -16\sqrt{x}$$

Perform the indicated operations.

(a) $3\sqrt{75} - \sqrt{18} + 2\sqrt{27}$

(b) $8\sqrt{w} - 3\sqrt{25w} + 5\sqrt{4w}$

Answers: (a) $21\sqrt{3} - 3\sqrt{2}$
(b) $3\sqrt{w}$

To obtain like terms, it may be necessary to rationalize denominators.

| **EXAMPLE 4**   RATIONALIZING BEFORE ADDING | **PRACTICE EXERCISE 4** |

Add or subtract.

(a) $3\sqrt{2} + \dfrac{5}{\sqrt{2}} = 3\sqrt{2} + \dfrac{5\sqrt{2}}{\sqrt{2}\ \sqrt{2}}$ Rationalize denominator

$$= 3\sqrt{2} + \frac{5\sqrt{2}}{2} \qquad \sqrt{2}\ \sqrt{2} = 2$$

$$= 3\boxed{\sqrt{2}} + \frac{5}{2}\boxed{\sqrt{2}} \qquad \frac{5\sqrt{2}}{2} = \frac{5 \cdot \sqrt{2}}{2 \cdot 1} = \frac{5}{2} \cdot \frac{\sqrt{2}}{1} = \frac{5}{2}\sqrt{2}$$

$$= \left(3 + \frac{5}{2}\right)\boxed{\sqrt{2}} \qquad \text{Distributive law}$$

$$= \left(\frac{6}{2} + \frac{5}{2}\right)\sqrt{2} = \frac{11}{2}\sqrt{2} = \frac{11\sqrt{2}}{2}$$

Add or subtract.

(a) $5\sqrt{3} + \dfrac{1}{\sqrt{3}}$

(b) $\dfrac{6}{\sqrt{5}} - 4\sqrt{5} = \dfrac{6\sqrt{5}}{\sqrt{5}\,\sqrt{5}} - 4\sqrt{5}$ Rationalize the denominator

$= \dfrac{6\sqrt{5}}{5} - 4\sqrt{5}$

$= \dfrac{6}{5}\sqrt{5} - 4\sqrt{5}$ $\dfrac{6\sqrt{5}}{5} = \dfrac{6\cdot\sqrt{5}}{5\cdot 1} = \dfrac{6}{5}\cdot\dfrac{\sqrt{5}}{1} = \dfrac{6}{5}\sqrt{5}$

$= \left(\dfrac{6}{5} - 4\right)\sqrt{5}$ Distributive law

$= \left(\dfrac{6}{5} - \dfrac{20}{5}\right)\sqrt{5}$

$= -\dfrac{14}{5}\sqrt{5} = -\dfrac{14\sqrt{5}}{5}$

(b) $\dfrac{9}{\sqrt{7}} - 6\sqrt{7}$

Answers: (a) $\dfrac{16\sqrt{3}}{3}$

(b) $-\dfrac{33\sqrt{7}}{7}$

9.4 EXERCISES A

Add or subtract.

1. $9\sqrt{2} + \sqrt{2}$

2. $-4\sqrt{5} + 3\sqrt{5}$

3. $-18\sqrt{xy} - 7\sqrt{xy}$

4. $3\sqrt{2} + 2\sqrt{3}$

5 $5\sqrt{12} - 3\sqrt{12}$

6. $\sqrt{50} + \sqrt{98}$

7. $\sqrt{18} - \sqrt{8}$

8. $6\sqrt{2} + 3\sqrt{8}$

9. $5\sqrt{3} - \sqrt{27}$

10 $4\sqrt{50} + 7\sqrt{18}$

11. $3\sqrt{18} - 5\sqrt{12}$

12. $-18\sqrt{7} - 2\sqrt{28}$

13. $15\sqrt{45} + 4\sqrt{20}$

14 $\sqrt{147} - 2\sqrt{75}$

15. $6\sqrt{44} + 2\sqrt{99}$

16. $\sqrt{121} - \sqrt{144}$

17. $2\sqrt{75} - 3\sqrt{125}$

18 $2\sqrt{4x} + 7\sqrt{9x}$

19. $-3\sqrt{25y} - 2\sqrt{9y}$

20. $8\sqrt{50y} + 2\sqrt{18y}$

21 $x\sqrt{y^3} + y\sqrt{x^2 y}$

Perform the indicated operations.

22 $\sqrt{18} + \sqrt{50} + \sqrt{72}$

23. $\sqrt{27} - \sqrt{75} - \sqrt{108}$

24. $5\sqrt{20} + 2\sqrt{45} - 9\sqrt{80}$

25 $3\sqrt{75} + 2\sqrt{12} - 5\sqrt{48}$

26. $-3\sqrt{24} - 5\sqrt{150} - 4\sqrt{54}$

27. $4\sqrt{99} - 7\sqrt{44} + 5\sqrt{52}$

28. $4\sqrt{98} - 8\sqrt{72} + 5\sqrt{32}$

29 $4\sqrt{300} - 2\sqrt{500} + 4\sqrt{125}$

30 $\sqrt{4x} + \sqrt{16x} - \sqrt{25x}$

31. $3\sqrt{5x} + 2\sqrt{20x} - 8\sqrt{45x}$

32 $9\sqrt{100a^2b^2} - 4\sqrt{36a^2b^2} - 10\sqrt{a^2b^2}$

33. $2\sqrt{25ab^3} - b\sqrt{36ab} - 5\sqrt{49ab^3}$

Rationalize the denominator and then add or subtract.

34 $\sqrt{3} + \dfrac{1}{\sqrt{3}}$

35. $2\sqrt{5} - \dfrac{3}{\sqrt{5}}$

36. $\dfrac{7}{\sqrt{2}} + 6\sqrt{2}$

37. $3\sqrt{7} - \dfrac{1}{\sqrt{7}}$

38 $3\sqrt{7} - \sqrt{\dfrac{1}{7}}$

39 $-3\sqrt{5} - \dfrac{9}{\sqrt{5}}$

40. $\dfrac{\sqrt{2}}{\sqrt{3}} + 2\sqrt{6}$

41 $\dfrac{\sqrt{2}}{\sqrt{10}} - \sqrt{5}$

42. $\sqrt{\dfrac{2}{3}} - 2\sqrt{6}$

Suppose we are given the formula $F = \sqrt{f} + 2\sqrt{g}$, relating the three quantities, F, f, and g.

43. Find F when $f = 8$ and $g = 18$.

44. Find F when $f = 60$ and $g = 15$.

45. Find F when $f = 0$ and $g = 7$.

46. Find F when $f = -4$ and $g = 9$.

FOR REVIEW

Simplify.

47. $3\sqrt{10}\,\sqrt{40}$

48. $\sqrt{3a^2}\,\sqrt{15a}$

49. $\dfrac{\sqrt{150x}}{\sqrt{3x}}$

50. $\dfrac{\sqrt{12x^2}}{\sqrt{5y}}$

Exercises 51–54 review material from Sections 6.3 and 6.4. They will help you prepare for the next section. Multiply the polynomials.

51. $(x - y)(x + 7y)$ **52.** $(x - 3y)^2$ **53.** $(x + 4y)(x - 4y)$ **54.** $(3x - 7y)(3x + 7y)$

ANSWERS: 1. $10\sqrt{2}$ 2. $-\sqrt{5}$ 3. $-25\sqrt{xy}$ 4. $3\sqrt{2} + 2\sqrt{3}$ (cannot be simplified)   5. $4\sqrt{3}$ 6. $12\sqrt{2}$ 7. $\sqrt{2}$
8. $12\sqrt{2}$ 9. $2\sqrt{3}$ 10. $41\sqrt{2}$ 11. $9\sqrt{2} - 10\sqrt{3}$ 12. $-22\sqrt{7}$ 13. $53\sqrt{5}$ 14. $-3\sqrt{3}$ 15. $18\sqrt{11}$ 16. -1
17. $10\sqrt{3} - 15\sqrt{5}$ 18. $25\sqrt{x}$ 19. $-21\sqrt{y}$ 20. $46\sqrt{2y}$ 21. $2xy\sqrt{y}$ 22. $14\sqrt{2}$ 23. $-8\sqrt{3}$ 24. $-20\sqrt{5}$
25. $-\sqrt{3}$ 26. $-43\sqrt{6}$ 27. $-2\sqrt{11} + 10\sqrt{13}$ 28. 0 29. $40\sqrt{3}$ 30. $\sqrt{x}$ 31. $-17\sqrt{5x}$ 32. $56ab$ 33. $-31b\sqrt{ab}$
34. $\dfrac{4\sqrt{3}}{3}$ 35. $\dfrac{7\sqrt{5}}{5}$ 36. $\dfrac{19\sqrt{2}}{2}$ 37. $\dfrac{20\sqrt{7}}{7}$ 38. $\dfrac{20\sqrt{7}}{7}$ 39. $\dfrac{-24\sqrt{5}}{5}$ 40. $\dfrac{7\sqrt{6}}{3}$ 41. $\dfrac{-4\sqrt{5}}{5}$ 42. $\dfrac{-5\sqrt{6}}{3}$
43. $8\sqrt{2}$ 44. $4\sqrt{15}$ 45. $2\sqrt{7}$ 46. no value for F when f is negative   47. 60 48. $3a\sqrt{5a}$ 49. $5\sqrt{2}$ 50. $\dfrac{2x\sqrt{15y}}{5y}$
51. $x^2 + 6xy - 7y^2$ 52. $x^2 - 6xy + 9y^2$ 53. $x^2 - 16y^2$ 54. $9x^2 - 49y^2$

9.4 EXERCISES B

Add or subtract.

1. $5\sqrt{3} - 9\sqrt{3}$

2. $-5\sqrt{7} - 9\sqrt{7}$

3. $-9\sqrt{xy} - 3\sqrt{xy}$

4. $5\sqrt{5} + 3\sqrt{3}$

5. $2\sqrt{27} - 4\sqrt{27}$

6. $\sqrt{48} + \sqrt{75}$

7. $\sqrt{45} - \sqrt{20}$

8. $7\sqrt{3} + 2\sqrt{12}$

9. $4\sqrt{2} - \sqrt{50}$

10. $8\sqrt{32} + 3\sqrt{50}$

11. $6\sqrt{27} - 4\sqrt{18}$

12. $-8\sqrt{5} + 2\sqrt{45}$

13. $4\sqrt{28} + 8\sqrt{63}$

14. $3\sqrt{75} - 5\sqrt{147}$

15. $-8\sqrt{99} + 8\sqrt{44}$

16. $\sqrt{169} - \sqrt{121}$

17. $5\sqrt{45} - 2\sqrt{48}$

18. $4\sqrt{25x} + 2\sqrt{16x}$

19. $-4\sqrt{4y} - 9\sqrt{25y}$

20. $6\sqrt{8y} + 5\sqrt{32y}$

21. $y\sqrt{x^3} + x\sqrt{xy^2}$

22. $\sqrt{8} + \sqrt{72} + \sqrt{98}$

23. $\sqrt{12} - \sqrt{27} - \sqrt{147}$

24. $7\sqrt{48} - 4\sqrt{75} - 6\sqrt{12}$

25. $3\sqrt{125} + 4\sqrt{80} - 6\sqrt{5}$

26. $-2\sqrt{6} - 2\sqrt{54} - 5\sqrt{150}$

27. $5\sqrt{28} - 2\sqrt{63} + 4\sqrt{99}$

28. $5\sqrt{147} - 2\sqrt{108} + 7\sqrt{48}$

29. $7\sqrt{700} - 4\sqrt{175} + 8\sqrt{44}$

30. $\sqrt{25x} + \sqrt{36x} - \sqrt{49x}$

31. $5\sqrt{3x} + 6\sqrt{27x} - 7\sqrt{75x}$

32. $5\sqrt{25a^2b^2} - 3\sqrt{16a^2b^2} - 20\sqrt{a^2b^2}$

33. $2a\sqrt{16ab} + \sqrt{4a^3b} - 3\sqrt{25a^3b}$

Rationalize the denominator and then add or subtract.

34. $\sqrt{2} + \dfrac{1}{\sqrt{2}}$

35. $3\sqrt{5} - \dfrac{4}{\sqrt{5}}$

36. $\dfrac{8}{\sqrt{3}} + 2\sqrt{3}$

37. $4\sqrt{5} - \dfrac{1}{\sqrt{5}}$

38. $4\sqrt{5} - \sqrt{\dfrac{1}{5}}$

39. $-4\sqrt{7} - \dfrac{3}{\sqrt{7}}$

40. $\dfrac{\sqrt{3}}{\sqrt{2}} + 5\sqrt{6}$

41. $\sqrt{\dfrac{3}{2}} - 5\sqrt{6}$

42. $\dfrac{\sqrt{5}}{\sqrt{10}} - \sqrt{2}$

Suppose we have the formula $P = \sqrt{n} - 3\sqrt{p}$, relating the three quantities P, n, and p.

43. Find P when $n = 49$ and $p = 45$.

44. Find P when $n = 52$ and $p = 117$.

45. Find P when $n = -9$ and $p = 14$.

46. Find P when $n = 11$ and $p = 0$.

FOR REVIEW

Simplify.

47. $5\sqrt{48}\,\sqrt{3}$

48. $\sqrt{8a^2}\,\sqrt{10a}$

49. $\dfrac{\sqrt{150x}}{\sqrt{2x}}$

50. $\dfrac{\sqrt{18y^2}}{\sqrt{7x}}$

Exercises 51–54 review material from Sections 6.3 and 6.4. They will help you prepare for the next section. Multiply the polynomials.

51. $(2x - y)(x + 4y)$

52. $(4x - y)^2$

53. $(2x - y)(2x + y)$

54. $(5x + 2y)(5x - 2y)$

9.4 EXERCISES C

Add or subtract.

1. $\sqrt{12x^4 - 4x^2y^2} - \sqrt{27x^4 - 9x^2y^2}$

2. $\sqrt{2x^2 - 20xy + 50y^2} + \sqrt{8x^2 - 80xy + 200y^2}$
[Answer: $3(x - 5y)\sqrt{2}$]

3. $3\sqrt[3]{40x^4y^4} - 2x\sqrt[3]{5xy^4}$ [*Hint:* $\sqrt[3]{a^3} = a$]

4. $-6ab\sqrt[3]{27a^5b^3} + 10\sqrt[3]{27a^8b^6}$
[Answer: $12a^2b^2\sqrt[3]{a^2}$]

9.5 SUMMARY OF TECHNIQUES AND RATIONALIZING DENOMINATORS

STUDENT GUIDEPOSTS

1 Simplifying Radical Expressions

2 Multiplying Binomial Radical Expressions

3 Rationalizing Binomial Denominators

1 SIMPLIFYING RADICAL EXPRESSIONS

Below is a summary of the simplifying techniques from the previous sections.

To Simplify a Radical Expression

1. Combine all like radicals using the distributive laws.
2. When needed, use the rules of multiplication and division,

$$\sqrt{a}\,\sqrt{b} = \sqrt{ab} \quad \text{and} \quad \frac{\sqrt{a}}{\sqrt{b}} = \sqrt{\frac{a}{b}}.$$

3. Remove all perfect squares from under the radicals.
4. Rationalize all denominators.

The following table illustrates the various techniques.

Simplification	*Type*
$3\sqrt{2} - \sqrt{2} = 2\sqrt{2}$	Combining like radicals
$\sqrt{2}\,\sqrt{3} = \sqrt{6}$	Using multiplication rule
$\dfrac{\sqrt{12}}{\sqrt{3}} = \sqrt{\dfrac{12}{3}} = \sqrt{4} = 2$	Using division rule and removing perfect squares
$\sqrt{9x^2y} = 3x\sqrt{y}$	Removing perfect squares
$\dfrac{3}{\sqrt{2}} = \dfrac{3\sqrt{2}}{\sqrt{2}\,\sqrt{2}} = \dfrac{3\sqrt{2}}{2}$	Rationalizing the denominator

To multiply radical expressions, we often use one or more of the simplifying techniques, as in the next example.

EXAMPLE 1   MULTIPLYING USING THE DISTRIBUTIVE LAW

Multiply.

(a) $\sqrt{5}(\sqrt{15} + \sqrt{5}) = \sqrt{5}\,\sqrt{15} + \sqrt{5}\,\sqrt{5}$ Distributive law

$= \sqrt{5 \cdot 15} + \sqrt{5 \cdot 5}$ $\sqrt{a}\,\sqrt{b} = \sqrt{ab}$

$= \sqrt{5 \cdot 5 \cdot 3} + \sqrt{5 \cdot 5}$ Factor

$= \sqrt{5^2}\,\sqrt{3} + \sqrt{5^2}$ 5^2 is a perfect square

$= 5\sqrt{3} + 5$

PRACTICE EXERCISE 1

Multiply.

(a) $\sqrt{2}(\sqrt{6} + 2\sqrt{2})$

(b) $\sqrt{3}(\sqrt{27} - \sqrt{12}) = \sqrt{3}\sqrt{27} - \sqrt{3}\sqrt{12}$ Distributive law

$= \sqrt{3 \cdot 3^3} - \sqrt{3 \cdot 3 \cdot 4}$ $\sqrt{a}\sqrt{b} = \sqrt{ab}$

$= \sqrt{3^4} - \sqrt{3^2 \cdot 2^2}$ $3^4, 3^2,$ and 2^2 are perfect squares

$= 3^2 - 3 \cdot 2$

$= 9 - 6 = 3$

(c) $\sqrt{7}\left(5\sqrt{7} - \dfrac{8}{\sqrt{7}}\right) = \sqrt{7}(5\sqrt{7}) - \sqrt{7}\left(\dfrac{8}{\sqrt{7}}\right)$ Distributive law

$= 5(\sqrt{7})^2 - \dfrac{\sqrt{7}}{\sqrt{7}} \cdot 8$ Commutative and associative laws

$= 5 \cdot 7 - 1 \cdot 8$ $(\sqrt{7})^2 = \sqrt{7}\sqrt{7} = 7$

$= 35 - 8 = 27$

(b) $\sqrt{5}(\sqrt{125} - \sqrt{20})$

(c) $\sqrt{3}\left(2\sqrt{3} - \dfrac{1}{\sqrt{3}}\right)$

Answers: (a) $2\sqrt{3} + 4$ (b) 15
(c) 5

---

② MULTIPLYING BINOMIAL RADICAL EXPRESSIONS

In Chapter 6 we used the FOIL method (*F*—First terms, *O*—Outside terms, *I*—Inside terms, and *L*—Last terms) to multiply binomials. Recall that, for example, to multiply $x + 2y$ and $2x - y$ we proceed as follows.

$$(x + 2y)(2x - y) = 2x \cdot x - x \cdot y + 2x \cdot 2y - 2y \cdot y$$

The same rule can be used to multiply binomial radical expressions.

$$(\sqrt{3} + 2\sqrt{2})(2\sqrt{3} - \sqrt{2}) = 2\sqrt{3} \cdot \sqrt{3} - \sqrt{3} \cdot \sqrt{2} + 2\sqrt{2} \cdot 2\sqrt{3} - 2\sqrt{2} \cdot \sqrt{2}$$

$$= 2 \cdot 3 - \sqrt{6} + 4\sqrt{6} - 2 \cdot 2$$

$$= 6 - \sqrt{6} + 4\sqrt{6} - 4$$

$$= 2 + 3\sqrt{6}$$

We multiplied the first terms Ⓕ, the outside terms Ⓞ, the inside terms Ⓘ, and the last terms Ⓛ, and then used the product rule and collected like terms.

| **EXAMPLE 2** Using FOIL | **PRACTICE EXERCISE 2** |

Multiply.

(a) $(\sqrt{3} - \sqrt{2})(\sqrt{3} + 5\sqrt{2})$

$= \sqrt{3}\sqrt{3} + \sqrt{3}(5\sqrt{2}) - \sqrt{2}\sqrt{3} - \sqrt{2}(5\sqrt{2})$

$= 3 + 5\sqrt{6} - \sqrt{6} - 5(2)$

$= 3 - 10 + (5 - 1)\sqrt{6}$

$= -7 + 4\sqrt{6}$

Multiply.

(a) $(\sqrt{7} - 2\sqrt{3})(\sqrt{7} + \sqrt{3})$

(b) $(\sqrt{3} - \sqrt{2})(\sqrt{3} + \sqrt{2})$
$$= \sqrt{3}\,\sqrt{3} + \sqrt{3}\,\sqrt{2} - \sqrt{2}\,\sqrt{3} - \sqrt{2}\,\sqrt{2}$$
$$= 3 + \sqrt{6} - \sqrt{6} - 2$$
$$= 3 - 2$$
$$= 1$$

(b) $(\sqrt{11} - 2\sqrt{2})(\sqrt{11} + 2\sqrt{2})$

Answers: (a) $1 - \sqrt{21}$ (b) 3

③ RATIONALIZING BINOMIAL DENOMINATORS

The product in Example 2(**b**) is the rational number 1. This is a special case of a more general result involving products of expressions of the form

$$(a - b)(a + b) = a^2 - b^2.$$

Notice that

$$(\sqrt{3} - \sqrt{2})(\sqrt{3} + \sqrt{2}) = (\sqrt{3})^2 - (\sqrt{2})^2 \qquad \sqrt{3} = a \text{ and } \sqrt{2} = b$$
$$= 3 - 2 = 1.$$

Any time that a sum and a difference like those above are multiplied, the result will be a rational number free of radicals. This observation leads to a way to rationalize the denominator of an expression with a binomial in the denominator. For example, we can rationalize the denominator of $\dfrac{1}{\sqrt{3} - \sqrt{2}}$ by multiplying the numerator and denominator by $\sqrt{3} + \sqrt{2}$.

$$\frac{1}{\sqrt{3} - \sqrt{2}} = \frac{1(\sqrt{3} + \sqrt{2})}{(\sqrt{3} - \sqrt{2})(\sqrt{3} + \sqrt{2})}$$

$$= \frac{\sqrt{3} + \sqrt{2}}{(\sqrt{3})^2 - (\sqrt{2})^2} = \frac{\sqrt{3} + \sqrt{2}}{3 - 2}$$

$$= \frac{\sqrt{3} + \sqrt{2}}{1} = \sqrt{3} + \sqrt{2}$$

Had the denominator been $\sqrt{3} + \sqrt{2}$, we would have rationalized by multiplying both numerator and denominator by $\sqrt{3} - \sqrt{2}$. The binomials $\sqrt{3} + \sqrt{2}$ and $\sqrt{3} - \sqrt{2}$ are called **conjugates** as are binomials of the form $\sqrt{3} + 2$ and $\sqrt{3} - 2$.

To Rationalize a Binomial Denominator

Multiply both numerator and denominator of the fraction by the conjugate of the denominator and simplify.

| **EXAMPLE 3** RATIONALIZING DENOMINATORS | **PRACTICE EXERCISE 3** |

Rationalize the denominators.

(a) $\dfrac{2}{\sqrt{7} + \sqrt{5}}$

$= \dfrac{2(\sqrt{7} - \sqrt{5})}{(\sqrt{7} + \sqrt{5})(\sqrt{7} - \sqrt{5})}$ Multiply numerator and denominator by $\sqrt{7} - \sqrt{5}$, the conjugate of $\sqrt{7} + \sqrt{5}$

$= \dfrac{2(\sqrt{7} - \sqrt{5})}{(\sqrt{7})^2 - (\sqrt{5})^2}$ $(a + b)(a - b) = a^2 - b^2$

$= \dfrac{2(\sqrt{7} - \sqrt{5})}{7 - 5}$ $(\sqrt{7})^2 = 7$ and $(\sqrt{5})^2 = 5$

$= \dfrac{2(\sqrt{7} - \sqrt{5})}{2}$

$= \sqrt{7} - \sqrt{5}$

(b) $\dfrac{\sqrt{3}}{\sqrt{3} - 2} = \dfrac{\sqrt{3}(\sqrt{3} + 2)}{(\sqrt{3} - 2)(\sqrt{3} + 2)}$ Multiply numerator and denominator by $\sqrt{3} + 2$, the conjugate of $\sqrt{3} - 2$

$= \dfrac{\sqrt{3}(\sqrt{3} + 2)}{(\sqrt{3})^2 - (2)^2}$ $(a - b)(a + b) = a^2 - b^2$

$= \dfrac{\sqrt{3}(\sqrt{3} + 2)}{3 - 4}$ $(\sqrt{3})^2 = 3$ and $(2)^2 = 4$

$= \dfrac{\sqrt{3}(\sqrt{3} + 2)}{-1}$

$= -\sqrt{3}(\sqrt{3} + 2)$ $\dfrac{a}{-1} = -a$

$= -\sqrt{3} \cdot \sqrt{3} - \sqrt{3} \cdot 2$ Distributive law

$= -3 - 2\sqrt{3}$

Rationalize the denominators.

(a) $\dfrac{4}{\sqrt{7} - \sqrt{3}}$

(b) $\dfrac{\sqrt{5}}{\sqrt{5} + 2}$

Answers: (a) $\sqrt{7} + \sqrt{3}$
(b) $5 - 2\sqrt{5}$

9.5 EXERCISES A

Refer to the summary of techniques and simplify.

1. $\sqrt{20} - \sqrt{45}$

2. $\sqrt{2}\,\sqrt{32}$

3. $\sqrt{3x}\,\sqrt{12x}$

4. $\dfrac{\sqrt{18xy^2}}{\sqrt{2x}}$

5 $\dfrac{\sqrt{25x^2}}{\sqrt{3y}}$

6. $\dfrac{3}{\sqrt{2}} + \dfrac{1}{\sqrt{2}}$

7. $\sqrt{147} - \sqrt{108}$

8. $\sqrt{3}\,\sqrt{12} + \sqrt{5}\,\sqrt{20}$

9. $\sqrt{2}\,\sqrt{10} - \sqrt{15}\,\sqrt{3}$

10. $\dfrac{\sqrt{2}}{\sqrt{50}} - \dfrac{\sqrt{3}}{\sqrt{75}}$

11. $\dfrac{\sqrt{98}}{\sqrt{2}} + \dfrac{\sqrt{125}}{\sqrt{5}}$

12. $\dfrac{\sqrt{3}}{\sqrt{2}} - \dfrac{\sqrt{5}}{\sqrt{2}}$

13. $\sqrt{3x}\,\sqrt{6x} + \sqrt{5x}\,\sqrt{10x}$

14. $\sqrt{x}\,\sqrt{y} + \dfrac{\sqrt{xy^2}}{\sqrt{y}}$

15. $\sqrt{5x} - \dfrac{\sqrt{x}}{\sqrt{5}}$

16. $6\sqrt{2}\,\sqrt{6} + 2\sqrt{5}\,\sqrt{15} - 3\sqrt{12}\,\sqrt{25}$

17. $\sqrt{x^3}\,\sqrt{y} - 3x\sqrt{x}\,\sqrt{y} + \dfrac{4x\sqrt{xy^2}}{\sqrt{y}}$

Multiply and simplify.

18. $\sqrt{3}(\sqrt{3} + \sqrt{2})$

19. $\sqrt{5}(\sqrt{125} - \sqrt{45})$

20. $\sqrt{2}(\sqrt{27} - \sqrt{12})$

21. $\sqrt{7}(\sqrt{5} - \sqrt{125})$

22. $\sqrt{5}(\sqrt{5} + 1)$

23. $\sqrt{6}(\sqrt{2} - 1)$

24. $\sqrt{3}\left(\sqrt{3} - \dfrac{1}{\sqrt{3}}\right)$

25. $\sqrt{7}\left(\dfrac{1}{\sqrt{7}} + \sqrt{7}\right)$

26 $\sqrt{15}\left(\dfrac{\sqrt{3}}{\sqrt{5}} + \dfrac{1}{\sqrt{15}}\right)$

27. $\sqrt{8}\left(\sqrt{2} - \dfrac{1}{\sqrt{8}}\right)$

28 $\sqrt{2}\left(\sqrt{6} + \dfrac{\sqrt{2}}{\sqrt{3}}\right)$

29. $\sqrt{5}\left(\sqrt{10} - \dfrac{\sqrt{5}}{\sqrt{2}}\right)$

30. $(\sqrt{5} + 1)(\sqrt{5} + 1)$

31. $(\sqrt{5} + 1)(\sqrt{5} - 1)$

32. $(\sqrt{2} - \sqrt{5})(\sqrt{2} - \sqrt{5})$

33. $(\sqrt{2} - \sqrt{5})(\sqrt{2} + \sqrt{5})$

34. $(2\sqrt{3} + 1)(\sqrt{3} + 1)$

35 $(\sqrt{2} - 2)(\sqrt{2} + 1)$

36. $(2\sqrt{5} + \sqrt{7})(2\sqrt{5} - \sqrt{7})$

37 $(3\sqrt{3} - \sqrt{2})(2\sqrt{3} + \sqrt{2})$

Rationalize the denominator.

38. $\dfrac{1}{\sqrt{2} - 1}$

39. $\dfrac{2}{\sqrt{5} - \sqrt{3}}$

40. $\dfrac{2}{\sqrt{7} + \sqrt{5}}$

41. $\dfrac{3}{\sqrt{5} - 2}$

42. $\dfrac{6}{\sqrt{5} + 2}$

43 $\dfrac{-8}{\sqrt{3} - \sqrt{7}}$

44 $\dfrac{2\sqrt{3}}{\sqrt{3}-1}$

45. $\dfrac{\sqrt{2}}{\sqrt{2}+1}$

46. $\dfrac{2\sqrt{5}}{\sqrt{5}-\sqrt{3}}$

47 $\dfrac{x-1}{\sqrt{x}-1}$

FOR REVIEW

Add or subtract.

48. $2\sqrt{75}-6\sqrt{48}+9\sqrt{12}$

49. $6\sqrt{20x}-2\sqrt{45x}-\sqrt{5x}$

50. An economist might use the formula $E = 3\sqrt{c}+\sqrt{e}$, relating the three quantities E, c, and e. Find E when $c = 98$ and $e = 50$.

Exercises 51–56 review material from Chapter 3 and Section 7.5 to help you prepare for the next section. Solve each equation.

51. $7x = 42$

52. $x + 8 = 31$

53. $2(x - 7) = x + 4$

54. $x^2 - 1 = x(x + 1) + 2$

55. $x^2 - 3x - 4 = 0$

56. $x^2 + 2x = 0$

ANSWERS: 1. $-\sqrt{5}$ 2. 8 3. $6x$ 4. $3y$ 5. $\dfrac{5x\sqrt{3y}}{3y}$ 6. $2\sqrt{2}$ 7. $\sqrt{3}$ 8. 16 9. $-\sqrt{5}$ 10. 0 11. 12
12. $\dfrac{\sqrt{6}-\sqrt{10}}{2}$ 13. $8x\sqrt{2}$ 14. $2\sqrt{xy}$ 15. $\dfrac{4\sqrt{5x}}{5}$ 16. $-8\sqrt{3}$ 17. $2x\sqrt{xy}$ 18. $3+\sqrt{6}$ 19. 10 20. $\sqrt{6}$
21. $-4\sqrt{35}$ 22. $5+\sqrt{5}$ 23. $2\sqrt{3}-\sqrt{6}$ 24. 2 25. 8 26. 4 27. 3 28. $\dfrac{8\sqrt{3}}{3}$ 29. $\dfrac{5\sqrt{2}}{2}$ 30. $6+2\sqrt{5}$ 31. 4
32. $7-2\sqrt{10}$ 33. -3 34. $7+3\sqrt{3}$ 35. $-\sqrt{2}$ 36. 13 37. $16+\sqrt{6}$ 38. $\sqrt{2}+1$ 39. $\sqrt{5}+\sqrt{3}$ 40. $\sqrt{7}-\sqrt{5}$
41. $3\sqrt{5}+6$ 42. $6\sqrt{5}-12$ 43. $2\sqrt{3}+2\sqrt{7}$ 44. $3+\sqrt{3}$ 45. $2-\sqrt{2}$ 46. $5+\sqrt{15}$ 47. $\sqrt{x}+1$ 48. $4\sqrt{3}$
49. $5\sqrt{5x}$ 50. $26\sqrt{2}$ 51. 6 52. 23 53. 18 54. -3 55. 4 *or* -1 56. 0 *or* -2

9.5 EXERCISES B

Refer to the summary of techniques and simplify.

1. $\sqrt{98} + \sqrt{50}$

2. $\sqrt{5y}\,\sqrt{75y}$

3. $\dfrac{\sqrt{108}}{\sqrt{3}}$

4. $\dfrac{\sqrt{50x^2 y}}{\sqrt{2y}}$

5. $\dfrac{\sqrt{4y^2}}{\sqrt{3x}}$

6. $\dfrac{5}{\sqrt{3}} + \dfrac{1}{\sqrt{3}}$

7. $2\sqrt{175} - 3\sqrt{28}$

8. $\sqrt{2}\,\sqrt{8} + \sqrt{3}\,\sqrt{27}$

9. $\sqrt{5}\,\sqrt{10} - \sqrt{6}\,\sqrt{3}$

10. $\dfrac{\sqrt{3}}{\sqrt{75}} + \dfrac{\sqrt{5}}{\sqrt{125}}$

11. $\dfrac{\sqrt{108}}{\sqrt{3}} + \dfrac{\sqrt{147}}{\sqrt{3}}$

12. $\dfrac{\sqrt{2}}{\sqrt{5}} - \dfrac{\sqrt{3}}{\sqrt{5}}$

13. $\sqrt{2x}\,\sqrt{6x} + \sqrt{15x}\,\sqrt{5x}$

14. $\dfrac{\sqrt{x^2 y}}{\sqrt{x}} - \sqrt{x}\,\sqrt{y}$

15. $\dfrac{\sqrt{x}}{\sqrt{7}} + \sqrt{7x}$

16. $3\sqrt{5}\,\sqrt{15} + 5\sqrt{27} - 6\sqrt{6}\,\sqrt{8}$

17. $\dfrac{3y\sqrt{x^2 y}}{\sqrt{y}} + 3x\sqrt{y^2} - \dfrac{5x\sqrt{y^3}}{\sqrt{y}}$

Multiply and simplify.

18. $\sqrt{2}(\sqrt{3} + \sqrt{2})$

19. $\sqrt{3}(\sqrt{27} - \sqrt{12})$

20. $\sqrt{5}(\sqrt{7} - \sqrt{28})$

21. $\sqrt{5}(\sqrt{12} - \sqrt{48})$

22. $\sqrt{7}(1 - \sqrt{7})$

23. $\sqrt{10}(\sqrt{5} + 1)$

24. $\sqrt{5}\left(\sqrt{5} - \dfrac{1}{\sqrt{5}}\right)$

25. $\sqrt{11}\left(\sqrt{11} + \dfrac{1}{\sqrt{11}}\right)$

26. $\sqrt{18}\left(\sqrt{2} - \dfrac{1}{\sqrt{18}}\right)$

27. $\sqrt{21}\left(\dfrac{\sqrt{3}}{\sqrt{7}} + \dfrac{1}{\sqrt{21}}\right)$

28. $\sqrt{3}\left(\sqrt{6} + \dfrac{\sqrt{3}}{\sqrt{2}}\right)$

29. $\sqrt{2}\left(\sqrt{10} - \dfrac{\sqrt{2}}{\sqrt{5}}\right)$

30. $(\sqrt{7} + 1)(\sqrt{7} + 1)$

31. $(\sqrt{7} + 1)(\sqrt{7} - 1)$

32. $(\sqrt{3} - \sqrt{7})(\sqrt{3} - \sqrt{7})$

33. $(\sqrt{3} - \sqrt{7})(\sqrt{3} + \sqrt{7})$

34. $(3\sqrt{5} + 1)(\sqrt{5} + 1)$

35. $(\sqrt{3} - 3)(\sqrt{3} + 1)$

36. $(2\sqrt{2} - \sqrt{3})(3\sqrt{2} + \sqrt{3})$

37. $(2\sqrt{3} + \sqrt{5})(2\sqrt{3} - \sqrt{5})$

Rationalize the denominator.

38. $\dfrac{2}{\sqrt{3} - 1}$

39. $\dfrac{3}{\sqrt{5} - \sqrt{2}}$

40. $\dfrac{4}{\sqrt{7} + \sqrt{3}}$

41. $\dfrac{-2}{\sqrt{7} - 3}$

42. $\dfrac{6}{\sqrt{5} - \sqrt{7}}$

43. $\dfrac{9}{\sqrt{7} + 2}$

44. $\dfrac{2\sqrt{3}}{\sqrt{3} + 1}$

45. $\dfrac{4\sqrt{5}}{\sqrt{5} - 1}$

46. $\dfrac{3\sqrt{5}}{\sqrt{5} - \sqrt{2}}$

47. $\dfrac{1 - x}{1 - \sqrt{x}}$

FOR REVIEW

Add or subtract.

48. $5\sqrt{125} + 3\sqrt{5} - 8\sqrt{45}$

49. $7\sqrt{12x} - 3\sqrt{48x} - 2\sqrt{75x}$

50. A meteorologist may use the formula $W = 5\sqrt{f} - \sqrt{g}$, relating the three quantities W, f, and g. Find W when $f = 72$ and $g = 8$.

Exercises 51–56 review material from Chapter 3 and Section 7.5 to help you prepare for the next section. Solve each equation.

51. $6x = 54$

52. $4 + x = 15$

53. $5(x - 3) = 2x - 6$

54. $4 - x^2 = x(3 - x) - 2$

55. $x^2 + x - 6 = 0$

56. $2x^2 - x = 0$

9.5 EXERCISES C

Multiply and simplify.

1. $(\sqrt{ax} - \sqrt{by})(\sqrt{ax} + \sqrt{by})$

2. $(\sqrt{x} - \sqrt{y})(x + \sqrt{xy} + y)$
   [Answer: $x\sqrt{x} - y\sqrt{y}$]

Rationalize the denominator and simplify.

3. $\dfrac{x\sqrt{y} - y\sqrt{x}}{\sqrt{x} - \sqrt{y}}$

4. $\dfrac{x^2}{\sqrt{4 - x^2}} + \sqrt{4 - x^2}$ $\left[\text{Answer: } \dfrac{4\sqrt{4 - x^2}}{4 - x^2}\right]$

9.6 SOLVING RADICAL EQUATIONS

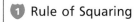

━━━━━━━━━━━ **STUDENT GUIDEPOSTS** ━━━━━━━━━━━

| ❶ Rule of Squaring | ❷ Solving Equations Involving Radicals |

❶ RULE OF SQUARING

Sometimes an equation contains a radical expression with a variable in the radicand, such as $\sqrt{x} = 5$. We use the following rule to solve such equations.

Rule of Squaring
If $a = b$,   then $a^2 = b^2$.

❷ SOLVING EQUATIONS INVOLVING RADICALS

When $\sqrt{x} = 5$, then the rule of squaring gives $(\sqrt{x})^2 = 5^2$ or $x = 25$. When this rule is applied to an equation with a variable, the resulting equation may have more solutions than the original. For instance, 2 is the only solution to the equation $x = 2$, but the equation formed by squaring both sides, $x^2 = 4$, has two solutions, 2 and -2. We call -2 an **extraneous root** in this case. Thus, when the rule of squaring is used, any possible solutions *must be* checked in the original equation, and extraneous roots must be discarded.

To Solve an Equation Involving Radicals

1. If only one radical is present, isolate this radical on one side of the equation and proceed to 3.
2. If two radicals are present, isolate one of the radicals on one side of the equation.
3. Square both sides to obtain an equation without the isolated radical expression(s).
4. Solve the resulting equation.
5. Check all possible solutions in the original equation.

EXAMPLE 1   SOLVING A RADICAL EQUATION

Solve.

$\sqrt{3x} - 5 = 7$

$\sqrt{3x} = 12$ Add 5 to both sides to isolate radical on the left

$(\sqrt{3x})^2 = (12)^2$ Use the rule of squaring

$3x = 144$ Squaring

$x = 48$ Divide both sides by 3

Check: $\sqrt{3(48)} - 5 \stackrel{?}{=} 7$

$\sqrt{144} - 5 \stackrel{?}{=} 7$

$12 - 5 \stackrel{?}{=} 7$

$7 = 7$

The solution is 48.

PRACTICE EXERCISE 1

Solve. $\sqrt{5y} + 2 = 12$

Answer: 20

EXAMPLE 2   SOLVING A RADICAL EQUATION

Solve.

$\sqrt{x + 3} + 4 = 11$

$\sqrt{x + 3} = 11 - 4$ Isolate the radical

$\sqrt{x + 3} = 7$

$(\sqrt{x + 3})^2 = 7^2$ Rule of squaring

$x + 3 = 49$

$x = 46$

Check: $\sqrt{46 + 3} + 4 \stackrel{?}{=} 11$

$\sqrt{49} + 4 \stackrel{?}{=} 11$

$7 + 4 \stackrel{?}{=} 11$

$11 = 11$

The solution is 46.

PRACTICE EXERCISE 2

Solve. $\sqrt{z - 2} - 1 = 3$

Answer: 18

EXAMPLE 3 EQUATION WITH TWO RADICALS	PRACTICE EXERCISE 3

Solve.

$$3\sqrt{2x - 5} - \sqrt{x + 23} = 0$$

$$3\sqrt{2x - 5} = \sqrt{x + 23} \qquad \text{Isolate the radicals}$$

$$(3\sqrt{2x - 5})^2 = (\sqrt{x + 23})^2$$

$$9(2x - 5) = x + 23 \qquad \text{Be sure to square the 3 on the left}$$

$$18x - 45 = x + 23$$

$$17x = 68$$

$$x = 4$$

Check: $\quad 3\sqrt{2 \cdot 4 - 5} - \sqrt{4 + 23} \stackrel{?}{=} 0$

$$3\sqrt{3} - \sqrt{27} \stackrel{?}{=} 0$$

$$3\sqrt{3} - 3\sqrt{3} = 0$$

The solution is 4.

Solve. $2\sqrt{3y + 1} - \sqrt{y + 15} = 0$

Answer: 1

EXAMPLE 4 EQUATION WITH NO SOLUTION	PRACTICE EXERCISE 4

Solve.

$$\sqrt{x^2 - 5} - x + 5 = 0$$

$$\sqrt{x^2 - 5} = x - 5 \qquad \text{Isolate the radical}$$

$$(\sqrt{x^2 - 5})^2 = (x - 5)^2 \qquad \text{Do \textit{not} square the right side}$$
$$\qquad\qquad\qquad\qquad \text{as } x^2 - 25$$

$$x^2 - 5 = x^2 - 10x + 25$$

$$-5 = -10x + 25 \qquad \text{Subtract } x^2 \text{ from both sides}$$

$$-30 = -10x$$

$$x = \frac{-30}{-10} = 3$$

Check: $\quad \sqrt{3^2 - 5} - 3 + 5 \stackrel{?}{=} 0$

$$\sqrt{9 - 5} - 3 + 5 \stackrel{?}{=} 0$$

$$\sqrt{4} - 3 + 5 \stackrel{?}{=} 0$$

$$2 - 3 + 5 \stackrel{?}{=} 0 \qquad \sqrt{4} \text{ is 2 \textit{not} } -2$$

$$4 \neq 0 \qquad\qquad \text{3 does not check}$$

The equation has no solution.

Solve. $\sqrt{y^2 + 9} - y + 1 = 0$

Answer: **no solution** (-4 **does not check**)

⚠ CAUTION ⚠

Remember that the radical represents the principal (nonnegative) root. Thus, in Example 4, we cannot replace $\sqrt{4}$ with -2 in the check.

Sometimes the equation that results when the radical is eliminated must be solved by factoring and using the zero-product rule. We see this in Example 5.

| EXAMPLE 5   USING THE ZERO-PRODUCT RULE | PRACTICE EXERCISE 5 |

Solve.

$$\sqrt{2x + 2} - x + 3 = 0$$

$\sqrt{2x + 2} = x - 3$ Isolate the radical

$(\sqrt{2x + 2})^2 = (x - 3)^2$

$2x + 2 = x^2 - 6x + 9$ $(a - b)^2 = a^2 - 2ab + b^2$

$0 = x^2 - 8x + 7$ Subtract $2x$ and 2

$0 = (x - 1)(x - 7)$ Factor

$x - 1 = 0$ or $x - 7 = 0$ Zero-product rule

$x = 1$ $x = 7$

Check: $\sqrt{2(1) + 2} - 1 + 3 \overset{?}{=} 0$ $\sqrt{2(7) + 2} - 7 + 3 \overset{?}{=} 0$

$\sqrt{4} - 1 + 3 \overset{?}{=} 0$ $\sqrt{16} - 7 + 3 \overset{?}{=} 0$

$2 - 1 + 3 \overset{?}{=} 0$ $4 - 7 + 3 \overset{?}{=} 0$

$4 \neq 0$ $0 = 0$

1 does not check.   7 does check.

The only solution is 7.

Solve. $\sqrt{17 - 4x} - x - 1 = 0$

Answer: 2

9.6   EXERCISES A

Solve the following equations.

1. $\sqrt{x} = 4$

② $3\sqrt{2x} - 9 = 0$

3. $\sqrt{4y} - 6 = 2$

4. $\sqrt{y + 2} = 6$

5. $\sqrt{4x - 3} = 5$

⑥ $3\sqrt{a - 3} + 6 = 0$

7. $4\sqrt{2a - 1} - 2 = 0$

8. $3\sqrt{x} = \sqrt{x + 16}$

9. $\sqrt{y + 3} = \sqrt{y + 5}$

10. $3\sqrt{y + 3} = \sqrt{y + 35}$

11. $\sqrt{3a + 2} - \sqrt{a + 8} = 0$

12. $\sqrt{5a - 3} - \sqrt{2a + 1} = 0$

13 $3\sqrt{x-1} - \sqrt{x+31} = 0$

14. $\sqrt{2x+1} = \sqrt{x+10}$

15. $3\sqrt{y+7} = 4\sqrt{y}$

16. $2\sqrt{2x+5} - 3\sqrt{3x-2} = 0$

17 $\sqrt{a^2-5} + a - 5 = 0$

18. $\sqrt{a^2+2} - a - 2 = 0$

19. $-\sqrt{x^2-12} + x + 6 = 0$

20 $\sqrt{x^2+3} + x = 6$

21. $\sqrt{x^2+16} - x + 8 = 0$

22. $\sqrt{x^2-8} + 3 = x$

23. $\sqrt{x+2} = x$

24. $\sqrt{x+1} = 1 - x$

25 $\sqrt{x-3} - x + 5 = 0$

26. $\sqrt{x+4} - x + 2 = 0$

27. $\sqrt{x + 15} - x - 3 = 0$

28 $\sqrt{x + 10} + x - 2 = 0$

FOR REVIEW

Simplify.

29. $\sqrt{5}(\sqrt{15} - \sqrt{12})$

30. $(\sqrt{3} + \sqrt{5})(2\sqrt{3} - \sqrt{5})$

31. $\dfrac{4}{\sqrt{11} - \sqrt{7}}$

32. $\dfrac{3\sqrt{5}}{\sqrt{5} - \sqrt{2}}$

33. If $c = \sqrt{a^2 + b^2}$, find c when $a = 4$ and $b = 3$.

34. If $b = \sqrt{c^2 - a^2}$, find b when $a = 5$ and $c = 13$.

ANSWERS: 1. 16   2. $\frac{9}{2}$ 3. 16   4. 34   5. 7   6. no solution   7. $\frac{5}{8}$ 8. 2   9. no solution   10. 1   11. 3   12. $\frac{4}{3}$ 13. 5   14. 9   15. 9   16. 2   17. 3   18. $-\frac{1}{2}$ 19. -4 20. $\frac{11}{4}$ 21. no solution   22. $\frac{17}{6}$ 23. 2   24. 0   25. 7   26. 5   27. 1   28. -1 29. $5\sqrt{3} - 2\sqrt{15}$ 30. $1 + \sqrt{15}$ 31. $\sqrt{11} + \sqrt{7}$ 32. $5 + \sqrt{10}$ 33. 5   34. 12

9.6   EXERCISES B

Solve the following equations.

1. $\sqrt{x} = 9$

2. $5\sqrt{3x} - 10 = 0$

3. $\sqrt{9y} - 6 = 9$

4. $\sqrt{y + 6} = 2$

5. $\sqrt{8x + 2} = 2$

6. $5\sqrt{a - 2} + 10 = 0$

7. $3\sqrt{3a + 1} - 6 = 0$

8. $5\sqrt{x} = \sqrt{x + 3}$

9. $\sqrt{2y + 1} = \sqrt{2y + 3}$

10. $3\sqrt{y - 2} = \sqrt{y + 12}$

11. $\sqrt{5a + 3} - \sqrt{a + 4} = 0$

12. $\sqrt{7a - 10} - \sqrt{4a - 5} = 0$

13. $2\sqrt{x - 5} - \sqrt{x + 22} = 0$

14. $\sqrt{3x + 5} = \sqrt{x + 15}$

15. $3\sqrt{y + 2} = 5\sqrt{y}$

16. $3\sqrt{2x - 4} - 2\sqrt{3x + 3} = 0$

17. $\sqrt{a^2 + 5} - a + 5 = 0$

18. $\sqrt{a^2 - 8} - a - 4 = 0$

19. $-\sqrt{x^2 - 20} + x + 10 = 0$

20. $\sqrt{x^2 + 7} - x + 7 = 0$

21. $\sqrt{x^2 - 15} + x = 5$

22. $\sqrt{x^2 + 3} - 3 = x$

23. $\sqrt{x + 6} = x$

24. $\sqrt{x + 1} = x - 1$

25. $\sqrt{x + 3} - x + 3 = 0$

26. $\sqrt{x + 6} - x - 4 = 0$

27. $\sqrt{x + 21} + x + 1 = 0$

28. $\sqrt{x + 5} + x - 1 = 0$

FOR REVIEW

Simplify.

29. $\sqrt{7}(\sqrt{14} - \sqrt{21})$

30. $(\sqrt{5} - \sqrt{3})(\sqrt{5} - 2\sqrt{3})$

31. $\dfrac{4}{\sqrt{11} + \sqrt{7}}$

32. $\dfrac{4\sqrt{7}}{\sqrt{7} - \sqrt{5}}$

33. If $c = \sqrt{a^2 + b^2}$, find c if $a = 6$ and $b = 8$.

34. If $a = \sqrt{c^2 - b^2}$, find a when $c = 7$ and $b = 5$.

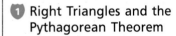

9.6 EXERCISES C

Solve.

1. $\sqrt{x + 5} - \sqrt{x - 3} = 2$
   [*Hint:*   You will need to square twice.]

2. $\sqrt{x - 3} + \sqrt{x + 2} = 5$

9.7 APPLICATIONS USING RADICALS

 ═════════════════════ STUDENT GUIDEPOSTS ═════════════════════

❶ Right Triangles and the
   Pythagorean Theorem

❷ 30°–60° and 45°–45° Right
   Triangles

❶ RIGHT TRIANGLES AND THE PYTHAGOREAN THEOREM

Many applied problems use radicals. For example, finding one side of a right triangle when the other two sides are given requires taking a square root. Remember that a **right triangle** is a triangle with a 90° (right) angle. The side opposite the 90° angle is the **hypotenuse** of the right triangle and the remaining two sides are its **legs.** We will agree to label the legs a and b and the hypotenuse c as in Figure 9.2. The next well-known theorem is named after the Greek mathematician Pythagoras.

Pythagorean Theorem
In a right triangle with legs a and b and hypotenuse c.
$$a^2 + b^2 = c^2.$$

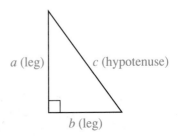

Figure 9.2 $a^2 + b^2 = c^2$

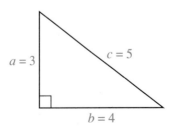

Figure 9.3

EXAMPLE 1 FINDING THE HYPOTENUSE

If the legs of a right triangle are 3 and 4, find the hypotenuse. See Figure 9.3.

By the Pythagorean theorem,

$$\begin{aligned} c^2 &= a^2 + b^2 \\ &= 3^2 + 4^2 && a = 3 \text{ and } b = 4 \\ &= 9 + 16 = 25. \end{aligned}$$

We need to solve $c^2 = 25$.

$$\begin{aligned} c^2 &= 25 \\ c^2 - 25 &= 0 && \text{Subtract 25} \\ c^2 - 5^2 &= 0 && 25 = 5^2 \\ (c + 5)(c - 5) &= 0 && \text{Factor using } a^2 - b^2 = (a + b)(a - b) \end{aligned}$$

$$c + 5 = 0 \quad \text{or} \quad c - 5 = 0$$
$$c = -5 \qquad\qquad c = 5$$

Since we want the length of the hypotenuse, we discard -5; $c = 5$. Notice that

$$\begin{aligned} c^2 &= 25 \\ c &= \sqrt{25} = 5. \end{aligned}$$

That is, if we take the principal square root of both sides of the equation, the result is 5. This technique is useful when the equation cannot be factored.

PRACTICE EXERCISE 1

If the legs of a right triangle are 9 and 12, find the hypotenuse.

Answer: 15

EXAMPLE 2 FINDING A LEG OF THE TRIANGLE

If $c = 7$ and $a = 3$, find b. See Figure 9.4.

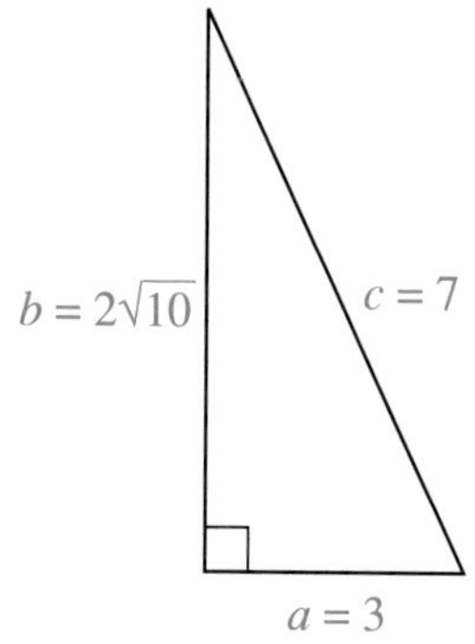

Figure 9.4

$$\begin{aligned} a^2 + b^2 &= c^2 && \text{Pythagorean theorem} \\ 3^2 + b^2 &= 7^2 && a = 3 \text{ and } c = 7 \\ 9 + b^2 &= 49 \\ b^2 &= 40 \\ b &= \sqrt{40} && \text{Take the principal square root on both sides} \\ b &= \sqrt{4 \cdot 10} = 2\sqrt{10} \end{aligned}$$

PRACTICE EXERCISE 2

If $c = 9$ and $b = 3$, find a.

Answer: $6\sqrt{2}$

| EXAMPLE 3 APPLICATION OF RADICALS | PRACTICE EXERCISE 3 |

The base of a 20-ft ladder used at a construction site is 8 ft from the base of a wall. How far is the top of the ladder from the ground? See Figure 9.5.

$$a^2 + b^2 = c^2 \quad \text{Pythagorean theorem}$$
$$a^2 + 8^2 = 20^2 \quad b = 8 \text{ and } c = 20$$
$$a^2 + 64 = 400$$
$$a^2 = 336$$
$$a = \sqrt{336} \quad \text{Take the principal square}$$
$$= \sqrt{16 \cdot 21} \quad \text{root on both sides}$$
$$= 4\sqrt{21}$$

20 ft

a

8 ft

Figure 9.5

The top of the ladder is $4\sqrt{21}$ ft from the ground. It is usually practical to give an approximate solution. Using a calculator, we find that $\sqrt{21} \approx 4.58$, so that $4\sqrt{21} \approx 18.3$. Thus, the ladder reaches about 18.3 ft up the wall.

A guy wire from the top of a pole to the ground measures 45 ft. If the pole is 30 ft in length, how far is the wire anchored from the base of the pole?

Answer: $15\sqrt{5} \approx 33.5$ ft

Applications from the business world, such as the next one, use square roots.

| EXAMPLE 4 APPLICATION TO BUSINESS | PRACTICE EXERCISE 4 |

A manufacturer of novelty items has found that his total daily expenses for production are given by the equation

$$e = 10\sqrt{n} + 25,$$

where e represents total expenses and n is the number of items produced.

(a) Find the total expenses when no items are produced. This is called **overhead.**

$$e = 10\sqrt{n} + 25 \quad \text{Expense equation}$$
$$= 10\sqrt{0} + 25 \quad \text{Substitute 0 for } n$$
$$= 10\sqrt{25}$$
$$= 10 \cdot 5 = 50$$

Overhead is $50 per day.

A manufacturer of puzzles has found that his daily expenses for production can be estimated by $e = 6\sqrt{n} + 49$, where e represents total expenses and n is the number of puzzles produced.

(a) Find the total expenses when no puzzles are produced.

(b) Find the total expenses when 600 items are produced.

$$e = 10\sqrt{n + 25} \qquad \text{Expense equation}$$
$$= 10\sqrt{600 + 25} \qquad \text{Substitute 600 for } n$$
$$= 10\sqrt{625}$$
$$= 10(25) = 250 \qquad (25)^2 = 625$$

Total expenses are $250 per day when 600 items are produced.

(c) How many items are made when the cost is $120 per day?

$$e = 10\sqrt{n + 25} \qquad \text{Expense equation}$$
$$120 = 10\sqrt{n + 25} \qquad \text{We solve for } n \text{ when } e \text{ is } 120$$
$$12 = \sqrt{n + 25} \qquad \text{To simplify, divide both sides by 10}$$
$$(12)^2 = (\sqrt{n + 25})^2 \qquad \text{Square both sides}$$
$$144 = n + 25$$
$$119 = n$$

Thus, 119 items are made on the day when expenses total $120.

(b) Find the total expense of producing 95 puzzles.

(c) How many puzzles were made on a day when the expenses were $78.

Answers: (a) $42   (b) $72
(c) 120

EXAMPLE 5   APPLICATION TO AERONAUTICS

If an airplane is flying above the earth, an equation that gives the approximate distance to the horizon in terms of the altitude of the plane is

$$d = \sqrt{8000h},$$

where d is the distance to the horizon in miles and h is the altitude (distance above the earth) in miles. See Figure 9.6.

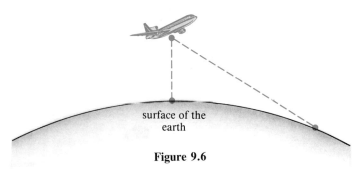

surface of the
earth

Figure 9.6

(a) Give the distance to the horizon when viewed from a plane at an altitude of 5 miles.

$$d = \sqrt{8000h} \qquad \text{Horizon equation}$$
$$d = \sqrt{(8000)(5)} \qquad h = 5 \text{ miles}$$
$$d = \sqrt{40000}$$
$$= \sqrt{(200)^2} \qquad (200)^2 = 40,000$$
$$= 200$$

The horizon is approximately 200 miles away.

PRACTICE EXERCISE 5

Use the equation $d = \sqrt{8000h}$ in the following:

(a) Give the distance to the horizon when viewed from a balloon at an altitude of 1 mile.

(b) Find the altitude of a plane when the horizon is 50 miles away.

$$d = \sqrt{8000h} \quad \text{Horizon equation}$$
$$50 = \sqrt{8000h} \quad d = 50 \text{ miles, solve for } h$$
$$(50)^2 = (\sqrt{8000h})^2 \quad \text{Square both sides}$$
$$2500 = 8000h$$
$$\frac{2500}{8000} = h$$
$$\frac{5}{16} = h$$

The plane is about $\frac{5}{16}$ of a mile (about 1650 ft) above the earth.

(b) Find the altitude of a balloon when the horizon is 150 miles away.

Answers: (a) $40\sqrt{5} \approx 89.4$ mi
(b) approximately 2.8 mi

❷ 30°–60° AND 45°–45° RIGHT TRIANGLES

To conclude this section we look at two special right triangles which have several applications in mathematics. A right triangle in which the two acute angles have measure 30° and 60° is called a **30°–60° right triangle.** The sides in such a triangle satisfy the following property.

> ### Property of 30°–60° Right Triangle
>
> In a 30°–60° right triangle, the length of the leg opposite the 30° angle is one-half the length of the hypotenuse.

Consider the right triangle in Figure 9.7. Since the hypotenuse c is 2, a must be 1 (one-half of c). We can find b using one of the equivalent forms of the Pythagorean theorem.

$$b^2 = c^2 - a^2$$
$$= 2^2 - 1^2$$
$$= 4 - 1$$
$$= 3$$
$$b = \sqrt{3} \quad \text{Take the principal square root of both sides}$$

In general, if a is the side opposite the 30° angle in a right triangle, $c = 2a$ and $b = \sqrt{3}a$.

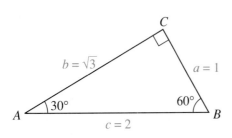

Figure 9.7 30°–60° Right Triangle

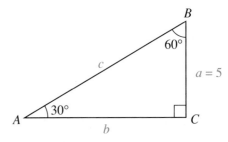

Figure 9.8

| EXAMPLE 6   SOLVING A 30°–60° RIGHT TRIANGLE | PRACTICE EXERCISE 6 |

In right $\triangle ABC$, $\angle C$ is the right angle, $\angle A$ has measure 30°, and $a = 5$. Find c and b.

First sketch the triangle as in Figure 9.8. Since $a = 5$ is one-half the hypotenuse, c must be 10. Also $b = \sqrt{3}a = \sqrt{3}(5) = 5\sqrt{3}$. We reach the same result using the Pythagorean theorem.

$$b^2 = c^2 - a^2$$
$$= 10^2 - 5^2$$
$$= 100 - 25 = 75$$
$$b = \sqrt{75} = \sqrt{25 \cdot 3} = 5\sqrt{3}$$

In right $\triangle ABC$, $\angle C$ is the right angle, $\angle B$ has measure 60°, and $c = 6$. Find a and b.

Answer: $a = 3$, $b = 3\sqrt{3}$

A right triangle with both acute angles measuring 45° is called a **45°–45° right triangle**. Since a 45°–45° right triangle is isosceles, the legs have equal length, and the Pythagorean theorem can be used to find the length of the hypotenuse. In Figure 9.9, since a and b are both 1, and

$$c^2 = a^2 + b^2$$
$$= 1^2 + 1^2 = 1 + 1 = 2,$$

we have that $c = \sqrt{2}$.

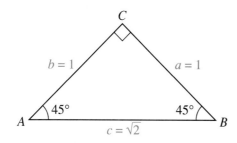

Figure 9.9   45°–45° Right Triangle

In general, if the legs are of length a then the hypotenuse is $\sqrt{2}a$.

9.7 EXERCISES A

Find the unknown leg or hypotenuse.

1. $a = 12$, $b = 5$

2. $a = 7$, $c = 10$

3. $b = 5$, $c = 9$

4. $a = 6$, $b = 8$

5. $a = 1$, $c = 5$

6. $b = 10$, $c = 20$

Solve.

7. A ladder 8 m long is placed on a building. If the base is 3 m from the building, how high up the side of the building is the top of the ladder?

8. The diagonal of a square is 12 centimeters. Find the length of the sides. [*Hint:* $a^2 + a^2 = 12^2$.]

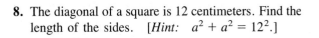

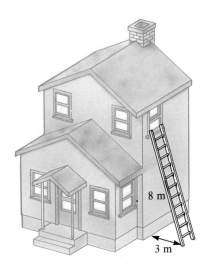

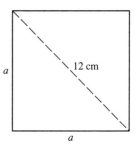

9. Find the length of the diagonal of a rectangle if the sides are 20 inches and 30 inches.

10 How long must a wire be to reach from the top of a 40-foot telephone pole to a point on the ground 30 feet from the base of the pole?

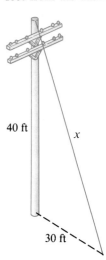

11. Three times the square root of a number is 6. Find the number.

12. Twice the square root of a number is 6 less than 12. Find the number.

13 The square root of 5 more than a number is twice the square root of the number. Find the number.

14. Twice the square root of one more than a number is the same as the square root of 13 more than the number. Find the number.

A manufacturer finds that the total cost per day of producing appliances is given by the equation $c = 100\sqrt{n} + 36$ where c is the total cost and n is the number produced. Use this information for Exercises 15–18.

15. What is c when no appliances are produced (overhead cost)?

16 What is the cost when 64 appliances are produced?

17. What is the cost when 39 appliances are produced?

18. How many appliances were made on a day when the cost was $2000?

A retailer finds that the total cost per day of selling sportcoats is given by the equation $c = 10\sqrt{n} + 100$, where c is the total cost and n is the number sold. Use this information in Exercises 19–22.

19. Find c when no sportcoats are sold (overhead cost).

20. Find the cost when 64 sportcoats are sold.

21. Find the number of coats sold on a day when the cost was $240.

22 Find the number of coats sold on a day when the cost was $50.

The distance in miles to the horizon from a point h miles above the earth is given by $d = \sqrt{8000h}$. Use this information in Exercises 23–26.

23. Find d when the altitude is 20 miles.

24. Find the distance to the horizon when the altitude is $\frac{1}{2}$ mile.

25. Find the altitude when the distance to the horizon is 120 miles.

26. Find the altitude when the distance to the horizon is 20 miles.

Use the Pythagorean theorem to find the length of the missing legs or hypotenuse in each triangle.

27

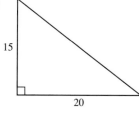

28.

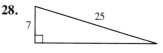

29.

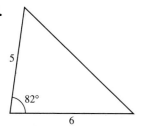

30.

Use the properties of 30°–60° or 45°–45° right triangles to find the missing legs or hypotenuse.

31.

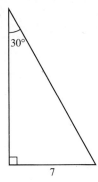

32.

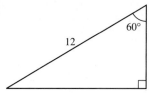

33.

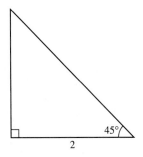

34.

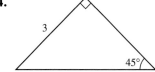

FOR REVIEW

Solve.

35. $\sqrt{3x + 1} = 5$

36. $\sqrt{4x - 5} - \sqrt{2x + 9} = 0$

37. $2\sqrt{y^2 + 1} - 2y - 1 = 0$

38. $2\sqrt{3y + 1} - \sqrt{15y - 11} = 0$

ANSWERS: 1. 13   2. $\sqrt{51}$ 3. $2\sqrt{14}$ 4. 10   5. $2\sqrt{6}$ 6. $10\sqrt{3}$ 7. $\sqrt{55}$ m (approximately 7.4 m)   8. $6\sqrt{2}$ cm (approximately 8.5 cm)   9. $10\sqrt{13}$ in (approximately 36.1 in)   10. 50 ft   11. 4   12. 9   13. $\frac{5}{3}$ 14. 3   15. $600   16. $1000   17. $500\sqrt{3}$ (approximately $866)   18. 364   19. $100   20. $180   21. 196   22. no solution   23. 400 mi   24. $20\sqrt{10}$ mi (approximately 63.2 mi)   25. 1.8 mi   26. 0.05 mi (264 ft)   27. 25   28. 24   29. The Pythagorean theorem does not apply since the triangle is not a right triangle.   30. 5   31. hypotenuse 14; leg $7\sqrt{3}$ 32. 6; $6\sqrt{3}$ 33. hypotenuse $2\sqrt{2}$; leg 2   34. hypotenuse $3\sqrt{2}$; leg 3   35. 8   36. 7   37. $\frac{3}{4}$ 38. 5

9.7   EXERCISES B

Find the unknown leg or hypotenuse.

1. $a = 4$, $b = 4$

2. $a = 5$, $c = 8$

3. $b = 14$, $c = 20$

4. $a = 9$, $b = 12$

5. $a = 3$, $c = 7$

6. $b = 20$, $c = 30$

Solve.

7. How long would a ladder need to be to reach the top of a 10-m building if the base of the ladder is 4 m from the building?

8. The diagonal of a square is 10 cm. Find the length of the sides.

9. Find the length of the diagonal of a rectangular pasture if the sides are 40 yards and 60 yards.

10. A wire 50 feet long is attached to the top of a tower and to the ground 20 feet from the base of the tower. How tall is the tower?

11. Four times the square root of a number is 20. Find the number.

12. Twice the square root of a number is 5 more than 15. Find the number.

13. The square root of 8 more than a number is 3 times the square root of the number. Find the number.

14. Twice the square root of 4 less than a number is the same as the square root of 1 less than the number. Find the number.

A manufacturer finds that the total cost per day of producing dresses is given by the equation $c = 20\sqrt{n + 16}$, where c is the total cost and n is the number produced. Use this information in Exercises 15–18.

15. What is c when no dresses are produced (overhead cost)?

16. What is the cost when 84 dresses are produced?

17. What is the cost when 24 dresses are produced?

18. How many dresses were made on a day when the cost was $360?

A retailer finds that the total cost per day of selling widgets is given by the equation $c = 50\sqrt{n} + 200$, where c is the total cost and n is the number sold. Use this information for Exercises 19–22.

19. Find c when no widgets are sold (overhead cost).

20. Find the cost when 81 widgets are sold.

21. Find the number of widgets sold on a day when the cost was $950.

22. Find the number of widgets sold on a day when the cost was $150.

The distance d in miles to the horizon from a point h miles above the earth is given by $d = 10\sqrt{80h}$. Use this information in Exercises 23–26.

23. Find d when the altitude is $\frac{1}{5}$ mile.

24. Find the distance to the horizon when the altitude is 10 miles.

25. Find the altitude when the distance to the horizon is 320 miles.

26. Find the altitude when the distance to the horizon is 8 miles.

Use the Pythagorean theorem to find the length of the missing legs or hypotenuse in each triangle.

27.
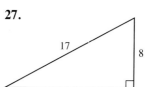

28.
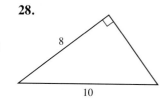

29.
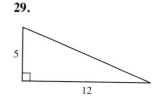

30.
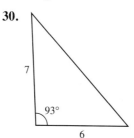

Use the properties of 30°–60° or 45°–45° right triangles to find the missing legs or hypotenuse.

31.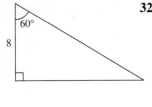

32.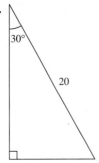

33.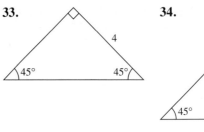

34.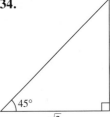

FOR REVIEW

Solve.

35. $\sqrt{2x - 3} = 3$

36. $\sqrt{3x + 2} - \sqrt{2x + 3} = 0$

37. $\sqrt{y^2 + 2} - y - 1 = 0$

38. $\sqrt{2y + 1} + \sqrt{y + 1} = 0$

9.7 EXERCISES C

Solve.

1. An airplane flew 120 km north from an airport and then 200 km west. How far was the plane from the airport?

2. A scientist uses the formula $a = \sqrt{1 + \frac{b}{x}}$. Solve this equation for b and find b when $a = 3$ and $x = \frac{1}{2}$. [Answer: $b = x(a^2 - 1)$; 4]

CHAPTER 9 REVIEW

KEY WORDS

9.1 A **perfect square** is the square of an integer or rational number.

If $a^2 = x$, then a is a **square root** of x.

The **principal square root** is the nonnegative square root of a positive real number.

In $\sqrt{x}$ the symbol $\sqrt{}$ is called a **radical** and x is the **radicand.**

9.2 The process of removing radicals from the denominator is called **rationalizing the denominator.**

9.4 **Like radicals** have the same radicand.

Unlike radicals have different radicands.

9.5 The expressions $\sqrt{a} + \sqrt{b}$ and $\sqrt{a} - \sqrt{b}$ are **conjugates** of each other.

9.6 An **extraneous root** is a "solution" obtained which is not a solution to the original problem.

9.7 A **right triangle** is a triangle with a 90° angle.

The **hypotenuse** is the side opposite the right angle and the **legs** are the other sides of a right triangle.

KEY CONCEPTS

9.1 The radical alone indicates the principal (non-negative) square root of a number. Thus $\sqrt{9} = 3$, not -3.

9.2, **1.** If $a \geq 0$ and $b \geq 0$, then
9.3
$$\sqrt{ab} = \sqrt{a}\,\sqrt{b}.$$ Multiplication rule

2. If $a \geq 0$ and $b > 0$, then
$$\sqrt{\frac{a}{b}} = \frac{\sqrt{a}}{\sqrt{b}}.$$ Division rule

3. When removing a perfect square from under a radical, multiply all factors; do not add. For example,
$$3\sqrt{4a} = 3\sqrt{4}\,\sqrt{a} = 3 \cdot 2\sqrt{a} =$$
$$6\sqrt{a}, \textit{ not } (3 + 2)\sqrt{a}, \text{ which is } 5\sqrt{a}.$$

9.4 $\sqrt{a + b} \neq \sqrt{a} + \sqrt{b}$ and $\sqrt{a - b} \neq \sqrt{a} - \sqrt{b}.$

9.5 To rationalize a binomial denominator, multiply both numerator and denominator of the fraction by the conjugate of the denominator and simplify.

9.6 **1.** To solve a radical equation, isolate a radical on one side of the equation and square both sides.

2. Check all solutions to a radical equation in the *original* equation.

9.7 **1.** Pythagorean theorem: In a right triangle with legs a and b and hypotenuse c, $a^2 + b^2 = c^2$.

2. In the 30°–60° triangle below,
$$c = 2a \text{ and } b = \sqrt{3}a.$$

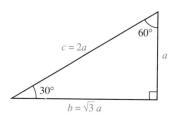

3. In the 45°–45° triangle below,
$$c = \sqrt{2}a \text{ and } b = a.$$

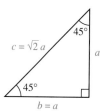

REVIEW EXERCISES

Part I

9.1 *Give the number of square roots of each number.*

1. 81   **2.** -81 **3.** 0

Evaluate each square root. Assume all variables and expressions are nonnegative.

4. $\sqrt{16}$ **5.** $\sqrt{9^2}$ **6.** $\sqrt{(-9)^2}$

7. $\pm\sqrt{144}$ **8.** $\sqrt{\dfrac{25}{16}}$ **9.** $\sqrt{-16}$

10. $\sqrt{0}$ **11.** $\sqrt{4a^2}$ **12.** $\sqrt{x^2 + 2x + 1}$

9.2 *Simplify each of the radicals.*

13. $\sqrt{50}$ **14.** $5\sqrt{8}$ **15.** $\sqrt{3^4}$

16. $\sqrt{16x^3}$ **17.** $\sqrt{8y^2}$ **18.** $\sqrt{50x^2y^3}$

19. $\sqrt{\dfrac{32}{81}}$ **20.** $\sqrt{\dfrac{50}{8}}$ **21.** $\sqrt{\dfrac{16y^2}{x^2}}$

22. $\sqrt{\dfrac{x^2y^3}{81}}$ **23.** $\dfrac{6}{\sqrt{7}}$ **24.** $\sqrt{\dfrac{50x^2}{y}}$

9.3 *Multiply and simplify.*

25. $\sqrt{8}\ \sqrt{18}$ **26.** $\sqrt{7}\ \sqrt{7}$ **27.** $3\sqrt{5}\ \sqrt{7}$

28. $\sqrt{12y}\ \sqrt{3y}$ **29.** $\sqrt{3xy}\ \sqrt{2xy^2}$ **30.** $\sqrt{5x}\ \sqrt{5y}\ \sqrt{5xy}$

Divide and simplify.

31. $\dfrac{\sqrt{99}}{\sqrt{11}}$ **32.** $\dfrac{\sqrt{50}}{\sqrt{8}}$ **33.** $\dfrac{3\sqrt{35}}{\sqrt{20}}$

34. $\dfrac{\sqrt{25x^2}}{\sqrt{2x^2}}$

35. $\dfrac{\sqrt{18x^2y}}{\sqrt{8y}}$

36. $\dfrac{\sqrt{25x^2}}{\sqrt{3y}}$

9.4 *Add or subtract.*

37. $5\sqrt{7} + 12\sqrt{7}$

38. $2\sqrt{5} - 7\sqrt{5}$

39. $7\sqrt{5} - 3\sqrt{3}$

40. $3\sqrt{8} + 7\sqrt{8}$

41. $\sqrt{49} - \sqrt{144}$

42. $3\sqrt{16x} + 4\sqrt{36x}$

43. $\sqrt{50} - \sqrt{98} + \sqrt{32}$

44. $2\sqrt{28} - 5\sqrt{112} + 3\sqrt{63}$

45. $3\sqrt{12x} + 2\sqrt{27x} - \sqrt{48x}$

46. $5\sqrt{48x^2} - \sqrt{27x^2} - \sqrt{3x^2}$

47. $\sqrt{5} + \dfrac{1}{\sqrt{5}}$

48. $\sqrt{\dfrac{3}{2}} - \sqrt{6}$

9.5 *Perform the indicated operations and simplify.*

49. $\sqrt{5}\ \sqrt{125} + \sqrt{3}\ \sqrt{27}$

50. $\dfrac{\sqrt{5}}{\sqrt{2}} - \dfrac{\sqrt{3}}{\sqrt{2}}$

51. $\sqrt{5}(\sqrt{3} - \sqrt{5})$

52. $\sqrt{7}\left(\sqrt{7} - \dfrac{1}{\sqrt{7}}\right)$ **53.** $(\sqrt{7} + 1)(\sqrt{7} + 1)$ **54.** $(\sqrt{7} + 1)(\sqrt{7} - 1)$

55. $(\sqrt{2} + \sqrt{3})(2\sqrt{2} - \sqrt{3})$ **56.** $(\sqrt{5} + 3)(\sqrt{5} - 1)$

Rationalize the denominator.

57. $\dfrac{2}{\sqrt{3} + 1}$ **58.** $\dfrac{-4}{\sqrt{7} - \sqrt{5}}$ **59.** $\dfrac{\sqrt{3}}{\sqrt{3} - \sqrt{2}}$ **60.** $\dfrac{\sqrt{7}}{\sqrt{7} - 1}$

9.6 *Solve.*

61. $2\sqrt{x} = \sqrt{x + 2}$ **62.** $\sqrt{7a - 1} - \sqrt{5a + 3} = 0$

63. $\sqrt{x^2 - 13} - x + 1 = 0$ **64.** $\sqrt{y + 5} + y - 1 = 0$

9.7 *Find the unknown leg or hypotenuse.*

65. $a = 6$, $b = 12$ **66.** $a = 5$, $c = 13$

Solve.

67. Find the length of a diagonal of a rectangular table with sides 8 feet and 12 feet.

68. Twice the square root of 4 less than a number is equal to the square root of 8 more than the number. Find the number.

A manufacturer finds that the total cost per day of producing wabbles is given by the equation $c = 500\sqrt{n + 144}$, *where c is the total cost and n is the number produced. Use this information in Exercises 69–70.*

69. Find the cost when 25 wabbles are produced.

70. How many wabbles were produced on a day when the cost was $10,000?

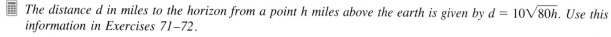 *The distance d in miles to the horizon from a point h miles above the earth is given by* $d = 10\sqrt{80h}$. *Use this information in Exercises 71–72.*

71. Find the distance to the horizon when the altitude is 0.05 miles.

72. Find the altitude when the distance to the horizon is 80 miles.

73. In a 30°–60° right triangle, a is the shorter leg and $a = 10$. Find b and c.

74. In a 45°–45° right triangle, $a = 7$. Find b and c.

Part II

In Exercises 75–86, perform the indicated operations and simplify. Rationalize all denominators.

75. $(\sqrt{6} - \sqrt{2})(\sqrt{6} + \sqrt{2})$

76. $\sqrt{x} + \dfrac{x}{\sqrt{x}}$

77. $\sqrt{5xy}\ \sqrt{10x^2y}$

78. $\sqrt{y^7}$

79. $\sqrt{\dfrac{14x^2y^3}{7xy}}$

80. $\dfrac{3}{\sqrt{5}}$

81. $\dfrac{5}{\sqrt{7} - \sqrt{2}}$

82. $6x\sqrt{8y^2} - 4xy\sqrt{2}$

83. $(\sqrt{2} - \sqrt{5})(3\sqrt{2} + \sqrt{5})$

84. $\sqrt{11}\left(\sqrt{11} + \dfrac{1}{\sqrt{11}}\right)$

85. $\dfrac{\sqrt{32x^2y^5}}{\sqrt{2xy^3}}$

86. $\sqrt{169} - \sqrt{196}$

The pressure *p* in pounds per square foot of a wind blowing against a flat surface is related to the velocity of the wind by the formula

$$v = \sqrt{\frac{p}{0.003}}$$

where *v* is in miles per hour. Use this information in Exercises 87–90.

87. A pressure gauge on a building registers a wind pressure of 10.8 lb/ft². What is the velocity of the wind?

88. During a storm, the wind pressure against the side of a mobile home measured 14.7 lb/ft². What was the velocity of the wind?

89. What is the pressure of the wind blowing 50 mph against the side of a bridge?

90. If the wind is blowing 15 mph, what pressure does it exert against the side of an apartment building?

91. $\sqrt{3x + 1} - 5 = 2$

92. $\sqrt{y - 1} - y + 3 = 0$

93. If *a* is the shorter of the two legs of a 30°–60° right triangle and *a* = 12 cm, find *b* and *c*.

94. If the hypotenuse of a 45°–45° right triangle is $7\sqrt{2}$ ft, find the length of each leg.

ANSWERS: 1. 2   2. 0   3. 1   4. 4   5. 9   6. 9   7. ±12 (12 and −12)   8. $\frac{5}{4}$ 9. meaningless   10. 0   11. 2*a*

12. *x* + 1   13. $5\sqrt{2}$ 14. $10\sqrt{2}$ 15. 9   16. $4x\sqrt{x}$ 17. $2y\sqrt{2}$ 18. $5xy\sqrt{2y}$ 19. $\frac{4\sqrt{2}}{9}$ 20. $\frac{5}{2}$ 21. $\frac{4y}{x}$ 22. $\frac{xy\sqrt{y}}{9}$

23. $\frac{6\sqrt{7}}{7}$ 24. $\frac{5x\sqrt{2y}}{y}$ 25. 12   26. 7   27. $3\sqrt{35}$ 28. 6*y*   29. $xy\sqrt{6y}$ 30. $5xy\sqrt{5}$ 31. 3   32. $\frac{5}{2}$ 33. $\frac{3\sqrt{7}}{2}$

34. $\frac{5\sqrt{2}}{2}$ 35. $\frac{3x}{2}$ 36. $\frac{5x\sqrt{3y}}{3y}$ 37. $17\sqrt{7}$ 38. $-5\sqrt{5}$ 39. $7\sqrt{5} - 3\sqrt{3}$ 40. $20\sqrt{2}$ 41. −5   42. $36\sqrt{x}$ 43. $2\sqrt{2}$

44. $-7\sqrt{7}$ 45. $8\sqrt{3x}$ 46. $16x\sqrt{3}$ 47. $\frac{6\sqrt{5}}{5}$ 48. $-\frac{\sqrt{6}}{2}$ 49. 34   50. $\frac{\sqrt{10} - \sqrt{6}}{2}$ 51. $\sqrt{15} - 5$ 52. 6

53. $8 + 2\sqrt{7}$ 54. 6   55. $1 + \sqrt{6}$ 56. $2 + 2\sqrt{5}$ 57. $\sqrt{3} - 1$ 58. $-2\sqrt{7} - 2\sqrt{5}$ 59. $3 + \sqrt{6}$ 60. $\frac{7 + \sqrt{7}}{6}$ 61. $\frac{2}{3}$

62. 2   63. 7   64. −1   65. $6\sqrt{5}$ 66. 12   67. $4\sqrt{13}$ ft (approximately 14.4 ft)   68. 8   69. $6500   70. 256

71. 20 mi   72. 0.8 mi   73. $b = 10\sqrt{3}$; $c = 20$ 74. $b = 7$; $c = 7\sqrt{2}$ 75. 4   76. $2\sqrt{x}$ 77. $5xy\sqrt{2x}$ 78. $y^3\sqrt{y}$

79. $y\sqrt{2x}$ 80. $\frac{3\sqrt{5}}{5}$ 81. $\sqrt{7} + \sqrt{2}$ 82. $8xy\sqrt{2}$ 83. $1 - 2\sqrt{10}$ 84. 12   85. $4y\sqrt{x}$ 86. −1   87. 60 mph

88. 70 mph   89. 7.5 lb/ft²   90. 0.675 lb/ft²   91. 16   92. 5   93. $b = 12\sqrt{3}$ cm; $c = 24$ cm   94. 7 ft

1. How many square roots does 36 have?

1. _____

Evaluate each square root. Assume all variables and expressions are nonnegative.

2. $\sqrt{(-5)^2}$

2. _____

3. $\pm\sqrt{121}$

3. _____

4. $\sqrt{\dfrac{81}{16}}$

4. _____

5. $\sqrt{16x^2}$

5. _____

6. $3\sqrt{48}$

6. _____

7. $\sqrt{\dfrac{8x}{y^2}}$

7. _____

8. $\sqrt{\dfrac{98x^2}{y}}$

8. _____

Multiply or divide and simplify.

9. $2\sqrt{6}\ \sqrt{24}$

9. _____

10. $\dfrac{\sqrt{27x^3}}{\sqrt{3x}}$

10. _____

Perform the indicated operations and simplify.

11. $\sqrt{20} + \sqrt{80}$

11. _____

12. $3\sqrt{18} - 5\sqrt{50}$

12. _____

13. $6\sqrt{12x} + 3\sqrt{3x} - \sqrt{75x}$

13. _____

14. $\sqrt{11} + \dfrac{1}{\sqrt{11}}$

14. _____

15. $\sqrt{2}\ \sqrt{8} + \sqrt{3}\ \sqrt{12}$

15. _____

16. $(\sqrt{5} + \sqrt{3})(\sqrt{5} - \sqrt{3})$

16. _____

17. $(\sqrt{2} + \sqrt{7})(2\sqrt{2} - \sqrt{7})$

17. _____

Rationalize the denominator.

18. $\dfrac{\sqrt{5}}{\sqrt{5} - 1}$

18. _____

Solve.

19. $\sqrt{x + 6} = \sqrt{3x - 1}$

19. _____

20. $\sqrt{x^2 + 7} + x - 1 = 0$

20. _____

21. Find the length of the diagonal of a rectangle with sides 6 cm and 8 cm.

21. _____

22. Given $d = 10\sqrt{80h}$.
   (a) Find d when h is $\frac{1}{16}$.

22(a) _____

   (b) Find h when d is 400.

22(b) _____

23. In a right triangle, $\angle A = 30°$ and $a = 20$. Find b and c.

23. _____

24. In a right triangle, $\angle B = 45°$ and the hypotenuse is 8. Find the length of each leg.

24. _____

500

Quadratic Equations

10.1 FACTORING AND TAKING ROOTS

STUDENT GUIDEPOSTS

1 Quadratic Equations
2 Solving by Factoring
3 Solving by Taking Roots

1 QUADRATIC EQUATIONS

Most of the equations we have solved have had the variable raised only to the first power. The only exception to this was in Section 5.5 where we briefly introduced the idea of a *quadratic equation*. An equation that can be expressed in the form

$$ax^2 + bx + c = 0,$$

where a, b, and c are constant real numbers and $a \neq 0$, is called a **quadratic equation.** (If a were 0, the term in which the variable is squared would be missing and the equation would not be quadratic.) We call $ax^2 + bx + c = 0$ the **general form** of a quadratic equation.

EXAMPLE 1 IDENTIFYING CONSTANTS IN GENERAL FORM	PRACTICE EXERCISE 1

Write each equation in general form and identify each constant a, b, and c.
(a) $3x^2 + 2x - 6 = 0$ $(3x^2 + 2x + (-6) = 0)$
This equation is already in general form with $a = 3$, $b = 2$ and $c = -6$. Note that $c = -6$ and not 6. In the general form of an equation, all signs are $+$ so any negative sign goes with the constant a, b, or c.

Write each equation in general form and identify each constant a, b, and c.

(a) $5x^2 - 4x + 8 = 0$

(b) $2y^2 = 3y + 1$
Collect all terms on one side of the equation to obtain

$$2y^2 - 3y - 1 = 0$$

so that $a = 2$, $b = -3$, and $c = 1$.

(b) $6y^2 = 1 - y$

(c) $y^2 = 3y + 1$
Collect all terms on the left side.

$$y^2 - 3y - 1 = 0$$

Then $a = 1$, $b = -3$, and $c = -1$. If we collect all terms on the right side we obtain

$$0 = -y^2 + 3y + 1$$

so that $a = -1$, $b = 3$, and $c = 1$. Both of these are correct since the equations are equivalent. However, it is common practice to write the equation so that the coefficient of the squared term is positive.

(d) $2x + 1 = 3 - 4x + 2$
Since there is no term in which the variable is squared, this is not a quadratic equation.

(e) $2y - 5y^2 = 3y^2 + 2y + 1$
Collect all terms to obtain

$$8y^2 + 1 = 0 \quad \text{or} \quad 8y^2 + 0 \cdot y + 1 = 0$$

so that $a = 8$, $b = 0$, and $c = 1$.

(f) $4 = 6x + 12x^2$
Collect all terms on one side.

$$12x^2 + 6x - 4 = 0$$

This is a general form of the equation. However, we should simplify by factoring out 2 and multiplying both sides by $\frac{1}{2}$. That is,

$$2(6x^2 + 3x - 2) = 0 \quad \text{or} \quad 6x^2 + 3x - 2 = 0.$$

After removing common factors, we have $a = 6$, $b = 3$, and $c = -2$.

(c) $-4y + 3 = -2y^2$

(d) $6 - 3x = 2 - (x - 5)$

(e) $3y + 7 + 7y^2 = 8y^2 - 3y + 2$

(f) $5 + 20x = 10x^2$

Answers:
(a) $5x^2 - 4x + 8 = 0$; $a = 5$, $b = -4$, $c = 8$
(b) $6y^2 + y - 1 = 0$; $a = 6$, $b = 1$, $c = -1$
(c) $2y^2 - 4y + 3 = 0$; $a = 2$, $b = -4$, $c = 3$
(d) not a quadratic equation
(e) $y^2 - 6y - 5 = 0$; $a = 1$, $b = -6$, $c = -5$
(f) $2x^2 - 4x - 1 = 0$; $a = 2$, $b = -4$, $c = -1$

② SOLVING BY FACTORING

Recall that an equation of the type

$$(x - 2)(x + 1) = 0$$

is solved by using the zero-product rule, which states that if the product of two numbers is zero then one or both of the numbers must be zero. Using this rule we set each factor equal to zero and solve.

$$x - 2 = 0 \quad \text{or} \quad x + 1 = 0$$
$$x = 2 \quad \text{or} \quad x = -1$$

Since $(x - 2)(x + 1) = x^2 - x - 2$, in effect we have solved the quadratic equation

$$x^2 - x - 2 = 0.$$

To Solve a Quadratic Equation by Factoring

1. Write the equation in general form.
2. Divide out any common numerical factor from each term.
3. Factor the left side into a product of two factors.
4. Use the zero-product rule to obtain two equations whose solutions are the solutions to the original quadratic equation.

EXAMPLE 2 **SOLVING A QUADRATIC EQUATION BY FACTORING**

Solve $2x^2 - x - 1 = 0$.

$$(2x + 1)(x - 1) = 0 \qquad \text{Factor}$$

$$2x + 1 = 0 \quad \text{or} \quad x - 1 = 0 \qquad \text{Zero-product rule}$$

$$2x = -1 \qquad\qquad x = 1$$

$$x = -\frac{1}{2}$$

Check: $\quad 2\left(-\frac{1}{2}\right)^2 - \left(-\frac{1}{2}\right) - 1 \stackrel{?}{=} 0 \qquad 2\,(1)^2 - (1) - 1 \stackrel{?}{=} 0$

$$2\left(\frac{1}{4}\right) + \frac{1}{2} - 1 \stackrel{?}{=} 0 \qquad\qquad 2 - 1 - 1 \stackrel{?}{=} 0$$

$$\frac{2}{4} + \frac{1}{2} - 1 \stackrel{?}{=} 0 \qquad\qquad 2 - 2 = 0$$

$$1 - 1 = 0$$

The solutions are 1 and $-\frac{1}{2}$.

PRACTICE EXERCISE 2

Solve $3x^2 + x - 2 = 0$.

Answer: $\frac{2}{3}$, -1

EXAMPLE 3 **REMOVING COMMON FACTOR FIRST**

Solve $2x^2 - 2x = 12$.

$$2x^2 - 2x - 12 = 0 \qquad \text{Write in general form}$$

$$2(x^2 - x - 6) = 0 \qquad \text{2 is a common factor}$$

$$x^2 - x - 6 = 0 \qquad \text{Divide both sides by 2}$$

$$(x - 3)(x + 2) = 0 \qquad \text{Factor}$$

$$x - 3 = 0 \quad \text{or} \quad x + 2 = 0 \qquad \text{Zero-product rule}$$

$$x = 3 \qquad\qquad x = -2$$

The solutions are 3 and -2. Check.

PRACTICE EXERCISE 3

Solve $4x^2 + 18x = 10$.

Answer: $\frac{1}{2}$, -5

EXAMPLE 4 **ONE SOLUTION ZERO**

Solve $5x^2 = 3x$.

$$5x^2 - 3x = 0 \qquad \text{Write in general form}$$

$$x(5x - 3) = 0 \qquad \text{Factor}$$

$$x = 0 \quad \text{or} \quad 5x - 3 = 0 \qquad \text{Zero-product rule}$$

$$5x = 3$$

$$x = \frac{3}{5}$$

The solutions are 0 and $\frac{3}{5}$. Check.

PRACTICE EXERCISE 4

Solve $7x = 14x^2$.

Answer: 0, $\frac{1}{2}$

Whenever we have a quadratic equation in general form with $c = 0$, that is, $ax^2 + bx = 0$, one factor is always x so zero is one solution to the equation.

//////////////////////| **CAUTION** |//////////////////////

Never divide both sides of an equation by an expression involving the variable. In Example 4 if we had divided by x we would have lost the solution 0.

//////////////

EXAMPLE 5 A MORE COMPLEX EQUATION	**PRACTICE EXERCISE 5**

Solve $5z(z - 6) - 8z = 2z$.

$$5z^2 - 30z - 8z = 2z$$
$$5z^2 - 38z = 2z$$
$$5z^2 - 40z = 0 \quad \text{Write in general form}$$
$$5(z^2 - 8z) = 0$$
$$z^2 - 8z = 0 \quad \text{Divide out the common factor of 5}$$
$$z(z - 8) = 0 \quad \text{Factor}$$
$$z = 0 \quad \text{or} \quad z - 8 = 0$$
$$z = 8$$

The solutions are 0 and 8. Check.

Solve $y(y - 3) - 7 = 2y + 7$.

Answer: $7, -2$

---

③ SOLVING BY TAKING ROOTS

A quadratic equation with $b = 0$, that is, $ax^2 + c = 0$, can be solved using a different technique. Consider the equation

$$x^2 - 4 = 0$$

or

$$x^2 = 4.$$

What number squared is 4? Clearly $2^2 = 4$ and $(-2)^2 = 4$ so that x is either 2 or -2. By taking the square root of both sides of the equation and using the symbol $\pm$ to indicate both the positive (principal) square root and the negative square root, we have

$$x = \pm\sqrt{4}$$
$$x = \pm 2.$$

Remember that $x = \pm 2$ is simply a shorthand way of writing $x = +2 \ or \ x = -2$.

To Solve a Quadratic Equation of the Form $ax^2 + c = 0$

1. Subtract c from both sides, obtaining $ax^2 = -c$.

2. Divide both sides by a, obtaining $x^2 = -\dfrac{c}{a}$, where $-\dfrac{c}{a} > 0$ if a and c have different signs.

3. Take the square root of both sides, obtaining $x = \pm\sqrt{-\dfrac{c}{a}}$.

CAUTION

This method gives real-number solutions only when the radicand $-\frac{c}{a}$ is positive. For this to happen a and c must have opposite signs.

EXAMPLE 6 SOLVING A QUADRATIC EQUATION BY TAKING ROOTS

Solve $3x^2 - 75 = 0$.

$$3x^2 = 75 \qquad \text{Add 75 to both sides}$$

$$x^2 = \frac{75}{3} \qquad \text{Divide both sides by 3}$$

$$x^2 = 25$$

$$x = \pm\sqrt{25} \qquad \text{Take square root of both sides}$$

$$= \pm 5$$

Check: $3(+5)^2 - 75 \stackrel{?}{=} 0 \qquad 3(-5)^2 - 75 \stackrel{?}{=} 0$

$\qquad\quad 3(25) - 75 \stackrel{?}{=} 0 \qquad\quad 3(25) - 75 \stackrel{?}{=} 0$

$\qquad\quad 75 - 75 = 0 \qquad\qquad 75 - 75 = 0$

The solutions are 5 and -5.

PRACTICE EXERCISE 6

Solve $64 - 4x^2 = 0$.

Answer: ± 4

EXAMPLE 7 AN EQUATION WITH IRRATIONAL SOLUTIONS

Solve $5x^2 - 40 = 0$.

$$5x^2 = 40 \qquad \text{Add 40}$$

$$x^2 = \frac{40}{5} \qquad \text{Divide by 5}$$

$$x = \pm\sqrt{8} \qquad \text{Take square root of both sides}$$

$$x = \pm\sqrt{4 \cdot 2} = \pm 2\sqrt{2}$$

The solutions are $2\sqrt{2}$ and $-2\sqrt{2}$. Check.

PRACTICE EXERCISE 7

Solve $8x^2 - 96 = 0$.

Answer: $\pm 2\sqrt{3}$

EXAMPLE 8 AN EQUATION WITH NO SOLUTION

Solve $x^2 + 5 = 0$.

$$x^2 = -5 \qquad \text{Subtract 5}$$

Recall from Section 9.1 that no real number squared can be negative. Hence, there are no real-number solutions to the equation $x^2 = -5$, nor to any equation that has x^2 equal to a negative number.

PRACTICE EXERCISE 8

Solve $x^2 + 17 = 0$.

Answer: no real number solutions

| **EXAMPLE 9** EQUATION IN FORM FOR TAKING ROOTS | **PRACTICE EXERCISE 9** |

Solve $(y + 3)^2 = 16$.

Although not in the form $ax^2 + c = 0$, we can use the same basic technique of taking the square root of both sides to solve.

$$y + 3 = \pm\sqrt{16}$$
$$y + 3 = \pm 4$$
$$y = -3 \pm 4 \qquad \text{Subtract 3 from both sides}$$
$$y = \begin{cases} -3 + 4 = 1 & \text{Use } + \text{ to obtain solution 1} \\ -3 - 4 = -7 & \text{Use } - \text{ to obtain solution } -7 \end{cases}$$

It is easy to substitute 1 and -7 into the original equation to see that they check and are the solutions.

Solve $(x + 7)^2 = 49$.

Answer: 0, -14

10.1 EXERCISES A

Write each equation in general quadratic form and identify the constants a, b, and c.

1. $2x^2 + x = 5$

2. $y - y^2 = 5$

3. $x^2 = 3x$

4. $y^2 = -7y - 12$

5. $2y^2 - y = 3 + y - y^2$

6. $2(x + x^2) = 3x - 1$

7. $x^2 + 1 = 0$

8. $\frac{1}{2}y^2 - y = 3 - y$

9. $35y - 2 = 0$

10. $2(x^2 + x) + x^3 = x^3 - 5$

11. $2y^3 + y^2 + 1 = 0$

12. $3x = 3(x + 1)$

13. Given the quadratic equation $-x^2 + 2x - 3 = 0$, we might conclude that $a = -1$, $b = 2$, and $c = -3$. Would $a = 1$, $b = -2$, and $c = 3$ be incorrect? Explain.

Solve.

14. $x^2 - 3x - 28 = 0$

15. $y^2 + 5y + 6 = 0$

16. $x^2 = 2x + 35$

17. $2z^2 - 5x = 3$

18. $6x^2 - 2x - 8 = 0$

19 $4z^2 = 7z$

20. $12x^2 + 6x = 0$

21. $y^2 - 10y = -25$

22. $z^2 - 6 = z$

23. $3(y^2 + 8) = 24 - 15y$

24. $z(8 - 2z) = 6$

25. $x(3x + 20) = 7$

26 $(3z - 1)(2z + 1) = 3(2z + 1)$

27. $(2y - 3)(y + 1) = 4(2y - 3)$

28 $4x^2 - 9 = 3$

29. $x^2 = 16$

30. $z^2 - 81 = 0$

31 $3x^2 - 48 = -48$

32. $2y^2 - 128 = 0$

33. $y^2 - \dfrac{9}{4} = 0$

34. $-3z^2 = -27$

35 $2x^2 - 150 = 0$

36. $4y^2 = 72$

37 $2x^2 + 50 = 0$

38. $-3y^2 = -6$

39. $3 - y^2 = y^2 - 3$

40. $z^2 + 2z = 2z - 4$

41 $(x - 2)^2 = 9$ **42.** $(y + 1)^2 = 49$ **43.** $(z - 1)^2 = 5$

44. The square of a number, decreased by 2, is equal to the negative of the number. Find the number.

45. The square of 2 more than a number is 64. Find the number.

The amount of money A that will result if a principal P is invested at r percent interest compounded annually for 2 years is given by $A = P(1 + r)^2$. Use this formula to find the interest rate for the given values of A and P in Exercises 46–47. Give the answer to the nearest tenth of a percent if appropriate.

46. $A = \$2645,\ P = \$2000$ **47.** $A = \$1765,\ P = \$1500$

Tell what is wrong with each of the "solutions" in Exercises 48–49.

48. $x^2 - x = 3(x - 1)$
  $x\cancel{(x - 1)} = 3\cancel{(x - 1)}$
      $x = 3$

49. $x^2 - x = 2$
  $x(x - 1) = 2$
  $x = 2$ or $x - 1 = 2$
        $x = 3$

FOR REVIEW

The following exercises review material from Sections 6.4 and 7.4. They will help you prepare for the next section. Find the products in Exercises 50–52.

50. $(x + 6)^2$ **51.** $(x - 10)^2$ **52.** $\left(x + \dfrac{1}{9}\right)^2$

Factor the trinomials in Exercises 53–55.

53. $x^2 - 4x + 4$ **54.** $x^2 + 10x + 25$ **55.** $x^2 + x + \dfrac{1}{4}$

Simplify the square roots in Exercises 56–58.

56. $\sqrt{8}$ **57.** $\sqrt{\dfrac{49}{4}}$ **58.** $\sqrt{\dfrac{7}{4}}$

10.1 EXERCISES B

Write each equation in general form and identify the constants a, b, and c.

1. $3x^2 + x = 7$

2. $y - y^2 = 4$

3. $z^2 = 9z$

4. $y^2 = -10y - 18$

5. $3z^2 - z = 5 + z - z^2$

6. $4(z + z^2) = 2z + 1$

7. $w^2 + 5 = 0$

8. $\dfrac{1}{3}x^2 - x = 5 - x$

9. $4x - 8 = 0$

10. $3(y^2 + y) - y^3 = 6 - y^3$

11. $5x^3 = x^2 + 7$

12. $3(w - 3) = 4w$

13. Given the quadratic equation $-y^2 + y - 7 = 0$, we might conclude that $a = -1$, $b = 1$, and $c = -7$. Would $a = 1$, $b = -1$, and $c = 7$ be incorrect? Explain.

Solve.

14. $x^2 + x - 6 = 0$

15. $y^2 + 3y + 2 = 0$

16. $x^2 = 2x + 24$

17. $2z^2 + 3 = 7z$

18. $3x^2 - 2x - 5 = 0$

19. $2z^2 = 3z$

20. $3x^2 - 18x = 0$

21. $y^2 - 8y + 16 = 0$

22. $z^2 - 20 = z$

23. $3(y^2 + 4y) = -6y$

24. $z(z - 3) = -2$

25. $x(2x + 9) = 5$

26. $(z - 2)(z + 1) = 3(z + 1)$

27. $5x^2 - 10 = 5$

28. $y(2y - 1) = 7(1 - 2y)$

29. $x^2 = 49$

30. $z^2 - 36 = 0$

31. $5x^2 - 7 = -7$

32. $2y^2 - 98 = 0$

33. $y^2 - \dfrac{25}{4} = 0$

34. $-8z^2 = -32$

35. $2x^2 - 100 = 0$

36. $3y^2 = 60$

37. $3x^2 + 75 = 0$

38. $-6y^2 = -18$

39. $7 - x^2 = x^2 - 7$

40. $z^2 + 5z = 5z - 1$

41. $(x - 5)^2 = 4$

42. $(y + 2)^2 = 1$

43. $(z - 3)^2 = 3$

44. The square of a number, less 10 is equal to three times the number. Find the number.

45. The square of 5 more than a number is 9. Find the number.

The amount of money A that will result if a principal P is invested at r percent interest compounded annually for 2 years is given by $A = P(1 + r)^2$. Use this formula to find the interest rate for the given values of A and P in Exercises 46–47. Give the answer to the nearest tenth of a percent if appropriate.

46. $A = \$1210,\ P = \$1000$

47. $A = \$2785,\ P = \$2400$

Tell what is wrong with each of the "solutions" in Exercises 48–49.

48. $x^2 - 2x = 0$

$x(x - 2) = 0$

$\dfrac{x(x - 2)}{x} = \dfrac{0}{x}$

$x - 2 = 0$

$x = 2$

49. $x(x - 5) = 6$

$x = 6$ or $x - 5 = 6$

$x = 11$

FOR REVIEW

The following exercises review material from Sections 6.4 and 7.4. They will help you prepare for the next section. Find the products in Exercises 50–52.

50. $(x + 3)^2$

51. $(x - 7)^2$

52. $\left(x - \dfrac{1}{6}\right)^2$

Factor the trinomials in Exercises 53–55.

53. $x^2 + 22x + 121$

54. $x^2 - 20x + 100$

55. $x^2 + \dfrac{2}{3}x + \dfrac{1}{9}$

Simplify the square roots in Exercises 56–58.

56. $\sqrt{32}$

57. $\sqrt{\dfrac{144}{25}}$

58. $\sqrt{\dfrac{11}{9}}$

10.1 EXERCISES C

Solve.

1. $6x^2 + 47x - 8 = 0$

2. $(2x - 1)(x + 5) =$
$(4x + 3)(x + 5)$ [Answer: 5, −2]

3. $(x^2 - 5)^2 = 8$

10.2 SOLVING BY COMPLETING THE SQUARE

1 Completing the Square   **2** Solving by Completing the Square

1 COMPLETING THE SQUARE

A trinomial $x^2 + bx + c$ is a perfect square trinomial if it factors into the square of a binomial. Notice that the trinomials in the left column below are perfect square trinomials. Also notice in the right column that for each perfect square trinomial, the constant term, c, is the square of one half the coefficient of the x term, b.

$$x^2 + 6x + 9 = (x + 3)^2 \qquad 9 = \left(\tfrac{1}{2} \cdot 6\right)^2$$
$$x^2 - 10x + 25 = (x - 5)^2 \qquad 25 = \left(\tfrac{1}{2} \cdot (-10)\right)^2$$
$$x^2 - x + \frac{1}{4} = \left(x - \frac{1}{2}\right)^2 \qquad \tfrac{1}{4} = \left(\tfrac{1}{2} \cdot (-1)\right)^2$$

In general, whenever we are given the first two terms of a trinomial, $x^2 + bx$, we can add $\left(\tfrac{1}{2} \cdot b\right)^2$ to it to get a perfect square trinomial. This process, called **completing the square,** can be used only when the coefficient in the squared term is 1.

EXAMPLE 1 COMPLETING THE SQUARE

What must be added to complete the square?

(a) $x^2 + 12x +$ _____

Add $\left(\dfrac{1}{2} \cdot 12\right)^2 = 6^2 = 36$.

(b) $x^2 - 20x +$ _____

Add $\left(\dfrac{1}{2} \cdot (-20)\right)^2 = (-10)^2 = 100$.

(c) $x^2 + \dfrac{2}{3}x +$ _____

Add $\left(\dfrac{1}{2} \cdot \dfrac{2}{3}\right)^2 = \left(\dfrac{1}{3}\right)^2 = \dfrac{1}{9}$.

PRACTICE EXERCISES 1

What must be added to complete the square?

(a) $x^2 + 14x +$ _____

(b) $x^2 - 22x +$ _____

(c) $x^2 - \dfrac{2}{3}x +$ _____

Answers: (a) 49   (b) 121   (c) $\tfrac{1}{9}$

2 SOLVING BY COMPLETING THE SQUARE

We can solve quadratic equations by completing the square (a method that will have applications in the next algebra course). For example, the equation

$$x^2 + 3x - 10 = 0$$

which can be solved by factoring,

$$(x + 5)(x - 2) = 0$$
$$x = -5, 2$$

can also be solved by completing the square. To do so, first isolate the constant term on the right side. (Leave space as indicated.)

$$x^2 + 3x \quad = 10$$

To complete the square on the left side, add the square of half the coefficient of x.

$$\left(\frac{1}{2} \cdot 3\right)^2 = \left(\frac{3}{2}\right)^2 = \frac{9}{4}$$

But if $\frac{9}{4}$ is added on the left, to keep equality we must add $\frac{9}{4}$ on the right.

$$x^2 + 3x + \frac{9}{4} = 10 + \frac{9}{4}$$

$$\left(x + \frac{3}{2}\right)^2 = \frac{49}{4}$$

Take the square root of each side, remembering to write both the positive and negative root.

$$x + \frac{3}{2} = \pm \sqrt{\frac{49}{4}} = \pm \frac{7}{2}$$

$$x = -\frac{3}{2} \pm \frac{7}{2} \qquad \text{Subtract } \tfrac{3}{2} \text{ from both sides}$$

$$x = -\frac{3}{2} + \frac{7}{2} \quad \text{or} \quad x = -\frac{3}{2} - \frac{7}{2}$$

$$x = \frac{4}{2} = 2 \qquad\qquad x = \frac{-10}{2} = -5$$

The solutions are the same as before, 2 and -5.

To Solve a Quadratic Equation by Completing the Square

1. Isolate the constant term on the right side of the equation.
2. If the coefficient of the squared item is 1, proceed to step 4.
3. If the coefficient of the squared term is not 1, divide each term by that coefficient.
4. Complete the square on the left side and add the same number to the right side.
5. Factor (check by multiplying the factors) and take the square root of both sides (positive and negative roots).
6. Use both the positive root and the negative root to obtain the solutions to the original equation.

EXAMPLE 2   SOLVING BY COMPLETING THE SQUARE

Solve $y^2 - 6y + 8 = 0$ by completing the square.

$$y^2 - 6y \quad\;\; = -8 \qquad \text{Isolate the constant, } b = -6$$

$$y^2 - 6y + 9 = -8 + 9 \qquad \text{Add 9 to both sides; } \left(\tfrac{-6}{2}\right)^2 = (-3)^2 = 9$$

$$(y - 3)^2 = 1$$

$$y - 3 = \pm\sqrt{1} \qquad \text{Take the square root of both sides}$$

$$y - 3 = \pm 1 \qquad \sqrt{1} = 1$$

$$y = 3 \pm 1$$

$$y = 3 + 1 \quad \text{or} \quad y = 3 - 1$$

$$y = 4 \qquad\qquad y = 2$$

The solutions are 2 and 4.

PRACTICE EXERCISE 2

Solve $y^2 - 8y + 12 = 0$ by completing the square.

Answer: 2, 6

As mentioned earlier, we can use the technique above to complete the square when the coefficient of the squared item is 1. If this is not the case, we need to divide each term of the equation by the coefficient of the squared term before we can use the method.

EXAMPLE 3   MAKING THE COEFFICIENT OF x^2 ONE

Solve $2x^2 + 2x - 3 = 0$ by completing the square.

$$2x^2 + 2x \qquad = 3 \qquad \text{Isolate the constant}$$

$$x^2 + x \qquad = \frac{3}{2} \qquad \text{Divide through by 2; } b = 1$$

$$x^2 + x + \frac{1}{4} = \frac{3}{2} + \frac{1}{4} \qquad \text{Complete the square by adding } \tfrac{1}{4} \text{ to both sides}$$

$$\left(x + \frac{1}{2}\right)^2 = \frac{7}{4} \qquad \tfrac{3}{2} + \tfrac{1}{4} = \tfrac{6}{4} + \tfrac{1}{4} = \tfrac{7}{4}$$

$$x + \frac{1}{2} = \pm\sqrt{\frac{7}{4}} = \pm\frac{\sqrt{7}}{\sqrt{4}} = \pm\frac{\sqrt{7}}{2}$$

$$x + \frac{1}{2} = \frac{\sqrt{7}}{2} \qquad \text{or} \qquad x + \frac{1}{2} = -\frac{\sqrt{7}}{2}$$

$$x = -\frac{1}{2} + \frac{\sqrt{7}}{2} \qquad\qquad x = -\frac{1}{2} - \frac{\sqrt{7}}{2}$$

$$x = \frac{-1 + \sqrt{7}}{2} \qquad\qquad x = \frac{-1 - \sqrt{7}}{2}$$

The solutions are $\dfrac{-1 \pm \sqrt{7}}{2}$.

PRACTICE EXERCISE 3

Solve $3x^2 - 2x - 6 = 0$ by completing the square.

Answer: $\dfrac{1 \pm \sqrt{19}}{3}$

10.2 EXERCISES A

What must be added to complete the square?

1. $x^2 - 8x + \underline{\hphantom{xxx}}$

2. $y^2 - 18y + \underline{\hphantom{xxx}}$

3. $z^2 + 7z + \underline{\hphantom{xxx}}$

4. $x^2 - x + \underline{\hphantom{xxx}}$

5. $y^2 + 9y + \underline{\hphantom{xxx}}$

6 $z^2 + \dfrac{1}{3}z + \underline{\hphantom{xxx}}$

7. $x^2 - \dfrac{1}{2}x + \underline{\hphantom{xxx}}$

8. $y^2 - \dfrac{4}{3}y + \underline{\hphantom{xxx}}$

9. $z^2 + z + \underline{\hphantom{xxx}}$

Solve by completing the square.

10. $x^2 + 8x + 15 = 0$

11. $y^2 + 10y + 21 = 0$

12. $z^2 - 2z - 8 = 0$

13 $x^2 - 2x - 1 = 0$

14. $y^2 + 4y = 96$

15 $3z^2 + 12z = 135$

16 $2x^2 + x - 1 = 0$

17. $2y^2 - 5y = -2$

18. $z^2 + 16z + 50 = 0$

19. $x^2 - 10x + 22 = 0$

20 $y^2 + 2y + 5 = 0$

21. $2z^2 - z - 3 = 0$

Solve by any method.

22. $x^2 - 5x = -4$

23 $5y^2 - 4y - 33 = 0$

24. $z^2 + 6z = 16$

25. $2x^2 - 5x - 3 = 0$

26. $2y^2 - 3y = 1$

27. $z^2 + z = z + 10$

28. A free-falling object, starting from rest, will fall d feet in t seconds according to the formula $d = 16t^2$.
  (a) How far will the object fall in 1 second?
  (b) How far will the object fall in 5 seconds?
  (c) How many seconds does it take for the object to fall 64 feet?

FOR REVIEW

The following exercises review material from Chapter 1 and Sections 8.4, 9.2, 9.4, and 9.7. They will help prepare for the next section. Evaluate each expression in Exercises 29–34 if $a = 2$, $b = -4$, and $c = 1$.

29. $-b$

30. b^2

31. $4ac$

32. $b^2 - 4ac$

33. $\sqrt{b^2 - 4ac}$

34. $-b + \sqrt{b^2 - 4ac}$

Solve each equation in Exercises 35–36.

35. $\dfrac{1}{x-1} + \dfrac{2}{x+1} = \dfrac{5}{x^2-1}$

36. $\sqrt{x^2-3} - x = 1$

The Pythagorean theorem states that $a^2 + b^2 = c^2$ where a and b are legs of a right triangle and c is the hypotenuse. In Exercises 37–38, find the missing leg or hypotenuse.

37. $a = 6$ and $b = 4$

38. $b = \dfrac{1}{2}$ and $c = \dfrac{3}{2}$

ANSWERS: 1. 16 2. 81 3. $\frac{49}{4}$ 4. $\frac{1}{4}$ 5. $\frac{81}{4}$ 6. $\frac{1}{36}$ 7. $\frac{1}{16}$ 8. $\frac{4}{9}$ 9. $\frac{1}{4}$ 10. $-3, -5$ 11. $-3, -7$ 12. $4, -2$ 13. $1 \pm \sqrt{2}$ 14. $-12, 8$ 15. $5, -9$ 16. $\frac{1}{2}, -1$ 17. $2, \frac{1}{2}$ 18. $-8 \pm \sqrt{14}$ 19. $5 \pm \sqrt{3}$ 20. no solution (square root of a negative number does not exist) 21. $\frac{3}{2}, -1$ 22. $4, 1$ (best to factor) 23. $3, -\frac{11}{5}$ (factor) 24. $2, -8$ (factor) 25. $3, -\frac{1}{2}$ (factor) 26. $\frac{3 \pm \sqrt{17}}{4}$ (complete the square) 27. $\pm\sqrt{10}$ (take roots) 28. (a) 16 feet (b) 400 feet (c) 2 seconds 29. 4 30. 16 31. 8 32. 8 33. $2\sqrt{2}$ 34. $4 + 2\sqrt{2}$ 35. 2 36. no solution (-2 does not check) 37. $2\sqrt{13}$ 38. $\sqrt{2}$

10.2 EXERCISES B

What must be added to complete the square?

1. $x^2 + 24x +$ _____

2. $y^2 - 4y +$ _____

3. $z^2 + 3z +$ _____

4. $x^2 + x +$ _____

5. $y^2 - 11y +$ _____

6. $z^2 + \dfrac{1}{2}z +$ _____

7. $x^2 - \dfrac{1}{3}x +$ _____

8. $y^2 + \dfrac{2}{3}y +$ _____

9. $z^2 - z +$ _____

Solve by completing the square.

10. $x^2 + 6x + 5 = 0$

11. $y^2 + 3y - 18 = 0$

12. $z^2 - 3z - 10 = 0$

13. $x^2 - 4x - 1 = 0$

14. $y^2 = 8y + 20$

15. $z^2 + 5z = 24$

16. $2x^2 - 3x + 1 = 0$

17. $2y^2 - 5y = 3$

18. $z^2 - z = 3$

19. $2x^2 + 6x + 1 = 0$

20. $y^2 + 2y = -9$

21. $z^2 + 15 = 8z$

Solve by any method.

22. $x^2 + 8x - 9 = 0$

23. $3y^2 + 11y - 4 = 0$

24. $z^2 - 14z + 49 = 0$

25. $2x^2 - 3x - 2 = 0$

26. $3y^2 - 6 = 2y$

27. $z^2 - z = 8 - z$

28. A free falling object, starting at rest, will fall d feet in t seconds according to the formula $d = 16t^2$.

   (a) How far will the object fall in 2 seconds?

   (b) How far will the object fall in 10 seconds?

   (c) How many seconds does it take for the object to fall 144 feet?

FOR REVIEW

The following exercises review material from Chapter 1 and Sections 8.4, 9.2, 9.4, and 9.7. They will help prepare for the next section. Evaluate each expression in Exercises 29–34 if $a = 3$, $b = -6$ and $c = 2$.

29. $-b$

30. b^2

31. $4ac$

32. $b^2 - 4ac$

33. $\sqrt{b^2 - 4ac}$

34. $-b + \sqrt{b^2 - 4ac}$

Solve each equation in Exercises 35–36.

35. $\dfrac{3}{x - 2} + \dfrac{4}{x + 2} = \dfrac{-1}{x^2 - 4}$

36. $\sqrt{x^2 + 5} - x = 1$

The Pythagorean theorem states that $a^2 + b^2 = c^2$ where a and b are legs of a right triangle and c is the hypotenuse. In Exercises 37–38, find the missing leg or hypotenuse.

37. $a = 9$ and $b = 2$

38. $a = \dfrac{1}{3}$ and $c = \dfrac{4}{3}$

10.2 EXERCISES C

Solve by completing the square.

1. $5x^2 - 2x - 10 = 0$

2. $x^2 + bx + c = 0$

$$\left[\text{Answer:} \quad x = \frac{-b \pm \sqrt{b^2 - 4ac}}{2} \right]$$

10.3 SOLVING BY THE QUADRATIC FORMULA

=== STUDENT GUIDEPOSTS ===

1 Quadratic Formula	**3** The Discriminant
2 Using the Quadratic Formula	**4** Applications Requiring the Quadratic Formula

1 QUADRATIC FORMULA

Factoring and taking roots are usually faster and easier methods than the method of completing the square. Completing the square becomes tedious since it repeats the same process time and time again. To avoid such repetition, we can work through the process once in a general case and memorize the result. This is how the *quadratic formula* is developed. Starting with a quadratic equation in general form

$$ax^2 + bx + c = 0 \quad (a \neq 0)$$

we complete the square.

$$ax^2 + bx = -c \qquad \text{Isolate the constant term}$$

$$\frac{\cancel{a}x^2}{\cancel{a}} + \frac{b}{a}x = -\frac{c}{a} \qquad \text{Divide by the coefficient of } x^2$$

$$x^2 + \frac{b}{a}x + \left(\frac{b}{2a}\right)^2 = -\frac{c}{a} + \left(\frac{b}{2a}\right)^2 \qquad \text{Add } \left(\frac{1}{2}\cdot\frac{b}{a}\right)^2 = \left(\frac{b}{2a}\right)^2$$

$$\left(x + \frac{b}{2a}\right)^2 = -\frac{c}{a} + \frac{b^2}{4a^2} \qquad \text{Factor}$$

$$\left(x + \frac{b}{2a}\right)^2 = -\frac{4ac}{4a^2} + \frac{b^2}{4a^2} \qquad \text{Find a common denominator}$$

$$= \frac{b^2 - 4ac}{4a^2} \qquad \text{Subtract the fractions}$$

$$x + \frac{b}{2a} = \pm\sqrt{\frac{b^2 - 4ac}{4a^2}} \qquad \text{Take square roots}$$

$$= \frac{\pm\sqrt{b^2 - 4ac}}{2a}$$

$$x = -\frac{b}{2a} \pm \frac{\sqrt{b^2 - 4ac}}{2a} \qquad \text{Subtract } \frac{b}{2a}$$

$$x = \frac{-b \pm \sqrt{b^2 - 4ac}}{2a} \qquad \text{Note that } 2a \text{ is the denominator of the entire expression}$$

② USING THE QUADRATIC FORMULA

This last formula is called the **quadratic formula** and it must be memorized. To use it to solve a quadratic equation, we identify the constants, a, b, and c and substitute into the quadratic formula.

To Solve a Quadratic Equation Using the Quadratic Formula

1. Write the equation in general form ($ax^2 + bx + c = 0$).
2. Identify the constants a, b, and c.
3. Substitute the values for a, b, and c into the quadratic formula,

$$x = \frac{-b \pm \sqrt{b^2 - 4ac}}{2a}.$$

4. Simplify the numerical expression to obtain the solutions.

EXAMPLE 1 USING THE QUADRATIC FORMULA

Solve $x^2 - 5x + 6 = 0$ using the quadratic formula.

We have $a = 1$, $b = -5$ (not 5), and $c = 6$.

$$x = \frac{-b \pm \sqrt{b^2 - 4ac}}{2a} = \frac{-(-5) \pm \sqrt{(-5)^2 - 4(1)(6)}}{2(1)} \quad \text{Substitute}$$

$$= \frac{5 \pm \sqrt{25 - 24}}{2} \quad \text{Watch all signs}$$

$$= \frac{5 \pm \sqrt{1}}{2} = \frac{5 \pm 1}{2} = \begin{cases} \dfrac{5 + 1}{2} = \dfrac{6}{2} = 3 \\ \dfrac{5 - 1}{2} = \dfrac{4}{2} = 2 \end{cases}$$

The solutions are 2 and 3.

PRACTICE EXERCISE 1

Solve $x^2 + 6x + 9 = 0$ using the quadratic formula.

Answer: -3

EXAMPLE 2 IRRATIONAL SOLUTIONS

Solve $3x^2 - 5 = -4x$ using the quadratic formula.

First we write the equation in general form, $3x^2 + 4x - 5 = 0$, and identify $a = 3$, $b = 4$, and $c = -5$ (not 5), then substitute.

$$x = \frac{-b \pm \sqrt{b^2 - 4ac}}{2a} = \frac{-(4) \pm \sqrt{(4)^2 - 4(3)(-5)}}{2(3)}$$

$$= \frac{-4 \pm \sqrt{16 + 60}}{6} \quad \text{Watch the signs}$$

$$= \frac{-4 \pm \sqrt{76}}{6} = \frac{-4 \pm \sqrt{4 \cdot 19}}{6}$$

$$= \frac{-4 \pm 2\sqrt{19}}{6} = \frac{2(-2 \pm \sqrt{19})}{2 \cdot 3}$$

$$= \frac{-2 \pm \sqrt{19}}{3}$$

The solutions are $\dfrac{-2 \pm \sqrt{19}}{3}$.

PRACTICE EXERCISE 2

Solve $7x - 2 = 2x^2$ using the quadratic formula.

Answer: $\dfrac{7 \pm \sqrt{33}}{4}$

To Solve a Quadratic Equation

1. First try factoring or taking roots.
2. If the first two methods do not work, go directly to the quadratic formula.

Notice that the quadratic equation in Example 1 could be solved much more quickly by factoring than by using the quadratic formula.

$$x^2 - 5x + 6 = 0$$
$$(x - 2)(x - 3) = 0$$
$$x - 2 = 0 \quad \text{or} \quad x - 3 = 0$$
$$x = 2 \quad \text{or} \quad x = 3$$

❸ THE DISCRIMINANT

In a quadratic equation $ax^2 + bx + c = 0$, the number $b^2 - 4ac$, called the **discriminant**, can be used to find out whether the equation has solutions. Notice that the discriminant is the number under the radical sign in the quadratic formula $x = \dfrac{-b \pm \sqrt{b^2 - 4ac}}{2a}$. Since the square root of a negative number is not a real number, a quadratic equation has no real-number solutions when the discriminant is a negative number. Consider

$$x^2 + x + 1 = 0.$$

Here $a = 1$, $b = 1$, and $c = 1$, so $b^2 - 4ac = (1) - 4(1)(1) = 1 - 4 = -3$. Since -3 is a negative number, $\sqrt{-3}$ is not a real number and $x^2 + x + 1 = 0$ has no real-number solutions.

It is helpful to clear all fractions first when solving quadratic equations with fractional coefficients. To solve

$$\frac{1}{3}x^2 - x + \frac{1}{3} = 0,$$

it would be difficult to use $a = \frac{1}{3}$, $b = -1$, and $c = \frac{1}{3}$ in the quadratic formula. However, if we multiply through by the LCD 3, we obtain the equivalent equation

$$x^2 - 3x + 1 = 0,$$

in which $a = 1$, $b = -3$, and $c = 1$. Avoiding fractions makes the arithmetic simpler.

❹ APPLICATIONS REQUIRING THE QUADRATIC FORMULA

In many applications of quadratic equations, solutions containing radicals must be estimated by using an approximate value for the radical.

| EXAMPLE 3 APPLICATION OF THE QUADRATIC FORMULA | PRACTICE EXERCISE 3 |

A large wheat field is in the shape of a right triangle with hypotenuse 5 km long and one leg 4 km longer than the other. Find the measure of each leg (side) of the field.

The triangular field is shown in Figure 10.1. Remember the Pythagorean theorem: the sum of the squares of the legs of a right triangle is equal to the square of the hypotenuse.

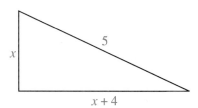

Figure 10.1

Let x = the measure of one leg,
$x + 4$ = the measure of the other leg.
Use the Pythagorean theorem.

Meg and Joe need a water line the length of their garden. If the garden is a rectangle with diagonal 8 meters and the length is 2 meters longer than the width, approximately how much pipe will be needed?

$$x^2 + (x + 4)^2 = 5^2$$
$$x^2 + x^2 + 8x + 16 = 25 \qquad \text{Don't forget the } 8x$$
$$2x^2 + 8x - 9 = 0$$

Since we cannot factor, use the quadratic formula.

$$x = \frac{-b \pm \sqrt{b^2 - 4ac}}{2a}$$

$$= \frac{-8 \pm \sqrt{64 - 4(2)(-9)}}{2(2)} \qquad a = 2, b = 8, \text{ and } c = -9$$

$$= \frac{-8 \pm \sqrt{64 + 72}}{4}$$

$$= \frac{-8 \pm \sqrt{136}}{4}$$

$$= \frac{-8 \pm \sqrt{4 \cdot 34}}{4}$$

$$= \frac{-8 \pm 2\sqrt{34}}{4} \qquad \sqrt{4 \cdot 34} = \sqrt{4}\,\sqrt{34} = 2\sqrt{34}$$

$$= \frac{2[-4 \pm \sqrt{34}]}{2 \cdot 2} \qquad \text{Factor and divide out 2}$$

$$= \frac{-4 \pm \sqrt{34}}{2}$$

Since $\dfrac{-4 - \sqrt{34}}{2}$ is a negative number, we can discard it (a length must be positive or zero). Thus, the length of one leg is exactly $\dfrac{-4 + \sqrt{34}}{2}$ km. Using a calculator to get a decimal approximation, we find $\sqrt{34} \approx 5.8$. Thus,

$$\frac{-4 + \sqrt{34}}{2} \approx \frac{-4 + 5.8}{2} = \frac{1.8}{2} = 0.9,$$

and the lengths of the sides of the field are approximately 0.9 km and 4.9 km ($x = 0.9$, $x + 4 = 4.9$). As a check,

$$(0.9)^2 + (4.9)^2 = 0.81 + 24.01 = 24.82 \approx 25 = 5^2.$$

Answer: approximately 6.6 m

Unless you are working an applied problem in which an approximate value would make more sense, or unless instructed otherwise, always leave answers in radical form.

10.3 EXERCISES A

Solve using the quadratic formula.

1. $x^2 + x - 20 = 0$ **2.** $y^2 - 3y = 10$ **3.** $z^2 + 8 = 9z$

4. $x^2 - 9 = 0$

5 $0 = 7y - 2 + 15y^2$

6. $5z - 2 = 3z^2$

7 $5x^2 - x = 1$

8. $2y^2 - 4y + 1 = 0$

9. $4z^2 + 9z + 3 = 0$

10. $x^2 + 2x = 5$

11 $z^2 = z - 5$

12. $y^2 - 6y = 1$

13. $x^2 - 3x - 9 = 0$

14 $y(y + 1) + 2y = 4$

15. $(z + 1)^2 = 5 + 3z$

16. $x^2 - 2x + 5 = 0$

17. $4y^2 - 1 = 0$

18 $(2z - 1)(z - 2) - 11 = 2(z + 4) - 8$

Solve using any method.

19. $2x^2 - 7x + 3 = 0$

20 $3z^2 + 10z - 1 = 0$

21. $3y^2 - 12 = 0$

22. $3x^2 - 2 = -5x$

23. $x^2 + x - 1 = 0$

24. $-x^2 - x + 1 = 0$

25. How do the solutions to Exercises 23 and 24 compare? Explain.

26 $\dfrac{2}{3}x^2 - \dfrac{1}{3}x - 1 = 0$

27. $\dfrac{1}{2}x^2 + x - 1 = 0$

28. $\dfrac{1}{4}x^2 + \dfrac{1}{2}x - \dfrac{3}{4} = 0$

29. Write the solution to Exercise 7 using a decimal approximation.

30. Write the solution to Exercise 8 using a decimal approximation.

Solve.

31. The hypotenuse of a right triangle is 3 cm long, and one leg is 1 cm more than the other. Find the approximate measure of each leg.

32 Winnie has a square garden measuring 4 yards on each side. She wishes to place a picket fence diagonally across the garden. What is the approximate length of the fence?

4 yd   4 yd

33. If an object is thrown upward with initial velocity of 128 feet per second, its height, h, above the ground in t seconds is given by $h = 128t - 16t^2$. How long will it take for the object to reach a height of 240 feet?

34. Show that the sum of the solutions to the equation $y^2 - 6y - 1 = 0$ is equal to $-\frac{b}{a}$. Note that $a = 1$ and $b = -6$.

FOR REVIEW

35. Solve $x^2 + x = 13 + x$ by taking roots.

36. Solve $2y^2 + 2y = 3$ by completing the square.

Exercises 37–38 review material from Sections 8.5 and 9.6 to prepare for the next section. Solve each equation.

37. $\dfrac{1}{x} + \dfrac{2}{x-3} = \dfrac{4}{x^2 - 3x}$

38. $\sqrt{x^2 - 11} - x + 1 = 0$

ANSWERS: 1. 4, −5   2. 5, −2   3. 1, 8   4. 3, −3   5. $\frac{1}{5}$, $-\frac{2}{3}$ 6. 1, $\frac{2}{3}$ 7. $\dfrac{1 \pm \sqrt{21}}{10}$ 8. $\dfrac{2 \pm \sqrt{2}}{2}$ 9. $\dfrac{-9 \pm \sqrt{33}}{8}$

10. $-1 \pm \sqrt{6}$ 11. no solution   12. $3 \pm \sqrt{10}$ 13. $\dfrac{3 \pm 3\sqrt{5}}{2}$ 14. 1, −4   15. $\dfrac{1 \pm \sqrt{17}}{2}$ 16. no solution   17. $\frac{1}{2}$, $-\frac{1}{2}$

18. $\frac{9}{2}$, −1   19. 3, $\frac{1}{2}$ 20. $\dfrac{-5 \pm 2\sqrt{7}}{3}$ 21. 2, −2   22. $\frac{1}{3}$, −2   23. $\dfrac{-1 \pm \sqrt{5}}{2}$ 24. $\dfrac{1 \pm \sqrt{5}}{-2}$ 25. Same (Multiply

numerator and denominator of $\dfrac{-1 \pm \sqrt{5}}{2}$ by −1 to obtain $\dfrac{1 \pm \sqrt{5}}{-2}$). The equations are equivalent (multiply both sides

of 23 by −1 to get 24).   26. $\frac{3}{2}$, −1   27. $-1 \pm \sqrt{3}$ 28. 1, −3   29. 0.56, −0.36 (using $\sqrt{21} \approx 4.6$)   30. 1.7, 0.3

(using $\sqrt{2} \approx 1.4$)   31. 1.6 cm, 2.6 cm   32. 5.7 yards   33. 3 seconds (on the way up) then again at 5 seconds (on the

way down)   34. The solutions are $3 + \sqrt{10}$ and $3 - \sqrt{10}$. Their sum is $(3 + \sqrt{10}) + (3 - \sqrt{10}) = 6 = -\frac{-6}{1} = -\frac{b}{a}$.

35. $\pm\sqrt{13}$ 36. $\dfrac{-1 \pm \sqrt{7}}{2}$ 37. $\frac{7}{3}$ 38. 6

10.3 EXERCISES B

Solve using the quadratic formula.

1. $x^2 - 4x - 21 = 0$

2. $y^2 - 10y + 9 = 0$

3. $z^2 + 5 = 6z$

4. $x^2 - 49 = 0$

5. $0 = 20y^2 - y - 1$

6. $7z - 3 = 4z^2$

7. $2x^2 - x = 2$

8. $y^2 + 5y + 5 = 0$

9. $3z^2 + 3 = 7z$

10. $x^2 - 4x = 1$

11. $3y^2 + 2y = 7$

12. $z^2 = z - 9$

13. $x^2 - 5x + 3 = 0$

14. $y(y + 3) + 2y = 6$

15. $3(z - 1)^2 = 3z$

16. $x^2 - 3x + 7 = 0$

17. $6y^2 + 7y = 3$

18. $(z + 1)^2 = 5(1 + z)$

Solve using any method.

19. $2x^2 - x - 1 = 0$

20. $7y^2 - 49 = 0$

21. $z^2 - 5z - 2 = 0$

22. $2x^2 = \frac{1}{2}(27x + 7)$

23. $x^2 + 6x - 1 = 0$

24. $-x^2 - 6x + 1 = 0$

25. How do the solutions to Exercises 23 and 24 compare? Explain.

26. $x^2 - \frac{1}{2}x - 5 = 0$

27. $\frac{1}{3}x^2 + x + \frac{1}{6} = 0$

28. $\frac{1}{11}x^2 + \frac{9}{11}x - 2 = 0$

29. Write the solution to Exercise 7 using a decimal approximation.

30. Write the solution to Exercise 8 using a decimal approximation.

31. The hypotenuse of a right triangle is 7 cm long, and one leg is 3 cm more than the other. Find the approximate measure of each leg.

32. A baseball diamond is a square 90 ft on each side. How far is it from home plate directly across to second base?

33. If an object is thrown upward with initial velocity of 64 feet per second its height, h, above the ground in t seconds is given by $h = 64t - 16t^2$. How long will it take for the object to reach a height of 48 feet?

34. Show that the product of the solutions to the equation $y^2 - 6y - 1 = 0$ is equal to $\frac{c}{a}$. Note that $a = 1$ and $c = -1$.

FOR REVIEW

35. Solve $2y^2 - 50 = 0$ by taking roots.

36. Solve $2x^2 - 6x + 3 = 0$ by the method of completing the square.

Exercises 37–38 review material from Sections 8.5 and 9.6 to prepare for the next section. Solve each equation.

37. $\frac{3}{x + 5} - \frac{2}{x} = \frac{5}{x^2 + 5x}$

38. $\sqrt{x^2 + 7} - x = 1$

10.3 EXERCISES C

Solve for x using the quadratic formula.

1. $0.3x^2 - 0.1x - 1.6 = 0$ [*Hint:* Multiply by 10.]   **2.** $x^2 + 3xy + y^2 = 0$ $\left[\text{Answer: } \dfrac{-y(3 \pm \sqrt{5})}{2}\right]$

Use the quadratic formula and your calculator to solve each quadratic equation. Give answers to two decimal places.

3. $\sqrt{2}x^2 - 3x - 7.1 = 0$
[Answer: 3.54, −1.42]

4. $3x^2 + \sqrt{5}x - \dfrac{1}{2} = 0$

10.4 SOLVING FRACTIONAL AND RADICAL EQUATIONS

STUDENT GUIDEPOSTS

1 Fractional Equations   **2** Radical Equations

1 FRACTIONAL EQUATIONS

In Chapter 8 we learned that to solve fractional equations we multiply both sides by the LCD. The fractional equations we look at in this section will result in quadratic equations when both sides are multiplied by the LCD. Remember that we always need to check for invalid answers when solving fractional equations.

EXAMPLE 1   SOLVING A FRACTIONAL EQUATION

Solve $\dfrac{6}{x} - x = 5$.

The LCD is x.

$$x\left[\dfrac{6}{x} - x\right] = \boxed{x} \cdot 5 \quad \text{Multiply by LCD}$$

$$\cancel{x} \cdot \dfrac{6}{\cancel{x}} - x \cdot x = 5x$$

$$6 - x^2 = 5x$$

$$x^2 + 5x - 6 = 0$$

$$(x + 6)(x - 1) = 0$$

$$x + 6 = 0 \quad \text{or} \quad x - 1 = 0$$

$$x = -6 \qquad x = 1$$

Check: $\dfrac{6}{(-6)} - (-6) \overset{?}{=} 5 \qquad \dfrac{6}{(1)} - (1) \overset{?}{=} 5$

$$-1 + 6 = 5 \qquad\qquad 6 - 1 = 5$$

The solutions are −6 and 1.

PRACTICE EXERCISE 1

Solve $x - \dfrac{11}{x} = \dfrac{10}{x} + 4$.

Answer: 7, −3

EXAMPLE 2 SOLVING A FRACTIONAL EQUATION

Solve $\dfrac{12}{x^2 - 4} - \dfrac{3}{x - 2} = -1$.

The LCD $= (x - 2)(x + 2)$.

$(x - 2)(x + 2) \left[\dfrac{12}{x^2 - 4} - \dfrac{3}{x - 2} \right] = (x - 2)(x + 2)(-1)$ Multiply both sides by LCD

$(x-2)(x+2) \cdot \dfrac{12}{(x-2)(x+2)} = (x-2)(x+2) \cdot \dfrac{3}{(x-2)}$ Distribute

$= (x - 2)(x + 2)(-1)$

$12 - 3(x + 2) = (x^2 - 4)(-1)$ Watch parentheses

$12 - 3x - 6 = -x^2 + 4$ Watch signs

$x^2 - 3x + 2 = 0$ Collect terms

$(x - 1)(x - 2) = 0$ Factor

$x - 1 = 0 \quad \text{or} \quad x - 2 = 0$

$x = 1 \qquad\qquad x = 2$

Check: $\dfrac{12}{(1)^2 - 4} - \dfrac{3}{(1) - 2} \overset{?}{=} -1 \qquad \dfrac{12}{(2)^2 - 4} - \dfrac{3}{(2) - 2} \overset{?}{=} -1$

$\dfrac{12}{-3} - \dfrac{3}{-1} \overset{?}{=} -1 \qquad\qquad \dfrac{2}{4 - 4} - \dfrac{3}{2 - 2} \overset{?}{=} -1$

$-4 + 3 = -1 \qquad\qquad\qquad \dfrac{12}{0} - \dfrac{3}{0} \neq -1$

Cannot divide by 0

The only solution is 1. Notice that 2 makes a denominator zero in the original equation. In general, any number for which a fraction is undefined must be discarded as a solution.

PRACTICE EXERCISE 2

Solve $\dfrac{3}{x + 3} + \dfrac{4}{x^2 - 9} = 1$.

Answer: 4, −1

EXAMPLE 3 SOLVING A FRACTIONAL EQUATION

Solve $\dfrac{y + 1}{y} = \dfrac{-2}{y - 2}$.

The LCD $= y(y - 2)$.

$y(y - 2) \left[\dfrac{y + 1}{y} \right] = y(y - 2) \left[\dfrac{-2}{y - 2} \right]$ Multiply by LCD

$(y - 2)(y + 1) = -2y$

$y^2 - y - 2 = -2y$

$y^2 + y - 2 = 0$

$(y + 2)(y - 1) = 0$

$y + 2 = 0 \quad \text{or} \quad y - 1 = 0$

$y = -2 \qquad\qquad y = 1$

PRACTICE EXERCISE 3

Solve $\dfrac{x^2}{x - 10} = \dfrac{3}{x - 10}$. Start by multiplying both sides by the LCD, $x - 10$.

Check: $\dfrac{(-2) + 1}{(-2)} \overset{?}{=} \dfrac{-2}{(-2) - 2}$ $\dfrac{(1) + 1}{(1)} \overset{?}{=} \dfrac{-2}{(1) - 2}$

$$\dfrac{-1}{-2} \overset{?}{=} \dfrac{-2}{-4} \qquad\qquad \dfrac{2}{1} \overset{?}{=} \dfrac{-2}{-1}$$

$$\dfrac{1}{2} = \dfrac{1}{2} \qquad\qquad\qquad 2 = 2$$

The solutions are -2 and 1.   Answer: $\pm\sqrt{3}$

② RADICAL EQUATIONS

With radical equations in Chapter 9, we squared both sides of the equation to eliminate a radical. Now we consider other radical equations that become quadratic equations when both sides are squared. We always need to check our answers when solving radical equations, since answers that do not satisfy the *original* equation may be introduced by the process of squaring.

EXAMPLE 4 SOLVING A RADICAL EQUATION	PRACTICE EXERCISE 4

Solve $\sqrt{12 - x} = x$.

$$\begin{aligned}
(\sqrt{12 - x})^2 &= x^2 && \text{Indicate the square of both sides} \\
12 - x &= x^2 && \text{Square both sides} \\
x^2 + x - 12 &= 0 && \text{Collect terms} \\
(x - 3)(x + 4) &= 0 && \text{Factor}
\end{aligned}$$

$x - 3 = 0$ or $x + 4 = 0$

    $x = 3$ $x = -4$

Check: $\sqrt{12 - (3)} \overset{?}{=} 3$ $\sqrt{12 - (-4)} \overset{?}{=} (-4)$

           $\sqrt{9} \overset{?}{=} 3$ $\sqrt{16} \ne -4$ ($\sqrt{16} = 4$,

            $3 = 3$ *not* -4.)

The only solution is 3.

Solve $x = \sqrt{8x + 20}$.

Answer: 10 (-2 does not check)

EXAMPLE 5 SOLVING A RADICAL EQUATION	PRACTICE EXERCISE 5

Solve $x + 2 = \sqrt{x + 8}$.

$$\begin{aligned}
(x + 2)^2 &= (\sqrt{x + 8})^2 && \text{Indicate the square of both sides} \\
x^2 + 4x + 4 &= x + 8 && \text{Square both sides, } (x + 2)^2 \ne (x^2 + 4) \\
x^2 + 3x - 4 &= 0 && \text{Collect terms} \\
(x + 4)(x - 1) &= 0 && \text{Factor}
\end{aligned}$$

$x + 4 = 0$ or $x - 1 = 0$

    $x = -4$ $x = 1$

Check: $(-4) + 2 \overset{?}{=} \sqrt{(-4) + 8}$ $(1) + 2 \overset{?}{=} \sqrt{(1) + 8}$

            $-2 \ne \sqrt{4}$ $3 \overset{?}{=} \sqrt{9}$

                                  $3 = 3$

The only solution is 1.

Solve $\sqrt{13 - 6x} = x - 3$.

Answer: no solution

| **EXAMPLE 6** **SOLVING A RADICAL EQUATION** | **PRACTICE EXERCISE 6** |

Solve $x = 4\sqrt{x + 1} - 4$.

$$x + 4 = 4\sqrt{x + 1}$$ Add 4 to both sides to isolate the radical

$$(x + 4)^2 = (4\sqrt{x + 1})^2$$ Indicate the square of both sides

$$x^2 + 8x + 16 = 16(x + 1)$$ Do not forget the middle term on left and to square 4 on right

$$x^2 + 8x + 16 = 16x + 16$$ Distribute the 16

$$x^2 - 8x = 0$$ Collect terms

$$x(x - 8) = 0$$ Factor out x

$$x = 0 \quad \text{or} \quad x - 8 = 0$$

$$x = 8$$

Check: $(0) \overset{?}{=} 4\sqrt{(0) + 1} - 4$ $(8) \overset{?}{=} 4\sqrt{(8) + 1} - 4$

$0 \overset{?}{=} 4\sqrt{1} - 4$ $8 \overset{?}{=} 4\sqrt{9} - 4$

$0 \overset{?}{=} 4 - 4$ $8 \overset{?}{=} 4 \cdot 3 - 4$

$0 = 0$ $8 \overset{?}{=} 12 - 4$

                         $8 = 8$

The solutions are 0 and 8.

Solve $\sqrt{x + 5} - x + 1 = 0$.

Answer: 4

10.4 EXERCISES A

Solve.

1. $\dfrac{3}{x - 4} = 1 + \dfrac{5}{x + 4}$

2. $\dfrac{y - 3}{y} = \dfrac{-4}{y + 1}$

3. $\dfrac{3}{1 + z} + \dfrac{2}{1 - z} = -1$

4. $\dfrac{a^2}{a + 1} - \dfrac{9}{a + 1} = 0$

5. $\dfrac{1}{x + 2} + \dfrac{x}{x - 2} = \dfrac{1}{2}$

6. $\dfrac{60}{y + 3} = \dfrac{60}{y} - 1$

7 $\dfrac{5}{z + 3} - \dfrac{1}{z + 3} = \dfrac{z + 3}{z + 2}$

8 $\dfrac{x - 1}{x} = \dfrac{x}{x + 1}$

9. $\dfrac{16}{a + 2} = 1 + \dfrac{2}{a - 4}$

10. $\dfrac{1}{x-3} + \dfrac{1}{x+3} = \dfrac{1}{x^2-9}$

11. $\dfrac{x^2}{2x+1} - \dfrac{5}{2x+1} = 0$

[*Hint:* Multiply through by $2x+1$.]

12 $\dfrac{x+2}{2} = \dfrac{x+5}{x+2}$

13. $x = \sqrt{3x+10}$

14. $\sqrt{5y+6} - y = 0$

15. $z - 7 = \sqrt{z-5}$

16. $\sqrt{15-a} = a - 3$

17 $1 + 2\sqrt{x-1} = x$

18. $y = 1 + 6\sqrt{y-9}$

19. $\sqrt{1-2a} = a - 1$

20. $\sqrt{2x+7} - 4 = x$

21. $\sqrt{z^2+2} = z + 1$

22 $3\sqrt{y+1} - y - 1 = 0$

23. $\sqrt{5-4x} = 2 - x$

24 $\sqrt{2x^2-5} = x$

25 A number increased by its reciprocal is the same as $\frac{5}{2}$. Find the number.

26. The principal square root of 2 more than a number is equal to the number itself. Find the number.

FOR REVIEW

Solve.

27. $z^2 - 8z + 8 = 0$

28. $x^2 - \dfrac{2}{3}x - 1 = 0$

29. $2y^2 + 3y - 2 = 0$

30. $5z^2 + 2z + 1 = 0$

31. It can be shown that a polygon (many-sided geometric figure) with x sides has a total of n diagonals where $n = \frac{1}{2}x^2 - \frac{3}{2}x$. A hexagon has 6 sides; how many diagonals does it have? Draw a hexagon and verify your answer by actual count.

32. Use the formula in Exercise 31 to find the number of sides that a polygon has if it is known to have 20 diagonals. Draw a polygon with the number of sides that you determine and count its diagonals.

ANSWERS: 1. 6, −8   2. 1, −3   3. −3, 2   4. 3, −3   5. 0, −6   6. 12, −15   7. −1   8. no solution   9. 6, 10   10. $\frac{1}{2}$ 11. $\pm\sqrt{5}$ 12. $-1 \pm \sqrt{7}$ 13. 5   14. 6   15. 9   16. 6   17. 5, 1   18. 25, 13   19. no solution   20. −3   21. $\frac{1}{2}$ 22. −1, 8   23. 1, −1   24. $\sqrt{5}$ 25. 2, $\frac{1}{2}$ 26. 2   27. $4 \pm 2\sqrt{2}$ 28. $\dfrac{1 \pm \sqrt{10}}{3}$ 29. $\frac{1}{2}$, −2   30. no solution   31. 9   32. 8

10.4 EXERCISES B

Solve.

1. $\dfrac{3}{x-1} = 1 + \dfrac{2}{x+1}$

2. $\dfrac{y}{1+y} = \dfrac{3-y}{4}$

3. $\dfrac{1}{x+2} + \dfrac{1}{x-2} = \dfrac{3}{8}$

4. $\dfrac{a^2}{a+5} - \dfrac{4}{a+5} = 0$

5. $\dfrac{6}{x+3} + \dfrac{x}{x-3} = 1$

6. $\dfrac{32}{y+5} = \dfrac{6}{y} + 2$

7. $\dfrac{4}{z+1} - \dfrac{1}{z+1} = \dfrac{z-1}{z+3}$

8. $x + \dfrac{2}{x-2} = \dfrac{1}{x-2}$

9. $\dfrac{z+2}{z} = \dfrac{1}{z+2}$

10. $\dfrac{3}{a-1} + 1 = \dfrac{6}{a^2-1}$

11. $\dfrac{x^2}{1+3x} - \dfrac{11}{1+3x} = 0$

12. $\dfrac{x+3}{x+1} = \dfrac{x+1}{3}$

13. $x = \sqrt{2x + 24}$

14. $\sqrt{4y + 5} - y = 0$

15. $z + 3 = \sqrt{12z + 9}$

16. $\sqrt{5y + 21} = y + 3$

17. $\sqrt{2x - 1} = x - 2$

18. $z = \sqrt{z + 7} - 1$

19. $\sqrt{7 - 3a} = 1 + a$

20. $2\sqrt{3 - x} - 5 = x$

21. $\sqrt{y^2 + 8} = y - 2$

22. $2\sqrt{y + 15} + y - 9 = 0$

23. $\sqrt{4x + 13} = x + 2$

24. $\sqrt{5x^2 - 11} = 2x$

25. A number decreased by its reciprocal is the same as $\frac{15}{4}$. Find the number.

26. The principal square root of 8 more than a number is three times the number. Find the number.

FOR REVIEW

Solve.

27. $x^2 - 6x + 6 = 0$

28. $y^2 - \frac{1}{3}y - 1 = 0$

29. $5z^2 + 14z - 3 = 0$

30. $4x^2 + x + 2 = 0$

31. It can be shown that a polygon with x sides has a total of n diagonals where $n = \frac{1}{2}x^2 - \frac{3}{2}x$. An octagon has 8 sides; how many diagonals does it have? Draw an octagon and verify your answer by actual count.

32. Use the formula in Exercise 31 to find the number of sides that a polygon has if it is known to have 5 diagonals. Draw a polygon with the number of sides that you determine and count its diagonals.

10.4 EXERCISES C

Solve.

1. $\dfrac{3}{x^2 - 5x + 6} - \dfrac{2}{x^2 + 3x - 10} = \dfrac{x + 2}{x^2 + 2x - 15}$

$\left[\text{Answer:} \quad \dfrac{1 \pm \sqrt{101}}{2} \right]$

2. $\sqrt{2x - 2} - \sqrt{x + 6} = -1$

10.5 MORE APPLICATIONS OF QUADRATIC EQUATIONS

═══ STUDENT GUIDEPOSTS ═══

① Basic Applications

② Geometry Applications

③ Work Applications

④ Motion Applications

① BASIC APPLICATIONS

Many applied problems can be solved using quadratic equations. We illustrated this in the preceding sections using several types of problems, and now we consider a wider variety of these applications.

EXAMPLE 1   AGE PROBLEM

Murphy's age in 3 years will be four times the square of his age now. How old is Murphy?

Let $\quad x$ = Murphy's present age,

$\quad x + 3$ = Murphy's age in 3 years,

$\quad 4x^2$ = four times the square of his present age.

The last two expressions are equal, so we solve the following equation.

$$4x^2 = x + 3$$
$$4x^2 - x - 3 = 0$$
$$(4x + 3)(x - 1) = 0$$
$$4x + 3 = 0 \quad \text{or} \quad x - 1 = 0$$
$$4x = -3 \qquad\qquad x = 1$$
$$x = -\frac{3}{4}$$

Since $-\frac{3}{4}$ could not be a person's age, $x = 1$ is the only possible solution.

Check: $\quad 4(\mathbf{1})^2 \stackrel{?}{=} \mathbf{1} + 3$

$\qquad\qquad 4 = 4$

Murphy is 1 year old.

PRACTICE EXERCISE 1

The square of Roberta's age plus the square of her age in 5 years is 97. How old is Roberta?

Answer: 4 years old

EXAMPLE 2   CONSECUTIVE INTEGERS

Twice the product of two consecutive positive even integers is 160. Find the integers.

Let $\quad x$ = the first positive even integer,

$\quad x + 2$ = the next consecutive even integer.

Twice the product of these two integers is 160, so we should solve the following equation.

$$2x(x + 2) = 160$$
$$x(x + 2) = 80 \quad \text{Multiply both sides by } \tfrac{1}{2}$$
$$x^2 + 2x = 80$$
$$x^2 + 2x - 80 = 0$$
$$(x - 8)(x + 10) = 0$$
$$x - 8 = 0 \quad \text{or} \quad x + 10 = 0$$
$$x = 8 \qquad\qquad x = -10$$

Since x must be positive, 8 is the only possible solution.

Check: $\quad 2 \cdot \mathbf{8}\,(\mathbf{8} + 2) \stackrel{?}{=} 160$

$\qquad\qquad 2 \cdot 8(10) \stackrel{?}{=} 160$

$\qquad\qquad 160 = 160$

The numbers are 8 and 10. $(x + 2 = 10.)$

PRACTICE EXERCISE 2

Four times the product of two consecutive negative odd integers is 780. Find the integers.

Answer: $-15, -13$

② GEOMETRY APPLICATIONS

When working a geometry problem remember to sketch a figure as shown in the next two examples.

EXAMPLE 3   RECTANGLE PROBLEM

Find the length and width of a rectangle if the length is 3 cm more than the width and the area is 180 cm^2.

Make a sketch as in Figure 10.2.

Let x = width of the rectangle,
   $x + 3$ = length of the rectangle.

Use the formula for the area of a rectangle, $A = l \cdot w$, to get the following equation.

$$x(x + 3) = 180$$
$$x^2 + 3x = 180$$
$$x^2 + 3x - 180 = 0$$
$$(x + 15)(x - 12) = 0$$
$$x + 15 = 0 \quad \text{or} \quad x - 12 = 0$$
$$x = -15 \qquad\qquad x = 12$$
$$(x + 3 = 15)$$

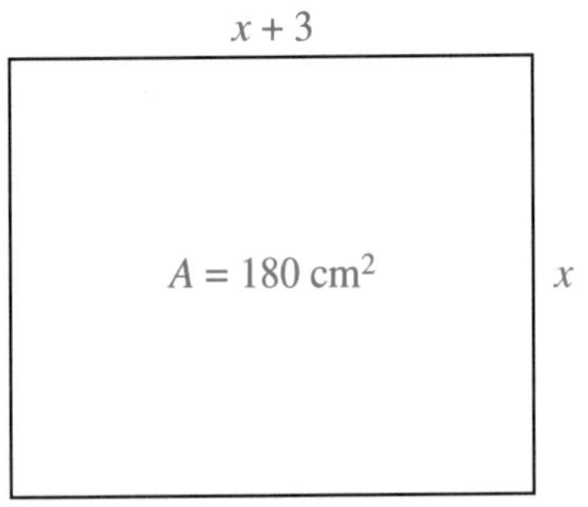

$x + 3$

$A = 180 \text{ cm}^2$ x

Figure 10.2

Since -15 could not be the width of a rectangle, 12 is the only possible solution.

Check: $12 (12 + 3) \stackrel{?}{=} 180$
              $12 \cdot 15 \stackrel{?}{=} 180$
                 $180 = 180$

The width is 12 cm and the length is 15 cm.

PRACTICE EXERCISE 3

Find the base and height of a triangle if the base is 2 cm more than the height and the area is 24 cm^2.

Answer: 8 cm, 6 cm

EXAMPLE 4 AREA AND PERIMETER

The number of square inches in the area of a square is 12 more than the number of inches in its perimeter. Find the length of a side.

A sketch of the square is shown in Figure 10.3.

Let x = length of a side (all sides are the same length, x),

x^2 = area of the square ($A = l \cdot w = x \cdot x$),

$4x$ = perimeter of the square ($P = 2 \cdot l + 2 \cdot w = 2 \cdot x + 2 \cdot x = 4x$).

Since the area is 12 more than the perimeter, adding 12 to the perimeter, $4x$, will give us the area, x^2.

$$x^2 = 4x + 12$$
$$x^2 - 4x - 12 = 0$$
$$(x + 2)(x - 6) = 6$$
$$x + 2 = 0 \quad \text{or} \quad x - 6 = 0$$
$$x = -2 \qquad\qquad x = 6$$

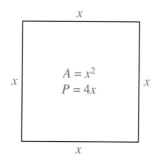

Figure 10.3

Since the length of a side cannot be negative, 6 is the only possible solution. Each side is 6 in.

PRACTICE EXERCISE 4

A box is 12 inches high and the volume is 420 in³. Find the length and width if the length is 2 in more than the width.

Let x = width of box,
$x + 2$ = length of box.
$12x(x + 2) =$ _____

Answer: 7 inches, 5 inches

---

③ **WORK APPLICATIONS**

Another variation of the work problems introduced in Section 8.5 results in a quadratic equation.

EXAMPLE 5 WORK PROBLEM

When each works alone, Ralph can do a job in 3 hours less time than Bert. When they work together, it takes 2 hours. How long does it take each to do the job by himself?

Since 2 = number of hrs to do the job together,

then $\dfrac{1}{2}$ = amount done together in 1 hr.

PRACTICE EXERCISE 5

Gloria can do a job in 2 hours less time than Evelyn. How long would it take each working alone, if together they can do the job in 5 hours?

Let x = number of hrs for Bert to do the job,

$\dfrac{1}{x}$ = amount done by Bert in 1 hr.

Then $x - 3$ = number of hrs for Ralph to do the job,

$\dfrac{1}{(x - 3)}$ = amount done by Ralph in 1 hr.

(Note: $\dfrac{1}{(x - 3)} \neq \dfrac{1}{x} - \dfrac{1}{3}$. To see this, substitute $x = 2$.)

We need to solve the following equation.

$$\dfrac{1}{x} + \dfrac{1}{(x - 3)} = \dfrac{1}{2}$$ (amount by Bert) +
(amount by Ralph) =
(amount together)

$$2x(x - 3)\left(\dfrac{1}{x} + \dfrac{1}{(x - 3)}\right) = 2x(x - 3) \cdot \dfrac{1}{2}$$ The LCD is $2x(x - 3)$

$$2x(x - 3)\dfrac{1}{x} + 2x(x - 3)\dfrac{1}{(x - 3)} = x(x - 3)$$

$$2(x - 3) + 2x = x^2 - 3x$$

$$2x - 6 + 2x = x^2 - 3x$$

$$0 = x^2 - 7x + 6$$

$$0 = (x - 6)(x - 1)$$

$$x - 6 = 0 \quad \text{or} \quad x - 1 = 0$$

$$x = 6 \qquad\qquad x = 1$$

If Bert did the job in 1 hr, Ralph would do it in $1 - 3 = -2$ hr, which makes no sense. Thus, Bert would take 6 hr and Ralph 3 hr.

Answer: Gloria approximately 9.1 hr; Evelyn approximately 11.1 hr

---

④ MOTION APPLICATIONS

The last example in this section can be classified as a motion or rate problem.

EXAMPLE 6   MOTION PROBLEM

A backpacker can hike 10 miles up a mountain and then return in a total time of 7 hours. Her hiking rate uphill is 3 mph slower than her rate downhill. At what rate does she hike up the mountain?

Let x = the rate hiking up the mountain.
Then $x + 3$ = the rate hiking down the mountain.

We use the distance formula $d = rt$, solved for $t = \dfrac{d}{r}$, to express the two times. Since the distance hiked uphill and the return distance downhill are both 10, we have

$$\dfrac{10}{x}$$ = the time spent hiking up the mountain,

$$\dfrac{10}{x + 3}$$ = the time spent hiking down the mountain.

Since the total time of the hike is 7 hours, we need to solve the following equation.

PRACTICE EXERCISE 6

A small plane flies 720 mi with the wind and returns against the wind in a total time of 15 hours. If the speed of the wind is 20 mph, what is the speed of the plane in still air?

$$\frac{10}{x} + \frac{10}{x + 3} = 7 \qquad \text{LCD} = x(x + 3)$$

$$x(x + 3)\left[\frac{10}{x} + \frac{10}{x + 3}\right] = 7x(x + 3) \qquad \begin{array}{l}\text{Multiply both sides} \\ \text{by the LCD}\end{array}$$

$$x(x + 3)\frac{10}{x} + x(x + 3)\frac{10}{x + 3} = 7x(x + 3) \qquad \begin{array}{l}\text{Use the distributive} \\ \text{law}\end{array}$$

$$10(x + 3) + 10x = 7x^2 + 21x$$

$$10x + 30 + 10x = 7x^2 + 21x$$

$$20x + 30 = 7x^2 + 21x$$

$$0 = 7x^2 + x - 30 \qquad \begin{array}{l}\text{Collect like terms} \\ \text{in the quadratic} \\ \text{equation}\end{array}$$

$$0 = (7x + 15)(x - 2) \qquad \text{Factor}$$

$$7x + 15 = 0 \qquad \text{or} \quad x - 2 = 0$$

$$x = \frac{-15}{7} \qquad \text{or} \qquad x = 2$$

Since $\frac{-15}{7}$ cannot represent a rate in this problem, we can discard it. Thus, her rate up the mountain is 2 mph and down the mountain is 5 mph.

Answer: 100 mph

CAUTION

It is important when working any word problem to be neat and complete. Do not try to take shortcuts, especially when writing down the pertinent information. Writing complete descriptions of the variables can eliminate time-consuming errors.

10.5 EXERCISES A

Solve. (Some problems have been started.)

1. If five is added to the square of a number the result is 41. Find the number.

Let x = the desired number,
  $x^2 + 5$ = five added to the square of
      the number.

2. If the square of Mary's age is decreased by 44, the result is 100. How old is Mary?

Let x = Mary's age,
  $x^2 - 44$ = the square of Mary's age
      decreased by 44.

3. If the square of a number is decreased by 10 the result is three times the number. Find the number.

Let x = the number.

4. Twice the square of Ernie's age, less 9, is the same as seventeen times his age. How old is Ernie?

Let x = Ernie's age.

5. The product of 1 more than a number and 1 less than the number is 99. Find the number.

Let x = the desired number,
$x + 1$ = 1 more than the number,
$x - 1$ = 1 less than the number.

6. Sam's present age times his age in five years is 84. How old is Sam?

7. The product of two positive consecutive integers is 120. Find the integers.

Let x = the first even integer,
$x + 2$ = the next even integer.

8 Three times the product of two positive consecutive odd integers is 297. Find the integers.

9. Adding 4 to the square of Marvin's age is the same as subtracting 3 from eight times his age. How old is Marvin?

10. The sum of the squares of two consecutive even positive integers is 100. Find the integers.

11. One number is 6 larger than another. The square of the larger is 96 more than the square of the smaller. Find the numbers.

12. Find the length and width of a rectangle if the length is 4 cm longer than the width and the area is 140 cm².

Let x = width of rectangle,
$x + 4$ = length of rectangle.

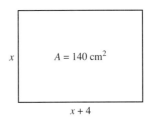

13. The number of square inches in the area of a square is 21 more than the number of inches in its perimeter. Find the length of a side.

Let x = length of sides of square.

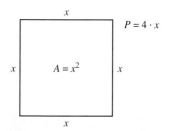

14. If the hypotenuse of a right triangle is 13 feet long and one leg is 7 feet longer than the other, find the measure of each leg.

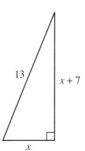

15. If the sides of a square are lengthened by 2 cm, its area becomes 169 cm². Find the length of a side.

Let x = length of sides of square,
    $x + 2$ = length of sides increased by 2 cm.

16. The area of a triangle is 24 ft² and the base is $\frac{1}{3}$ as long as the height. Find the base and height.

17. A box is 6 inches high. The length is 8 inches longer than the width and the volume is 1080 in³. Find the width and length.

18. The area of a circle is 1256 cm². Find the radius. (Use $\pi \approx 3.14$.)

19. The area of a parallelogram is 55 ft². If the base is 1 ft greater than twice the height, find the base and height.

20. The perimeter of a rectangle is 62 inches and the length is 3 inches more than the width. Find the dimensions.

21. It takes one pipe 6 hours longer to fill a tank than it takes a second pipe. Working together they fill the tank in 4 hours. How long would it take each working alone to fill the tank?

22. It takes Joe 9 hours longer to grade a set of tests than Dan. If they work together they can grade them in 20 hours. How long would it take each to grade the tests if they worked alone?

23. A picture frame is 24 in by 18 in as in the sketch below. If the area of the picture itself is 216 in², what is the width of the frame?

Let x = the width of the frame.
$(24 - 2x)(18 - 2x) = 216$

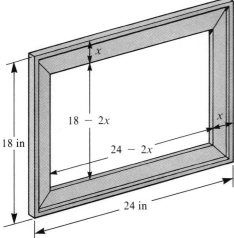

24. The pressure p, in lb per sq ft, of a wind blowing v mph can be approximated by the equation $p = 0.003v^2$. What is the approximate velocity of the wind when it creates a pressure of 1.875 lb per sq ft against the side of a building?

25. The amount of money A that will result if a principal P is invested at r percent integer compounded annually for 2 years is given by $A = P(1 + r)^2$. If $1000 grows to $1210 in 2 years using this formula, what is the interest rate?

26 A boat travels 48 miles upstream and then returns in a total time of 10 hours. If the speed of the stream is 2 mph, what is the speed of the boat in still water?

27. Mary Conners can walk 8 miles up a mountain and then return in a total time of 6 hours. Her speed downhill is 2 mph faster than her speed uphill. What is her speed uphill?

FOR REVIEW

Solve.

28. $\dfrac{x^2}{x-3} = \dfrac{9}{x-3} + 10$

29. $x = \sqrt{7x + 18} - 4$

30. The square root of Neil's age in 6 years is the same as one-fifth of his present age. How old is Neil?

ANSWERS: 1. 6, −6   2. 12 years old   3. 5, −2   4. 9 years old   5. 10, −10   6. 7 years old   7. 10, 12   8. 9, 11
9. 1 year old or 7 years old   10. 6, 8   11. 5, 11   12. 14 cm, 10 cm   13. 7 in   14. 5 ft, 12 ft   15. 11 cm   16. 12 ft,
4 ft   17. 10 in, 18 in   18. 20 cm   19. 11 ft, 5 ft   20. 17 in, 14 in   21. 6 hr, 12 hr   22. 36 hr, 45 hr   23. 3 in
24. 25 mph   25. 10%   26. 10 mph   27. 2 mph   28. 7   29. 1, −2   30. 30 years old

---

10.5 EXERCISES B

Solve.

1. If 4 is added to the square of a number, the result is 85. Find the number.

2. If the square of Rosemary's age is decreased by 24, the result is 300. How old is Rosemary?

3. If the square of a number is decreased by 32, the result is four times the number. Find the number.

4. Twice the square of Arlo's age, less 22, is the same as twenty times his age. How old is Arlo?

5. The product of 1 more than a number and 1 less than the number is 224. Find the number.

6. Troy's present age times his age in seven years is 260. How old is Troy?

7. The product of two positive consecutive even integers is 168. Find the integers.

8. Twice the product of two positive consecutive odd integers is 126. Find the integers.

9. Subtracting 100 from the square of Raul's age is the same as adding 60 to twelve times his age. How old is Raul?

10. The sum of the squares of two consecutive even positive integers is 164. Find the integers.

11. One number is 5 more than another. The square of the larger exceeds the square of the smaller by 95. Find the number.

12. Find the length and width of a rectangle if the length is 7 meters more than the width and the area is 198 m².

13. The number of square inches in the area of a square is 5 more than the number of inches in its perimeter. Find the length of a side.

14. If the hypotenuse of a right triangle is 26 cm long and one leg is 14 cm longer than the other, find the measure of each leg.

15. If the sides of a square are lengthened by 2 feet, its area becomes 100 ft². Find the length of a side.

16. The area of a triangle is 18 m² and the base is $\frac{1}{4}$ the height. Find the base and height.

17. A box is 8 yards high. The length is 4 yards longer than the width and the volume is 360 yd³. Find the width and length.

18. The area of a circle is 200.96 cm². Find the radius. (Use $\pi \approx 3.14$.)

19. The area of a parallelogram is 175 ft². If the base is 4 feet more than three times the height, find the base and height.

20. The perimeter of a rectangular garden is 28 meters and the length is 4 meters more than the width. Find the dimensions of the garden.

21. It takes Jeff 16 hours longer to repair his car than it takes his father, who is a mechanic, to do the same job. If they could do the job together in 6 hours, how long would it take each if they worked alone?

22. Graydon, the registered cheese cutter at Perko's Delicatessen, is training a new assistant, Burford. It takes Burford 24 hours longer to process the Tillamook cheese shipment than it takes Graydon. If together they could process the cheese in 5 hours, how long would it take each, working alone, to cut and display the cheese?

23. A rectangular backyard is to have a sidewalk placed entirely around its perimeter in such a way that 875 ft² of lawn area are enclosed inside the walk. If the dimensions of the yard are 40 ft by 30 ft, what is the width of the sidewalk?

24. The pressure p, in lb per sq ft, of a wind blowing v miles per hour can be approximated by the equation $p = 0.003v^2$. What is the approximate wind velocity when a pressure of 2.7 lb per sq ft is exerted against the side of a camper?

25. The amount of money A that will result if a principal P is invested at r percent interest compounded annually for 2 years is given by $A = P(1 + r)^2$. If $2000 grows to $2645 in 2 years using this formula, what is the interest rate?

26. Pat Marx swims 4 miles downstream and then returns in a total time of 3 hours. If the speed of the stream is 1 mph, what is Pat's speed in still water?

27. Chuck Little rode a bicycle with the wind for 18 miles. He then returned against the wind and the total time of his trip was 5 hours. If his speed with the wind was 3 mph faster than against the wind, what was his speed against the wind?

FOR REVIEW

Solve.

28. $\dfrac{z^2}{z - 2} = \dfrac{4}{z - 2} + 5$

29. $\sqrt{8y - 7} = y$

30. The square root of 6 more than a number is the same as 6 less than the number. Find the number.

10.5 EXERCISES C

Solve.

1. Two boats leave an island with one heading south and the other west. After 4 hours they are 100 miles apart. What is the speed of each boat if one travels 5 mph faster than the other? [Answer: 15 mph, 20 mph]

2. A boat requires one hour longer to go 80 miles upstream than to make the return trip downstream. What is the speed of the boat in still water if the speed of the stream is 2 mph?

10.6 SOLVING FORMULAS

═══ STUDENT GUIDEPOSTS ═══

1 Linear Equations   **2** Quadratic and Radical Equations

Many times, equations involve a variable and constants represented by letters instead of numerical constants. Formulas are excellent examples of these types of equations. For instance, we have used such formulas as

$$A = lw \qquad P = 2l + 2w \qquad c^2 = a^2 + b^2 \qquad c = \pi d.$$

To solve a given formula for a particular letter, remember that the letters play the same role as numbers and all of our equation-solving rules apply. If you have trouble solving for a letter in a formula, it may be helpful to make up a similar equation with numbers instead of letters, and pattern your solution steps after the procedure you use to solve the new equation. We will demonstrate this technique in the examples.

1 LINEAR EQUATIONS

Recall that solving a formula for a particular variable is a process of isolating the variable on one side of the equation.

EXAMPLE 1 ADDITION-SUBTRACTION RULE	PRACTICE EXERCISE 1

Solve $a + x = b$ for x.

    Formula   *Similar numerical equation*

$$a + x = b \qquad\qquad\qquad 3 + x = 7$$
$$a - a + x = b - a \quad \text{Subtract } a \quad 3 - 3 + x = 7 - 3 \quad \text{Subtract } 3$$
$$0 + x = b - a \qquad\qquad\quad 0 + x = 7 - 3$$
$$x = b - a \qquad\qquad\qquad\quad x = 4$$

The solution is $x = b - a$.

Solve $2y + z = x$ for z.

Answer: $z = x - 2y$

EXAMPLE 2 MULTIPLICATION-DIVISION RULE

Solve $cx = d$ for x.

Formula	*Similar numerical equation*

$$cx = d \qquad\qquad 5x = 8$$

$$\frac{1}{c} \cdot cx = \frac{1}{c} \cdot d \quad \text{Multiply by } \tfrac{1}{c} \qquad \frac{1}{5} \cdot 5x = \frac{1}{5} \cdot 8 \quad \text{Multiply by } \tfrac{1}{5}$$

$$1 \cdot x = \frac{d}{c} \qquad\qquad 1 \cdot x = \frac{8}{5}$$

$$x = \frac{d}{c} \qquad\qquad x = \frac{8}{5}$$

The solution is $x = \dfrac{d}{c}$.

PRACTICE EXERCISE 2

Solve $5uv = w$ for u.

Answer: $u = \dfrac{w}{5v}$

EXAMPLE 3 COMBINATION OF RULES

Solve $ax + b = c$ for x.

Formula	*Similar numerical equation*

$$ax + b = c \qquad\qquad 3x + 5 = 20$$

$$ax + b - b = c - b \quad \text{Subtract } b \qquad 3x + 5 - 5 = 20 - 5 \quad \text{Subtract 5}$$

$$ax = c - b \qquad\qquad 3x = 15$$

$$\frac{1}{a} \cdot ax = \frac{1}{a}(c - b) \ \text{Multiply by } \tfrac{1}{a} \qquad \frac{1}{3} \cdot 3x = \frac{1}{3} \cdot 15 \ \text{Multiply by } \tfrac{1}{3}$$

$$x = \frac{c - b}{a} \qquad\qquad x = \frac{15}{3} = 5$$

The solution is $x = \dfrac{c - b}{a}$.

PRACTICE EXERCISE 3

Solve $mn - k = 6$ for n.

Answer: $n = \dfrac{k + 6}{m}$

EXAMPLE 4 FORMULA FOR THE AREA OF RECTANGLE

Solve $A = lw$ for l.

Formula	*Similar numerical equation*

$$A = lw \qquad\qquad 12 = l \cdot 4$$

$$A \cdot \frac{1}{w} = lw \cdot \frac{1}{w} \quad \text{Multiply by } \tfrac{1}{w} \qquad 12 \cdot \frac{1}{4} = l \cdot 4 \cdot \frac{1}{4} \quad \text{Multiply by } \tfrac{1}{4}$$

$$\frac{A}{w} = l \cdot 1 \qquad\qquad \frac{12}{4} = l \cdot 1$$

$$\frac{A}{w} = l \qquad\qquad 3 = l$$

The solution is $l = \dfrac{A}{w}$.

PRACTICE EXERCISE 4

Solve $PV = RT$ for R.

Answer: $R = \dfrac{PV}{T}$

EXAMPLE 5 **DIVIDING BY TWO VARIABLES**	**PRACTICE EXERCISE 5**

Solve $a = bcx$ for x.

Solve $3xyz = uv$ for x.

$$a = bcx$$

$$\frac{1}{bc} \cdot a = \frac{1}{bc} \cdot bcx \qquad \text{Multiply by } \frac{1}{bc}, \text{ the reciprocal of the coefficient of } x$$

$$\frac{a}{bc} = 1 \cdot x$$

$$\frac{a}{bc} = x$$

The solution is $x = \dfrac{a}{bc}$.

Answer: $x = \dfrac{uv}{3yz}$

② QUADRATIC AND RADICAL EQUATIONS

Some formulas involve quadratic equations or radical equations. We shall assume that the variables and constants are chosen so that division by zero and negative numbers under a radical sign are avoided.

EXAMPLE 6 **QUADRATIC EQUATION**	**PRACTICE EXERCISE 6**

Solve $a = bx^2$ for x.

Solve $uy^2 - v = 0$ for y.

Formula		*Similar numerical equation*	

$$a = bx^2 \qquad\qquad\qquad 50 = 2x^2$$

$$\frac{1}{b} \cdot a = \frac{1}{b} \cdot bx^2 \quad \text{Multiply by } \frac{1}{b} \qquad \frac{1}{2} \cdot 50 = \frac{1}{2} \cdot 2x^2 \quad \text{Multiply by } \frac{1}{2}$$

$$\frac{a}{b} = x^2 \qquad\qquad\qquad 25 = x^2$$

$$\pm\sqrt{\frac{a}{b}} = x \qquad \begin{array}{l}\text{Take square root} \\ \text{of both sides}\end{array} \qquad \pm 5 = \pm\sqrt{25} = x$$

The solutions are $x = \pm\sqrt{\dfrac{a}{b}}$.

Answer: $y = \pm\sqrt{\dfrac{v}{u}}$

EXAMPLE 7 **RADICAL EQUATION**	**PRACTICE EXERCISE 7**

Solve $a = \sqrt{\dfrac{x}{b}}$ for x.

Solve $m = \sqrt{\dfrac{2k}{n}}$ for n.

Formula	Similar numerical equation

$$a = \sqrt{\dfrac{x}{b}}$$

$$5 = \sqrt{\dfrac{x}{2}}$$

$$(a)^2 = \left(\sqrt{\dfrac{x}{b}}\right)^2 \quad \text{Square both sides} \quad (5)^2 = \left(\sqrt{\dfrac{x}{2}}\right)^2$$

$$a^2 = \dfrac{x}{b}$$

$$25 = \dfrac{x}{2}$$

$$b \cdot a^2 = b \cdot \dfrac{x}{b} \quad \text{Multiply by } b \quad 2 \cdot 25 = 2 \cdot \dfrac{x}{2}$$

$$ba^2 = x$$

$$50 = x$$

The solution is $x = ba^2$.

Answer: $n = \dfrac{2k}{m^2}$

10.6   EXERCISES A

1. Solve $g = x + h$ for x.

2. Solve $t - x = k$ for x.

3. Solve $bx + d = e$ for x.

4 Solve $g - ax = m$ for x.

5. Solve $d = rt$ for t.

6. Solve $d = rt$ for r.

7. Solve $E = IR$ for I.

8. Solve $I = prt$ for t.

9. Solve $I = prt$ for r.

10. Solve $I = prt$ for p.

11 Solve $A = \dfrac{1}{2}bh$ for b.

12. Solve $A = \dfrac{1}{2}bh$ for h.

13. Solve $r = \dfrac{d}{t}$ for d.

14 Solve $r = \dfrac{d}{t}$ for t.

15. Solve $A = \pi r^2$ for r.

16 Solve $S = \dfrac{1}{2}gt^2$ for t.

17. Solve $E = mc^2$ for m.

18. Solve $E = mc^2$ for c.

19. Solve $A = \frac{4}{3}\pi r^2$ for r.

20 Solve $U = \sqrt{\frac{a}{c}}$ for c.

21. Solve $U = \sqrt{\frac{a}{c}}$ for a.

22. Solve $\sqrt{x + a} = b$ for x.

FOR REVIEW

Solve.

23. The product of 2 less than a number and 3 more than the number is 176. Find the number.

24. The hypotenuse of a right triangle is 5 cm and one leg is 1 cm longer than the other. Find the measure of each leg.

25. The area of a triangle is 24 ft², and the base is 3 times as long as the height. Find the base and height.

26. Laura takes 2 days to type a manuscript and Holly takes 8 days. How long would it take if they worked together?

Exercises 27–28 review material from Chapter 4 to help you prepare for the next section. Graph each equation.

27. $y + 2x - 4 = 0$

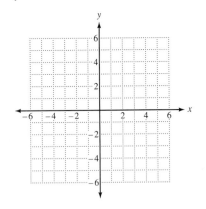

28. $y = 3x + 6$

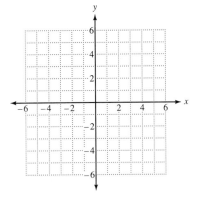

ANSWERS: 1. $g - h$ 2. $t - k$ 3. $\dfrac{e - d}{b}$ 4. $\dfrac{g - m}{a}$ 5. $\dfrac{d}{r}$ 6. $\dfrac{d}{t}$ 7. $\dfrac{E}{R}$ 8. $\dfrac{I}{pr}$ 9. $\dfrac{I}{pt}$ 10. $\dfrac{I}{rt}$ 11. $\dfrac{2A}{h}$

12. $\dfrac{2A}{b}$ 13. rt 14. $\dfrac{d}{r}$ 15. $\pm\sqrt{\dfrac{A}{\pi}}$ 16. $\pm\sqrt{\dfrac{2S}{g}}$ 17. $\dfrac{E}{c^2}$ 18. $\pm\sqrt{\dfrac{E}{m}}$ 19. $\pm\sqrt{\dfrac{3A}{4\pi}}$ 20. $\dfrac{a}{U^2}$ 21. cU^2 22. $b^2 - a$

23. $-14, 13$ 24. 3 cm, 4 cm 25. 12 ft, 4 ft 26. $\frac{8}{5}$ days

27.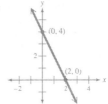

28.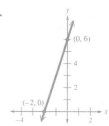

10.6 EXERCISES B

1. Solve $x + k = w$ for x.

2. Solve $t = m - x$ for x.

3. Solve $ax + w = m$ for x.

4. Solve $p - bx = w$ for x.

5. Solve $A = bh$ for h.

6. Solve $A = bh$ for b.

7. Solve $T = cn$ for n.

8. Solve $V = lwh$ for h.

9. Solve $V = lwh$ for w.

10. Solve $V = lwh$ for l.

11. Solve $W = \dfrac{1}{3}pq$ for p.

12. Solve $W = \dfrac{1}{3}pq$ for q.

13. Solve $a = \dfrac{b}{m}$ for b.

14. Solve $a = \dfrac{b}{m}$ for m.

15. Solve $V = \pi r^2 h$ for h.

16. Solve $V = \pi r^2 h$ for r.

17. Solve $S = a\pi r^2$ for r.

18. Solve $S = a\pi r^2$ for a.

19. Solve $W = \dfrac{5}{4}am^2$ for m.

20. Solve $T = \sqrt{\dfrac{a}{w}}$ for a.

21. Solve $T = \sqrt{\dfrac{a}{w}}$ for w.

22. Solve $\sqrt{x + t} = p$ for x.

Solve.

23. Of two positive numbers, one is 4 larger than the other and the sum of the squares of the two is 136. Find the numbers.

24. The height of a triangle is 8 inches more than the base. If the area is 90 in^2, find the height and base.

25. The length of a rectangle is 4 cm more than the width and the area is 96 cm². Find the dimensions.

26. The amount of money A that will result if a principal P is invested at r percent interest compounded annually for 2 years is given by $A = P(1 + r)^2$. If $5000 grows to $6498 in 2 years using this formula, what is the interest rate?

Exercises 27–28 review material from Chapter 3 to help you prepare for the next section. Graph each equation.

27. $2y - x - 6 = 0$

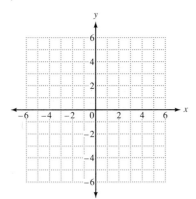

28. $y = -3x - 3$

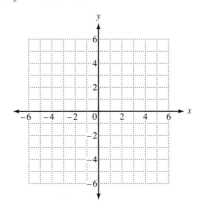

10.6 EXERCISES C

1. Solve $\dfrac{u^2}{1 + v^2} = 2$ for positive v.

2. Solve $\dfrac{x}{\sqrt{2x + y}} = 1$ for x.

[Answer: $x = 1 \pm \sqrt{1 + y}$]

10.7 GRAPHING QUADRATIC EQUATIONS

STUDENT GUIDEPOSTS

1 Parabolas

2 Graphing Quadratic Equations

3 Graphs for $a > 0$ and $a < 0$

4 Vertex of a Parabola

1 PARABOLAS

In Chapter 3 we learned how to graph a linear equation

$$ax + by + c = 0 \qquad \text{General form}$$
$$\text{or} \quad y = mx + b \qquad \text{Slope-intercept form}$$

by plotting the intercepts (or the intercept and one additional point if the intercepts are both (0, 0)) and drawing the straight line through them. The graph of every **quadratic equation in two variables** of the form

$$y = ax^2 + bx + c \quad (a \neq 0)$$

is a **parabola,** a U-shaped curve similar to the one shown in Figure 10.4. Applications of parabolas or surfaces in the shape of a parabola are numerous in science, engineering, business, and architecture.

② GRAPHING QUADRATIC EQUATIONS

To graph a quadratic equation, we choose several values for x, calculate the corresponding y-values, and plot the resulting points. A table of values helps us keep a record of the points.

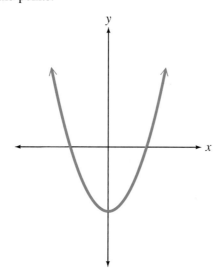

Figure 10.4

| **EXAMPLE 1**   GRAPHING A PARABOLA | **PRACTICE EXERCISE 1** |

Graph $y = x^2$.

Find y-values for the following values of x.

If $x = 0$, $y = x^2 = 0^2 = 0.$ If $x = -1$, $y = x^2 = (-1)^2 = 1.$

If $x = 1$, $y = x^2 = 1^2 = 1.$ If $x = -2$, $y = x^2 = (-2)^2 = 4.$

If $x = 2$, $y = x^2 = 2^2 = 4.$

Graph $y = x^2 - 1$.

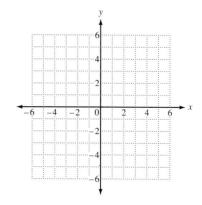

The completed table is shown below. Plotting the ordered pairs, we obtain the parabola in Figure 10.5.

x	y
0	0
1	1
2	4
−1	1
−2	4

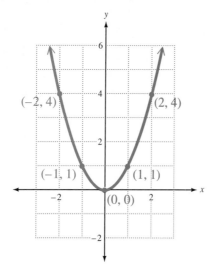

Figure 10.5

Answer: same as Figure 10.5 except all points down one unit

EXAMPLE 2   GRAPHING A PARABOLA

Graph $y = -x^2$.

Be careful when finding values for y in this case. For example, if $x = 1$, $y = -x^2 = -(1)^2 = -1$. The completed table appears beside the graph given in Figure 10.6.

x	y
0	0
1	−1
2	−4
−1	−1
−2	−4

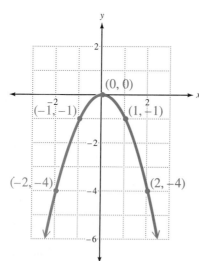

Figure 10.6

PRACTICE EXERCISE 2

Graph $y = -x^2 + 2$.

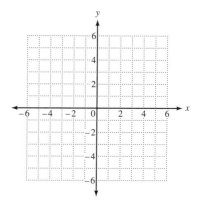

Answer: same as Figure 10.6 except all points up two units

❸ GRAPHS FOR $a > 0$ AND $a < 0$

Notice that the parabola in Example 1 opens up while the one in Example 2 opens down. In general, the direction in which a parabola opens depends on the sign of the coefficient of the x^2-term, a. If $a > 0$ (as in Example 1) the parabola opens up, and if $a < 0$ (as in Example 2), the parabola opens down.

Direction a Parabola Opens

The graph of $y = ax^2 + bx + c$ $(a \neq 0)$ is a parabola which

1. opens up when the coefficient of x^2 is positive,

2. opens down when the coefficient of x^2 is negative.

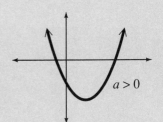 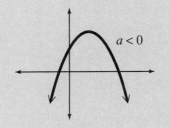

EXAMPLE 3	GRAPH OPENING UP

Graph $y = x^2 - 2x - 3$.

   We know that the parabola opens up since the coefficient of x^2 is positive. From the values computed in the table, the graph is plotted in Figure 10.7.

x	y
0	-3
1	-4
2	-3
3	0
-1	0

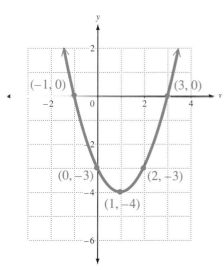

Figure 10.7

PRACTICE EXERCISE 3

Graph $y = -x^2 + 2x + 3$.

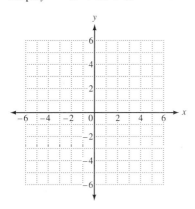

Answer: like Figure 10.7 except opening down

EXAMPLE 4   GRAPH OPENING DOWN

Graph $y = -x^2 + 2x$.

Use care when substituting for x. The graph is in Figure 10.8.

x	y
0	0
1	1
2	0
3	−3
−1	−3

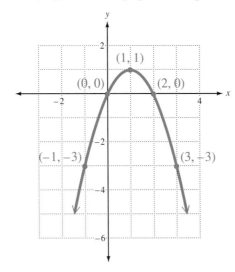

Figure 10.8

PRACTICE EXERCISE 4

Graph $y = x^2 - 2x$.

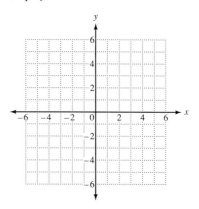

Answer: like Figure 10.8 except opening up

---

④ VERTEX OF A PARABOLA

Consider the following equation.

$$y = (x - 1)^2 - 4$$
$$= x^2 - 2x + 1 - 4$$
$$= x^2 - 2x - 3$$

Thus, $y = (x - 1)^2 - 4$ and $y = x^2 - 2x - 3$ have the same graph, given in Figure 10.7. Notice that $(1, -4)$ is the low point or **vertex** of the parabola. If the equation of a parabola is written in the form

$$y = (x - \boxed{h})^2 + \boxed{k},$$

then $\boxed{(h, k)}$ is the vertex and the graph is a parabola opening up.

A similar rule holds for parabolas opening down. Consider

$$y = -(x - 1)^2 + 1$$
$$= -(x^2 - 2x + 1) + 1$$
$$= -x^2 + 2x - 1 + 1$$
$$= -x^2 + 2x.$$

Thus, $y = -(x - 1)^2 + 1$ and $y = -x^2 + 2x$ have the same graph, given in Figure 10.8, with vertex $(1, 1)$. If the equation of a parabola is written in the form

$$y = -(x - \boxed{h})^2 + \boxed{k},$$

then $\boxed{(h, k)}$ is the vertex, and the graph is a parabola opening down.

| EXAMPLE 5 FINDING THE VERTEX | PRACTICE EXERCISE 5 |

Find the vertex of each parabola.

(a) $y = (x + 3)^2 - 5$

We must have the form $y = (x - h)^2 + k$.

$$y = (x + 3)^2 - 5$$
$$= [x - (-3)]^2 + (-5)$$

Thus, $h = -3$ and $k = -5$. The vertex is $(-3, -5)$.

(b) $y = -(x + 1)^2 + 7$

Write in the form $y = -(x - h)^2 + k$.

$$y = -(x + 1)^2 + 7$$
$$= -[x - (-1)]^2 + 7$$

Thus, $h = -1$ and $k = 7$. The vertex is $(-1, 7)$.

(c) $y = (x + 5)^2$

Write in the form $y = (x - h)^2 + k$.

$$y = (x + 5)^2$$
$$= [x - (-5)]^2 + 0$$

Thus, $h = -5$ and $k = 0$. The vertex is $(-5, 0)$.

Find the vertex of each parabola.

(a) $y = (x - 4)^2 - 5$

(b) $y = -(x + 2)^2 + 4$

(c) $y = -(x + 8)^2$

Answers: (a) $(4, -5)$
(b) $(-2, 4)$ (c) $(-8, 0)$

10.7 EXERCISES A

Before graphing, tell whether the parabola opens up or down. Then graph the equation.

1. $y = x^2 + 1$

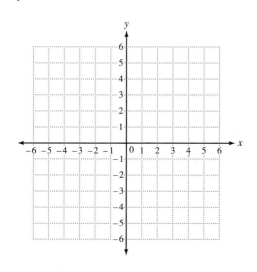

2. $y = -x^2 + 1$

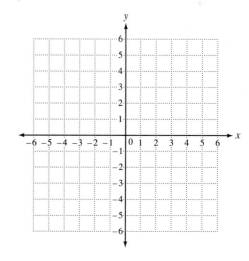

3. $y = x^2 - 2$

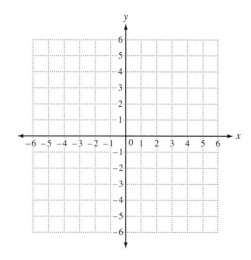

4. $y = -x^2 + 2$

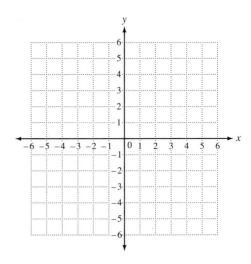

5. $y = x^2 + 2x + 1$

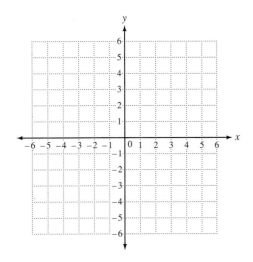

6. $y = -x^2 - 2x - 1$

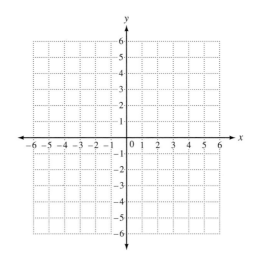

7. $y = x^2 + 2x$

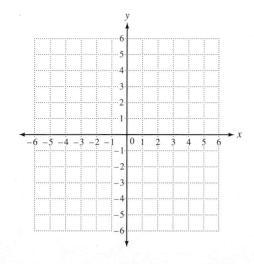

8. $y = -x^2 - 2x$

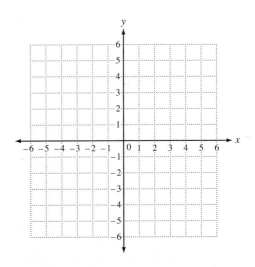

9. $y = x^2 - 3x + 2$

10. $y = -x^2 + 3x - 2$

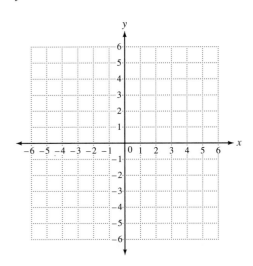

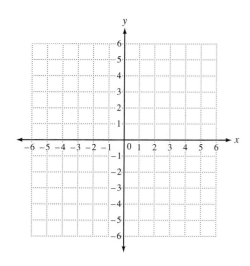

Give the vertex of the parabola.

11. $y = (x - 4)^2 + 2$

12. $y = -(x - 1)^2 + 5$

13. $y = (x + 3)^2 - 2$

14. $y = -(x + 10)^2 - 9$

15 $y = x^2 - 3$

16. $y = -x^2 - 3$

FOR REVIEW

17. Solve $p = ax + t$ for x.

18. Solve $M = \dfrac{1}{3}wt^2$ for t.

19. Solve $A = \dfrac{v}{2w}$ for v.

20. Solve $A = \dfrac{v}{2w}$ for w.

21. Solve $S = \sqrt{\dfrac{m}{r}}$ for m.

22. Solve $\sqrt{3x + u} = a$ for x.

ANSWERS:

1.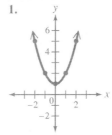

2.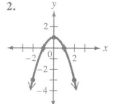

3.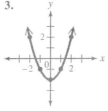

4.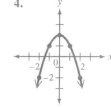

5.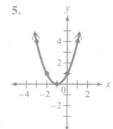

6.
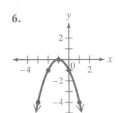
7.
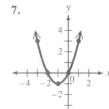
8.
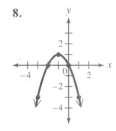
9.
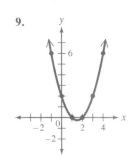
10.
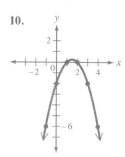

11. (4, 2) 12. (1, 5) 13. (−3, −2) 14. (−10, −9) 15. (0, −3) 16. (0, −3) 17. $\dfrac{p-t}{a}$ 18. $\pm\sqrt{\dfrac{3M}{w}}$ 19. $2Aw$

20. $\dfrac{v}{2A}$ 21. S^2r 22. $\dfrac{a^2-u}{3}$

10.7 EXERCISES B

Before graphing, tell whether the parabola opens up or down. Then graph the equation.

1. $y = x^2 + 2$

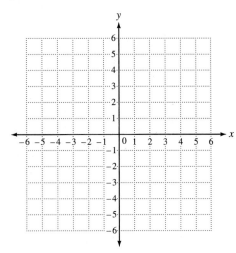

2. $y = -x^2 - 1$

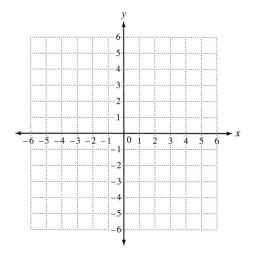

3. $y = x^2 - 1$

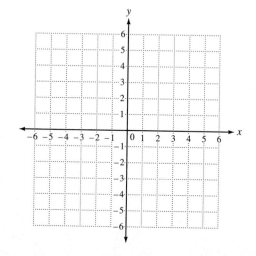

4. $y = -x^2 + 3$

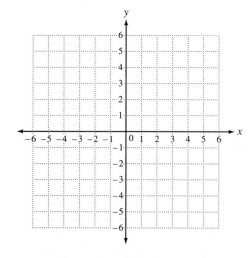

5. $y = x^2 - 2x + 1$

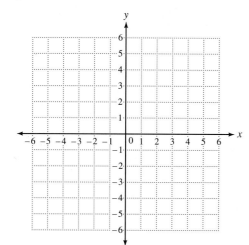

6. $y = -x^2 + 2x - 1$

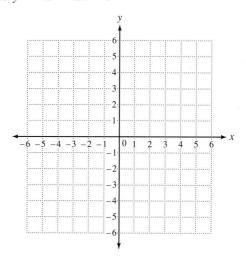

7. $y = x^2 + 4x$

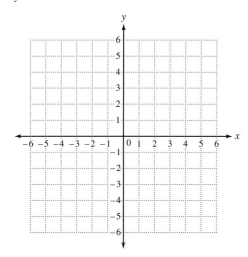

8. $y = -x^2 - 4x$

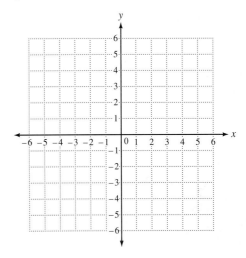

9. $y = x^2 + 3x + 2$

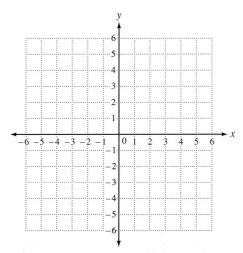

10. $y = -x^2 - 3x - 2$

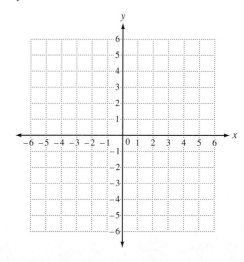

Give the vertex of the parabola.

11. $y = (x - 7)^2 + 1$

12. $y = -(x - 4)^2 + 8$

13. $y = (x + 12)^2 - 8$

14. $y = -(x + 9)^2 - 14$

15. $y = (x + 1)^2$

16. $y = -x^2 + 8$

FOR REVIEW

17. Solve $q = bx - w$ for x.

18. Solve $V = \dfrac{4}{3}am^2$ for m.

19. Solve $R = \dfrac{I}{2E}$ for I.

20. Solve $R = \dfrac{I}{2E}$ for E.

21. Solve $W = \sqrt{\dfrac{a}{n}}$ for n.

22. Solve $\sqrt{a + 2x} = z$ for x.

10.7 EXERCISES C

Find the vertex of each parabola by completing the square.

1. $y = x^2 + 10x + 20$ [Answer: $(-5, -5)$]

2. $y = -x^2 + 6x - 4$

CHAPTER 10 REVIEW

KEY WORDS

10.1 A **quadratic equation** is an equation that can be written in the form $ax^2 + bx + c = 0$, $a \neq 0$.

10.2 To **complete the square** on $x^2 + bx$ add $\left(\frac{1}{2}b\right)^2$ to make the expression a perfect square.

10.3 The **discriminant** of $ax^2 + bx + c = 0$ is $b^2 - 4ac$.

10.7 A **quadratic equation in two variables** is an equation of the form $y = ax^2 + bx + c$, $a \neq 0$.

A **parabola** is the graph of a quadratic equation in two variables.

The **vertex** is the low point of a parabola opening up and the high point of a parabola opening down.

KEY CONCEPTS

10.1, The best ways to solve a quadratic equation are
10.2 by factoring and taking roots. If these methods do not work, use the quadratic formula instead of completing the square.

10.3 The quadratic formula for solving $ax^2 + bx + c = 0$ is

$$x = \frac{-b \pm \sqrt{b^2 - 4ac}}{2a}.$$

Make sure that an equation is put into the above general form before trying to identify the constants a, b, and c. Also, the entire numerator $-b \pm \sqrt{b^2 - 4ac}$, is divided by $2a$, *not* just the radical term.

10.4 **1.** To solve a fractional equation, multiply both sides by the LCD of all fractions. Be sure to check all possible solutions in the original equations and exclude those that make any denominator zero.

2. To solve a radical equation, isolate a radical and square both sides of the equation. If a binomial occurs on one side, don't forget the middle term when squaring. For example,

$$x + 4 = 3\sqrt{x + 1} \quad \text{becomes}$$
$$x^2 + 8x + 16 = 9(x + 1).$$

Be sure to square 3 also. Finally, check all possible answers in the *original* equa-

tion and remember that the radical only represents the positive (principal) square root.

10.5 Be precise and write out all details (including complete descriptions of the variable) when solving word problems.

10.6 Solving a similar numerical equation can often help when solving a formula for a particular variable.

10.7 The graph of a quadratic equation $y = ax^2 + bx + c$ ($a \neq 0$) is a parabola that opens up if $a > 0$ and down if $a < 0$.

REVIEW EXERCISES

Part I

10.1 *Solve the following quadratic equations.*

1. $2x^2 - 32 = 0$

2. $-3z^2 = -15$

3. $x^2 + 6x - 72 = 0$

4. $2y^2 + y = 21$

5. $(z - 5)^2 = 25$

6. $21x^2 - 4x - 32 = 0$

10.2 *What must be added to complete the square?*

7. $x^2 + x +$ _____

8. $y^2 + \dfrac{1}{4}y +$ _____

9. Solve $x^2 - 3x + 1 = 0$ by completing the square.

10.3 *Solve.*

10. $2y^2 + y = 5$

11. $2z^2 + 11z = -(10 + z)$

12. $x^2 + 2x - 5 = 0$

13. $2y^2 + y + 5 = 0$

10.4 *Solve.*

14. $\dfrac{z + 2}{-z} = \dfrac{1}{z + 2}$

15. $\dfrac{3}{1 + x} = -1 - \dfrac{2}{1 - x}$

16. $x = 1 + \sqrt{4 - 4x}$

17. $4 = 4\sqrt{x + 1} - x$

10.5 **18.** The product of two positive consecutive integers is 420. Find the integers.

19. Mike's present age times his age in 7 years is 30. Find his present age.

20. The number of square inches in a square is 3 less than the number of inches in its perimeter. Find the length of its sides.

21. Find the number whose square is 18 more than three times the number.

22. The pressure p, in lb per sq ft, of a wind blowing v mph can be approximated by the equation $p = 0.003v^2$. What is the approximate wind velocity when a pressure of 3.675 lb per sq ft is exerted against the side of a skyscraper?

23. Use $A = P(1 + r)^2$ to find the interest rate if $3000 grows to $3630 in 2 years.

10.6 **24.** Solve $c = \dfrac{1}{3}dh$ for h.   **25.** Solve $a - x = 2b$ for x.   **26.** Solve $D = au^2$ for u.

27. Solve $D = au^2$ for a.   **28.** Solve $g = \sqrt{\dfrac{a}{d}}$ for a.   **29.** Solve $g = \sqrt{\dfrac{a}{d}}$ for d.

10.7 **30.** Tell whether the parabola with the given equation opens up or down.
     (a) $y = 3x^2 + x - 1$ **(b)** $y = -2x^2 + 5$

Graph the given equations.

31. $y = x^2 - 3$ **32.** $y = x^2 - 6x + 8$

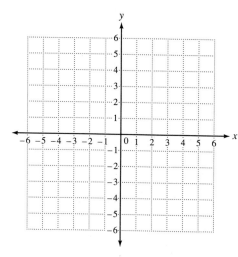 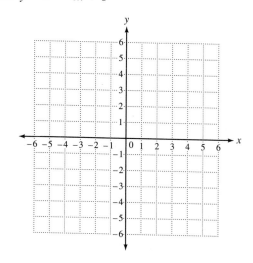

Give the vertex of the parabola.

33. $y = -(x + 4)^2 - 3$ **34.** $y = x^2 + 6$

Part II

Solve.

35. Solve $a = cu^2$ for u.   **36.** Solve $2hr^2 = 3V$ for r.

37. The square of a number, less 7, is equal to 9. Find the number.   **38.** Twice the square root of 2 more than a number is the same as 1 less than the number. Find the number.

39. Lennie has a garden in the shape of a square 12 yards on a side. He wishes to lay a water pipe from one corner diagonally across to the other. What is the approximate length of this pipe?

40. If it takes Bill 12 hours longer to paint a room than Peter and together they can complete the job in 8 hours, how long would it take each to do it alone?

41. $\dfrac{5}{x-2} + 1 = \dfrac{2}{x-3}$

42. $x^2 - 6x + 2 = 0$

43. $1 - x^2 = 7x - 5$

44. $\sqrt{x+1} + 1 = x$

45. $9x^2 = 1$

46. $2x^2 = 9x + 5$

Give the vertex of the parabola.

47. $y = -x^2 - 8$

48. $y = -(x-2)^2 + 5$

What must be added to complete the square?

49. $x^2 - 14x +$ ____

50. $y^2 + \dfrac{1}{3}x +$ ____

ANSWERS: 1. ± 4 2. $\pm\sqrt{5}$ 3. $6, -12$ 4. $3, -\frac{7}{2}$ 5. $0, 10$ 6. $\frac{4}{3}, -\frac{8}{7}$ 7. $\frac{1}{4}$ 8. $\frac{1}{64}$ 9. $\dfrac{3 \pm \sqrt{5}}{2}$

10. $\dfrac{-1 \pm \sqrt{41}}{4}$ 11. $-1, -5$ 12. $-1 \pm \sqrt{6}$ 13. no solution   14. $-1, -4$ 15. $2, -3$ 16. 1   17. $0, 8$

18. $20, 21$ 19. 3 years old   20. 1 in or 3 in   21. $6, -3$ 22. 35 mph   23. 10%   24. $\dfrac{3c}{d}$ 25. $a - 2b$ 26. $\pm\sqrt{\dfrac{D}{a}}$

27. $\dfrac{D}{u^2}$ 28. dg^2 29. $\dfrac{a}{g^2}$ 30. (a) up (b) down

31.

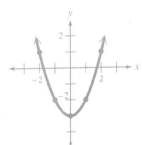

32.

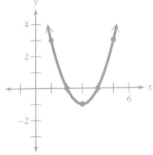

33. $(-4, -3)$ 34. $(0, 6)$ 35. $\pm\sqrt{\dfrac{a}{c}}$ 36. $\pm\sqrt{\dfrac{3V}{2h}}$ 37. $4, -4$ 38. 7   39. 16.8 yd (using $\sqrt{2} \approx 1.4$)   40. Peter: 12 hr,

Bill: 24 hr   41. $1 \pm \sqrt{6}$ 42. $3 \pm \sqrt{7}$ 43. $\dfrac{-7 \pm \sqrt{73}}{2}$ 44. 3   45. $\pm\frac{1}{3}$ 46. $-\frac{1}{2}, 5$ 47. $(0, -8)$ 48. $(2, 5)$ 49. 49

50. $\frac{1}{36}$

Solve the following quadratic equations.

1. $x^2 + 13x + 36 = 0$

1. _____

2. $3z^2 - z = 27 - z$

2. _____

3. $y(y - 2) = 3y$

3. _____

4. $x^2 - 4x + 1 = 0$

4. _____

5. What must be added to complete the square?

$y^2 + \dfrac{1}{2}y +$ _____

5. _____

Solve.

6. $\dfrac{2}{x - 1} + 2 = \dfrac{3}{x - 2}$

6. _____

7. $4 = x + \sqrt{2 + x}$

7. _____

Solve.

8. The product of two positive consecutive even integers is 120. Find the integers.

8. _____

9. It takes 8 hours less time to fill a pond using a large pipe than it takes using a smaller pipe. If when used together they fill the pond in 3 hours, how long would it take each to fill it alone?

9. _____

10. A boat travels 70 miles upstream and then returns in a total time of 12 hours. If the speed of the stream is 2 mph, what is the speed of the boat in still water?

10. _____

11. Solve $w = \sqrt{\dfrac{b}{m}}$ for m.

11. _____

12. You are given the equation $y = x^2 + 2$.
  (a) Does the graph of this parabola open up or down?
  (b) What is the vertex of the parabola?
  (c) Graph the equation.

12(a) _____
12(b) _____
12(c)

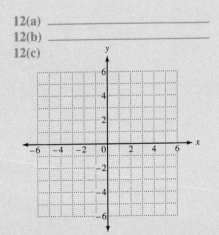

Geometry Appendix

A.1 LINES AND ANGLES

1 SEGMENTS, LINES, AND RAYS

The material in this Geometry Appendix is included for students who need a review of some of the concepts of geometry.

A point is a precise location in space often symbolized by a dot and labeled with a capital letter, such as the point A shown below.

·A

Figure A.1 displays three geometric figures. An arrowhead indicates that the curve continues in that direction without end.

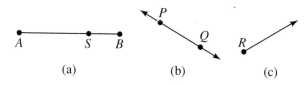

$$\text{(a)} \qquad \text{(b)} \qquad \text{(c)}$$

Figure A.1

A **segment** is a geometric figure made up of two **endpoints** and all points between them. The segment with endpoints A and B in Figure A.1(a) can be denoted by either $\overline{AB}$ or $\overline{BA}$. The point S is said to be **on the segment** $\overline{AB}$. Figure A.1(b) represents a straight **line**, which consists of the points on the segment $\overline{PQ}$ together with all points beyond P and Q. We denote this line by $\overleftrightarrow{PQ}$ or $\overleftrightarrow{QP}$.

Any two distinct points determine one and only one line. The two distinct points E and F determine the unique line $\overleftrightarrow{EF}$ shown in Figure A.2. For convenience, we often use a small letter such as l to represent a line.

Figure A.2

A **ray** is that part of a line lying on one side of a point P on the line, and includes the point P. A ray is shown in color in Figure A.3. By selecting another point Q on the ray, we have a way to denote the ray by $\overrightarrow{PQ}$. Unlike the notations used for segments and lines, the rays $\overrightarrow{PQ}$ and $\overrightarrow{QP}$ are different. A ray has only one endpoint and it is always written first. Taken together, the rays $\overrightarrow{PQ}$ and $\overrightarrow{QP}$ make up the line $\overleftrightarrow{PQ}$. Another example of a ray is shown in Figure A.1(c).

Figure A.3

EXAMPLE 1 SEGMENTS, LINES, AND RAYS

Use Figure A.4 to answer each question.

Figure A.4

(a) Is point C on $\overline{AB}$?   No

(b) Is point C on $\overleftrightarrow{AB}$?   Yes

(c) Is point C on $\overrightarrow{BA}$?   No

(d) What are the endpoints of $\overline{CA}$? A and C

(e) What are the endpoints of $\overrightarrow{CA}$?   Only C

(f) What are the endpoints of $\overleftrightarrow{CA}$?   There are no endpoints.

(g) Are $\overline{AB}$ and $\overline{BA}$ the same?   Yes

(h) Are $\overrightarrow{AB}$ and $\overrightarrow{BA}$ the same?   No

---

PRACTICE EXERCISE 1

Draw two points X and Y and place point Z on $\overleftrightarrow{XY}$ but not on $\overline{XY}$. Use this figure to answer the following questions.

(a) Is Z on $\overrightarrow{YX}$?

(b) Is Z on $\overline{YX}$?

(c) Is Y on $\overrightarrow{ZX}$?

(d) What are the endpoints of $\overline{XY}$?

(e) What are the endpoints of $\overrightarrow{XY}$?

(f) What are the endpoints of $\overleftrightarrow{ZX}$?

(g) Are $\overline{YZ}$ and $\overline{ZY}$ the same?

(h) Are $\overrightarrow{ZX}$ and $\overrightarrow{XZ}$ the same?

---

② COINCIDING, INTERSECTING, AND PARALLEL LINES

A **plane** is a surface with the property that any two points in it can be joined by a line, all points of which are also contained in the surface. Intuitively, a plane can be thought of as any flat surface such as a blackboard or a desktop. If two lines in a plane represent the same line, they are **coinciding lines.** When two lines share exactly one common point, they are **intersecting lines.** When two distinct lines in a plane do not intersect, the lines are called **parallel lines.** These three situations are illustrated in Figure A.5.

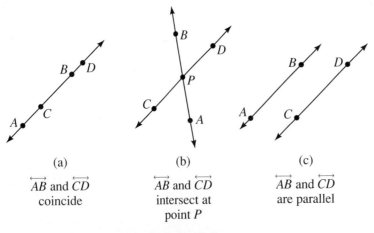

(a)
$\overleftrightarrow{AB}$ and $\overleftrightarrow{CD}$
coincide

(b)
$\overleftrightarrow{AB}$ and $\overleftrightarrow{CD}$
intersect at
point P

(c)
$\overleftrightarrow{AB}$ and $\overleftrightarrow{CD}$
are parallel

Figure A.5

Two segments, two rays, or a segment and a ray can also be called **parallel** if they are on parallel lines. Thus, in Figure A.5(c) $\overline{AB}$ is parallel to $\overleftrightarrow{CD}$, $\overrightarrow{AB}$ is parallel to $\overrightarrow{DC}$, and $\overline{AB}$ is parallel to $\overrightarrow{CD}$.

❸ ANGLES

An **angle** is a geometric figure consisting of two rays which share a common endpoint, called the **vertex** of the angle. In Figure A.6, the two rays $\overrightarrow{AC}$ and $\overrightarrow{AB}$ form the angle denoted by $\angle BAC$, $\angle CAB$, or simply $\angle A$. The rays $\overrightarrow{AC}$ and $\overrightarrow{AB}$ are the **sides** of $\angle A$.

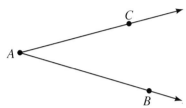

Figure A.6

We are familiar with measuring the length of segments by some suitable unit of measure such as inch, centimeter, foot, or meter. In order to measure an angle, we need a measuring unit. The most common unit is the degree (°). An angle with measure 0° is formed by two coinciding rays such as $\overrightarrow{AB}$ and $\overrightarrow{AC}$ in Figure A.7. As the ray $\overrightarrow{AB}$ rotates in a counterclockwise direction from ray $\overrightarrow{AC}$, the two rays form larger and larger angles as shown in Figure A.8.

Figure A.7

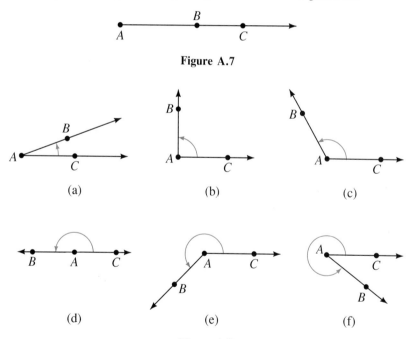

Figure A.8

If $\overrightarrow{AB}$ is allowed to rotate completely around until it coincides with ray $\overrightarrow{AC}$ again, the resulting angle is said to measure 360°. Thus, an angle of measure 1° is formed by making $\frac{1}{360}$ of a complete rotation, as in Figure A.9.

Figure A.9

4 CLASSIFYING ANGLES

When two rays forming an angle are in opposite directions, such as in Figure A.8(d), the resulting angle is a **straight angle** and has measure 180°. When a ray $\overrightarrow{AB}$ is rotated through one-fourth of a complete revolution from $\overrightarrow{AC}$, as in Figure A.8(b), the resulting angle is a **right angle** and has measure 90°. Right angles are often marked with a square corner at the vertex, such as ⌐__ . Angles measuring between 0° and 90°, such as the angle in Figure A.8(a), are **acute angles.** Angles measuring between 90° and 180°, such as the angle in Figure A.8(c), are **obtuse angles.**

Two angles with the same measure are **equal** or **congruent.** Two angles whose measures total 90° are **complementary,** and two angles whose measures total 180° are **supplementary.**

When two lines intersect, four angles are formed. The intersecting lines $\overleftrightarrow{AB}$ and $\overleftrightarrow{CD}$ in Figure A.10 form the four angles $\angle APD$, $\angle DPB$, $\angle BPC$, and $\angle CPA$. $\angle APC$ and $\angle DPB$ are **vertical angles** to each other. The same is true for $\angle APD$ and $\angle BPC$.

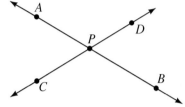

Figure A.10

5 PERPENDICULAR LINES

Vertical angles have the same measure. Thus, in Figure A.10, $\angle APC$ and $\angle DPB$ are equal, as are $\angle APD$ and $\angle BPC$. When the four angles formed by two intersecting lines all have measure 90°, that is, when they are all right angles, the lines are **perpendicular.**

EXAMPLE 2   LINES AND ANGLES

The following statements refer to Figure A.11.

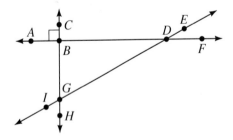

Figure A.11

(a) $\overleftrightarrow{AB}$ and $\overleftrightarrow{DF}$ are coinciding lines.

(b) $\overleftrightarrow{AF}$ and $\overleftrightarrow{IE}$ are intersecting lines.

(c) $\overrightarrow{GI}$ and $\overrightarrow{GH}$ are the sides of $\angle IGH$.

(d) $\angle IGD$ is a straight angle.

(e) $\angle ABC$ is a right angle.

(f) $\angle IGB$ and $\angle BGD$ are supplementary.

PRACTICE EXERCISE 2

Use the following figure and answer true or false in (a)–(k).

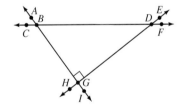

(a) $\overleftrightarrow{HD}$ and $\overleftrightarrow{GE}$ are coinciding lines.

(b) $\overleftrightarrow{AI}$ and $\overleftrightarrow{HE}$ are parallel lines.

(c) $\overrightarrow{AB}$ and $\overrightarrow{DB}$ are the sides of $\angle ABD$.

(d) $\angle BGI$ is a straight angle.

(e) $\angle BGD$ is a right angle.

(f) $\angle BDE$ and $\angle EDF$ are complementary.

(g) ∠*EDF* is an acute angle.

(h) ∠*EDB* is an obtuse angle.

(i) ∠*IGH* and ∠*BGD* are vertical angles.

(j) ∠*EDB* and ∠*FDG* are equal (they are vertical angles).

(k) $\overleftrightarrow{AF}$ and $\overleftrightarrow{CH}$ are perpendicular.

(g) ∠*ABD* is an acute angle.

(h) ∠*BDE* is an obtuse angle.

(i) ∠*ABC* and ∠*CBG* are vertical angles.

(j) ∠*DGI* and ∠*IGH* are equal.

(k) $\overleftrightarrow{CB}$ and $\overleftrightarrow{HG}$ are intersecting lines.

Answers: (a) true   (b) false
(c) false (d) true   (e) true
(f) false (g) false (h) true
(i) false (j) true   (k) true

⑥ PARALLEL LINES CUT BY A TRANSVERSAL

A line that intersects each of two parallel lines is called a **transversal.** The two lines are said to be ''cut'' by the transversal and several angles are formed. In Figure A.12, transversal $\overleftrightarrow{AD}$ cuts parallel lines $\overleftrightarrow{GF}$ and $\overleftrightarrow{HE}$ forming four **interior angles,** ∠*GBC,* ∠*FBC,* ∠*BCH,* and ∠*BCE,* and four **exterior angles,** ∠*ABG,* ∠*ABF,* ∠*HCD,* and ∠*ECD.* Since ∠*GBC* and ∠*BCE* are on alternate sides of the transversal, they are called **alternate interior angles.** Similarly, ∠*FBC* and ∠*BCH* are alternate interior angles. In Figure A.12, ∠*ABF* and ∠*BCE* are **corresponding angles.** Three other pairs of corresponding angles are ∠*ABG* and ∠*BCH,* ∠*GBC* and ∠*HCD,* and ∠*FBC* and ∠*ECD.*

Alternate interior angles always have the same measure. Thus, referring to Figure A.12 again, ∠*FBC* and ∠*BCH* have the same measure, as do ∠*GBC* and ∠*BCE.* Also, corresponding angles always have the same measure so that, for example, ∠*ABF* has the same measure as ∠*BCF* in Figure A.12.

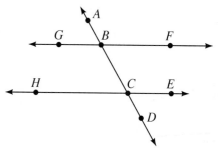

Figure A.12

EXAMPLE 3 TRANSVERSAL CUTTING PARALLEL LINES

The following statements refer to Figure A.13 in which $\overleftrightarrow{AB}$ is parallel to $\overleftrightarrow{DE}.$

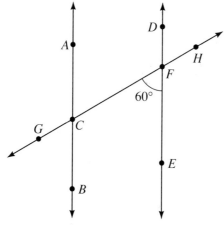

Figure A.13

(a) $\overleftrightarrow{GH}$ is a transversal.

(b) ∠*DFC* is an interior angle.

(c) ∠*ACG* is an exterior angle.

PRACTICE EXERCISE 3

Use the following figure in which $\overleftrightarrow{PS}$ is parallel to $\overleftrightarrow{WU}$ and answer true or false in (a)–(i)

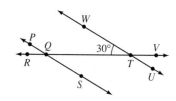

(a) $\overleftrightarrow{QT}$ is a transversal.

(b) ∠*VTU* is an interior angle.

(c) ∠*RQS* is an exterior angle.

(d) ∠*EFC* and ∠*ACF* are alternate interior angles.

(e) ∠*DFC* has the same measure as ∠*FCB*.

(f) ∠*HFE* has the same measure as ∠*FCB*.

(g) ∠*DFC* has measure 120°.

(h) ∠*ACG* has measure 120°.

(i) ∠*GCB* has measure 60°.

(d) ∠*WTV* and ∠*QTU* are alternate interior angles.

(e) ∠*UTQ* and ∠*PQT* have the same measure.

(f) ∠*VTU* and ∠*PQR* have the same measure.

(g) ∠*WTV* has measure 150°.

(h) ∠*RQS* has measure 30°.

(i) ∠*PQT* has measure 150°.

Answers: (a) true   (b) false
(c) true   (d) false   (e) true
(f) true   (g) true   (h) false
(i) true

A.1 EXERCISES A

Use the figure below to answer each question in Exercises 1–16.

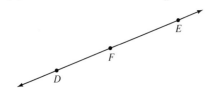

1. Is point *E* on $\overline{DF}$?

2. Is point *E* on $\overrightarrow{FD}$?

3. Is point *E* on $\overleftrightarrow{DF}$?

4. Is *E* on $\overrightarrow{DF}$?

5. What are the endpoints of $\overrightarrow{FE}$?

6. What are the endpoints of $\overleftrightarrow{FE}$?

7. What are the endpoints of $\overline{FE}$?

8. Are $\overline{DE}$ and $\overline{ED}$ the same?

9. Are $\overrightarrow{DE}$ and $\overrightarrow{ED}$ the same?

10. Are $\overleftrightarrow{DE}$ and $\overleftrightarrow{ED}$ the same?

11. Do $\overleftrightarrow{DE}$ and $\overleftrightarrow{DF}$ coincide?

12. What is the measure of ∠*DFE*?

13. What is the measure of ∠*EDF*?

14. What is the vertex of ∠*EFD*?

15. Is ∠*EFD* a right angle?

16 Are ∠*EDF* and ∠*DFE* supplementary?

Use the figure below to answer true *or* false *in Exercises 17–30. If the statement is false, tell why.*

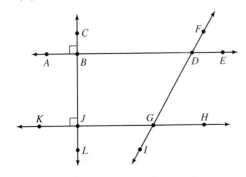

17. ∠*DBJ* is a right angle.

18. ∠*BDG* is an acute angle.

19. ∠*DGH* and ∠*JGI* are obtuse angles.

20. ∠*FDB* and ∠*EDG* have the same measure.

21. $\overleftrightarrow{AE}$ is parallel to $\overleftrightarrow{KH}$.

22. $\overleftrightarrow{CB}$ is perpendicular to $\overleftrightarrow{GH}$.

23. The measure of $\angle GJL$ is 90°.

24. $\overrightarrow{GD}$ and $\overrightarrow{JG}$ are the sides of $\angle JGD$.

25. $\angle IGJ$ and $\angle JGF$ are supplementary.

26. $\overline{BE}$ is parallel to $\overrightarrow{JK}$.

27. $\overleftrightarrow{IF}$ is a transversal which cuts parallel lines $\overleftrightarrow{AE}$ and $\overleftrightarrow{KH}$.

28. $\angle FDE$ and $\angle BDG$ are corresponding angles.

29. $\angle EDG$ and $\angle DGJ$ are alternate interior angles.

30. $\angle BDG$ and $\angle DGH$ have the same measure.

ANSWERS: 1. no 2. no 3. yes 4. yes 5. only the point F 6. There are no endpoints. 7. F and E 8. yes 9. no 10. yes 11. yes 12. 180° 13. 0° 14. F 15. no 16. yes 17. true 18. true 19. false (acute) 20. true 21. true 22. true 23. true 24. false ($\overrightarrow{GD}$ and $\overrightarrow{GJ}$) 25. true 26. true 27. true 28. false ($\angle FDE$ and $\angle DGH$) 29. true 30. true

A.1 EXERCISES B

Use the figure at the right to answer each question in Exercises 1–16.

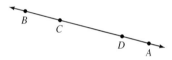

1. Is point C on $\overline{AB}$?

2. Is point B on $\overrightarrow{DC}$?

3. Is point A on $\overleftrightarrow{BD}$?

4. Is A on $\overrightarrow{DC}$?

5. What are the endpoints of $\overrightarrow{DC}$?

6. What are the endpoints of $\overleftrightarrow{DC}$?

7. What are the endpoints of $\overline{DC}$?

8. Are $\overline{BD}$ and $\overline{DB}$ the same?

9. Are $\overrightarrow{BD}$ and $\overrightarrow{DB}$ the same?

10. Are $\overleftrightarrow{BD}$ and $\overleftrightarrow{DB}$ the same?

11. Do $\overleftrightarrow{BA}$ and $\overleftrightarrow{DC}$ coincide?

12. What is the measure of $\angle DBC$?

13. What is the measure of $\angle DCB$?

14. What is the vertex of $\angle DCB$?

15. Is $\angle ADC$ a straight angle?

16. Are $\angle BCD$ and $\angle CBD$ complementary?

Use the figure below to answer true or false *in Exercises 17–30. If the statement is false, tell why.*

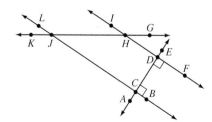

17. $\angle HDC$ is a straight angle.

18. $\angle JHD$ is an obtuse angle.

19. $\angle IHJ$ and $\angle GHD$ are vertical angles.

20. $\angle LJH$ and $\angle KJC$ have the same measure.

21. $\overleftrightarrow{IF}$ is parallel to $\overleftrightarrow{LJ}$.

22. $\overleftrightarrow{AC}$ is perpendicular to $\overleftrightarrow{IH}$.

23. The measure of $\angle JCA$ is 180°.

24. $\overrightarrow{HG}$ and $\overrightarrow{HF}$ are the sides of $\angle GHF$.

25. $\angle JHD$ and $\angle DHG$ are complementary.

26. $\overleftrightarrow{CL}$ is perpendicular to $\overleftrightarrow{IF}$.

27. $\overleftrightarrow{KG}$ is a transversal which cuts parallel lines $\overleftrightarrow{IF}$ and $\overleftrightarrow{LB}$.

28. $\angle GHD$ and $\angle HJC$ are corresponding angles.

29. $\angle DHJ$ and $\angle HJC$ are alternate interior angles.

30. $\angle IHJ$ and $\angle LJK$ have the same measure.

A.1 EXERCISES C

Use the figure below to answer the questions in Exercises 1–8. Lines $\overleftrightarrow{BG}$ and $\overleftrightarrow{MI}$ are parallel.

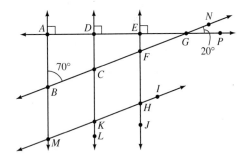

1. What is the measure of $\angle BCK$?

2. What is the measure of $\angle EFG$?

3. What is the measure of $\angle FHI$?

4. What is the measure of $\angle HKL$?

5. What is the measure of $\angle EGF$?

6. What is the sum of the measures of $\angle GEF$, $\angle EFG$, and $\angle FGE$?

7. What is the sum of the measures of $\angle GDC$, $\angle DCG$, and $\angle CGD$?

8. What is the sum of the measures of $\angle CBM$, $\angle BMK$, $\angle MKC$, and $\angle KCB$?

A.2 TRIANGLES

STUDENT GUIDEPOSTS

① Triangles	③ Congruent Triangles
② Classifying Triangles	④ Similar Triangles

① TRIANGLES

In this section we present an introduction to the study of triangles and their properties. Consider three noncollinear points A, B, and C and the corresponding segments $\overline{AB}$, $\overline{BC}$, and $\overline{CA}$, as shown in Figure A.14. The resulting geometric figure is a **triangle** with sides $\overline{AB}$, $\overline{BC}$, and $\overline{CA}$, **vertices** (plural of **vertex**) A, B, and C, and is denoted by ΔABC (Δ represents the word "triangle").

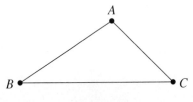

Figure A.14

② CLASSIFYING TRIANGLES

Triangles are often classified by their angles, as shown in Figure A.15. They can also be classified by their sides, as in Figure A.16. We can also describe triangles with a combination of these terms, as in Figure A.17.

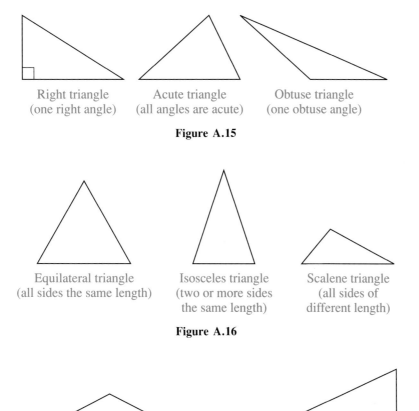

Right triangle
(one right angle)

Acute triangle
(all angles are acute)

Obtuse triangle
(one obtuse angle)

Figure A.15

Equilateral triangle
(all sides the same length)

Isosceles triangle
(two or more sides
the same length)

Scalene triangle
(all sides of
different length)

Figure A.16

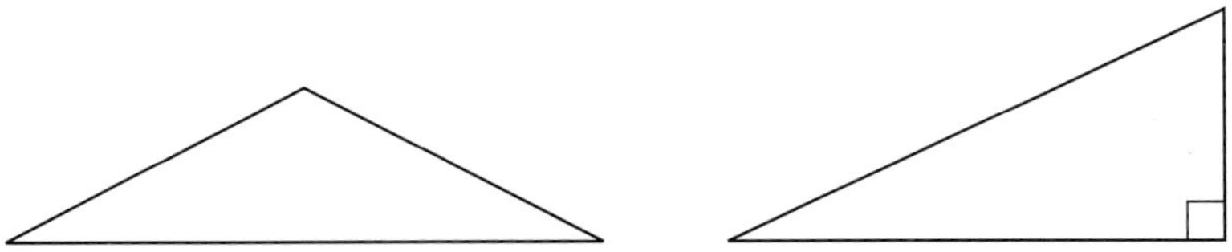

Obtuse isosceles triangle

Right scalene triangle

Figure A.17

In Chapter 9, after the study of radicals, we discuss some of the properties of isosceles right triangles and right triangles with acute angles of 30° and 60°. Here we restrict ourselves to some general properties of triangles.

Sum of Angles

The sum of the measures of the angles in any triangle is 180°.

EXAMPLE 1 **ANGLES OF TRIANGLE**

Consider $\triangle ABC$ in Figure A.18. If $\angle A$ has measure $35°$ and $\angle B$ has measure $70°$, find the measure of $\angle C$.

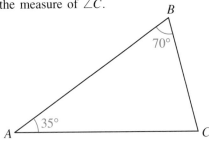

Figure A.18

The sum of the measures of $\angle A$ and $\angle B$ is $70° + 35° = 105°$. For the sum of the three measures to equal $180°$, $\angle C$ must measure $180° - 105° = 75°$.

In right $\triangle ABC$, one acute angle has measure $42°$. What is the measure of the other acute angle?

Answer: $48°$

❸ CONGRUENT TRIANGLES

Two triangles are **congruent** if they can be made to coincide by placing one on top of the other, either directly or by flipping one of them over. In Figure A.19, $\triangle ABC$ and $\triangle DEF$ are congruent, since $\triangle ABC$ could be placed on top of $\triangle DEF$. $\triangle ABC$ and $\triangle GHI$ are also congruent, since if $\triangle GHI$ were flipped over it could be made to coincide with $\triangle ABC$. On the other hand, $\triangle ABC$ and $\triangle JKL$ are not congruent, since they cannot be made to coincide.

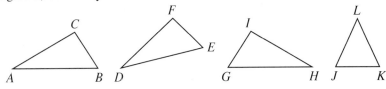

Figure A.19

More precisely, two triangles are congruent if the three sides of one triangle are equal in length to the three sides of the other, and the measures of the three angles of one are equal to the measures of the three angles of the other. Fortunately, in practical situations we do not need to establish equality of all six parts to show that two triangles are congruent. Three ways to show congruence are summarized as follows.

Congruent Triangles

Two triangles are congruent when one of the following can be shown.

1. Each of the three sides of one triangle is equal in length to a side of the other triangle (congruence by side-side-side or SSS).

2. Two sides of one triangle are equal in length to two sides of the other triangle, and the angle formed by these sides in one triangle is equal in measure to the angle formed by the corresponding sides in the other triangle (congruence by side-angle-side or SAS).

3. Two angles of one triangle are equal in measure to two angles of the other triangle, and the side between these two angles in one triangle is equal in length to the corresponding side in the other triangle (congruence by angle-side-angle or ASA).

EXAMPLE 2   CONGRUENT TRIANGLES

State why the triangles in Figure A.20 are congruent.

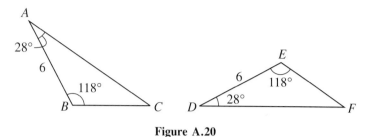

Figure A.20

Since the measures of $\overline{AB}$ and $\overline{DE}$ are both 6, and $\overline{AB}$ and $\overline{DE}$ are located between the corresponding equal angles, $\angle A$ and $\angle D$, and $\angle B$ and $\angle E$, $\triangle ABC$ is congruent to $\triangle DEF$ by ASA.

State why the given triangles are congruent.

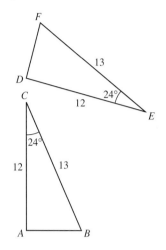

Answer: The triangles are congruent by SAS.

4 SIMILAR TRIANGLES

Congruent triangles have the same size and shape. Triangles that have the same shape but may differ in size are called **similar.** Congruent triangles are always similar, but similar triangles need not be congruent. For example, the triangles in Figure A.21 are similar but clearly not congruent.

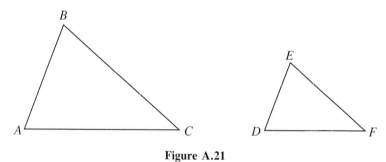

Figure A.21

To determine if two triangles are similar, we need compare only the angles.

Similar Triangles

Two triangles are similar when the angles of one are equal in measure to the angles of the other.

Since the sum of the angles of any triangle equals 180°, whenever two angles of one triangle are equal to two angles of a second, the remaining angles are also equal, making the triangles similar.

EXAMPLE 3 SIMILAR TRIANGLES	PRACTICE EXERCISE 3

State why the triangles in Figure A.22 are similar.

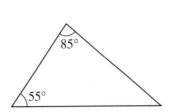

Figure A.22

State why the triangles are similar.

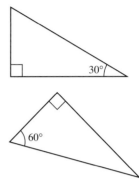

Since $85° + 40° = 125°$ and $180° - 125° = 55°$, the angles of the tri-angles are equal. Thus, the triangles are similar.

Answer: Both triangles are right triangles. Since $90° + 30° = 120°$ and $180° - 120° = 60°$, the angles of the triangles are equal.

In Section A.3 we will discuss perimeter and area of a triangle.

A.2   EXERCISES A

1. $\triangle ABC$ is given.

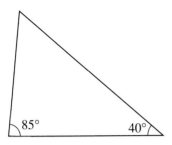

(a) What is the measure of $\angle C$?
(b) Classify the triangle by its angles.
(c) Classify the triangle by its sides.

2. $\triangle DEF$ is given.

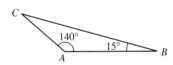

(a) What is the measure of $\angle F$?
(b) Classify the triangle by its angles.
(c) Classify the triangle by its sides.

Decide whether the given triangles are congruent. If so, state why.

3.

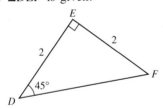

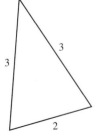

4.

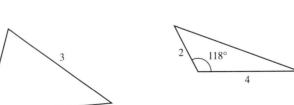

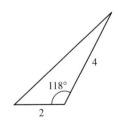

5.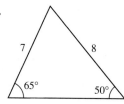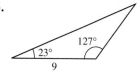

Decide whether the given triangles are similar.

7. 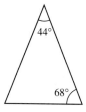   **8.**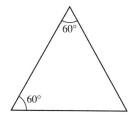

9. 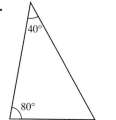   **10.**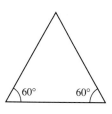

11. Is it possible for an isosceles triangle to be a right triangle?

12. Are all isosceles triangles equilateral?

FOR REVIEW

Answer true *or* false *in Exercises 13–16. If the statement is false, tell why.*

13. A straight angle has measure 180°.

14. Two angles whose measures total 180° are complementary.

15. Vertical angles have the same measure.

16. Alternate interior angles are always complementary.

ANSWERS: 1. (a) 25° (b) obtuse triangle (c) scalene triangle   2. (a) 45° (b) right triangle (c) isosceles triangle
3. congruent by SSS   4. congruent by SAS   5. The triangles are similar but not congruent.   6. congruent by ASA
7. They are similar.   8. They are similar.   9. They are not similar.   10. They are similar.   11. yes   12. no
13. true   14. false (supplementary)   15. true   16. false (equal)

A.2   EXERCISES B

1. $\triangle ABC$ is given.

(a) What is the measure of $\angle C$?
(b) Classify the triangle by its angles.
(c) Classify the triangle by its sides.

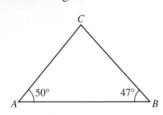

2. $\triangle DEF$ is given.

(a) What is the measure of $\angle F$?
(b) Classify the triangle by its angles.
(c) Classify the triangle by its sides.

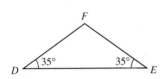

Decide whether the given triangles are congruent. If so, state why.

3.

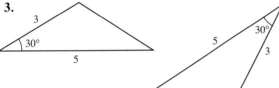

4.

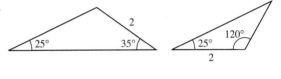

5.

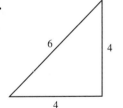

6.

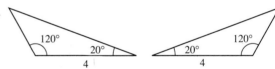

Decide whether the given triangles are similar.

7.

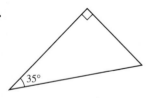

8.

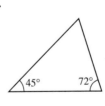
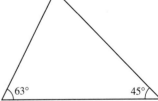

9.

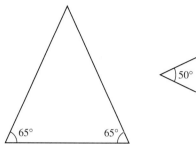

10.

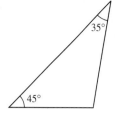

11. Is it possible for an equilateral triangle to be a right triangle?

12. Are all equilateral triangles isosceles?

FOR REVIEW

Answer true *or* false *in Exercises 13–16. If the statement is false, tell why.*

13. An acute angle has measure 90°.

14. Two angles whose measures total 360° are complementary.

15. Corresponding angles always have the same measure.

16. Interior angles on the same side of a transversal are supplementary.

A.2 EXERCISES C

Use the figure below to answer the questions in Exercises 1–6.

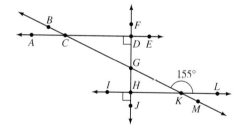

1. What is the measure of ∠GKH?

2. What is the measure of ∠DGC?

3. What is the measure of ∠BCA?

4. Is △DGC similar to △GHK?

5. Is △DGC congruent to △GHK?

6. What is the measure of ∠DGK?

A.3 PERIMETER AND AREA

1 RECTANGLES

In this section we review formulas for finding the perimeter and area of six geometric figures. These formulas provide problem-solving tools for numerous applications in algebra. We begin with the **rectangle,** a four-sided figure having four right angles and parallel opposite sides of equal length. Figure A.23 shows a rectangle with **length** l and **width** w.

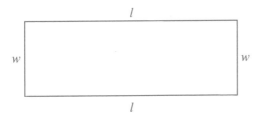

Rectangle

Figure A.23

The **perimeter** of a geometric figure is the distance around the figure. Thus, the perimeter of a rectangle is

$$l + w + l + w = l + l + w + w$$
$$= 2l + 2w.$$

Perimeter of a Rectangle
The perimeter P of a rectangle is twice the length plus twice the width.
$P = 2l + 2w$

The perimeter of the rectangle in Figure A.24 is given by

$$P = 2l + 2w$$
$$= 2(\textbf{5 cm}) + 2(\textbf{3 cm})$$
$$= 10 \text{ cm} + 6 \text{ cm} = 16 \text{ cm}.$$

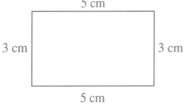

Figure A.24

EXAMPLE 1   PERIMETER OF RECTANGLE

Find the length of a fence needed to enclose a rectangular pasture that is 72 m wide and 90 m long.

    The perimeter of the pasture is the length of the fence needed.

$$P = 2l + 2w$$
$$= 2(90 \text{ m}) + 2(72 \text{ m})$$
$$= 180 \text{ m} + 144 \text{ m}$$
$$= 324 \text{ m}$$

Thus, 324 m of fence are needed.

PRACTICE EXERCISE 1

A farmer needs to fence a rectangular pasture which is 520 ft long and 470 ft wide. How long must the fence be?

Answer: **1980 ft**

2 SQUARES

A special type of rectangle, called a **square,** is shown in Figure A.25. All its sides have the same length. If s is the length of a side, then the perimeter of a square is

$$P = 2l + 2w$$
$$= 2s + 2s$$
$$= 4s.$$

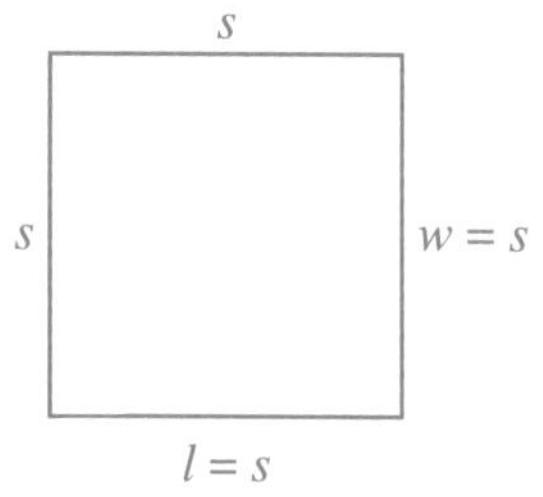

Figure A.25

Perimeter of a Square

The perimeter of a square is four times the length of a side.

$$P = 4s$$

EXAMPLE 2   PERIMETER OF A SQUARE

Find the perimeter of a square whose sides are 15.3 m. See Figure A.26.

$$P = 4s$$
$$= 4(\mathbf{15.3 \text{ m}})$$
$$= 61.2 \text{ m}$$

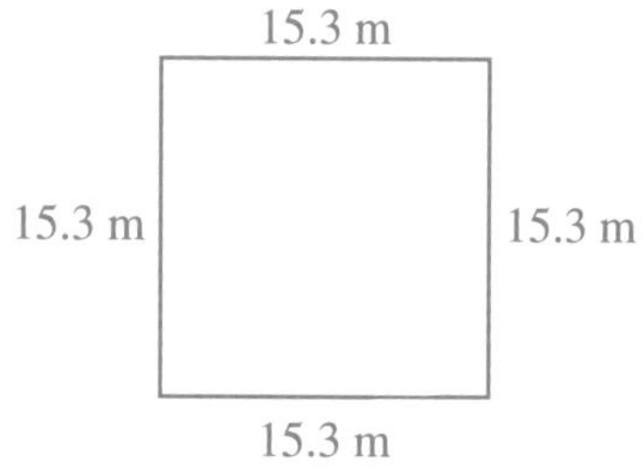

Figure A.26

PRACTICE EXERCISE 2

A square has side $3\frac{3}{4}$ ft. Find the perimeter.

Answer: **15 ft**

❸ SQUARE UNITS OF MEASURE FOR AREA

In Figure A.27, the area of a square which has side 1 cm is one **square centimeter** and is written 1 cm^2. Likewise, if the side is 1 in, the area is one **square inch,** 1 in^2. Similarly, if some other unit of length is used, the area is 1 square unit, or 1 unit^2.

Area = A = 1 square centimeter (sq cm)
        = 1 cm^2

Area = A = 1 square inch (sq in)
        = 1 in^2

Figure A.27

Areas of rectangles are given in square units. The rectangle in Figure A.28 has area 10 cm^2 since 10 of the 1-cm^2 squares are contained in it. In general, this is what is meant by *area*. Notice that 10 cm^2 can be found by multiplying 5 cm by 2 cm.

$$10 \text{ cm}^2 = 5 \text{ cm} \cdot 2 \text{ cm}$$

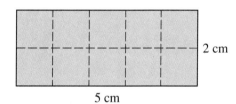

5 cm

Figure A.28

Area of a Rectangle

The area A of a rectangle is the length times the width.

$$A = lw$$

EXAMPLE 3 AREA OF A RECTANGLE

Find the area of the rectangle with width 6 m and length 12 m, shown in Figure A.29. (Sometimes we say "the rectangle that is 6 m by 12 m.")

$A = lw$
  $= \textbf{(12 m)(6 m)}$
  $= (12)(6) \text{ m}^2$
  $= 72 \text{ m}^2$

6 m

12 m

The area is 72 square meters.

Figure A.29

PRACTICE EXERCISE 3

Find the area of a rectangle that is 17 cm by 20 cm.

Since all sides of a square have the same measure s, the area of a square is

$$lw = ss = s^2.$$

Area of a Square

The area of a square is the square of a side.

$$A = s^2$$

EXAMPLE 4 AREA OF A COMPOSITE REGION

Find the area of the region in Figure A.30.

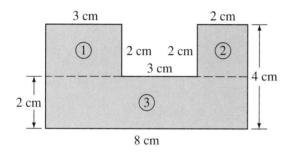

Figure A.30

The region can be divided into two rectangles and a square.

① is 2 cm by 3 cm   Area ① = 3 cm × 2 cm = 6 cm²
② is 2 cm by 2 cm   Area ② = 2 cm × 2 cm = 4 cm²
③ is 2 cm by 8 cm   Area ③ = 8 cm × 2 cm = 16 cm²

The total area is the sum of the three areas.

$$A = 6 \text{ cm}^2 + 4 \text{ cm}^2 + 16 \text{ cm}^2 = 26 \text{ cm}^2.$$

PRACTICE EXERCISE 4

Find the area of the region shown.

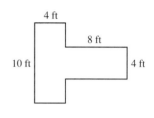

Answer: 72 ft²

---

④ PARALLELOGRAMS

A four-sided figure whose opposite sides are parallel is a **parallelogram.** A rectangle is a special case of this type of figure. Several parallelograms are shown in Figure A.31.

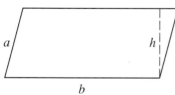

b = base a = side h = height

Parallelograms

Figure A.31

Perimeter of a Parallelogram

The perimeter of a parallelogram with side a and base b is given by

$$P = 2a + 2b.$$

EXAMPLE 5 PERIMETER OF A PARALLELOGRAM	PRACTICE EXERCISE 5

Find the perimeter of the parallelogram shown in Figure A.32.

$P = 2(10.5 \text{ m}) + 2(6.5 \text{ m})$

$= 34 \text{ m}$

6.5 cm

10.5 cm

Figure A.32

Find the perimeter of a parallelogram with base $7\frac{1}{2}$ ft and side $4\frac{1}{2}$ ft.

Answer: 24 ft

To discover how to find the area of a parallelogram, cut off one end and put it on the other end. This forms a rectangle (Figure A.33) with length $l = b$ and width $w = h$. Thus, the area of the parallelogram (which is the area of the rectangle) is the base times the height.

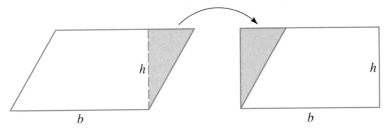

h h

b b

Figure A.33

Area of a Parallelogram

The area of a parallelogram with base b and height h is

$$A = bh.$$

EXAMPLE 6 AREA OF A PARALLELOGRAM	PRACTICE EXERCISE 6

Find the area of the parallelogram shown in Figure A.34, with height 15.2 cm and base 20.5 cm.

$A = bh$

$= (20.5 \text{ cm})(15.2 \text{ cm})$

$= (20.5)(15.2) \text{ cm}^2$

$= 311.6 \text{ cm}^2$

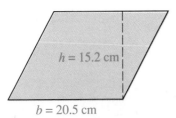

$h = 15.2 \text{ cm}$

$b = 20.5 \text{ cm}$

Figure A.34

Find the area of a parallelogram with base 10.8 mm and height 12.5 mm.

Answer: 135 mm^2

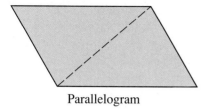

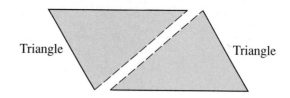

Figure A.35

⑤ TRIANGLES

If the parallelogram in Figure A.35 is cut along the dotted line, the area of each of the two triangles formed is one-half the area of the parallelogram. Thus, a triangle can be made into a parallelogram by adding to it a triangle of the same size. This is shown in Figure A.36, where triangle B is the same size and shape as the original triangle A. Since the heights of the triangle and the parallelogram are the same, the area of the triangle must be one-half the area of the parallelogram.

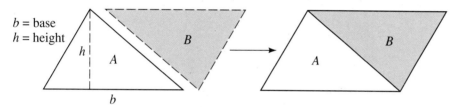

$b = $ base
$h = $ height

Figure A.36

Area of a Triangle

The area of a triangle with base b and height h is

$$A = \frac{1}{2}bh.$$

EXAMPLE 7 APPLICATION OF AREA

A wall to be painted is in the shape shown in Figure A.37.

(a) How many square feet must be painted?
First find the area of the rectangle.

$$A = lw = 10 \text{ ft} \cdot 8 \text{ ft} = 80 \text{ ft}^2$$

The base of the triangle is 10 ft and the height is 6 feet.

$$A = \frac{1}{2}bh = \frac{1}{2}(10 \text{ ft})(6 \text{ ft}) = 30 \text{ ft}^2$$

The total area to be covered is

$$80 \text{ ft}^2 + 30 \text{ ft}^2 = 110 \text{ ft}^2.$$

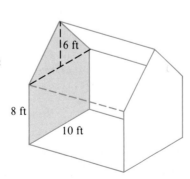

Figure A.37

PRACTICE EXERCISE 7

The wall shown in the figure is to be painted.

1.5 m

4 m

8 m

(a) What is the area to be painted?

(b) How much paint will it take if one gallon covers 220 ft²? Multiply 110 ft² by $\dfrac{1 \text{ gal}}{220 \text{ ft}^2}$.

$$\frac{110 \text{ ft}^2}{1} \cdot \frac{1 \text{ gal}}{220 \text{ ft}^2} = \frac{110 \text{ gal}}{2(110)} = \frac{1}{2} \text{ gal}$$

It will take $\frac{1}{2}$ gallon to cover the area.

(b) If one liter of paint covers 19 m², how many liters of paint are needed?

Answers: (a) 38 m²   (b) 2 L

⑥ CIRCLES

A **circle** is a figure consisting of all points located the same distance r from a fixed point O called its **center.** In Figure A.38 the segment $\overline{OA}$ or r is a **radius** of the circle. The segment $\overline{DE}$, with length d, is a **diameter** of the circle. The distance around the circle is the **circumference** C of the circle. Since the diameter is twice the radius, we can write

$$d = 2r \quad \text{and} \quad r = \frac{d}{2}.$$

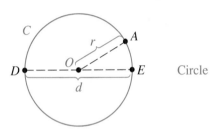

Circle

Figure A.38

EXAMPLE 8 DIAMETER OF A CIRCLE

Find the diameter of the circle with radius 2.3 cm, shown in Figure A.39.

$$d = 2r = 2(\textbf{2.3 cm})$$
$$= 4.6 \text{ cm}$$

Figure A.39

PRACTICE EXERCISE 8

Find the radius of a circle with diameter 22.4 meters.

Answer: 11.2 m

If the circumference, C, of any circle is divided by its diameter, d, the result is a constant called π (the Greek letter *pi*). Suppose, for example, that we measure the circle formed by the top of a can. If the circumference C is 6.2 inches, and the diameter d is 2.0 inches, then

$$\frac{C}{d} = \frac{6.2 \text{ in}}{2.0 \text{ in}} = 3.1.$$

For any circle measured, $\frac{C}{d}$ will always be close to 3.1, because $\frac{C}{d}$ is always equal to π. This number has been calculated to many decimal places, but it is usually approximated as 3.14 or $\frac{22}{7}$.

Since $\frac{C}{d} = \pi$, we have $C = \pi d$.

Circumference of a Circle

The circumference of a circle with diameter d and radius r is

$$C = \pi d = 2\pi r.$$

EXAMPLE 9   CIRCUMFERENCE OF A CIRCLE

Find the circumference of a circle with diameter 32 mm. Use 3.14 for π to give an approximation of the actual value.

$$C = \pi d = \pi(32 \text{ mm}) = 32\pi \text{ mm}$$

The approximate value of C is

$$C \approx 3.14(32 \text{ mm})$$
$$\approx 100.5 \text{ mm}. \qquad \text{Rounded to the nearest tenth}$$

PRACTICE EXERCISE 9

Find the circumference of a circle with radius 63 inches. Use $\frac{22}{7}$ for π.

Answer: 396 in

Area of a Circle

The area A of a circle is given by

$$A = \pi r^2.$$

EXAMPLE 10   APPLICATION OF AREA OF A CIRCLE

In the machine part shown in Figure A.40, each circular hole has radius 3 cm. Find the area of metal which is left. Use 3.14 for π.

First calculate the area of the rectangle.

$$A = lw$$
$$= (16 \text{ cm})(14 \text{ cm})$$
$$= 224 \text{ cm}^2$$

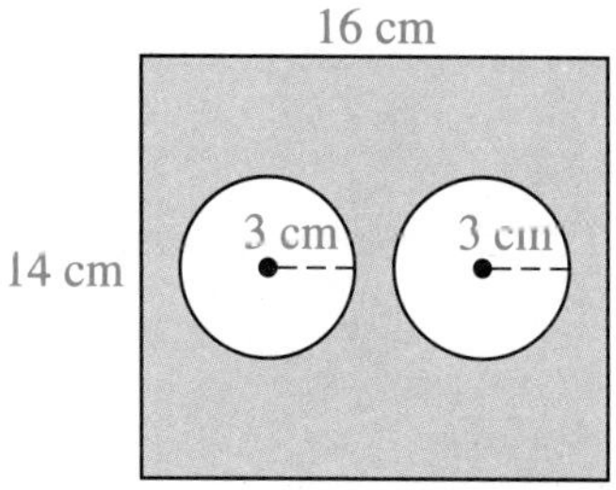

Figure A.40

Now find the area of each circle.

$$A = \pi r^2 \approx 3.14(3 \text{ cm})^2$$
$$= (3.14)(9) \text{ cm}^2 = 28.26 \text{ cm}^2$$

Thus, the two circles have a combined area of

$$2(28.26 \text{ cm}^2) = 56.52 \text{ cm}^2.$$

The area of metal is the area of the rectangle minus the area of the circles.

$$\begin{array}{r} 224.00 \\ -56.52 \\ \hline 167.48 \end{array}$$

The area left is 167.48 cm^2.

PRACTICE EXERCISE 10

A triangle with base 10 inches and height 5 inches has three holes of radius 1 inch drilled through it. What is the area left in the triangle after the holes are drilled? Use 3.14 for π.

Answer: 15.58 in^2

A.3 EXERCISES A

Find the perimeter of the rectangle (or square) with the given width and length.

1. 6 ft by 11 ft

2. 2.5 m by 3.6 m

3. 102 mm by 30.5 mm

4. 6.8 mi by $9\frac{1}{2}$ mi

5. 7.2 cm by 7.2 cm

6. $22\frac{1}{2}$ mi by $22\frac{1}{2}$ mi

Find the area of the rectangle (or square) with the given width and length.

7. 6 ft by 11 ft

8. 2.5 m by 3.6 m

9. 102 mm by 30.5 mm

10. 6.8 mi by $9\frac{1}{2}$ mi

11. 7.2 cm by 7.2 cm

12. $22\frac{1}{2}$ mi by $22\frac{1}{2}$ mi

Solve.

13. Find the perimeter and area of the given figure.

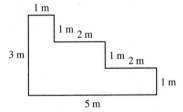

14. Fencing costs $3.20 per yard. If a pasture 62 yd by 205 yd is to be fenced, how much will it cost?

15. A rose garden is 7.2 m by 10.8 m. If a square with side 2.5 m is used in the middle of the garden for a fountain, what area remains for roses?

16 The area shown is to be carpeted. How much will it cost if carpet costs $16.50 per square yard?

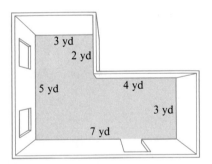

Find the area of each figure.

17.

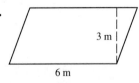

18.

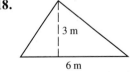

19.

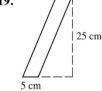

20.

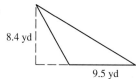

21.

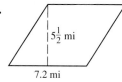

22.

Find the perimeter of each figure.

23.

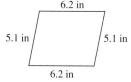

24.

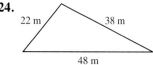

25 Find the area of the figure.

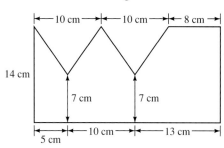

26. The recreation area sketched in the figure is to be carpeted. How much will it cost if carpet is $14.50 per square yard?

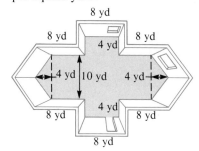

Find the diameter of the circle with the given radius.

27. $r = 11$ in

28. $r = 5.85$ cm

29. $r = \dfrac{3}{4}$ ft

Find the radius of the circle with the given diameter.

30. $d = 24$ m

31. $d = \dfrac{5}{2}$ in

32. $d = 7.24$ mm

Find the circumference and area of the circle. Leave the answer using π.

33. $r = 7$ yd

34. $r = \dfrac{7}{2}$ ft

35 $r = 7.7$ mi

Find the circumference and area of the circle. Use 3.14 for π.

36. $r = 6.8$ km

37. $d = 92$ in

38 $r = 0.05$ cm

Solve.

39 How much more area is there for water to pass through in a $\frac{3}{4}$-in diameter water hose than there is in a $\frac{1}{2}$-in diameter water hose? Use 3.14 for π.

40. A 12-inch diameter pizza costs \$4.50. A 16-inch diameter pizza costs \$7.50. Which pizza costs less per square inch?

FOR REVIEW

41. Can an isosceles triangle be obtuse?

42. If two triangles are similar, can one be acute and the other be obtuse?

ANSWERS: (*Some answers are rounded.*) 1. 34 ft   2. 12.2 m   3. 265 mm   4. 32.6 mi   5. 28.8 cm   6. 90 mi
7. 66 ft^2 8. 9.0 m^2 9. 3111 mm^2 10. 64.6 mi^2 11. 51.84 cm^2 12. 506.25 mi^2 13. 16 m; 9 m^2 14. \$1708.80
15. 71.51 m^2 16. \$445.50   17. 18 m^2 18. 9 m^2 19. 125 cm^2 20. 39.9 yd^2 21. 39.6 mi^2 22. 27.2 ft^2 23. 22.6 in
24. 108 m   25. 322 cm^2 26. \$4988   27. 22 in   28. 11.7 cm   29. $\frac{3}{2}$ ft   30. 12 m   31. $\frac{5}{4}$ in   32. 3.62 mm
33. 14π yd; 49π yd^2 34. 7π ft; $\frac{49}{4}\pi$ ft^2 35. 15.4π mi; 59.29π mi^2 36. 42.7 km; 145.2 km^2
37. 288.9 in; 6644.2 in^2 38. 0.314 cm; 0.00785 cm^2 39. 0.246 in^2 40. 12-in: \$0.0398 per in^2; 16-in: \$0.0373 per in^2;
16-in pizza costs less   41. yes   42. no

A.3   EXERCISES B

Find the perimeter of the rectangle (or square) with the given width and length.

1. 10 ft by 17 ft

2. 8.2 m by 9.5 m

3. 2500 mm by 45.5 mm

4. 2.2 mi by $6\frac{1}{2}$ mi

5. 40.2 cm by 40.2 cm

6. $10\frac{1}{2}$ mi by $10\frac{1}{2}$ mi

Find the area of the rectangle (or square) with the given width and length.

7. 10 ft by 17 ft

8. 8.2 m by 9.5 m

9. 2500 mm by 45.5 mm

10. 2.2 mi by $6\frac{1}{2}$ mi

11. 40.2 cm by 40.2 cm

12. $10\frac{1}{2}$ mi by $10\frac{1}{2}$ mi

Solve.

13. The play area shown needs to have a new fence around it. How long will the fence need to be?

14. A rancher needs to fence an area that is 2500 yd by 4200 yd. If fencing costs \$2.60 per yard, how much will it cost?

15. A flower garden is in the shape of a square which is 22.5 m on a side. If a walk 1.5 m wide is put across the garden parallel to one of the sides, how much area is left for flowers?

16. The area shown is covered with carpeting costing $24.50 per square yard. How much did it cost?

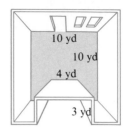

10 yd
10 yd
4 yd
3 yd

Find the area of each figure.

17.

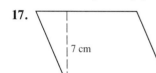

7 cm
10 cm

18.

7 cm
10 cm

19.

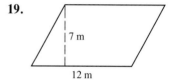

7 m
12 m

20.

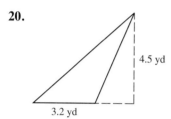

4.5 yd
3.2 yd

21.

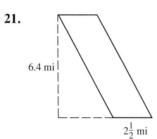

6.4 mi
$2\frac{1}{2}$ mi

22.

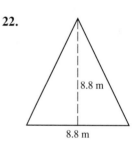

8.8 m
8.8 m

Find the perimeter of each figure.

23.

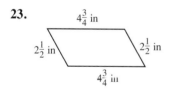

$4\frac{3}{4}$ in
$2\frac{1}{2}$ in $2\frac{1}{2}$ in
$4\frac{3}{4}$ in

24.

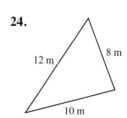

8 m
12 m
10 m

25. Find the area of the figure.

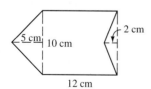

5 cm   10 cm   2 cm
12 cm

26. The patio area shown is to be covered with outdoor carpeting. What will be the total cost if the carpeting can be installed for $22.50 per square meter?

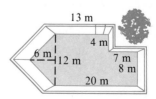

13 m
4 m
6 m   12 m   7 m
8 m
20 m

Find the diameter of the circle with the given radius.

27. $r = 28$ cm

28. $r = 2.75$ ft

29. $r = \dfrac{7}{6}$ in

Find the radius of the circle with the given diameter.

30. $d = 120$ mm

31. $d = \dfrac{2}{3}$ in

32. $d = 28.6$ m

Find the circumference and area of the circle. Leave the answer using π.

33. $r = 14$ m

34. $r = \dfrac{14}{11}$ cm

35. $r = 5.67$ yd

Find the circumference and area of the circle. Use 3.14 for π.

36. $d = 32.8$ ft

37. $r = 420$ mm

38. $d = 1.64$ in

Solve.

39. A circular garden has radius 9 m. If a 1-m-wide circular walk is put around it, what is the area of the walk? Use 3.14 for π. [*Hint:* Find the area of a circle with 10-m radius and subtract the area of the garden.]

40. A 14-inch diameter pizza costs $6.00 and a 16-inch diameter pizza costs $8.00. Which pizza costs less per square inch? Use 3.14 for π.

FOR REVIEW

41. Can an isosceles triangle and a right triangle be congruent?

42. If two angles of a triangle are 45° and 55°, is the triangle a right triangle?

A.3 EXERCISES C

Solve.

1. A patio is in the shape of a trapezoid with bases 8.1 yd and 6.7 yd and height 5.8 yd. Assuming no waste, to the nearest cent how much will it cost to cover the patio with outdoor carpeting which costs $2.15 a square foot? ($A = \frac{1}{2}(b_1 + b_2)\, h$ where b_1 and b_2 are bases and h is the height.)
[Answer: $830.50]

2. A sheet of metal is in the shape of a triangle with base 1500 cm and height 800 cm, and has a piece removed from the center of it in the shape of a parallelogram with base 125 cm and height 85 cm. How much will it cost to cover the metal on both sides with rustproofing paint which costs $0.45 per square meter?

3. Find the area (to the nearest hundredth) of the metal left on the machine part shown which has three circular holes drilled in it. Use 3.14 for π.
[Answer: 1257.77 cm²]

4. Find the exterior perimeter (to the nearest tenth) of the machine part in Exercise 3.

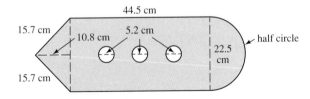

A.4 VOLUME AND SURFACE AREA

═══════════════ STUDENT GUIDEPOSTS ═══════════════

1 Volume   **4** Surface Area
2 Rectangular Solids   **5** Cylinders
3 Cubes   **6** Spheres

1 VOLUME

To measure length we use units like 1 cm and 1 inch. When we talk about area we use square units like 1 cm^2 and 1 in^2. To find volume we will use cubic units like 1 cm^3 and 1 in^3.

The **volume** of a cube that is 1 cm by 1 cm by 1 cm is one **cubic centimeter** and is written 1 cm^3. Similarly, a cube that is 1 in by 1 in by 1 in is one **cubic inch,** 1 in^3. See Figure A.41. If some other unit were used for measuring, the volume of the cube would be 1 cubic unit, or 1 unit^3.

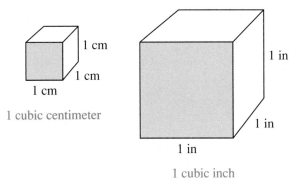

1 cubic centimeter

1 cubic inch

Figure A.41

2 RECTANGULAR SOLIDS

The cubic unit of measure is used for the volume of solids. The **rectangular solid** in Figure A.42(a) has volume 20 cm^3 since there are 10 cm^3 on the bottom layer and 10 cm^3 on the top layer. In Figure A.42(b) there is one more layer (10 cm^3) than in Figure A.42(a), so the volume is 30 cm^3. To find these volumes without a diagram, multiply the area of the base by the height.

$$20 \text{ cm}^3 = \underbrace{5 \text{ cm} \cdot 2 \text{ cm}}_{\text{area of base}} \cdot \underbrace{2 \text{ cm}}_{\text{height}}$$

$$30 \text{ cm}^3 = \underbrace{5 \text{ cm} \cdot 2 \text{ cm}}_{\text{area of base}} \cdot \underbrace{3 \text{ cm}}_{\text{height}}$$

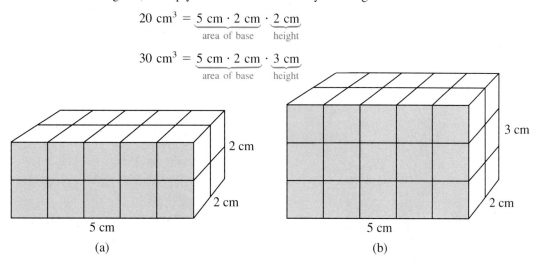

(a)   (b)

Figure A.42

Volume of a Rectangular Solid

The volume V of a rectangular solid is the area of the base lw times the height h.

$$V = lwh$$

EXAMPLE 1   VOLUME OF A RECTANGULAR SOLID

A highway construction truck has a bed which is 5.5 yd long, 2.5 yd wide, and 2 yd deep. How many cubic yards of gravel will it hold?
   Find the volume of the truck bed.

$$V = lwh$$
$$= (5.5 \text{ yd})(2.5 \text{ yd})(2 \text{ yd})$$
$$= (5.5)(2.5)(2) \text{ yd}^3$$
$$= 27.5 \text{ yd}^3$$

The truck will hold 27.5 yd³ of gravel.

PRACTICE EXERCISE 1

A trailer for hauling cotton is 18 ft by 8 ft by 10 ft. What volume of cotton will it hold?

Answer: 1440 ft³

---

❸ CUBES

A **cube** is a rectangular solid for which $l = w = h$. We talk about *edges* of the cube, and call the length of each edge e.

Volume of a Cube

The volume V of a cube is the cube of the edge.

$$V = e^3 \quad (e^3 \text{ is } e \text{ times } e \text{ times } e)$$

EXAMPLE 2   VOLUME OF A CUBE

Find the volume of the cube in Figure A.43 that has an edge of 1.5 ft.

$$V = (\textbf{1.5 ft})^3$$
$$= (1.5 \text{ ft})(1.5 \text{ ft})(1.5 \text{ ft})$$
$$= 3.375 \text{ ft}^3$$

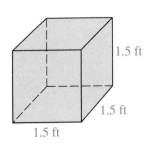

1.5 ft

1.5 ft

1.5 ft

Figure A.43

PRACTICE EXERCISE 2

Find the volume of a cube which has an edge of $3\frac{1}{2}$ inches.

Answer: $42\frac{7}{8}$ in³

---

❹ SURFACE AREA

Another problem involving a rectangular solid is finding the area of the surface. If we can do this, we can find the amount of material needed to make a rectangular tank or the amount of paint needed to paint a building. The **surface area** of an object is the area of all its surfaces. Consider the solid in Figure A.44, which was used earlier to count cubic centimeters. Now, we count square centimeters on its surface. There are

$$(5 \text{ cm})(3 \text{ cm}) = 15 \text{ cm}^2$$

on the front face. But the back is just like the front, so there are

$$2(15 \text{ cm}^2) = 30 \text{ cm}^2 \qquad 2lh$$

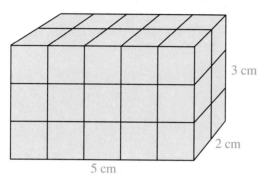

Figure A.44

on the front and back together. On the two ends there are

$$2(2 \text{ cm})(3 \text{ cm}) = 12 \text{ cm}^2. \qquad 2wh$$

On the bottom there are

$$(5 \text{ cm})(2 \text{ cm}) = 10 \text{ cm}^2,$$

and the same number on the top. The total of the top and bottom is

$$2(10 \text{ cm}^2) = 20 \text{ cm}^2. \qquad 2lw$$

The total surface area is the sum of all surfaces.

Area of front and back:	30 cm^2	2*lh*
Area of two ends:	12 cm^2	2*wh*
Area of bottom and top:	20 cm^2	2*lw*
Total surface area:	62 cm^2	

Surface Area of a Rectangular Solid

The total surface area A of a rectangular solid is

$$A = 2lh + 2wh + 2lw.$$

EXAMPLE 3   SURFACE AREA OF A RECTANGULAR SOLID

Find the total surface area of a rectangular solid 6 m by 3 m by 7 m.

$$A = 2lh + 2wh + 2lw$$
$$= 2(6 \text{ m})(7 \text{ m}) + 2(3 \text{ m})(7 \text{ m}) + 2(6 \text{ m})(3 \text{ m})$$
$$= 2(6)(7)\text{m}^2 + 2(3)(7)\text{m}^2 + 2(6)(3)\text{m}^2$$
$$= 84 \text{ m}^2 + 42 \text{ m}^2 + 36 \text{ m}^2 = 162 \text{ m}^2$$

PRACTICE EXERCISE 3

Find the surface area of a rectangular solid 12 ft by 16 ft by 20 ft.

Answer: 1504 ft^2

| EXAMPLE 4 APPLICATION OF SURFACE AREA | PRACTICE EXERCISE 4 |

The four vertical sides of a storage building are to be painted. One gallon of paint will cover 150 ft^2 of area. How much paint will be used if the building is 16 ft by 12 ft by 8 ft?

The area of the front and back is twice the length times the height.

$$2(16 \text{ ft})(8 \text{ ft}) = 256 \text{ ft}^2$$

The area of the two ends is twice the width times the height.

$$2(12 \text{ ft})(8 \text{ ft}) = 192 \text{ ft}^2$$

The total area is the sum of these two areas.

Area of front and back: 256 ft^2
Area of two ends: 192 ft^2
Total area: 448 ft^2

Since the paint covers 150 ft^2 per gal, the unit fraction is $\dfrac{1 \text{ gal}}{150 \text{ ft}^2}$.

$$\frac{448 \text{ ft}^2}{1} \times \frac{1 \text{ gal}}{150 \text{ ft}^2} = \frac{448}{150} \text{ gal}$$

$$\approx 2.99 \text{ gal} \quad \text{To the nearest hundredth}$$

It will take about 3 gal to paint the building.

The outside walls of a workshop are to be covered with siding that costs $1.50 per square foot. How much will it cost if the building is 16 ft wide, 18 ft long, and 8 ft high?

Answer: $816

Surface Area of a Cube

The surface area of a cube with edge e is

$$A = 6e^2.$$

| EXAMPLE 5 SURFACE AREA OF A CUBE | PRACTICE EXERCISE 5 |

Find the surface area of a cube with edge 14 cm.

$$A = 6e^2$$
$$= 6(14 \text{ cm})^2$$
$$= 6(196) \text{ cm}^2$$
$$= 1176 \text{ cm}^2$$

Find the surface area of a cube with edge 200 mm.

Answer: 240,000 mm^2

---

5 CYLINDERS

A **cylinder** (see Figure A.45) is a solid with circular ends of the same radius. For example, a can of beans is in the shape of a cylinder.

Remember that the volume of a rectangular solid is the area of the base times the height. The same is true for a cylinder. Since the base is a circle with area πr^2, we have the following formula.

Volume of a Cylinder

The volume V of a cylinder is the area of the base, πr^2, times the height, h.

$$V = \pi r^2 h$$

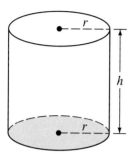

Figure A.45

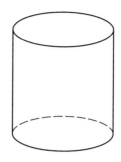

Figure A.46

EXAMPLE 6 VOLUME OF A CYLINDER

Find the volume of the cylinder in Figure A.46 which is 8 cm high and has a base of radius 3 cm. Use 3.14 for π.

Since $r = 3$ cm and $h = 8$ cm,

$$\begin{aligned}
V &= \pi r^2 h \\
&\approx 3.14(\quad)^2(\quad) \\
&= (3.14)(9)(8) \text{ cm}^3 = 226.08 \text{ cm}^3.
\end{aligned}$$

PRACTICE EXERCISE 6

Find the volume of a cylinder with radius of the base 6 cm and height 20 cm. Use 3.14 for π.

Answer: 2260.8 cm^3

EXAMPLE 7 APPLICATION OF VOLUME

A can of blueberry pie filling has diameter 3 in and height 4.5 in. How many cans of filling are needed to fill a 9-in diameter pie pan 1 in deep?

First find the volume of pie filling in one can.

$$\begin{aligned}
V &= \pi r^2 h \\
&\approx 3.14(1.5 \text{ in})^2(4.5 \text{ in}) \\
&= (3.14)(2.25)(4.5) \text{ in}^3 \\
&\approx 31.8 \text{ in}^3
\end{aligned}$$

Now find the volume of pie filling in a 9-in diameter (4.5-in radius) pie pan.

$$\begin{aligned}
V &= \pi r^2 h \\
&\approx 3.14(4.5 \text{ in})^2(\quad) \\
&= (3.14)(20.25)(1) \text{ in}^3 \\
&\approx 63.6 \text{ in}^3
\end{aligned}$$

Since one can has 31.8 in^3 of filling, multiply by the unit fraction $\dfrac{1 \text{ can}}{31.8 \text{ in}^3}$.

$$\frac{63.6 \text{ in}^3}{1} \times \frac{1 \text{ can}}{31.8 \text{ in}^3} = \frac{63.6}{31.8} \text{ can} = 2 \text{ cans}$$

It will take 2 cans of filling for the pie.

PRACTICE EXERCISE 7

Water is stored in a cylindrical tank with diameter of the base 12 m and height 10 m. How many water trucks having cylindrical tanks with diameter 1.6 m and length 5 m can be filled from the storage tank?

Answer: 112.5

To see how to find the surface area of a cylinder, consider a can without top or bottom. See Figure A.47(a). If it is cut along the seam and pressed flat, as in Figure A.47(b), a rectangle is formed. The length of the rectangle is the circumference of the circle and the width is the height. Thus, the surface area of the side of the can is $2\pi rh$.

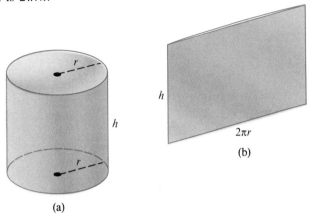

(a)

(b)

Figure A.47

Since the area of the top of the can is πr^2, and the area of the bottom is also πr^2, we have the following.

Surface Area of a Cylinder

The surface area A of a cylinder is

$$A = 2\pi rh + 2\pi r^2.$$

EXAMPLE 8   SURFACE AREA OF A CYLINDER

Find the surface area of a cylinder with radius 7 ft and height 9 ft.

$$A = 2\pi rh + 2\pi r^2$$
$$\approx 2(3.14)(7\text{ ft})(9\text{ ft}) + 2(3.14)(7\text{ ft})^2$$
$$= 2(3.14)(7)(9)\text{ ft}^2 + 2(3.14)(49)\text{ ft}^2$$
$$= 703.36\text{ ft}^2$$

PRACTICE EXERCISE 8

Find the surface area of a cylinder with radius 16 ft and height 4 ft.

Answer: 2009.6 ft^2

⑥ SPHERES

A **sphere** (Figure A.48) is a solid in the shape of a ball. The radius of a sphere is the distance from its center to the surface. The diameter is twice the radius. If a sphere is cut through the center, the cut surface is a circle.

Sphere

Figure A.48

Figure A.49

Volume of a Sphere

The volume V of a sphere is

$$V = \frac{4}{3}\pi r^3.$$

EXAMPLE 9   VOLUME OF A SPHERE

Find the volume of the sphere in Figure A.49 with diameter 18 cm.
If the diameter is 18 cm, then the radius r is $\frac{18}{2}$ cm or 9 cm.

$$V = \frac{4}{3}\pi r^3 \approx \frac{4}{3}(3.14)(9 \text{ cm})^3$$

$$= \frac{4}{3}(3.14)729 \text{ cm}^3$$

$$= 3052.08 \text{ cm}^3$$

PRACTICE EXERCISE 9

Finding the volume of a sphere with diameter 10 ft.

Answer: 523.3 ft^3

Surface Area of a Sphere

The surface area A of a sphere is

$$A = 4\pi r^2.$$

EXAMPLE 10   APPLICATION OF SURFACE AREA

A spherical tank with radius 3.2 m is to be covered with insulation. If the insulation costs \$2.50 per square meter, how much will the material cost?

$$A = 4\pi r^2$$

$$\approx 4(3.14)(3.2 \text{ m})^2$$

$$= 4(3.14)(10.24) \text{ m}^2$$

$$\approx 128.61 \text{ m}^2 \quad \text{To the nearest hundredth}$$

Since the insulation costs \$2.50 per square meter, the cost is

$$\frac{128.61 \text{ m}^2}{1} \times \frac{\$2.50}{1 \text{ m}^2} \approx \$321.53. \quad \text{To the nearest cent}$$

PRACTICE EXERCISE 10

A spherical tank with diameter 6.4 m is to be prepared for painting. If the labor costs are \$4.50 per square meter, how much will the job cost?

Answer: \$578.76

A.4 EXERCISES A

Find the volume of the rectangular solid.

1. 10 m by 5 m by 6 m

2. 2.3 cm by 1.2 cm by 4.5 cm

3 $9\frac{1}{2}$ in by 3.4 in by $8\frac{3}{4}$ in

Find the volume of the cube with the given edge length.

4. 8 m

5. 10.2 cm

6. $1\frac{3}{5}$ in

Find the total surface area of each rectangular solid.

7. 6 ft by 2 ft by 5 ft

8. $1\frac{1}{2}$ in by $\frac{1}{2}$ in by 4 in

9. 10.2 yd by $8\frac{1}{2}$ yd by 4.5 yd

10 15.5 cm by 6.4 cm by 20.2 cm

Find the surface area of the cube with the given edge.

11. 5 ft

12. 21 mm

13. $1\frac{1}{2}$ yd

Solve.

14. Frank's truck is to be used to carry topsoil. If the truck bed is 2 yd by 1.5 yd by 0.8 yd, how many cubic yards of topsoil will it hold?

15 A tank is in the shape of a cube. How many grams of water will it hold if the edge of the cube is 22 cm? Note that 1 cm^3 of water weighs 1 g.

16 How many square centimeters of glass did it take to make the bottom and sides of the tank in Exercise 15?

17. Elizabeth owns a rectangular building that is 30 ft by 12 ft by 8 ft. How much will it cost to put siding on the building if the siding costs $0.75 per square foot?

Find the volume of each cylinder.

18. $r = 6$ cm, $h = 5$ cm

19 $d = 24.4$ in, $h = 30$ in

20. $d = 11.4$ m, $h = 4.4$ m

Find the surface area of each cylinder.

21. $r = 6$ cm, $h = 5$ cm

22 $d = 24.4$ in, $h = 30$ in

23. $d = 11.4$ m, $h = 4.4$ m

Find the volume of each sphere.

24. $r = 12$ mm

25 $r = \dfrac{3}{4}$ in

26. $d = 5.6$ yd

Find the surface area of each sphere.

27. $r = 12$ mm

28. $r = \dfrac{3}{4}$ in

29 $d = 5.6$ yd

Solve.

30 A can of cherry pie filling has diameter 8 cm and height 12 cm. How many cans (to the nearest tenth) are needed to fill a 20-cm diameter pie pan 3 cm deep?

31. A cylindrical storage tank has radius 3.8 ft and height 9.8 ft. How many gallons (to the nearest tenth) of paint are needed to paint the tank (including top and bottom) if one gallon covers 150 ft²?

32. A spherical tank has radius 8.6 m. If a rust-preventing material costs $1.50 per square meter, what is the cost to rustproof the tank?

33. A spherical tank with radius 32 cm is filled with gasoline. How many liters are in the tank? Note that 1 cm³ = 0.001 L.

FOR REVIEW

Find the area of metal left on each machine part shown which has circular holes drilled in it. Use 3.14 for π.

34.

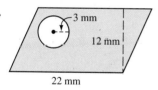

35

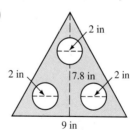

A.4 EXERCISES B

Find the volume of the rectangular solid.

1. 6 m by 12 m by 15 m

2. 9.2 cm by 1.5 cm by 7.5 cm

3. $6\frac{1}{2}$ in by 5.5 in by $3\frac{1}{5}$ in

Find the volume of the cube with the given edge length.

4. 12 m

5. 0.1 cm

6. $2\frac{1}{6}$ in

Find the total surface area of each rectangular solid.

7. 3 ft by 5 ft by 12 ft

8. $4\frac{1}{2}$ in by $\frac{1}{4}$ in by 6 in

9. 8.6 yd by $4\frac{1}{5}$ yd by 7.5 yd

10. 12.2 cm by 100 cm by 6.5 cm

Find the surface area of the cube with the given edge.

11. 17 ft

12. 0.01 mm

13. $4\frac{1}{6}$ yd

Solve.

14. A large earth mover has a bed which is 5 yd by 4 yd by 10 yd. How many cubic yards of dirt will it hold?

15. A water tank is 22.5 cm by 10 cm by 5.2 cm. How many grams of water will it hold? Note that 1 cm^3 of water weighs 1 g.

16. A box which is a cube with edge 8.5 dm is to be covered with foil. How much foil is required?

17. A rectangular tank 4 m by 6 m by 5 m (5 m is the height) needs to be insulated on the sides and top. If insulation costs $2.15 per square meter, how much will it cost?

Find the volume of each cylinder.

18. $r = 9$ cm, $h = 15$ cm

19. $d = 6.8$ m, $h = 10.5$ m

20. $d = 20.4$ in, $h = 8.5$ in

Find the surface area of each cylinder.

21. $r = 9$ cm, $h = 15$ cm

22. $d = 6.8$ m, $h = 10.5$ m

23. $d = 20.4$ in, $h = 8.5$ in

Find the volume of each sphere.

24. $r = 30$ mm

25. $r = \frac{2}{5}$ in

26. $d = 9.8$ yd

Find the surface area of each sphere.

27. $r = 30$ mm

28. $r = \frac{2}{5}$ in

29. $d = 9.8$ yd

Solve.

30. Mercury is stored in a cylinder of radius 16 cm and height 20 cm. How many cylindrical tubes with radius 0.5 cm and height 100 cm can be filled from a full supply?

31. A cylindrical tank is to be made from sheet metal. If the tank is to have radius 2.2 ft and height 12 ft, what will the metal cost if it is $16.50 per square foot?

32. A spherical tank with diameter 36 ft is to be insulated. If insulation costs $0.50 per square foot, how much will the job cost?

33. How many liters of chemical can be stored in a sphere with radius 650 mm? Note that 1 mm^3 = 0.000001 L.

FOR REVIEW

Find the area of metal left on each machine part shown which has circular holes drilled in it. Use 3.14 for π.

34.

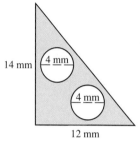

35.

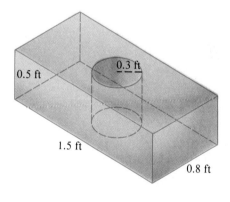

A.4 EXERCISES C

1. Find the volume of the wood remaining when the rectangular block shown at right has a hole bored through it. Use 3.14 for π and give the answer rounded to the nearest hundredth.
[Answer: 0.46 ft^3]

2. Find the total surface area of the wood remaining in the block in Exercise 1.

GEOMETRY APPENDIX REVIEW

KEY WORDS

A.1 A **segment** is a geometric figure made up of two endpoints and all points between them.

A **line** consists of all the points on a segment together with all the points beyond the endpoints of the segment.

A **ray** is that part of a line lying on one side of point P on the line, and includes the point P.

A **plane** is a surface with the property that any two points in it can be joined by a line, all points of which are also contained in the surface.

Parallel lines are distinct lines in a plane that do not intersect.

An **angle** is a geometric figure consisting of two rays which share a common endpoint called the **vertex** of the angle.

A **straight** angle has measure 180°.

A **right** angle has measure 90°.

A **acute** angle has measure between 0° and 90°.

A **obtuse** angle has measure between 90° and 180°.

Two angles are **complementary** if their measures total 90°.

Two angles are **supplementary** if their measures total 180°

Vertical angles are nonadjacent angles formed by intersecting lines.

When the four angles formed by two intersecting lines all have measure 90°, the lines are **perpendicular.**

A **transversal** is a line that intersects each of two parallel lines.

A.2 A **triangle** is a three-sided geometric figure.

Two triangles are **congruent** if they can be made to coincide by placing one on top of the other.

Two triangles are **similar** if the angles of one are equal to the angles of the other.

A.3 A **rectangle** is a four-sided figure having four right angles and parallel opposite sides of equal length.

The **perimeter** of a geometric figure is the distance around the figure.

A **square** is a rectangle with all sides the same length.

A **parallelogram** is a four-sided figure whose opposite sides are **parallel.**

A **circle** is a figure consisting of all points located the same distance from a fixed point called the **center** of the circle.

The **circumference** of a circle is the distance around the circle.

A.4 A **cube** is a rectangular solid for which the length equals the width equals the height.

A **cylinder** is a solid with circular ends of the same radius.

A **sphere** is a solid in the shape of a ball.

KEY CONCEPTS

A.1 1. Angles are measured in degrees (°).
2. Vertical angles are equal in measure.
3. Alternate interior angles are equal in measure.
4. Corresponding angles are equal in measure.

A.2 1. Similar triangles have all corresponding angles equal.
2. Triangles can be shown to be congruent by SSS, SAS, or ASA.

A.3 Rectangle: $P = 2l + 2w$, $A = lw$
Square: $P = 4s$, $A = s^2$
Parallelogram: $A = bh$
Triangle: $A = \frac{1}{2}bh$
Circle: $d = 2r$, $C = \pi d = 2\pi r$, $A = \pi r^2$

A.4 Rectangular solid: $V = lwh$,
$A = 2lh + 2wh + 2lw$
Cube: $V = e^3$, $A = 6e^2$
Cylinder: $V = \pi r^2 h$, $A = 2\pi rh + 2\pi r^2$
Sphere: $V = \frac{4}{3}\pi r^3$, $A = 4\pi r^2$

REVIEW EXERCISES

Part I

Use the figure to answer true *or* false *in Exercises 1–10. If the statement is false, tell why.*

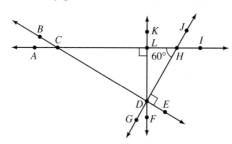

A.1,
A.2

1. ∠*CDH* is a straight angle.

2. ∠*LDB* has measure 60°.

3. $\overleftrightarrow{FK}$ is perpendicular to $\overleftrightarrow{AC}$.

4. The measure of ∠*CDE* is 180°.

5. ∠*LCD* and ∠*LHD* are supplementary.

6. ∠*BCA* has measure 30°.

7. $\overrightarrow{CD}$ is parallel to $\overline{HI}$.

8. ∠*LHD* is similar to △*CLD*.

9. △*CDH* is congruent to △*CLD*.

10. △*CDH* is a right triangle.

A.3 *Find the perimeter and area of the rectangle or square.*

11. 9 cm by 20 cm

12. $6\frac{1}{2}$ ft by 4.2 ft

13. 3.2 mi by 3.2 mi

Find the area of each figure.

14.

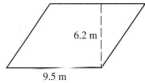

6.2 m

9.5 m

15.

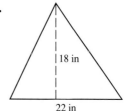

18 in

22 in

Find the perimeter of each figure.

16.

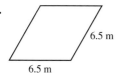

6.5 m

6.5 m

17.

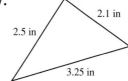

2.1 in

2.5 in

3.25 in

18. What will it cost to carpet the area shown if carpet costs $22.50 per square yard?

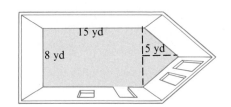

Find the circumference and area of each circle. Use 3.14 for π and round answers to the nearest tenth.

19. $r = 6.8$ km

20. $d = 22$ in

21. $r = 8\frac{1}{2}$ ft

A.4 *Find the volume of each rectangular solid.*

22. 2.2 ft by 1.5 ft by 6.4 ft

23. 52 cm by 22 cm by 8 cm

Find the volume of each cube with the given edge.

24. 16 m

25. 8.2 in

Find the surface area of each rectangular solid.

26. 2.2 ft by 1.5 ft by 6.4 ft

27. 53 cm by 22 cm by 8 cm

Find the surface area of each cube with the given edge.

28. 16 m

29. 8.2 in

Find the volume of each cylinder. Use 3.14 for π.

30. $r = 8$ cm, $h = 12$ cm

31. $d = 2.2$ in, $h = 1.5$ in

Find the surface area of each cylinder.

32. $r = 8$ cm, $h = 12$ cm

33. $d = 2.2$ in, $h = 1.5$ in

Find the volume of each sphere.

34. $r = 10$ ft

35. $d = 8.8$ m

Find the surface area of each sphere.

36. $r = 10$ ft

37. $d = 8.8$ m

Part II

38. How many grams of water will a tank 10 cm by 8 cm by 2.5 cm hold?

39. How many square centimeters of glass did it take to make the sides and bottom of the tank in Exercise 38?

40. The area shown is to be carpeted. How much will it cost if carpet is $19.50 per square yard?

41. A 12-inch diameter pizza costs $5.00 and a 16-in diameter pizza costs $8.00. Which pizza costs less per square inch?

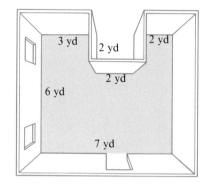

42. Are the given triangles congruent? Explain.

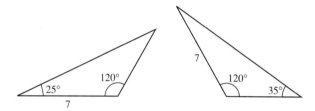

43. A spherical tank with radius 12 is to be insulated. If insulation costs $0.50 per square foot, how much will it cost to insulate the tank?

ANSWERS: 1. false (right angle) 2. true 3. true 4. true 5. false (complementary) 6. true 7. false ($\overleftrightarrow{CD}$ and $\overrightarrow{HI}$ intersect at C) 8. true 9. false (only one side equal) 10. true 11. 58 cm; 180 cm^2 12. 21.4 ft; 27.3 ft^2 13. 12.8 mi; 10.24 mi^2 14. 58.9 m^2 15. 198 in^2 16. 26 m 17. 7.85 in 18. $3150 19. 42.7 km; 145.2 km^2 20. 69.1 in; 379.9 in^2 21. 53.4 ft; 226.9 ft^2 22. 21.12 ft^3 23. 9152 cm^3 24. 4096 m^3 25. 551.368 in^3 26. 53.96 ft^2 27. 3472 cm^2 28. 1536 m^2 29. 403.44 in^2 30. 2411.52 cm^3 31. 5.7 in^3 32. 1004.8 cm^2 33. 17.96 in^2 34. 4186.7 ft^3 35. 356.6 m^3 36. 1256 ft^2 37. 243.2 m^2 38. 200 g 39. 170 cm^2 40. $741 41. 16-in costs less (12-in: $0.044 per square inch, 16-in: $0.040 per square inch) 42. yes; congruent by ASA 43. $904.32

Given that $\overleftrightarrow{AD}$ is parallel to $\overleftrightarrow{HE}$, use the figure to answer true *or* false in problems 1–4. If the statement is false, tell why.

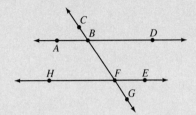

1. $\angle ABD$ is a straight angle.

1. _____

2. $\angle BFE$ and $\angle BFH$ are complementary.

2. _____

3. $\overrightarrow{AB}$ is parallel to $\overleftrightarrow{FG}$.

3. _____

4. $\angle EFG$ and $\angle ABC$ are equal.

4. _____

5. Are the given triangles congruent?

5. _____

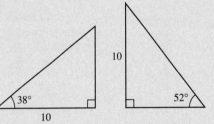

6. Find the area of the given figure.

6. _____

7. Find the perimeter of the given figure.

7. _____

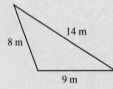

8. Find the volume of a cylinder with radius 8 cm and height 10 cm.

8. _____

9. Find the surface area of a sphere with radius 20 inches.

9. _____

Solve.

10. Gravel costs $1.50 per cubic meter. How much would a truckload cost if the bed of the truck is 4.2 m by 5.0 m by 2.5 m?

10. _____

CONTINUED

Problems 11–20 refer to the figure below. Answer true *or* false. *If the statement is false, tell why.*

11. ∠*FBE* is an obtuse angle.

12. ∠*MKF* is a right angle.

13. $\overrightarrow{EG}$ is parallel to $\overrightarrow{HM}$.

14. $\overleftrightarrow{AE}$ is perpendicular to $\overline{JH}$.

15. ∠*IJH* has measure 41°.

16. ∠*FBE* has measure 41°.

17. ∠*BFE* and ∠*BFK* are complementary.

18. Δ*KJF* is a right triangle.

19. ∠*CBD* has measure 41°.

20. Δ*BFE* is similar to Δ*JKF*.

21. The area shown is to be carpeted. How much will it cost if carpet is $22.50 per square yard?

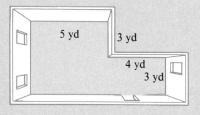

11. _____

12. _____

13. _____

14. _____

15. _____

16. _____

17. _____

18. _____

19. _____

20. _____

21. _____

22. Find the circumference of a circle with radius 2.6 ft. Use 3.14 for π.

22. _____

23. Find the area of a circle with diameter 9.2 m. Use 3.14 for π.

23. _____

24. Find **(a)** the volume and **(b)** the surface area of a cube with edge 3.6 ft.

24. (a) _____

(b) _____

25. A 14-in diameter pizza costs $7.50 and a 16-in pizza costs $8.50. Which costs less per square inch?

25. _____

CHAPTER 1

1. Reduce $\frac{242}{220}$ to lowest terms.

2. Bill received 28 of the 42 votes cast. What fractional part of the votes did he receive? Reduce the fraction to lowest terms.

Perform the indicated operations.

3. $\frac{5}{11} \cdot \frac{33}{35}$

4. $\frac{20}{39} \div \frac{6}{13}$

5. $\frac{2}{15} + \frac{8}{3}$

6. $\frac{11}{14} - \frac{3}{4}$

7. $29.705 + 3.01$

8. $17.6 - 9.72$

9. $(0.0049)(52.8)$

10. $2.86 \div 0.052$

11. Change $\frac{43}{18}$ to a mixed number.

12. Change $8\frac{3}{13}$ to an improper fraction.

13. There are 475 light bulbs in a shipment. If $\frac{1}{25}$ of them are defective, how many are defective?

14. Harry bought $2\frac{3}{8}$ pounds of one candy and $3\frac{11}{16}$ pounds of another. What total weight of candy did he buy?

15. Convert $\frac{3}{8}$ to a percent.

16. Convert 2.72 to a fraction.

17. The tax rate in Upstate, NY, is 6%. What is the tax on a $38.50 purchase?

18. Write $yyyyyy$ in exponential notation.

19. Evaluate $x^2 + 4(x - y)$ for $x = 5$ and $y = 3$.

20. If $A = P(1 + r)^t$, find A when P is $2000, the interest is 8% compounded annually, and t is 2 years.

CHAPTER 2

21. Find the reciprocal of $-\frac{3}{4}$.

22. Find $|-8|$.

Place the correct symbol, $=$, $<$, or $>$, between the pairs of numbers.

23. $-8 \quad -10$

24. $\frac{8}{31} \quad \frac{4}{17}$

Perform the indicated operations.

25. $-4 - (-16)$

26. $14 + (-24)$

27. $\left(-\frac{5}{9}\right) \cdot \left(-\frac{3}{20}\right)$

28. $(7.5) \div (-2.5)$

Evaluate.

29. $2 \cdot 4 + (8 - 12) - 2(3 - 1)$

30. $-8 - [-(-8)]$

Factor.

31. $2y + 12$

32. $-14x - 21$

Collect like terms.

33. $-8a + 3b - 4a - 5 - b$

34. $7y - (2y - 5) + 3$

35. Evaluate $2u^2 - v^2$ when $u = -2$ and $v = 4$.

36. Write $-5x^4$ without exponents.

37. Evaluate. $-\sqrt{144}$

38. Evaluate. $\sqrt{\dfrac{12}{75}}$

CHAPTER 3

Solve.

39. $x + 10 = 19$

40. $16x = 4$

41. $x - \dfrac{5}{3} = \dfrac{7}{3}$

42. $-5.5x = 33$

43. $\dfrac{2}{3}x = 26$

44. $\dfrac{x}{\frac{3}{4}} = 12$

45. $4x + 8 = 7x - 4$

46. $-3x - 2 = -5x + 9$

47. $5(x + 2) - 2(x - 10) = 0$

48. $9x - (2x + 3) = 4$

49. The sum of two consecutive even integers is 134. What are the integers?

50. A suit is put on sale at a 20% discount. If the original price was $160, what is the sale price?

51. The area of a triangle is 85 in^2 and the base is 10 inches. What is the height?

52. Two trains leave Portland, one traveling north and the other south. If one is traveling 30 mph faster than the other, how fast is each traveling if they are 360 miles apart after 3 hours?

Graph on the number line.

53. $3x + 8 = 5$

54. $4(x - 2) \geq 4 - (x - 3)$

Solve the inequality.

55. $4x + 9 \geq 7x - 3$

56. $2x - (3x - 2) < 9x - 8$

CHAPTER 4

57. The point with coordinates $(-3, -5)$ is located in which quadrant?

58. Complete the ordered pair $(5, \;\;\;)$ so that it is a solution to $2x + 3y = 10$.

59. Cost, c, is related to number of items produced, n, by $c = 100n + 50$. Find c when n is 18.

60. True or false: $5 + 3y = 2x$ is a linear equation.

61. Find the slope of the line through the points $(7, -1)$ and $(2, -5)$. Determine the general form of the equation of the line.

62. Write $3x - 4y = 12$ in slope-intercept form and give the slope and the y-intercept.

63. Find the general form of the equation of the line with slope $-\frac{4}{5}$ and y-intercept $(0, 7)$.

64. If l_1 passes through $(1, 2)$ and $(-1, 6)$ and l_2 passes through $(0, -3)$ and $(6, 0)$, are the lines parallel, perpendicular, or neither?

Graph in the given Cartesian coordinate system.

65. $y = -3x + 6$

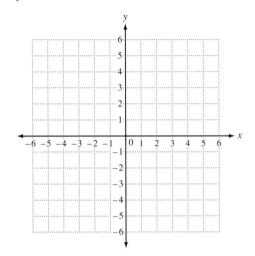

66. $2x - 3y + 6 \leq 0$

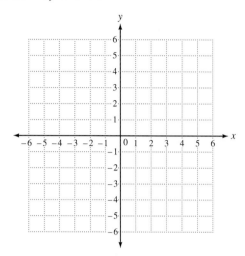

CHAPTER 5

67. Without graphing, tell whether the graphs are intersecting, parallel, or coinciding.
$3x + 4y = 7$
$9x + 12y = 5$

68. If the lines in a system of equations are intersecting, the system has _____ solution(s).

Solve the system.

69. $3x - y = -2$
   $-5x + 2y = 5$

70. $2x + 3y = 8$
   $4x + 7y = 16$

71. $2x - 3y = 12$
   $3x + 9 = 0$

72. $4x - 1 = 0$
   $3y + 2 = 0$

73. A boat travels 40 miles upstream in 4 hours and returns to the starting point in 2 hours. What is the speed of the stream?

74. Pam Johnson earned $1180 in one year on an investment of $12,000. Part of the money was invested at 9% and part at 11% simple interest. How much was invested at 11%?

75. Twice one number plus a second is 13. The first minus the second is 5. Find the numbers.

76. A candy shop has one candy that sells for 95¢ per lb and a second that sells for 65¢ per lb. How much of each should be used to make 60 lb of candy selling for 75¢ per lb?

CHAPTER 6

Simplify and write without negative exponents.

77. $3x^2x^3$

78. $\dfrac{y^6}{y^2}$

79. $(-3a^3)^2$

80. a^{-2}

81. Write 36,200,000 in scientific notation.

82. Write 7.2×10^{-4} using standard notation.

Evaluate using scientific notation.

83. $(0.0000021)(3,000,000)$

84. $\dfrac{0.000055}{11,000,000}$

85. Is $-8x^3y$ a monomial, binomial, or trinomial?

86. Give the degree of $3x^4 - 6x^5 + 5x - 2$.

87. Write in descending order.
$3x - 4x^3 - 6x^5 + 4x^2 + x^7$

88. Profit in dollars when x coats are sold is given by $2x^2 - 5$. What is the profit when 6 coats are sold?

Add.

89. $(-5a + 2) + (6a - 1)$

90. $(2x^2y^2 - 3xy + 2) + (-6x^2y^2 - xy - 5)$

Subtract.

91. $(9y - 3) - (-2y + 5)$

92. $(3x^2 - 2x + 5) - (4x^2 + 9x - 6)$

93. $(7a - 6) - (-a + 1) - (3a - 5)$

94. $(6x^2y - 4xy) - (4xy^2 + 2xy) - (2x^2y + xy^2)$

Multiply.

95. $-3a^2(4a - 2)$

96. $(3x + 4y)(9x - 2y)$

97. $(a - 2)(3a^2 + 2a - 5)$

98. $(5x + 7)^2$

99. $(2x + 5y)(2x - 5y)$

100. $(x^2 - 2)^2$

Divide.

101. $\dfrac{6y^9 - 2y^6 + 10y^3}{2y^2}$

102. $(2x^3 - 27x + 20) \div (x + 4)$

CHAPTER 7

Factor by removing the greatest common factor.

103. $24x - 16$

104. $6x^3y^2 - 3x^2y^3 + 9x^2y^2$

Factor by grouping.

105. $3x^3 + 2x^2 + 6x + 4$

106. $a^2b^2 - a^2 + 3b^2 - 3$

Factor.

107. $x^2 + 13x + 40$

108. $y^2 - 18y + 81$

109. $6x^2 - 24y^2$

110. $x^2 + xy - 20y^2$

111. $2x^2 - 9x - 5$

112. $9x^2 + 12xy + 4y^2$

Solve.

113. $(2x - 1)(x + 8) = 0$

114. $x^2 - 8x - 20 = 0$

115. The length of a rectangle is four times the width. If the area is 144 cm^2, find the dimensions.

116. Profit is given by $P = n^2 - 2n + 3$, where n is the number of items sold. How many items were sold when the profit was $38?

CHAPTER 8

117. What values of the variable must be excluded in $\dfrac{x}{(x + 1)(x - 8)}$?

118. Reduce $\dfrac{a^2 - b^2}{(a - b)^2}$ to lowest terms.

Perform the indicated operations.

119. $\dfrac{x^2 - x - 2}{x^2 - 2x - 3} \cdot \dfrac{x^2 - 3x}{x + 2}$

120. $\dfrac{y^2 - 1}{y^2 + 2y} \div \dfrac{y - 1}{y + 2}$

121. $\dfrac{5}{y^2 - 4y + 3} + \dfrac{7}{y^2 + y - 2}$

122. $\dfrac{x + 6}{x^2 - 4} - \dfrac{x}{x - 2}$

123. Simplify. $\dfrac{x - \dfrac{1}{x}}{1 - \dfrac{1}{x}}$

124. Solve. $\dfrac{1}{y + 2} + \dfrac{1}{y - 2} = \dfrac{1}{y^2 - 4}$

125. A car goes 205 miles on 15 gallons of gasoline. How many miles can the car go on 75 gallons of gasoline?

126. The denominator of a fraction is 5 more than the numerator. If 2 is added to both the numerator and the denominator, the result is $\frac{1}{2}$. Find the fraction.

127. The cost c of nuts varies directly as the weight w of the nuts. If 12 lb of nuts cost \$30, how much would 80 lb cost?

128. The volume V of a gas varies inversely as the pressure P. If V is 16 in^3 when the pressure is 3 pounds per square inch (psi), what is the volume when $P = 12$ psi?

CHAPTER 9

Evaluate. Assume all variables and expressions are nonnegative.

129. $\sqrt{(-16)^2}$

130. $\pm\sqrt{169}$

131. $\sqrt{81x^2}$

132. $5\sqrt{8x^2y^2}$

133. $\sqrt{\dfrac{27x^2}{4y^2}}$

134. $\sqrt{\dfrac{49y^2}{x}}$

Perform the indicated operations and simplify.

135. $5\sqrt{15}\,\sqrt{45}$

136. $\dfrac{\sqrt{24x^3y}}{\sqrt{3xy}}$

137. $\sqrt{98} + \sqrt{50}$

138. $3\sqrt{45} - 2\sqrt{80} + \sqrt{125}$

139. $\sqrt{28x} + 3\sqrt{7x} - 2\sqrt{63x}$

140. $\sqrt{3} - \dfrac{2}{\sqrt{3}}$

141. $(\sqrt{7} - 2\sqrt{2})(\sqrt{7} + \sqrt{2})$

142. $\dfrac{\sqrt{6}}{\sqrt{3} - \sqrt{2}}$

Solve.

143. $\sqrt{2x - 1} - \sqrt{3x - 7} = 0$

144. $\sqrt{x^2 + 3} + 1 - x = 0$

145. One side of a rectangle is 12 m long and the diagonal is 13 m. Find the length of the other side.

146. Three times the square root of 5 more than a number is equal to 15. Find the number.

147. In a 30°–60° right triangle, a is the side opposite the 30° angle. If $a = 22$, find b and c.

148. In a 45°–45° right triangle, $b = 45$ cm. Find a and c.

CHAPTER 10

Solve.

149. $x^2 - 12x + 35 = 0$

150. $x^2 + 14x + 49 = 0$

151. $2x^2 - 11x + 5 = 0$

152. $x^2 - 3x - 1 = 0$

153. $3y^2 - 5 = y(y + 3)$

154. $(z - 2)(z + 2) = 5z$

155. $\sqrt{7 - 2x} = x - 2$

156. $\dfrac{x + 2}{x} = \dfrac{2}{x - 1}$

157. The number of square inches in a square is 3 less than the perimeter of the square. Find the length of a side.

158. Al can do a job in 3 hours less time than Joe. Together they can do the job in 2 hours. How long would it take each to do the job.

159. Solve $a = \sqrt{\dfrac{2b}{c}}$ for b.

160. Graph $y = -x^2 + 1$.

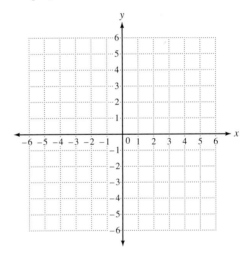

65.

66.

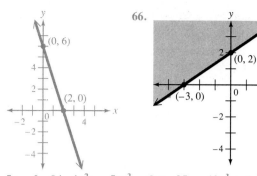

67. parallel   68. exactly one   69. (1, 5)   70. (4, 0)
71. $(-3, -6)$ 72. $\left(\frac{1}{4}, -\frac{2}{3}\right)$ 73. 5 mph   74. $5000
75. (6, 1)   76. 20 lb of 95¢ candy and 40 lb of 65¢ candy

77. $3x^5$ 78. y^4 79. $9a^6$ 80. $\dfrac{1}{a^2}$ 81. 3.62×10^7

82. 0.00072   83. $6.3 \times 10^0 = 6.3$ 84. 5×10^{-12}
85. monomial   86. 5   87. $x^7 - 6x^5 - 4x^3 + 4x^2 + 3x$
88. $67   89. $a + 1$ 90. $-4x^2y^2 - 4xy - 3$
91. $11y - 8$ 92. $-x^2 - 11x + 11$

93. $5a - 2$ 94. $4x^2y - 5xy^2 - 6xy$ 95. $-12a^3 + 6a^2$ 96. $27x^2 + 30xy - 8y^2$ 97. $3a^3 - 4a^2 - 9a + 10$
98. $25x^2 + 70x + 49$ 99. $4x^2 - 25y^2$ 100. $x^4 - 4x^2 + 4$ 101. $3y^7 - y^4 + 5y$ 102. $2x^2 - 8x + 5$ 103. $8(3x - 2)$
104. $3x^2y^2(2x - y + 3)$ 105. $(x^2 + 2)(3x + 2)$ 106. $(a^2 + 3)(b + 1)(b - 1)$ 107. $(x + 8)(x + 5)$ 108. $(y - 9)^2$
109. $6(x + 2y)(x - 2y)$ 110. $(x - 4y)(x + 5y)$ 111. $(2x + 1)(x - 5)$ 112. $(3x + 2y)^2$ 113. $\frac{1}{2}, -8$ 114. $-2, 10$

115. 6 cm, 24 cm   116. 7   117. $-1, 8$ 118. $\dfrac{a + b}{a - b}$ 119. $\dfrac{x(x - 2)}{x + 2}$ 120. $\dfrac{y + 1}{y}$ 121. $\dfrac{12y - 11}{(y + 2)(y - 3)(y - 1)}$

122. $\dfrac{-x - 3}{x + 2}$ 123. $x + 1$ 124. $\frac{1}{2}$ 125. 1025 mi   126. $\frac{3}{8}$ 127. $200   128. 4 in^3 129. 16   130. ± 13 131. $9x$

132. $10xy\sqrt{2}$ 133. $\dfrac{3x\sqrt{3}}{2y}$ 134. $\dfrac{7y\sqrt{x}}{x}$ 135. $75\sqrt{3}$ 136. $2x\sqrt{2}$ 137. $12\sqrt{2}$ 138. $6\sqrt{5}$ 139. $-\sqrt{7x}$ 140. $\dfrac{\sqrt{3}}{3}$

141. $3 - \sqrt{14}$ 142. $3\sqrt{2} + 2\sqrt{3}$ 143. 6   144. no solution   145. 5 m   146. 20   147. $b = 22\sqrt{3}, c = 44$

148. $a = 45$ cm, $c = 45\sqrt{2}$ cm   149. 5, 7   150. -7 151. $\frac{1}{2}, 5$ 152. $\dfrac{3 \pm \sqrt{13}}{2}$ 153. $\frac{5}{2}, -1$ 154. $\dfrac{5 \pm \sqrt{41}}{2}$

155. 3   156. $2, -1$ 157. 3 in, 1 in   158. Al needs 3 hr and Joe 6 hr   159. $b = \dfrac{a^2 c}{2}$

160.

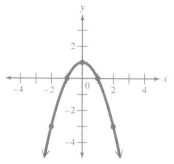

TABLE OF SQUARES AND SQUARE ROOTS

n	n^2	$\sqrt{n}$	$\sqrt{10n}$	n	n^2	$\sqrt{n}$	$\sqrt{10n}$
1	1	1.000	3.162	51	2601	7.141	22.583
2	4	1.414	4.472	52	2704	7.211	22.804
3	9	1.732	5.477	53	2809	7.280	23.022
4	16	2.000	6.325	54	2916	7.348	23.238
5	25	2.236	7.071	55	3025	7.416	23.452
6	36	2.449	7.746	56	3136	7.483	23.664
7	49	2.646	8.367	57	3249	7.550	23.875
8	64	2.828	8.944	58	3364	7.616	24.083
9	81	3.000	9.487	59	3481	7.681	24.290
10	100	3.162	10.000	60	3600	7.746	24.495
11	121	3.317	10.488	61	3721	7.810	24.698
12	144	3.464	10.954	62	3844	7.874	24.900
13	169	3.606	11.402	63	3969	7.937	25.100
14	196	3.742	11.832	64	4096	8.000	25.298
15	225	3.873	12.247	65	4225	8.062	25.495
16	256	4.000	12.649	66	4356	8.124	25.690
17	289	4.123	13.038	67	4489	8.185	25.884
18	324	4.243	13.416	68	4624	8.246	26.077
19	361	4.359	13.784	69	4761	8.307	26.268
20	400	4.472	14.142	70	4900	8.367	26.458
21	441	4.583	14.491	71	5041	8.426	26.646
22	484	4.690	14.832	72	5184	8.485	26.833
23	529	4.796	15.166	73	5329	8.544	27.019
24	576	4.899	15.492	74	5476	8.602	27.203
25	625	5.000	15.811	75	5625	8.660	27.386
26	676	5.099	16.125	76	5776	8.718	27.568
27	729	5.196	16.432	77	5929	8.775	27.749
28	784	5.292	16.733	78	6084	8.832	27.928
29	841	5.385	17.029	79	6241	8.888	28.107
30	900	5.477	17.321	80	6400	8.944	28.284
31	961	5.568	17.607	81	6561	9.000	28.460
32	1024	5.657	17.889	82	6724	9.055	28.636
33	1089	5.745	18.166	83	6889	9.110	28.810
34	1156	5.831	18.439	84	7056	9.165	28.983
35	1225	5.916	18.708	85	7225	9.220	29.155
36	1296	6.000	18.974	86	7396	9.274	29.326
37	1369	6.083	19.235	87	7569	9.327	29.496
38	1444	6.164	19.494	88	7744	9.381	29.665
39	1521	6.245	19.748	89	7921	9.434	29.833
40	1600	6.325	20.000	90	8100	9.487	30.000
41	1681	6.403	20.248	91	8281	9.539	30.166
42	1764	6.481	20.494	92	8464	9.592	30.332
43	1849	6.557	20.736	93	8649	9.644	30.496
44	1936	6.633	20.976	94	8836	9.695	30.659
45	2025	6.708	21.213	95	9025	9.747	30.822
46	2116	6.782	21.448	96	9216	9.798	30.984
47	2209	6.856	21.679	97	9409	9.849	31.145
48	2304	6.928	21.909	98	9604	9.899	31.305
49	2401	7.000	22.136	99	9801	9.950	31.464
50	2500	7.071	22.361	100	10000	10.000	31.623

Chapter 1

1. $\frac{23}{25}$ 2. $\frac{24}{75}$ 3. 24 4. $\frac{11}{15}$ 5. $\frac{3}{11}$ 6. $\frac{9}{4}$ 7. $\frac{29}{18}$ 8. $\frac{1}{24}$ 9. $5\frac{3}{5}$ 10. $\frac{44}{15}$ 11. 432 12. $4\frac{7}{10}$ liters 13. 17.117
14. 18.859 15. 2.1252 16. 44 17. 20% 18. $\frac{63}{20}$ 19. \$1.08 20. a^3 21. 1 22. 75 23. 225 24. 125 25. 35
26. 5 27. \$594.05

Chapter 2

1. false 2. 13 3. true 4. $<$ 5. 141° 6. -22 7. 8 8. $-\frac{1}{8}$ 9. 3 10. 5 11. -18 12. -19 13. -6
14. $-3(y - 5)$ 15. $5 - b - 2y$ 16. 3 17. 15 18. $-12y + 6$ 19. 11 20. $-\frac{5}{4}$ 21. true 22. false

Chapter 3

1. 15 2. 8 3. $\frac{1}{2}$ 4. -2 5. 36 6. 4 7. -3 8. 13 9. 3 10. 83, 85 11. \$62.10 12. 40 cm 13. 48 mph,
57 mph 14. $x \geq -3$ 15. $x < \frac{1}{2}$ 16. 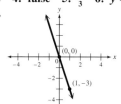 17.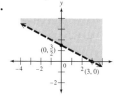

Chapter 4

1. IV 2. $\left(-3, \frac{2}{5}\right)$ 3. 400 items 4. false 5. $\frac{1}{3}$ 6. $y = -\frac{2}{5}x + 4$; $-\frac{2}{5}$; $(0, 4)$ 7. $6x + y - 19 = 0$ 8. perpendicular
9. 10. 11. 12.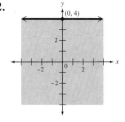

Chapter 5

1. intersecting lines 2. no 3. $(3, -1)$ 4. no solution 5. infinitely many 6. 3, 15 7. 24°, 156° 8. 30 nickels,
22 dimes 9. 450 mph 10. 40 lb candy, 60 lb nuts

Chapter 6

1. 5^2a^3 2. $2y^7$ 3. a^4 4. $-8x^6$ 5. $\frac{1}{z^3}$ 6. $\frac{y^7}{x^8}$ 7. $-\frac{1}{64}$ 8. 3×10^{-2} 9. binomial 10. 4 11. $-6y^5 + 4y^3 - 6y +$
6 12. $-4x^2y^2 + 5x^2y - 4x^2$ 13. \$285 14. $-2x + 10$ 15. $3x^4 + 10x^3 - 3x^2 + 7$ 16. $7a^2b^2 + 6ab + 2$ 17. $-7x + 7$
18. $-4x^4 - x^3 + 6x^2 + 1$ 19. $4a^2b - 4ab^2 - ab$ 20. $-12x^4 + 16x^3 - 4x^2$ 21. $x^2 - 3x - 40$ 22. $8a^2 + 22a - 21$
23. $3y^3 + 22y^2 - 15y + 8$ 24. $4x^2 - y^2$ 25. $9x^2 - 24xy + 16y^2$ 26. $4x^2 + 20x + 25$ 27. $6x^3 - 5x^2 + 6x + 8$
28. $x^4 - 9$ 29. $5x^2 - 2x - 3$ 30. $3x^2 - 2x + 4$

Chapter 7

1. $4(5x + 3)$ 2. $7(5y - 1)$ 3. $6x^2(4x^2 - 2x + 3)$ 4. $5x^3y^2(y + 8)$ 5. $(y^2 + 2)(7x^2 - 1)$ 6. $(x + 7)(x + 8)$
7. $(x - 8)^2$ 8. $3(y + 5)(y - 5)$ 9. $(x + 2)(x - 10)$ 10. $(2x - 3)(x + 8)$ 11. $(3x + y)(x + 5y)$ 12. $\frac{5}{4}$, $-\frac{3}{2}$
13. 6, -5 14. 1 or -4 15. 4 cm, 6 cm 16. (a) \$332 (b) 12

Chapter 8

1. $0, -5$ 2. yes 3. $\dfrac{1}{a}$ 4. $\dfrac{5(x + 4)}{x}$ 5. $\dfrac{1}{y - 3}$ 6. $\dfrac{1}{a - b}$ 7. $\dfrac{2x}{x + 2}$ 8. $\dfrac{y - 5}{(y - 1)(y + 1)(y - 2)}$ 9. no solution
10. $\frac{10}{13}$ 11. 37 gallons 12. \$13.80 13. $\frac{15}{8}$ days 14. $\dfrac{1}{1 + y}$ 15. 5

Chapter 9

1. 2 2. 5 3. ± 11 4. $\frac{9}{4}$ 5. $4x$ 6. $12\sqrt{3}$ 7. $\dfrac{2\sqrt{2x}}{y}$ 8. $\dfrac{7x\sqrt{2y}}{y}$ 9. 24 10. $3x$ 11. $6\sqrt{5}$ 12. $-16\sqrt{2}$
13. $10\sqrt{3x}$
14. $\dfrac{12\sqrt{11}}{11}$ 15. 10 16. 2 17. $-3 + \sqrt{14}$ 18. $\dfrac{5 + \sqrt{5}}{4}$ 19. $\frac{7}{2}$ 20. -3 21. 10 cm 22. (a) $10\sqrt{5}$ (b) 20
23. $b = 20\sqrt{3}$, $c = 40$ 24. $4\sqrt{2}$

Chapter 10

1. $-4, -9$ 2. $3, -3$ 3. $0, 5$ 4. $2 \pm \sqrt{3}$ 5. $\frac{1}{16}$ 6. $3, \frac{1}{2}$ 7. 2 8. 10, 12 9. large pipe: 4 hours, small pipe:
12 hours 10. 12 mph 11. $m = \dfrac{b}{w^2}$ 12. (a) up (b) $(0, 2)$ (c)

Geometry Appendix

1. true 2. false (supplementary) 3. false ($\overrightarrow{AB}$ and $\overleftrightarrow{FG}$ intersect at B) 4. true 5. yes 6. 12 ft^2 7. 31 m
8. 2009.6 cm^3 9. 5024 in^2 10. \$78.75 11. false (acute) 12. true 13. true 14. true 15. true 16. true
17. false (supplementary) 18. true 19. true 20. true 21. \$945 22. 16.3 ft 23. 66.4 m^2 24. (a) 46.7 ft^3
(b) 77.8 ft^2 25. 16-in (14 in: \$0.049 per square inch; 16 in: \$0.042 per square inch)

SOLUTIONS TO SELECTED EXERCISES

All exercises are selected from the A set of exercises.

1.2

8. $4\frac{2}{5} \cdot \frac{3}{11} = \frac{22}{5} \cdot \frac{3}{11} = \frac{2 \cdot 11}{5} \cdot \frac{3}{11}$

$= \frac{2 \cdot \cancel{11} \cdot 3}{5 \cdot \cancel{11}} = \frac{6}{5}$

14. $3\frac{1}{5} \div \frac{2}{5} = \frac{16}{5} \div \frac{2}{5} = \frac{16}{5} \cdot \frac{5}{2} =$

$\frac{\cancel{2} \cdot 2 \cdot 2 \cdot 2 \cdot \cancel{5}}{\cancel{5} \cdot \cancel{2}} =$

$\frac{2 \cdot 2 \cdot 2}{1} = \frac{8}{1} = 8$

43. $\frac{5}{6}$ of 72 means $\frac{5}{6} \cdot 72$.

$\frac{5}{6} \cdot 72 = \frac{5}{6} \cdot \frac{72}{1} = \frac{5}{2 \cdot 3} \cdot \frac{2 \cdot 2 \cdot 2 \cdot 3 \cdot 3}{1} =$

$\frac{5 \cdot \cancel{2} \cdot 2 \cdot 2 \cdot \cancel{3} \cdot 3}{\cancel{2} \cdot \cancel{3}} = \frac{5 \cdot 2 \cdot 2 \cdot 3}{1} = 60$ girls

$72 - 60 = 12$ boys

46. Divide $12\frac{1}{2}$ by $\frac{3}{4}$.

$12\frac{1}{2} \div \frac{3}{4} = \frac{25}{2} \div \frac{3}{4} = \frac{25}{2} \cdot \frac{4}{3} = \frac{5 \cdot 5}{2} \cdot \frac{2 \cdot 2}{3} =$

$\frac{5 \cdot 5 \cdot \cancel{2} \cdot 2}{\cancel{2} \cdot 3} = \frac{50}{3} = 16\frac{2}{3}$ gallons

1.3

41. $7\frac{3}{4} + 9\frac{1}{8} = \frac{31}{4} + \frac{73}{8} = \frac{31 \cdot 2}{2 \cdot 2 \cdot 2} + \frac{73}{2 \cdot 2 \cdot 2}$

$= \frac{62 + 73}{2 \cdot 2 \cdot 2} = \frac{135}{8}$ **Total distance**

$\frac{135}{8} \cdot \frac{1}{2} = \frac{135}{16} = 8\frac{7}{16}$ mi **One half distance**

1.4

37. $\frac{15}{20} = \frac{3 \cdot \cancel{5}}{4 \cdot \cancel{5}} = \frac{3}{4}$ **Fraction of shots**

$\frac{3}{4} = 0.75 = 75\%$ **Percent of shots**

39. Find 55% of 5000.
$(0.55)(5000) = 2750$ votes

1.5

51. $x - [a(b + 1) - c] = 12 - [2(3 + 1) - 5] =$
$12 - [2 \cdot 4 - 5] = 12 - [8 - 5] = 12 - 3 = 9$

60. $A = 2\pi rh + 2\pi r^2 = 2(3.14)(2)(10) + 2(3.14)(2)^2 =$
$125.60 + 25.12 = 150.72$ cm^2

2.1

26. Given $\frac{17}{43}$ and $\frac{28}{79}$. The cross product is $17 \cdot 79 = 1343$, and the second cross product is $43 \cdot 28 = 1204$. Since $1343 > 1204$, we have $\frac{17}{43} > \frac{28}{79}$.

2.2

51. The difference in temperature was from $+35°$ to $-13°$. That is, we must find $35 - (-13) = 35 + 13 = 48°$. Thus, the total change in temperature was a decrease of $48°$.

2.3

44. Since 20 penalty points must be deducted from the total 3 times, we have that $3(-20) = -60$ represents the deduction. Thus, $423 + (-60) = 363$ points is her total after losing the penalty points.

50. $3 - \frac{2 + (-4)}{1 - 5} = 3 - \frac{-2}{-4}$

$= 3 - \frac{1}{2} = \frac{6}{2} - \frac{1}{2} = \frac{5}{2}$

63. When $a = -1$, $b = 2$, and $c = -4$,

$2[3(a + 1) + 2(4 + c)]$

$= 2[3((-1) + 1) + 2(4 + (-4)]$

$= 2[3(0) + 2(0)] = 2[0 + 0] = 2[0] = 0.$

75. -3000 represents the amount owed. $+1200$ represents the amount repaid, and -650 is the amount again borrowed. We must compute $(-3000) + (1200) + (-650) = -2450$. Thus, the Brisebois family owes \$2450, so that -2450 represents the status of their account.

2.4

37. $\frac{1}{2}x - \frac{2}{3}y - \frac{5}{2}x - \frac{1}{3}y$

$= \frac{1}{2}x - \frac{5}{2}x - \frac{2}{3}y - \frac{1}{3}y$

$$= \left(\frac{1}{2} - \frac{5}{2}\right)x - \left(\frac{2}{3} + \frac{1}{3}\right)y$$

$$= -2x - y$$

43. $2x - (-x + 1) + 3$

$$= 2x + x - 1 + 3$$

$$= 3x + 2$$

47. $a - [3a - (1 - 2a)] = a - [3a - 1 + 2a] =$

$a - [5a - 1] = a - 5a + 1 = -4a + 1$

55. When $x = -3$, $-2x^2 = -2(-3)^2 = -2 \cdot 9 = -18$. Note that we square -3 before multiplying the result by -2.

56. When $x = -3$, $(-2x)^2 = ((-2)(-3))^2 = (6)^2 = 36$. Note that we first multiply -2 by -3, obtaining 6, then square the result. Compare Exercise 55 with Exercise 56.

68. When $x = -3$ and $y = -1$, $|x - y| = |(-3) - (-1)| = |(-3) + 1| = |-2| = 2$. Notice that we first simplify inside the absolute value bars.

2.5

17. Always reduce fractions when they appear under a radical. Thus,

$$\sqrt{\frac{8}{18}} = \sqrt{\frac{2 \cdot 4}{2 \cdot 9}} = \sqrt{\frac{4}{9}} = \frac{2}{3}.$$

29. Since $3^2 = 9 < 15 < 16 = 4^2$, we have $\sqrt{3^2} = 3 < \sqrt{15} < 4 = \sqrt{4^2}$.

Thus, $\sqrt{15}$ is between 3 and 4.

3.1

28. Change $2\frac{1}{2}$ to $\frac{5}{2}$ and subtract $\frac{5}{2}$ from both sides of the equation.

$$x + 2\frac{1}{2} = 5$$

$$x + \frac{5}{2} = 5$$

$$x + \frac{5}{2} - \frac{5}{2} = 5 - \frac{5}{2}$$

$$x + 0 = \frac{10}{2} - \frac{5}{2}$$

$$x = \frac{5}{2} = 2\frac{1}{2}$$

3.2

15. Since $-\frac{1}{8}$ is the divisor of z we must multiply both sides by $-\frac{1}{8}$.

$$\frac{z}{-\frac{1}{8}} = 16$$

$$\left(-\frac{1}{8}\right)\frac{z}{-\frac{1}{8}} = 16\left(-\frac{1}{8}\right)$$

$$\frac{\left(-\frac{1}{8}\right)}{\left(-\frac{1}{8}\right)}z = -\frac{16}{8} \qquad \frac{-\frac{1}{8}}{-\frac{1}{8}} = 1$$

$$1 \cdot z = -2$$

$$z = -2$$

16. First multiply both sides by 3.

$$\frac{2x}{3} = 4$$

$$(3)\frac{2x}{3} = 4(3)$$

$$\frac{3}{3} \cdot 2x = 12$$

$$2x = 12$$

$$\frac{2x}{2} = \frac{12}{2} \qquad \textbf{Now divide by 2}$$

$$x = 6$$

22. First change $2\frac{1}{2}$ to $\frac{5}{2}$.

$$2\frac{1}{2}x = 10$$

$$\frac{5}{2}x = 10$$

$$\frac{2}{5} \cdot \frac{5}{2}x = 10 \cdot \frac{2}{5} \qquad \textbf{The product } \frac{2}{5} \cdot \frac{5}{2} \textbf{ is 1}$$

$$1 \cdot x = \frac{10}{1} \cdot \frac{2}{5}$$

$$x = 4 \qquad \textbf{10 ÷ 5 = 2 and 2 · 2 = 4}$$

3.3

11. Collect like terms on the left side.

$$-\frac{1}{7}y + y = 18$$

$$\left(-\frac{1}{7} + 1\right)y = 18$$

$$\left(-\frac{1}{7} + \frac{7}{7}\right)y = 18$$

$$\frac{6}{7}y = 18$$

$$\frac{7}{6} \cdot \frac{6}{7}y = 18 \cdot \frac{7}{6} \qquad \frac{7}{6} \textbf{ is the reciprocal of } \frac{6}{7}$$

$$1 \cdot y = \frac{18}{1} \cdot \frac{7}{6}$$

$$y = 21 \qquad \begin{array}{l}\textbf{18 ÷ 6 = 3} \\ \textbf{and 3 · 7 = 21}\end{array}$$

14. $-7 + 21y + 23 = 16$

$\quad\quad 21y + 16 = 16$ **Collect like terms**

$\quad\quad\quad 21y = 0$ **Subtract 16 from both sides**

$\quad\quad\quad\quad y = 0$ **Divide both sides by 21**

17. $6y - 4y + 1 = 12 + 2y - 11$

$\quad\quad 2y + 1 = 1 + 2y$ **Collect like terms**

Since this is an identity, every real number is a solution.

19. Clear fractions by multiplying through by 2 and collect like terms on both sides.

$$3y + \frac{5}{2}y + \frac{3}{2} = \frac{1}{2}y + \frac{5}{2}y$$

$$2\left(3y + \frac{5}{2}y + \frac{3}{2}\right) = 2\left(\frac{1}{2}y + \frac{5}{2}y\right)$$

$6y + 5y + 3 = y + 5y$ **Multiply each term by 2**

$\quad\quad 11y + 3 = 6y$ **Collect like terms**

$\quad\quad\quad 11y = 6y - 3$ **Subtract 3 from both sides**

$\quad 11y - 6y = -3$ **Subtract 6y from both sides**

$\quad\quad\quad 5y = -3$

$\quad\quad\quad\quad y = -\frac{3}{5}$ **Divide both sides by 5**

29. $5(2 - 3x) = 15 - (x + 7)$

$10 - 15x = 15 - x - 7$ **Clear parentheses**

$10 - 15x = 8 - x$ **Collect like terms**

$\quad\quad\quad 10 = 8 + 14x$ **Add 15x to both sides**

$\quad\quad\quad\quad 2 = 14x$ **Subtract 8 from both sides**

$\quad\quad\frac{2}{14} = x$ **Divide both sides by 14**

$\quad\quad\frac{1}{7} = x$

36. The fractions could be cleared by multiplying both sides by the LCD of 2 and 3, which is 6. However, in this exercise the fractions are cleared when we use the distributive law.

$$\frac{1}{3}(6x - 9) = \frac{1}{2}(8x - 4)$$

$$\frac{1}{3} \cdot 6x - \frac{1}{3} \cdot 9 = \frac{1}{2} \cdot 8x - \frac{1}{2} \cdot 4$$

$2x - 3 = 4x - 2$

$2x = 4x - 2 + 3$

$2x - 4x = 1$

$-2x = 1$

$\dfrac{-2x}{-2} = \dfrac{1}{-2}$

$x = -\dfrac{1}{2}$

37. $5(x + 1) - 4x = x - 5$

$5x + 5 - 4x = x - 5$ **Clear parentheses**

$\quad\quad x + 5 = x - 5$ **Collect like terms**

$\quad\quad\quad 5 = -5$ **Subtract x from both sides**

Since we obtain a contradiction, the equation has no solution.

38. Clear all parentheses first.

$$3(2 - 4x) = 4(2x - 1) - 2(1 + x)$$

$6 - 12x = 8x - 4 - 2 - 2x$

$6 - 12x = 8x - 2x - 4 - 2$ **Collect like terms**

$-12x = 6x - 6 - 6$

$-12x - 6x = -12$

$-18x = -12$

$\dfrac{-18x}{-18} = \dfrac{-12}{-18}$

$x = \dfrac{2}{3}$

41. $-4(-x + 1) + 3(2x + 3) - 7x = -10$

$4x - 4 + 6x + 9 - 7x = -10$ **Clear parentheses**

$\quad\quad\quad 3x + 5 = -10$ **Collect like terms**

$\quad\quad\quad 3x = -15$ **Subtract 5 from both sides**

$\quad\quad\quad x = -5$ **Divide both sides by 3**

3.5

17. Write out complete details when solving a word problem.

Let x = length of one piece of rope,

$x + 7$ = length of other piece of rope (it is 7 feet longer than the first).

$x + (x + 7) = 17$ **The sum of the lengths must be 17**

$\quad\quad 2x + 7 = 17$

$\quad\quad 2x = 10$

$\quad\quad x = 5$

$\quad\quad x + 7 = 12$

Thus, one length is 5 ft and the other is 12 ft.

18. Let x = Tony's age,

$\frac{1}{4}x$ = Angela's age (If Tony is 4 times as old as Angela, then Angela is $\frac{1}{4}$ as old as Tony.),

$2x$ = Theresa's age (If Tony is $\frac{1}{2}$ as old as Theresa, then Angela is 2 times as old as Tony.).

$x + \dfrac{1}{4}x + 2x = 39$ **The sum of their ages is 39**

$4\left(x + \dfrac{1}{4}x + 2x\right) = 4(39)$ **Clear fractions**

$$4x + x + 8x = 156$$
$$13x = 156$$
$$\frac{13}{13}x = \frac{156}{13}$$
$$x = 12 \qquad \text{Tony's age}$$
$$\frac{1}{4}x = 3 \qquad \text{Angela's age}$$
$$2x = 24 \qquad \text{Theresa's age}$$

32. Write out complete details. Let x = amount that Percy had invested. We are given that the amount of interest was \$31.50 and the percent interest was 6%.

(% interest) · (amount invested) = (amount of interest)

$$0.06 \qquad x \qquad = \qquad 31.50$$
$$0.06x = 31.50$$
$$\frac{0.06}{0.06}x = \frac{31.50}{0.06}$$
$$x = 525$$

The amount invested was \$525.

34. Let x = the original price of the dress, $0.30x$ = amount of discount.

(original price) − (amount of discount) = (new price)

$$x \qquad - \qquad 0.30x \qquad = \qquad 51.45$$
$$(1 - 0.30)x = 51.45$$
$$0.7x = 51.45$$
$$x = \frac{51.45}{0.7} = 73.50$$

The original price of the dress was \$73.50.

38. x = sum invested. $0.045x$ = amount of interest.

(sum invested) + (amount of interest) = (amt. in one year)

$$x \qquad + \qquad 0.045x \qquad = \qquad 1254$$
$$(1 + 0.045)x = 1254$$
$$1.045x = 1254$$
$$x = \frac{1254}{1.045} = 1200$$

The sum invested was \$1200.

3.6

14. The volume of a rectangular solid is $V = lwh$. Since we have $V = 4725$ m^3, $w = 15$ m, and $h = 15$ m, substitute and solve for l.

$$4725 = l(15)(15)$$
$$4725 = l(225)$$
$$\frac{4725}{225} = l$$
$$21 = l$$

Thus, the length is 21 m.

17. Let x = number of feet that the water rises. The volume of the cube is $10 \cdot 10 \cdot 10 = 1000$ ft^3. Thus, 1000 ft^3 of water will be the increase in the level of the tank. But this volume is represented by $40 \cdot 50 \cdot x$.

$$(40)(50)x = 1000$$
$$2000x = 1000$$
$$x = \frac{1000}{2000} = \frac{1}{2} = 0.5$$

The level of the tank will increase 0.5 ft.

19. Let x = numer of meters that the level rises. From Exercise 18 the volume increase is 33.5 m^3 (the volume of the sphere).

$$(10)(8)x = 33.5$$
$$80x = 33.5$$
$$x = \frac{33.5}{80} \approx 0.4$$

Thus, the level increased about 0.4 m.

20. Let x = cost in dollars to enclose the pasture. The perimeter of the pasture is $P = 2l + 2w = 2(3) + 2(2) = 6 + 4 = 10$ mi. The length in feet is $10(5280) = 52,800$. Now multiply the number of feet (52,800 ft) by the cost per foot in dollars, \$0.35.
$x = (52,800)(0.35) = 18,480$
The total cost is \$18,480.

25. The distance one travels in 8 hours is $(8)(22) = 176$ nautical miles. The distance the second travels in 8 hours is $(8)(17) = 136$ nautical miles. Thus, in 8 hours they will be $176 - 136 = 40$ nautical miles apart.

Supplementary Applied Problems

5. Let x = the length of the pole. We can obtain an equation to use by setting the length of the pole equal to the sum of the lengths.

(total length) = (length in sand) + (length in water) + (length in air)

$$x \qquad = \qquad \frac{1}{4}x \qquad + \qquad 9 \qquad + \qquad \frac{3}{8}x$$
$$x = \frac{1}{4}x + 9 + \frac{3}{8}x$$
$$8x = 2x + 72 + 3x \qquad \text{Multiply by the LCD 8}$$
$$8x = 5x + 72$$
$$3x = 72$$
$$x = \frac{72}{3} = 24$$

The pole is 24 ft long.

10. Let x = measure of first angle of the triangle,
$x + 30$ = measure of second angle of the triangle,
$4(x + 30)$ = measure of third angle of the triangle (if the second is $\frac{1}{4}$ the third, then the third is 4 times the second).

$$x + (x + 30) + 4(x + 30) = 180 \quad \textbf{Sum of angles}$$
$$x + x + 30 + 4x + 120 = 180 \quad \textbf{is 180}$$
$$6x + 150 = 180$$
$$6x = 30$$
$$x = 5$$
$$x + 30 = 35$$
$$4(x + 30) = 140$$

The first angle is 5°, the second is 35°, and the third is 140°.

13. Let x = width of rectangle,
   $2x$ = length of rectangle,
   $x + 8$ = width increased by 8,
   $2x - 8$ = length decreased by 8.
For a square the length and width are equal.

$$x + 8 = 2x - 8$$
$$x = 2x - 16$$
$$-x = -16$$
$$x = 16$$
$$2x = 32$$

Thus, the width is 16 ft and the length is 32 ft.

19. x = population in 1980.
$0.20x$ = amount of decrease.

(population in 1980) − (amt. of decrease) = (pop. in 1990)

$$x \qquad - \qquad 0.20x \qquad = \qquad 740$$
$$(1 - 0.20)x = 740$$
$$0.8x = 740$$
$$x = \frac{740}{0.8} = 925$$

There were 925 people in Deserted in 1980.

20. x = amount invested at 14% interest,

$10,000 - x$ = amount invested at 11% interest
(If x dollars are invested at 14%, the amount left to invest at 11% is $10,000 - x$),

$0.14x$ = amount of interest earned on 14% investment,

$0.11(10,000 - x)$ = amount of interest earned on 11% investment.

$$0.14x + 0.11(10,000 - x) = 1310$$
$$14x + 11(10,000 - x) = 131,000 \quad \textbf{Multiply by}$$
$$14x + 110,000 - 11x = 131,000 \quad \textbf{100 to clear}$$
$$\textbf{decimals}$$
$$3x = 21,000$$
$$x = 7000$$
$$10,000 - x = 3000$$

Thus, $7000 was invested at 14% and $3000 at 11%.

24. x = score Becky must make to have 90 average. The average of her 4 scores must be 90.

$$\frac{96 + 78 + 91 + x}{4} = 90$$

$$\frac{265 + x}{4} = 90$$
$$4\left(\frac{265 + x}{4}\right) = 90(4)$$
$$265 + x = 360$$
$$x = 95$$

Becky must make at least 95 on her fourth test to get an A.

3.7

27. $2a + 1 > a - 3$
$$2a > a - 4$$
$$a > -4$$

33. Multiply both sides by −4 to make the coefficient of y 1. When multiplying by a negative number we must reverse the inequality.

$$-\frac{1}{4}y \geq 2$$
$$-4\left(-\frac{1}{4}\right)y \leq 2(-4)$$
$$y \leq -8$$

45. $3(2x + 3) - (3x + 2) < 12$
$$6x + 9 - 3x - 2 < 12$$
$$3x + 7 < 12$$
$$3x < 5$$
$$x < \frac{5}{3}$$

50. Multiply by the LCD, 8, to clear the fractions. Since 8 is positive the inequality stays the same.

$$\frac{3}{4}z - \frac{3}{8} < \frac{3}{2} + \frac{1}{8}z$$
$$8\left(\frac{3}{4}z - \frac{3}{8}\right) < 8\left(\frac{3}{2} + \frac{1}{8}z\right)$$
$$6z - 3 < 12 + z$$
$$5z < 15$$
$$z < \frac{15}{5} \qquad \textbf{5 is positive}$$
$$z < 3$$

57. x = the number,
   $3x$ = the product of the number and 3,
   $x - 8$ = the number less 8.

$$3x \geq x - 8$$
$$2x \geq -8$$
$$x \geq \frac{-8}{2}$$
$$x \geq -4$$

60. Let x = number of cars he must sell in August. The average number of cars in the three month period is given by $\dfrac{47 + 62 + x}{3}$.

To win the trip, $\dfrac{47 + 62 + x}{3} \geq 50$.

$$47 + 62 + x \geq 150 \quad \textbf{Multiply both sides by 3}$$
$$109 + x \geq 150$$
$$x \geq 41$$

Thus, Darrell must sell at least 41 cars in August.

17. (a) Substitute 0 for x in $x - 5 = 0$.

$$0 - 5 = 0$$
$$-5 = 0$$

This is a contradiction so x cannot be 0. In fact, x must always be 5.

(b) Substitute 0 for y in $x - 5 = 0$. But y does not occur in the equation. Thus, if y is 0, x is still 5, so that $(5, 0)$ is the completed ordered pair.

(c) Substitute 5 for x in $x - 5 = 0$.

$$5 - 5 = 0$$
$$0 = 0$$

Thus, x is 5 and y can be any number. That is, $(5, \text{any number})$ is a solution to the equation.

(d) As in part (b), y can be -10 but x is still 5. Thus, $(5, -10)$ is the completed ordered pair.

22. Using the equation $y = 55x$, we must complete each pair.

(a) For $(1, \ )$: Substitute 1 for x to obtain $y = 55 \cdot 1 = 55$. Thus, $(1, 55)$ is the completed pair and the car travels 55 miles in 1 hour.

(b) For $(5, \ )$: Substitute 5 for x to obtain $y = 55 \cdot 5 = 275$. Thus, $(5, 275)$ is the completed pair and the car travels 275 miles in 5 hours.

(c) For $(10, \ )$: Substitute 10 for x to obtain $y = 55 \cdot 10 = 550$. The completed pair is $(10, 550)$ and the car travels 550 mi in 10 hr.

4.3

14. The three points will all lie on the same straight line if the slopes of the lines through $(2, 3)$ and $(0, 2)$ and through $(0, 2)$ and $(-2, 1)$ are equal. The slope of the line through $(2, 3)$ and $(0, 2)$ is

$$\frac{3 - 2}{2 - 0} = \frac{1}{2},$$

and the slope of the line through $(0, 2)$ and $(-2, 1)$ is

$$\frac{2 - 1}{0 - (-2)} = \frac{1}{2}.$$

Thus, the three points all lie on the same straight line.

16. First choose two points on $5x + 1 = 0$. Since all points on this line have x-coordinate $-\frac{1}{5}$ ($5x + 1 = 0$ means $5x = -1$ or $x = -\frac{1}{5}$), two such points are $\left(-\frac{1}{5}, 1\right)$ and

$\left(-\frac{1}{5}, 2\right)$. Substitute into the slope formula.

$$m = \frac{y_2 - y_1}{x_2 - x_1} = \frac{2 - 1}{\left(-\frac{1}{5}\right) - \left(-\frac{1}{5}\right)} = \frac{1}{0}$$

Since m is undefined, the slope is undefined (the line is parallel to the y-axis).

4.4

15. Substitute $-\frac{1}{2}$ for m and $(3, -2)$ for (x_1, y_1) in

$$y - y_1 = m(x - x_1).$$
$$y - (-2) = -\frac{1}{2}(x - 3)$$
$$2(y + 2) = -(x - 3) \quad \textbf{Clear fractions}$$
$$2y + 4 = -x + 3$$
$$x + 2y + 1 = 0 \quad \textbf{General form}$$

18. First find the slope.

$$m = \frac{y_2 - y_1}{x_2 - x_1} = \frac{-2 - (-1)}{3 - (-1)}$$
$$= \frac{-2 + 1}{3 + 1}$$
$$= \frac{-1}{4}$$

Then substitute into the slope-intercept form using $(3, -2)$ for (x_1, x_2) and $-\frac{1}{4}$ for m.

$$y - y_1 = m(x - x_1).$$
$$y - (-2) = -\frac{1}{4}(x - 3)$$
$$4(y + 2) = -(x - 3) \quad \textbf{Multiply both sides by 4}$$
$$4y + 8 = -x + 3$$
$$x + 4y + 5 = 0 \quad \textbf{General form}$$

19. We are given that

$$y = mx + b$$

where b = overhead cost and m = variable cost. Overhead cost = b = \$25 and variable cost = m = \$10. Thus,

$$y = mx + b$$
$$= 10x + 25.$$

5.1

7. Solve for y.

$$2x - 7y + 1 = 0 \qquad -6x + 21y + 3 = 0$$
$$-7y = -2x - 1 \qquad 21y = 6x - 3$$
$$y = \frac{2}{7}x + \frac{1}{7} \qquad y = \frac{6}{21}x - \frac{3}{21}$$
$$y = \frac{2}{7}x - \frac{1}{7}$$

Since the coefficients of x are both $\frac{2}{7}$, the lines are either parallel or coinciding (they have the same slope).

Since the constants are unequal, the y-intercepts are different so the lines do not coincide. Thus, the lines are parallel.

10. Since the line $2y = 8$ is the same as $y = 4$, its graph is a horizontal line with y-intercept $(0, 4)$. Also, the line $2x = 8$ is the same as $x = 4$, and its graph is a vertical line with x-intercept $(4, 0)$. Since a horizontal line and a vertical line are perpendicular, they are intersecting.

14. To determine whether $(-3, 2)$ is a solution to the system, substitute into both equations.

$$2x + 3y = 0 \qquad\qquad x + 8y = 19$$
$$2(-3) + 3(2) \overset{?}{=} 0 \qquad (-3) + 8(2) \overset{?}{=} 19$$
$$-6 + 6 \overset{?}{=} 0 \qquad\qquad -3 + 16 \overset{?}{=} 19$$
$$0 = 0 \qquad\qquad\qquad 13 \neq 19$$

Thus, although $(-3, 2)$ is a solution to the first equation, it is not a solution to the system since it does not solve the second equation.

5.2

5. First determine the number of solutions by writing each equation in slope-intercept form.

$$x + 3y = 4 \qquad\qquad -2x - 6y = -8$$
$$3y = -x + 4 \qquad\qquad -6y = 2x - 8$$
$$y = -\frac{1}{3}x + \frac{4}{3} \qquad\qquad y = -\frac{2}{6}x + \frac{8}{6}$$
$$\qquad\qquad\qquad\qquad = -\frac{1}{3}x + \frac{4}{3}$$

Since the equations are the same, the lines are coinciding so there will be infinitely many solutions. We might also have observed sooner that if the second equation is multiplied by $-\frac{1}{2}$ on both sides the result is the first equation. Thus, obviously the two equations are the same.

8. The system will have exactly one solution since the two lines have slope -1 and $-\frac{1}{3}$. Graph each equation by finding its intercepts.
$$x + y = 3 \qquad x + 3y = -1$$

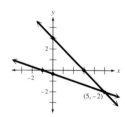

x	y
0	3
3	0

x	y
0	$-\frac{1}{3}$
-1	0

The point of intersection appears to have coordinates $(5, -2)$. Check by substitution.

$$x + y = 3 \qquad\qquad x + 3y = -1$$
$$5 + (-2) \overset{?}{=} 3 \qquad 5 + 3(-2) \overset{?}{=} -1$$
$$3 = 3 \qquad\qquad -1 = -1$$

Thus, the solution is the ordered pair $(5, -2)$.

13. Write in slope-intercept form.

$$x + 4y = 1 \qquad\qquad -x - 4y = 1$$
$$4y = -x + 1 \qquad\qquad -4y = x + 1$$
$$y = -\frac{1}{4}x + \frac{1}{4} \qquad\qquad y = -\frac{1}{4}x - \frac{1}{4}$$

Although the slopes are the same $\left(\text{both are } -\frac{1}{4}\right)$, the y-intercepts are different. Thus, the lines are parallel, and the system has no solution.

5.3

7. $3x - 5y = 19$
$2x - 4y = 16$

Simplify by multiplying the second equation by $\frac{1}{2}$.

$$x - 2y = 8$$

Solve for x.

$$x = 2y + 8.$$

Substitute into the first equation.

$$3(2y + 8) - 5y = 19$$
$$6y + 24 - 5y = 19$$
$$y + 24 = 19$$
$$y = -5$$

Substitute -5 for y in $x - 2y = 8$.

$$x - 2(-5) = 8$$
$$x + 10 = 8$$
$$x = -2$$

Thus, the solution is the ordered pair $(-2, -5)$.

10. $3x - 3y = 1$
$x - y = -1$

Solve the second equation for x, $x = y - 1$, and substitute into the first.

$$3(y - 1) - 3y = 1$$
$$3y - 3 - 3y = 1$$
$$-3 = 1$$

Since a contradiction is obtained, the system has no solution.

11. $2x + 2y = -6$
$-x - y = 3$

Solve the second equation for y.

$$-y = x + 3$$
$$y = -x - 3$$

Substitute into the first equation.

$$2x + 2(-x - 3) = -6$$
$$2x - 2x - 6 = -6$$
$$-6 = -6$$

Since an identity is obtained, the system has infinitely many solutions.

14. $3x + 5y = 30$
$5x + 3y = 34$

Solving for either variable in either equation will result in fractions. Perhaps it is best to solve for x in the first.

$3x = -5y + 30$

$x = -\dfrac{5}{3}y + 10$

Substitute into the second.

$5\left(-\dfrac{5}{3}y + 10\right) + 3y = 34$

$-\dfrac{25}{3}y + 50 + 3y = 34$

$-25y + 150 + 9y = 102$ **Multiply through by 3**
        $-16y = -48$ **to clear the fraction**

          $y = 3$

Substitute 3 for y in the second equation.

$5x + 3(3) = 34$

      $5x = 25$

       $x = 5$

Thus, the solution pair is $(5, 3)$.

5.4

5. $6x + 5y = 11$
$3x - 7y = -4$

Multiply the second equation by -2 and add.

$\begin{array}{r} 6x + 5y = 11 \\ -6x + 14y = 8 \\ \hline 19y = 19 \\ y = 1 \end{array}$

Substitute 1 for y in the first equation and solve for x.

$6x + 5(1) = 11$

  $6x + 5 = 11$

      $6x = 6$

       $x = 1$

Thus, the solution is the pair $(1, 1)$.

10. $2x + 3y = -4$
$1 + 2y = -3x$

First rewrite the system.

$2x + 3y = -4$

$3x + 2y = -1$

Multiply the first equation by 3 and the second equation by -2 and add.

$\begin{array}{r} 6x + 9y = -12 \\ -6x - 4y = 2 \\ \hline 5y = -10 \\ y = -2 \end{array}$

Now substitute -2 for y in the first equation.

$2x + 3(-2) = -4$

  $2x - 6 = -4$

      $2x = 2$

       $x = 1$

Thus, the solution is $(1, -2)$.

11. $3x + 7y = -21$
$7x + 3y = -9$

If we multiply the first equation by -7 and the second by 3, the coefficients of x will be -21 and 21. Then by adding, we eliminate x and obtain an equation in the one variable y.

$\begin{array}{r} -21x - 49y = 147 \\ 21x + 9y = -27 \\ \hline -40y = 120 \\ y = -3 \end{array}$

Substitute -3 for y in the first equation.

$3x + 7(-3) = -21$

      $3x = 0$

       $x = 0$

Thus, the solution is the pair $(0, -3)$.

16. $3x + 3y = 3$
$4x + 4y = -3$

Multiply the first equation by -4, the second by 3, and add.

$\begin{array}{r} -12x - 12y = -12 \\ 12x + 12y = -9 \\ \hline 0 = -21 \end{array}$

Since a contradiction is obtained, the system has no solution.

17. $2x + 3y = -4$
$5x + 7y = -10$

Multiply the first equation by -5, the second by 2, and add to eliminate x.

$\begin{array}{r} -10x - 15y = 20 \\ 10x + 14y = -20 \\ \hline -y = 0 \\ y = 0 \end{array}$

Substitute 0 for y in the first equation.

$2x + 3(0) = -4$

      $2x = -4$

       $x = -2$

Thus, the solution is $(-2, 0)$.

5.5

11. Let x = number of pounds of 90¢ candy,
     y = number of pounds of \$1.50 candy.
Since there are 30 pounds in the mixture,

$$x + y = 30.$$

The value of the mixture is
$$90x + 150y = 110(30).$$
Notice that the units were changed to cents, and don't forget to multiply 110¢ by 30 pounds on the right side. Solve the first equation for x and substitute.
$$x = 30 - y$$
$$90(30 - y) + 150y = 100(30)$$
$$9(30 - y) + 15y = 11(30) \quad \textbf{Divide through}$$
$$\textbf{by 10}$$
$$270 - 9y + 15y = 330$$
$$6y = 60$$
$$y = 10$$
Then $x = 30 - y = 30 - 10 = 20$. Thus, there are 20 lb of the 90¢ candy and 10 lb of the $1.50 candy.

17. Let $x =$ Terry's average speed riding in still air,
$y =$ average wind speed,
$x + y =$ Terry's average speed riding with the wind,
$x - y =$ Terry's average speed riding against the wind.

Use the distance formula, $d = rt$, twice noting that both distances are 40 miles.
$$40 = (x - y)4 \quad \textbf{Distance against the wind}$$
$$40 = (x + y)2 \quad \textbf{Distance with the wind}$$
Thus, dividing the first equation by 4 and the second by 2, we have the system:
$$\begin{array}{r} x - y = 10 \\ \underline{x + y = 20} \\ 2x = 30 \\ x = 15. \end{array}$$
Substitute 15 for x in the first equation.
$$15 - y = 10$$
$$-y = -5$$
$$y = 5$$
Thus, Terry's rate riding is 15 mph, and the wind speed is 5 mph.

19. Let $x =$ amount to be invested at 10% interest,
$y =$ amount to be invested at 12% interest.
Then $x + y = 10,000$. The total to invest is $10,000. The total interest is the sum of the interests earned on each part.
$$(0.10)x + (0.12)y = 1160$$
Multiply through by 100 to clear all decimals,
$$10x + 12y = 116,000,$$
then divide through by 2 to decrease the coefficients.
$$5x + 6y = 58,000$$
Solve the first equation for x and substitute.
$$5(10,000 - y) + 6y = 58,000$$
$$50,000 - 5y + 6y = 58,000$$
$$y = 8000$$
Then $x = 10,000 - y = 10,000 - 8000 = 2000$. Thus, $2000 is invested at 10% and $8000 is invested at 12%.

21. Let $x =$ daily rate charged,
$y =$ mileage rate charged.
The system of equations to solve is:
$$3x + 400y = 82$$
$$5x + 500y = 120.$$
Divide the second equation by 5 and solve for x.
$$x + 100y = 24$$
$$x = 24 - 100y.$$
Substitute into the first equation.
$$3(24 - 100y) + 400y = 82$$
$$72 - 300y + 400y = 82$$
$$100y = 10$$
$$y = 0.10$$
Then $x = 24 - 100y = 24 - 100(0.10) = 24 - 10 = 14$. Thus, the charges are $14 per day and 10¢ per mile.

6.1

27. $\left(\dfrac{2y}{x^3}\right)^{-2} = \dfrac{(2y)^{-2}}{(x^3)^{-2}} \quad \left(\dfrac{a}{b}\right)^n = \dfrac{a^n}{b^n}$

$\phantom{\left(\dfrac{2y}{x^3}\right)^{-2}} = \dfrac{2^{-2}y^{-2}}{(x^3)^{-2}} \quad (ab)^n = a^n b^n$

$\phantom{\left(\dfrac{2y}{x^3}\right)^{-2}} = \dfrac{2^{-2}y^{-2}}{x^{-6}} \quad (a^m)^n = a^{mn}$

$\phantom{\left(\dfrac{2y}{x^3}\right)^{-2}} = \dfrac{x^6}{2^2 y^2}$

$\phantom{\left(\dfrac{2y}{x^3}\right)^{-2}} = \dfrac{x^6}{4y^2}$

30. $\dfrac{a^{-2}b^2}{a^4 b^{-3}} = a^{-2-4}b^{2-(-3)} \quad \dfrac{a^m}{a^n} = a^{m-n}$

$\phantom{\dfrac{a^{-2}b^2}{a^4 b^{-3}}} = a^{-6}b^5$

$\phantom{\dfrac{a^{-2}b^2}{a^4 b^{-3}}} = \dfrac{1}{a^6} \cdot b^5 \quad a^{-n} = \dfrac{1}{a^n}$

$\phantom{\dfrac{a^{-2}b^2}{a^4 b^{-3}}} = \dfrac{b^5}{a^6}$

51. When $a = -2$, $a^{-2} = (-2)^{-2} = \dfrac{1}{(-2)^2} = \dfrac{1}{4}$.

52. When $a = -2$, $-2a = -2(-2) = 4$.

53. When $a = -2$, $-a^2 = -(-2)^2 = -(4) = -4$. Note the difference between Exercises 51, 52, and 53.

54. $a^{-2} + b^{-2} = (-2)^{-2} + (3)^{-2}$ **Substitute -2 for a and 3 for b**

$\phantom{a^{-2} + b^{-2}} = \dfrac{1}{(-2)^2} + \dfrac{1}{3^2} \quad a^{-n} = \dfrac{1}{a^n}$

$\phantom{a^{-2} + b^{-2}} = \dfrac{1}{4} + \dfrac{1}{9}$

$\phantom{a^{-2} + b^{-2}} = \dfrac{13}{36}$

55. $(a + b)^{-2} = ((-2) + 3)^{-2}$ **Substitute −2 for a**
          $= (1)^{-2}$ **and 3 for b**

          $= \dfrac{1}{1^2} = 1$

Note the difference between Exercises 54 and 55.

59. $a^{-1}b^{-1} = (-2)^{-1}(3)^{-1}$

          $= \left(\dfrac{1}{-2}\right)\left(\dfrac{1}{3}\right)$ $a^{-1} = \dfrac{1}{a^1} = \dfrac{1}{a}$

          $= -\dfrac{1}{6}$

60. $(ab)^{-1} = ((-2)(3))^{-1}$

          $= (-6)^{-1}$

          $= \dfrac{1}{(-6)^1}$

          $= -\dfrac{1}{6}$

6.2

19. $(0.0000022)(300) = (2.2 \times 10^{-6})(3.0 \times 10^2) =$
$(2.2)(3.0) \times (10^{-6})(10^2) = 6.6 \times 10^{-4}$

6.3

28. $-8x^{10} + x^5 - 2x^{10} - 7x^5 + 1 - x^{10} + 3$

      $= (-8x^{10} - 2x^{10} - x^{10}) + (x^5 - 7x^5) + (1 + 3)$

      $= (-8 - 2 - 1)x^{10} + (1 - 7)x^5 + (1 + 3)$

      $= -11x^{10} - 6x^5 + 4$

35. $-4x^2y + 2xy^2 + x^2 - 3x^2y$

      $= (-4x^2 - 3x^2y) + 2xy^2 + x^2$

      $= (-4 - 3)x^2y + 2xy^2 + x^2$

      $= -7x^2y + 2xy^2 + x^2$

38. $7y^2 - 2y - 5 = 7(-2)^2 - 2(-2) - 5$

                $= 7 \cdot 4 + 4 - 5$

                $= 28 + 4 - 5 = 27$

41. We need to evaluate $0.05x + 15.5$ when $x = 440$.

$0.05(440) + 15.5 = 22 + 15.5 = 37.5$

Thus, the cost is $37.50 to manufacture 440 bolts.

6.4

10. $(-4a^3 + 7a^4 + 3a + 2) + (5 - 3a + 7a^3) +$
      $(17a^4 - 5 + 12a^3)$

      $= (7a^4 - 4a^3 + 3a + 2) + (7a^3 - 3a + 5) +$
      $(17a^4 + 12a^3 - 5)$

      $= 7a^4 + 17a^4 - 4a^3 + 7a^3 + 12a^3 + 3a - 3a +$
      $2 + 5 - 5$

      $= (7 + 17)a^4 + (-4 + 7 + 12)a^3 + (3 - 3)a +$
      $(2 + 5 - 5)$

          $= 24a^4 + 15a^3 + 0 \cdot a + 2$

          $= 24a^4 + 15a^3 + 2$

16. $0.03y^3 - 0.75y^2 - 3y + 2$
         $-0.15y^3 \qquad\qquad + 5y - 0.3$
         $\underline{\;0.21y^3 - 0.13y^2 \qquad\quad + 0.6\;}$

          $0.09y^3 - 0.88y^2 + 2y + 2.3$

Simply add the coefficients of like terms (down the
columns) to obtain the coefficients in the sum.

23. $(56x - 93x^3 + 21x^4 + 32x^5) - (3x^4 - 7x^5 + 15x - 32)$

      $= (32x^5 + 21x^4 - 93x^3 + 56x) -$
      $(-7x^5 + 3x^4 + 15x - 32)$

      $= 32x^5 + 21x^4 - 93x^3 + 56x + 7x^5 - 3x^4 -$
      $15x + 32$

      $= 32x^5 + 7x^5 + 21x^4 - 3x^4 - 93x^3 + 56x -$
      $15x + 32$

      $= (32 + 7)x^5 + (21 - 3)x^4 - 93x^3 +$
      $(56 - 15)x + 32$

      $= 39x^5 + 18x^4 - 93x^3 + 41x + 32$

28. $(8y^{10} - y^8) - (3y^{12} + 2y^{10} - y^8)$

      $= 8y^{10} - y^8 - 3y^{12} - 2y^{10} + y^8$ **Clear parentheses**

      $= -3y^{12} + 8y^{10} - 2y^{10} - y^8 + y^8$

      $= -3y^{12} + (8 - 2)y^{10} + (-1 + 1)y^8$ **Distributive**
                                                    laws
      $= -3y^{12} + 6y^{10}$

31. $(9x^4 + 3x^3 + 8x) + (3x^4 + x^3 - 7x^2) -$
      $(12x^4 - 3x^2 + x)$

      $= 9x^4 + 3x^3 + 8x + 3x^4 + x^3 - 7x^2 - 12x^4 + 3x^2 - x$

      $= (9 + 3 - 12)x^4 + (3 + 1)x^3 + (-7 + 3)x^2 +$
      $(8 - 1)x$

      $= 0 \cdot x^4 + 4x^3 - 4x^2 + 7x$

      $= 4x^3 - 4x^2 + 7x$

38. $(-2a^2b + ab - 4ab^2) - (6a^2b + 4ab^2)$

      $= -2a^2b + ab - 4ab^2 - 6a^2b - 4ab^2$

      $= (-2 - 6)a^2b + ab + (-4 - 4)ab^2)$

      $= -8a^2b + ab - 8ab^2$

40. $(6x^2y - xy) - (3x^2y - 7xy^2) - (4xy - 5xy^2)$

      $= 6x^2y - xy - (3x^2y + 7xy^2 - 4xy + 5xy^2$
                                             Clear parentheses

      $= 6x^2y - 3x^2y - xy - 4xy7xy^2 + 5xy^2$

      $= (6 - 3)x^2y + (-1 - 4)xy + (7 + 5)xy^2$
                                             Distributive laws

      $= 3x^2y - 5xy + 12xy^2$

6.5

16. $-10x^2(x^5 - 6x^3 + 7x^2)$

      $= (-10x^2)(x^5) - (-10x^2)(6x^3) +$ **Distributive**
      $(-10x^2)(7x^2)$ **laws**

$= -10x^2x^5 + 60x^2x^3 - 70x^2x^2$ **Multiply coefficients**

$= -10x^7 + 60x^5 - 70x^4$ $a^m a^n = a^{m+n}$

23. $(x^2 + 4x - 2)(2x^2 - x + 3)$

$= (x^2 + 4x - 2)(2x^2) + (x^2 + 4x - 2)(-x) + (x^2 + 4x - 2)(3)$

$= (x^2)(2x^2) + (4x)(2x^2) + (-2)(2x^2) + (x^2)(-x) + (4x)(-x) + (-2)(-x) + (x^2)(3) + (4x)(3) + (-2)(3)$

$= 2x^4 + 8x^3 - 4x^2 - x^3 - 4x^2 + 2x + 3x^2 + 12x - 6$

$= 2x^4 + (8 - 1)x^3 + (-4 + 3)x^2 + (2 + 12)x - 6$

$= 2x^4 + 7x^3 - x^2 + 14x - 6$

37. $(5x - 2)(3x + 4)$

$= (5x)(3x) + (5x)(4) + (-2)(3x) + (-2)(4)$

$= 15x^2 + 20x - 6x - 8$

$= 15x^2 + 14x - 8$

41. $(2z^2 + 1)(z^2 - 2)$

$= (2z^2)(z^2) + (2z^2)(-2) + (1)(z^2) + (1)(-2)$

$= 2z^4 - 4z^2 + z^2 - 2$

$= 2z^4 - 3z^2 - 2$ **Collect like terms**

37.

$$\begin{array}{r} 0.3x^2 + 0.2 \\ \underline{0.5x\ \ \ - 0.7} \\ 0.15x^3 \qquad\qquad + 0.10x \\ \underline{\quad - 0.21x^2 \qquad\qquad - 0.14} \\ 0.15x^3 - 0.21x^2 + 0.10x - 0.14 \end{array}$$

55.

$$\begin{array}{r} 3y^2 + 5y\ \ - 6 \\ \underline{y^2 - 3y\ \ + 2} \\ 3y^4 + 5y^3 -\ \ 6y^2 \\ -9y^3 - 15y^2 + 18y \\ \underline{\qquad\qquad 6y^2 + 10y - 12} \\ 3y^4 - 4y^3 - 15y^2 + 28y - 12 \end{array}$$

59. $(2a - 3b)(5a + b)$

$= (2a)(5a) + (2a)(b) + (-3b)(5a) + (-3b)(b)$

$= 10a^2 + 2ab - 15ab - 3b^2$

$= 10a^2 - 13ab - 3b^2$

65. To find the products $(x - 1)(x + 1)(x + 2)$, we multiply the first two factors then the result times the third.

$(x - 1)(x + 1) = x^2 + x - x - 1$ **FOIL**

$\qquad\qquad\qquad = x^2 - 1$

Now multiply by $(x + 2)$.

$(x^2 - 1)(x + 2) = x^3 + 2x^2 - x - x - 2$ **FOIL**

Then $(x - 1)(x + 1)(x + 2) = x^3 + 2x^2 - x - 2$.

67. Use the formula $A = lw$ and substitute $(2x + 1)$ for l and $(3x - 2)$ for w.

$a = (2x + 1)(3x - 2)$

$= (2x)(3x) + (2x)(-2) + (1)(3x) + (1)(-2)$ **FOIL**

$= 6x^2 - 4x + 3x - 2$

$= 6x^2 - x - 2$ **Collect like terms**

Thus, the area is $(6x^2 - x - 2)$ ft^2.

6.6

13. $(4y - 9)(4y + 9)$

$= (4y)^2 - (9)^2$ $(a - b)(a + b) = a^2 - b^2$

$= 4^2y^2 - 81 = 16y^2 - 81$

31. $(0.7y - 3)^2 = (0.7y)^2 - 2(0.7y)(3) + (3)^2$

$= (0.7)^2y^2 - 6(0.7)y + 9 = 0.49y^2 - 4.2y + 9$

37. $(2x^2 - y)^2 = (2x^2)^2 - 2(2x^2)(y) + (y)^2$

$= 2^2(x^2)^2 - 4x^2y + y^2 = 4x^4 - 4x^2y + y^2$

39. $(a^2 + 2b)^2$

$= (a^2)^2 + 2(a^2)(2b) + (2b)^2$ **Use a^2 for a and $2b$ for b**

$= a^4 + 4a^2b + 4b^2$ **in $(a + b)^2 = a^2 + 2ab + b^2$**

40. $(x + 1)^2 - (x - 1)^2$

$= (x^2 + (2)(x)(1) + 1^2) - (x^2 - (2)(x)(1) + 1^2)$

Use both perfect square formulas and be sure to enclose the products in parentheses.

$= (x^2 + 2x + 1) - (x^2 - 2x + 1)$

$= x^2 + 2x + 1 - x^2 + 2x - 1$

$= 4x$

Remember to use parentheses when subtracting in order to avoid a common sign error.

44. Substitute $(2a + 7)$ for l and $(a - 1)$ for w in $A = lw$.

$A = (2a + 7)(a - 1)$

$= 2a^2 - 2a + 7a - 7$ **FOIL**

$= 2a^2 + 5a - 7$ **Collect like terms**

Thus, the area is $(2a^2 + 5a - 7)$ mi^2.

6.7

7. $(3a^{12} - 9a^6 + 27a^5 + 81a^4) \div 9a^2$

$= \dfrac{3a^{12} - 9a^6 + 27a^5 + 81a^4}{9a^2}$

$= \dfrac{3a^{12}}{9a^2} - \dfrac{9a^6}{9a^2} + \dfrac{27a^5}{9a^2} + \dfrac{81a^4}{9a^2}$

$= \dfrac{1}{3}a^{12-2} - 1 \cdot a^{6-2} + 3a^{5-2} + 9a^{4-2}$

$= \dfrac{1}{3}a^{10} - a^4 + 3a^3 + 9a^2$

10. $\dfrac{-8x^3 + 6x^2 - 4x}{0.2x} = \dfrac{-8x^3}{0.2x} + \dfrac{6x^2}{0.2x} - \dfrac{4x}{0.2x}$

$= -40x^{3-1} + 30x^{2-1} - 20x^{1-1}$ $\dfrac{8}{0.2} = \dfrac{80}{2} = 40$

$= -40x^2 + 30x - 20x^0$

$= -40x^2 + 30x - 20$ $x^0 = 1$

13. $\dfrac{-5a^2b + 3ab - 2a}{-a} = \dfrac{-5a^2b}{-a} + \dfrac{3ab}{-a} - \dfrac{2a}{-a} =$

$5a^{2-1}b - 3a^{1-1}b + 2a^{1-1} = 5ab - 3a^0b + 2a^0 =$

$5ab - 3b + 2$

22. $(6 + 8y - y^2) \div (4 - y)$

$$= \frac{6 + 8y - y^2}{4 - y}$$

$$= \frac{(-1)(6 + 8y - y^2)}{(-1)(4 - y)}, \quad \text{To make } y \text{ in denominator positive}$$

$$= \frac{-6 - 8y + y^2}{-4 + y}$$

$$= \frac{y^2 - 8y - 6}{y - 4} \quad \text{Descending order}$$

$$\begin{array}{r} y - 4 \\ y - 4 \overline{) y^2 - 8y - 6} \\ \underline{y^2 - 4y} \\ -4y - 6 \\ \underline{-4y + 16} \\ -22 \end{array}$$ The answer is $y - 4 - \dfrac{22}{y - 4}$.

23. $\begin{array}{r} 3a - 8 \\ a - 5 \overline{) 3a^2 - 23a + 40} \\ \underline{3a^2 - 15a} \\ -8a + 40 \\ \underline{-8a + 40} \\ 0 \end{array}$

Remember to subtract at this step then bring down the 40
Remember again to subtract

Thus, the quotient is $3a - 8$.

28. $\begin{array}{r} y^4 - 2y^3 + 4y^2 - 8y + 16 \\ y + 2 \overline{) y^5 \qquad\qquad\qquad\qquad + 32} \\ \underline{y^5 + 2y^4} \\ -2y^4 \\ \underline{-2y^4 - 4y^3} \\ 4y^3 \\ \underline{4y^3 + 8y^2} \\ -8y^2 \\ \underline{-8y^2 - 16y} \\ 16y + 32 \\ \underline{16y + 32} \\ 0 \end{array}$

30. $\begin{array}{r} x^2 + 1 \\ x^2 - 1 \overline{) x^4 \qquad\quad - 1} \\ \underline{x^4 - x^2} \\ +x^2 - 1 \\ \underline{x^2 - 1} \\ 0 \end{array}$ **Subtract, don't add**

7.1

14. $36y^4 = 2^2 \cdot 3^2 \cdot y^4$

$6y^3 = 2 \cdot 3 \cdot y^3$

$42y^5 = 2 \cdot 3 \cdot 7 \cdot y^5$

The factors 2, 3, and y are in all three monomials. The lowest power of 2 and 3 is 1 and the lowest power of y is 3. Thus, the GCF is $2 \cdot 3 \cdot y^3 = 6y^3$.

27. The only common factor of $3a(a + 2)$ and $5(a + 2)$ is $(a + 2)$. Thus, the GCF is $(a + 2)$.

29. Since 5 is the GCF of 5, -15, and 10, and since ab^2 is the GCF of a^3b^3, a^2b^2, and ab^2, the GCF of $5a^3b^3 - 15a^2b^2 + 10ab^2$ is $5ab^2$.

41. $-6y^{10} - 8y^8 - 4y^5$

$$= (-2y^5) \cdot 3y^5 + (-2y^5) \cdot 4y^3 + (-2y^5) \cdot 2$$

$$= -2y^5(3y^5 + 4y^3 + 2)$$

47. $a^2(a + 2) + 3(a + 2)$

$$= (a + 2) \cdot a^2 + (a + 2) \cdot 3 \quad \text{GCF is } (a + 2)$$

$$= (a + 2)(a^2 + 3)$$

50. $x^2a + x^2b + y^2a + y^2b = (x^2a + x^2b) + (y^2a + y^2b)$

$$= x^2(a + b) + y^2(a + b)$$

$$= (x^2 + y^2)(a + b)$$

56. $-x^2y - x^2 - 3y - 3$

$$= (-1)(x^2y + x^2 + 3y + 3) \quad \text{Factor } -1 \text{ from all terms}$$

$$= (-1)[x^2(y + 1) + 3(y + 1)] \quad \text{Factor } x^2 \text{ from first two and 3 from second two}$$

$$= (-1)[(x^2 + 3)(y + 1)] \quad (y + 1) \text{ is common to the two terms in brackets}$$

$$= -(x^2 + 3)(y + 1)$$

7.2

10. $y^2 + 5y - 24 = (y + \underline{\quad})(y + \underline{\quad})$

$b = 5$ and $c = -24$

Factors of $c = -24$	Sum of factors
12, -2	10
-2, 12	-10
8, -3	5

With c negative, we know that the factors must have opposite signs. Since we have found factors of c whose sum is $b = 5$, we can stop the table.

$y^2 + 5y - 24 = (y + 8)(y - 3)$

14. $x^2 - 2x - 63 = (x + \underline{\quad})(x + \underline{\quad})$

$b = -2$ and $c = -63$

Factors of $c = -63$	Sum of factors
21, -3	18
-21, 3	-18
9, -7	2
-9, 7	-2

With c negative, we only try factors with opposite signs. Since $b = -2$ the factors are -9 and 7.

$x^2 - 2x - 63 = (x - 9)(x + 7)$

20. $u^2 - 9uv + 20v^2 = (u + \underline{\quad}v)(u + \underline{\quad}v)$

$b = -9$ and $c = 20$

Factors of $c = 20$	Sum of factors
-10, -2	-12
-5, -4	-9

With c positive and b negative, we only try factors that are both negative. Thus, $u^2 - 9uv + 20v^2 = (u - 5v)(u - 4v)$.

35. $u^2 - 22uv + 121v^2 = (u + \underline{\quad}v)(u + \underline{\quad}v)$

$b = -22$ and $c = 121$

Factors of $c = 121$	Sum of factors
-121, -1	-122
-11, -11	-2

With c positive and b negative, we only try factors that are both negative. Thus, $u^2 - 22uv + 121v^2 = (u - 11v)(u - 11v)$.

10. For $6z^2 - 13z - 28$ we have $a = 6$, $b + -13$, and $c = -28$. The factors of $c = -28$ will have opposite signs.

Factors of $c = 6$ Factors of $c = -28$
6, 1   28, -1 and -28, 1
3, 2   14, -2 and -14, 2
                  7, -4 and -7, 4

$6z^2 - 13z - 28 = (\underline{}z + \underline{})(\underline{}z + \underline{})$

$\overset{?}{=} (6z - 1)(z + 28)$ **Does not work**

$\overset{?}{=} (3z - 2)(2z + 14)$ **There would have to be a common factor of 2 for the 14, 2 factors to work**

$\overset{?}{=} (3z - 4)(2z + 7)$ **Does not work**

$\overset{?}{=} (3z + 4)(2z - 7)$ **This works**

$6z^2 - 13z - 28 = (3z + 4)(2z - 7)$

17. $-45x^2 + 150x - 125$

$= (-5)9x^2 + (-5)(-30x) + (-5)(25)$

$= -5(9x^2 - 30x + 25)$

Now factor $9x^2 - 30x + 25$.

Factors of $a = 9$ Factors of $c = 25$
9, 1 -25, -1
3, 3 -5, -5

$9x^2 - 30x + 25 = (\underline{}x + \underline{})(\underline{}x + \underline{})$

$\overset{?}{=} (9x - 1)(x - 25)$ **Does not work**

$\overset{?}{=} (3x - 5)(3x - 5)$ **This works**

$9x^2 - 30x + 25 = (3x - 5)(3x - 5)$

We must include the common factor in the final answer.
$-45x^2 + 150x - 125 = -5(3x - 5)(3x - 5)$

22. $2x^2 + 7xy + 5y^2$ has $a = 2$, $b = 7$, and $c = 5$.

Factors of $a = 2$ Factors of $c = 5$
2, 1   5, 1

$2x^2 + 7xy + 5y^2 = (\underline{}x + \underline{}y)(\underline{}x + \underline{}y)$

$\overset{?}{=} (2x + y)(x + 5y)$ **Does not work**

$\overset{?}{=} (2x + 5y)(x + y)$ **This works**

$2x^2 + 7xy + 5y^2 = (2x + 5y)(x + y)$

31. $4u^2 - v^2$ has $a = 4$, $b = 0$, and $c = -1$.

Factors of $a = 4$ Factors of $c = -1$
4, 1   1, -1
2, 2

$4u^2 - v^2 = (\underline{}u + \underline{}v)(\underline{}u + \underline{}v)$

$\overset{?}{=} (4u + v)(u - v)$ **Does not work**

$\overset{?}{=} (2u + v)(2u - v)$ **This works**

$4u^2 - v^2 = (2u + v)(2u - v)$

11. $9u^2 + 6u + 1$

$= (3u)^2 + 6u + 1^2$ **$(3u)^2$ and 1^2 are perfect squares**

$= (3u)^2 + 2 \cdot 3u \cdot 1 + 1^2$ **$3u = a$ and $1 = b$**

$= (3u + 1)^2$ **$a^2 + 2ab + b^2 = (a + b)^2$**

19. $-12y^2 + 60y - 75$

$= (-3)4y^2 + (-3)(-20y) + (-3)(25)$ **Common factor first**

$= -3(4y^2 - 20y + 25)$

$= -3[(2y)^2 - 2 \cdot 2y \cdot 5 + (5)^2]$ **$2y = a$ and $5 = b$**

$= -3(2y - 5)^2$

28. $25x^2 - 10xy + y^2 = (5x)^2 - 2 \cdot 5x \cdot y + (y)^2$

$= (5x - y)^2$

30. $u^4 - v^4 = (u^2)^2 - (v^2)^2$

$= (u^2 + v^2)(u^2 - v^2)$

$= (u^2 + v^2)(u + v)(u - v)$

41. In $4y^2 - 16y + 15$ the constants are $a = 4$, $b = -16$, and $c = 15$.

Factors of $a = 4$ Factors of $c = 15$
4, 1 -15, -1
2, 2 -5, -3

Since $b = -16$, the most likely combinations are 4, 1 with -5, -3 or 2, 2 with -5, -3.

$4y^2 - 16y + 15 = (\underline{}y + \underline{})(\underline{}y + \underline{})$

$\overset{?}{=} (4y - 5)(y - 3)$ **Does not work**

$\overset{?}{=} (2y - 5)(2y - 3)$ **This works**

$4y^2 - 16y + 15 = (2y - 5)(2y - 3)$

43. $3x^9 - 147x^3 = (3x^3)x^6 - (3x^3)49$ **Common factor is $3x^3$**

$= 3x^3(x^6 - 49)$

$= 3x^3[(x^3)^3 - (7)^2]$ **$x^3 = a$ and $7 = b$**

$= 3x^3(x^3 + 7)(x^3 - 7)$

49. $100x^5 - 60x^4y + 9x^3y^2$

$= (x^3)100x^2 - (x^3)60xy + x^3(9y^2)$

$= x^3(100x^2 - 60xy + 9y^2)$

$= x^3[(10x)^2 - 2(10x)(3y) + (3y)^2]$

$= x^3(10x - 3y)^2$

7.5

13. $x(3x + 7) = 0$

$x = 0$ or $3x + 7 = 0$ **Zero-product rule; do not divide both sides by x**

                    $3x = -7$

$x = 0$ or $x = -\dfrac{7}{3}$

Thus, the solutions are 0 and $-\frac{7}{3}$. Remember that whenever one factor is x, 0 will always be one solution.

23. $4u^2 - 8u = 0$

$(4u) \cdot u - (4u) \cdot 2 = 0$

              $4u(u - 2) = 0$

$u = 0$ or $u - 2 = 0$

                   $u = 2$

The solutions are 0 and 2. Do not forget the zero solution.

26. $(3x - 5) - (x + 7) = 0$

Do not try to use the zero-product rule on a problem like this. It involves a zero-difference, not a zero-product. Clear parentheses and solve.

$3x - 5 - x - 7 = 0$ Watch the signs

$2x - 12 = 0$ Collect like terms

$2x = 12$

$x = 6$

Thus, the solution is 6.

27.

$y(y + 3) = 10$

$y^2 + 3y = 10$

$y^2 + 3y - 10 = 0$

$(y - 2)(y + 5) = 0$

$y - 2 = 0$ or $y + 5 = 0$

$y = 2$ $y = -5$

The solutions are 2 and −5.

7.6

5.

Let $n = $ first integer,

$n + 1 = $ next consecutive integer,

$n(n + 1) = $ product of consecutive integers.

$n(n + 1) = 240$

$n^2 + n = 240$

$n^2 + n - 240 = 0$

$(n - 15)(n + 16) = 0$

$n - 15 = 0$ or $n + 16 = 0$

$n = 15$ $n = -16$

$n + 1 = 16$ $n + 1 = -15$

Thus, 15 and 16 form one solution and −16 and −15 the other.

8. Let $n = $ the first even positive integer,

$n + 2 = $ the next consecutive even positive integer.

The equation to solve is

$n^2 + (n + 2)^2 = 100$

$n^2 + n^2 + 4n + 4 = 100$

$2n^2 + 4n - 96 = 0$

$n^2 + 2n - 48 = 0$ Divide out common factor 2

$(n + 8)(n - 6) = 0$ Factor

$n + 8 = 0$ or $n - 6 = 0$ Zero-product rule

$n = -8$ $n = 6$

Since n must be positive, we discard −8. Thus, the integers are 6 and 8 (since $n + 2 = 8$).

10. Let $x = $ length of one side,

$x^2 = $ area of square,

$4x = $ perimeter of square.

If we add 4 to the area, the value is the same as the perimeter.

$x^2 + 4 = 4x$

$x^2 - 4x + 4 = 0$

$(x - 2)^2 = 0$

$x - 2 = 0$ or $x - 2 = 0$

$x = 2$ $x = 2$

The side is 2 cm.

15. We need to find P when n is 30.

$P = n^2 - 3n - 60$

$= (30)^2 - 3(30) - 60$ Substitute 30 for n

$= 900 - 90 - 60$

$= 750$

Thus, the profit is $750 when 30 appliances are sold.

17. We need to find n when P is 120.

$P = n^2 - 3n - 60$

$120 = n^2 - 3n - 60$ Substitute 120 for P

$0 = n^2 - 3n - 180$

$0 = (n + 12)(n - 15)$

$n + 12 = 0$ or $n - 15 = 0$ Zero-product rule

$n = -12$ $n = 15$

Since the number of appliances cannot be negative, we discard −12. Thus, 15 were sold when the profit was $120.

19. We need to find N when n is 7.

$N = \frac{1}{2}n(n - 1)$

$= \frac{1}{2}(7)(7 - 1)$ Substitute 7 for n

$= \frac{1}{2}(7)(6) = 21$

Thus, 21 committees of two can be formed.

21. We need to find n when N is 10.

$N = \frac{1}{2}n(n - 1)$

$10 = \frac{1}{2}n(n - 1)$

$20 = n(n - 1)$ Clear fraction

$20 = n^2 - n$

$0 = n^2 - n - 20$

$0 = (n + 4)(n - 5)$

$n + 4 = 0$ or $n - 5 = 0$

$n = -4$ $n = 5$

Since n cannot be negative, there are 5 members of the organization.

8.1

7. $\dfrac{2x+7}{x^2+2x+1}$ We need to solve $x^2+2x+1=0$. First factor, then use the zero-product rule.

$$x^2+2x+1=0$$
$$(x+1)(x+1)=0$$
$$x+1=0 \quad \text{or} \quad x+1=0$$
$$x=-1 \qquad\qquad x=-1$$

Thus, -1 is the only number to be excluded.

16. $\dfrac{2}{x^2}$ and $\dfrac{2+x}{x^2+x}$ are not equivalent since one cannot be obtained from the other by multiplying both numerator and denominator by the same expression. Notice that the second fraction can be formed from the first by adding x to both numerator and denominator, but this process does not make the fraction equivalent. To see this, substitute 1 for x in both fractions.

17. $\dfrac{x-1}{3x-1}$ and $\dfrac{x}{3x}$ are not equivalent.

To obtain one from the other we would have to add or subtract the same number in both the numerator and denominator, and this process (unlike multiplying or dividing) does not yield equivalent fractions.

29. $\dfrac{x^2-3x-10}{x^2-6x+5}=\dfrac{(x-5)(x+2)}{(x-5)(x-1)}=\dfrac{x+2}{x-1}$

32. $\dfrac{x^2-9}{3-x}=\dfrac{(x-3)(x+3)}{3-x}=\dfrac{x-3}{3-x}\cdot(x+3)=$

$(-1)(x+3)=-(x+3)$

36. $\dfrac{x^2-7xy+6y^2}{x^2-4xy-12y^2}$

$=\dfrac{(x-y)(x-6y)}{(x+2y)(x-6y)}$ **Factor numerator and denominator**

$=\dfrac{x-y}{x+2y}$ **Divide out factors $(x-6y)$**

Do not cancel the x's since they are terms, not factors, of the numerator and denominator.

8.2

10. $\dfrac{z^2-z-20}{z^2+7z+12}\cdot\dfrac{z+3}{z^2-25}$

$=\dfrac{(z-5)(z+4)}{(z+3)(z+4)}\cdot\dfrac{(z+3)}{(z-5)(z+5)}$

$=\dfrac{(z-5)(z+4)(z+3)}{(z+3)(z+4)(z-5)(z+5)}=\dfrac{1}{z+5}$

14. $\dfrac{2x^2-5xy+3y^2}{x^2-y^2}\cdot(x^2+2xy+y^2)$

$=\dfrac{(2x-3y)(x-y)}{(x-y)(x+y)}\cdot\dfrac{(x+y)(x+y)}{1}$

$=\dfrac{(2x-3y)(x-y)(x+y)(x+y)}{(x-y)(x+y)}$

$=(2x-3y)(x+y)$

22. The reciprocal of $z+2$ is $\frac{1}{z+2}$ and *not* $\frac{1}{z}+\frac{1}{2}$; a common error.

29. $(x+6)\div\dfrac{x^2-36}{x^2-6x}$

$=\dfrac{(x+6)}{1}\cdot\dfrac{x^2-6x}{x^2-36}$ **Multiply by the reciprocal**

$=\dfrac{(x+6)}{1}\cdot\dfrac{x(x-6)}{(x+6)(x-6)}$ **Factor**

$=\dfrac{(x+6)x(x-6)}{(x+6)(x-6)}$ **Divide out common factors**

$=x$

34. $\dfrac{x^2+16x+64}{2x^2-128}\div\dfrac{3x^2+30x+48}{x^2-6x-16}$

$=\dfrac{x^2+16x+64}{2x^2-128}\cdot\dfrac{x^2-6x-16}{3x^2+30x+48}$

$=\dfrac{(x+8)(x+8)}{2(x-8)(x+8)}\cdot\dfrac{(x-8)(x+2)}{3(x+8)(x+2)}$

$=\dfrac{(x+8)(x+8)(x-8)(x+2)}{2(x-8)(x+8)\cdot 3\cdot(x+8)(x+2)}$

$=\dfrac{1}{2\cdot 3}=\dfrac{1}{6}$

35. $\dfrac{x^2-5xy+6y^2}{x^2-4y^2}\div(x^2-2xy-3y^2)$

$=\dfrac{x^2-5xy+6y^2}{x^2-4y^2}\cdot\dfrac{1}{x^2-2xy-3y^2}$

$=\dfrac{(x-2y)(x-3y)}{(x-2y)(x+2y)}\cdot\dfrac{1}{(x-3y)(x+y)}$

$=\dfrac{1}{(x+2y)(x+y)}$

38. $\dfrac{a^2-y^2}{a^2-ay}\cdot\dfrac{2a^2+ay}{a^2-4y^2}\div\dfrac{a+y}{a+2y}$

$=\dfrac{(a-y)(a+y)}{a(a-y)}\cdot\dfrac{a(2a+y)}{(a-2y)(a+2y)}\cdot\dfrac{(a+2y)}{a+y}$

$=\dfrac{(a-y)(a+y)a(2a+y)(a+2y)}{a(a-y)(a-2y)(a+2y)(a+y)}$

$=\dfrac{2a+y}{a-2y}$

8.3

5. $\dfrac{2y}{y+2}-\dfrac{y+1}{y+2}=\dfrac{2y-(y+1)}{y+2}$ **Use parentheses**

$=\dfrac{2y-y-1}{y+2}=\dfrac{y-1}{y+2}$

10. $\dfrac{z}{2}+\dfrac{3z-1}{-2}$

$=\dfrac{z}{2}+\dfrac{(-1)(3z-1)}{(-1)(-2)}$

$$= \frac{z}{2} + \frac{1 - 3z}{2}$$

$$= \frac{z + (1 - 3z)}{2}$$

$$= \frac{z + 1 - 3z}{2} = \frac{-2z + 1}{2}$$

14. $\dfrac{2a}{a - 1} + \dfrac{3a}{1 - a} = \dfrac{2a}{a - 1} + \dfrac{3a}{(-1)(a - 1)} =$

$$\frac{2a}{a - 1} + \frac{-3a}{a - 1} = \frac{2a - 3a}{a - 1} = \frac{-a}{a - 1}$$

20. $\dfrac{2x}{x^2 + x - 6} + \dfrac{x - 3}{6 - x - x^3}$

$$= \frac{2x}{x^2 + x - 6} + \frac{x - 3}{(-1)(-6 + x + x^2)}$$

$$= \frac{2x}{x^2 + x - 6} + \frac{(-1)(x - 3)}{x^2 + x - 6}$$

$$= \frac{2x + (-1)(x - 3)}{x^2 + x - 6}$$

$$= \frac{2x - x + 3}{x^2 + x - 6} = \frac{\cancel{(x + 3)}}{\cancel{(x + 3)}(x - 2)} = \frac{1}{x - 2}$$

24. $\dfrac{x + y}{x - y} - \dfrac{x + y}{y - x}$

$$= \frac{x + y}{x - y} - \frac{(-1)(x + y)}{(-1)(y - x)}$$

$$= \frac{x + y}{x - y} + \frac{x + y}{x - y}$$

$$= \frac{(x + y) + (x + y)}{x - y}$$

$$= \frac{2x + 2y}{x - y} = \frac{2(x + y)}{x - y}$$

25. $\dfrac{5x}{x + 1} + \dfrac{2x - 1}{x + 1} - \dfrac{3x}{x + 1} = \dfrac{5x + 2x - 1 - 3x}{x + 1}$

$$= \frac{4x - 1}{x + 1}$$

8.4

10. $\dfrac{1}{x^2 + 2x + 1}$ and $\dfrac{2}{x^2 - 1}$

$x^2 + 2x + 1 = (x + 1)(x + 1)$ **The LCD consists of one $(x - 1)$ and two $(x + 1)$'s**

$$x^2 - 1 = (x + 1)(x - 1)$$

Thus, LCD $= (x - 1)(x + 1)^2$.

14. $\dfrac{5x}{x + y}, \dfrac{7y}{2x + 2y},$ and $\dfrac{2xy}{3x + 3y}$

The factored denominators are $x + y$, $2(x + y)$, and $3(x + y)$. The LCD must consist of one $(x + y)$, one 2, and one 3. Thus, LCD $= 2 \cdot 3 \cdot (x + y) = 6(x + y)$.

23. $\dfrac{4y}{y^2 - 36} - \dfrac{4}{y + 6}$

$$= \frac{4y}{(y - 6)(y + 6)} - \frac{4(y - 6)}{(y + 6)(y - 6)}$$

$$= \frac{4y - 4(y - 6)}{(y - 6)(y + 6)}$$

$$= \frac{4y - 4y + 24}{(y - 6)(y + 6)} = \frac{24}{(y - 6)(y + 6)}$$

31. $\dfrac{3}{a^2 - a - 12} - \dfrac{2}{a^2 - 9}$

$$= \frac{3}{(a - 4)(a + 3)} - \frac{2}{(a - 3)(a + 3)}$$

$$= \frac{3(a - 3)}{(a - 4)(a + 3)(a - 3)} - \frac{2(a - 4)}{(a - 3)(a + 3)(a - 4)}$$

$$= \frac{3(a - 3) - 2(a - 4)}{(a - 4)(a + 3)(a - 3)}$$

$$= \frac{3a - 9 - 2a + 8}{(a - 4)(a + 3)(a - 3)} = \frac{a - 1}{(a - 4)(a + 3)(a - 3)}$$

35. $\dfrac{3}{x + 1} + \dfrac{5}{x - 1} - \dfrac{10}{x^2 - 1}$

$$= \frac{3}{x + 1} + \frac{5}{x - 1} - \frac{10}{(x - 1)(x + 1)}$$
Factor denominators, the LCD $= (x - 1)(x + 1)$

$$= \frac{3(x - 1)}{(x - 1)(x + 1)} + \frac{5(x + 1)}{(x - 1)(x + 1)}$$

$$- \frac{10}{(x - 1)(x + 1)}$$ **Supply missing factors**

$$= \frac{3(x - 1) + 5(x + 1) - 10}{(x - 1)(x + 1)}$$ **Add and subtract numerators over LCD.**

$$= \frac{3x - 3 + 5x + 5 - 10}{(x - 1)(x + 1)}$$

$$= \frac{8x - 8}{(x - 1)(x + 1)}$$ **Collect like terms**

$$= \frac{8\cancel{(x - 1)}}{\cancel{(x - 1)}(x + 1)} = \frac{8}{x + 1}$$ **Divide out common factor**

37. $\dfrac{2y}{y + 5} - \dfrac{3y}{y - 2} - \dfrac{2y^2}{y^2 + 3y - 10}$

$$= \frac{2y}{y + 5} - \frac{3y}{y - 2} - \frac{2y^2}{(y + 5)(y - 2)}$$

$$= \frac{2y(y - 2)}{(y + 5)(y - 2)} - \frac{3y(y + 5)}{(y - 2)(y + 5)}$$

$$- \frac{2y^2}{(y + 5)(y - 2)}$$

$$= \frac{2y(y - 2) - 3y(y + 5) - 2y^2}{(y + 5)(y - 2)}$$

$$= \frac{2y^2 - 4y - 3y^2 - 15y - 2y^2}{(y + 5)(y - 2)}$$

$$= \frac{-3y^2 - 19y}{(y + 5)(y - 2)} = \frac{-y(3y + 19)}{(y + 5)(y - 2)}$$

8.5

4. $\dfrac{1}{z-3}+\dfrac{3}{z-5}=0$

$\dfrac{1}{z-3}=-\dfrac{3}{z-5}$ Subtract $\dfrac{3}{z-5}$ from both sides

$(1)(z-5)=(-3)(z-3)$ **Multiply by LCD**

$z-5=-3z+9$ **Clear parentheses**

$4z=14$

$z=\dfrac{14}{4}=\dfrac{7}{2}$

Since $\frac{7}{2}$ will check in the original equation, it is the solution.

8. $\dfrac{z}{z+1}-1=\dfrac{1}{z}$

Multiply through by the LCD, $z(z+1)$.

$z(z+1)\dfrac{z}{z+1}-z(z+1)(1)=z(z+1)\dfrac{1}{z}$

$z^2-(z^2+z)=z+1$

Use parentheses in this step to avoid a sign error

$z^2-z^2-z=z+1$

$-2z=1$

$z=-\dfrac{1}{2}$

The solution is $-\frac{1}{2}$, which does indeed check in the original equation.

11. $\dfrac{4}{z-3}+\dfrac{2z}{z^2-9}=\dfrac{1}{z+3}$

The LCD $=(z-3)(z+3)$; so multiply both sides by it.

$\left[\dfrac{4}{z-3}+\dfrac{2z}{(z-3)(z+3)}\right](z-3)(z+3)$

$=\dfrac{1}{z+3}(z-3)(z+3)$

$\dfrac{4}{(z-3)}(z-3)(z+3)+$

$\dfrac{2z}{(z-3)(z+3)}(z-3)(z+3)$

$=\dfrac{1}{(z+3)}(z-3)(z+3)$

$4(z+3)+2z=z-3$

$4z+12+2z=z-3$

$6z+12=z-3$

$5z=-15$

$z=-3$

However, if -3 is substituted into the original equation, we obtain zero in two denominators. That is, -3 must be excluded as a replacement for z. Thus, the equation has no solution.

16. $\dfrac{3}{x-3}+\dfrac{2}{x+3}=-1$

The LCD is $(x-3)(x+3)$. Multiply both sides by it, and do not forget to multiply on the right side.

$\left[\dfrac{3}{x-3}+\dfrac{2}{x+3}\right](x-3)(x+3)$

$=(-1)(x-3)(x+3)$

$\dfrac{3}{(x-3)}(x-3)(x+3)+\dfrac{2}{(x+3)}(x-3)(x+3)=$

$(-1)(x^2-9)$

$3(x+3)+2(x-3)=-x^2+9$

$3x+9+2x-6=-x^2+9$

$5x+3=-x^2+9$

$x^2+5x-6=0$

$(x+6)(x-1)=0$

$x+6=0$ or $x-1=0$ **Zero-product rule**

$x=-6$ $x=1$

Both -6 and 1 are solutions since both check in the original equation.

23. Let $x=$ the desired number.
The equation to solve is

$\dfrac{5+x}{7+x}=\dfrac{4}{5}.$

$5(5+x)=4(7+x)$ **Multiply by LCD**

$25+5x=28+4x$ **Clear parentheses**

$x=3$

25. Let $x=$ the numerator of the fraction,
$x+4=$ the denominator of the fraction.

Then the fraction is $\dfrac{x}{x+4}.$

Now we increase *both* the numerator and the denominator by 1, and set the result equal to $\frac{2}{3}$.

$\dfrac{x+1}{x+4+1}=\dfrac{2}{3}$

$\dfrac{x+1}{x+5}=\dfrac{2}{3}$ **Multiply both sides by the LCD $=3(x+5)$**

$3(x+5)\cdot\dfrac{(x+1)}{(x+5)}=3(x+5)\cdot\dfrac{2}{3}$

$3(x+1)=(x+5)2$

$3x+3=2x+10$

$x=7$

Thus, the fraction (which was what we were asked to find) is $\frac{7}{11}$.

28. Let $n=$ the desired number,
$n-4=4$ less than the number,

$\dfrac{1}{n-4}=$ the reciprocal of 4 less than the number,

$\dfrac{1}{n}=$ the reciprocal of the desired number.

The equation to solve is

$$\frac{1}{n-4} = 3 \cdot \frac{1}{n}$$

$$\frac{1}{n-4} = \frac{3}{n}.$$

Multiplying both sides by $n(n-4)$ results in the equation

$$n = 3(n-4)$$
$$n = 3n - 12$$
$$-2n = -12$$
$$n = 6.$$

Thus, the desired number is 6, and 6 does satisfy the description of the problem.

30. Let t = time required for Bob's father to paint house,
12 = time required for Bob to paint house,
9 = time required for Bob and his father to paint the house working together.

Then $\dfrac{1}{12} + \dfrac{1}{t} = \dfrac{1}{9}.$

That is, the amount Bob does in 1 day plus the amount his father does in 1 day equals the amount they do together in 1 day. Multiply through by $36t$, the LCD.

$$36t \cdot \frac{1}{12} + 36t \cdot \frac{1}{t} = 36t \cdot \frac{1}{9}$$

$$3t + 36 = 4t$$

$$36 = t$$

It would take Bob's father 36 days to paint the house.

32. Let x = time for Max to repair the car working alone,
$4x$ = time for Irv to repair the car working alone,

$$\frac{1}{x} = \text{amount done by Max in 1 hr,}$$

$$\frac{1}{4x} = \text{amount done by Irv in 1 hr,}$$

$$\frac{1}{8} = \text{amount done together in 1 hr.}$$

The equation to solve is

$$\frac{1}{x} + \frac{1}{4x} = \frac{1}{8}.$$

Multiply through by the LCD, $8x$.

$$8x\left(\frac{1}{x}\right) + 8x\left(\frac{1}{4x}\right) = 8x\left(\frac{1}{8}\right)$$

$$8 + 2 = x$$

$$10 = x$$

It will take Max 10 hr and Irv 40 hr to repair the car working alone.

8.6

11. $\dfrac{y-3}{y+2} = \dfrac{y-2}{y+3}$

$$(y-3)(y+3) = (y+2)(y-2) \quad \textbf{Product of means}$$
$$y^2 - 9 = y^2 - 4 \qquad\qquad \textbf{equals product}$$
$$-9 = -4 \qquad\qquad\qquad \textbf{of extremes}$$

Since we obtain a contradiction, the proportion has no solutions.

17. Let x = the number of defective tires expected in a sample of 1288 tires. Then

$$\frac{6}{184} = \frac{x}{1288}.$$

$$6 \cdot 1288 = 184x$$

$$7728 = 184x$$

$$42 = x$$

You would expect 42 tires to be defective.

26. Let x = approximate number of fish in the lake. The ratio of total fish to the number tagged should be about the same as the ratio of the number of fish caught in the sample to the number of fish in the sample that were tagged.

That is, we must solve

$$\frac{x}{100} = \frac{70}{14}. \quad \begin{array}{l}\textbf{Product of means equals} \\ \textbf{product of extremes}\end{array}$$

$$14x = (70)(100)$$

$$x = 500$$

There are about 500 fish in the lake.

30. $c = kw$ **This is the direct variation equation**

$480 = k \cdot 12$ **Substitute values for c and w, and let c be in cents to avoid decimals**

$40 = k$ **The value of the constant k is 40**

$c = 40w$ **This is the variation equation using the known value for the constant**

$c = 40 \cdot 66 = 2640$ **Substitute the new value of w**

Thus, the cost of 66 lb of peaches would be $26.40.

34. The variation equation is

$$a = kt,$$

where a is the amount spent on recreation and t is total income. Then

$$1920 = k(16,000)$$

$$\frac{1920}{16,000} = k$$

$$0.12 = k.$$

The variation equation is $a = 0.12t$.

Next find a when his total income is increased by $5000; that is, when $t = 21,000$.

$$a = 0.12(21,000)$$

$$a = 2520$$

Hence, the man would spend about $2520 on recreation under these conditions.

37. The variation equation is

$$t = \frac{d}{r}$$

where t is the time, d is the fixed distance (constant), and r is the rate. Then

$$6 = \frac{d}{500}$$

$$6(500) = d$$

$$3000 = d.$$

Thus, the variation equation is

$$t = \frac{3000}{r}.$$

When $r = 1200$ mph we have

$$t = \frac{3000}{1200} = 2.5 \text{ hours.}$$

8.7

11. $\dfrac{\frac{1}{2x} - 1}{\frac{1}{x} - 2}$ The LCD of all fractions is $2x$.
Multiply both numerator and denominator by $2x$.

$$\frac{\left[\frac{1}{2x} - 1\right]2x}{\left[\frac{1}{x} - 2\right]2x} = \frac{\frac{1}{2x} \cdot 2x - 1 \cdot 2x}{\frac{1}{x} \cdot 2x - 2 \cdot 2x} = \frac{1 - 2x}{2 - 4x} =$$

$$\frac{\cancel{(1 - 2x)}}{2\cancel{(1 - 2x)}} = \frac{1}{2}$$

15. $\dfrac{z - 2 - \frac{3}{z}}{1 + \frac{1}{z}} = \dfrac{\left[z - 2 - \frac{3}{z}\right]z}{\left[1 + \frac{1}{z}\right]z} = \dfrac{z^2 - 2z - 3}{z + 1}$

$$= \frac{(z - 3)\cancel{(z + 1)}}{\cancel{(z + 1)}} = z - 3$$

17. $\dfrac{x - 3 + \frac{2}{x}}{x - 4 + \frac{3}{x}} = \dfrac{x\left(x - 3 + \frac{2}{x}\right)}{x\left(x - 4 + \frac{3}{x}\right)}$ The LCD is x

$$= \frac{x^2 - 3x + 2}{x^2 - 4x + 3}$$

$$= \frac{\cancel{(x - 1)}(x - 2)}{\cancel{(x - 1)}(x - 3)}$$

$$= \frac{x - 2}{x - 3}$$

19. $V = \dfrac{3}{\frac{1}{S_1} + \frac{1}{S_2}}$

$$= \frac{3(S_1 S_2)}{\left(\frac{1}{S_1} + \frac{1}{S_2}\right)(S_1 S_2)}$$ $S_1 S_2$ is the LCD

$$= \frac{3S_1 S_2}{\frac{S_1 S_2}{S_1} + \frac{S_1 S_2}{S_2}} = \frac{3S_1 S_2}{S_2 + S_1} = \frac{3S_1 S_2}{S_1 + S_2}$$

If $S_1 = 2$ and $S_2 = 5$, then

$$V = \frac{3S_1 S_2}{S_1 + S_2} = \frac{3(2)(5)}{2 + 5} = \frac{30}{7}.$$

9.1

14. $\sqrt{\dfrac{1000}{10}} = \sqrt{100}$ Reduce the fraction

$$= 10$$

19. $\sqrt{49a^2} = \sqrt{7^2 a^2}$ Factor 49
$$= \sqrt{(7a)^2} \quad a^n b^n = (ab)^n$$
$$= 7a$$

23. $\sqrt{x^2 + 10x + 25} = \sqrt{(x + 5)^2}$ Factor the radicand
$$= x + 5$$

Do not make the mistake of trying to take the square roots term-by-term.

25. We are given that $p = \sqrt{w}$ and that w is 25. Then
$$p = \sqrt{w} = \sqrt{25} = 5.$$
The productivity is 5.

28. With $p = \sqrt{w}$ and $p = 4$ we have
$$p = \sqrt{w}$$
$$4 = \sqrt{w}$$
$$(4)^2 = (\sqrt{w})^2$$
$$16 = w.$$

9.2

28. $\sqrt{\dfrac{x^3 y^3}{49}} = \dfrac{\sqrt{x^3 y^3}}{\sqrt{49}} = \dfrac{\sqrt{x^2 \cdot x \cdot y^2 \cdot y}}{7} = \dfrac{\sqrt{x^2}\sqrt{y^2}\sqrt{xy}}{7}$

$$= \frac{xy\sqrt{xy}}{7}$$

31. $\sqrt{\dfrac{9}{5}} = \dfrac{\sqrt{9}}{\sqrt{5}} = \dfrac{3}{\sqrt{5}}$

$$= \frac{3\sqrt{5}}{\sqrt{5}\,\sqrt{5}} \quad \text{Rationalize the denominator}$$

$$= \frac{3\sqrt{5}}{5}$$

34. $\sqrt{\dfrac{36y^2}{x}} = \dfrac{\sqrt{36y^2}}{\sqrt{x}} = \dfrac{\sqrt{36}\sqrt{y^2}}{\sqrt{x}} = \dfrac{6y}{\sqrt{x}} =$

$$\frac{6y\sqrt{x}}{\sqrt{x}\,\sqrt{x}} = \frac{6y\sqrt{x}}{x}$$

35. $\sqrt{\dfrac{25x^3}{y^3 z^4}} = \sqrt{\dfrac{25x^2 x}{y^2 y (z^2)^2}}$ Identify perfect squares

$$= \frac{5x}{yz^2}\sqrt{\frac{x}{y}} \quad \text{Remove perfect squares}$$

$$= \frac{5x}{yz^2} \sqrt{\frac{xy}{y^2}}$$

Multiply numerator and denominator of radicand by y to obtain a perfect square denominator

$$= \frac{5x}{yz^2} \cdot \frac{\sqrt{xy}}{y}$$

$$= \frac{5x\sqrt{xy}}{y^2z^2}$$

9.3

14. $\sqrt{3a^2} \sqrt{12a} = \sqrt{3a^2 \cdot 12a} = \sqrt{3^2 \cdot 2^2 \cdot a^2 \cdot a} =$
$\sqrt{3^2} \sqrt{2^2} \sqrt{a^2} \sqrt{a} = 3 \cdot 2 \cdot a\sqrt{a} = 6a\sqrt{a}$

17. $\sqrt{6x^2y} \sqrt{2xy^2} = \sqrt{6x^2y \cdot 2xy^2} =$
$\sqrt{2^2 \cdot 3 \cdot x^2 \cdot y \cdot x \cdot y^2} = \sqrt{2^2 x^2 y^2 \cdot 3xy} =$
$\sqrt{2^2} \sqrt{x^2} \sqrt{y^2}\sqrt{3xy} = 2xy\sqrt{3xy}$

19. $\sqrt{5x} \sqrt{15xy} \sqrt{3y} = \sqrt{5x \cdot 15xy \cdot 3y} = \sqrt{5^2 \cdot 3^2 \cdot x^2y^2}$
$= \sqrt{5^2}\sqrt{3^2} \cdot \sqrt{x^2} \cdot \sqrt{y^2} = 5 \cdot 3 \cdot xy = 15xy$

23. $\dfrac{\sqrt{98}}{\sqrt{18}} = \sqrt{\dfrac{98}{18}} \quad \dfrac{\sqrt{a}}{\sqrt{b}} = \sqrt{\dfrac{a}{b}}$

$\quad = \sqrt{\dfrac{\cancel{2} \cdot 49}{\cancel{2} \cdot 9}}$ Factor, look for perfect squares, cancel the 2's

$\quad = \dfrac{\sqrt{49}}{\sqrt{9}} = \dfrac{7}{3}$

26. $\dfrac{5\sqrt{15}}{\sqrt{75}} = 5\sqrt{\dfrac{15}{75}}$

$\quad = 5\sqrt{\dfrac{\cancel{3} \cdot 5}{\cancel{3} \cdot 25}}$

$\quad = 5\sqrt{\dfrac{5}{25}}$ Do not cancel 5's since 25 is a perfect square

$\quad = 5\dfrac{\sqrt{5}}{\sqrt{25}} = \dfrac{\cancel{5}\sqrt{5}}{\cancel{5}} = \sqrt{5}$

31. $\dfrac{\sqrt{3y}}{\sqrt{4y}} = \sqrt{\dfrac{3\cancel{y}}{4\cancel{y}}} = \sqrt{\dfrac{3}{4}} = \dfrac{\sqrt{3}}{\sqrt{4}} = \dfrac{\sqrt{3}}{2}$

32. $\dfrac{\sqrt{27a^3}}{\sqrt{3}} = \sqrt{\dfrac{27a^3}{3}} \quad \dfrac{\sqrt{a}}{\sqrt{b}} = \sqrt{\dfrac{a}{b}}$

$\quad = \sqrt{9a^2a}$

$\quad = 3a\sqrt{a}$

37. $\dfrac{5\sqrt{2xy}}{\sqrt{25x}} = 5\sqrt{\dfrac{2xy}{25x}} = 5\sqrt{\dfrac{2y}{25}} = 5\dfrac{\sqrt{2y}}{\sqrt{25}} =$

$\dfrac{\cancel{5}\sqrt{2y}}{\cancel{5}} = \sqrt{2y}$

38. $\dfrac{\sqrt{18x}}{\sqrt{2y}} = \sqrt{\dfrac{18x}{2y}} = \sqrt{\dfrac{9x}{y}} = \dfrac{\sqrt{9}\sqrt{x}}{\sqrt{y}} =$

$\dfrac{3\sqrt{x}\sqrt{y}}{\sqrt{y}\sqrt{y}} = \dfrac{3\sqrt{xy}}{y}$

9.4

5. $5\sqrt{12} - 3\sqrt{12} = (5-3)\sqrt{12}$ Distributive law
$\quad = 2\sqrt{12}$
$\quad = 2\sqrt{4 \cdot 3}$ Look for perfect squares
$\quad = 2 \cdot 2\sqrt{3}$
$\quad = 4\sqrt{3}$

10. $4\sqrt{50} + 7\sqrt{18} = 4\sqrt{25 \cdot 2} + 7\sqrt{9 \cdot 2} =$
$4\sqrt{25}\,\sqrt{2} + 7\sqrt{9}\,\sqrt{2} = 4 \cdot 5\sqrt{2} + 7 \cdot 3\sqrt{2} =$
$20\sqrt{2} + 21\sqrt{2} = (20 + 21)\sqrt{2} = 41\sqrt{2}$

14. $\sqrt{147} - 2\sqrt{75} = \sqrt{49 \cdot 3} - 2\sqrt{25 \cdot 3}$
$\quad = 7\sqrt{3} - 2 \cdot 5\sqrt{3}$
$\quad = 7\sqrt{3} - 10\sqrt{3}$
$\quad = (7 - 10)\sqrt{3}$
$\quad = -3\sqrt{3}$

18. $2\sqrt{4x} + 7\sqrt{9x} = 2\sqrt{4}\,\sqrt{x} + 7\sqrt{9}\,\sqrt{x}$
$\quad = 2 \cdot 2\sqrt{x} + 7 \cdot 3\sqrt{x} = 4\sqrt{x} + 21\sqrt{x}$
$\quad = (4 + 21)\sqrt{x} = 25\sqrt{x}$

21. $x\sqrt{y^3} + y\sqrt{x^2y} = x\sqrt{y^2 \cdot y} + y\sqrt{x^2y}$
$\quad = x\sqrt{y^2}\,\sqrt{y} + y\sqrt{x^2}\,\sqrt{y} = xy\sqrt{y} + yx\sqrt{y}$
$\quad = (xy + xy)\sqrt{y} = 2xy\sqrt{y}$

22. $\sqrt{18} + \sqrt{50} + \sqrt{72}$
$\quad = \sqrt{9 \cdot 2} + \sqrt{25 \cdot 2} + \sqrt{36 \cdot 2}$
$\quad = \sqrt{9}\,\sqrt{2} + \sqrt{25}\,\sqrt{2} + \sqrt{36}\,\sqrt{2}$
$\quad = 3\sqrt{2} + 5\sqrt{2} + 6\sqrt{2} = (3 + 5 + 6)\sqrt{2}$
$\quad = 14\sqrt{2}$

25. $3\sqrt{75} + 2\sqrt{12} - 5\sqrt{48}$
$\quad = 3\sqrt{25 \cdot 3} + 2\sqrt{4 \cdot 3} - 5\sqrt{16 \cdot 3}$
$\quad = 3\sqrt{25}\,\sqrt{3} + 2\sqrt{4}\,\sqrt{3} - 5\sqrt{16}\,\sqrt{3}$
$\quad = 3 \cdot 5\sqrt{3} + 2 \cdot 2\sqrt{3} - 5 \cdot 4\sqrt{3}$
$\quad = 15\sqrt{3} + 4\sqrt{3} - 20\sqrt{3} = -\sqrt{3}$

29. $4\sqrt{300} - 2\sqrt{500} + 4\sqrt{125}$
$\quad = 4\sqrt{100 \cdot 3} - 2\sqrt{100 \cdot 5} + 4\sqrt{25 \cdot 5}$ Look for perfect squares
$\quad = 4 \cdot 10\sqrt{3} - 2 \cdot 10\sqrt{5} + 4 \cdot 5\sqrt{5}$
$\quad = 40\sqrt{3} - 20\sqrt{5} + 20\sqrt{5}$
$\quad = 40\sqrt{3}$

30. $\sqrt{4x} + \sqrt{16x} - \sqrt{25x}$
$\quad = \sqrt{4}\,\sqrt{x} + \sqrt{16}\,\sqrt{x} - \sqrt{25}\,\sqrt{x}$
$\quad = 2\sqrt{x} + 4\sqrt{x} - 5\sqrt{x} = \sqrt{x}$

32. $9\sqrt{100a^2b^2} - 4\sqrt{36a^2b^2} - 10\sqrt{a^2b^2}$
$\quad = 9\sqrt{100}\,\sqrt{a^2}\,\sqrt{b^2} - 4\sqrt{36}\,\sqrt{a^2}\,\sqrt{b^2} -$
$\quad \quad 10\sqrt{a^2}\sqrt{b^2}$
$\quad = 9 \cdot 10 \cdot ab - 4 \cdot 6 \cdot ab - 10 \cdot ab$
$\quad = 90ab - 24ab - 10ab = 56ab$

34. $\sqrt{3} + \dfrac{1}{\sqrt{3}} = \sqrt{3} + \dfrac{\sqrt{3}}{\sqrt{3}\,\sqrt{3}}$

$\quad = \sqrt{3} + \dfrac{\sqrt{3}}{3} = \left(1 + \dfrac{1}{3}\right)\sqrt{3} = \left(\dfrac{3}{3} + \dfrac{1}{3}\right)\sqrt{3}$

$\quad = \dfrac{4}{3}\sqrt{3} = \dfrac{4\sqrt{3}}{3}$

38. $3\sqrt{7} - \sqrt{\dfrac{1}{7}} = 3\sqrt{7} - \dfrac{1}{\sqrt{7}}$

$\quad = 3\sqrt{7} - \dfrac{1\cdot\sqrt{7}}{\sqrt{7}\,\sqrt{7}}$ **Rationalize**

$\quad = 3\sqrt{7} - \dfrac{\sqrt{7}}{7}$

$\quad = \left(3 - \dfrac{1}{7}\right)\sqrt{7}$ **Distributive law**

$\quad = \left(\dfrac{21}{7} - \dfrac{1}{7}\right)\sqrt{7} = \dfrac{20\sqrt{7}}{7}$

39. $-3\sqrt{5} - \dfrac{9}{\sqrt{5}} = -3\sqrt{5} - \dfrac{9\sqrt{5}}{\sqrt{5}\,\sqrt{5}}$

$\quad = -3\sqrt{5} - \dfrac{9\sqrt{5}}{5} = \left(-3 - \dfrac{9}{5}\right)\sqrt{5}$

$\quad = \left(-\dfrac{15}{5} - \dfrac{9}{5}\right)\sqrt{5} = -\dfrac{24\sqrt{5}}{5}$

41. $\dfrac{\sqrt{2}}{\sqrt{10}} - \sqrt{5} = \sqrt{\dfrac{2}{10}} - \sqrt{5} = \sqrt{\dfrac{1}{5}} - \sqrt{5}$

$\quad = \dfrac{1}{\sqrt{5}} - \sqrt{5} = \dfrac{\sqrt{5}}{\sqrt{5}\,\sqrt{5}} - \sqrt{5} = \dfrac{\sqrt{5}}{5} - \sqrt{5}$

$\quad = \left(\dfrac{1}{5} - 1\right)\sqrt{5} = -\dfrac{4\sqrt{5}}{5}$

9.5

5. $\dfrac{\sqrt{25x^2}}{\sqrt{3y}} = \dfrac{\sqrt{25}\,\sqrt{x^2}}{\sqrt{3y}} = \dfrac{5x}{\sqrt{3y}} \cdot \dfrac{\sqrt{3y}}{\sqrt{3y}} = \dfrac{5x\sqrt{3y}}{3y}$

10. $\dfrac{\sqrt{2}}{\sqrt{50}} - \dfrac{\sqrt{3}}{\sqrt{75}} = \sqrt{\dfrac{2}{50}} - \sqrt{\dfrac{3}{75}}$

$\quad = \sqrt{\dfrac{\cancel{2}}{25\cdot\cancel{2}}} - \sqrt{\dfrac{\cancel{3}}{25\cdot\cancel{3}}} = \sqrt{\dfrac{1}{25}} - \sqrt{\dfrac{1}{25}} = 0$

14. $\sqrt{x}\,\sqrt{y} + \dfrac{\sqrt{xy^2}}{\sqrt{y}} = \sqrt{xy} + \sqrt{\dfrac{xy^2}{y}}$

$\quad = \sqrt{xy} + \sqrt{xy} = 2\sqrt{xy}$

17. $\sqrt{x^3}\,\sqrt{y} - 3x\sqrt{x}\,\sqrt{y} + \dfrac{4x\sqrt{xy^2}}{\sqrt{y}}$

$\quad = \sqrt{x^3 y} - 3x\sqrt{xy} + 4x\sqrt{\dfrac{xy^2}{y}}$

$\quad = \sqrt{x^2 xy} - 3x\sqrt{xy} + 4x\sqrt{xy}$

$\quad = x\sqrt{xy} - 3x\sqrt{xy} + 4x\sqrt{xy}$

$\quad = (x - 3x + 4x)\sqrt{xy} = 2x\sqrt{xy}$

20. $\sqrt{2}(\sqrt{27} - \sqrt{12}) = \sqrt{2}\,\sqrt{27} - \sqrt{2}\,\sqrt{12}$

$\quad = \sqrt{2\cdot27} - \sqrt{2\cdot12}$

$\quad = \sqrt{9\cdot6} - \sqrt{4\cdot6}$

$\quad = 3\sqrt{6} - 2\sqrt{6}$

$\quad = \sqrt{6}$

26. $\sqrt{15}\left(\dfrac{\sqrt{3}}{\sqrt{5}} + \dfrac{1}{\sqrt{15}}\right) = \dfrac{\sqrt{15}\,\sqrt{3}}{\sqrt{5}} + \dfrac{\sqrt{15}}{\sqrt{15}}$

$\quad = \dfrac{\sqrt{5\cdot3\cdot3}}{\sqrt{5}} + 1 = \dfrac{\cancel{\sqrt{5}}\,\sqrt{9}}{\cancel{\sqrt{5}}} + 1 = 3 + 1 = 4$

28. $\sqrt{2}\left(\sqrt{6} + \dfrac{\sqrt{2}}{\sqrt{3}}\right) = \sqrt{2}\,\sqrt{6} + \dfrac{\sqrt{2}\,\sqrt{2}}{\sqrt{3}}$

$\quad = \sqrt{2\cdot2\cdot3} + \dfrac{2}{\sqrt{3}} = \sqrt{4}\,\sqrt{3} + \dfrac{2}{\sqrt{3}}$

$\quad = 2\sqrt{3} + \dfrac{2}{\sqrt{3}} \cdot \dfrac{\sqrt{3}}{\sqrt{3}} = 2\sqrt{3} + \dfrac{2\sqrt{3}}{3}$

$\quad = \left(2 + \dfrac{2}{3}\right)\sqrt{3} = \dfrac{8\sqrt{3}}{3}$

35. $(\sqrt{2} - 2)(\sqrt{2} + 1)$

$\quad = \sqrt{2}\cdot\sqrt{2} + \sqrt{2}\cdot1 + (-2)\sqrt{2} + (-2)\cdot1$

$\quad = 2 + \sqrt{2} - 2\sqrt{2} - 2$

$\quad = 2 - 2 + \sqrt{2} - 2\sqrt{2} = -\sqrt{2}$

37. $(3\sqrt{3} - \sqrt{2})(2\sqrt{3} + \sqrt{2})$

$\quad = (3\sqrt{3})(2\sqrt{3}) + (3\sqrt{3})(\sqrt{2}) + (-\sqrt{2})(2\sqrt{3}) +$

$\quad (-\sqrt{2})(\sqrt{2}) = 6\cdot3 + 3\sqrt{6} - 2\sqrt{6} - 2$

$\quad = 18 - 2 + 3\sqrt{6} - 2\sqrt{6} = 16 + \sqrt{6}$

43. $\dfrac{-8}{\sqrt{3} - \sqrt{7}} = \dfrac{-8(\sqrt{3} + \sqrt{7})}{(\sqrt{3} - \sqrt{7})(\sqrt{3} + \sqrt{7})}$

$\quad = \dfrac{-8(\sqrt{3} + \sqrt{7})}{(\sqrt{3})^2 - (\sqrt{7})^2} = \dfrac{-8(\sqrt{3} + \sqrt{7})}{3 - 7}$

$\quad = \dfrac{-8(\sqrt{3} + \sqrt{7})}{-4} = 2(\sqrt{3} + \sqrt{7}) = 2\sqrt{3} + 2\sqrt{7}$

44. $\dfrac{2\sqrt{3}}{\sqrt{3} - 1} = \dfrac{2\sqrt{3}(\sqrt{3} + 1)}{(\sqrt{3} - 1)(\sqrt{3} + 1)}$ **Rationalize**

$\quad = \dfrac{2\sqrt{3}(\sqrt{3} + 1)}{3 - 1}$

$\quad = \dfrac{\cancel{2}\sqrt{3}(\sqrt{3} + 1)}{\cancel{2}}$

$\quad = 3 + \sqrt{3}$

47. $\dfrac{x - 1}{\sqrt{x} - 1} = \dfrac{(x - 1)(\sqrt{x} + 1)}{(\sqrt{x} - 1)(\sqrt{x} + 1)}$

$\quad = \dfrac{(x - 1)(\sqrt{x} + 1)}{(\sqrt{x})^2 - (1)^2}$

$\quad = \dfrac{\cancel{(x - 1)}(\sqrt{x} + 1)}{\cancel{x - 1}} = \sqrt{x} + 1$

9.6

2. $3\sqrt{2x} - 9 = 0$

$\quad\quad 3\sqrt{2x} = 9$ **Isolate the radical**

$\quad\quad\quad \sqrt{2x} = 3$ **Simplify**

$\quad\quad (\sqrt{2x})^2 = (3)^2$ **Square both sides**

$\quad\quad\quad\quad 2x = 9$

$\quad\quad\quad\quad x = \dfrac{9}{2}$

Check: $3\sqrt{2 \cdot \dfrac{9}{2}} - 9 = 3\sqrt{9} - 9 = 3 \cdot 3 - 9 = 0$

6. $3\sqrt{a - 3} + 6 = 0$

$\quad\quad 3\sqrt{a - 3} = -6$ **Isolate the radical**

$\quad\quad\quad \sqrt{a - 3} = -2$ **Divide by 3**

$\quad\quad (\sqrt{a - 3})^2 = (-2)^2$ **Square both sides**

$\quad\quad\quad\quad a - 3 = 4$

$\quad\quad\quad\quad\quad a = 7$

Check: $3\sqrt{a - 3} + 6 = 3\sqrt{7 - 3} + 6$

$\quad\quad\quad\quad = 3\sqrt{4} + 6 = 3 \cdot 2 + 6 \neq 0$

Since 7 does not check, there is no solution.

13. $3\sqrt{x - 1} - \sqrt{x + 31} = 0$

$\quad\quad 3\sqrt{x - 1} = \sqrt{x + 31}$ **Isolate the radicals**

$\quad\quad (3\sqrt{x - 1})^2 = (\sqrt{x + 31})^2$ **Square both sides**

$\quad\quad 9(x - 1) = x + 31$ **Square the 3 also**

$\quad\quad 9x - 9 = x + 31$

$\quad\quad\quad 8x = 40$

$\quad\quad\quad\quad x = 5$

Check: $3\sqrt{5 - 1} - \sqrt{5 + 31} = 3\sqrt{4} - \sqrt{36} =$

$\quad\quad\quad 3 \cdot 2 - 6 = 0$

17. $\sqrt{a^2 - 5} + a - 5 = 0$

$\quad\quad \sqrt{a^2 - 5} = 5 - a$ **Subtract a and add 5**

$\quad\quad (\sqrt{a^2 - 5})^2 = (5 - a)^2$

$\quad\quad a^2 - 5 = 25 - 10a + a^2$ **Do not forget the −10a**

$\quad\quad\quad -5 = 25 - 10a$ **Subtract a²**

$\quad\quad\quad 10a = 30$

$\quad\quad\quad\quad a = 3$

Check: $\sqrt{3^2 - 5} + 3 - 5 = \sqrt{9 - 5} - 2 =$

$\quad\quad\quad \sqrt{4} - 2 = 2 - 2 = 0$

20. $\sqrt{x^2 + 3} + x = 6$

$\quad\quad \sqrt{x^2 + 3} = 6 - x$

$\quad\quad (\sqrt{x^2 + 3})^2 = (6 - x)^2$

$\quad\quad x^2 + 3 = 36 - 12x + x^2$

$\quad\quad\quad 3 = 36 - 12x$

$\quad\quad\quad 12x = 33$

$\quad\quad\quad\quad x = \dfrac{33}{12} = \dfrac{11}{4}$

Check: $\sqrt{\left(\dfrac{11}{4}\right)^2 + 3} + \dfrac{11}{4} = \sqrt{\dfrac{121}{16} + \dfrac{48}{16}} +$

$\dfrac{11}{4} = \sqrt{\dfrac{169}{16}} + \dfrac{11}{4} = \dfrac{13}{4} + \dfrac{11}{4} = \dfrac{24}{4} = 6$

25. $\sqrt{x - 3} - x + 5 = 0$

$\quad\quad \sqrt{x - 3} = x - 5$

$\quad\quad (\sqrt{x - 3})^2 = (x - 5)^2$

$\quad\quad x - 3 = x^2 - 10x + 25$

$\quad\quad 0 = x^2 - 11x + 28$

$\quad\quad 0 = (x - 7)(x - 4)$

$\quad\quad x - 7 = 0$ or $x - 4 = 0$

$\quad\quad\quad x = 7$ or $x = 4$

Check:

$\quad \sqrt{7 - 3} - 7 + 5 = \sqrt{4} - 2 = 2 - 2 = 0$

$\quad \sqrt{4 - 3} - 4 + 5 = \sqrt{1} + 1 = 1 + 1 \neq 0$

Since 7 checks and 4 does not, the only solution is 7.

28. $\sqrt{x + 10} + x - 2 = 0$

$\quad\quad \sqrt{x + 10} = 2 - x$ **Isolate radical**

$\quad\quad (\sqrt{x + 10})^2 = (2 - x)^2$ **Square both sides**

$\quad\quad x + 10 = 4 - 4x + x^2$ **Don't forget the middle term −4x**

$\quad\quad 0 = x^2 - 5x - 6$

$\quad\quad 0 = (x - 6)(x + 1)$

$\quad\quad x - 6 = 0$ or $x + 1 = 0$

$\quad\quad\quad x = 6$ $x = -1$

Check: $x = 6$ Check: $x = -1$

$\sqrt{6 + 10} + 6 - 2 \overset{?}{=} 0$ $\sqrt{-1 + 10} + (-1) - 2 \overset{?}{=} 0$

$\sqrt{16} + 6 - 2 \overset{?}{=} 0$ $\sqrt{9} - 1 - 2 \overset{?}{=} 0$

$4 + 6 - 2 \neq 0$ $3 - 1 - 2 = 0$

Since 6 does not check, the only solution is −1.

9.7

10. Let x = length of wire needed to reach top. Use the Pythagorean theorem.

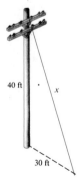

$x^2 = (40)^2 + (30)^2$

$x^2 = 1600 + 900$

$$x^2 = 2500$$

$$x = \sqrt{2500} = 50$$

The wire must be 50 ft long.

13.   Let x = the number,
$x + 5$ = five more than the number,
$\sqrt{x + 5}$ = square root of 5 more than the number,
$2\sqrt{x}$ = twice the square root of the number.

$$\sqrt{x + 5} = 2\sqrt{x}$$

$$(\sqrt{x + 5})^2 = (2\sqrt{x})^2$$

$$x + 5 = 4x$$

$$5 = 3x$$

$$\frac{5}{3} = x$$

Check: $\sqrt{\dfrac{5}{3} + 5} = \sqrt{\dfrac{5}{3} + \dfrac{15}{3}} = \sqrt{\dfrac{20}{3}} =$

$$\sqrt{\frac{4 \cdot 5}{3}} = \sqrt{4}\sqrt{\frac{5}{3}} = 2\sqrt{\frac{5}{3}}$$

16. We are given that $n = 64$ and asked to find c.

$$c = 100\sqrt{n + 36} = 100\sqrt{64 + 36} =$$

$$100\sqrt{100} = 100(10) = 1000$$

The cost is $1000 when 64 appliances are produced.

22. We need to find n when c is 50.

$$c = 10\sqrt{n} + 100$$

$$50 = 10\sqrt{n} + 100 \quad \textbf{Substitute 50 for } c$$

$$-50 = 10\sqrt{n}$$

$$-5 = \sqrt{n}$$

At this point we can stop since we know that $\sqrt{n}$ must be nonnegative and hence cannot be -5. Thus, there is no value of n for $c = 50$.

27. We need to find c, the hypotenuse, given the two legs $a = 15$ and $b = 20$. By the Pythagorean theorem,

$$c^2 = a^2 + b^2.$$

$$= 15^2 + 20^2$$

$$= 225 + 400 = 625$$

$$c = \sqrt{625} = 25$$

10.1

19. $$4z^2 = 7z$$

$$4z^2 - 7z = 0$$

$$z(4z - 7) = 0$$

$$z = 0 \quad \text{or} \quad 4z - 7 = 0$$

$$4z = 7$$

$$z = \frac{7}{4}$$

The solutions are 0 and $\frac{7}{4}$.

26. $(3z - 1)(2z + 1) = 3(2z + 1)$

$$6z^2 + z - 1 = 6z + 3$$

$$6z^2 - 5z - 4 = 0$$

$$(3z - 4)(2z + 1) = 0$$

$$3z - 4 = 0 \quad \text{or} \quad 2z + 1 = 0$$

$$3z = 4 \qquad\qquad 2z = -1$$

$$z = \frac{4}{3} \qquad\qquad z = -\frac{1}{2}$$

The solutions are $\frac{4}{3}$ and $-\frac{1}{2}$.

28. $4x^2 - 9 = 3$

$$4x^2 = 12$$

$$x^2 = \frac{12}{4}$$

$$x^2 = 3$$

$$x = \pm\sqrt{3}$$

The solutions are $\sqrt{3}$ and $-\sqrt{3}$.

31. $3x^2 - 48 = -48$

$$3x^2 = 0 \qquad \textbf{Add 48 to both sides}$$

$$x^2 = 0 \qquad \textbf{Divide both sides by 3}$$

$$x = \pm\sqrt{0} \qquad \textbf{Take square root of both sides}$$

$$x = \pm 0 = 0 \quad \textbf{+0 and −0 are both 0}$$

Thus, 0 is the only solution.

35. $2x^2 - 150 = 0$

$$2x^2 = 150$$

$$x^2 = 75$$

$$x = \pm\sqrt{75} = \pm\sqrt{25 \cdot 3} = \pm 5\sqrt{3}$$

The two solutions are $5\sqrt{3}$ and $-5\sqrt{3}$.

37. $2x^2 + 50 = 0$

$$2x^2 = -50$$

$$x^2 = -25$$

Since $\sqrt{-25}$ is not a real number, there are no real solutions to this equation.

41. $(x - 2)^2 = 9$

$$(x - 2) = \pm\sqrt{9} \quad \textbf{Take square root}$$

$$x - 2 = \pm 3 \qquad\quad \textbf{of both sides}$$

$$x = 2 \pm 3 \quad \textbf{Add 2 to both sides}$$

$$x = 2 + 3 = 5 \quad \text{or} \quad x = 2 - 3 = -1$$

The solutions are 5 and -1.

10.2

6. $z^2 + \dfrac{1}{3}z +$ _____

$$\left[\frac{1}{2} \text{ of } \frac{1}{3}\right]^2 = \left[\frac{1}{2} \cdot \frac{1}{3}\right]^2 = \left[\frac{1}{6}\right]^2 = \frac{1}{36}$$

We need to add $\frac{1}{36}$ to complete the square.

13. $x^2 - 2x - 1 = 0$

$x^2 - 2x \qquad = 1$

$x^2 - 2x + 1 = 1 + 1 \qquad 1 = \left[\frac{1}{2} \text{ coefficient of } x\right]^2$

$(x - 1)^2 = 2$

$x - 1 = \pm\sqrt{2}$ **Take square root of both sides**

$x = 1 \pm \sqrt{2}$ **Add 1 to both sides**

15. $3z^2 + 12z = 135$

$3z^2 + 12z - 135 = 0$

$z^2 + 4z - 45 = 0$ **Divide through by 3**

$z^2 + 4z \qquad = 45$

$z^2 + 4z + 4 = 45 + 4$ $\left[\frac{1}{2} \text{ of } 4\right]^2 = [2]^2 = 4$

$(z + 2)^2 = 49$

$z + 2 = \pm\sqrt{49} = \pm 7$

$z = -2 \pm 7$

$z = -2 + 7 = 5$ or $z = -2 - 7 = -9$

The solutions are 5 and -9.

16. $2x^2 + x - 1 = 0$

$x^2 + \frac{1}{2}x - \frac{1}{2} = 0$ **Divide through by 2 to make coefficient of x^2-term equal to 1**

$x^2 + \frac{1}{2}x + \frac{1}{16} = \frac{1}{2} + \frac{1}{16}$ $\left[\frac{1}{2} \text{ of } \frac{1}{2}\right]^2 = \left[\frac{1}{4}\right]^2 = \frac{1}{16}$

$\left(x + \frac{1}{4}\right)^2 = \frac{9}{16}$

$x + \frac{1}{4} = \pm\sqrt{\frac{9}{16}} = \pm\frac{3}{4}$

$x = -\frac{1}{4} \pm \frac{3}{4}$

$x = -\frac{1}{4} + \frac{3}{4} = \frac{2}{4} = \frac{1}{2}$ or

$x = -\frac{1}{4} - \frac{3}{4} = -1$

The solutions are $\frac{1}{2}$ and -1.

20. $y^2 + 2y + 5 = 0$

$y^2 + 2y \qquad = -5$

$y^2 + 2y + 1 = -5 + 1$

$(y + 1)^2 = -4$

But since $\sqrt{-4}$ is not a real number, there are no real solutions.

23. $5y^2 - 4y - 33 = 0$

We first try to factor.

$(5y + 11)(y - 3) = 0$

$5y + 11 = 0$ or $y - 3 = 0$ **Zero-product rule**

$5y = -11 \qquad\qquad y = 3$

$y = -\frac{11}{5}$

The solutions are $-\frac{11}{5}$ and 3.

10.3

5. $0 = 7y - 2 + 15y^2$

$15y^2 + 7y - 2 = 0$ **Write in general form**

$y = \dfrac{-7 \pm \sqrt{49 - 4(15)(-2)}}{2 \cdot 15}$ $a = 15, b = 7, c = -2$

$= \dfrac{-7 \pm \sqrt{49 + 120}}{30} = \dfrac{-7 \pm \sqrt{169}}{30} = \dfrac{-7 \pm 13}{30}$

$y = \dfrac{-7 + 13}{30} = \dfrac{6}{30} = \dfrac{1}{5}$ or

$y = \dfrac{-7 - 13}{30} = \dfrac{-20}{30} = -\dfrac{2}{3}$

The solutions are $\frac{1}{5}$ and $-\frac{2}{3}$.

7. $5x^2 - x = 1$

$5x^2 - x - 1 = 0$ **General form, $a = 5, b = -1, c = -1$**

$x = \dfrac{-b \pm \sqrt{b^2 - 4ac}}{2a}$

$= \dfrac{-(-1) \pm \sqrt{(-1)^2 - 4(5)(-1)}}{2(5)}$

$= \dfrac{1 \pm \sqrt{1 - (-20)}}{10}$

$= \dfrac{1 \pm \sqrt{21}}{10}$

11. $z^2 = z - 5$

$z^2 - z + 5 = 0$ **General form, $a = 1, b = -1, c = 5$**

$z = \dfrac{-b \pm \sqrt{b^2 - 4ac}}{2a}$

$= \dfrac{-(-1) \pm \sqrt{(-1)^2 - 4(1)(5)}}{2(1)}$

$= \dfrac{1 \pm \sqrt{1 - 20}}{2}$

$= \dfrac{1 \pm \sqrt{-19}}{2}$

Since $\sqrt{-19}$ is not a real number, the equation has no real solution.

14. $y(y + 1) + 2y = 4$

$y^2 + y + 2y = 4$

$y^2 + 3y - 4 = 0$

$$y = \frac{-3 \pm \sqrt{9 - 4(1)(-4)}}{2 \cdot 1} \qquad a = 1,\ b = 3,$$
$$c = -4$$
$$= \frac{-3 \pm \sqrt{9 + 16}}{2} = \frac{-3 \pm \sqrt{25}}{2} = \frac{-3 \pm 5}{2}$$

$$y = \frac{-3 + 5}{2} = \frac{2}{2} = 1 \quad \text{or}$$

$$y = \frac{-3 - 5}{2} = \frac{-8}{2} = -4$$

The solutions are 1 and -4.

18. $(2z - 1)(z - 2) - 11 = 2(z + 4) - 8$

$$2z^2 - 5z + 2 - 11 = 2z + 8 - 8$$

$$2z^2 - 7z - 9 = 0$$

$$z = \frac{7 \pm \sqrt{49 - 4(2)(-9)}}{2 \cdot 2} \qquad a = 2,\ b = -7,$$
$$c = -9$$
$$= \frac{7 \pm \sqrt{49 + 72}}{4} = \frac{7 \pm \sqrt{121}}{4} = \frac{7 \pm 11}{4}$$

$$z = \frac{7 + 11}{4} = \frac{18}{4} = \frac{9}{2} \quad \text{or}$$

$$z = \frac{7 - 11}{4} = \frac{-4}{4} = -1$$

The solutions are $\frac{9}{2}$ and -1.

20. $3z^2 + 10z - 1 = 0$

$$z = \frac{-10 \pm \sqrt{100 - 4(3)(-1)}}{2 \cdot 3} \qquad a = 3,\ b = 10,$$
$$c = -1$$
$$= \frac{-10 \pm \sqrt{100 + 12}}{6} = \frac{-10 \pm \sqrt{112}}{6}$$

$$= \frac{-10 \pm \sqrt{16 \cdot 7}}{6}$$

$$= \frac{-10 \pm 4\sqrt{7}}{6} = \frac{\cancel{2}(-5 \pm 2\sqrt{7})}{\cancel{2} \cdot 3}$$

$$= \frac{-5 \pm 2\sqrt{7}}{3}$$

The solutions are $\frac{-5 + 2\sqrt{7}}{3}$ and $\frac{-5 - 2\sqrt{7}}{3}$.

26. $\frac{2}{3}x^2 - \frac{1}{3}x - 1 = 0$

$2x^2 - x - 3 = 0$ **Multiply by 3 to clear fractions**

$(2x - 3)(x + 1) = 0$

$2x - 3 = 0 \quad \text{or} \quad x + 1 = 0$

$\qquad 2x = 3 \qquad\qquad x = -1$

$\qquad\quad x = \frac{3}{2}$

The solutions are $\frac{3}{2}$ and -1.

32. Let x = length of the fence.
By the Pythagorean theorem,

$$x^2 = 4^2 + 4^2$$

$$x^2 = 16 + 16$$

$$x^2 = 16 \cdot 2$$

$$x = \pm\sqrt{16 \cdot 2} = \pm 4\sqrt{2}.$$

4 yd   4 yd

Since we are interested in the length of the fence, we discard the negative solution, $-4\sqrt{2}$. Thus, the length of the fence is $4\sqrt{2}$ yd, or approximately 5.7 yards (using $\sqrt{2} \approx 1.414$).

10.4

7. $\frac{5}{z + 3} - \frac{1}{z + 3} = \frac{z + 3}{z + 2}$ **LCD = $(z + 3)(z + 2)$**

$$\left[\frac{5}{z + 3} - \frac{1}{z + 3}\right](z + 3)(z + 2)$$

$$= \frac{z + 3}{\cancel{z + 2}}(z + 3)\cancel{(z + 2)} \quad \textbf{Multiply by LCD}$$

$$\frac{5}{\cancel{z + 3}}\cancel{(z + 3)}(z + 2) - \frac{1}{\cancel{z + 3}}\cancel{(z + 3)}(z + 2)$$

$$= (z + 3)(z + 3)$$

$$5(z + 2) - (z + 2) = z^2 + 6z + 9$$

$$5z + 10 - z - 2 = z^2 + 6z + 9$$

$$0 = z^2 + 2z + 1$$

$$0 = (z + 1)(z + 1)$$

Thus, $z + 1 = 0$ or $z = -1$, which does indeed check in the original equation.

8. $\frac{x - 1}{x} = \frac{x}{x + 1}$ **LCD = $x(x + 1)$**

$$\frac{x - 1}{\cancel{x}}\cancel{x}(x + 1) = \frac{x}{\cancel{x + 1}}x\cancel{(x + 1)} \quad \textbf{Multiply by LCD}$$

$$x^2 - 1 = x^2$$

$$-1 = 0$$

Since we obtain a contradiction, there is no solution.

12. $\qquad \frac{x + 2}{2} = \frac{x + 5}{x + 2}$

$(x + 2)(x + 2) = 2(x + 5)$ **Cross-product equation**

$$x^2 + 4x + 4 = 2x + 10$$

$$x^2 + 2x - 6 = 0$$

Since this will not factor, we use the quadratic formula with $a = 1$, $b = 2$, and $c = -6$.

$$x = \frac{-b \pm \sqrt{b^2 - 4ac}}{2a}$$

$$= \frac{-2 \pm \sqrt{2^2 - 4(1)(-6)}}{2(1)}$$

$$= \frac{-2 \pm \sqrt{4 + 24}}{2}$$

$$= \frac{-2 \pm \sqrt{28}}{2}$$

$$= \frac{-2 \pm 2\sqrt{7}}{2} \qquad \sqrt{28} = \sqrt{4 \cdot 7} = 2\sqrt{7}$$

$$= \frac{2(-1 \pm \sqrt{7})}{2}$$

$$= -1 \pm \sqrt{7}$$

The solutions are $-1 \pm \sqrt{7}$.

17. $1 + 2\sqrt{x - 1} = x$

$\qquad 2\sqrt{x - 1} = x - 1$ **Isolate the radical**

$\qquad (2\sqrt{x - 1})^2 = (x - 1)^2$ **Square both sides**

$\qquad 4(x - 1) = x^2 - 2x + 1$ **Remember to square 2 on the left**

$\qquad 4x - 4 = x^2 - 2x + 1$

$\qquad 0 = x^2 - 6x + 5$

$\qquad 0 = (x - 1)(x - 5)$

$x - 1 = 0 \quad \text{or} \quad x - 5 = 0$

$\quad x = 1 \quad \text{or} \qquad x = 5$

Since both 1 and 5 check in the original equation, they are the solutions.

22. $3\sqrt{y + 1} - y - 1 = 0$

$\qquad 3\sqrt{y + 1} = y + 1$ **Isolate the radical**

$\qquad (3\sqrt{y + 1})^2 = (y + 1)^2$ **Square both sides**

$\qquad 9(y + 1) = y^2 + 2y + 1$

$\qquad 9y + 9 = y^2 + 2y + 1$

$\qquad 0 = y^2 - 7y - 8$

$\qquad 0 = (y - 8)(y + 1)$

$y - 8 = 0 \quad \text{or} \quad y + 1 = 0$

$\quad y = 8 \qquad\qquad y = -1$

The solutions are 8 and -1, which do check.

24. $\sqrt{2x^2 - 5} = x$

$\qquad (\sqrt{2x^2 - 5})^2 = x^2$ **Square both sides**

$\qquad 2x^2 - 5 = x^2$

$\qquad x^2 = 5$

$\qquad x = \pm\sqrt{5}$

But if we check $-\sqrt{5}$, we have

$$\sqrt{2(-\sqrt{5})^2 - 5} \stackrel{?}{=} -\sqrt{5}$$

$$\sqrt{5} \neq -\sqrt{5}.$$

Since $-\sqrt{5}$ does not check, it must be discarded.

Thus, $\sqrt{5}$ is the only solution that checks.

25. Let n = the desired number,

$\dfrac{1}{n}$ = reciprocal of the number,

$n + \dfrac{1}{n}$ = the number increased by its reciprocal.

$$n + \frac{1}{n} = \frac{5}{2} \quad \textbf{LCD = 2n}$$

$$2n\left[n + \frac{1}{n}\right] = 2n \cdot \frac{5}{2}$$

$$2n^2 + 2 = 5n$$

$$2n^2 - 5n + 2 = 0$$

$$(2n - 1)(n - 2) = 0$$

$$2n - 1 = 0 \quad \text{or} \quad n - 2 = 0$$

$$2n = 1 \quad \text{or} \qquad n = 2$$

$$n = \frac{1}{2}$$

Thus, the solutions are $\frac{1}{2}$ and 2.

10.5

8. Let x = a positive odd integer,

$x + 2$ = the next consecutive positive odd integer.

$$3x(x + 2) = 297$$

$$x(x + 2) = 99 \quad \textbf{Divide both sides by 3}$$

$$x^2 + 2x - 99 = 0$$

$$(x - 9)(x + 11) = 0$$

$$x - 9 = 0 \quad \text{or} \quad x + 11 = 0$$

$$x = 9 \quad \text{or} \qquad x = -11$$

Since x must be positive, we discard -11. Thus, the two integers are 9 and 11.

16. Let x = height of a triangle.

$\frac{1}{3}x$ = base of the triangle.

Since the formula for the area of a triangle is $A = \frac{1}{2}bh = \frac{1}{2}$(base)(height), we must solve:

$$24 = \frac{1}{2}\left(\frac{1}{3}x\right)(x)$$

$$24 = \frac{1}{6}x^2$$

$$144 = x^2 \quad \textbf{Multiply both sides by 6}$$

$$\pm\sqrt{144} = x$$

$$\pm 12 = x.$$

Since the height of a triangle must be positive we discard -12. The height is 12 ft and the base is 4 ft.

19. The area of a parallelogram is given by $A = bh$.

Let x = height of parallelogram,

$2x + 1$ = base of parallelogram.

The equation to solve is

$$55 = x(2x + 1).$$

$$55 = 2x^2 + x$$

$$0 = 2x^2 + x - 55$$

$$0 = (2x + 11)(x - 5)$$

$$2x + 11 = 0 \qquad \text{or} \quad x - 5 = 0$$

$$x = -\frac{11}{2} \qquad\qquad x = 5$$

Since the height cannot be negative, we discard $-\frac{11}{2}$. Thus, the height is 5 ft and the base is 11 ft (since $2x + 1 = 2(5) + 1 = 11$).

22. Let $x =$ the number of hrs required for Dan to grade tests,

$x + 9 =$ number of hrs required for Joe to grade tests.

$$\frac{1}{x} + \frac{1}{x + 9} = \frac{1}{20} \qquad \begin{array}{l}\textbf{Number each grades in 1 hr}\\ \textbf{is } \frac{1}{20} \textbf{ of the total}\end{array}$$

$$20x(x + 9)\frac{1}{x} + 20x(x + 9)\frac{1}{(x + 9)} = 20x(x + 9)\frac{1}{20}$$

$$20x + 180 + 20x = x^2 + 9x$$

$$0 = x^2 - 31x - 180$$

$$0 = (x + 5)(x - 36)$$

$$x + 5 = 0 \qquad \text{or} \quad x - 36 = 0$$

$$x = -5 \quad \text{or} \qquad\quad x = 36$$

Discard -5. Thus, Dan takes 36 hrs and Joe takes 45 hrs to grade the tests.

25.
$$A = P(1 + r)^2$$

$$1210 = 1000(1 + r)^2$$

$$\frac{1210}{1000} = (1 + r)^2$$

$$\pm\sqrt{1.21} = 1 + r$$

$$\pm 1.1 = 1 + r$$

Thus, $r = -1 \pm 1.1$, or $r = -2.1$ and $r = 0.1$. Since the negative value for r must be discarded, $r = 0.1$ or $r = 10\%$.

26. Let $x =$ speed of boat in still water,
$x + 2 =$ speed of boat downstream,
$x - 2 =$ speed of boat upstream.
We use the distance formula $d = rt$, solved for t, to obtain:

$$\frac{48}{x - 2} = \text{time of travel upstream,}$$

$$\frac{48}{x + 2} = \text{time of travel downstream.}$$

Since the total time is 10 hr, we need to solve

$$\frac{48}{x - 2} + \frac{48}{x + 2} = 10.$$

Multiply both sides by the LCD, $(x - 2)(x + 2)$ to obtain

$$48(x + 2) + 48(x - 2) = 10(x - 2)(x + 2).$$

$$48x + 96 + 48x - 96 = 10x^2 - 40$$

$$0 = 10x^2 - 96x - 40$$

$$0 = 5x^2 - 48x - 20$$

$$0 = (5x + 2)(x - 10)$$

$$5x + 2 = 0 \qquad \text{or} \quad x - 10 = 0$$

$$x = -\frac{2}{5} \qquad\qquad x = 10$$

We can discard $-\frac{2}{5}$, so the speed of the boat in still water is 10 mph.

10.6

4. $g - ax = m$

$$-ax = m - g \quad \text{Subtract } g \text{ from both sides}$$

$$x = \frac{m - g}{-a} \quad \text{Divide both sides by } -a$$

$$x = \frac{g - m}{a} \quad \begin{array}{l}\textbf{Multiply numerator and}\\ \textbf{denominator by } -1\end{array}$$

11. Solve $A = \frac{1}{2}bh$ for b.

$$2A = bh \quad \text{Multiply both sides by 2}$$

$$\frac{2A}{h} = b \quad \text{Divide both sides by } h$$

14. $r = \dfrac{d}{t}$

$$rt = d \quad \text{Multiply both sides by } t$$

$$t = \frac{d}{r} \quad \text{Divide both sides by } r$$

16. Solve $S = \frac{1}{2}gt^2$ for t.

$$2S = gt^2 \quad \text{Multiply by 2}$$

$$\frac{2S}{g} = t^2 \quad \text{Divide by } g$$

$$\pm\sqrt{\frac{2S}{g}} = t \quad \text{Take square root of both sides}$$

20. Solve $U = \sqrt{\dfrac{a}{c}}$ for c.

$$U^2 = \frac{a}{c} \quad \text{Square both sides}$$

$$U^2 c = a \quad \text{Multiply both sides by } c$$

$$c = \frac{a}{U^2} \quad \text{Divide both sides by } U^2$$

10.7

15. To determine the vertex of the parabola with equation

$$y = x^2 - 3$$

we observe that the equation can be written in the form

$$y = (x - 0)^2 - 3.$$

Then it is clear that $h = 0$ and $k = -3$, so the vertex is $(0, -3)$.

A.1

16. Since $\angle EDF = 0°$ and $\angle DFE = 180°$, the sum of their measures is 180°, making them supplementary.

A.3

16. Divide the area into two rectangles. One is 7 yd long and 3 yd wide and the other is 3 yd long and 2 yd wide.

$$7 \text{ yd} \times 3 \text{ yd} = 21 \text{ yd}^2$$
$$3 \text{ yd} \times 2 \text{ yd} = 6 \text{ yd}^2$$
$$21 \text{ yd}^2 + 6 \text{ yd}^2 = 27 \text{ yd}^2$$
$$\frac{\$16.50}{1 \text{ yd}^2} \times \frac{27 \text{ yd}^2}{1} = (\$16.50)(27)$$
$$= \$445.50$$

25. First find the area of the rectangle which is 28 cm long and 14 cm wide.

$$A = (28 \text{ cm})(14 \text{ cm}) = 392 \text{ cm}^2$$

Now find the area of the two triangles and subtract from the area of the rectangle. The base of each triangle is 10 cm and the height is 14 cm − 7 cm = 7 cm.

$$A = \frac{1}{2}(10 \text{ cm})(7 \text{ cm}) = 35 \text{cm}^2$$
$$2A = 70 \text{ cm}^2$$

Thus, the area is 392 cm² − 70 cm² = 322 cm².

35. $c = 2\pi r$
$$= 2\pi(7.7 \text{ mi})$$
$$= 15.4\pi \text{ mi}$$
$A = \pi r^2$
$$= \pi(7.7 \text{ mi})^2$$
$$= 59.29\pi \text{ mi}^2$$

38. $c = 2\pi r$
$$= 2\pi(0.05 \text{ cm})$$
$$\approx 2(3.14)(0.05) \text{ cm}$$
$$= 0.314 \text{ cm}$$
$A = \pi r^2$
$$= \pi(0.05 \text{ cm})^2$$
$$\approx (3.14)(0.0025) \text{ cm}^2$$
$$= 0.00785 \text{ cm}^2$$

39. Calculate the area for each hose and subtract. The radius of the first hose is $\frac{3}{8}$ in and the radius of the second hose is $\frac{1}{4}$ in.

$$A = \pi r^2$$
$$\approx 3.14\left(\frac{3}{8} \text{ in}\right)^2$$
$$\approx 0.442 \text{ in}^2$$

$$A = \pi r^2$$
$$\approx 3.14\left(\frac{1}{4} \text{ in}\right)^2$$
$$\approx 0.196 \text{ in}^2$$
$$0.442 \text{ in}^2 - 0.196 \text{ in}^2 = 0.246 \text{ in}^2$$

A.4

3. Convert each number to decimals:
9.5 in by 3.4 in by 8.75 in.
$$V = (9.5 \text{ in})(3.4 \text{ in})(8.75 \text{ in})$$
$$= 282.625 \text{ in}^3$$

10. $A = 2lh + 2wh + 2lw$
$$= 2(15.5 \text{ cm})(20.2 \text{ cm}) + 2(6.4 \text{ cm})(20.2 \text{ cm}) + 2(15.5 \text{ cm})(6.4 \text{ cm})$$
$$= 1083.16 \text{ cm}^2$$

15. $V = e^3$
$$= (22 \text{ cm})^3$$
$$= 10{,}648 \text{ cm}^3$$
Since water weighs $1\frac{\text{gm}}{\text{cm}^3}$ the answer is 10,648 g.

16. Area of bottom = $(22 \text{ cm})^2 = 484 \text{ cm}^2$
Area of 4 sides = $4(22 \text{ cm})^2 = 4(484)\text{cm}^2$
$$= 1936 \text{ cm}^2$$
Total required = 484 cm² + 1936 cm²
$$= 2420 \text{ cm}^2$$

19. $V = \pi r^2 h$
$$= \pi(12.2 \text{ in})^2(30 \text{ in}) \quad \text{Radius is } \tfrac{1}{2} \text{ the diameter}$$
$$\approx (3.14)(12.2)^2(30)\text{in}^2$$
$$\approx 14{,}020.7 \text{ in}^3$$

22. $A = 2\pi rh + 2\pi r^2$
$$\approx 2(3.14)(12.2 \text{ in})(30 \text{ in}) + 2(3.14)(12.2 \text{ in})^2$$
$$\approx 3233.2 \text{ in}^2$$

25. $V = \frac{4}{3}\pi r^3$
$$= \frac{4}{3}\pi\left(\frac{3}{4} \text{ in}\right)^3$$
$$\approx \left(\frac{4}{3}\right)(3.14)\left(\frac{3}{4}\right)^3 \text{ in}^3$$
$$\approx 1.766 \text{ in}^3$$

29. $A = 4\pi r^2$
$$= 4\pi(2.8 \text{ yd})^2$$
$$\approx 4(3.14)(2.8)^2 \text{ yd}^2$$
$$\approx 98.47 \text{ yd}^2$$

30. Calculate the volume of the can and the volume of the pie pan and compare the two.

$$V = \pi r^2 h$$
$$= \pi(4 \text{ cm})^2(12 \text{ cm})$$
$$\approx (3.14)(4)^2(12) \text{ cm}^3$$
$$= 602.88 \text{ cm}^3$$

$$V = \pi r^2 h$$
$$= \pi(10 \text{ cm})^2(3 \text{ cm})$$
$$\approx (3.14)(10)^2(3) \text{ cm}^3$$
$$= 942 \text{ cm}^3$$

$$\frac{942 \text{ cm}^3}{1} \times \frac{1 \text{ can}}{602.88 \text{ cm}^3} = 1.6 \text{ cans}$$

35. Find the area of the triangle and subtract 3 times the area of one hole.

$$A = \frac{1}{2}bh$$
$$= \frac{1}{2}(9 \text{ in})(7.8 \text{ in})$$
$$= 35.1 \text{ in}^2$$

$$A = \pi r^2$$
$$= \pi(1 \text{ in})^2 \quad \textbf{Radius is } \frac{1}{2}\textbf{(2 in)} = \textbf{1 in}$$
$$\approx 3.14 \text{ in}^2$$

The area of the three circles is

$$3(3.14 \text{ in}^2) = 9.42 \text{ in}^2.$$
$$35.1 \text{ in}^2 - 9.42 \text{ in}^2 = 25.68 \text{ in}^2$$

INDEX